EVALUATION COMPLEXITY OF ALGORITHMS FOR NONCONVEX OPTIMIZATION

MOS-SIAM Series on Optimization

This series is published jointly by the Mathematical Optimization Society and the Society for Industrial and Applied Mathematics. It includes research monographs, books on applications, textbooks at all levels, and tutorials. Besides being of high scientific quality, books in the series must advance the understanding and practice of optimization. They must also be written clearly and at an appropriate level for the intended audience.

Series Volumes

Cartis, Coralia, Gould, Nicholas I. M., and Toint, Philippe L., *Evaluation Complexity of Algorithms for Nonconvex Optimization: Theory, Computation, and Perspectives*

Cui, Ying and Pang, Jong-Shi, *Modern Nonconvex Nondifferentiable Optimization*

Shapiro, Alexander, Dentcheva, Darinka, and Ruszczyński, Andrzej, *Lectures on Stochastic Programming: Modeling and Theory, Third Edition*

Delfour, Michel C., *Introduction to Optimization and Hadamard Semidifferential Calculus, Second Edition*

Campi, Marco C. and Garatti, Simone, *Introduction to the Scenario Approach*

Beck, Amir, *First-Order Methods in Optimization*

Terlaky, Tamás, Anjos, Miguel F., and Ahmed, Shabbir, editors, *Advances and Trends in Optimization with Engineering Applications*

Todd, Michael J., *Minimum-Volume Ellipsoids: Theory and Algorithms*

Bienstock, Daniel, *Electrical Transmission System Cascades and Vulnerability: An Operations Research Viewpoint*

Koch, Thorsten, Hiller, Benjamin, Pfetsch, Marc E., and Schewe, Lars, editors, *Evaluating Gas Network Capacities*

Corberán, Ángel and Laporte, Gilbert, *Arc Routing: Problems, Methods, and Applications*

Toth, Paolo and Vigo, Daniele, *Vehicle Routing: Problems, Methods, and Applications, Second Edition*

Beck, Amir, *Introduction to Nonlinear Optimization: Theory, Algorithms, and Applications with MATLAB*

Attouch, Hedy, Buttazzo, Giuseppe, and Michaille, Gérard, *Variational Analysis in Sobolev and BV Spaces: Applications to PDEs and Optimization, Second Edition*

Shapiro, Alexander, Dentcheva, Darinka, and Ruszczyński, Andrzej, *Lectures on Stochastic Programming: Modeling and Theory, Second Edition*

Locatelli, Marco and Schoen, Fabio, *Global Optimization: Theory, Algorithms, and Applications*

De Loera, Jesús A., Hemmecke, Raymond, and Köppe, Matthias, *Algebraic and Geometric Ideas in the Theory of Discrete Optimization*

Blekherman, Grigoriy, Parrilo, Pablo A., and Thomas, Rekha R., editors, *Semidefinite Optimization and Convex Algebraic Geometry*

Delfour, M. C., *Introduction to Optimization and Semidifferential Calculus*

Ulbrich, Michael, *Semismooth Newton Methods for Variational Inequalities and Constrained Optimization Problems in Function Spaces*

Biegler, Lorenz T., *Nonlinear Programming: Concepts, Algorithms, and Applications to Chemical Processes*

Shapiro, Alexander, Dentcheva, Darinka, and Ruszczyński, Andrzej, *Lectures on Stochastic Programming: Modeling and Theory*

Conn, Andrew R., Scheinberg, Katya, and Vicente, Luis N., *Introduction to Derivative-Free Optimization*

Ferris, Michael C., Mangasarian, Olvi L., and Wright, Stephen J., *Linear Programming with MATLAB*

Attouch, Hedy, Buttazzo, Giuseppe, and Michaille, Gérard, *Variational Analysis in Sobolev and BV Spaces: Applications to PDEs and Optimization*

Wallace, Stein W. and Ziemba, William T., editors, *Applications of Stochastic Programming*

Grötschel, Martin, editor, *The Sharpest Cut: The Impact of Manfred Padberg and His Work*

Renegar, James, *A Mathematical View of Interior-Point Methods in Convex Optimization*

Ben-Tal, Aharon and Nemirovski, Arkadi, *Lectures on Modern Convex Optimization: Analysis, Algorithms, and Engineering Applications*

Conn, Andrew R., Gould, Nicholas I. M., and Toint, Phillippe L., *Trust-Region Methods*

EVALUATION COMPLEXITY OF ALGORITHMS FOR NONCONVEX OPTIMIZATION

THEORY, COMPUTATION, AND PERSPECTIVES

Coralia Cartis
University of Oxford
Oxford, United Kingdom

Nicholas I. M. Gould
STFC-Rutherford Appleton Laboratory
Harwell, United Kingdom

Philippe L. Toint
University of Namur
Namur, Belgium

siam.
Society for Industrial and Applied Mathematics
Philadelphia

Mathematical Optimization Society
Mathematical Optimization Society
Philadelphia

10 9 8 7 6 5 4 3 2 1

Illustrations: Front cover, *The Root of Complexity*, linocut by Philippe Toint, 2020.
Title page, *Nonconvex Landscape*, Northumberland, England, by Nick Gould, 2004.

Publications Director	Kivmars H. Bowling
Executive Editor	Elizabeth Greenspan
Acquisitions Editor	Paula Callaghan
Developmental Editor	Mellisa Pascale
Managing Editor	Kelly Thomas
Production Editor	Louis R. Primus
Copy Editor	Julia Cochrane
Production Manager	Donna Witzleben
Production Coordinator	Cally A. Shrader
Compositor	Cheryl Hufnagle
Graphic Designer	Doug Smock

Library of Congress Cataloging-in-Publication Data

Names: Cartis, C. (Coralia), author. | Gould, Nicholas I. M., author. | Toint, Ph. L. (Philippe L.), author.
Title: Evaluation complexity of algorithms for nonconvex optimization : theory, computation, and perspectives / Coralia Cartis, Nicholas I.M. Gould, Philippe L. Toint.
Description: Philadelphia : Society for Industrial and Applied Mathematics, [2022] | Includes bibliographical references and indexes. | Summary: "This is the go-to book for those interested in solving nonconvex problems"-- Provided by publisher.
Identifiers: LCCN 2021058246 (print) | LCCN 2021058247 (ebook) | ISBN 9781611976984 (hardcover) | ISBN 9781611976991 (ebook)
Subjects: LCSH: Mathematical optimization. | Nonconvex programming.
Classification: LCC QA402.5 .C368 2022 (print) | LCC QA402.5 (ebook) | DDC 519.7–dc23/eng20220215
LC record available at *https://lccn.loc.gov/2021058246*
LC ebook record available at *https://lccn.loc.gov/2021058247*

To our families
and
the memory of our friend Andy Conn

Contents

Preface

Do you know the difference between an optimist and a pessimist? The former believes we live in the best possible world, and the latter is afraid that the former might be right.... In that sense, this is an optimistic book in that it shows that apparently daunting tasks (such as minimizing nonlinear nonconvex functions subject to nonlinear equality and inequality constraints) may not be completely intractable, even in the worst-possible case. It is also a pessimistic book because the complexity results obtained are often relatively poor from the practical point of view while being theoretically sharp.

The developments presented are unquestionably biased toward previous work of its authors, while opening doors and giving pointers to others. They provide a fresh and often surprising view on traditional methods in optimization. Here is a brief list of some remarkable facts, even for seasoned optimizers:

The worst-case evaluation complexity of convexly constrained nonlinear optimization is of the same order as that of unconstrained optimization.

Newton's method, an old trusted friend, and many of its variants can be as slow in the worst case as the steepest-descent algorithm in its worst case (a well-known loser).

Regularization with second-order models (a relative newcomer) is the optimal second-order algorithm in terms of worst-case evaluation complexity and is significantly faster in this respect than most other standard methods.

Linesearch methods cannot be guaranteed to converge when searching for approximate minimizers of order above three.

Fast global rates of convergence to approximate minimizers of arbitrary order can be achieved by algorithms using function values or derivatives of low degree only.

Of course, these statements require qualifications, assumptions, and proofs. These, in turn, when seen as a whole, constitute a complex yet structured construction, a hall with many pillars and perspectives.

Interested? Read on...

Acknowledgments

It is a real pleasure to acknowledge the many useful and friendly discussions we have had with our coauthors Stefania Bellavia, Ernesto Birgin, Nicolas Boumal, Xiaojun Chen, John-Lennon Gardenghi, Serge Gratton, Gianmarco Gurioli, Sadok Jerad, Marius Lange, José Mario Martínez, Benedetta Morini, Margherita Porcelli, Tyrone Rees, Daniel Robinson, Phillipe Sampaio, Sandra Santos, Katya Scheinberg, Jennifer Scott, Ehouarn Simon, Valeria Simoncini, Sue Thorne, and Hung Wang. They have undoubtedly contributed to our progress and understanding of the topic of this book. They have also made the pursuit of research in this domain a humanly enriching process.

We also most appreciate the extremely fruitful discussions we have held over the years with Richard Byrd, Frank Curtis, Roger Fletcher, Geovani Grapiglia, Andreas Griewank, Raphael Hauser, Michal Kočvara, Adrian Lewis, Ken McKinnon, Yurii Nesterov, Jorge Nocedal, Dominique Orban, Michael Overton, Mike Powell, Nick Trefethen, Luis Nunes Vicente, Andreas Wächter, and Steve Wright on and around the subject of evaluation complexity, and for the different perspectives these interactions have provided. Thanks also go to Paula Callaghan, Cheryl Hufnagle, Kathleen LeBlanc, Mellisa Pascale, Lou Primus, and the team at SIAM for their sage advice and unwavering patience during the publication process.

Most importantly, we are deeply grateful to Jared, Penny, and Claire, our spouses, who have tolerated the many hours we spent writing this book and have supported our scientific efforts for so many years while at the same time bringing us back to real life when necessary.

On a more personal note, Cora would like to thank her Romanian family (Livia and Ioan, Dan and Kristie) for their selfless provision of childcare, for their endless love and prayers, and their constant inspiration to do good ... and many thanks to her daughter Ana Sofia—for always bringing a smile! The Mathematical Institute at the University of Oxford and the Turing Institute for Data Science deserve many thanks for supporting our collaboration and many productive visits. Balliol College, through its Oliver Smithies Programme, its friendly environment, and its sabbatical leave award, has, figuratively and literally, nourished and sustained our joint work and plans: thank you! Cora is also grateful to her doctoral students and postdoctoral researchers for useful discussions and for their patience.

Nick would like to thank his colleagues in the Scientific Computing Department at the Rutherford Appleton Laboratory for their long-lasting support, and both the Engineering and Physical Sciences Research Council (EPSRC) and UK Research and Innovation (UKRI) for funding much of the research that resulted in papers quoted here. He is also eternally grateful to his dogs, friends, and extended family—you know who you all are—for keeping him close to an even keel. Oh, and to the Royal Oak in Wantage, of course.

Philippe would like to thank his daughters and their families for their continued refreshing presence. He is indebted to the Department of Mathematics at the University of Namur for its lasting and friendly support, before and after retirement. He also acknowledges the kind support provided on several occasions by the University of Oxford, Balliol College, the Leverhulme Trust, the Hong Kong Polytechnic University, the ENSEEIHT (Toulouse), and the University of Florence. He finally thanks J. S. Bach, J. Brahms, C. Monteverdi, O. Messiaen, M. Davis, C. Corea, J. Surman, and many other music composers for proposing such a wonderful and creative universe, and L. Chauvier, his engraving teacher, for her patience, humour, and understanding.

Book Website

Corrections, updates, and additional material will be available from

`https://bookstore.siam.org/mo30/bonus.`

Basic Notation

$[i:j], [i:j)$	sets of integers $\{i,\ldots,j\}$ and $\{i,\ldots,j-1\}$, respectively
$[v]_i$	ith component of a vector of real numbers v
I	identity matrix (of appropriate order)
e_i	ith column of I
$\langle\cdot,\cdot\rangle$	Euclidean inner product
$\|\cdot\|$	Euclidean norm or subordinate tensor norm
$\mathcal{O}(\cdot), o(\cdot)$	asymptotic "no-larger-than" and "no-smaller-than" orders of their arguments
k	iteration counter; as a subscript indicates a quantity at iteration k
v_k	kth iterated version of a vector v
n	number of optimization variables (parameters)
m	number of explicit residuals or constraints
$\mathbb{R}$, $\mathbb{R}^n$, $\mathbb{R}^{m\times n}$	spaces of real scalars, n-dimensional vectors, and m by n matrices
x, x_k	optimization variables (in $\mathbb{R}^n$)
$f(x)$	objective function
$\nabla_x^\ell f(x)$	derivative tensor of f of order ℓ; $\nabla_x^0 f(x) = f(x)$, $\nabla_x^1 f(x) =$ gradient, $\nabla_x^2 f(x) =$ Hessian matrix
$L_{f,j}$	Lipschitz constant for the jth derivative of f over an appropriate domain
$f_{\rm low}$	lower bound on f in an appropriate domain
s, s_k	step from x or x_k
$T_{f,j}(s)$	jth-order Taylor-series approximation of $f(x+s)$ about x
$m_k(s)$	model of $f(x+s)$ at $x = x_k$
Δ, Δ_k	trust-region radii
$\mathcal{S}, \mathcal{U}$	sets of indices of successful and unsuccessful iterations
σ, σ_k	regularization parameters
$r(x)$	residual function (in $\mathbb{R}^m$)
$\Phi_\ell(x)$	merit function $\|r(x)\|^\ell/\ell$ for residuals
$\mathcal{C}, \mathcal{F}$	convex and feasible sets, respectively
$P_\mathcal{C}(x)$	projection of x into $\mathcal{C}$
$c(x)$	constraint function (in $\mathbb{R}^m$)
y, y_k	Lagrange multipliers (in $\mathbb{R}^m$)
$\mathcal{L}(x,y)$	Lagrangian function $f(x) - \langle y, c(x)\rangle$
$\ker^\ell[T]$	the kernel $\{v \in \mathbb{R}^n \mid T[v]^\ell = 0\}$ of the cubically symmetric ℓ-tensor T
$\phi_{f,q}^\delta(x)$	qth-order δ-measure of optimality of f at x
$\mathcal{B}$, $\mathcal{B}_\delta$	the unit ball $\{s \mid \|s\| \leq 1\}$ and the ball of radius δ, $\{s \mid \|s\| \leq \delta\}$
ϵ	vector of approximate optimality tolerances
κ	a constant, independent of ϵ, with a subscript that indicates the context

Chapter 1

Introduction

1.1 Nonlinear optimization and optimality conditions

1.1.1 Introduction

The focus of this book is on finite-dimensional continuous optimization problems, that is, on finding a small or large value of a function f that depends on parameters $x \in \mathbb{R}^n$ and where x is allowed to take any value from within a continuous domain of interest, the feasible set. Since there is no loss in doing so, we shall concentrate on minimization, because to maximize we simply need to minimize $-f$. We are especially interested in nonconvex problems, informally those problems that potentially admit local minimizers, since these lie at the heart of applications from many branches of science, engineering, economics, and planning. Although ideally we would like to find the smallest value (the global minimum), this unfortunately is beyond our current expertise and indeed there are strong theoretical reasons why the problem will remain intractable in general. Thus we shall content ourselves with local minimization, that is to say with trying to find a value of x, a local minimizer, for which all adjacent feasible points have a larger value of f. It is perhaps disappointing to discover that even this more modest goal may be beyond us, and we will have to lower our sights further. In particular, while we can easily identify, using calculus, conditions that must be satisfied at a local minimizer, these necessary "optimality" conditions need not be sufficient. Indeed, the same conditions might actually be satisfied by a maximizer or some other deviant point. Fortunately, the difference between (second-order) necessary and sufficient conditions is rather minor, so we would be unlucky in practice to be misled.

Thus our hunt will be to try to find a value (or values) of x that satisfies necessary local optimality conditions. We will consider these in detail in Section 1.1.2. Notice that there is then a tacit assumption that a calculus-based approach is warranted; this may not be the case if the functions that define our problem are not differentiable.

For some problems, any possible value of x is allowed. These are generally known as unconstrained optimization problems, and the least-squares fitting of a model to experimental data is a good example. But for the vast majority of examples that arise in practice, some values of x are clearly "off limits," and most real-life examples involve constraints. Sometimes these constraints are very simple—the values of the variables x are not allowed to be negative—while in other applications the constraints are formidable. Some restrictions are defined by physical equations $c(x) = 0$, while others permit a range of values $c(x) \geq 0$—we arbitrarily categorize constrained optimization problems by whether they are equality or inequality constrained, but in practice they may actually be both, in which case a hybrid approach may be necessary. There is actually a vast amassed knowledge of methods/algorithms[1] for solving optimization problems, often backed by a reassuring guarantee of convergence under reasonable assumptions.

1.1.2 Problem classes and optimality conditions

Necessary optimality conditions are fundamentally important in what we study in this book. They translate the statement that x_* is a local minimizer of f over some domain into a set of verifiable conditions that must be achieved. Sufficient optimality conditions assert that if a set of conditions occur, f must be a local minimizer within the domain of interest.

Almost everything we discuss in this book concerns continuous functions. If $\mathcal{S}$ is an open subset of $\mathbb{R}^n$, we let $C^0(\mathcal{S})$ be the set of functions that are continuous throughout $\mathcal{S}$; the set C^0 in which $\mathcal{S}$ is omitted consists of functions that are continuous everywhere. Likewise, the sets $C^1(\mathcal{S})$, $C^2(\mathcal{S})$, and more generally $C^p(\mathcal{S})$ contain functions that are continuously differentiable, twice continuously differentiable, and p-times continuously differentiable throughout $\mathcal{S}$, respectively.

1.1.2.1 Functions and their derivatives

At various stages, we will be concerned with an objective function $f : \mathbb{R}^n \to \mathbb{R}$ and a vector of constraint functions $c \colon \mathbb{R}^n \to \mathbb{R}^m$, where n and m are positive integers. Often f and c are assumed to be continuously differentiable over a set of interest, and we then denote the gradient and Jacobian by $\nabla_x^1 f(x)$ and $\nabla_x^1 c(x)$, respectively, where the component $[\nabla_x^1 f(x)]_j = \partial f(x)/\partial x_j$ denotes the usual first-order partial derivative of f with respect to x_j and $[\nabla_x^1 c(x)]_{i,j} = \partial c_i(x)/\partial x_j$ for $1 \le i \le m$ and $1 \le j \le n$. When f and c are required to be twice continuously differentiable, we denote their (symmetric) Hessian matrices by $\nabla_x^2 f(x)$ and $\nabla_x^2 c_i(x)$, $1 \le i \le m$, where here the components $[\nabla_x^2 f(x)]_{j_1,j_2} = \partial^2 f(x)/\partial x_{j_1}\partial x_{j_2}$ and $[\nabla_x^2 c_i(x)]_{j_1,j_2} = \partial^2 c_i(x)/\partial x_{j_1}\partial x_{j_2}$ for $1 \le j_1, j_2 \le n$. When constraints are present, we will encounter the Lagrangian function

$$\mathcal{L}(x,y) \stackrel{\text{def}}{=} f(x) - \sum_{i=1}^m y_i c_i(x), \tag{1.1.1}$$

along with its gradient

$$\nabla_x^1 \mathcal{L}(x,y) = \nabla_x^1 f(x) - \sum_{i=1}^m y_i \nabla_x^1 c_i(x) \equiv \nabla_x^1 f(x) - (\nabla_x^1 c(x))^T y$$

and its Hessian

$$\nabla_x^2 \mathcal{L}(x,y) = \nabla_x^2 f(x) - \sum_{i=1}^m y_i \nabla_x^2 c_i(x).$$

More generally, the pth derivative of f at x is denoted by $\nabla_x^p f(x)$ and is a cubical, symmetric tensor of order p whose components are $[\nabla_x^p f(x)]_{j_1,\ldots,j_p} = \partial^p f(x)/\partial x_{j_1}\ldots\partial x_{j_p}$ for $1 \le j_1,\ldots,j_p \le n$. For such a tensor, the product

$$\left[\nabla_x^p f(x)[v_1,\ldots,v_q]\right]_{j_{q+1},\ldots,j_p} \stackrel{\text{def}}{=} \sum_{j_1=1}^n \cdots \sum_{j_q=1}^n [\nabla_x^p f(x)]_{j_1,j_2,\ldots,j_p} [v_1]_{j_1} \cdots [v_q]_{j_q} \tag{1.1.2}$$

[1]We use the terms *methods* and *algorithms* interchangeably throughout, although computer science purists argue that an algorithm must terminate but a method might not.

is the result of applying $\nabla_x^p f(x)$ to the vectors $v_1, \ldots, v_q$, where $q \leq p$, for $1 \leq j_{q+1}, \ldots, j_p \leq n$, and is itself a cubical, symmetric tensor of dimension $p-q$. By convention, we write $\nabla_x^0 f(x) = f(x)$.

For a real, sufficiently differentiable univariate function (i.e., a function of a single variable) ϕ, $\phi^{(p)}$ denotes its pth derivative and $\phi^{(0)}$ is a synonym for ϕ.

1.1.2.2 Unconstrained problems

Unconstrained minimization will be the focus of much of this book. The aim is to find a point x_* that satisfies necessary local optimality conditions for the problem

$$\underset{x\in\mathbb{R}^n}{\text{minimize}} \quad f(x), \tag{1.1.3}$$

in which $f : \mathbb{R}^n \to \mathbb{R}$ is a continuous function. In the unconstrained case, both first- and second-order necessary and sufficient conditions will undoubtedly be familiar to our readers. The third-order necessary condition is less well known; indeed it is stated in terms of the third-order tensor product (1.1.2) for which

$$\nabla_x^3 f(x)[d]^3 \equiv \nabla_x^3 f(x)[d,d,d] = \sum_{i=1}^{n}\sum_{j=1}^{n}\sum_{k=1}^{n}[\nabla_x^3 f(x)]_{i,j,k}[d]_i[d]_j[d]_k$$

involving $\nabla_x^3 f(x_*)$; we will return to both this and to why even higher order conditions become increasingly computationally unwieldy in Section 12.1.

Theorem 1.1.1: Unconstrained optimality conditions

Let $f: \mathbb{R}^n \to \mathbb{R} \in C^1$, and suppose that $x_* \in \mathbb{R}^n$ is a local minimizer of f. Then

$$\nabla_x^1 f(x_*) = 0. \tag{1.1.4}$$

If $f \in C^2$, then $\nabla_x^2 f(x_*)$ is positive semidefinite, while if $f \in C^3$, $\nabla_x^3 f(x_*)[d]^3 = 0$ for all vectors d for which $\nabla_x^2 f(x_*)d = 0$. Conversely, if, for a given $x_* \in \mathbb{R}^n$, (1.1.4) holds and $\nabla_x^2 f(x_*)$ is positive definite, then x_* is a strict local minimizer of $f \in C^2$.

A *local minimizer* is a point x_* for which $f(x_*) \leq f(x)$ for all x in a small, open, feasible neighbourhood $\mathcal{N}$ of x_*. It is *strict*—sometimes called *strong*—if $f(x_*) < f(x)$ for all $x \neq x_* \in \mathcal{N}$. From a computational perspective, the first- and second-order necessary optimality conditions are often summarized as requiring that

$$\|\nabla_x^1 f(x_*)\| = 0 \quad \text{and} \quad \max\Big(0, -\lambda_{\min}[\nabla_x^2 f(x_*)]\Big) = 0, \tag{1.1.5}$$

where $\|\cdot\|$ is the norm corresponding to the Euclidean inner product $\langle\cdot,\cdot\rangle$ and $\lambda_{\min}$ is the leftmost eigenvalue[2] of its symmetric matrix argument (see Section A.5 of Appendix A).

1.1.2.3 Convexly constrained problems

A more realistic problem occurs when the minimization variables are restricted to a closed, convex set. A set $\mathcal{C}$ is *convex* if $x + t(y-x) \in \mathcal{C}$ for all $x, y \in \mathcal{C}$ and $t \in [0,1]$ and

[2]The second optimality condition in (1.1.5) is also frequently written as $\lambda_{\min}[\nabla_x^2 f(x_*)] \geq 0$.

closed if it contains all its limit points. The aim in convexly constrained minimization changes to finding a point x_* that satisfies necessary local optimality conditions for the problem

$$\underset{x \in \mathcal{C}}{\text{minimize}} \; f(x),$$

where $\mathcal{C} \subseteq \mathbb{R}^n$ is convex and closed.[3] Things are slightly more complicated when the unknowns are restricted to lie in a convex domain.

Theorem 1.1.2: Convexly constrained optimality conditions

Let $\mathcal{C}$ be a nonempty, closed, convex set; let $f : \mathbb{R}^n \to \mathbb{R} \in C^1(\mathcal{S})$, where the open set $\mathcal{S}$ contains $\mathcal{C}$; and suppose that x_* is a local minimizer of f within $\mathcal{C}$. Then

$$\langle d, \nabla^1_x f(x_*) \rangle \geq 0 \;\; \textit{for all} \;\; d \in \mathcal{C}(x_*), \tag{1.1.6}$$

where

$$\mathcal{C}(x) \stackrel{\text{def}}{=} \{ d \in \mathbb{R}^n \mid x + d \in \mathcal{C} \}. \tag{1.1.7}$$

If $f \in C^2(\mathcal{S})$, then

$$\langle d, \nabla^2_x f(x_*) d \rangle \geq 0 \;\; \textit{for all} \;\; d \in \mathcal{C}(x_*) \;\; \textit{for which} \;\; \langle d, \nabla^1_x f(x_*) \rangle = 0, \tag{1.1.8}$$

while if $f \in C^3(\mathcal{S})$, then

$$\begin{array}{c} \nabla^3_x f(x_*)[d]^3 \geq 0 \;\; \textit{for all} \;\; d \in \mathcal{C}(x_*) \;\; \textit{for which} \\ \langle d, \nabla^1_x f(x_*) \rangle = 0 \;\; \textit{and} \;\; \langle d, \nabla^2_x f(x_*) d \rangle = 0. \end{array} \tag{1.1.9}$$

Conversely, if, for a given $x_* \in \mathcal{C}$, (1.1.6) holds and

$$\langle d, \nabla^2_x f(x_*) d \rangle > 0 \;\; \textit{for all} \;\; d \neq 0 \in \mathcal{C}(x_*) \;\; \textit{for which} \;\; \langle d, \nabla^1_x f(x_*) \rangle = 0,$$

then x_* is a strict local minimizer of $f \in C^2(\mathcal{S})$.

Of course, Theorem 1.1.1 follows immediately from Theorem 1.1.2, since in that case $\mathcal{C}(x_*) = \mathbb{R}^n$. Conditions (1.1.6), (1.1.8), and (1.1.9) may superficially appear awkward to verify. However, it is clear that they are equivalent to

$$\underset{d \in \mathcal{C}(x_*),\; \|d\| \leq 1}{\text{minimum}} \; \langle d, \nabla^1_x f(x_*) \rangle = 0, \tag{1.1.10}$$

$$\underset{\substack{d \in \mathcal{C}(x_*),\; \|d\| \leq 1, \\ \langle d, \nabla^1_x f(x_*) \rangle = 0}}{\text{minimum}} \; \langle d, \nabla^2_x f(x_*) d \rangle = 0, \tag{1.1.11}$$

and

$$\underset{\substack{d \in \mathcal{C}(x_*),\; \|d\| \leq 1, \\ \langle d, \nabla^1_x f(x_*) \rangle = 0, \\ \langle d, \nabla^2_x f(x_*) d \rangle = 0}}{\text{minimum}} \; \nabla^3_x f(x_*)[d]^3 = 0, \tag{1.1.12}$$

where the extra restriction $\|d\| \leq 1$ simply ensures that the minima exist. The minimization problem in (1.1.10) is convex, and thus, in general, there are efficient methods to check this condition; indeed for simple convex sets such as simple bounds on the variables, checking is trivial. By contrast, the problems in (1.1.11) and (1.1.12) are nonconvex in

[3] The set $\mathcal{C}$ needs to be closed for the problem to be well defined.

general, and (1.1.11) is known to be (NP-)hard (see Section 1.2.3). A further useful alternative to (1.1.6) is that

$$\left\| P_{\mathcal{C}}[x_* - \nabla_x^1 f(x_*)] - x_* \right\| = 0,$$

where the projection $P_{\mathcal{C}}[x]$ of x into $\mathcal{C}$ is (uniquely)

$$P_{\mathcal{C}}[x] \stackrel{\text{def}}{=} \arg\min_{y \in \mathcal{C}} \|y - x\|_2.$$

It is important to notice that there is no requirement that $\mathcal{C}$ in the above be described algebraically, that is, as the intersection of a finite number of linear or convex inequality constraints. If it is, the conditions in the next two subsections may provide yet further alternatives.

1.1.2.4 Equality-constrained problems

Next in our hierarchy are problems whose constraints may be nonconvex. We focus first on the case where the feasible region is defined by a set of equations, and our task is to find a point x_* that satisfies necessary local optimality conditions for the equality-constrained minimization problem

$$\underset{x \in \mathbb{R}^n}{\text{minimize}} \;\; f(x) \;\; \text{subject to} \;\; c(x) = 0,$$

where $c : \mathbb{R}^n \to \mathbb{R}^m$ is a continuous vector-valued function.

There is a technical detail in what follows here and in the next subsection. This is that we require that the (boundary of the) feasible domain be sufficiently smooth to be represented locally by linear approximations. We call this a *constraint qualification*, and its effect is to rule out cusps and other nasty pathologies. We do not wish to go into detail here since there are seemingly hundreds of constraint qualifications of varying strength and sophistication, with more arising every year [280, 340]. Fortunately for many convex regions of interest, such as those that arise as the intersection of a finite number of linear equations or inequalities, the constraint qualification is automatic.

Theorem 1.1.3: Equality-constrained optimality conditions

Let $\mathcal{S}$ be an open set containing the feasible region $\{x \mid c(x) = 0\}$, let $f : \mathbb{R}^n \to \mathbb{R}$ and $c : \mathbb{R}^n \to \mathbb{R}^m \in C^1(\mathcal{S})$, suppose that $x_* \in \mathbb{R}^n$ is a local minimizer of $f(x)$ subject to $c(x) = 0$, and suppose that a constraint qualification is satisfied at x_*. Then there exist Lagrange multipliers $y_* \in \mathbb{R}^m$ for which

$$\nabla_x^1 \mathcal{L}(x_*, y_*) = 0 \;\; \text{and} \;\; c(x_*) = 0. \tag{1.1.13}$$

If $f, c \in C^2(\mathcal{S})$, then

$$\langle d, \nabla_x^2 \mathcal{L}(x_*, y_*) d \rangle \geq 0 \;\; \text{for all} \;\; d \;\; \text{such that} \;\; \nabla_x^1 c(x_*) d = 0. \tag{1.1.14}$$

Conversely, if, for given $x_* \in \mathbb{R}^n$, $y_* \in \mathbb{R}^m$, and $f, c \in C^2(\mathcal{S})$, (1.1.13) holds and

$$\langle d, \nabla_x^2 \mathcal{L}(x_*, y_*) d \rangle > 0 \;\; \text{for all} \;\; d \;\; \text{such that} \;\; \nabla_x^1 c(x_*) d = 0, \tag{1.1.15}$$

then x_* is a strict local minimizer of $f(x)$ subject to $c(x) = 0$.

Notice that although the conditions (1.1.14) and (1.1.15) might appear rather foreboding, they are fully equivalent to requiring that the leftmost eigenvalue of the "reduced" Hessian $N_*^T \nabla_x^2 \mathcal{L}(x_*, y_*) N_*$ be nonnegative and strictly positive, respectively, where N_* is any full-rank matrix whose columns span the nullspace of $J(x_*)$.

1.1.2.5 Inequality-constrained problems

The most general problem that we will examine involves potentially nonconvex inequality constraints, and for this we seek to find a point x_* that satisfies necessary local optimality conditions for the inequality-constrained minimization problem

$$\underset{x \in \mathbb{R}^n}{\text{minimize}} \;\; f(x) \;\; \text{subject to} \;\; c(x) \geq 0.$$

Notice that this formulation introduces a hitherto unseen combinatorial aspect, namely, that at any feasible point some of the constraints may be irrelevant. We define the *active set* at the feasible point x as

$$\mathcal{A}(x) \stackrel{\text{def}}{=} \{1 \leq i \leq m \mid c_i(x) = 0\}.$$

The set of constraints whose indices lie in the active set are said to be *active*, while the remainder are *inactive*.

Theorem 1.1.4: Inequality-constrained optimality conditions

Let $\mathcal{S}$ be an open set containing the feasible region $\{x \mid c(x) \geq 0\}$, let $f\colon \mathbb{R}^n \to \mathbb{R}$ and $c\colon \mathbb{R}^n \to \mathbb{R}^m \in C^1(\mathcal{S})$, suppose that $x_* \in \mathbb{R}^n$ is a local minimizer of $f(x)$ subject to $c(x) \geq 0$, and suppose that a constraint qualification is satisfied at x_*. Then there exist Lagrange multipliers $y_* \in \mathbb{R}^m$ for which

$$\nabla_x^1 \mathcal{L}(x_*, y_*) = 0, \;\; c(x_*) \geq 0, \;\; y_* \geq 0, \;\; \text{and} \;\; \langle y_*, c(x_*) \rangle = 0. \tag{1.1.16}$$

If $f, c \in C^2(\mathcal{S})$, then

$$\langle d, \nabla_x^2 \mathcal{L}(x_*, y_*) d \rangle \geq 0 \;\; \text{for all} \;\; d \in \mathcal{N}_+, \tag{1.1.17}$$

where

$$\mathcal{N}_+ = \left\{ d \in \mathbb{R}^n \;\middle|\; \begin{array}{l} \langle d, \nabla_x^1 c_i(x_*) \rangle = 0 \text{ for all } i \in \mathcal{A}(x_*) \text{ with } [y_*]_i > 0 \text{ and} \\ \langle d, \nabla_x^1 c_i(x_*) \rangle \geq 0 \text{ for all } i \in \mathcal{A}(x_*) \text{ with } [y_*]_i = 0 \end{array} \right\}.$$

Conversely, if, for given $x_* \in \mathbb{R}^n$, $y_* \in \mathbb{R}^m$, and $f, c \in C^2(\mathcal{S})$, (1.1.16) holds and

$$\langle d, \nabla_x^2 \mathcal{L}(x_*, y_*) d \rangle > 0 \;\; \text{for all} \;\; d \in \mathcal{N}_+, \tag{1.1.18}$$

then x_* is a strict local minimizer of $f(x)$ subject to $c(x) \geq 0$.

Unfortunately, unlike for the equality-constrained case, verifying (1.1.17) and (1.1.18) does not reduce to checking eigenvalues because of the constraints $\langle d, \nabla_x^1 c_i(x_*) \rangle \geq 0$, which must hold for all $i \in \mathcal{A}(x_*)$ when $[y_*]_i = 0$ in $\mathcal{N}_+$. The necessary condition (1.1.17) holds if

$$\underset{d \in \mathcal{N}_+, \, \|d\| \leq 1}{\text{minimum}} \; \langle d, \nabla_x^2 \mathcal{L}(x_*, y_*) d \rangle \geq 0, \tag{1.1.19}$$

but solving the quadratic program in (1.1.19) is regrettably, once again, (NP-)hard (see Section 1.2.3).

1.1.2.6 Criticality measures

Now that we have surveyed a variety of criticality conditions that must necessarily hold at local minimizers of a problem under consideration, our attention immediately turns to ways of recognizing them. A key concept is what we shall call criticality measures. A qth-order *criticality measure* $\omega_q(x)$ is a continuous, nonnegative function that is zero if x satisfies the qth-order criticality condition for the problem in hand. For example, for the differentiable unconstrained optimization problem (1.1.3), the measures

$$\omega_1(x) = \|\nabla_x f(x)\| \text{ and } \omega_2(x) = \max\Big(0, -\lambda_{\min}[\nabla_x^2 f(x)]\Big)$$

are suitable because of (1.1.5), and we shall see further examples[4] as the book progresses. Continuity is important because we would like to say that the relevant measure is small whenever we are close to a critical point.

1.1.3 Algorithms

Armed with our optimality conditions (and their associated measures), we soon recognize that we are unlikely to be able to satisfy them generally by analytic reasoning alone, and thus we have to resort to numerical solution. We proceed by iteration. Given any "arbitrary" estimate x_0 of a local minimizer, we design an algorithm that builds a sequence of, one hopes, improving estimates $\{x_k\}$, $k = 1, 2, \ldots$. The intention is that by theoretical means we can establish that a limit point of the generated sequence satisfies our optimality conditions, and that for some sufficiently large k, x_k is "close" to locally optimal. A typical method proceeds from x_k to x_{k+1} by finding a promising step s_k and then assessing whether $x_k + s_k$ is "better" than x_k. If it is, $x_{k+1} = x_k + s_k$, but otherwise $x_{k+1} = x_k$ and a refined step must be sought; variants are possible, and we will consider this in detail in Chapter 2 and beyond. To assess the quality of $x_k + s_k$ relative to x_k normally requires that we evaluate problem functions and/or derivatives at $x_k + s_k$ and compare these to those already evaluated at x_k.

Notes and references for Section 1.1

Optimality conditions and algorithms are covered in detail in many textbooks [19, 62, 147, 168, 307]. Criticality measures are an important ingredient in the convergence analysis of trust-region methods and other algorithms [131, Sect. 8.1 and what follows].

1.2 What is evaluation-complexity analysis?

1.2.1 Traditional convergence questions

Traditionally, the analysis of algorithms for nonconvex optimization has focused on two aspects, global and asymptotic convergence. Starting from an arbitrary point, does the algorithm converge to a point, or points, at which necessary optimality conditions hold under sensible, stated assumptions, and, if so, ultimately how fast is this convergence? The former is a reasonable requirement of anything worthy of being named an algorithm for its problem, and given competing convergent methods, the latter is often used to distinguish and rank them.

[4] With different notation (other than $\omega_q(x)$) for optimality measures.

Such a taxonomy is undoubtedly crude for a number of reasons. To start, different algorithms have different requirements—some may require second derivatives of the defining problem functions, while others might work with function values alone—and consequently might operate under different assumptions. Secondly, what constitutes an iteration can differ enormously. An iteration for a method that relies on function and gradient values alone typically performs vector operations, while those that use Hessians often require the solution of linear systems. A third consideration is that most modern algorithms rely to some extent on parallelism, either explicitly or via compiler code optimization. Finally, many problems exhibit considerable external (or perhaps hidden, internal) structure, and efficiency will depend to a large degree on how this is exploited, if at all.

1.2.2 Worst-case complexity

A rival convergence preoccupation has arisen more recently which, to a certain extent, subsumes global and asymptotic convergence questions. Simplistically, this considers a given algorithm, with prespecified termination conditions, applied to a generic problem class, and asks the question: given any instance of the problem in its class, what is the worst that the algorithm might do? This is often referred to as *worst-case complexity*. Of course, this leads us to the question of how we measure "worst."

Broadly, we might consider two possibilities. The first is to say that any iterative algorithm comprises a sequence of computations, and thus we might consider the largest number of computations that might be required; the far-from-trivial question of what a computation is then needs to be addressed. We refer to this as *computational complexity*, and we give a brief overview of this in the next section.

Computational complexity is fine for problems for which the data is readily available, but what of problems that involve the evaluation of functions and derivatives for which the algorithm has no control and cannot know the associated computational costs? This is the domain of *evaluation complexity*,[5] and the central question here addresses, as the name suggests, how many evaluations of the functions (and their derivatives if needed) that define the problem are required before the algorithm terminates. Notice that this measure of complexity implicitly assumes that evaluations are the dominant cost. For some problems, for example, those for which an evaluation might require an expensive simulation, this may be justified, while for others, ignoring other costs in the iteration is perhaps unwarranted.

The focus of this book is on evaluation complexity, and, while recognizing the defects mentioned above, we will make no further apologies for its possible shortcomings as we proceed. We should also mention that there is some interest in *average-case complexity*, in which the question relates to how algorithms behave on "average" problems from their class. Although such an analysis may be realistic for some problems that are amenable to computational-complexity analysis, it is harder to say what an average nonlinear problem might be, and thus we say no more here.

1.2.3 Worst-case computational complexity

> Very roughly,[6] a *polynomial* algorithm for solving a problem is one in which the worst-case running time can be shown to be a polynomial function of

[5] This is also frequently referred to as *oracle complexity*.

[6] This section is shamelessly lifted from the second and third authors' previous book, *Trust-Region Methods* [131, p. 171], since we couldn't think of a better way to describe what was written there.

the size of the data. For instance, there are polynomial algorithms for solving systems of linear equations and for solving linear and convex quadratic optimization problems. At the other extreme, some algorithms have a worst-case running time that is an exponential (or worse) function of the data; the simplex method for linear programming with all common pivoting rules is exponential. Between these two extremes lies a large class of problems that are called NP. The class *NP* belongs to the set of what are called decision problems. *Decision* problems are problems for which an answer "yes" or "no" must be generated. (For instance, given any A and b, we might be asked whether there is an x for which $Ax \leq b$.) The class NP comprises precisely those decision problems for which a "yes" answer for a particular problem instance may be checked in polynomial time (e.g., given that we are told that the particular system $Ax \leq b$ has a solution, we need to verify this in a polynomial number of steps). While the class P of all decision problems that have polynomial solution algorithms lies in NP, it is not known if the two classes are the same. A subset of NP, the set of *NP-complete* problems, are those decision problems for which any other problem in NP may be transformed to the given problem by a polynomial-time algorithm. The class NP-complete contains such famous "hard" problems as integer linear programming and the traveling salesperson problem. Significantly, unless P = NP, no problem in NP-complete is solvable in polynomial time. Not every problem lies in NP, but many of them have the property of all NP problems that they may be reduced to the given problem by a polynomial-time algorithm. These problems are known as *NP-hard* problems. The example of most interest to us is the nonconvex quadratic programming problem, instances of which we have seen in (1.1.11) and (1.1.19). Of course, all NP-complete problems are NP-hard, and in fact there is a decision variant of the nonconvex quadratic programming problem that lies in NP-complete.

A few words of caution may be necessary. Just because an algorithm is polynomial does not mean that it is an effective method in practice (e.g., the ellipsoidal method for linear programming is polynomial but extremely poor in almost every case), nor does it mean that an exponential algorithm is bad except in the worst case (the simplex method for linear programming has an excellent average-case behaviour). Thus, although complexity theory plays an important role in the design of algorithms, it should not be regarded as the only goal. Just because a problem is NP-hard does not mean that people do not want to solve it but merely implies that our algorithm will typically rely on heuristics and may behave badly in the (fortunately infrequent) worst case.

1.2.4 Worst-case evaluation complexity

As we have already seen, an algorithm for general nonlinear optimization seeks an improving sequence of iterates $\{x_k\}_{k\geq 0}$ by evaluating problem functions (and possibly derivatives) at each x_k and computing the next x_{k+1} on the basis of such information. If we make no assumption about the problem under consideration aside from its generic class (unconstrained, convexly constrained, etc.), the information obtained at x_k is completely local and predicts nothing about the problem in any neighbourhood. Thus, inevitably we need to assume that the problem functions we are considering possess additional properties, such as continuity or differentiability, so that information obtained at x_k is able to predict what might happen at x_{k+1}.

The evaluation of problem functions at a given x_k is frequently viewed as consulting a (black-box) *oracle*. The oracle only returns pointwise information and tells nothing globally about the problem. It is a *pth-degree oracle* if it provides derivatives of order p; a zeroth-degree oracle provides function (and possibly constraint) values alone, while first- and second-order oracles return gradients (or Jacobians) and Hessians, respectively. It is a *pth-degree-product oracle* if it gives the derivative-tensor product (1.1.2) for a given sufficiently differentiable function f, the slope $\langle s, \nabla_x^1 f(x) \rangle$ and Hessian-vector product $\nabla_x^2 f(x) s$ for given s being commonly used examples. A *pth-order algorithm* is one that uses oracles (or product oracles) of degrees up to p, but note that not every iteration necessarily uses each order.

Given our algorithm, we now consider one or more conditions—or *goals*—that we would like the iterates to satisfy simultaneously. These can be explicit—for example, when is the norm of the gradient first smaller than 0.1?—or implicit—when is the objective function within a tenth of its (unknown) optimal value? Our quest is to determine how many evaluations are required to guarantee that the desired goals are fulfilled by our algorithm. But, since we haven't specified the problem functions, we are concerned with those problems that make this bound as large as possible.

We have to be careful. In order to derive an evaluation-complexity bound for our algorithm, the particular line of analysis that we employ may be unduly pessimistic; inevitably, as we will see, such analyses rely on a sequence of inequalities, some of which may not be sharp, and perhaps a different approach might have produced a better bound. Thus, it is equally useful, if possible, to demonstrate an example for which the predicted worst-case bound is close to attained (modulo less important constants).

This, then, is the theme of our book. For a given algorithm designed for a specific problem class and with a series of desired goals, on the one hand, we aim to find a bound on the number of evaluations needed simultaneously to satisfy all of the goals, but, on the other, we aim to provide an example that matches as well as possible the proposed worst-case evaluation bound. This is, of course, an algorithm-centric question.

Throughout the book, we will denote this needed number of evaluations of the problem functions by $\#^{\mathrm{F}}_{\mathrm{algo}}$ (note the superscript F) and the associated number of evaluations of their derivatives by $\#^{\mathrm{D}}_{\mathrm{algo}}$ (note the superscript D).

A related, equally vital question relates to the algorithm itself. Namely, is the proposed pth-order algorithm itself suboptimal among all pth-order algorithms designed for the class of problems under consideration? Indeed, what is the best pth-order algorithm? The answer depends, of course, on what the stated goals are. We have surprises in store here!

In some simple cases in which the internal cost of an iteration is constant, evaluation complexity is equivalent to *iteration complexity*, for which the aim is to count the number of iterations, denoted by $\#^{\mathrm{I}}_{\mathrm{algo}}$, required to satisfy the stated goals. The bounds that we shall derive will normally count iterations indirectly but more generally separate the counts for each of the oracles consulted.

We should mention one issue that is frequently addressed in computational complexity but generally ignored in the evaluation case. That is, we tacitly assumed that the computations in the algorithms we examine are performed perfectly. This is certainly not the case for finite-precision floating-point computation—even carefully rounded results vary slightly from their exact counterparts. This is a limitation on most of what we have to say. Nonetheless, it is an aspiration for future analyses.

One other issue of note is the relationship between global convergence of an algorithm and its evaluation- or iteration-complexity analysis. As we have already hinted, the latter asks the question of how many evaluations or iterations are needed to achieve one or more predefined goals. This is usually stated in terms of quantities that have to be zero for us to judge the algorithm to have converged globally. In evaluation- or iteration-complexity analysis, we ask instead how many evaluations/iterations are needed to ensure that these quantities are small, or more precisely when they are all smaller than their particular specified (small) tolerances $\epsilon > 0$. So long as the algorithm itself does not depend on ϵ, global convergence is then a consequence of a finite complexity bound for a given ϵ, since we may then simply pick a sequence of such ϵ's whose limit is zero and we know that the algorithm will terminate finitely for each element of the ϵ sequence. This may not be the case for algorithms that depend on ϵ. However, it is still possible for this last class of algorithms to use a "meta-algorithm" consisting in running the algorithm for a succession of smaller and smaller ϵ's, each time starting the new minimization where the previous one terminated. While this defines a satisfyingly convergent process, it is of a slightly different nature. We will encounter algorithms of both types in this book.

All the algorithms considered in this book are deterministic. This is a deliberate choice made to circumscribe a manageable topic but does not in any way deny the real interest, both practical and theoretical, of methods involving stochastic elements such as sampling strategies.

1.2.5 Worst-case evaluation complexity for convex optimization

Although convex minimization is not the focus of our book, it has had a profound effect on the development of the theory that underpins general evaluation complexity. In this section we briefly summarize the topic and some of its major results.

A real-valued function f is *convex* over a convex domain $\mathcal{C}$ if

$$f(\alpha x + (1-\alpha)y) \leq \alpha f(x) + (1-\alpha) f(y)$$

for all $x, y \in \mathcal{C}$ and $\alpha \in [0,1]$. Any local minimizer of such a function within $\mathcal{C}$ is also a global minimizer. If additionally $f \in C^1(\mathcal{S})$ for some open set $\mathcal{S} \supseteq \mathcal{C}$, then

$$f(x) \geq f(y) + \langle x - y, \nabla_x^1 f(y) \rangle \tag{1.2.1}$$

for all $x, y \in \mathcal{C}$, while if $f \in C^2(\mathcal{S})$, $H(x)$ is positive semidefinite over the interior of $\mathcal{C}$. Conversely, if the Hessian of f is positive semidefinite over the interior of $\mathcal{C}$, then f is convex there. The convex function $f \in C^1(\mathcal{S})$ is *strongly* convex if

$$f(x) \geq f(y) + \langle x - y, \nabla_x^1 f(y) \rangle + \tfrac{1}{2}\lambda_{\rm L} \|x - y\|^2 \tag{1.2.2}$$

for some $\lambda_{\rm L} > 0$ and all $x, y \in \mathcal{C}$. Such a function has a unique global minimizer within $\mathcal{C}$, and, if additionally $C^2(\mathcal{S})$, all the eigenvalues of the Hessian of f are at least $\lambda_{\rm L}$ (and thus the Hessian is positive definite). If all the eigenvalues of the Hessian of f are at least $\lambda_{\rm L} > 0$ over the interior of $\mathcal{C}$, then f is strongly convex there.

The aim of convex minimization is to find the smallest value of a convex objective function $f : \mathcal{S} \to \mathbb{R}$ within a closed, convex set $\mathcal{C} \subseteq \mathcal{S}$. Descent-based algorithms for the problem are designed to construct a sequence of iterates $\{x_k\}_{k\geq 0} \in \mathcal{C}$ such that $f(x_{k+1}) \leq f(x_k)$, $k \geq 0$, and for which

$$\lim_{k\to\infty} f(x_k) = f(x_*),$$

where x_* is a global minimizer. Given $\epsilon > 0$, such algorithms will be deemed to have succeeded as soon as the gap

$$\Gamma_k \stackrel{\text{def}}{=} f(x_k) - f(x_*) \leq \epsilon, \tag{1.2.3}$$

and the goal is to determine how large k might be as a function of ϵ, and other problem-specific data and algorithmic parameters, in the worst case. Notice that (1.2.3) may not be a practical termination test since $f(x_*)$ is generally unknown; we shall return to this shortly.

Throughout our book, we adopt a common, very convenient shorthand. Suppose that a and b are positive real numbers. Then we declare that the quantities $a = \mathcal{O}(b)$ if $a \leq \kappa b$ for some $\kappa > 0$, and say informally that a is of *order* b. In addition, b and c are the *same order* if both $b = \mathcal{O}(c)$ and $c = \mathcal{O}(b)$, i.e., they differ by a multiplicative constant. We also say that $a = o(b)$ if $\lim_{b\to 0} a/b = 0$ [246, Thm. 3.3.18]. At various stages in our development, we shall also talk of the pth derivative of a generic (convex or otherwise) function $h \in \mathcal{C}^p(\mathcal{S})$ being *Lipschitz continuous* over the convex set $\mathcal{C}$ within the open domain $\mathcal{S}$. By this we mean that

$$\|\nabla_x^p h(x) - \nabla^p h_x(y)\| \leq L_{h,p}\|x - y\|$$

for some *Lipschitz constant* $L_{h,p} \geq 0$ and all $x, y \in \mathcal{C}$. In many places we shall actually assume that $L_{h,p} \geq 1$, but since if this is not the case we can simply replace $L_{h,p}$ by 1 there is no loss of generality in doing so, except perhaps exceptionally when $\nabla_x^p h(x)$ is constant throughout $\mathcal{C}$.

It is known [299, Thm. 2.1.6] that there is a convex C^1 function f with a Lipschitz-continuous gradient $\nabla_x^1 f$ for which any first-order algorithm whose iterate x_k lies in the space spanned by the initial point x_0 and the k gradients $\{\nabla_x^1 f(x_i)\}_{i=0}^{k-1}$ requires at least $\mathcal{O}(\epsilon^{-1/2})$ evaluations of $\nabla_x^1 f$ to guarantee (1.2.3); the constant hidden by the order notation involves the gradient Lipschitz constant and an upper bound on the distance from x_0 to the (set of) global minimizer(s). Although many well-known first-order algorithms for unconstrained minimization, including the steepest-descent method with a variety of stepsizes (see Chapter 2), can be shown to require at most $\mathcal{O}(\epsilon^{-1})$ evaluations [299, Thm. 2.1.14], there are so-called optimal-gradient methods that match the best possible $\mathcal{O}(\epsilon^{-1/2})$ bound [299, Sect. 2.2]. Thus, remarkably in the convex case, the case is closed for first-order algorithms (aside possibly from improving noncrucial constants). If higher-order derivatives are available, things may improve. In particular, there is a convex C^p function with pth-order Lipschitz derivative for which any pth-order algorithm requires at least $\mathcal{O}(\epsilon^{-2/(3p+1)})$ calls to its oracle to ensure (1.2.3) [11]. This is significant since for $p = 2$, there is once again an optimal algorithm [289] that requires at most $\mathcal{O}(\epsilon^{-2/7})$ oracle calls, thus matching the indicated lower evaluation-complexity bound. It is now known that there are pth-order optimal algorithms that generalize [289], for which the same is true for $p > 2$ [65, 176, 177, 250].

Things improve significantly in the strongly convex case. For this, an example with a Lipschitz-continuous gradient is known to require at least $\mathcal{O}(|\log \epsilon^{-1}|)$ evaluations of $\nabla_x^1 f$ to guarantee (1.2.3) and additionally $\|x_k - x_*\|^2 \leq \epsilon$ [299, Thm. 2.1.12]. Now the order notation additionally hides a dependence on λ_{L}. Fortunately, optimal first-order methods (but now using what are referred to as gradient maps to cope with any convex constraints) match the order of this lower bound [299, Thm. 2.2.8]. For C^2 Lipschitz functions, there is an example for which an improved $\mathcal{O}(|\log|\log \epsilon^{-1}||)$ evaluations are necessary, and there is a matching optimal algorithm [289] for which this bound also suffices [11]. Similar bounds are also possible for C^p Lipschitz strongly convex functions and generalizations of these [176].

Finally, we comment on a computationally viable alternative to the stopping rule (1.2.3). We know from Theorem 1.1.2 that (1.1.10) must hold at any minimizer x_*, and thus it is reasonable to aim for the first iterate x_k for which

$$\underset{d \in \mathcal{C}(x_k),\ \|d\| \leq 1}{\text{minimum}} \langle d, \nabla_x^1 f(x_k) \rangle \geq -\epsilon, \tag{1.2.4}$$

where $\mathcal{C}(x)$ is defined as in (1.1.7). Since $x_* \in \mathcal{C}(x_k)$ and we are considering algorithms that ensure $x_k \to x_*$, eventually $\|x_k - x_*\| \leq 1$. Once this occurs, we have

$$\langle x_* - x_k, \nabla_x^1 f(x_k) \rangle \geq \underset{d \in \mathcal{C}(x_k),\ \|d\| \leq 1}{\text{minimum}} \langle d, \nabla_x^1 f(x_k) \rangle,$$

and, combining this with (1.2.1) and (1.2.4), we find that

$$\Gamma_k = f(x_k) - f(x_*) \leq -\langle x_* - x_k, \nabla_x^1 f(x_k) \rangle \leq \epsilon,$$

that is, that (1.2.3) holds. The advantage of using (1.2.4) is, of course, that the minimization is computed at x_k and, unlike (1.2.3), does not require knowledge of $f(x_*)$. Notice, in the unconstrained case in which $\mathcal{C}(x_k) = \mathbb{R}^n$, that (1.2.4) reduces to requiring that

$$\|\nabla_x^1 f(x_k)\| \leq \epsilon. \tag{1.2.5}$$

In the worst case, it can be shown that most sensible first-order methods require at most $\mathcal{O}(\epsilon^{-1})$ gradient evaluations to satisfy (1.2.5) for convex f but that there are "optimal" gradient methods for which the bound drops to $\mathcal{O}(\epsilon^{-1/2}|\log \epsilon^{-1}|)$ evaluations and to $\mathcal{O}(\epsilon^{-1/3}|\log \epsilon^{-1}|)$ Hessian and gradient evaluations for some second-order methods [301].

1.2.6 Entering the nonconvex jungle...

This then leads to our main topic, minimization problems for which the objective or the constraints (if any) are nonconvex, i.e., not convex. We have seen a lower bound on the worst-case evaluation complexity of $\mathcal{O}(\epsilon^{-1/2})$ evaluations for convex problems to find a point that satisfies (1.2.3), and this then must apply to the more general nonconvex case. We have also mentioned that there are "optimal" algorithms that match this lower bound, but these make considerable use of convexity. More importantly, nonconvex problems may have multiple local minima, not to mention critical points of varying order. We are not trying to find the global minimizer; indeed it can be shown [299, Thm. 1.1.1] that to do so may require $\mathcal{O}(\epsilon^{-n})$ evaluations to satisfy (1.2.3) for C^0 Lipschitz functions within a bounded domain using any algorithm, which is a truly fearsome prospect. Things may be almost as bad for C^p Lipschitz functions within a compact domain, where the lower bound "improves" to $\mathcal{O}(\epsilon^{-n/p})$ evaluations [247], and even if the function has no nonglobal critical points, it may nonetheless require $\mathcal{O}(\epsilon^{-(n-1)/p})$ evaluations [296, Sect. 1.6.2].

Now that we have explored the background and described our benchmarks, we are just about ready to explore the nonconvex world. Before we do, we need to make a comment about the influence of problem dimension. Without a doubt, dimension is a crucial aspect of computational complexity—typical results state, for example, that $\mathcal{O}(n^p)$ operations are required, where n is an intrinsic size of the problem. For evaluation complexity, things are less clear. If a problem depends on n independent[7] minimization variables, it

[7] Taken in a very broad sense—we mean here that the value of each individual variable must be accounted for when computing the function value.

is inconceivable that an evaluation will require fewer than n operations, but this is not the concern of evaluation complexity, since it is only the number of evaluations that count. It is certainly the case that dimension may appear, as it did in the awful $\mathcal{O}(\epsilon^{-n/p})$ bounds in the previous paragraph. Nonetheless, for many of the bounds we shall encounter later, using higher-order algorithms for problems with more smoothness, the effects of dimension are less transparent; in cases where dimension dependence is explicit, we will highlight it. Typically, though, these bounds will depend on the maximum possible decrease in an objective function, various algorithmic parameters, and problem Lipschitz constants. Although Lipschitz constants are usually benign, occasionally dimensionality lurks, sometimes devastatingly. For example, consider the function

$$f(x) = \left(\sum_{i=1}^{n} x_i\right)^{\exp(n)}$$

of n variables, whose gradient is

$$\nabla_x^1 f(x) = \exp(n) \left(\sum_{i=1}^{n} x_i\right)^{\exp(n)-1} e,$$

where exp is the exponential function and e is the vector of ones. But then, for instance,

$$\nabla_x^1 f(e_1) - \nabla_x^1 f(0) = \exp(n) e,$$

where e_1 is the first coordinate vector, has Euclidean norm $\exp(n)\sqrt{n}$, while $\|e_1 - 0\| = 1$. Thus the Euclidean-norm Lipschitz constant of the gradient of f is at least $\exp(n)\sqrt{n}$ on the Euclidean unit ball. The existence of such examples is why we refrain from calling complexity bounds in which Lipschitz constants appear but the dimension n does not appear explicitly "dimensionless," since we feel this could mislead the reader in view of the potential hidden dependence of the latter on the former.

Notes and references for Section 1.2

The seminal books by Nemirovsky and Yudin [296] and Nesterov [299, 303] provide a wealth of fascinating insights into the convex world. The introductory chapters in [296,299] also offer many thought-provoking comments relating to the nonconvex case and are thoroughly recommended. Excellent introductions to computational complexity, P versus NP, and related topics are given by Garey and Johnson [173], Papadimitriou and Steiglitz [313], and Vavasis [358].

1.3 Commonly used concepts and notation

Our intention in this book is to introduce concepts and notation as we need them and to delegate many details to appendices. Indeed, we assume readers have some familiarity with real analysis and linear and multilinear algebra, but they may wish to consult Appendix A to get a feel for the background required. Nonetheless, because some notation is so central to what we shall say, we briefly summarize this aspect before we get started.

The ith component of a vector v will be written as $[v]_i$, or more succinctly v_i, so long as no conflict with iteration subscripts occurs. The i,jth entry of a matrix A is $[A]_{i,j}$, or more concisely $a_{i,j}$. We let the $n \times n$ identity matrix be I_n, or more simply I when the dimension is clear, and denote its ith column by e_i. We strive to denote scalars by lowercase Greek (α) letters, vectors by lowercase Roman letters (x), matrices and tensors by capital roman letters (H), and sets and their boundaries by capital calligraphic letters ($\mathcal{S}$, $\partial\mathcal{S}$), but there will be exceptions to conform to well-established practice.

The natural numbers with and without 0 will be denoted by $\mathbb{N}^0$ and $\mathbb{N}$, respectively. We shall use the shorthand $[m : n]$ for the set of integers $\{m, \ldots, n\}$ between m and n, while $[m : n)$ is the same set without n. The set of real numbers is $\mathbb{R}$, and the real coordinate space of n dimensions is $\mathbb{R}^n$.

Unless otherwise specified, we will always use the Euclidean inner product $\langle u, v \rangle \stackrel{\text{def}}{=} \sum_{i=1}^n u_i v_i$ and its associated vector norm $\|v\| = \|v\|_2 \stackrel{\text{def}}{=} \sqrt{\langle v, v \rangle}$ for given vectors $u, v \in \mathbb{R}^n$, along with the subordinate matrix norm $\|A\| = \|A\|_2 \stackrel{\text{def}}{=} \sigma_{\max}(A)$, the largest singular value of the matrix A.

We will adopt the convention that for a function of interest $r(x)$ that we wish to evaluate at an iterate x_k, we will write $r_k \stackrel{\text{def}}{=} r(x_k)$. Thus, for example,

$$f_k \stackrel{\text{def}}{=} f(x_k) \quad \text{and} \quad c_k \stackrel{\text{def}}{=} c(x_k).$$

Finally, $\lfloor a \rfloor$ and $\lceil a \rceil$ are the lower and upper integer roundings of $a \in \mathbb{R}$, respectively, while the indicator function $\delta_{[\text{expression}]}$ takes the value 1 if the logical expression is true and 0 if it is false.

Additional notation and background material are provided in the first sections of Appendix A (p. 429).

1.4 Assumptions and organization

We shall use a variety of assumptions on the problem functions involved as well as on components of the algorithms under study. Since we believe the reader might appreciate not having to flip pages and return to the complete assumption statement every time it is mentioned, we have tried to denote the assumptions by symbols that are mnemonic of what is meant. Assumptions will therefore be denoted by a two- or three-part string.

1. The first part indicates the object the assumption is about. It can be a symbol referring to a function (**f**) or a set ($\mathcal{C}$).

2. The second part indicates what is assumed about the first part. When considering functions, we will typically assume some smoothness and will denote it using strings of the form **DpL** or **DqpL** indicating that the derivatives of order q (or 1) to p are Lipschitz continuous. Other strings may be used, but we will always attempt to provide a mnemonic for describing what is assumed.

3. The optional third part indicates where the assumption is supposed to hold, typically the neighbourhood of some set $\mathcal{S}$ like the feasible set of constrained problems. When the third part is not specified, the assumption is meant to apply on the whole of $\mathbb{R}^n$.

For example, assumption **f.D1pL.F** says that the function f has Lipschitz-continuous derivatives of orders one to p in a neighbourhood (made more explicit in the full statement of the assumption) of the set $\mathcal{F}$.

An important comment must be made here. In our theoretical derivations, we tend to make global assumptions that do not depend on the sequence of iterates generated by the algorithm under study—we assume, for instance, global Lipschitz continuity of first and/or second derivatives. Our choice to do so is based on two considerations.

The first is that algorithm-dependent assumptions are much more difficult to verify (or, if an a priori perspective is considered, to anticipate). The second is to avoid undue technicality in the discussions themselves. We are fully aware, however, that such global assumptions may very often be significantly weakened without qualitatively affecting the complexity results obtained by restricting the assumptions to a suitable neighbourhood of the tree/path of iterates generated by the iterates of the algorithm under study. For example, we could have assumed that the gradient is Lipschitz continuous merely on the path of successive search directions of a linesearch-based algorithm. Thus, provided that the iterates remain in parts of $\mathbb{R}^n$ where the relevant derivatives exist or behave well, the complexity results we establish under global assumptions can nevertheless be expected to hold despite the failure of these global assumptions. A typical but important example is the minimization of polynomials of degree higher than two, where one expects the derivatives of degree higher than two to grow at infinity and therefore global Lipschitz continuity of the derivatives of degree one and higher to fail. However, the complexity results obtained under global assumptions nevertheless apply if the iterates remain in a bounded domain, as is often the case, particularly if the objective function has bounded level sets.

Our intention is to state and prove central results as we proceed. Some proofs will occur in the main text, while others, particularly the more technical ones that might interrupt the flow of our argument, will be delegated to Appendix B; this appendix is laid out to refer to the chapters in which the results in the main text are stated. Since we do not wish to overload our readers, some less-central results will be given without proof, but we shall always endeavour to provide pointers to the literature. As we have already mentioned, some assumptions we use will be weaker than absolutely necessary, and this is part of a more general strategy of aiming to privilege clarity of exposition at the expense of sometimes looser choices of constants in the complexity bounds.

The book is divided into two main parts. In the first, we focus on algorithms for optimization where one seeks a low-order (i.e., first-, second-, or third-order) approximate minimizer. We start by considering unconstrained problems. Although we accept that such problems arise less frequently than those involving constraints in practice, nonlinear fitting notwithstanding, this choice is deliberate since it allows us to introduce basic ideas within a simple framework. In Chapters 2 to 4, we examine what one might call "traditional" evaluation complexity, because it highlights methods that are in use today, as well as their ability to find first-, second-, and even third-order critical points. Chapter 5 considers how problem structure, such as least-norm fitting, partial separability, or measure domination, can be incorporated in the same framework. We then apply the lessons learned to more practical problems. These include problems involving convex (Chapter 6) and, later, general constraints (Chapter 7); we distinguish the two because the former frequently do not require an oracle, while the latter will. We conclude our analysis of algorithms designed to find low-order approximate minimizers in Chapters 8 to 11 with a discussion on methods for solving the subproblems arising in the course of the various minimization algorithms. We show that, at least for first- and second-degree algorithms, these subproblems can be solved efficiently.

Things become more adventurous in the second part of the book, where we consider both higher-order models and higher-order criticality. The latter is especially interesting, since it leads us to reexamine what a critical point is. We start by considering unconstrained problems (Chapter 12), and subsequently turn (Chapter 13) to discussing extensions that permit inexact function and derivative values. Finally, Chapter 14 considers both composite and (simple and general) constrained problems in the context of high-order models and higher-order criticality.

Part I

Evaluation Complexity for Low-Order Optimality

The first part of our book concerns the verification of approximate criticality of orders one to three. This somewhat traditional point of view will appeal to many practitioners, but as we see later we shall aspire to grander things.

Previous page: From the combook 15, by Philippe Toint, 2020

Part IA

Evaluation Complexity for Unconstrained Problems

We start by considering problems without constraints. This setting has the advantage of simplifying exposition while still covering an important part of numerical optimization in practice.

Previous page: Nonconvex seascape, Tromvik, Norway, by Nick Gould, 2005

Chapter 2

Methods Using First Derivatives

Chapter summary

This chapter explores the worst-case complexity of obtaining an ϵ-approximate first-order minimizer of a nonconvex, unconstrained optimization problem, that is, a point at which the norm of the gradient of the objective function is below a specified threshold ϵ. Naturally this presupposes that the gradient exists, but, as we see first, this alone is not strong enough to derive useful complexity results. Consequently, we strengthen our assumptions to require that the gradient be Lipschitz continuous. This then enables us to investigate the worst-case complexity of three well-known classes of first-order algorithms: linesearch methods, trust-region schemes, and adaptive-regularization algorithms. We show that to obtain an ϵ-approximate first-order minimizer with first-order algorithms, that is, algorithms that only use oracles of degree at most one, requires at most $\mathcal{O}(\epsilon^{-2})$ evaluations of the objective function and its gradient, irrespective of the class of algorithms considered. Moreover, we show that such a bound is sharp for functions that do not have differentiable gradients.

2.1 A requirement

Having surveyed the principle of evaluation complexity in Section 1.2, we start our investigations by considering optimization methods applied to the general smooth (possibly nonconvex) unconstrained problem

$$\underset{x \in \mathbb{R}^n}{\text{minimize}} \quad f(x), \tag{2.1.1}$$

where f is a real-valued objective function from $\mathbb{R}^n$ into $\mathbb{R}$. It is natural to consider the widest possible class of such optimization problems for which first derivatives might be used, that is, those problems whose objective functions satisfy the following requirements.

f.Bb

$f(x)$ is bounded below on $\mathbb{R}^n$; that is, there exists a constant f_{low} such that, for all $x \in \mathbb{R}^n$,

$$f(x) \geq f_{\text{low}}.$$

f.C1 $f(x)$ is continuously differentiable in $\mathbb{R}^n$.

Since the second of these assumptions says that our objective function has a continuous gradient, our goal in this chapter is to design algorithms to find *ϵ-approximate first-order minimizers*, that is, points x_ϵ for which

$$\|\nabla_x^1 f(x_\epsilon)\| \leq \epsilon \tag{2.1.2}$$

for a prespecified $\epsilon > 0$, and to characterize the evaluation complexity of doing so. Notice the terminology: "first-order" reflects the fact that the test involves the gradient but no higher-order derivatives, "ϵ-approximate" indicates the required accuracy of the approximation, and "minimizer" relates to Theorem 1.1.1, for which necessarily $\nabla_x^1 f(x_*) = 0$ at a minimizer. Of course (2.1.2) will also hold in the neighbourhood of other critical points of f, and to distinguish these we will need higher-order derivatives.

Unfortunately, it turns out that the assumptions above are too weak for worst-case analysis because it is easy to construct, for any algorithm, a one-dimensional example on which the algorithm will fail in any prescribed number of iterations.

Example 2.1.1 To see this, consider any finite sequence of test points $x_0, \ldots, x_m$, and assume, without loss of generality, that they are reordered by increasing value and that $x_{k+1} \leq x_k + 1$ (this last property can easily be ensured by introducing additional intermediate test points if necessary). We then build a continuously differentiable function such that

$$f(x_k) = 0 \quad \text{for all} \quad k \geq 0 \tag{2.1.3}$$

and

$$f'(x_0) = -1 \quad \text{and} \quad f'(x_k) = 1 \quad \text{for all} \quad k \geq 1 \tag{2.1.4}$$

using Hermite interpolation (see Appendix A.9) with polynomials of degree three of the form $p_k(t) = c_0 + c_1 t + c_2 t^2 + c_3 t^3$ on each interval $[x_k, x_{k+1}]$, where $t \in [0, T_k]$ with $T_k = x_{k+1} - x_k \leq 1$. It is then not difficult to verify that the conditions (2.1.3)–(2.1.4) give

$$c_0 = 0, \quad c_1 = 1, \quad c_2 = -3/T_k, \quad \text{and} \quad c_3 = 2/T_k^2$$

for $k \geq 0$, and thus

$$p_k(t) = t - 3t^2/T_k + 2t^3/T_k^2$$

for $t \in [0, T_k]$, while, for $k = 0$,

$$c_0 = 0, \quad c_1 = 1, \quad c_2 = 1/T_k, \quad \text{and} \quad c_3 = 0,$$

which gives that $p_0(t) = -t + t^2/T_k$ for $t \in [0, T_0]$. Using these elements, we may now construct

$$f(x) = \begin{cases} 0 & \text{for } x \leq x_0 - 1, \\ (x - x_0 + 1)^2 - (x - x_0 + 1)^3 & \text{for } x \in [x_0 - 1, x_0], \\ p_k(x - x_k) & \text{for } x \in [x_k, x_{k+1}] \\ & \quad (k = 0, \ldots, m-1), \\ (x - x_m) - 2(x - x_m)^2 + (x - x_m)^3 & \text{for } x \in [x_m, x_m + 1], \text{ and} \\ 0 & \text{for } x \geq x_m + 1, \end{cases}$$

which is continuously differentiable (satisfies **f.C1**) and is such that $|f(x)| \leq 6$ for all x (and hence satisfies **f.Bb**). Figure 2.1.1 shows an example of such a function for 7 test points in the unit interval with $x_0 = 0$ and the rest of the sequence generated (before reordering) by a typical bisection process, namely, $x_k = (1/2)^{k-1}$ for $k \geq 1$.

Observe now that any first-order method starting at $x_0 = 0$ and using this sequence of test points (in any order) will fail to find a point x_k with $|f'(x_k)| < 1$ in m iterations because it will discover that $f(x_k) = f(x_0) = 0$ and $f'(x_k) = 1$ for all $1 \leq k \leq m$. Since m is arbitrary (but finite), this leads to the conclusion that no finite sequence of test points will guarantee that the norm of the gradient falls below any prescribed accuracy threshold.

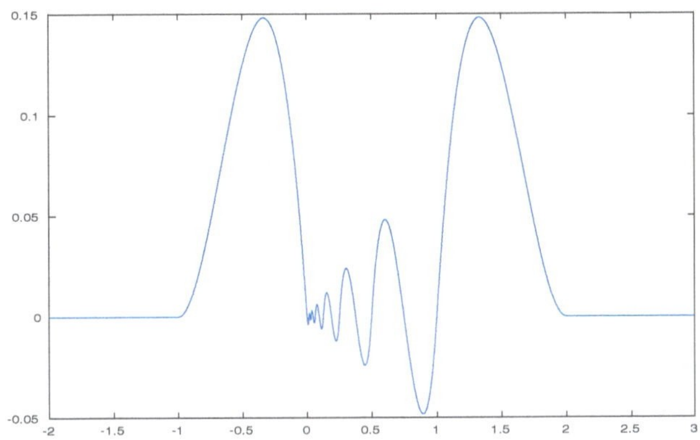

Figure 2.1.1. The shape of $f(x)$ for 7 tests points in [0,1] corresponding to the sequence $x_k = (\frac{1}{2})^{k-1}$ ($k \geq 1$, before reordering).

In fact, this short analysis shows that **f.Bb** and **f.C1** are insufficient to quantify worst-case complexity bounds, essentially because they do not prevent the gradient from oscillating infinitely fast (see Figure 2.1.2). In particular, the quantity

$$\frac{\|\nabla_x^1 f(x_0) - \nabla_x^1 f(\min_{k=1,\ldots,m} x_k)\|}{\|x_0 - \min_{k=1,\ldots,m} x_k\|} = \frac{2}{\|x_0 - \min_{k=1,\ldots,m} x_k\|}$$

becomes arbitrarily large as the number of test points grows.

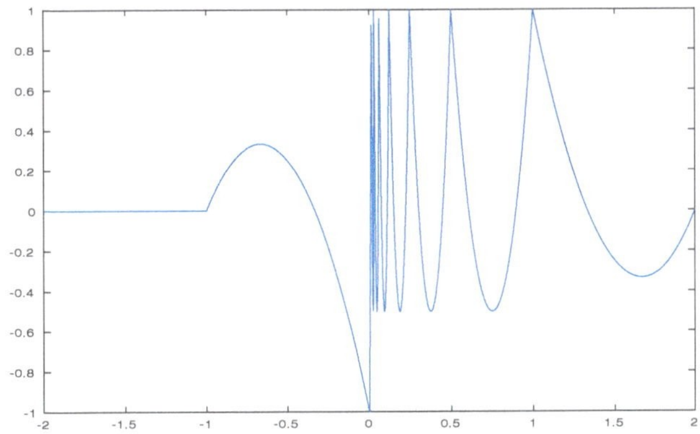

Figure 2.1.2. The fast oscillations of the gradient $f'(x)$ for 7 test points in [0,1] corresponding to the sequence $x_k = (\frac{1}{2})^{k-1}$ ($k \geq 1$, before reordering).

If we wish to quantify the worst-case complexity of first-order algorithms, we are thus led to be slightly more restrictive, and we require an additional property of the gradient of the objective function that prevents such rapid growth.

f.gL

$f(x)$ is continuously differentiable in $\mathbb{R}^n$, and its gradient $\nabla_x^1 f(x)$ is Lipschitz continuous on $\mathbb{R}^n$; that is, there exists a constant $L_{f,1} \geq 1$ such that

$$\|\nabla_x^1 f(x) - \nabla_x^1 f(y)\| \leq L_{f,1}\|x - y\| \tag{2.1.5}$$

for all $x, y \in \mathbb{R}^n$.

Note that if the gradient's Lipschitz constant $L_{f,1}$ is zero, then f is linear and either is constant, in which case optimization is meaningless, or has a minimizer at infinity (violating **f.Bb**). Thus our assumption that $L_{f,1} \geq 1$ can be made without loss of generality. As we will show in the remainder of this chapter, the modified set of assumptions (**f.Bb** and **f.gL**) is now sufficient for deriving relevant worst-case evaluation-complexity bounds, which depend on the Lipschitz constant $L_{f,1}$.

Notes and references for Section 2.1

This section and the presented example are adapted from Vavasis [359] for the unconstrained case.

2.2 Linesearch algorithms

Since the first optimization method using gradients that comes to mind is the steepest-descent algorithm, we begin our analysis there.

2.2.1 The steepest-descent algorithm with backtracking-Armijo linesearch

The steepest-descent algorithm is an iterative method intended to find approximate first-order minimizers of the smooth unconstrained minimization problem (2.1.1). Given the iterate x_k at iteration k, the steepest-descent method searches for an approximate minimizer of $f(x_k - t\nabla_x^1 f(x_k))$ considered as a function of the nonnegative real parameter t. The name of the method derives from the fact that $-\nabla_x^1 f(x_k)$ is the direction in $\mathbb{R}^n$ with the most negative slope of f at x_k measured in the Euclidean norm.

The first step in our development is to say precisely what we mean by finding an approximate minimizer along the steepest-descent direction. A common approach is to use a *linesearch* to find a point $x_k - t_k\nabla_x^1 f(x_k)$ that provides a good reduction in f. In what follows, we consider the well-known *backtracking-Armijo* technique, in which $t_k = 2^{-i_k} t_{\rm ini}$, where $t_{\rm ini} > 0$ is given and i_k is the smallest nonnegative integer for which

$$f(x_k - t_k\nabla_x^1 f(x_k)) \leq f(x_k) - \alpha t_k\|\nabla_x^1 f(x_k)\|^2 \tag{2.2.1}$$

for a given $\alpha \in (0, 1)$. The intention is that such a step provides a significant reduction in f and is not too short. Crucially, as we shall shortly show, such a step always exists.

Our algorithm may be stated formally as follows.

Algorithm 2.2.1: Steepest-descent algorithm with backtracking-Armijo linesearch (SDBA)

Step 0: Initialization. An initial point x_0 is given, as well as an accuracy level $\epsilon > 0$. Constants $t_{\rm ini}$ and α satisfying

$$t_{\rm ini} > 0 \;\text{ and }\; 0 < \alpha < 1 \tag{2.2.2}$$

are also given. Compute $f(x_0)$, and set $k = 0$.

Step 1: Test for termination. Compute $\nabla_x^1 f(x_k)$. If $\|\nabla_x^1 f(x_k)\| \le \epsilon$, terminate with the ϵ-approximate first-order minimizer $x_\epsilon = x_k$.

Step 2: Linesearch: initialization.
Set $t_k = t_{\rm ini}$ and $i_k = 0$, and compute $f(x_k - t_k \nabla_x^1 f(x_k))$.

Step 3: Linesearch: acceptance or backtracking.

1. If (2.2.1) holds, go to Step 4.
2. Else set $t_k \leftarrow \frac{1}{2} t_k$ and $i_k \leftarrow i_k + 1$, compute $f(x_k - t_k \nabla_x^1 f(x_k))$, and return to Step 3.

Step 4: Compute the new iterate and gradient.
Set $x_{k+1} = x_k - t_k \nabla_x^1 f(x_k)$, increment k by one, and return to Step 1.

Note that $f(x_{k+1})$ is computed in Step 2 or 3 and thus need not be computed again in Step 4. The choice of the initial candidate for t_k in Step 2 simply reflects that no a priori estimate of the steplength is truly reliable in a purely first-order method unless one has additional information beyond that assumed here. In practice, one often chooses $t_{\rm ini} = 1$ or initializes t_k to the steplength computed at the previous iteration, for lack of a better choice. Our analysis in fact allows for any initialization of t_k provided that it is uniformly bounded away from zero and infinity. A typical value for the other constant in (2.2.2) is $\alpha = \frac{1}{1000}$. Note that we have chosen to halve t_k in Step 3.2 purely for convenience of exposition, and one could choose a different contraction factor without altering the substance of what follows.

Before we analyze the algorithm, we must show that (2.2.1) is well defined.

Lemma 2.2.1: Lower bound on the stepsize

Suppose that **f.gL** *holds. Then*

$$f(x_k - t\nabla_x^1 f(x_k)) \le f(x_k) - \alpha t \|\nabla_x^1 f(x_k)\|^2 \tag{2.2.3}$$

holds for all

$$t \in [0, t_{\rm low}], \;\; \textit{where}\;\; t_{\rm low} \stackrel{\rm def}{=} \frac{2(1-\alpha)}{L_{f,1}}. \tag{2.2.4}$$

Proof. Since **f.gL** holds, Taylor's theorem (e.g., Corollary A.8.4, p. 437, with $p = 1$) gives that

$$f(x_k - t\nabla_x^1 f(x_k)) \le f(x_k) - t\|\nabla_x^1 f(x_k)\|^2 + \tfrac{1}{2} L_{f,1} \|t\nabla_x^1 f(x_k)\|^2. \tag{2.2.5}$$

It follows from (2.2.4) that

$$\tfrac{1}{2}L_{f,1}t \leq (1-\alpha),$$

and combining this with (2.2.5) gives (2.2.3). □

Hence for every iteration k, there must be a smallest index $i_k \geq 0$ for which $2^{-i_k}t_{\text{ini}} \leq t_{\text{low}}$, and thus (2.2.1) is satisfied. Indeed, this provides the bounds

$$i_k \leq \left\lceil \log_2\left(\frac{t_{\text{ini}}}{t_{\text{low}}}\right)\right\rceil = \left\lceil \log_2\left(\frac{L_{f,1}t_{\text{ini}}}{2(1-\alpha)}\right)\right\rceil \tag{2.2.6}$$

and

$$t_k \geq \tfrac{1}{2}t_{\text{low}}. \tag{2.2.7}$$

2.2.2 Complexity of finding approximate first-order minimizers

We next turn to estimating the worst-case evaluation complexity of the SDBA algorithm (p. 25) for the task of finding an ϵ-approximate first-order minimizer.

Theorem 2.2.2: Complexity of the steepest-descent algorithm with a backtracking-Armijo linesearch

Suppose that **f.Bb** *and* **f.gL** *hold. Then there is a positive constant*

$$\kappa_{\text{SDBA}} \stackrel{\text{def}}{=} \frac{L_{f,1}}{\alpha(1-\alpha)}$$

for which the SDBA *algorithm (p. 25) requires at most*

$$\#^{F}_{\text{SBDA}} = \left\lceil \log_2\left(\frac{L_{f,1}t_{\text{ini}}}{2(1-\alpha)}\right)\right\rceil \kappa_{\text{SDBA}}\frac{f(x_0)-f_{\text{low}}}{\epsilon^2} = \mathcal{O}\left(\epsilon^{-2}\right) \tag{2.2.8}$$

evaluations of the objective function and at most

$$\#^{D}_{\text{SBDA}} \stackrel{\text{def}}{=} \kappa_{\text{SDBA}}\frac{f(x_0)-f_{\text{low}}}{\epsilon^2} + 1 = \mathcal{O}\left(\epsilon^{-2}\right) \tag{2.2.9}$$

gradient evaluations to determine an iterate x_ϵ for which $\|\nabla^1_x f(x_\epsilon)\| \leq \epsilon$.

We note here and hereafter that evaluation bounds such as (2.2.8) and (2.2.9) should be interpreted as the smallest integer not exceeding the stated number quoted. We could have used the $\lfloor\cdot\rfloor$ notation[8] to be precise, but this would lead to additional notational clutter.

Proof. It follows from (2.2.1), **f.gL**, (2.2.7), and Step 4 of the algorithm that

$$f(x_k) - f(x_{k+1}) \geq \alpha t_k \|\nabla^1_x f(x_k)\|^2 \geq \tfrac{1}{2}\alpha t_{\text{low}}\|\nabla^1_x f(x_k)\|^2 \tag{2.2.10}$$

for all $k \geq 0$. Suppose that

$$\|\nabla^1_x f(x_k)\| \geq \epsilon \tag{2.2.11}$$

[8] Around the entire right-hand side of (2.2.8).

for the first k_ϵ iterations. Then **f.Bb**, (2.2.10), and (2.2.11) imply that

$$\begin{aligned} f(x_0) - f_{\text{low}} \geq f(x_0) - f(x_{k_\epsilon}) &= \sum_{k=0}^{k_\epsilon - 1} f(x_k) - f(x_{k+1}) \\ &\geq \tfrac{1}{2} k_\epsilon \alpha t_{\text{low}} \epsilon^2 \end{aligned} \tag{2.2.12}$$

and hence that

$$k_\epsilon \leq \kappa_{\text{SDBA}} \frac{f(x_0) - f_{\text{low}}}{\epsilon^2}, \quad \text{where } \kappa_{\text{SDBA}} = \frac{2}{\alpha t_{\text{low}}} = \frac{L_{f,1}}{\alpha(1-\alpha)}, \tag{2.2.13}$$

using (2.2.4). Since each iteration requires the evaluation of a single gradient and i_k function values, and because the last iteration requires one further gradient evaluation, the inequalities (2.2.6) and (2.2.13) provide the bounds (2.2.8) and (2.2.9). □

We pause briefly to reexamine the steps in the proof above, since the same basic pattern will occur many times throughout the book. First, in (2.2.10), we have a lower bound on the decrease in objective function value $f(x_k) - f(x_{k+1})$ during a particular iteration, expressed in terms of the criticality measure, in this case $\|\nabla_x^1 f(x_k)\|$, that we wish to examine. Then, in (2.2.11), we assume that this measure is large enough that the algorithm has not terminated; these together give a nonzero lower bound on the decrease in the objective function during the iteration in question. Next, in (2.2.12), we use a "telescoping-sum" argument to bound the maximum objective decrease possible over all such iterations. Finally, in (2.2.13), the telescoped sum leads to an upper bound on the total number of iterations for which (2.2.11) is possible, after which the algorithm must have terminated. It is then simply a matter of calculating how many function values and how many gradients are required during each iteration.

2.2.2.1 Sharpness

Now that we have a worst-case bound for the number of evaluations necessary to achieve a specified level of criticality, the obvious next question is whether such a bound is realistic or simply the consequence of an imprecise analysis. That is, is there actually a function for which Algorithm 2.2.1 requires $\mathcal{O}(\epsilon^{-2})$ evaluations? We now show that, yes indeed, the evaluation-complexity bound we have just derived is sharp in order.

Theorem 2.2.3: Slow first-order convergence of the steepest-descent algorithm

The SDBA *algorithm applied to minimize a function f satisfying* **f.Bb** *and* **f.gL** *may require as many as ϵ^{-2} evaluations of $f(x)$ and $\nabla_x^1 f(x)$ to produce an iterate x_ϵ such that $\|\nabla_x^1 f(x_\epsilon)\| \leq \epsilon$.*

Proof. To prove this statement, we will consider an arbitrary $\epsilon \in (0,1)$ and an arbitrary choice of the constants α and $t_{\text{ini}} \geq \alpha$ in the SDBA algorithm and construct an objective function f from $\mathbb{R}$ to $\mathbb{R}$ satisfying the assumptions **f.gL** and **f.Bb** such that the SDBA algorithm takes precisely

$$k_\epsilon = \epsilon^{-2} \tag{2.2.14}$$

function and gradient evaluations to obtain an iterate x_{k_ϵ} such that $\|\nabla_x^1 f(x_{k_\epsilon})\| \leq \epsilon$. In what will become a standard procedure in the rest of this book, we will first construct a sequence of iterates $\{x_k\}$ for which $f(x_k) = f_k$ and $\nabla_x^1 f(x_k) = g_k$ for associated sequences of function and gradient values $\{f_k\}$ and $\{g_k\}$ and then apply Hermite interpolation (Theorem A.9.2 on page 441) to exhibit the function f itself.

We start by defining, for all $k \in [0 : k_\epsilon]$,

$$\omega_k \stackrel{\text{def}}{=} \epsilon\,\frac{k_\epsilon - k}{k_\epsilon} \in [0, \epsilon] \tag{2.2.15}$$

and

$$g_k \stackrel{\text{def}}{=} -(\epsilon + \omega_k). \tag{2.2.16}$$

We note immediately that

$$|g_k| > \epsilon \;\text{ for all }\; k \in [0 : k_\epsilon - 1] \;\text{ and }\; |g_{k_\epsilon}| = \epsilon. \tag{2.2.17}$$

We then define

$$x_0 = 0 \;\text{ and }\; x_{k+1} = x_k - t_{\text{ini}} g_k \;\text{ for all }\; k \in [0 : k_\epsilon - 1] \tag{2.2.18}$$

as well as

$$f_0 = 4\,t_{\text{ini}} \;\text{ and }\; f_{k+1} = f_k - t_{\text{ini}} |g_k|^2 \;\text{ for all }\; k \in [0 : k_\epsilon - 1]. \tag{2.2.19}$$

Observe that the sequences $\{\omega_k\}$ and $\{f_k\}$ are decreasing and that

$$f_{k_\epsilon} = f_0 - t_{\text{ini}} \sum_{k=0}^{k_\epsilon - 1} |g_k|^2 = f_0 - t_{\text{ini}} \sum_{k=0}^{k_\epsilon - 1} |\epsilon + \omega_k|^2 \geq f_0 - t_{\text{ini}}\, k_\epsilon\, (2\epsilon)^2 = f_0 - 4\,t_{\text{ini}},$$

where we used (2.2.16) to deduce the second equality, (2.2.15) to deduce the following inequality, and (2.2.14) to obtain the last equality. Hence (2.2.19) implies that

$$f_k \in [0, 4\,t_{\text{ini}}] \;\text{ for all }\; k \in [0 : k_\epsilon]. \tag{2.2.20}$$

In addition, $s_k = -t_{\text{ini}} g_k > 0$ for $k \in [0 : k_\epsilon - 1]$. Finally,

$$|f_{k+1} - f_k + t_{\text{ini}} |g_k|^2| = 0 \;\text{ and }\; |g_{k+1} - g_k| = -\omega_{k+1} + \omega_k = \epsilon^3 \leq \frac{1}{t_{\text{ini}}} s_k. \tag{2.2.21}$$

These last bounds and (2.2.20) allow us to use Theorem A.9.2 (p. 441) for $p = 1$ and

$$\kappa_f = \max\left[\frac{1}{t_{\text{ini}}}, 4t_{\text{ini}}, 2\right] \tag{2.2.22}$$

(the second term in the max bounding $|f_k|$ because of (2.2.20) and the third bounding $|g_k|$ because of (2.2.15), (2.2.16), and $\epsilon \leq 1$). We then deduce that there exists a continuously differentiable function f from $\mathbb{R}$ to $\mathbb{R}$ with Lipschitz-continuous gradient (i.e., satisfying **f.gL**) such that, for $k \in [0 : k_\epsilon]$,

$$f(x_k) = f_k \quad\text{and}\quad \nabla_x^1 f(x_k) = g_k.$$

Moreover, the range of f is a constant independent of ϵ, hence guaranteeing **f.Bb**. Figure 2.2.1 on the facing page shows the form[9] of the function $f(x)$ over the interval

[9]For the choices $t_{\text{ini}} = \frac{1}{10}$ and $\alpha = \frac{1}{1000}$.

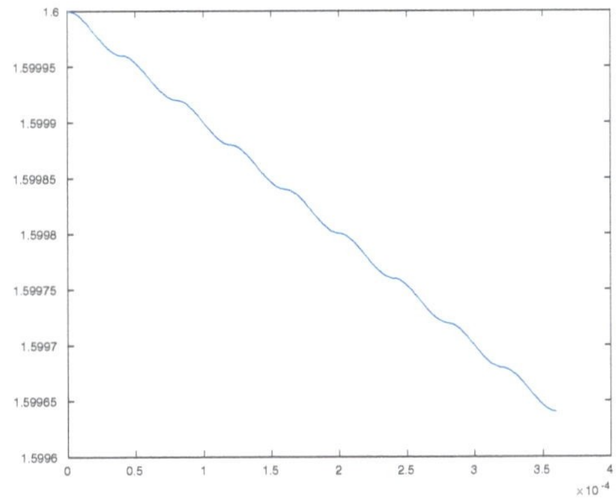

Figure 2.2.1. The form of the objective function f over $[x_0, x_{10}]$.

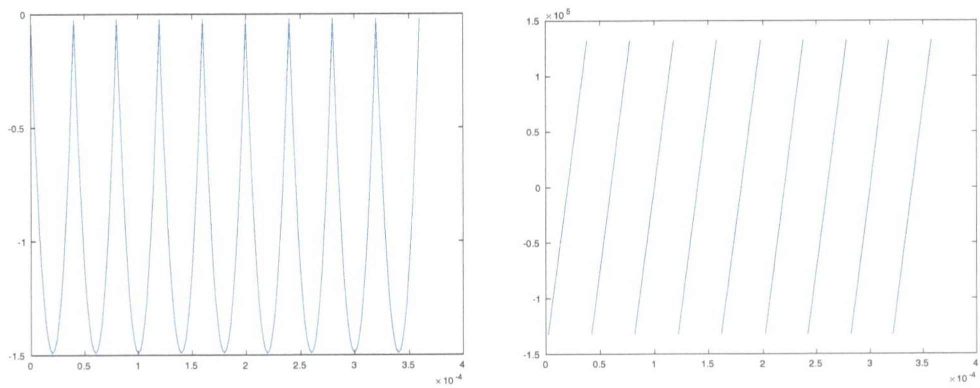

Figure 2.2.2. The form of the gradient (left) and Hessian (right) of f over $[x_0, x_{10}]$.

$[x_0, x_{10}]$, while Figure 2.2.2 shows the forms of its gradient (left) and Hessian (right) over the same interval. Note that the gradient is continuous and the Hessian is discontinuous but bounded, implying **f.gL**. The function is nonconvex.

The fact that the second part of (2.2.19) and $t_{\rm ini} \geq \alpha$ ensure that (2.2.1) holds for $k \in [0 : k_\epsilon - 1]$ then implies that the sequence $\{x_k\}$ (and hence the sequences $\{f_k\}$ and $\{g_k\}$) can be seen as generated by the SDBA algorithm applied to f, starting from $x_0 = 0$. In view of (2.2.17), we thus deduce that the SDBA algorithm applied to f starting from the origin needs exactly $k_\epsilon = \epsilon^{-2}$ iterations (in which the objective function and its gradient are evaluated once) to produce an iterate x_ϵ such that $\|\nabla_x^1 f(x_\epsilon)\| \leq \epsilon$. This implies the desired conclusion.[10] □

It is important to notice that while Theorem 2.2.3 says that there is a function satisfying **f.gL** for which the convergence of Algorithm 2.2.1 is as slow at the worst-case bound predicted in Theorem 2.2.2—indeed, we provide such an example—it does *not* imply that the same is true if we further restrict the problem class by requiring more smoothness, for instance, by applying the algorithm to functions that are twice

[10]The fact that the example derived in Theorem 2.2.3 is unidimensional is not restrictive, since it is always possible to make the value of its objective function and gradient independent of all dimensions but one.

continuously differentiable with Lipschitz-continuous Hessians. This question is covered in Section 12.2.2.4 (see the discussion following Theorem 12.2.17) for a much more general class of algorithms, where we will show that allowing more smoothness without modifying the algorithm to exploit it does not result in better evaluation complexity.

2.2.3 Other linesearch algorithms

2.2.3.1 Flexible Armijo

We have chosen to present the evaluation-complexity analysis for a version of steepest descent using the backtracking-Armijo linesearch, mostly because this is probably the simplest version of the algorithm. But other choices are obviously possible.

The first of these is a variant of backtracking Armijo that allows increasing as well as decreasing stepsizes. In particular, if (2.2.1) is satisfied when $t_k = t_{\rm ini}$, the step may be doubled until (2.2.1) fails or a prescribed number, $i_{\rm max}$, of doublings occurs, whichever happens first. Since in such cases $t_k \geq t_{\rm ini}$, and thus $f(x_k)-f(x_{k+1}) \geq \frac{1}{2}\alpha t_{\rm ini}\|\nabla_x^1 f(x_k)\|^2$ rather than (2.2.10), an $\mathcal{O}(\epsilon^{-2})$ evaluation complexity may be established just as in the proof of Theorem 2.2.2 if $i_{\rm max}$ is finite.[11]

2.2.3.2 Goldstein–Armijo

A well-regarded variant is the more involved *Goldstein–Armijo* scheme where t_k is chosen to satisfy two simultaneous inequalities. The first is the Armijo rule (2.2.1) that we have already discussed, while the second ensures that steps are not unduly short by imposing the bound

$$f(x_k - t_k\nabla_x^1 f(x_k)) \geq f(x_k) - \beta t_k\|\nabla_x^1 f(x_k)\|^2 \tag{2.2.23}$$

for given $\beta \in (\alpha, 1)$. The point $x_k - t_k\nabla_x^1 f(x_k)$ is then chosen as the next iterate x_{k+1}. In practice, a bracket $[t^{\rm low}, t^{\rm up}]$ is first sought to contain a suitable t_k, that is, such that (2.2.1) holds and (2.2.23) fails at $x_k - t^{\rm low}\nabla_x^1 f(x_k)$, while (2.2.23) holds and (2.2.1) fails at $x_k - t^{\rm up}\nabla_x^1 f(x_k)$. The desired t_k is then computed by bisection within this bracket.[12] This too yields the same $\mathcal{O}(\epsilon^{-2})$ worst-case evaluation-complexity bound, and the sharpness example exhibited in Theorem 2.2.3 still applies so long as $t_{\rm ini} \in [\alpha, \beta]$.

2.2.3.3 Nesterov's approach

An alternative view is provided by Nesterov's [299] original analysis of steepest descent. This contribution is based on a different idea, using the overestimation property (A.19) in Corollary A.8.4 (p. 437) to define a convex upper bound

$$\phi(t) \stackrel{\rm def}{=} f(x_k) - t\|\nabla_x^1 f(x_k)\|^2 + \tfrac{1}{2}t^2 L_{f,1}\|\nabla_x^1 f(x_k)\|^2 \geq f(x_k - t\nabla_x^1 f(x_k)), \tag{2.2.24}$$

which is valid for all $t \geq 0$ (see Corollary A.8.4 for $p = 1$). Assuming that $L_{f,1}$ is known, $\phi(t)$ can then be minimized, yielding a stepsize $t_k = 1/L_{f,1}$. Of course, $L_{f,1}$ is typically unavailable in practice, but [299] shows that this covers different linesearch schemes,

[11] If $i_{\rm max}$ is infinite, the linesearch will nevertheless terminate in a finite number of doublings, since after i doublings without termination

$$f(x_0) - f_{\rm low} \geq f(x_k) - f(x_k - t_{\rm ini}2^i\nabla_x^1 f(x_k)) \geq 2^i t_{\rm ini}\alpha\epsilon^2,$$

which is impossible if $i \geq \lceil \log_2\left((f(x_0) - f_{\rm low})/t_{\rm ini}\alpha\epsilon^2\right)\rceil$.

[12] It can be shown that at each bisection point it is impossible for both (2.2.1) and (2.2.23) to fail, because $\beta < \alpha$, and thus the bisection process is well defined.

including constant or suitably monotonically decreasing stepsizes, exact line minimization (see next paragraph), or even the Goldstein–Armijo technique. (This, however, doesn't take into account the fact that more than a single function evaluation may be necessary in the last two cases.) The argument in [299] is based on the observation that, in this context,

$$f(x_k) - f(x_k - t_k \nabla_x^1 f(x_k)) \geq f(x_k) - \phi(t_k) = \frac{1}{2L_{f,1}} \|\nabla_x^1 f(x_k)\|^2, \tag{2.2.25}$$

providing sufficient descent (as is obtained from (2.2.1) above) for a single iteration (rather than a single function evaluation). This procedure is therefore adequate for analyzing *iteration* worst-case complexity, as opposed to *function-evaluation* complexity. Unsurprisingly, it produces an $\mathcal{O}(\epsilon^{-2})$ worst-case iteration-complexity bound.

2.2.3.4 Exact linesearch

From the theoretical viewpoint, it is also interesting to see if our above results remain true if exact linesearch is used, that is if Steps 2 and 3 of the SDBA algorithm (p. 25) are replaced by

$$t_k = \arg\min_{t \geq 0} f(x_k - t\nabla_x^1 f(x_k)), \tag{2.2.26}$$

where, for example, ties are broken by choosing the first minimizer of $f(x_k - t\nabla_x^1 f(x_k))$ if there is more than one. The resulting SDEX algorithm is of course most often numerically unrealistic, except for special functions such as quadratics, where the minimizer can be determined analytically. But it remains an ideal that numerically sounder techniques attempt to imitate, justifying our curiosity. To analyze its complexity, we use (2.2.24) in a manner very similar to Nesterov's [299] analysis and obtain the following upper bound.

Theorem 2.2.4: Complexity of the steepest-descent algorithm with an exact linesearch

Suppose that **f.Bb** *and* **f.gL** *hold. Then there exists a positive constant* κ_{SDEX} *such that the* SDEX *algorithm requires at most*

$$\#^I_{\text{SDEX}} \stackrel{\text{def}}{=} \kappa_{\text{SDEX}} \frac{f(x_0) - f_{\text{low}}}{\epsilon^2} = \mathcal{O}\left(\epsilon^{-2}\right) \tag{2.2.27}$$

iterations to produce an iterate x_ϵ *such that* $\|\nabla_x^1 f(x_\epsilon)\| \leq \epsilon$.

Proof. We first note that **f.C1**, Taylor's expansion at x_k, and **f.gL** give that, for each $k \geq 0$,

$$f(x_k) - f(x_k - t\nabla_x^1 f(x_k)) \geq f(x_k) - f(x_k) + t\|\nabla_x^1 f(x_k)\|^2 - \tfrac{1}{2}t^2 L_{f,1}\|\nabla_x^1 f(x_k)\|^2$$

for any $t \geq 0$. Maximizing the right-hand side of this inequality with respect to t, we obtain that

$$f(x_k) - f\left(x_k - \frac{1}{L_{f,1}}\nabla_x^1 f(x_k)\right) \geq \frac{1}{2L_{f,1}}\|\nabla_x^1 f(x_k)\|^2 \geq \frac{\epsilon^2}{2L_{f,1}} \tag{2.2.28}$$

for each iteration k, as long as $\|\nabla_x^1 f(x_k)\| > \epsilon$.

But (2.2.26) ensures that the slope of $f(x_k - t\nabla_x^1 f(x_k))$ must be zero at $x_{k+1} = x_k - t_k \nabla_x^1 f(x_k)$, giving that, for all k,

$$\begin{aligned}
0 &= \langle \nabla_x^1 f(x_k), \nabla_x^1 f(x_{k+1}) \rangle \\
&= \|\nabla_x^1 f(x_k)\|^2 + \langle \nabla_x^1 f(x_k), \nabla_x^1 f(x_k - t_k \nabla_x^1 f(x_k)) - \nabla_x^1 f(x_k) \rangle \\
&\geq \|\nabla_x^1 f(x_k)\|^2 (1 - L_{f,1} t_k),
\end{aligned}$$

where we use the Cauchy–Schwarz inequality and **f.gL**. This implies that t_k, the argument of the (first) minimum in (2.2.26), is such that $t_k \geq 1/L_{f,1}$ and therefore, because of (2.2.28), that, for each k,

$$f(x_k) - f(x_k - t_k \nabla_x^1 f(x_k)) \geq f(x_k) - f\left(x_k - \frac{1}{L_{f,1}} \nabla_x^1 f(x_k)\right) \geq \frac{\epsilon^2}{2L_{f,1}}$$

as long as $\|\nabla_x^1 f(x_k)\| > \epsilon$. Thus a maximum number of

$$\left\lfloor \frac{2L_{f,1}(f(x_0) - f_{\text{low}})}{\epsilon^2} \right\rfloor$$

such iterations may take place before x_k is found such that $\|\nabla_x^1 f(x_k)\| \leq \epsilon$. This gives (2.2.27) with

$$\kappa_{\text{SDEX}} \stackrel{\text{def}}{=} 2L_{f,1}. \qquad \square$$

Observe that we have bounded the *number of iterations* possibly needed to compute an ϵ-approximate first-order minimizer but that this is merely a very optimistic lower bound on the number of evaluations of f. Indeed, each iteration may require an (unquantified) number of such evaluations to find the line minimizer (2.2.26). As it turns out, this bound is also sharp.

Theorem 2.2.5: Slow first-order convergence of the steepest-descent algorithm

The SDEX *algorithm applied to minimize a function f satisfying* ***f.Bb*** *and* ***f.gL*** *may require as many as ϵ^{-2} iterations of $f(x)$ and $\nabla_x^1 f(x)$ to produce an iterate x_ϵ such that $\|\nabla_x^1 f(x_\epsilon)\| \leq \epsilon$.*

Proof. As for Theorem 2.2.3, an example is constructed such that the SDEX algorithm takes exactly ϵ^{-2} iterations to satisfy the termination criterion. However, since minimization is exact at every iteration and, by definition,

$$\langle \nabla_x^1 f(x_{k+1}), \nabla_x^1 f(x_k) \rangle = 0,$$

a univariate example is now impossible. The more complicated two-dimensional example of slow convergence is constructed and verified in Appendix B (p. 445). The contour lines of the resulting objective function and the path of iterates are shown in Figure 2.2.3. $\square$

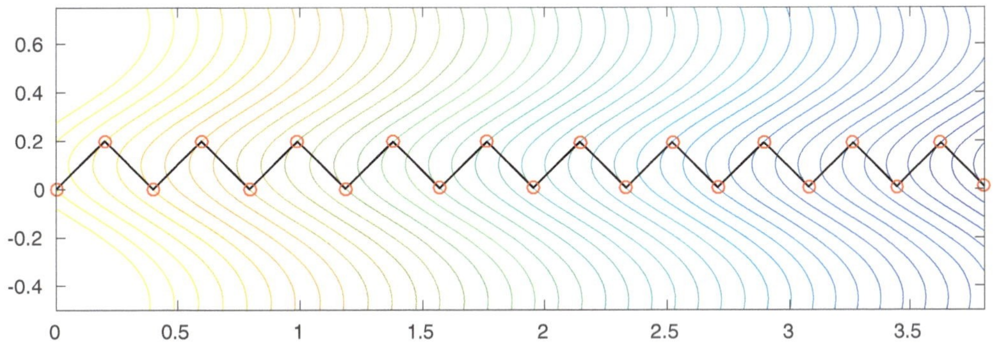

Figure 2.2.3. The contour lines of the objective function on which the SDEX algorithm started at the origin converges slowly (Theorem 2.2.5) with $\epsilon = 0.1$ and the associated path of iterates generated by the first 20 iterations.

2.2.3.5 Nonmonotone gradient-related algorithms

We also note that the analysis of Section 2.2.2 can be extended to the class of nonmonotone gradient-related algorithms. This class of methods is interesting because of its generality (it covers many possible linesearch variants) and its often excellent behaviour in practice. In this class, the new iterate is found by first selecting a direction d_k whose angle with the steepest descent is controlled by the condition

$$\langle \nabla_x^1 f(x_k), d_k \rangle \leq -\kappa_1 \|\nabla_x^1 f(x_k)\|^2 \quad \text{and} \quad \|d_k\| \leq \kappa_2 \|\nabla_x^1 f(x_k)\|,$$

where κ_1 and κ_2 are positive constants independent of k. Once the direction is fixed, the new iterate $x_{k+1} = x_k + t_k d_k$ is found such that the nonmonotone Goldstein–Armijo conditions

$$f(x_k + t_k d_k) \leq \max_{0 \leq j \leq M} [f(x_{k-j})] + \alpha\, t_k \langle \nabla_x^1 f(x_k), d_k \rangle$$

and

$$f(x_k + t_k d_k) \geq \max_{0 \leq j \leq M} [f(x_{k-j})] + \beta\, t_k \langle \nabla_x^1 f(x_k), d_k \rangle$$

hold, where $M \geq 0$, $\alpha \in (0,1)$, and $\beta \in (\alpha, 1)$ are constants. It can be shown that a result similar to Theorem 2.2.2 still holds and thus that such methods also require at most $\mathcal{O}(\epsilon^{-2})$ evaluations of the objective function and its gradient to produce an ϵ-approximate first-order minimizer. Moreover, the sharpness example in Theorem 2.2.3 applies when using such methods. Furthermore, the Goldstein–Armijo method discussed in Section 2.2.3.2 is a special case, and thus any analysis for the nonmonotone case immediately applies to it.

Notes and references for Section 2.2

For an introduction to linesearch methods and the Goldstein–Armijo technique, see Dennis and Schnabel [147, p. 117ff]. The original proposal for the steepest-descent algorithm dates back to Cauchy [115], while the backtracking strategy in Section 2.2.1 is due to Armijo [12]. The discussion of Section 2.2.3.3 is obviously inspired by the iterationwise analysis in Nesterov [299, pp. 26–29]. At variance, the function-evaluation view is taken in the first part of Section 2.2.2, which is new, as is the associated example of slow convergence discussed in the proof of Theorem 2.2.3. A first analysis of the complexity of steepest descent with exact linesearch appeared in a technical report [92] and is more in line with Nesterov's approach. It also produces an $\mathcal{O}(\epsilon^{-2})$ complexity bound. This is of interest since the example of Theorem 2.2.3 (see Figure 2.2.1 in particular) could (erroneously) suggest that slow convergence might be alleviated by exact line minimization. The proof of Theorem 2.2.5 is adapted from [92]. Nonmonotone linesearches were first introduced by Grippo, Lampariello, and Lucidi [220] (see also [307, pp. 33 and 37]), while the complexity analysis for nonmonotone gradient-related methods is more recent [112, 204].

2.3 The trust-region method

Having examined steepest-descent linesearch algorithms for solving (2.1.1), we now turn to methods that use first-order information (i.e., the gradient) less directly. Specifically, a local *model* of the objective function is constructed using the gradient. This model is then (locally) optimized, with the intention of moving the next iterate toward an ϵ-approximate first-order minimizer. The first of these methods is the trust-region algorithm.

2.3.1 The TR1 algorithm

The basic *trust-region algorithm* is based on the idea that a (typically quadratic) model of the objective function is minimized, possibly approximately, in a *trust region* centered at the current iterate, that is, a region in which this model is believed to provide a sufficiently good approximation of the objective function.

More specifically, iteration $k \geq 0$ consists of the following steps. We first define a quadratic model at the current iterate x_k by

$$m_k(s) = f(x_k) + \langle \nabla_x^1 f(x_k), s \rangle + \tfrac{1}{2}\langle s, B_k s \rangle, \tag{2.3.1}$$

where B_k is a uniformly bounded symmetric matrix (see **m.HBa**), whose purpose is to approximate the objective function within a *trust region*, that is, the set of all points

$$\mathcal{B}_{\Delta_k} = \{ s \in \mathbb{R}^n \mid \|s\| \leq \Delta_k \}, \tag{2.3.2}$$

for some *trust-region radius* Δ_k. Observe that, at this stage, we do not require that the matrix B_k approximate the Hessian of f but insist that the gradient of the model $m_k(s)$ coincide with that of f at x_k, so that $\nabla_s^1 m_k(0) = \nabla_x^1 f(x_k)$.

We next seek a *trial step* s_k to a *trial point* $x_k + s_k \in \mathcal{B}_{\Delta_k}$ with the aim of reducing the model (while satisfying the bound $\|s_k\| \leq \Delta_k$). We ask that the reduction in the model at the trial point be at least equal to (a fraction of) the model reduction that can be achieved along the steepest-descent direction *within the trust region.*

Having determined the trial step, we now compute the objective function at $x_k + s_k$ and compare it to the value predicted by the model at this point, that is, $m_k(s_k)$. If the reduction predicted by the model is realized for the objective function, the trial point is accepted as the next iterate and the trust region is expanded or kept the same; such an iteration is said to be *successful.* If the model reduction turns out to be a poor prediction of the actual behaviour of the objective function, the trial point is rejected—the iteration is *unsuccessful*—and the trust region is contracted, with the hope that the model provides a better prediction within the smaller region.

More formally, the basic trust-region algorithm is described on the next page.

Algorithm 2.3.1: Trust-region algorithm with a first-order model (TR1)

Step 0: Initialization. An initial point x_0, an accuracy level $\epsilon > 0$, and an initial trust-region radius $\Delta_0 > 0$ are given. The constants η_1, η_2, γ_1, and γ_2 are also given and satisfy

$$0 < \eta_1 \le \eta_2 < 1 \quad \text{and} \quad 0 < \gamma_1 \le \gamma_2 < 1 < \gamma_3. \tag{2.3.3}$$

Compute $f(x_0)$ and $\nabla_x f(x_0)$, and set $k = 0$.

Step 1: Test for termination. If $\|\nabla_x^1 f(x_k)\| \le \epsilon$, terminate with the ϵ-approximate first-order minimizer $x_\epsilon = x_k$.

Step 2: Step computation. Compute a step $s_k \in \mathcal{B}_{\Delta_k}$ such that

$$m_k(s_k) \le m_k(s_k^{\mathrm{C}}), \tag{2.3.4}$$

where m_k is given by (2.3.1) and

$$s_k^{\mathrm{C}} \stackrel{\text{def}}{=} -t_k^{\mathrm{C}} \nabla_x^1 f(x_k) \quad \text{with} \quad t_k^{\mathrm{C}} \stackrel{\text{def}}{=} \mathop{\arg\min}_{t \ge 0,\, x_k - t\nabla_x^1 f(x_k) \in \mathcal{R}_k} m_k(-t\nabla_x^1 f(x_k)). \tag{2.3.5}$$

Step 3: Acceptance of the trial point. Compute $f(x_k + s_k)$ and define

$$\rho_k = \frac{f(x_k) - f(x_k + s_k)}{m_k(0) - m_k(s_k)}. \tag{2.3.6}$$

If $\rho_k \ge \eta_1$, then define $x_{k+1} = x_k + s_k$ and compute $\nabla_x f(x_{k+1})$. Otherwise, define $x_{k+1} = x_k$.

Step 4: Trust-region radius update. Set

$$\Delta_{k+1} \in \begin{cases} [\Delta_k, \gamma_3 \Delta_k] & \text{if } \rho_k \ge \eta_2, \\ [\gamma_2 \Delta_k, \Delta_k] & \text{if } \rho_k \in [\eta_1, \eta_2), \\ [\gamma_1 \Delta_k, \gamma_2 \Delta_k] & \text{if } \rho_k < \eta_1. \end{cases} \tag{2.3.7}$$

Increment k by one and go to Step 1.

Reasonable values for the constants in (2.3.3) might be, for example, $\eta_1 = 0.01$, $\eta_2 = 0.9$, $\gamma_1 = \gamma_2 = 0.5$, and $\gamma_3 = 2$. The step s_k^{C} in (2.3.5) is referred to as the *Cauchy step* and the point $x_k^{\mathrm{C}} \stackrel{\text{def}}{=} x_k + s_k^{\mathrm{C}}$ as the *Cauchy point*. We note that the condition (2.3.4) may be relaxed to require instead that merely

$$m_k(0) - m_k(s_k) \ge \kappa(m_k(0) - m_k(s_k^{\mathrm{C}})) \tag{2.3.8}$$

for some $\kappa \in (0, 1]$, without affecting the results below.

The key difference between this method and those considered in Section 2.2 is that here one minimizes a (possibly crude) approximate *quadratic model* of the objective function along the steepest-descent direction (see (2.3.5)) instead of considering the true objective function itself.

For convenience, here and for many algorithms to come, we define

$$\mathcal{S} \stackrel{\text{def}}{=} \{k \in \mathbb{N}^0 \mid \rho_k \geq \eta_1\} \quad \text{and} \quad \mathcal{U} \stackrel{\text{def}}{=} \mathbb{N}^0 \setminus \mathcal{S}, \tag{2.3.9}$$

the sets of *successful* and *unsuccessful* iterations, respectively, and

$$\mathcal{S}_k \stackrel{\text{def}}{=} \{j \in [0:k] \mid \rho_j \geq \eta_1\} \quad \text{and} \quad \mathcal{U}_k \stackrel{\text{def}}{=} [0:k] \setminus \mathcal{S}_k, \tag{2.3.10}$$

the corresponding sets up to iteration k. Notice that $x_{k+1} = x_k + s_k$ for $k \in \mathcal{S}$, while $x_{k+1} = x_k$ for $k \in \mathcal{U}$.

For future reference, we now state a property of the TR1 algorithm that solely depends on the mechanism (2.3.7) to update the trust-region radius.

Lemma 2.3.1: Successful and unsuccessful trust-region iterations

Suppose that the TR1 *algorithm is used and that* $\Delta_k \geq \Delta_{\min}$ *for some* $\Delta_{\min} > 0$. *Then*

$$k \leq |\mathcal{S}_k| \left(1 + \frac{\log \gamma_3}{|\log \gamma_2|}\right) + \frac{1}{|\log \gamma_2|} \left|\log\left(\frac{\Delta_{\min}}{\Delta_0}\right)\right|. \tag{2.3.11}$$

Proof. Observe that (2.3.7) and our assumption imply that

$$\Delta_{i+1} \leq \gamma_3 \Delta_i, \quad i \in \mathcal{S}_k, \quad \text{and} \quad \Delta_{i+1} \leq \gamma_2 \Delta_i, \quad i \in \mathcal{U}_k.$$

Using our assumption, we thus deduce inductively that

$$\Delta_{\min} \leq \Delta_k \leq \Delta_0 \gamma_3^{|\mathcal{S}_k|} \gamma_2^{|\mathcal{U}_k|},$$

which gives that

$$\gamma_3^{|\mathcal{S}_k|} \gamma_2^{|\mathcal{U}_k|} \geq \frac{\Delta_{\min}}{\Delta_0},$$

and we obtain inequality (2.3.11) by taking logarithms on both sides and recalling that $\gamma_2 \in (0,1)$ and that $k = |\mathcal{S}_k| + |\mathcal{U}_k|$. □

In other words, so long as the trust-region radius is bounded from below, the total number of iterations performed thus far is bounded in terms of the number of successful ones. Note that this lemma is independent of the specific choice of s_k, a property that we will use later.

2.3.2 Complexity of finding approximate first-order minimizers

A complete and general convergence analysis of the trust-region method is clearly beyond the scope of the present book, and we restrict our attention to simple formulations of the results that are necessary for conducting the worst-case complexity analysis. We start by revisiting the assumptions needed for this task. If we continue to require that the objective function be bounded below (**f.Bb**) and continuously differentiable with a Lipschitzian gradient (**f.gL**), we need to add a condition on the model Hessian B_k.

m.HBa	The model's Hessian matrix B_k is uniformly bounded above; that is, there exists a constant $\kappa_B \geq 0$ such that, for all $k \geq 0$, $$\|B_k\| \leq \kappa_B.$$

The first crucial result is well known in the trust-region literature as the *Cauchy decrease* property, and assesses the magnitude of the function decrease at each iteration.

Lemma 2.3.2: Model decrease at the Cauchy point

Suppose that $m_k(s)$ is quadratic. Then

$$m_k(0) - m_k(s_k^{\rm C}) \geq \tfrac{1}{2}\|\nabla_x^1 f(x_k)\| \min\left[\frac{\|\nabla_x^1 f(x_k)\|}{1+\|B_k\|}, \Delta_k\right] \tag{2.3.12}$$

and the ratio (2.3.6) is well defined as long as $\|\nabla_x^1 f(x_k)\| \neq 0$.

Proof. To prove that (2.3.12) holds, we first note that the quadratic nature of the model m_k (see (2.3.1)) allows us to distinguish three possibilities for the Cauchy step $s_k^{\rm C}$.

• The first case is when the model's curvature along the steepest-descent direction is positive, resulting in $m_k(-t\nabla_x^1 f(x_k))$ being a convex quadratic whose minimizer $x_k - t_k^* \nabla_x^1 f(x_k)$ with $t_k^* \stackrel{\rm def}{=} \|\nabla_x^1 f(x_k)\|^2 / \langle \nabla_x^1 f(x_k), B_k \nabla_x^1 f(x_k)\rangle$ lies in the interior of the trust region $\mathcal{B}_{\Delta_k}$. In this case, it is easy to verify that $t_k^{\rm C} = t_k^*$, and we then deduce (2.3.12) by substituting these values in (2.3.1) and by using the Cauchy–Schwarz inequality and the trivial relationship $\|B_k\| \leq 1 + \|B_k\|$.

• The second case is similar, except that the quadratic's minimizer now lies outside $\mathcal{B}_{\Delta_k}$, in which case we obtain that the constrained minimizer lies exactly on the trust-region boundary, yielding that

$$t_k^{\rm C} = \Delta_k / \|\nabla_x^1 f(x_k)\|. \tag{2.3.13}$$

We then use the property that, as is the case for all strongly convex quadratics,

$$m_k(-t\nabla_x^1 f(x_k)) \leq m_k(0) - \tfrac{1}{2} t \|\nabla_x^1 f(x_k)\|^2 \quad \text{for all} \quad t \in [0, t_k^*]$$

to deduce that

$$m_k(0) - m_k(s_k^{\rm C}) \geq \tfrac{1}{2}\|\nabla_x^1 f(x_k)\| \Delta_k,$$

again implying (2.3.12).

• The final case is when the model's curvature along the steepest-descent direction is nonpositive, implying that

$$m_k(-t\nabla_x^1 f(x_k)) \leq m_k(0) - t\|\nabla_x^1 f(x_k)\|^2 \leq m_k(0) - \tfrac{1}{2} t \|\nabla_x^1 f(x_k)\|^2 \text{ for all } t \geq 0. \tag{2.3.14}$$

Moreover, the constrained minimizer again lies on the trust-region boundary, giving (2.3.13). Substituting this identity into (2.3.14) also implies (2.3.12).

Now observe that (2.3.12) and the inequality $\|\nabla_x^1 f(x_k)\| > 0$ then give that

$$m_k(0) - m_k(s_k^{\rm C}) > 0$$

because (2.3.7) ensures that $\Delta_k > 0$ for all k. Hence the ratio (2.3.6) is well defined. □

In view of **m.HBa**, this result nearly allows us to derive a model decrease (translated to a function decrease at successful iterations) which is bounded below by a multiple of ϵ^2 in the case where $\|\nabla_x^1 f(x_k)\| \geq \epsilon$.

The following results, based on the Lipschitz-continuity assumption **f.gL**, provide the missing argument. The first establishes conditions that ensure that the trust-region radius increases.

Lemma 2.3.3: Condition for a successful iteration

Suppose that **f.gL** *and* **m.HBa** *hold and that*

$$\Delta_k < \frac{(1-\eta_2)\|\nabla_x^1 f(x_k)\|}{2(L_{f,1}+\kappa_B)}. \tag{2.3.15}$$

Then $\rho_k \geq \eta_2$, iteration k is successful, and $\Delta_{k+1} \geq \Delta_k$.

Proof. We obtain from (2.3.15), the bound $0 < \eta_2 < 1$ in (2.3.3), and $L_{f,1} \geq 1$ that

$$\Delta_k < \frac{\|\nabla_x^1 f(x_k)\|}{L_{f,1}+\kappa_B} \leq \frac{\|\nabla_x^1 f(x_k)\|}{1+\kappa_B}.$$

As a consequence, the mechanism of the algorithm and (2.3.12) immediately imply that

$$\begin{aligned} m_k(0) - m_k(s_k) &\geq m_k(0) - m_k(s_k^{\mathrm{C}}) \\ &\geq \tfrac{1}{2}\|\nabla_x^1 f(x_k)\| \min\left[\frac{\|\nabla_x^1 f(x_k)\|}{1+\kappa_B}, \Delta_k\right] \\ &= \tfrac{1}{2}\|\nabla_x^1 f(x_k)\|\Delta_k. \end{aligned} \tag{2.3.16}$$

By contrast, the mean-value theorem, (2.3.1), the Cauchy–Schwarz inequality, **f.gL**, and **m.HBa** yield that

$$|f(x_k+s_k)-m_k(s_k)| = |\langle \nabla_x^1 f(\xi_k s_k) - \nabla_x^1 f(x_k), s_k\rangle - \tfrac{1}{2}\langle s_k, B_k s_k\rangle| \leq (L_{f,1}+\kappa_B)\|s_k\|^2$$

for some $\xi_k \in [0,1]$. Combining this relation with (2.3.16) and using the bound $\|s_k\| \leq \Delta_k$, we then obtain that

$$|\rho_k - 1| = \left|\frac{f(x_k+s_k)-m_k(s_k)}{m_k(0)-m_k(s_k)}\right| \leq \frac{(L_{f,1}+\kappa_B)\Delta_k^2}{\frac{1}{2}\|\nabla_x^1 f(x_k)\|\Delta_k} \leq 1-\eta_2$$

because of (2.3.15). Hence $\rho_k \geq \eta_2 \geq \eta_1$, $k \in \mathcal{S}_k$, and the remaining conclusion follows from (2.3.7). □

The next result provides a uniform lower bound on Δ_k.

Lemma 2.3.4: Lower bound for the trust-region radius

Suppose that **f.gL** *and* **m.HBa** *hold. Then we have that*

$$\Delta_k \geq \frac{\gamma_1(1-\eta_2)}{2(L_{f,1}+\kappa_B)} \min\left[1, \frac{\Delta_0}{\|\nabla_x^1 f(x_0)\|}\right] \min_{i\in[0:k]} \|\nabla_x^1 f(x_i)\| \tag{2.3.17}$$

for all $k \geq 0$.

Proof. Observe first that, by construction, (2.3.17) holds for $k = 0$. Assume now that iteration $k \geq 1$ is the first such that (2.3.17) fails and let

$$\alpha \stackrel{\text{def}}{=} \frac{(1-\eta_2)}{2(L_{f,1}+\kappa_B)} \min\left[1, \frac{\Delta_0}{\|\nabla_x^1 f(x_0)\|}\right].$$

Since (2.3.7) ensures that $\gamma_1 \Delta_{k-1} \leq \Delta_k$, we deduce that

$$\frac{\Delta_{k-1}}{\min_{i\in[0:k]} \|\nabla_x^1 f(x_i)\|} < \alpha,$$

and thus, since $\|\nabla_x^1 f(x_{k-1})\| \geq \min_{i\in[0:k-1]} \|\nabla_x^1 f(x_i)\| \geq \min_{i\in[0:k]} \|\nabla_x^1 f(x_i)\|$, we have that

$$\frac{\Delta_{k-1}}{\|\nabla_x^1 f(x_{k-1})\|} \leq \frac{\Delta_{k-1}}{\min_{i\in[0:k-1]} \|\nabla_x^1 f(x_i)\|} \leq \frac{\Delta_{k-1}}{\min_{i\in[0:k]} \|\nabla_x^1 f(x_i)\|} < \alpha. \tag{2.3.18}$$

Hence Lemma 2.3.3 guarantees that $\Delta_{k-1} \leq \Delta_k$, and therefore, from the second inequality in (2.3.18) and the fact that (2.3.17) is violated at iteration k, that

$$\frac{\Delta_{k-1}}{\min_{i\in[0:k-1]} \|\nabla_x^1 f(x_i)\|} \leq \frac{\Delta_k}{\min_{i\in[0:k]} \|\nabla_x^1 f(x_i)\|} < \gamma_1 \alpha < \alpha.$$

As a consequence, (2.3.17) is also violated at iteration $k-1$. But this contradicts the assumption that iteration k is the first such that this inequality fails. This latter assumption is thus impossible, and no such iteration can exist. □

Lemma 2.3.5: Model decrease (TR1)

Suppose that **f.gL** *and* **m.HBa** *hold. Then we have that, for all* $k \geq 0$,

$$m_k(x_k) - m_k(x_k + s_k) \geq \kappa_{\text{decr}} \|\nabla_x^1 f(x_k)\| \min_{i\in[0:k]} \|\nabla_x^1 f(x_i)\|, \tag{2.3.19}$$

where

$$\kappa_{\text{decr}} \stackrel{\text{def}}{=} \tfrac{1}{2} \min\left[\frac{1}{1+\kappa_B}, \frac{\gamma_1(1-\eta_2)}{2(L_{f,1}+\kappa_B)} \min\left[1, \frac{\Delta_0}{\|\nabla_x^1 f(x_0)\|}\right]\right]. \tag{2.3.20}$$

Proof. The result directly follows from (2.3.4), (2.3.12), and (2.3.17). □

We are now in a position to bound the total number of successful iterations necessary to find an ϵ-approximate first-order minimizer using the TR1 algorithm.

Lemma 2.3.6: Bound on the number of successful TR1 iterations

Suppose that **f.Bb**, **f.gL**, *and* **m.HBa** *hold. Then there exists a positive constant* κ^S_{TR1} *such that, for any* $\epsilon > 0$, *the* TR1 *algorithm (p. 35) requires at most*

$$\#^{SI}_{\text{TR1}} \stackrel{\text{def}}{=} \kappa^S_{\text{TR1}} \frac{f(x_0) - f_{\text{low}}}{\epsilon^2}$$

successful iterations before an iterate x_ϵ *is computed for which* $\|\nabla_x^1 f(x_\epsilon)\| \leq \epsilon$.

Proof. First, recall that if $j \in \mathcal{U}_k$, $f(x_j) = f(x_{j+1})$ and $\nabla_x^1 f(x_j) = \nabla_x^1 f(x_{j+1})$. Suppose that $\|\nabla_x^1 f(x_j)\| > \epsilon$ for all $j \in \mathcal{S}_k$ and thus that $\|\nabla_x^1 f(x_j)\| > \epsilon$ for all $0 \leq j \leq k$. Combining **f.Bb**, a telescoping sum for which $f(x_j) = f(x_{j+1})$ when $j \in \mathcal{U}_k$, (2.3.6), the fact that $\rho_j \geq \eta_1$ for all $j \in \mathcal{S}_k$, and (2.3.19), we deduce that

$$\begin{aligned} f(x_0) - f_{\text{low}} &\geq f(x_0) - f(x_{k+1}) \\ &= \sum_{j \in \mathcal{S}_k} [f(x_j) - f(x_{j+1})] \\ &\geq \sum_{j \in \mathcal{S}_k} \eta_1 [m_k(0) - m_k(s_k)] \\ &> \eta_1 \kappa_{\text{decr}} |\mathcal{S}_k| \epsilon^2, \end{aligned}$$

where κ_{decr} is given by (2.3.20). This provides an upper bound on $|\mathcal{S}_k|$ which is independent of k, from which we obtain the desired conclusion with

$$\kappa_{\text{TR1}}^{\text{S}} \stackrel{\text{def}}{=} (\eta_1 \kappa_{\text{decr}})^{-1} = \frac{4(L_{f,1} + \kappa_B)}{\gamma_1 \eta_1 (1 - \eta_2)} \max\left[1, \frac{\|\nabla_x^1 f(x_0)\|}{\Delta_0}\right], \tag{2.3.21}$$

since $\gamma_1 < 1$, $\eta_2 \in (0,1)$, and $L_{f,1} \geq 1$. □

Finally, we may combine all the previous results with Lemma 2.3.1 to obtain an upper bound on the iteration (and function-evaluation) complexity of the trust-region method for the task of computing an ϵ-approximate first-order minimizer.

Theorem 2.3.7: Complexity of the first-order trust-region algorithm

Suppose that **f.Bb**, **f.gL**, *and* **m.HBa** *hold. Then there exist positive constants $\kappa_{\text{TR1}}^{\text{S}}$ (see Lemma 2.3.6), κ_{TR1}^{A}, κ_{TR1}^{B}, and κ_{TR1}^{C} such that the* TR1 *algorithm (p. 35) requires at most*

$$\#_{\text{TR1}}^{F} \stackrel{\text{def}}{=} \kappa_{\text{TR1}}^{A} \frac{f(x_0) - f_{\text{low}}}{\epsilon^2} + \kappa_{\text{TR1}}^{B} |\log \epsilon| + \kappa_{\text{TR1}}^{C} = \mathcal{O}\left(\epsilon^{-2}\right)$$

function evaluations and at most

$$\#_{\text{TR1}}^{D} \stackrel{\text{def}}{=} \kappa_{\text{TR1}}^{S} \frac{f(x_0) - f_{\text{low}}}{\epsilon^2} + 1 = \mathcal{O}\left(\epsilon^{-2}\right)$$

gradient evaluations to produce an iterate x_ϵ such that $\|\nabla_x^1 f(x_\epsilon)\| \leq \epsilon$.

Proof. The number of successful iterations needed to find an ϵ-approximate first-order minimizer is bounded above by Lemma 2.3.6. Since for all iterations before termination $\|\nabla_x^1 f(x_k)\| > \epsilon$, (2.3.17) implies that $\Delta_k \geq \Delta_{\min}$, where

$$\Delta_{\min} = \frac{\gamma_1(1 - \eta_2)}{2(L_{f,1} + \kappa_B)} \min\left[1, \frac{\Delta_0}{\|\nabla_x^1 f(x_0)\|}\right] \epsilon.$$

Thus the total number of iterations (including unsuccessful ones) can be bounded above using Lemma 2.3.1, yielding that at most

$$\kappa_{\text{TR1}}^{\text{S}} \frac{f(x_0) - f_{\text{low}}}{\epsilon^2} \left(1 + \frac{\log \gamma_3}{|\log \gamma_2|}\right) + \frac{1}{|\log \gamma_2|} \left|\log\left(\frac{\Delta_{\min}}{\Delta_0}\right)\right|$$

iterations are required. Since the TR1 algorithm uses at most one evaluation of f per iteration and at most one evaluation of its gradient per successful iteration (plus one evaluation of f and g at the final iteration), we deduce that the desired conclusion holds with

$$\kappa_{\text{TR1}}^{\text{A}} \stackrel{\text{def}}{=} \kappa_{\text{TR1}}^{\text{S}} \left(1 + \frac{\log \gamma_3}{|\log \gamma_2|}\right), \quad \kappa_{\text{TR1}}^{\text{B}} \stackrel{\text{def}}{=} \frac{1}{|\log \gamma_2|}, \tag{2.3.22}$$

and

$$\kappa_{\text{TR1}}^{\text{C}} \stackrel{\text{def}}{=} \frac{1}{|\log \gamma_2|} \left|\log\left(\frac{\gamma_1(1-\eta_2)}{2\Delta_0(L_{f,1}+\kappa_B)} \min\left[1, \frac{\Delta_0}{\|\nabla_x^1 f(x_0)\|}\right]\right)\right| + 1. \tag{2.3.23}$$

□

The worst-case complexity of the TR1 algorithm is thus of the same order as that of steepest descent (the term in $|\log \epsilon|$ is small compared to that in ϵ^{-2} as ϵ shrinks), which is not unexpected because we only requested a model decrease at least equal to that at the Cauchy point, itself clearly resulting from a move along $-\nabla_x^1 f(x_k)$.

If we now return to the one-dimensional example of Theorem 2.2.3, it is interesting that we can view the iterates and values given by (2.2.18)–(2.2.19) as the result of applying the TR1 algorithm from x_0 with $B_k = 2t_{\text{ini}}^{-1}$ and $\Delta_0 = 1$. All iterations are successful since the model and the objective-function decreases match exactly, implying that $\rho_k = 1$ for each k. We may thus define $\Delta_k = 1$ for all $k \geq 0$. As a consequence, we obtain the following sharpness result.

Theorem 2.3.8: Slow first-order convergence of the TR1 algorithm

The TR1 *algorithm (p. 35) applied to a function f satisfying* **f.Bb**, **f.gL**, *and* **m.HBa** *may require as many as ϵ^{-2} evaluations of $f(x)$ and $\nabla_x^1 f(x)$ to produce an iterate x_ϵ such that $\|\nabla_x^1 f(x_\epsilon)\| \leq \epsilon$.*

Notes and references for Section 2.3

The trust-region method analyzed here in the guise of the TR1 algorithm is reduced to its essential features, but it can be extended in many different ways. We refer the reader to the comprehensive book [131] for in-depth coverage of this domain and to [374] for a recent survey. We simply note that obvious extensions include nonquadratic objective-function models, alternative techniques for computing the Cauchy point, and iteration-dependent norms for the definition of the region $\mathcal{B}_{\Delta_k}$. The complexity result of Theorem 2.3.7 extends at least to these variants.

The first analysis of worst-case complexity for trust-region methods is, to the best of our knowledge, to be found in [210,212], where it is considered in the context of a multilevel version of the trust-region method.

The techniques of proof in this section are clearly inspired by [131]. In particular, Lemma 2.3.2 is a simplified version of Theorem 6.3.1 in this reference, while Lemma 2.3.4 is an adapted condensation of Theorems 6.4.2 and 6.4.3. Lemma 2.3.1 is inspired by developments in [84].

2.4 The adaptive-regularization method

The last class of methods we shall describe in this chapter are *adaptive-regularization algorithms*. As we will discover as the book evolves, these are highly significant. In particular, to date, they provide the lowest[13] worst-case evaluation-complexity bounds of all known algorithms for unconstrained problems, independent of the order of derivatives used.

Adaptive-regularization algorithms are closely related to trust-region methods in that they also use a *model* whose purpose is to approximate the objective function in a neighbourhood of the current iterate. While trust-region methods define this neighbourhood explicitly (by specifying a trust-region radius), adaptive-regularization methods define it implicitly by penalizing long steps. This is achieved by adding a *regularization term*, comprising a *regularization parameter*[14] multiplied by a specific power of the steplength, to an unregularized model of f. A trial point is then computed by (approximately) minimizing this *regularized* model. As for trust-region methods, the achieved objective-function reduction at the trial point is compared to that predicted by the *unregularized* model. The trial point is accepted as the next iterate if the ratio of actual to predicted reductions is large enough, and the regularization parameter allowed to decrease, thereby encouraging longer steps. Conversely, if this ratio is small, the trial point is rejected, and the regularization parameter increased, with the aim of shortening the next attempted step.

The rationale behind the regularization parameter update is given by the important overestimation property (A.19) in Corollary A.8.4 (p. 437): this property states that the regularized model must be an overestimate of the true objective function whenever the regularization parameter is large enough (and exceeds the associated Lipschitz constant). While this argument may seem somewhat fuzzy at this stage, in particular because we have not yet defined the unregularized and regularized models, the power of that estimation property will become clear as we proceed.

2.4.1 The AR1 algorithm

When solving the unconstrained minimization problem (2.1.1) using first-order information, the corresponding adaptive-regularization algorithm (AR1) once again proceeds by iteration. Given an iterate x_k, the method uses the *regularized model*

$$m_k(s) \stackrel{\text{def}}{=} f(x_k) + \langle \nabla_x^1 f(x_k), s\rangle + \tfrac{1}{2}\sigma_k\|s\|^2 = T_{f,1}(x_k, s) + \tfrac{1}{2}\sigma_k\|s\|^2, \tag{2.4.1}$$

where $T_{f,1}(x_k, s)$ is the first-order Taylor-series expansion of f around x_k (the *unregularized model*) and where $\frac{1}{2}\sigma_k\|s\|^2$ is the *regularization term* involving the regularization parameter $\sigma_k > 0$. A step s_k is then computed by minimizing $m_k(s)$. The remainder of the AR1 algorithm proceeds as we have outlined in the paragraph above. The complete AR1 algorithm is formally stated as follows.

Algorithm 2.4.1: Adaptive-regularization algorithm with a first-order model (AR1)

Step 0: Initialization. An initial point x_0, an accuracy level $\epsilon > 0$, and an initial regularization parameter $\sigma_0 > 0$ are given. The constants η_1, η_2, γ_1, γ_2, γ_3, and $\sigma_{\min}$ are also given and satisfy

$$\sigma_{\min} \in (0, \sigma_0], \quad 0 < \eta_1 \le \eta_2 < 1, \quad \text{and} \quad 0 < \gamma_1 < 1 < \gamma_2 < \gamma_3. \tag{2.4.2}$$

Compute $f(x_0)$ and $\nabla_x^1 f(x_0)$, and set $k = 0$.

---

[13] Recent competitors, often hybrids, have sometimes achieved as good bounds, but never better ones. We shall discuss these at the appropriate time.

[14] Sometimes referred to as the regularization *weight*.

Step 1: Test for termination. If $\|\nabla_x^1 f(x_k)\| \leq \epsilon$, terminate with the ϵ-approximate first-order minimizer $x_\epsilon = x_k$.

Step 2: Step computation. Let

$$s_k = \arg\min_{s \in \mathbb{R}^n} m_k(s).$$

Step 3: Acceptance of the trial point. Compute $f(x_k + s_k)$, and define

$$\rho_k = \frac{f(x_k) - f(x_k + s_k)}{T_{f,1}(x_k, 0) - T_{f,1}(x_k, s_k)}. \tag{2.4.3}$$

If $\rho_k \geq \eta_1$, then define $x_{k+1} = x_k + s_k$ and compute $\nabla_x^1 f(x_{k+1})$. Otherwise define $x_{k+1} = x_k$.

Step 4: Regularization parameter update. Set

$$\sigma_{k+1} \in \begin{cases} [\max(\sigma_{\min}, \gamma_1 \sigma_k), \sigma_k] & \text{if } \rho_k \geq \eta_2, \\ [\sigma_k, \gamma_2 \sigma_k] & \text{if } \rho_k \in [\eta_1, \eta_2), \\ [\gamma_2 \sigma_k, \gamma_3 \sigma_k] & \text{if } \rho_k < \eta_1. \end{cases} \tag{2.4.4}$$

Increment k by one and go to Step 1.

In this description, reasonable values for the constants of (2.4.2) might be $\eta_1 = 0.01$, $\eta_2 = 0.9$, $\gamma_1 = 2$, and $\gamma_2 = \gamma_3 = 0.5$.

For this very simple first-order case, it is easy to verify that $m_k(s)$ is strongly convex and that its (global) minimizer is obtained at

$$s_k = \arg\min_{s \in \mathbb{R}^n} m_k(s) = -\frac{\nabla_x^1 f(x_k)}{\sigma_k}. \tag{2.4.5}$$

Observe that (2.4.5) and the definition of $T_{f,1}(x_k, s)$ in (2.4.1) ensure that

$$T_{f,1}(x_k, 0) - T_{f,1}(x_k, s_k) = \frac{\|\nabla_x^1 f(x_k)\|^2}{\sigma_k} \tag{2.4.6}$$

and hence that we may rewrite (2.4.3) as

$$\rho_k = \sigma_k \frac{f(x_k) - f(x_k + s_k)}{\|\nabla_x^1 f(x_k)\|^2}. \tag{2.4.7}$$

Notice that (2.4.3) is identical to (2.3.6), and that the TR1 and AR1 algorithms are similar in many other aspects. However, the AR1 algorithm differs from the gradient-related linesearch methods of Section 2.2.3.5 and the TR1 algorithm in that its step is always a multiple of the steepest-descent direction because of (2.4.5). Indeed, it may not have escaped the reader's attention that the AR1 algorithm may also be viewed as a reformulation of the SDBA method where an attempt is made—via the minimization of the regularized model (2.4.1), which itself depends on the adaptive-regularization parameter σ_k—to determine a good value of t_{ini} at each iteration. Such similarities between linesearch and regularization methods are peculiar to first-order models; as we shall see in later chapters, the search directions resulting from higher-order regularized models almost always change direction as the regularization parameter increases, while a linesearch method persists in the same direction until it succeeds.

We now define the sets of successful and unsuccessful iterations as with the TR1 algorithm using (2.3.10). As was the case then, we bound the number of unsuccessful iterations by a multiple of the number of successful ones. The proof of the next lemma is very close in spirit to that of Lemma 2.3.1.

Lemma 2.4.1: Successful and unsuccessful adaptive-regularization iterations

Suppose that the AR1 *algorithm is used and that* $\sigma_k \leq \sigma_{\max}$ *for some* $\sigma_{\max} > 0$. *Then*

$$k \leq |\mathcal{S}_k| \left(1 + \frac{|\log \gamma_1|}{\log \gamma_2}\right) + \frac{1}{\log \gamma_2} \log\left(\frac{\sigma_{\max}}{\sigma_0}\right). \tag{2.4.8}$$

Proof. The regularization parameter update (2.4.4) gives that, for each k,

$$\gamma_1 \sigma_j \leq \max[\gamma_1 \sigma_j, \sigma_{\min}] \leq \sigma_{j+1}, \quad j \in \mathcal{S}_k, \quad \text{and} \quad \gamma_2 \sigma_j \leq \sigma_{j+1}, \quad j \in \mathcal{U}_k.$$

Thus we deduce inductively that

$$\sigma_0 \gamma_1^{|\mathcal{S}_k|} \gamma_2^{|\mathcal{U}_k|} \leq \sigma_k.$$

Therefore, using our assumption that $\sigma_k \leq \sigma_{\max}$, we deduce that

$$|\mathcal{S}_k| \log \gamma_1 + |\mathcal{U}_k| \log \gamma_2 \leq \log\left(\frac{\sigma_{\max}}{\sigma_0}\right).$$

The desired result (2.4.8) then follows from the equality $k = |\mathcal{S}_k| + |\mathcal{U}_k|$ and the inequality $\gamma_1 < 1$ given by (2.4.2). □

Observe that this result is completely independent of the form of the model $m_k(s)$ and depends entirely on the mechanism defined by (2.4.4).

2.4.2 Complexity of finding approximate first-order minimizers

The worst-case analysis of the AR1 algorithm is interesting because of its simple yet powerful nature, and this make it extendable in many useful ways. Its structure is the backbone of the most effective analysis tool proposed in this book.

The first step of the analysis is to show that the definition of the regularized model guarantees a quantifiable measure of the reduction obtained in the unregularized one, as given by (2.4.6). For this decrease to be bounded away from zero, we need to establish that the regularization parameters cannot grow to infinity.

Lemma 2.4.2: Upper bound on the regularization parameter

Suppose that **f.gL** *holds. Then, for all* $k \geq 0$,

$$\sigma_k \leq \sigma_{\max} \stackrel{\text{def}}{=} \max\left[\sigma_0, \frac{\gamma_3 L_{f,1}}{2(1-\eta_2)}\right]. \tag{2.4.9}$$

Proof. Suppose that

$$\sigma_k \geq \frac{L_{f,1}}{2(1-\eta_2)}. \tag{2.4.10}$$

Using successively (2.4.3), the overestimation property (A.19) in Corollary A.8.4 (p. 437) for the case $p = 1$, (2.4.6), (2.4.5), and (2.4.10), we may then deduce that

$$|\rho_k - 1| = \frac{|f(x_k + s_k) - T_{f,1}(x_k, s_k)|}{|T_{f,1}(x_k, 0) - T_{f,1}(x_k, s_k)|} \leq \frac{L_{f,1}}{2\sigma_k} \leq 1 - \eta_2$$

and hence that $\rho_k \geq \eta_2$. Thus iteration k is successful, and it follows from (2.4.4) that $\sigma_{k+1} \leq \sigma_k$. As a consequence, the mechanism of (2.4.4) in the algorithm ensures that (2.4.9) holds. □

The intuition behind this lemma (and behind all of its variants to come) is that, whenever the regularization parameter exceeds the relevant Lipschitz constant, the regularized model *overestimates* f because of Corollary A.8.4 (p. 437). Therefore any decrease obtained in this model is a lower bound on the decrease in f, ensuring that every iteration is successful, and consequently removing the need to increase the regularization parameter any further.

We may now focus on, at most, how many *successful* iterations may be needed to compute an ϵ-approximate first-order minimizer using the AR1 algorithm.

Lemma 2.4.3: Bound on the number of successful AR1 iterations

Suppose that **f.Bb** *and* **f.gL** *hold. Then there exists a positive constant* κ^{S}_{AR1} *such that the* AR1 *algorithm (p. 42) requires at most*

$$\#^{SI}_{\text{AR1}} \stackrel{\text{def}}{=} \kappa^{S}_{\text{AR1}} \frac{f(x_0) - f_{\text{low}}}{\epsilon^2} \tag{2.4.11}$$

successful iterations before an iterate x_ϵ *is computed for which* $\|\nabla^1_x f(x_\epsilon)\| \leq \epsilon$.

Proof. Just as in the proof of Lemma 2.3.6, suppose that $\|\nabla^1_x f(x_j)\| > \epsilon$ for all $j \in [1:k]$. Then, using **f.Bb**, a telescoping sum, the definition of $|\mathcal{S}_k|$ in (2.3.10), (2.4.7), the fact that $\rho_j \geq \eta_1$ for all $j \in \mathcal{S}_k$, and Lemma 2.4.2, we have that

$$\begin{aligned} f(x_0) - f_{\text{low}} &\geq f(x_0) - f(x_{k+1}) \\ &= \sum_{j \in \mathcal{S}_k} [f(x_j) - f(x_{j+1})] \\ &\geq \sum_{j \in \mathcal{S}_k} \frac{\rho_j \|\nabla^1_x f(x_j)\|^2}{\sigma_j} \\ &\geq \eta_1 \sum_{j \in \mathcal{S}_k} \frac{\|\nabla^1_x f(x_j)\|^2}{\sigma_j} \\ &\geq \frac{\eta_1}{\sigma_{\max}} \sum_{j \in \mathcal{S}_k} \|\nabla^1_x f(x_j)\|^2 \\ &> \frac{\eta_1}{\sigma_{\max}} |\mathcal{S}_k| \epsilon^2. \end{aligned}$$

This implies the bound

$$|\mathcal{S}_k| < \frac{\sigma_{\max}[f(x_0) - f_{\text{low}}]}{\eta_1 \epsilon^2},$$

which, in view of (2.4.9), gives the desired conclusion with

$$\kappa_{\text{AR1}}^{\text{S}} \stackrel{\text{def}}{=} \frac{1}{\eta_1} \max\left[\sigma_0, \frac{\gamma_3 L_{f,1}}{2(1-\eta_2)}\right].$$

□

Given Lemma 2.4.1, we are now ready to state the worst-case evaluation bound for the AR1 algorithm.

Theorem 2.4.4: Complexity of the first-order adaptive-regularization algorithm

Suppose that **f.Bb** *and* **f.gL** *hold. Then there exist positive constants* κ_{AR1}^{S} *(see Lemma 2.4.3),* κ_{AR1}^{A}*, and* κ_{AR1}^{C} *such that the* AR1 *algorithm (p. 42) requires at most*

$$\#_{\text{AR1}}^{F} \stackrel{\text{def}}{=} \kappa_{\text{AR1}}^{A} \frac{f(x_0) - f_{\text{low}}}{\epsilon^2} + \kappa_{\text{AR1}}^{C} = \mathcal{O}\left(\epsilon^{-2}\right) \tag{2.4.12}$$

function evaluations and at most

$$\#_{\text{AR1}}^{D} \stackrel{\text{def}}{=} \kappa_{\text{AR1}}^{S} \frac{f(x_0) - f_{\text{low}}}{\epsilon^2} + 1 = \mathcal{O}\left(\epsilon^{-2}\right)$$

gradient evaluations to compute an iterate x_ϵ *such that* $\|\nabla_x^1 f(x_\epsilon)\| \leq \epsilon$.

Proof. The number of successful iterations needed to find an ϵ-approximate first-order minimizer is bounded above by (2.4.11). The total number of iterations (including unsuccessful ones) can also be bounded above using Lemma 2.4.1 with $\sigma_{\max}$ given by Lemma 2.4.2, which yields that this number is at most

$$\kappa_{\text{AR1}}^{S} \frac{f(x_0) - f_{\text{low}}}{\epsilon^2} \left(1 + \frac{|\log \gamma_1|}{\log \gamma_2}\right) + \frac{1}{\log \gamma_2} \log\left(\frac{\sigma_{\max}}{\sigma_0}\right).$$

Since the AR1 algorithm uses at most one evaluation of f per iteration and at most one evaluation of its gradient per successful iteration (plus one evaluation of f and g at the final iteration), we deduce from (2.4.9) and Lemma 2.4.3 that the desired conclusion holds with

$$\kappa_{\text{AR1}}^{\text{A}} \stackrel{\text{def}}{=} \frac{1}{\eta_1} \max\left[\sigma_0, \frac{\gamma_3 L_{f,1}}{1-\eta_2}\right] \left(1 + \frac{|\log \gamma_1|}{\log \gamma_2}\right)$$

and

$$\kappa_{\text{AR1}}^{\text{C}} \stackrel{\text{def}}{=} \frac{1}{\log \gamma_2} \log\left(\max\left[1, \frac{\gamma_3 L_{f,1}}{\sigma_0(1-\eta_2)}\right]\right) + 1.$$

□

We may now exploit the one-dimensional example of Theorem 2.2.3 once more and verify that applying the AR1 algorithm to the objective function f defined in this section using $\sigma_0 = t_{\text{ini}}^{-1}$ again yields the slowly convergent sequence of iterates. This is because the unregularized model decrease always matches the achieved objective reduction and thus σ_k can be kept constant and equal to σ_0 for all k. As a consequence, we obtain the following sharpness result.

Theorem 2.4.5: Slow first-order convergence of the AR1 algorithm

*The AR1 algorithm (p. 42) applied to minimize a function f satisfying **f.Bb** and **f.gL** may require as many as ϵ^{-2} evaluations of $f(x)$ and $\nabla_x^1 f(x)$ to produce an iterate x_ϵ such that $\|\nabla_x^1 f(x_\epsilon)\| \leq \epsilon$.*

Thus far, the reader may be unconvinced by the hyperbole with which we started the section. There seems little to choose between linesearch methods such as the SDBA algorithm, trust-region ones like the TR1 algorithm, and the much-heralded AR1 regularization algorithm that we have just encountered. Each of these appears to require at most $\mathcal{O}(\epsilon^{-2})$ evaluations to find an ϵ-first-order approximate minimizer, and for each there are examples for which this pessimistic bound may be achieved. The most important lessons here are more how each of these methods is structured, and the different tricks at our disposal for analyzing them. Fear not—significant differences that justify our interest will start to emerge from the next chapter onward.

Notes and references for Section 2.4

The origins of the adaptive-regularization algorithm may, in our view, be traced back to two different contributions. The first [284], related specifically to the AR1 algorithm, is the proximal-point method, in which the step is obtained by minimizing a linear approximation of the (possibly nonsmooth) objective function augmented by a quadratic penalty term in the steplength (just as in AR1). The second is an unpublished paper [218] where an adaptive-regularization method for a second-order model was proposed. This method was subsequently rediscovered and analyzed [305, 364], with the important former paper providing the first worst-case complexity analysis for a version of the algorithm requiring exact model minimization, a requirement that has subsequently been relaxed [85]. More detailed coverage of the origins and development of the adaptive-regularization method can be found in [84].

Interestingly, the early versions of the adaptive-regularization algorithm used a definition of the ratio ρ_k which compared the achieved reduction in the objective function to that predicted by the *regularized* model. This resulted in a substantially more complicated analysis using the Cauchy-point decrease (see [85]). The statement of the AR1 algorithm (with the definition (2.4.3) using the *unregularized* model decrease as the denominator) and the resulting worst-case analysis are largely inspired by [45, 157, 158]. Lemma 2.3.1 is Theorem 2.1 in [85].

Although the analysis we have presented here uses the strong Lipschitz assumption **f.gL** for brevity, it has long been recognized that essentially identical results are possible if instead (2.1.5) is simply required to hold for all x and y in an open convex set containing the "tree of iterates" defined as $\cup_{k\geq 0}[x_k, x_k + s_k]$, where the x_k are the iterates generated by the algorithm and the s_k are the corresponding trial steps [85].

2.5 A final comment

As it turns out, Vavasis [359] has given an example of a bound-constrained nonconvex problem with $n = 2$ for which *any* method based on first-order information alone is guaranteed to take $\mathcal{O}(\epsilon^{-1/2})$ iterations—and function evaluations—to find an approximate minimizer. This bound is exceeded by all results presented in this chapter and corresponds to the known optimal complexity bound for first-order methods on *convex* problems (see Section 1.2.5). It is therefore unlikely, although not yet provably impossible, that it will be achievable for more general *nonconvex* cases.

Chapter 3

Methods Using Second Derivatives

Chapter summary

We next turn our attention to second-order algorithms, that is, methods which use oracles or product-oracles of degree at most two for problems whose Hessians exist. We might expect the additional information provided by the Hessian to ensure a better worst-case performance, but we quickly see that such a hope may be misplaced. Indeed we show that the celebrated Newton's method can converge as slowly as steepest descent in its worst case! And the same is true for a number of traditional second-order methods, such as the Goldfeld–Quandt–Trotter and second-order trust-region algorithms in most cases.

By stark contrast, the second-order regularization method will be shown to be substantially faster, since we will prove that it requires at most (and may require at least) $\mathcal{O}(\epsilon_1^{-3/2})$ *evaluations of the objective function and its first two derivatives to find an* ϵ_1*-approximate first-order minimizer. Moreover, convergence to* (ϵ_1, ϵ_2)*-approximate second-order minimizers will be achieved in at most* $\mathcal{O}(\max[\epsilon_1^{-3/2}, \epsilon_2^{-3}])$ *evaluations of the objective function and its first two derivatives.*

We then discuss a variety of algorithmic variants that produce regularization-like fast convergence, including crossovers between trust-region and regularization schemes, as well as alternative regularization techniques. We conclude the chapter by providing a general framework for methods that ensure fast convergence.

3.1 Slow convergence of Newton's method

What happens to the worst-case evaluation complexity for methods where we now allow the use of (true) second derivatives? The answer to this question is somewhat surprising. Indeed, *many second-order algorithms do* not *converge faster, in the worst case, than first-order ones, in their worst case.*

To substantiate this claim, we shall construct an example showing that Newton's method, the prototype of most second-order algorithms, may require at least $\mathcal{O}(\epsilon_1^{-2})$ evaluations of the objective function, as well as its gradient and Hessian, before finding an ϵ_1-approximate first-order minimizer. Note the use of ϵ_1 in the previous sentence, rather than ϵ (as in the previous chapter): this is because convergence to second-order points will involve both first- and second-order optimality tests, for which we will use different tolerances ϵ_1 and ϵ_2. The need for ϵ_1 and ϵ_2 will become clearer when we discuss the globalized version of Newton's method in Sections 3.2 and 3.3. We will also return to the question of choosing optimality goals in Section 12.1.

Of course the reader might reasonably contend that because the unfettered Newton iteration is known to diverge in many cases [172], it is hardly surprising that a poor worst-case bound could arise without precautions, even in favourable circumstances. However, this is not the point, but rather that many of the methods we shall consider later in this chapter reduce to Newton's method when close to a minimizer, indeed sometimes even globally, throughout the run of the algorithm, under suitable conditions such as the presence of positive curvature. It is thus helpful to consider what might happen under these supposedly favourable circumstances.

The standard Newton iteration for solving the unconstrained minimization problem uses the following assumption.

f.C2	f is twice continuously differentiable.

The method is outlined in Algorithm 3.1.1.

Algorithm 3.1.1: Newton's method for unconstrained minimization

Step 0: Initialization. A starting point $x_0 \in \mathbb{R}^n$ is given. Set $k = 0$.

Step 1: Test for termination. Compute $\nabla_x^1 f(x_k)$. If $\|\nabla_x^1 f(x_k)\| \leq \epsilon_1$, terminate with x_k as the ϵ_1-approximate first-order minimizer.

Step 2: Step computation. Compute $\nabla_x^2 f(x_k)$ and the step s_k as the solution of the linear system

$$\nabla_x^2 f(x_k) s_k = -\nabla_x^1 f(x_k). \tag{3.1.1}$$

Step 3: Acceptance of the trial point. Set $x_{k+1} = x_k + s_k$, increment k by one, and go to Step 1.

Of course, the method as described in this outline makes strong (favourable) assumptions: it presumes that the matrix $\nabla_x^2 f(x_k)$ is positive definite for each k and that $f(x_k + s_k)$ is sufficiently smaller than $f(x_k)$ without any specific globalization procedure, such as a linesearch or a shrinking trust region. However, since our example will ensure both of these properties, the above description is sufficient for our purpose. It is important, for what follows, that the step defined by (3.1.1) be equivalently defined by

$$s_k = \arg\min_{s \in \mathbb{R}^n} T_{f,2}(x_k, s) = \arg\min_{s \in \mathbb{R}^n} f(x_k) + \langle \nabla_x^1 f(x_k), s \rangle + \tfrac{1}{2} \langle s, \nabla_x^2 f(x_k) s \rangle \tag{3.1.2}$$

whenever $\nabla_x^2 f(x_k)$ is positive definite.[15]

The complexity result we are about to present is actually true for a more restrictive class of twice continuously differentiable functions, those with Lipschitz-continuous Hessians. These are defined as follows.

[15] A globalization technique, such as that in the TR2 algorithm discussed later in this chapter, is necessary when $\nabla_x^2 f(x_k)$ is not positive definite.

f.HL $f(x)$ is twice continuously differentiable in $\mathbb{R}^n$, and its Hessian $\nabla_x^2 f(x)$ is Lipschitz continuous on $\mathbb{R}^n$; that is, there exists a constant $L_{f,2} \geq 1$ such that

$$\|\nabla_x^2 f(x) - \nabla_x^2 f(y)\| \leq L_{f,2}\|x - y\| \tag{3.1.3}$$

for all $x, y \in \mathbb{R}^n$.

Of course **f.HL** subsumes **f.C2**, and **f.HL** holds if $f \in C^3$ with bounded third derivatives. The assumption that $L_{f,2} \geq 1$ is made without losing generality and simply saves us all from a sequence of expressions involving terms such as $\max[1, L_{f,2}]$.

Theorem 3.1.1: Slow first-order convergence of Newton's method

*Newton's method (p. 50) applied to minimize a function f satisfying **f.Bb**, **f.gL**, and **f.HL** may require more than $\lceil \epsilon_1^{-2} \rceil$ evaluations of $f(x)$, $\nabla_x^1 f(x)$, and $\nabla_x^2 f(x)$ to produce an iterate x_ϵ such that $\|\nabla_x^1 f(x_\epsilon)\| \leq \epsilon_1$.*

Proof. Our aim is to build a function $f : \mathbb{R}^2 \to \mathbb{R}$ such that, for any $\epsilon_1 \in (0,1)$, Newton's method takes at least ϵ_1^{-2} iterations to find an ϵ_1-approximate first-order minimizer x_ϵ for which

$$\|\nabla_x^1 f(x_\epsilon)\| \leq \epsilon_1,$$

when started from the origin.

Consider a hypothetical sequence of iterates $\{x_k\}_{k=0}^{\infty}$ such that $f(x_k) = f_k$, $\nabla_x^1 f(x_k) = g_k$, and $\nabla_x^2 f(x_k) = H_k$ are defined, for $k \geq 0$, by the relations

$$f_k = 2 - k\epsilon_1^2, \quad g_k = \begin{pmatrix} -\epsilon_1^2 f_k \\ -\epsilon_1 f_k \end{pmatrix}, \quad \text{and} \quad H_k = \begin{pmatrix} \epsilon_1^2 f_k^2 & 0 \\ 0 & f_k^2 \end{pmatrix}. \tag{3.1.4}$$

If this sequence of function, gradient, and Hessian values could be generated by Newton's method starting from the origin and applied to a twice continuously differentiable function with globally Lipschitz-continuous Hessian, then for $k \in [0 : k_\epsilon]$ with

$$k_\epsilon \stackrel{\text{def}}{=} \left\lceil \epsilon_1^{-2} \right\rceil,$$

it is easy to check that

$$f_k \in [1,2], \quad \|g_k\| = \epsilon_1 f_k \sqrt{1 + \epsilon_1^2} > \epsilon_1, \quad \text{and} \quad s_k = \frac{1}{f_k} \begin{pmatrix} 1 \\ \epsilon_1 \end{pmatrix}, \tag{3.1.5}$$

with the last equality resulting from (3.1.1). As a consequence, this Newton iteration would require at least k_ϵ iterations (and evaluations of f_k, g_k, and H_k) before terminating. In addition,

$$x_0 = \begin{pmatrix} 0 \\ 0 \end{pmatrix} \quad \text{and} \quad x_k = \sum_{j=0}^{k-1} s_j. \tag{3.1.6}$$

The next step in our construction is to build a smooth function with bounded second and third derivatives (which implies that its gradient and Hessian are Lipschitz continuous) interpolating (3.1.4) at (3.1.6). The idea is to exploit the large components of the step along the first dimension (see (3.1.5)) to define $f(x)$ and its first two derivatives to be identical to those of

$$q_k(x) \stackrel{\text{def}}{=} f_k + \langle g_k, x - x_k \rangle + \tfrac{1}{2}\langle x - x_k, H_k(x - x_k)\rangle \equiv T_{f,2}(x_k, x - x_k),$$

for which $q_k(x_k) = f_k$, $\nabla_x^1 q_k(x_k) = g_k$, and $\nabla_x^2 q_k(x_k) = H_k$, in a domain around x_k. More specifically, for $k \geq 1$, let

$$\sigma(\alpha) \stackrel{\text{def}}{=} \begin{cases} 1 & \text{if } 0 \leq \alpha \leq \frac{1}{6}, \\ 1 + 27(\alpha - \frac{1}{6})^3 \left[-10 + 45(\alpha - \frac{1}{6}) - 54(\alpha - \frac{1}{6})^2\right] & \text{if } \frac{1}{6} \leq \alpha \leq \frac{1}{2}, \\ 0 & \text{if } \alpha \geq \frac{1}{2}. \end{cases}$$

This piecewise polynomial is defined[16] to be identically equal to 1 near the origin and to decrease smoothly to zero (with bounded first, second, and third derivatives) between $\frac{1}{6}$ and $\frac{1}{2}$. Using this function, we may then define, for each $k \geq 0$, a *local support* function $\sigma(\|x - x_k\|)$ that is identically equal to one in a (circular) neighbourhood of x_k of radius $\frac{1}{6}$ and decreases smoothly (with bounded first, second, and third derivatives) to zero for all points whose distance to x_k exceeds $\frac{1}{2}$.

We are now in a position to define the objective function as

$$f_{SN}(x) = \sum_{k=0}^{k_\epsilon} \sigma(\|x - x_k\|)\, q_k(x) \tag{3.1.7}$$

for all x in $\mathbb{R}^2$. Note that the sum in (3.1.7) involves at most two nonzero terms for each x, because

$$[s_k]_1 = 1/f_k > \tfrac{1}{2},$$

where we used (3.1.5), and thus the distance between the first components of successive iterates (and thus also between the iterates themselves) exceeds $\frac{1}{2}$. This function, with its first two derivatives, obviously interpolates (3.1.4), is bounded below, and has bounded first, second, and third derivatives since the large values of $q_k(x)$ that occur in their expressions always occur far from x_k and are thus annihilated by the support function. In particular, f satisfies **f.Bb**, **f.gL**, and **f.HL**. □

The contour lines of $f_{SN}(x)$ as given by (3.1.7) are shown in Figure 3.1.1 on the next page, together with the iteration path. A perspective view is provided in Figure 3.1.2. In these figures, we see the path of iterates hopping from one "pass" to the next along the very slowly decreasing mountain ridge.

Although the reader might at first be concerned that restricting the example presented in Theorem 3.1.1 to two dimensions is somehow special, it is of course trivial to embed it in n dimensions. For instance, we might pick $f_n(x) = f([x]_1, [x]_2) + f_q([x]_3, \ldots, [x]_n)$, where f_q is a convex quadratic function of its argument. Such a function will behave precisely as f for all iterations $k \geq 1$ since Newton's method will minimize f_q in a single iteration.

Thus Newton's method can be as slow (in its worst case) as steepest descent (in its worst case)! Observe that the Hessians $\nabla_x^2 f(x_k)$ remain positive definite for all $k \geq 0$ but

[16] Using Hermite interpolation, with boundary conditions

$$\sigma\left(\tfrac{1}{6}\right) = 1, \quad \sigma'\left(\tfrac{1}{6}\right) = 0, \quad \sigma''\left(\tfrac{1}{6}\right) = 0, \quad \sigma\left(\tfrac{1}{2}\right) = 0, \quad \sigma'\left(\tfrac{1}{2}\right) = 0, \text{ and } \sigma''\left(\tfrac{1}{2}\right) = 0.$$

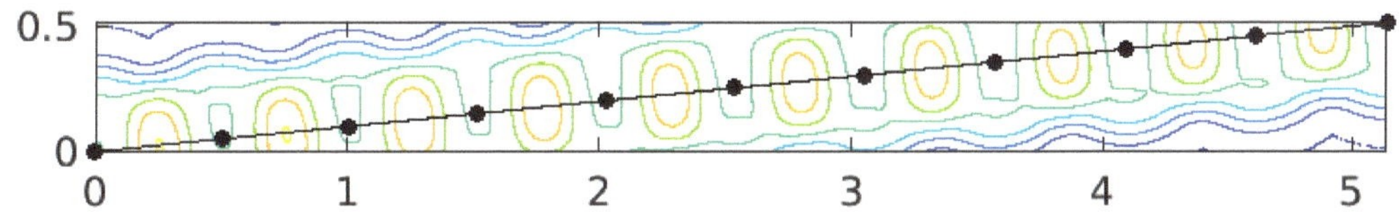

Figure 3.1.1. Contour lines of $f_{SN}(x)$ and the path of iterates for $k = 0, \dots, 10$.

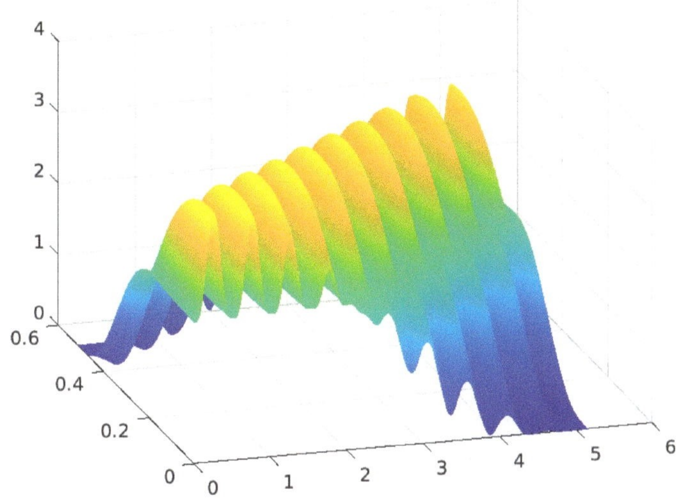

Figure 3.1.2. A view of $f_{SN}(x)$.

become singular in the limit. This is to be expected in the worst case, since otherwise the well-known locally quadratic rate of convergence of Newton's method would take over, providing very fast convergence instead. It is also possible to relax (3.1.2) to allow for an inexact minimization of the quadratic model by tolerating a nonzero but suitably small residual to the Newton's equation (3.1.1), using a condition of the form

$$\|\nabla_s^1 T_{f,2}(x_k, s_k)\| = \|\nabla_x^1 f(x_k) + \nabla_x^2 f(x_k) s_k\| \leq \theta \|\nabla_x^1 f(x_k)\|^\alpha \tag{3.1.8}$$

for some $\theta \in (0, 1]$ and $\alpha \geq 1$. Unsurprisingly, this relaxation does not improve the worst-case complexity. In the form of the algorithm of page 50, Newton's method lacks the ability to handle problems for which nonpositive curvature—that is, directions s for which the inequality $\langle s, \nabla_x^2 f(x_k) s \rangle \leq 0$—occurs. As a consequence, analyzing its worst-case behaviour for the task of finding approximate second-order minimizers is meaningless. We will return to this question later in this chapter, when we analyze globalized variants.

Notes and references for Section 3.1

Newton's method has a long and illustrious history as the prototypical rapidly converging local method [148, 172, 372]. The first indication that the method could be as slow as steepest descent in the worst case appeared in [83], where it was shown that the $\mathcal{O}(\epsilon_1^{-2})$ complexity bound was "essentially sharp" (meaning it could be approached arbitrarily closely for a variable objective function range; see the paper for details). This result, however, made explicit use of the relaxation of the Hessian Lipschitz-continuity assumption to the path of iterates, a hard to verify assumption. Such an assumption may be replaced by the more amenable requirement of

Lipschitz continuity of the Hessian on the whole space [94]. A stronger version of the result of [83] (now using a fixed objective function range) appeared later [106] within a somewhat more general context. In particular, the example discussed therein allows for an inexact minimization of the quadratic model as suggested by (3.1.8), thereby justifying our comment at the end of the section. The example presented here merges ideas of [94] and [106].

3.2 The trust-region method using second derivatives

3.2.1 The TR2 algorithm

As in Chapter 2, we next turn our attention to the most common variant of Newton's method using exact second derivatives that can be proved to be globally convergent: the second-order trust-region method TR2.

An obvious generalization of the TR1 algorithm would be to use the same framework, except that the step would then be computed as

$$s_k = \arg\min_{x_k+s\in\mathcal{R}_k} T_{f,2}(x_k,s). \tag{3.2.1}$$

We immediately note that there are efficient methods for computing the step (3.2.1), and details are given in Chapters 9 and 10 (see also [131, Chap. 7]). These methods typically involve factorizing the matrix $\nabla_x^2 f(x_k)+\lambda I$ for some suitable $\lambda\geq 0$, thereby ensuring the positive semidefiniteness of the modified Hessian. Such methods are based on the remarkable fact that s_k satisfies the linear system

$$(\nabla_x^2 f(x_k)+\lambda_k I)s_k = -\nabla_x^1 f(x_k), \tag{3.2.2}$$

where $\nabla_x^2 f(x_k)+\lambda_k I$ is positive semidefinite, $\lambda_k\geq 0$, and $\lambda_k(\|s_k\|^2-\Delta_k^2)=0$ (see Theorem 8.2.7 on p. 226).

While this is indeed possible, and sometimes done for problems small or sparse enough to make factorizations affordable, we will focus here on an algorithm in which an inexact step can be used. The conditions we impose on this step reflect both first- and second-order optimality requirements. The first-order condition has already been stated in (2.3.4) and (2.3.5) of the TR1 algorithm on pages 35 and 35, although now we need to use $m_k(s)=T_{f,2}(x_k,s)$ in the latter. In addition, we also require that a meaningful step be made along a significant direction of negative curvature—a direction s for which $\langle s,\nabla_x^2 f(x_k)s\rangle<0$—should such a direction occur. In particular, let u_k be a normalized eigenvector corresponding to the leftmost eigenvalue, $\lambda_{\min}[\nabla_x^2 f(x_k)]<0$, of the Hessian $\nabla_x^2 f(x_k)$, that is,

$$\nabla_x^2 f(x_k)u_k=\lambda_{\min}[\nabla_x^2 f(x_k)]u_k, \quad \|u_k\|=1, \quad \text{and} \quad \langle u_k,\nabla_x^1 f(x_k)\rangle\leq 0. \tag{3.2.3}$$

We then require that

$$T_{f,2}(x_k,s_k)\leq T_{f,2}(x_k,s_k^{\mathrm{E}}), \tag{3.2.4}$$

where

$$s_k^{\mathrm{E}} \stackrel{\text{def}}{=} t_k^{\mathrm{E}}u_k \quad \text{with} \quad t_k^{\mathrm{E}} \stackrel{\text{def}}{=} \arg\min_{t\geq 0,\, x_k-tu_k\in\mathcal{B}_{\Delta_k}} T_{f,2}(x_k,tu_k), \tag{3.2.5}$$

and, as we recall from (2.3.2),

$$\mathcal{B}_{\Delta_k}=\{s\in\mathbb{R}^n \mid \|s\|\leq\Delta_k\}.$$

If $\nabla_x^2 f(x_k)$ is positive semidefinite, i.e., if $\lambda_{\min}[\nabla_x^2 f(x_k)]\geq 0$, we simply define $s_k^{\mathrm{E}}=0$.

The point $x_k + s_k^{\mathrm{E}}$ is called the *eigenpoint* because it involves a move along an eigenvector, and (3.2.4) is referred to as the *eigenpoint decrease condition*. Obviously, the step to the global minimizer defined in (3.2.1) satisfies condition (3.2.4).

Since we now have (and use) second derivatives, it is possible to seek points that approximately satisfy the second-order necessary conditions given in Theorem 1.1.1. To this end, we say that x_ϵ is an *ϵ-approximate second-order minimizer* of f if

$$\|\nabla_x^1 f(x_\epsilon)\| \le \epsilon_1 \text{ and } \lambda_{\min}[\nabla_x^2 f(x_\epsilon)] \ge -\epsilon_2 \tag{3.2.6}$$

for given $\epsilon_1, \epsilon_2 > 0$. Of course this is an extension of (2.1.2), and later we may sometimes write the second condition in (3.2.6) as the entirely equivalent

$$\max\Big(0, -\lambda_{\min}[\nabla_x^2 f(x_\epsilon)]\Big) \le \epsilon_2.$$

In view of Theorem 1.1.1, if, by chance, in addition

$$\|\nabla_x^1 f(x_\epsilon)\| \le \epsilon_1 \text{ and } \lambda_{\min}[\nabla_x^2 f(x_\epsilon)] > \epsilon_2, \tag{3.2.7}$$

we say that x_ϵ is an *ϵ-approximate strict local minimizer*. Although our algorithm is only designed to guarantee (3.2.6), a user will undoubtedly be delighted should (3.2.7) occur, since then there is no need to worry about whether higher-order approximate optimality conditions are satisfied.

We formally state the TR2 algorithm as follows.

Algorithm 3.2.1: Trust-region algorithm with a second-order model (TR2)

Step 0: Initialization. An initial point x_0, accuracy thresholds $\epsilon_1, \epsilon_2 > 0$, and an initial trust-region radius $\Delta_0 > 0$ are given. The constants η_1, η_2, γ_1, and γ_2 are also given and satisfy

$$0 < \eta_1 \le \eta_2 < 1 \text{ and } 0 < \gamma_1 \le \gamma_2 < 1 < \gamma_3. \tag{3.2.8}$$

Compute $f(x_0)$, $\nabla_x^1 f(x_0)$, and $\nabla_x^2 f(x_0)$, and set $k = 0$.

Step 1: Test for termination. If $\epsilon_2 = \infty$ and

$$\|\nabla_x^1 f(x_k)\| \le \epsilon_1,$$

terminate with the (ϵ_1, ϵ_2)-approximate first-order minimizer $x_\epsilon = x_k$. Otherwise, if $\epsilon_2 < \infty$ and

$$\|\nabla_x^1 f(x_k)\| \le \epsilon_1 \text{ and } \lambda_{\min}[\nabla_x^2 f(x_k)] > \epsilon_2,$$

terminate with the (ϵ_1, ϵ_2)-approximate strict local minimizer $x_\epsilon = x_k$. Otherwise, if

$$\|\nabla_x^1 f(x_k)\| \le \epsilon_1 \text{ and } \lambda_{\min}[\nabla_x^2 f(x_k)] \ge -\epsilon_2, \tag{3.2.9}$$

terminate with the (ϵ_1, ϵ_2)-approximate second-order minimizer $x_\epsilon = x_k$.

Step 2: Step computation. Compute a step $s_k \in \mathcal{B}_{\Delta_k}$ such that

$$T_{f,2}(x_k, s_k) \le T_{f,2}(x_k, s_k^{\mathrm{C}}),$$

where $s_k^{\rm C}$ is defined by

$$s_k^{\rm C} \stackrel{\rm def}{=} -t_k^{\rm C}\nabla_x^1 f(x_k) \;\text{ with }\; t_k^{\rm C} \stackrel{\rm def}{=} \mathop{\arg\min}_{t\geq 0,\, x_k - t\nabla_x^1 f(x_k)\in\mathcal{R}_k} T_{f,2}(x_k, -t\nabla_x^1 f(x_k)),$$

and, if $\epsilon_2 < \infty$,

$$T_{f,2}(x_k, s_k) \leq T_{f,2}(x_k, s_k^{\rm E}), \tag{3.2.10}$$

where $s_k^{\rm E}$ is defined as in (3.2.5).

Step 3: Acceptance of the trial point. Compute $f(x_k + s_k)$ and define

$$\rho_k = \frac{f(x_k) - f(x_k + s_k)}{T_{f,2}(x_k, 0) - T_{f,2}(x_k, s_k)}. \tag{3.2.11}$$

If $\rho_k \geq \eta_1$, then define $x_{k+1} = x_k + s_k$ and compute $\nabla_x^1 f(x_{k+1})$ and $\nabla_x^2 f(x_{k+1})$. Otherwise, define $x_{k+1} = x_k$.

Step 4: Trust-region radius update. Set

$$\Delta_{k+1} \in \left\{ \begin{array}{ll} [\Delta_k, \gamma_3\Delta_k] & \text{if } \rho_k \geq \eta_2, \\ [\gamma_2\Delta_k, \Delta_k] & \text{if } \rho_k \in [\eta_1, \eta_2), \\ [\gamma_1\Delta_k, \gamma_2\Delta_k] & \text{if } \rho_k < \eta_1. \end{array} \right. \tag{3.2.12}$$

Increment k by one and go to Step 1.

As with TR1 and AR1, we define successful and unsuccessful iterations by (2.3.9)–(2.3.10) and note that Lemma 2.3.1 still applies to the TR2 algorithm because it uses the same updating rules for Δ_k as TR1. Note also that, because TR2 can equally be used to compute approximate first-order minimizers alone, we allow $\epsilon_2 = +\infty$, in which case the second parts of (3.2.9) and (3.2.10) are deactivated.

Notes and references for Subsection 3.2.1

The definitions of the eigenpoint and the eigenpoint decrease condition are taken from [131, Sect. 6.6.1]. Again, many variants of the TR2 algorithm can be found in the literature. Alternatives include relaxing condition (3.2.10) just as we did in (2.3.8), using global model minimization within the trust region, etc. Once more, we refer the reader to [131, Chaps. 6 and 10] for discussion of some of these variants and their theoretical properties.

3.2.2 Complexity of finding approximate first-order minimizers

In order to derive an upper bound on the number of evaluations of $f(x)$, $\nabla_x^1 f(x)$, and $\nabla_x^2 f(x)$ required to find an ϵ_1-approximate first-order minimizer, we first observe that the TR2 algorithm (with $\epsilon_2 = +\infty$) can be considered as a special case of TR1 where $B_k = \nabla_x^2 f(x_k)$, provided we replace **m.HBa** by the following assumption.

f.HBa

The Hessian matrix $\nabla_x^2 f(x)$ is uniformly bounded above; that is, there exists a constant $\kappa_H \geq 1$ such that

$$\|\nabla_x^2 f(x)\| \leq \kappa_H$$

for all $x \in \mathbb{R}^n$.

In particular, the first-order conditions on the step (the Cauchy decrease condition (2.3.5)) are identical for the TR1 and TR2 algorithms, and, therefore, since **f.HBa** implies **f.gL**, Lemmas 2.3.5 and 2.3.3 remain valid for the TR2 algorithm with

$$\kappa_B = \kappa_H = L_{f,1} \tag{3.2.13}$$

because of Theorem A.8.5 (p. 438)—once again, there is no loss of generality in assuming that $\kappa_H \geq 1$. As a consequence, the evaluation-complexity analysis of TR2 for the convergence to ϵ_1-approximate first-order minimizers is directly inherited from that of the TR1 algorithm. For the sake of completeness, we restate here the complexity result for TR2.

Theorem 3.2.1: First-order complexity of the TR2 algorithm

*Suppose that **f.Bb** and **f.HBa** hold. Then there exist positive constants κ^{S}_{TR2}, κ^{A}_{TR2}, κ^{B}_{TR2}, and κ^{C}_{TR2} for which the TR2 algorithm (p. 55) with $\epsilon_2 = \infty$ requires at most*

$$\#^{F}_{TR2,1} \stackrel{\text{def}}{=} \kappa^{A}_{TR2} \frac{f(x_0) - f_{\text{low}}}{\epsilon_1^2} + \kappa^{B}_{TR2} |\log \epsilon_1| + \kappa^{C}_{TR2} = \mathcal{O}\left(\epsilon_1^{-2}\right)$$

evaluations of f and at most

$$\#^{D}_{TR2,1} \stackrel{\text{def}}{=} \kappa^{S}_{TR2} \frac{f(x_0) - f_{\text{low}}}{\epsilon_1^2} + 1 = \mathcal{O}\left(\epsilon_1^{-2}\right)$$

evaluations of the gradient and Hessian to produce an iterate x_ϵ such that $\|\nabla^1_x f(x_\epsilon)\| \leq \epsilon_1$.

In this statement, the constants κ^{A}_{TR2}, κ^{B}_{TR2}, κ^{C}_{TR2}, and κ^{S}_{TR2} are now given by

$$\kappa^{A}_{TR2} \stackrel{\text{def}}{=} \kappa^{S}_{TR2} \left(1 + \frac{\log \gamma_3}{|\log \gamma_2|}\right), \qquad \kappa^{B}_{TR2} \stackrel{\text{def}}{=} \frac{1}{|\log \gamma_2|},$$

$$\kappa^{C}_{TR2} \stackrel{\text{def}}{=} \frac{1}{|\log \gamma_2|} \left| \log \left(\frac{\gamma_1(1-\eta_2)}{4\Delta_0 \kappa_H} \min\left[1, \frac{\Delta_0}{\|\nabla^1_x f(x_0)\|}\right] \right) \right| + 1,$$

and

$$\kappa^{S}_{TR2} \stackrel{\text{def}}{=} \frac{8\kappa_H}{\gamma_1 \eta_1 (1-\eta_2)} \max\left[1, \frac{\|\nabla^1_x f(x_0)\|}{\Delta_0}\right]$$

(see (2.3.22), (2.3.23), and (2.3.21)), where we have used the relationship $\kappa_B = \kappa_H = L_{f,1}$.

We pause here to reflect. As it stands, the evaluation-complexity bound given by Theorem 3.2.1 for TR2 is no better than that for the first-order trust-region method recorded in Theorem 2.3.7, despite using exact second derivatives. Might this simply be that our analysis is insufficient, and that a better bound is possible with more care? As we now see, the answer is no—the bound in Theorem 3.2.1 is accurate.

Theorem 3.2.2: Slow first-order convergence of the TR2 algorithm

*The TR2 algorithm (p. 55) applied to minimize a function f satisfying **f.Bb** and **f.HBa** may require more than $\lceil \epsilon_1^{-2} \rceil$ iterations and evaluations of $f(x)$, $\nabla_x^1 f(x)$, and $\nabla_x^2 f(x)$ to produce a point x_ϵ such that $\|\nabla_x^1 f(x_\epsilon)\| \leq \epsilon_1$.*

Proof. The desired conclusion follows if we show that the sequence of iterates in the example considered in the proof of Theorem 3.1.1 can also be generated by the TR2 algorithm. This is easily seen. It is enough to start the TR2 algorithm with $\Delta_0 = 2$. Suppose that $\Delta_i = 2$ for $i = 0, \dots, k-1$, and thus that s_i is permitted by the TR2 algorithm since $\|s_i\| < 2$ by construction. Consider iteration k. Once again, $\|s_k\| < 2$ by construction. The predicted value of the second-order model

$$T_{f,2}(x_k, s) = f_k + \langle g_k, s \rangle + \tfrac{1}{2}\langle s, H_k s \rangle$$

at s_k is therefore given by

$$T_{f,2}(x_k, s_k) = f_k + \langle g_k, s_k \rangle + \tfrac{1}{2}\langle s_k, H_k s_k \rangle = f_k - \epsilon_1^2 = f_{k+1},$$

where the last equality results from (3.1.4). Thus $\rho_k = 1$, iteration k is successful, and we may, once again, choose $\Delta_{k+1} = 2$. Therefore the entire sequence $\{x_k\}_{k \geq 0}$ as described in the proof of Theorem 3.1.1 may result from the TR2 algorithm, and thus from (3.1.5) we have that $\|g_k\| > \epsilon_1$ for all $k \leq \lceil \epsilon_1^{-2} \rceil$. $\square$

We did not specify the condition $\epsilon_2 = +\infty$ in this last theorem because (3.1.4) implies that both the second part of (3.2.9) and (3.2.10) hold at every iteration of the example, since $\nabla_x^2 f(x_k)$ is positive definite. Hence the example and Theorem 3.2.2 apply for the TR2 algorithm independent of the value of ϵ_2.

Notes and references for Subsection 3.2.2

The assumption **f.HBa** is very strong and stated in this form for simplicity. It is sufficient for Theorem 3.2.1 to hold that $\|\nabla_x^2 f(x_k)\| \leq \kappa_H$ for all k. This is much weaker but harder to check a priori. The analysis presented is essentially that discussed in [89], in which $T_{f,2}(x_k, s)$ may be replaced by a model of the form (2.3.1), where

$$\|(B_k - \nabla_x^2 f(x_k)) s_k\| \leq \kappa_{BH} \|s_k\|^2 \tag{3.2.14}$$

for some constant $\kappa_{BH} > 0$. Assumption **f.HBa** is then replaced by **m.HBa**. See also Section 13.1.4 (p. 337) and particularly (13.1.2) for a discussion of how much Hessian (and gradient) accuracy can be relaxed.

3.2.3 Complexity of finding approximate second-order minimizers

Since, as we have just seen, there is no gain from an evaluation-complexity perspective from using TR2 rather than TR1, we now turn to the main advantage of TR2, namely, its ability to converge to approximate second-order minimizers when $\epsilon_2 \in (0, 1]$. Our first step to show this is to deduce a lower bound on the model reduction resulting from the eigenpoint decrease condition (3.2.4) (equivalently (3.2.10)).

Lemma 3.2.3: Model decrease at the eigenpoint

Suppose that (3.2.3)–(3.2.5) hold and that $\lambda_{\min}[\nabla_x^2 f(x_k)] < 0$. *Then*

$$T_{f,2}(x_k, s_k^{\mathrm{E}}) \le T_{f,2}(x_k, 0) - \tfrac{1}{2}|\lambda_{\min}[\nabla_x^2 f(x_k)]|\Delta_k^2, \tag{3.2.15}$$

and the ratio (3.2.11) is well defined.

Proof. Using the relation $\nabla_x^2 f(x_k)u_k = \lambda_{\min}[\nabla_x^2 f(x_k)]u_k$, we obtain that

$$\begin{aligned} T_{f,2}(x_k, tu_k) &= T_{f,2}(x_k, 0) + t\langle \nabla_x^1 f(x_k), u_k\rangle + \tfrac{1}{2}t^2\langle u_k, \nabla_x^2 f(x_k)u_k\rangle \\ &= T_{f,2}(x_k, 0) + t\langle \nabla_x^1 f(x_k), u_k\rangle + \tfrac{1}{2}t^2\lambda_{\min}[\nabla_x^2 f(x_k)]. \end{aligned}$$

Because $\lambda_{\min}[\nabla_x^2 f(x_k)] < 0$, this quadratic is concave in t and its minimum for $t \ge 0, x_k - tu_k \in \mathcal{B}_{\Delta_k}$, is achieved on the boundary of the trust region. As a consequence, $t_k^{\mathrm{E}} = \Delta_k$ because u_k is normalized. We therefore obtain, using the inequality $\langle u_k, \nabla_x^1 f(x_k)\rangle \le 0$, that

$$\begin{aligned} T_{f,2}(x_k, s_k^{\mathrm{E}}) &\le T_{f,2}(x_k, 0) + \Delta_k\langle \nabla_x^1 f(x_k), u_k\rangle + \tfrac{1}{2}\lambda_{\min}[\nabla_x^2 f(x_k)]\Delta_k^2 \\ &\le T_{f,2}(x_k, 0) + \tfrac{1}{2}\lambda_{\min}[\nabla_x^2 f(x_k)]\Delta_k^2, \end{aligned}$$

which is the desired bound (3.2.15). The requirement (3.2.10) and (3.2.15) together give that

$$T_{f,2}(x_k, 0) - T_{f,2}(x_k, s_k) \ge T_{f,2}(x_k, 0) - T_{f,2}(x_k, s_k^{\mathrm{E}}) > \tfrac{1}{2}|\lambda_{\min}[\nabla_x^2 f(x_k)]|\Delta_k^2 > 0,$$

and thus the denominator in (3.2.11) is nonzero. □

As was the case for convergence to first-order points, we now deduce a condition for a successful iteration, but now based on the decrease at the eigenpoint rather than at the Cauchy point. This uses the full power of **f.HL**.

Lemma 3.2.4: Condition for a successful iteration (second order)

*Suppose that **f.HL** holds, that* $\lambda_{\min}[\nabla_x^2 f(x_k)] < 0$, *and that*

$$\Delta_k < \frac{3(1-\eta_2)|\lambda_{\min}[\nabla_x^2 f(x_k)]|}{L_{f,2}}. \tag{3.2.16}$$

Then $\rho_k \ge \eta_2$, *iteration* k *is successful, and* $\Delta_{k+1} \ge \Delta_k$.

Proof. Combining the definition of ρ_k in (3.2.11), the bound (A.19) in Corollary A.8.4 (p. 437) when $p = 2$, and the model decrease (3.2.15), we deduce that

$$|\rho_k - 1| \le \frac{\tfrac{1}{6}L_{f,2}\Delta_k^3}{\tfrac{1}{2}|\lambda_{\min}[\nabla_x^2 f(x_k)]|\Delta_k^2} \le \frac{L_{f,2}\Delta_k}{3|\lambda_{\min}[\nabla_x^2 f(x_k)]|} < 1 - \eta_2,$$

where we used (3.2.16) to deduce the last inequality. Hence the desired conclusion follows. □

We now combine the conditions for a successful iteration derived from the first- and second-order decreases to obtain an (ϵ_1, ϵ_2)-dependent lower bound on the trust-region radius.

Lemma 3.2.5: Lower bound for the trust-region radius (second order)

Suppose that **f.HBa** *and* **f.HL** *hold and that the* TR2 *algorithm with* $\epsilon_2 \in (0,1]$ *has not terminated at or before iteration* k_ϵ. *Then there exists a constant* $\kappa_\Delta \in (0,1)$ *such that, for all* $k \in [0 : k_\epsilon]$,

$$\Delta_k \geq \kappa_\Delta \min[\epsilon_1, \epsilon_2]. \tag{3.2.17}$$

Proof. Because the TR2 algorithm has not terminated at iteration k_ϵ, we have that

$$\|\nabla_x^1 f(x_k)\| > \epsilon_1 \quad \text{or} \quad \lambda_{\min}[\nabla_x^2 f(x_k)] < -\epsilon_2 \tag{3.2.18}$$

for each $k \in [0 : k_\epsilon]$. Suppose that there is a first index $k \in [0 : k_\epsilon]$ for which

$$\Delta_{k+1} < \frac{\gamma_1(1-\eta_2)\min[\epsilon_1, \epsilon_2]}{\max\left[4\kappa_H, \frac{1}{3}L_{f,2}\right]}. \tag{3.2.19}$$

If the first part of (3.2.18) holds, we have from (3.2.12) that

$$\Delta_k < \frac{1-\eta_2}{4\kappa_H}\epsilon_1 < \frac{1-\eta_2}{4\kappa_H}\|\nabla_x^1 f(x_k)\|,$$

and Lemma 2.3.3 with (3.2.13) guarantees that $\Delta_{k+1} \geq \Delta_k$. Alternatively, if the second part of (3.2.18) holds, (3.2.12) implies that

$$\Delta_k < \frac{1-\eta_2}{\frac{1}{3}L_{f,2}}\epsilon_2 < \frac{3(1-\eta_2)}{L_{f,2}}|\lambda_{\min}[\nabla_x^2 f(x_k)]|,$$

and Lemma 3.2.4 again ensures that $\Delta_{k+1} \geq \Delta_k$. This last conclusion therefore holds irrespective of which part of (3.2.18) holds at iteration $k-1$. But this is impossible since k is the first index such that (3.2.19) holds. Hence (3.2.19) cannot occur for $k \in [0 : k_\epsilon]$, and thus (3.2.17) holds for all $k \in [1 : k_\epsilon]$ with

$$\kappa_\Delta \stackrel{\text{def}}{=} \min\left[\frac{\gamma_1(1-\eta_2)}{\max\left[4\kappa_H, \frac{1}{3}L_{f,2}\right]}, \Delta_0\right] \in (0,1),$$

with the last inclusion resulting from (3.2.8). It also holds for $k = 0$ since trivially $\Delta_0 \geq \Delta_0 \min[\epsilon_1, \epsilon_2]$ because $\epsilon_2 \leq 1$, which completes the proof. □

We may now use the familiar telescoping-sum argument to deduce an upper bound on the number of evaluations required by TR2 to compute an approximate second-order minimizer.

Theorem 3.2.6: Second-order complexity of the TR2 algorithm

Suppose that **f.Bb**, **f.HBa**, *and* **f.HL** *hold. Then there exist positive constants* κ^T_{TR2}, κ^D_{TR2}, κ^E_{TR2}, *and* κ^F_{TR2} *such that, for any* $\epsilon_1 > 0, \epsilon_2 \in (0,1]$, *the* TR2 *algorithm (p. 55) requires at most*

$$\begin{aligned}\#^F_{\text{TR2,2}} &\stackrel{\text{def}}{=} \kappa^D_{\text{TR2}} \frac{f(x_0)-f_{\text{low}}}{\min[\epsilon_1^2\epsilon_2, \epsilon_2^3]} + \kappa^E_{\text{TR2}} |\log(\min[\epsilon_1, \epsilon_2])| + \kappa^F_{\text{TR2}} \\ &= \mathcal{O}\left(\max[\epsilon_1^{-2}\epsilon_2^{-1}, \epsilon_2^{-3}]\right)\end{aligned} \tag{3.2.20}$$

evaluations of f *and at most*

$$\#^D_{\text{TR2,2}} \stackrel{\text{def}}{=} \kappa^T_{\text{TR2}} \frac{f(x_0)-f_{\text{low}}}{\min[\epsilon_1^2\epsilon_2, \epsilon_2^3]} + 1 = \mathcal{O}\left(\max[\epsilon_1^{-2}\epsilon_2^{-1}, \epsilon_2^{-3}]\right) \tag{3.2.21}$$

evaluations of the gradient and Hessian to produce an iterate x_ϵ *such that* $\|\nabla_x^1 f(x_\epsilon)\| \leq \epsilon_1$ *and* $\lambda_{\min}[\nabla_x^2 f(x_k)] \geq -\epsilon_2$.

Proof. Suppose that the TR2 algorithm has not terminated at iteration k. As in the previous proof, this implies that (3.2.18) holds for all $k \in [0 : k_\epsilon]$. Combining this observation with (2.3.12) (p. 37), (3.2.15), and (3.2.17), we then deduce that, for all $j \in [0 : k]$,

$$\begin{aligned}&T_{f,2}(x_j, 0) - T_{f,2}(x_j, s_j) \\ &\quad \geq \min\left\{\tfrac{1}{2}\epsilon_1 \min\left[\frac{\epsilon_1}{1+\kappa_H}, \kappa_\Delta \min[\epsilon_1, \epsilon_2]\right], \tfrac{1}{2}\kappa_\Delta^2 \epsilon_2 \min[\epsilon_1, \epsilon_2]^2\right\} \\ &\quad > \tfrac{1}{2}\min\left[\frac{1}{1+\kappa_H}, \kappa_\Delta^2\right] \min[\epsilon_1^2\epsilon_2, \epsilon_2^3],\end{aligned} \tag{3.2.22}$$

where we used the inclusions $\kappa_\Delta \in (0,1)$ and $(\epsilon_1, \epsilon_2) \in (0,1]^2$ to deduce the last inequality. We then obtain, successively using **f.Bb**, (2.3.10), and (3.2.22), that

$$\begin{aligned}f(x_0) - f_{\text{low}} &\geq f(x_0) - f(x_{k+1}) \\ &= \sum_{j\in\mathcal{S}_k} [f(x_j) - f(x_{j+1})] \\ &\geq \eta_1 \sum_{j\in\mathcal{S}_k} [T_{f,2}(x_j, 0) - T_{f,2}(x_j, s_j)] \\ &= \tfrac{1}{2}\eta_1 \min\left[\frac{1}{1+\kappa_H}, \kappa_\Delta^2\right] \min[\epsilon_1^2\epsilon_2, \epsilon_2^3] |\mathcal{S}_k|.\end{aligned}$$

Hence

$$|\mathcal{S}_k| \leq \frac{f(x_0) - f_{\text{low}}}{\tfrac{1}{2}\eta_1 \min\left[\frac{1}{1+\kappa_H}, \kappa_\Delta^2\right] \min[\epsilon_1^2\epsilon_2, \epsilon_2^3]}.$$

This gives an upper bound on the number of successful iterations of the TR2 algorithm before termination and, since the gradient and Hessian are only evaluated (once) on such iterations, also gives (3.2.21) with

$$\kappa_{\mathsf{TR2}}^{\mathrm{T}} \stackrel{\mathrm{def}}{=} \left\{ \frac{1}{2}\eta_1 \min\left[\frac{1}{1+\kappa_H}, \left(\min\left[\frac{\gamma_1(1-\eta_2)}{\max\left[4\kappa_H, \frac{1}{3}L_{f,2}\right]}, \Delta_0 \right] \right)^2 \right] \right\}^{-1}.$$

This bound and Lemma 2.3.1 then yield (3.2.20) with

$$\kappa_{\mathsf{TR2}}^{\mathrm{D}} \stackrel{\mathrm{def}}{=} \kappa_{\mathsf{TR2}}^{\mathrm{T}} \left(1 + \frac{\log \gamma_3}{|\log \gamma_2|} \right), \quad \kappa_{\mathsf{TR2}}^{\mathrm{E}} \stackrel{\mathrm{def}}{=} \frac{1}{|\log \gamma_2|},$$

and

$$\kappa_{\mathsf{TR2}}^{\mathrm{F}} \stackrel{\mathrm{def}}{=} \frac{1}{|\log \gamma_2|} \left(\min\left[\frac{\gamma_1(1-\eta_2)}{\Delta_0 \max\left[4\kappa_H, \frac{1}{3}L_{f,2}\right]}, \Delta_0 \right] \right) + 1.$$

□

If one chooses $\epsilon_1 = \epsilon_2 = \epsilon$, the evaluation-complexity upper bounds for the TR2 algorithm given by Theorems 3.2.1 and 3.2.6 take the form

$$\mathcal{O}\left(\epsilon^{-3/2}\right) \quad \text{and} \quad \mathcal{O}\left(\epsilon^{-3}\right)$$

for finding ϵ-approximate first- and second-order minimizers, respectively. As the reader has come to expect from the similar first-order result in Theorem 3.2.2, we now show that the bound in $\mathcal{O}(\epsilon^{-3})$ in the second-order case is sharp.

Theorem 3.2.7: Slow second-order convergence of the TR2 algorithm

The TR2 algorithm (p. 55) applied to minimize a function f satisfying ***f.Bb***, ***f.HBa***, *and* ***f.HL*** *may require ϵ^{-3} iterations and evaluations of $f(x)$, $\nabla_x^1 f(x)$, and $\nabla_x^2 f(x)$ to produce a point x_ϵ such that $\|\nabla_x^1 f(x_\epsilon)\| \le \epsilon$ and $\lambda_{\min}[\nabla_x^2 f(x_k)] \ge -\epsilon$.*

Proof. We achieve this conclusion by setting $\epsilon_1 = \epsilon_2 = \epsilon \in (0,1]$ and building a sufficiently smooth function f_{STR2} from $\mathbb{R}$ to $\mathbb{R}$ such that, when applied to minimize this function (starting from the origin), the TR2 algorithm needs exactly ϵ^{-3} evaluations of f, $\nabla_x^1 f$, and $\nabla_x^2 f$ to produce an iterate such that (3.2.9) holds. Let us define

$$k_\epsilon = \lceil \epsilon^{-3} \rceil \tag{3.2.23}$$

and the sequences $\{f_k\}_{k=0}^{k_\epsilon}$, $\{g_k\}_{k=0}^{k_\epsilon}$, and $\{H_k\}_{k=0}^{k_\epsilon}$ by

$$f_0 = 1, \quad f_{k+1} \stackrel{\mathrm{def}}{=} f_k - \tfrac{1}{2}\epsilon^3 \left(2 - \frac{k}{k_\epsilon}\right), \quad g_k = 0, \text{ and } H_k = -\epsilon\left(2 - \frac{k}{k_\epsilon}\right). \tag{3.2.24}$$

As we have done before, these are intended to specify the objective function, gradient, and Hessian values at successive iterates x_k generated by the TR2 algorithm. We immediately observe that this algorithm will terminate exactly at iteration k_ϵ since $g_k = 0$ for all k, $\lambda_{\min}[H_k] = H_k < -\epsilon$ for $k \in [0 : k_\epsilon - 1]$, and $\lambda_{\min}[H_{k_\epsilon}] = -\epsilon$. Also note that $\{f_k\}_{k=0}^{k_\epsilon}$ is monotonically decreasing and that

$$f_{k_\epsilon} = 1 - \sum_{k=0}^{k_\epsilon - 1} \tfrac{1}{2}\epsilon^3 \left(2 - \frac{k}{k_\epsilon}\right) = 1 - \tfrac{1}{2}\epsilon^3 \left(2k_\epsilon - \frac{k_\epsilon - 1}{2}\right) > 1 - \epsilon^3 k_\epsilon \ge 0,$$

where we used the fact that (3.2.23) implies that $\epsilon^3 k_\epsilon \leq 1$ to deduce the last inequality. Hence

$$f_k \in [0,1] \quad \text{for} \quad k \in [0:k_\epsilon].$$

At the putative iterate x_k, we therefore have that

$$T_{f,2}(x_k, s) = f_k + \tfrac{1}{2}\langle s, H_k s\rangle,$$

which is a negative-definite quadratic. For such a model, the TR2 algorithm may choose the step as a move to the trust-region boundary, that is, $s_k = \Delta_k$. Suppose now that $\Delta_0 = \epsilon$. Then $s_1 = \epsilon$ and

$$T_{f,2}(x_0, 0) - T_{f,2}(x_0, s_0) = -\tfrac{1}{2}\langle s_0, H_0 s_0\rangle = -\tfrac{1}{2}\Delta_0^2 H_0 = \epsilon^3 = f_0 - f_1.$$

As a consequence, $\rho_0 = 1$, iteration 0 is successful, $x_1 = \epsilon$, and (3.2.12) allows the choice $\Delta_1 = \Delta_0 = \epsilon$. It is easy to verify that the same situation is reproduced at every iteration, and therefore that the sequences $\{f_k\}_{k=0}^{k_\epsilon}$, $\{g_k\}_{k=0}^{k_\epsilon}$, and $\{H_k\}_{k=0}^{k_\epsilon}$ defined by (3.2.24) and $\{x_k\}_{k=0}^{k_\epsilon}$, $\{s_k\}_{k=0}^{k_\epsilon}$, and $\{\Delta_k\}_{k=0}^{k_\epsilon}$ defined by

$$x_k = (k-1)\epsilon \quad \text{and} \quad s_k = \Delta_k = \epsilon \quad \text{for} \quad k \in [0:k_\epsilon] \tag{3.2.25}$$

can be interpreted as generated by the TR2 algorithm, provided that the values (3.2.24) can be interpolated by a sufficiently smooth function. Observing now that

$$f_{k+1} - T_{f,2}(x_k, s_k) = 0, \tag{3.2.26}$$

$$|g_{k+1} - T_{f,2}^{(1)}(x_k, s_k)| = |0 - H_k s_k| = \epsilon^2\left(2 - \frac{k}{k_\epsilon}\right) \leq 2 s_k^2,$$

and

$$|H_{k+1} - T_{f,2}^{(2)}(x_k, s_k)| = |H_{k+1} - H_k| = \frac{\epsilon}{k_\epsilon} \leq s_k, \tag{3.2.27}$$

we have that the required interpolation is guaranteed by Theorem A.9.2 (p. 441) with $\kappa_f = 2$ (see the bounds (3.2.26)–(3.2.27) above, (3.2.24), and (3.2.25)). We therefore deduce that the TR2 algorithm applied on the resulting interpolated function f_{STR2} will stop exactly at iteration k_ϵ. Moreover, this function has bounded range, gradient, and Hessian, as well as Lipschitz-continuous second derivatives. Hence assumptions **f.Bb**, **f.HBa**, and **f.HL** hold, and we may deduce the desired conclusion. □

Notes and references for Subsection 3.2.3

The second-order worst-case evaluation-complexity analysis for the trust region was proposed in [90], which we essentially followed in this section. If one chooses ϵ_2 to be of the order of the square root of ϵ_1, as in the local optimality measure of Nesterov and Polyak [305], the evaluation-complexity upper bound of Theorem 3.2.6 then becomes $\mathcal{O}(\epsilon_1^{-5/2})$ instead of $\mathcal{O}(\max[\epsilon_1^{-2}\epsilon_2^{-1}, \epsilon_2^{-3}])$. Note that Theorem 3.1.1 implies that this complexity may be as bad as $\mathcal{O}(\epsilon_1^{-2})$, since approximate second-order optimality requires first order, which is coherent with Theorem 3.2.6.

The example of slow convergence in the proof of Theorem 3.2.7 is new, although the same upper complexity bound was shown to be "essentially sharp" in [89].

We note that the complexity bound for second-order criticality of TR2 mixes first- and second-order steps and tolerances; a trust-region algorithm with decoupled first- and second-order steps, and ensuing decoupled second-order complexity bound, can be found in [211].

We conclude Sections 3.1 and 3.2 with a word of caution: while Newton's method and the second-order trust-region algorithms may not converge as fast as one would wish *in the worst case*, their practical performance is very often *excellent*. In particular, the TR2 algorithm and several of its variants using inexact minimization within the trust region are clearly among today's best computational unconstrained optimization methods, across a wide range of applications. As an example, the package TRU from the GALAHAD open-source software library [192] is based on the TR2 framework, but with many subproblem enhancements that will be revealed in Part II of our book.

3.3 Adaptive regularization using second derivatives

The previous section might suggest to the reader that, at least in the worst case, adding additional derivative information beyond the first order doesn't help. Fortunately, such a hasty conclusion is false. In this section, we describe another algorithm—which may also be considered a variant of Newton's method—whose evaluation complexity is significantly better than that of first-order methods.

3.3.1 The AR2 algorithm

The adaptive-regularization algorithm may be motivated in two different ways. The first is to consider the method as a variant of the second-order trust-region scheme, in which the steplength is controlled not explicitly (by requiring that $\|s_k\| \leq \Delta_k$), but implicitly instead. This is achieved by adding a term to the quadratic model $T_{f,2}(x_k, s)$ used in (3.1.2) to penalize long steps. Thus, rather than updating Δ_k from iteration to iteration, it is the strength of the penalization that is updated. However, this says little about which form the penalization should take.[17]

The second motivation is more interesting because it also suggests a suitable form for the penalization term. The reader might remember that we made significant use, in the analysis of the AR1 algorithm, of the overestimation property expressed by inequality (A.19) in Corollary A.8.4 (p. 437). In particular, the penalization in $\|s\|^2$ in (2.4.1) made direct reference to the error term in (A.19) for $p = 1$. Considering now second-order models, that is, using $p = 2$ in (A.19), immediately suggests defining the regularized second-order model as

$$m_k(s) \stackrel{\text{def}}{=} f(x_k) + \langle \nabla_x^1 f(x_k), s \rangle + \tfrac{1}{2}\langle s, \nabla_x^2 f(x_k) s \rangle + \tfrac{1}{6}\sigma_k \|s\|^3 = T_{f,2}(x_k, s) + \tfrac{1}{6}\sigma_k \|s\|^3, \tag{3.3.1}$$

whose gradient and Hessian are given by

$$\nabla_s^1 m_k(s) = \nabla_x^1 f(x_k) + \nabla_x^2 f(x_k) s + \frac{\sigma_k}{2}\|s\| s \tag{3.3.2}$$

and

$$\nabla_s^2 m_k(s) = \nabla_x^2 f(x_k) + \sigma_k \|s\| \left(I + \frac{s s^T}{\|s\|^2} \right).$$

This improved model, compared to (2.4.1), is the main difference between the first-order regularization algorithm AR1 and the new second-order one, which we call AR2. We immediately note that $m_k(s)$ is bounded below because any negative curvature present in

[17] Many forms of penalization have been proposed in the past, such as the Levenberg–Morrison–Marquardt algorithm or the proximal-point method, but without improving the worst-case evaluation complexity beyond that of Newton's method.

$\nabla_x^2 f(x_k)$ is dominated by the third-order regularization term for large $\|s\|$. However, computing the global minimizer of (3.3.1) no longer results from a simple analytic formula, as was the case for AR1, but more elaborate techniques are needed—they are covered in detail in Chapters 9 and 10, and remarkably a global minimizer s_k of the model $m_k(s)$ once again satisfies the linear system (3.2.2) for positive-semidefinite $\nabla_x^2 f(x_k)+\lambda_k I$, but now explicitly $\lambda_k = \frac{1}{2}\sigma_k\|s_k\|$ (see Theorem 8.2.8 on p. 226)—yielding exact or inexact minimizers of $m_k(s)$.

Because of this complication, we focus here on a version of the AR2 algorithm that allows for inexact model minimization. Our algorithm is formally described as follows.

Algorithm 3.3.1: Adaptive-regularization algorithm with a second-order model (AR2)

Step 0: Initialization. An initial point x_0; accuracy thresholds $\epsilon_1 \in (0,1]$, $\epsilon_2 > 0$; and an initial regularization parameter $\sigma_0 > 0$ are given. The constants η_1, η_2, γ_1, γ_2, γ_3, θ_1, θ_2, and $\sigma_{\min}$ are also given and satisfy

$$\begin{aligned}\sigma_{\min} \in (0,\sigma_0], \quad \theta_1 > 0, \quad \theta_2 > 0, \quad 0 < \eta_1 \leq \eta_2 < 1,\\ \text{and} \quad 0 < \gamma_1 < 1 < \gamma_2 < \gamma_3.\end{aligned} \tag{3.3.3}$$

Compute $f(x_0)$, $\nabla_x^1 f(x_0)$, and $\nabla_x^2 f(x_0)$, and set $k = 0$.

Step 1: Test for termination. If $\epsilon_2 < \infty$ and

$$\|\nabla_x^1 f(x_k)\| \leq \epsilon_1 \quad \text{and} \quad \lambda_{\min}[\nabla_x^2 f(x_k)] > \epsilon_2,$$

terminate with the (ϵ_1,ϵ_2)-approximate strict local minimizer $x_\epsilon = x_k$. Otherwise, if

$$\|\nabla_x^1 f(x_k)\| \leq \epsilon_1 \quad \text{and} \quad \lambda_{\min}[\nabla_x^2 f(x_k)] \geq -\epsilon_2, \tag{3.3.4}$$

terminate with the (ϵ_1,ϵ_2)-approximate second-order minimizer $x_\epsilon = x_k$.

Step 2: Step computation. Compute a step s_k such that $m_k(s_k) < m_k(0)$,

$$\|\nabla_s^1 m_k(s_k)\| \leq \tfrac{1}{2}\theta_1\|s_k\|^2, \tag{3.3.5}$$

and, if $\epsilon_2 < \infty$,

$$\lambda_{\min}[\nabla_s^2 m_k(s_k)] \geq -\theta_2\|s_k\|. \tag{3.3.6}$$

Step 3: Acceptance of the trial point. Compute $f(x_k+s_k)$ and define

$$\rho_k = \frac{f(x_k) - f(x_k+s_k)}{T_{f,2}(x_k,0) - T_{f,2}(x_k,s_k)}. \tag{3.3.7}$$

If $\rho_k \geq \eta_1$, then define $x_{k+1} = x_k + s_k$ and compute $\nabla_x^1 f(x_{k+1})$ and $\nabla_x^2 f(x_{k+1})$. Otherwise define $x_{k+1} = x_k$.

Step 4: Regularization parameter update. Set

$$\sigma_{k+1} \in \left\{ \begin{array}{ll} [\max(\sigma_{\min},\gamma_1\sigma_k),\sigma_k] & \text{if } \rho_k \geq \eta_2,\\ [\sigma_k,\gamma_2\sigma_k] & \text{if } \rho_k \in [\eta_1,\eta_2),\\ [\gamma_2\sigma_k,\gamma_3\sigma_k] & \text{if } \rho_k < \eta_1. \end{array}\right. \tag{3.3.8}$$

Increment k by one and go to Step 1.

The AR2 algorithm is very similar in structure to AR1. Indeed, since $\|\nabla_s^1 m_k(s)\|$ must be zero at a local minimizer of $m_k(s)$, the step computation in AR2 is a relaxed equivalent of the exact requirement (2.4.5); that is,

$$s_k = \arg\min_{s\in\mathbb{R}^n} m_k(s), \tag{3.3.9}$$

from AR1. Moreover, (3.3.7) is the same as (2.4.3) for the models under consideration. As for the TR2 algorithm, we have left open the possibility of using AR2 simply to find an approximate first-order minimizer by selecting $\epsilon_2 = +\infty$, which deactivates the second part of (3.3.4) and (3.3.6). As is now usual, we define the sets of *successful* and *unsuccessful* iterations by (2.3.9)–(2.3.10).

Notes and references for Subsection 3.3.1

The overestimation bound (A.19) in Corollary A.8.4 (p. 437) is a standard result of standard analysis; see for example Lemma 4.1.14 in Dennis and Schnabel [147]. However, as far as we are aware, Griewank, in an unpublished technical report [218], was the first to suggest using the model $m_k(s)$ *to compute a step* as a means to construct globally convergent affine-invariant variants of Newton's method. He introduced a slight variant of the model $m_k(s)$,

$$m_k^G(s) \stackrel{\text{def}}{=} f(x_k) + \langle \nabla_x^1 f(x_k), s\rangle + \tfrac{1}{2}\langle s, \nabla_x^2 f(x_k) s\rangle + \tfrac{1}{3}\sigma_k \|s\|_{G_k}^3,$$

to regularize the quadratic Newton model, especially in the presence of negative curvature; here $\sigma_k\|\cdot\|_{G_k}^3$ is chosen iteratively to ensure (A.19) while preserving affine invariance. He then proved global convergence to second-order minimizers of a method where the step is computed by finding any exact second-order minimizer of the model that provides descent. All such minimizers (including the global one) were characterized. He also outlined a proof of quadratic local convergence. Finally, he suggested minimizing m_k^G approximately (using a variant of the nonlinear conjugate-gradient (CG) method), considered the rate of convergence of this modified method, and gave some preliminary numerical results.

More recently, Nesterov and Polyak [305] independently considered a similar idea using the model $m_k(s)$, although from a different perspective. They were able to show that, if the step is computed by globally and exactly minimizing the cubic model and if the objective's Hessian is globally Lipschitz continuous, then the resulting algorithm has a better global iteration-complexity bound than that achieved by the steepest-descent method. They completed this remarkable result by showing that the model's global minimizer can be computed in a manner acceptable from the complexity point of view and proposed superior complexity bounds for the (star) convex, "gradient-dominated" (see Section 5.3), and other special cases. Global convergence to second-order critical points and an asymptotically quadratic rate of convergence were also proved for this method, but no numerical results were provided.

Even more recently and again independently, Weiser, Deuflhard, and Erdmann [364] also pursued the ideas, motivated (as was Griewank) by the design of an affine-invariant version of Newton's method. Their approach, directly evolved from techniques for convex problems, makes use of the cubic model $m_k^G(s)$, where G is positive definite and σ_k is an estimate of the global Lipschitz constant (the techniques for updating σ_k are similar to Griewank's). The proposed method does not consider global model minimization but rather uses approximate techniques for finding a local minimizer, such as Krylov-space techniques and nonlinear CGs. Again, global Lipschitz continuity is assumed, but no formal convergence or complexity analysis is presented. Limited but encouraging numerical experience is discussed.

A method named ACO (for Adaptive Cubic Overestimation) and later ARC (for Adaptive Regularization with Cubics), which combined provable complexity results, adaptive regularization, and inexact model minimization, were proposed in [85]. It differs, however, from the AR2 algorithm of page 65 in that it uses, like its precursors mentioned above, a different, more restrictive definition of a successful iteration. In these methods, the definition of the ratio ρ_k

mimics that traditionally used for the TR2 algorithm, that is,

$$\rho_k = \frac{f(x_k) - f(x_k + s_k)}{m_k(0) - m_k(s_k)} \tag{3.3.10}$$

(see (3.2.11)), and thus compares the achieved reduction to that predicted by the regularized model. In contrast, (3.3.7) compares the achieved reduction to that predicted by the (unregularized) Taylor's expansion. We will discuss below how this seemingly relatively minor change, introduced independently by Birgin et al. [45] and Dussault [157,158], contributes to simplifying the proof of the complexity bound for the AR2 algorithm. We also note that further variants of ARC have been analyzed by Bianconcini, Liuzzi, Morini, and Sciandrone—periodic performance evaluation during model solution is considered in [41], while a nonmonotone variant is presented in [42]. These variants preserve the complexity bound for AR2 that is analyzed in this section.

We conclude our comments on the AR2 algorithm by noting that alternatives to the step-computation truncation rules (3.3.5) and (3.3.6) are possible. Of all the early inexact rules that were suggested [84, Sect. 3.3], the choice

$$\|\nabla_s^1 m_k(s_k)\| \leq \theta_1 \|\nabla_s^1 m_k(0)\| \min(1, \|s_k\|)$$

for given $\theta_1 \in (0,1)$ provides the same order of first-order worst-case evaluation bound as what we shall shortly derive for the AR2 algorithm [85, Sect. 4]. More recently, a different pair of first- and second-order rules[18] have been proposed [215] in which s_k is required to satisfy $m_k(s_k) < m_k(0)$,

$$\|\nabla_s^1 T_{f,2}(x_k, s_k)\| \leq \tfrac{1}{2}\theta_1 \sigma_k \|s_k\|^2, \tag{3.3.11}$$

and, if $\epsilon_2 < \infty$,

$$\lambda_{\min}[\nabla_x^2 f(x_k)] + \theta_2 \omega(s_k) \sigma_k \|s_k\| \geq 0 \tag{3.3.12}$$

for given $\theta_1, \theta_2 > 1$, where

$$\omega(s) \stackrel{\text{def}}{=} \begin{cases} 1 + \dfrac{2}{\sqrt{3}} \sqrt{\max\left[0, 1 + 2\dfrac{\langle \nabla_s^1 T_{f,2}(x_k, s_k), u_a \rangle}{\sigma \|s\|^2}\right]} & \text{if } s \neq 0, \\ 1 & \text{otherwise.} \end{cases}$$

Here u_a is an approximate normalized eigenvector associated with $\lambda_{\min}[\nabla_x^2 f(x_k)]$ whose sign is chosen such that $\langle \nabla_s^1 T_{f,2}(x_k, s_k), u_a \rangle \leq 0$, which ensures that $\omega(s)$ is continuous at the origin (see [215] for further details). Such tests have the advantage that they do not rely on the smoothness of the norm used in the regularization term (and thus of the model m_k), which then allows the AR2 algorithm to be defined using an arbitrary regularization norm. Moreover, the test (3.3.11) indicates that the Newton step, for which $\nabla_s^1 T_{f,2}(x_k, s_k) = 0$, is always acceptable so long as $m_k(s_k) < m_k(0)$. Finally, the requirements (3.3.11) and (3.3.12) result in a further simplification of the complexity analysis for the (modified) AR2 algorithm, in particular in streamlined versions of Lemmas 3.3.3 and 3.3.8. This analysis shows that the worst-case evaluation complexity of the modified AR2 method is still $\mathcal{O}(\epsilon_1^{-3/2})$ for finding first-order approximate minimizers and $\mathcal{O}(\max[\epsilon_1^{-3/2}, \epsilon_2^{-3}])$ for second-order ones. Whether these modifications will turn out to be numerically advantageous is the subject of ongoing research.

3.3.2 Complexity of finding approximate first-order minimizers

The evaluation-complexity analysis for the convergence of AR2 to an ϵ_1-approximate first-order minimizer is remarkably simple and elegant. As we will discover throughout the book, it is also possible to generalize it in many other contexts.

[18]The term on the right-hand side of (3.3.11) may be replaced by $\frac{1}{2}\theta_1 \max[1, \sigma_k] \|s_k\|^2$ without detrimental theoretical consequences, and this is less restrictive for small σ_k.

We start by establishing an elementary lower bound on the Taylor-series model decrease.

Lemma 3.3.1: Decrease in the Taylor-series model

At every iteration k of the AR2 *algorithm, one has that*

$$T_{f,2}(x_k,0) - T_{f,2}(x_k,s_k) > \frac{\sigma_k}{6}\|s_k\|^3. \tag{3.3.13}$$

Proof. From (3.3.1), we have that

$$T_{f,2}(x_k,0) - T_{f,2}(x_k,s_k) = m_k(0) - m_k(s_k) + \frac{\sigma_k}{6}\|s_k\|^3,$$

and (3.3.13) follows from the requirement (in Step 2) that $m_k(s_k) < m_k(0)$. □

As was the case for the AR1 algorithm in Lemma 2.4.2, we now show that the regularization parameter σ_k must remain bounded.

Lemma 3.3.2: Upper bound on the regularization parameter

Suppose that **f.HL** *holds. Then, for all iterations $k \geq 0$ of the* AR2 *algorithm,*

$$\sigma_k \leq \sigma_{\max} \stackrel{\text{def}}{=} \max\left[\sigma_0, \frac{\gamma_3 L_{f,2}}{1-\eta_2}\right]. \tag{3.3.14}$$

Proof. Successively using (3.3.7), (A.19) in Corollary A.8.4 (p. 437) for $p = 2$, and (3.3.13), we obtain that

$$\begin{aligned}|\rho_k - 1| &= \frac{|f(x_k) - f(x_k+s_k) - T_{f,2}(x_k,0) + T_{f,2}(x_k,s_k)|}{T_{f,2}(x_k,0) - T_{f,2}(x_k,s_k)}\\ &= \frac{|f(x_k+s_k) - T_{f,2}(x_k,s_k)|}{T_{f,2}(x_k,0) - T_{f,2}(x_k,s_k)}\\ &\leq \frac{L_{f,2}}{\sigma_k}.\end{aligned}$$

Thus, if $\sigma_k \geq L_{f,2}/(1-\eta_2)$, $\rho_k \geq \eta_2$, iteration k is successful and (3.3.8) implies that $\sigma_{k+1} \leq \sigma_k$. The mechanism of the algorithm then implies that (3.3.14) must hold. □

Next, we establish a crucial lower bound on the step length.

Lemma 3.3.3: Lower bound on the length of the step

Suppose that **f.HL** *holds. Then, if iteration k of the* AR2 *algorithm is successful,*

$$\|s_k\|^2 \geq \frac{2}{L_{f,2} + \theta_1 + \sigma_{\max}}\|\nabla_x^1 f(x_{k+1})\|. \tag{3.3.15}$$

Proof. Combining a simple "add and subtract" argument with the triangle inequality yields that

$$\begin{aligned}\|\nabla_x^1 f(x_{k+1})\| &\leq \big\|\nabla_x^1 f(x_{k+1}) - \nabla_s^1 T_{f,2}(x_k,s_k) + \nabla_s^1 T_{f,2}(x_k,s_k) \\ &\quad + \tfrac{1}{2}\sigma_k\|s_k\|s_k - \tfrac{1}{2}\sigma_k\|s_k\|s_k\big\| \\ &\leq \|\nabla_x^1 f(x_{k+1}) - \nabla_s^1 T_{f,2}(x_k,s_k)\| \\ &\quad + \big\|\nabla_s^1 T_{f,2}(x_k,s_k) + \tfrac{1}{2}\sigma_k\|s_k\|s_k\big\| + \tfrac{1}{2}\sigma_k\|s_k\|^2.\end{aligned} \tag{3.3.16}$$

Since iteration k is successful, $\nabla_x^1 f(x_{k+1}) = \nabla_x^1 f(x_k+s_k)$ and we may use **f.HL** and (A.20) in Corollary A.8.4 (p. 437) for $j=1$ and $p=2$ to deduce that

$$\|\nabla_x^1 f(x_{k+1}) - \nabla_s^1 T_{f,2}(x_k,s_k)\| = \|\nabla_x^1 f(x_k+s_k) - \nabla_s^1 T_{f,2}(x_k,s_k)\| \leq \tfrac{1}{2}L_{f,2}\|s\|^2. \tag{3.3.17}$$

We also deduce from (3.3.1) and (3.3.5) that

$$\big\|\nabla_s^1 T_{f,2}(x_k,s_k) + \tfrac{1}{2}\sigma_k\|s_k\|s_k\big\| = \|\nabla_s^1 m_k(s_k)\| \leq \tfrac{1}{2}\theta_1\|s_k\|^2. \tag{3.3.18}$$

Substituting (3.3.17) and (3.3.18) into (3.3.16) and using (3.3.14) then immediately gives (3.3.15). □

We are now ready to derive an upper bound on the number of successful iterations of the AR2 algorithm with $\epsilon_2 = +\infty$ using the by now familiar telescoping-sum argument.

Lemma 3.3.4: Bound on the number of successful AR2 iterations

Suppose that **f.Bb** *and* **f.HL** *hold. Then there exists a positive constant κ_{AR2}^{s} such that, for any $\epsilon_1 \in (0,1]$, the AR2 algorithm (p. 65) with $\epsilon_2 = +\infty$ requires at most*

$$\#_{\text{AR2}}^{SI} \stackrel{\text{def}}{=} \kappa_{\text{AR2}}^{s}\frac{f(x_0)-f_{\text{low}}}{\epsilon_1^{3/2}} + 1 \tag{3.3.19}$$

successful iterations before an iterate x_ϵ is computed such that $\|\nabla_x^1 f(x_{\epsilon_1})\| \leq \epsilon$.

Proof. For any $k \geq 0$ such that the algorithm does not terminate at or before iteration k, we have that

$$\begin{aligned}f(x_0) - f_{\text{low}} &\geq f(x_0) - f(x_{k+1}) \\ &= \sum_{j\in\mathcal{S}_k}[f(x_j) - f(x_{j+1})] \\ &\geq \eta_1\sum_{j\in\mathcal{S}_k}[T_{f,2}(x_j,0) - T_{f,2}(x_j,s_j)],\end{aligned} \tag{3.3.20}$$

where we have used the mechanism of Step 3 for successful iterations (i.e., $k \in \mathcal{S}_k$; see (2.3.10)) to derive the last inequality. We now successively substitute (3.3.13) into (3.3.20), use the fact that $\sigma_k \geq \sigma_{\min}$ for all k because of (3.3.8), and obtain that

$$f(x_0) - f_{\text{low}} \geq \tfrac{1}{6}\eta_1\sigma_{\min}\sum_{j\in\mathcal{S}_k}\|s_j\|^3 \geq \frac{2^{3/2}\,\eta_1\sigma_{\min}}{6\,(L_{f,2}+\theta_1+\sigma_{\max})^{3/2}}\sum_{j\in\mathcal{S}_{k-1}}\|\nabla_x^1 f(x_{j+1})\|^{3/2},$$

where we used Lemma 3.3.3 to deduce the last inequality.

Since the algorithm has not terminated at or before iteration k, we also have that $\|\nabla_x^1 f(x_{j+1})\| \geq \epsilon_1$ for $j \in \mathcal{S}_{k-1}$, which allows us to write that

$$\begin{aligned} f(x_0) - f_{\text{low}} &\geq \frac{2^{3/2}\eta_1\sigma_{\min}}{6\left(L_{f,2}+\theta_1+\sigma_{\max}\right)^{3/2}} \sum_{j\in\mathcal{S}_{k-1}} \|\nabla_x^1 f(x_{j+1})\|^{3/2} \\ &\geq \frac{2^{3/2}\eta_1\sigma_{\min}}{6\left(L_{f,2}+\theta_1+\sigma_{\max}\right)^{3/2}}\, \epsilon_1^{3/2}\, |\mathcal{S}_{k-1}| \end{aligned}$$

(note the change from $\mathcal{S}_k$ to $\mathcal{S}_{k-1} \subseteq \mathcal{S}_k$ in the sum). As a consequence, since $|\mathcal{S}_{k-1}| \leq |\mathcal{S}_k| + 1$,

$$|\mathcal{S}_k| \leq \frac{6\left(L_{f,2}+\theta_1+\sigma_{\max}\right)^{3/2}}{2^{3/2}\eta_1\sigma_{\min}}\,\frac{f(x_0)-f_{\text{low}}}{\epsilon_1^{3/2}} + 1,$$

which implies (3.3.19) with

$$\kappa_{\text{AR2}}^{\text{S}} \stackrel{\text{def}}{=} \frac{6\left(L_{f,2}+\theta_1+\sigma_{\max}\right)^{3/2}}{2^{3/2}\eta_1\sigma_{\min}} = \frac{6\left(L_{f,2}+\theta_1+\max\left[\sigma_0, \frac{\gamma_3 L_{f,2}}{1-\eta_2}\right]\right)^{3/2}}{2^{3/2}\eta_1\sigma_{\min}},$$

where we substituted the value of $\sigma_{\max}$ given by (3.3.14). □

We may now conclude our analysis by calling upon Lemma 2.4.1 (p. 44), which is equally applicable to the AR2 algorithm since (3.3.8) is identical to (2.4.4).

Theorem 3.3.5: First-order complexity of the AR2 algorithm

Suppose that **f.Bb** *and* **f.HL** *hold. Then there exist positive constants κ_{AR2}^{S} (see Lemma 3.3.4), κ_{AR2}^{A}, κ_{AR2}^{B}, and κ_{AR2}^{C} such that, for any $\epsilon_1 \in (0,1]$, the* AR2 *algorithm (p. 65) with $\epsilon_2 = +\infty$ requires at most*

$$\#_{\text{AR1,1}}^{F} \stackrel{\text{def}}{=} \kappa_{\text{AR2}}^{A}\frac{f(x_0)-f_{\text{low}}}{\epsilon_1^{3/2}} + \kappa_{\text{AR2}}^{C} = \mathcal{O}\left(\epsilon_1^{-3/2}\right) \tag{3.3.21}$$

evaluations of f and at most

$$\#_{\text{AR2,2}}^{D} \stackrel{\text{def}}{=} \kappa_{\text{AR2}}^{S}\frac{f(x_0)-f_{\text{low}}}{\epsilon_1^{3/2}} + 2 = \mathcal{O}\left(\epsilon_1^{-3/2}\right) \tag{3.3.22}$$

evaluations of $\nabla_x^1 f(x)$ and $\nabla_x^2 f(x)$ to produce an iterate x_ϵ such that $\|\nabla_x^1 f(x_\epsilon)\| \leq \epsilon_1$.

Proof. The desired conclusion follows from combining the bound of Lemma 3.3.4 with Lemma 2.4.1 and noting that the objective function f is evaluated at every iteration, while the gradient and Hessian are only evaluated at successful ones. The constants derived from Lemma 2.4.1 are

$$\kappa_{\text{AR2}}^{\text{A}} \stackrel{\text{def}}{=} \kappa_{\text{AR2}}^{\text{S}}\left(1 + \frac{|\log\gamma_1|}{\log\gamma_2}\right)$$

and

$$\begin{aligned}\kappa_{\text{AR2}}^{\text{C}} &\stackrel{\text{def}}{=} \left(1 + \frac{|\log \gamma_1|}{\log \gamma_2}\right) + \frac{1}{\log \gamma_2} \log\left(\max\left[1, \frac{\gamma_3 L_{f,2}}{\sigma_0(1-\eta_2)}\right]\right) + 1 \\ &= \frac{1}{\log \gamma_2}\left\{|\log \gamma_1| + \log\left(\max\left[1, \frac{\gamma_3 L_{f,2}}{\sigma_0(1-\eta_2)}\right]\right)\right\} + 2,\end{aligned}$$

where we also used (3.3.14) and added 1 in the expression of $\kappa_{\text{AR2}}^{\text{C}}$ and (3.3.22) to take the final iteration into account. □

This complexity bound is significantly better than those in $\mathcal{O}(\epsilon_1^{-2})$ evaluations obtained for the TR2 and Newton algorithms. Observe also that there is no need to assume a bounded Hessian (**f.HBa**), by contrast with Theorem 3.2.1. The new bound is also better than that obtained for the AR1 algorithm (see (2.4.12) on p. 46), which is once again $\mathcal{O}(\epsilon_1^{-2})$ evaluations.

While it is clear that a bound of $\mathcal{O}(\epsilon_1^{-3/2})$, such as (3.3.21), is better than one of $\mathcal{O}(\epsilon_1^{-2})$ if ϵ_1 is allowed to become arbitrarily small, surprises may lurk in the constants *when ϵ_1 is fixed.* Indeed, the inequality

$$\kappa_{\text{AR1}}^{\text{A}} \frac{f(x_0) - f_{\text{low}}}{\epsilon_1^2} + \kappa_{\text{AR1}}^{\text{C}} \geq \kappa_{\text{AR2}}^{\text{A}} \frac{f(x_0) - f_{\text{low}}}{\epsilon_1^{3/2}} + \kappa_{\text{AR2}}^{\text{C}}$$

may not hold for all values of ϵ_1, since its validity depends on the relative sizes of the constituent constants compared to $\epsilon_1^{1/2}$. In particular, these constants increase with the values of the Lipschitz constants $L_{f,1}$ and $L_{f,2}$, and it is not uncommon for the value of the latter to significantly exceed that of the former. Thus caution is advisable when interpreting complexity bounds[19] for fixed ϵ_1.

True to habit, the evaluation-complexity bound of Theorem 3.3.5 can be shown to be sharp in order. We will not develop the specific example to prove this statement here but postpone the argument to Section 12.2.2.4, where it is presented in a more general context. However, we formally state the result here for completeness.

Theorem 3.3.6: Slow first-order convergence of the AR2 algorithm

*The AR2 algorithm (p. 65) applied to minimize a function f satisfying **f.Bb** and **f.HL** may require $\epsilon_1^{-3/2}$ iterations and evaluations of $f(x)$, $\nabla_x^1 f(x)$, and $\nabla_x^2 f(x)$ to produce a point x_ϵ such that $\|\nabla_x^1 f(x_\epsilon)\| \leq \epsilon_1$.*

Proof. See the paragraph following Corollary 12.2.16 (p. 316) in Section 12.2.2.4. □

Notes and references for Subsection 3.3.2

The improved complexity bound of Theorem 3.3.5 was first obtained by Nesterov and Polyak [305], albeit for a variant of the AR2 algorithm using exact model minimization (3.3.9). Both of these conditions were relaxed in [85] for their ACO algorithm, for which encouraging numerical results were also presented. However, the ACO algorithm still involves the definition of ρ_k given by (3.3.10) instead of (3.3.7). As a consequence, a more complicated argument is necessary to obtain a suitable formula for the guaranteed model decrease—the guaranteed decrease (3.3.13)

[19]Bounds are just bounds . . . fortunately, overwhelming practical experience suggests that using higher-degree models leads to faster convergence.

cannot be invoked in this case. This argument follows along the lines of the standard trust-region theory and uses the notion of a Cauchy point, in this case the minimizer of the regularized model in the steepest-descent direction. Unsurprisingly, establishing its improved complexity properties requires an assumption on the boundedness of the Hessians (**f.HBa**). The streamlined analysis presented here hinges on that of Birgin et al. [45]. As before, here and in the next subsection, **f.HL** may be relaxed to merely require that (3.1.3) hold for all x and y in an open convex set containing the iterates generated by the algorithm [85]. An example of "essentially slow" convergence of the ACO algorithm to approximate first-order minimizers is presented in [83].

As for the trust-region case, it is entirely possible to use an appropriate regularized second-order model

$$m_k(s) = f(x_k) + \langle \nabla_x^1 f(x_k), s\rangle + \tfrac{1}{2}\langle s, B_k s\rangle + \tfrac{1}{6}\sigma_k \|s\|^3$$

instead of (3.3.1) [89, 362]. So long as (3.2.14) holds for the approximate Hessian B_k, it is trivial to refine inequality (3.3.16) in the proof of Lemma 3.3.3 so that (3.3.15) holds for a slightly different iteration-independent constant $2/(L_{f,2} + \theta_1 + \kappa_{BH} + \sigma_{\max})$. This results in an $\mathcal{O}(\epsilon_1^{-3/2})$ first-order evaluation-complexity bound. Once again, though, this is a special case of general developments in Section 13.1.4 (p. 337) and is used specifically in finite difference–based second-order methods [93].

Variants of the step-termination condition (3.3.5) are also possible [84, 215]. We briefly discuss the latter here, in which (3.3.5) is replaced by the condition

$$\|\nabla_s^1 T_{f,2}(x_k, s_k)\| \le \tfrac{1}{2}\theta_1\sigma_k\|s_k\|^2, \tag{3.3.23}$$

where $\theta_1 > 1$. Note that this inequality must hold at a stationary point of $m_k(s)$ since $\nabla_x^1 T_{f,2}(x_k, s_k) = -\frac{1}{2}\sigma_k\|s_k\|s_k$ at such a point. We now outline why a modified AR2 algorithm that uses (3.3.23) instead of (3.3.5) retains an $\mathcal{O}(\epsilon_1^{3/2})$ worst-case evaluation-complexity bound to find an ϵ_1-approximate first-order minimizer. The key is clearly a modification of Lemma 3.3.3.

Lemma 3.3.7: Modified lower bound on the length of the step

*Suppose that **f.HL** holds. Then, if iteration k of the modified* AR2 *algorithm using (3.3.23) rather than (3.3.5) is successful,*

$$\|s_k\|^2 \ge \frac{2}{L_{f,2} + \theta_1\sigma_{\max}}\|\nabla_x^1 f(x_{k+1})\|. \tag{3.3.24}$$

Proof. The triangle inequality gives that

$$\|\nabla_x^1 f(x_{k+1})\| \le \|\nabla_x^1 f(x_{k+1}) - \nabla_s^1 T_{f,2}(x_k, s_k)\| + \|\nabla_s^1 T_{f,2}(x_k, s_k)\|.$$

The successful nature of iteration k implies that $\nabla_x^1 f(x_{k+1}) = \nabla_x^1 f(x_k + s_k)$, and thus we find that

$$\|\nabla_x^1 f(x_{k+1})\| \le \tfrac{1}{2}(L_{f,2} + \theta_1\sigma_{\max})\|s\|^2,$$

using (3.3.17), (3.3.23) and (3.3.14), which directly leads to (3.3.24). □

The proofs of Lemma 3.3.4 and Theorem 3.3.5 are then unmodified except for a change of constants from $L_{f,2} + \theta_1 + \sigma_{\max}$ to $L_{f,2} + \theta_1\sigma_{\max}$.

3.3.3 Complexity of finding approximate second-order minimizers

We conclude the evaluation-complexity analysis for the AR2 algorithm by investigating the convergence to an approximate second-order minimizer (i.e., when one chooses $\epsilon_2 \in (0, 1]$). Compared to our discussion in the previous section, the only change required is that the termination rule of the algorithm and the step acceptance condition now involve checking for sufficient small negative curvature (if any). Thus Lemmas 2.4.1, 3.3.1, and 3.3.2 remain valid. Moreover, Lemma 3.3.3 also remains valid because of (3.3.5), but we now extend it to take (3.3.6) into account.

Lemma 3.3.8: Lower bound on the length of the step (second order)

Suppose that **f.HL** *holds. Then, if iteration k of the* AR2 *algorithm with $\epsilon_2 \in (0,1]$ is successful and $\lambda_{\min}[\nabla_x^2 f(x_{k+1})] < 0$, one has that*

$$\|s_k\| \geq \frac{1}{L_{f,2} + \theta_2 + \sigma_{\max}} |\lambda_{\min}[\nabla_x^2 f(x_{k+1})]|. \tag{3.3.25}$$

Proof. Since $\min_z[a(z)+b(z)] \geq \min_z[a(z)] + \min_z[b(z)]$, we have that

$$\begin{aligned}
&\lambda_{\min}[\nabla_x^2 f(x_{k+1})] \\
&\quad = \min_{\|d\|=1} \langle \nabla_x^2 f(x_{k+1}) d, d\rangle \\
&\quad = \min_{\|d\|=1} \Big[\Big\langle (\nabla_x^2 f(x_{k+1}) - \nabla_s^2 T_{f,2}(x_k,s))d, d \Big\rangle \\
&\qquad\qquad\qquad - \tfrac{1}{6}\sigma_k \Big\langle \nabla_s^2(\|s_k\|^3)d, d\Big\rangle + \langle \nabla_s^2 m_k(s_k)d, d\rangle \Big] \\
&\quad \geq \min_{\|d\|=1} \Big\langle (\nabla_x^2 f(x_{k+1}) - \nabla_s^2 T_{f,2}(x_k,s))d, d \Big\rangle \\
&\qquad\qquad\qquad + \tfrac{1}{6}\sigma_k \min_{\|d\|=1} \Big\langle \nabla_s^2(-\|s_k\|^3)d, d\Big\rangle + \lambda_{\min}[\nabla_s^2 m_k(s_k)].
\end{aligned} \tag{3.3.26}$$

But using **f.HL** and (A.20) in Corollary A.8.4 (p. 437) with $p = j = 2$ yields that

$$\begin{aligned}
\min_{\|d\|=1} & \Big\langle (\nabla_x^2 f(x_{k+1}) - \nabla_s^2 T_{f,2}(x_k,s))d, d \Big\rangle \\
& \geq - \max_{\|d\|=1} \Big| \Big\langle (\nabla_x^2 f(x_{k+1}) - \nabla_s^2 T_{f,2}(x_k,s))d, d \Big\rangle \Big| \\
& = - \Big\| \nabla_x^2 f(x_{k+1}) - \nabla_s^2 T_{f,2}(x_k,s) \Big\| \\
& \geq -L_{f,2}\|s_k\|.
\end{aligned} \tag{3.3.27}$$

Moreover, direct calculation gives that

$$\begin{aligned}
\min_{\|d\|=1} \Big\langle \nabla_s^2(-\|s_k\|^3)d, d\Big\rangle &= - \max_{\|d\|=1} \langle \nabla_s^2(-\|s_k\|^3)d, d\rangle \\
&= - \max_{\|d\|=1} \Big\langle 3\Big[\frac{1}{\|s_k\|} s_k s_k^T + \|s_k\| I\Big] d, d\Big\rangle \\
&= -6\|s_k\|.
\end{aligned} \tag{3.3.28}$$

Thus, combining (3.3.6) and (3.3.26)–(3.3.28), we deduce that

$$\lambda_{\min}[\nabla_x^2 f(x_{k+1})] \geq -(L_{f,2} + \sigma_k + \theta_2)\|s_k\|.$$

This inequality, our assumption that $\lambda_{\min}[\nabla_x^2 f(x_{k+1})] < 0$, and (3.3.14) now imply (3.3.25). □

We then follow the familiar argument to prove our final complexity result for AR2.

Theorem 3.3.9: Second-order complexity of the AR2 algorithm

*Suppose that **f.Bb** and **f.HL** hold. Then there exist positive constants κ^T_{AR2}, κ^D_{AR2}, κ^E_{AR2}, and κ^F_{AR2} such that, for any $(\epsilon_1, \epsilon_2) \in (0,1]^2$, the AR2 algorithm (p. 65) requires at most*

$$\#^F_{\text{AR2,2}} \stackrel{\text{def}}{=} \kappa^D_{\text{AR2}} \frac{f(x_0) - f_{\text{low}}}{\min[\epsilon_1^{3/2}, \epsilon_2^3]} + \kappa^F_{\text{AR2}} = \mathcal{O}\left(\max[\epsilon_1^{-3/2}, \epsilon_2^{-3}]\right) \tag{3.3.29}$$

evaluations of f and at most

$$\#^D_{\text{AR2,2}} \stackrel{\text{def}}{=} \kappa^T_{\text{AR2}} \frac{f(x_0) - f_{\text{low}}}{\min[\epsilon_1^{3/2}, \epsilon_2^3]} + 1 = \mathcal{O}\left(\max[\epsilon_1^{-3/2}, \epsilon_2^{-3}]\right) + 2 \tag{3.3.30}$$

evaluations of the gradient and Hessian to produce an iterate x_ϵ such that $\|\nabla^1_x f(x_\epsilon)\| \le \epsilon_1$ and $\lambda_{\min}[\nabla^2_x f(x_\epsilon)] \ge -\epsilon_2$.

Proof. Using once again the telescoping-sum argument of Theorem 3.3.5, we have that, for any k such that the AR2 algorithm has not terminated before or at iteration $k+1$,

$$f(x_0) - f_{\text{low}} \ge f(x_0) - f(x_{k+1}) \ge \tfrac{1}{6}\eta_1 \sum_{j \in \mathcal{S}_k} \sigma_j \|s_j\|^3. \tag{3.3.31}$$

Since the algorithm has not terminated, we must have that either $\|\nabla^1_x f(x_{k+1})\| > \epsilon_1$ or $\lambda_{\min}[\nabla^2_x f(x_{k+1})] < -\epsilon_2$. As a consequence, we may combine Lemmas 3.3.3 and 3.3.8 to obtain that

$$\begin{aligned} \|s_j\| &\ge \min\left[\frac{\sqrt{2\epsilon_1}}{\sqrt{L_{f,2} + \theta_1 + \sigma_{\max}}}, \frac{\epsilon_2}{L_{f,2} + \theta_2 + \sigma_{\max}}\right] \\ &> \frac{\min[\sqrt{\epsilon_1}, \epsilon_2]}{L_{f,2} + \max[1, \theta_1, \theta_2] + \sigma_{\max}} \end{aligned}$$

for all $j \in \mathcal{S}_{k-1}$. Hence (3.3.31) and (3.3.8) give that

$$f(x_0) - f_{\text{low}} \ge \tfrac{1}{6}\eta_1 \sum_{j \in \mathcal{S}_{k-1}} \sigma_j \|s_j\|^3 \ge \frac{\eta_1 \sigma_{\min} |\mathcal{S}_{k-1}|}{6(L_{f,2} + \max[1, \theta_1, \theta_2] + \sigma_{\max})^3} \min[\epsilon_1^{3/2}, \epsilon_2^3]$$

and, therefore, using $|\mathcal{S}_k| \le |\mathcal{S}_{k-1}| + 1$,

$$|\mathcal{S}_k| \le \frac{6(L_{f,2} + \max[1, \theta_1, \theta_2] + \sigma_{\max})^3}{\eta_1 \sigma_{\min}} \times \frac{f(x_0) - f_{\text{low}}}{\min[\epsilon_1^{3/2}, \epsilon_2^3]} + 1,$$

which provides an upper bound on the number of successful iterations and, given that the gradient and Hessian are evaluated once at each such iteration, yields the bound (3.3.30) with

$$\kappa^T_{\text{AR2}} \stackrel{\text{def}}{=} \frac{6(L_{f,2} + \max[1, \theta_1, \theta_2] + \sigma_{\max})^3}{\eta_1 \sigma_{\min}}.$$

We may now apply Lemma 2.4.1 to deduce an upper bound on the total number of iterations and evaluations of f given by (3.3.29) with

$$\kappa^D_{\text{AR2}} \stackrel{\text{def}}{=} \kappa^T_{\text{AR2}} \left(1 + \frac{|\log \gamma_1|}{\log \gamma_2}\right)$$

and

$$\kappa^{\rm F}_{\sf AR2} \stackrel{\rm def}{=} \frac{1}{\log \gamma_2}\left\{|\log\gamma_1| + \log\left(\max\left[1, \frac{\gamma_3 L_{f,2}}{\sigma_0(1-\eta_2)}\right]\right)\right\} + 2,$$

where we also used (3.3.14) and added 1 to take the evaluations at the last iteration into account. □

This result is different from the equivalent result for the TR2 algorithm since $\epsilon_1^{-2}\epsilon_2^{-1}$ is not equal to $\epsilon_1^{3/2}$ in general, and also because, as was the case for Theorem 3.3.5, there is no need to assume **f.HBa**.

Again, this result is sharp—we record this statement formally here but postpone the proof until Section 12.2.2.4 (see the paragraph following Corollary 12.2.16 on p. 316).

Theorem 3.3.10: Slow second-order convergence of the AR2 algorithm

*The AR2 algorithm (p. 65) applied to minimize a function f satisfying **f.Bb** and **f.HL** may require $\max[\epsilon_1^{-3/2}, \epsilon_2^{-3}]$ iterations and evaluations of $f(x)$, $\nabla_x^1 f(x)$, and $\nabla_x^2 f(x)$ to produce a point x_ϵ such that $\|\nabla_x^1 f(x_\epsilon)\| \leq \epsilon_1$ and $\lambda_{\min}[\nabla_x^2 f(x_k)] \geq -\epsilon_2$.*

Notes and references for Subsection 3.3.3

The bound (3.3.29) was first shown by Nesterov and Polyak [305] for the special case where $\epsilon_2 = \sqrt{\epsilon_1}$, an exact step is computed by (3.3.9), and the Hessian Lipschitz constant $L_{f,2}$ is known explicitly, in which case the evaluation-complexity bound becomes $\mathcal{O}(\epsilon_1^{-3/2})$. Again these conditions were weakened for the ACO algorithm in [85], where it was in particular assumed that the step s_k satisfies the conditions

$$\langle \nabla_x^1 f(x_k), s_k\rangle + \langle B_k s_k, s_k\rangle + \sigma_k\|s_k\|^3 = 0 \;\text{ and }\; \langle B_k s_k, s_k\rangle + \sigma_k\|s_k\|^3 \geq 0.$$

The analysis presented here follows along the broad lines of [107].

Currently there are fewer implementations of adaptive-regularization methods than there are of the trust-region variety. One example, though, is the package ARC, again from the GALAHAD library [192], which uses the TR2 framework, but with many subproblem enhancements that will be revealed in Part II of this book.

3.4 More algorithms with $\mathcal{O}(\epsilon^{-3/2})$ complexity

Thus far in our journey of discovery, the *best-known* bound for the number of evaluations to find an ϵ_1-approximate first-order minimizer that we have exhibited is $\mathcal{O}(\epsilon_1^{-3/2})$, which is achieved by the AR2 algorithm. The same algorithm also achieves the best-known $\mathcal{O}(\max[\epsilon_1^{-3/2}, \epsilon_2^{-3}])$ bound for the number of evaluations to find an (ϵ_1, ϵ_2)-approximate second-order minimizer among the methods that we have described.

Leaving aside the question of whether there are methods that achieve better, or even optimal, bounds, it is of interest to know at least if the AR2 algorithm is alone. Encouraged by the improved complexity of the AR2 algorithm compared to its predecessors, complexity-efficient variants of more traditional approaches, such as those based on trust-region and quadratic regularization—Levenberg–Morrison–Marquardt-type methods—have been developed, often with the additional intention of promoting good performance in practice. This is the focus of this section.

3.4.1 TRϵ, a tolerance-dependent trust-region variant

The first variant was introduced by Curtis, Robinson, Royer, and Wright [136] and is a simple modification of the TR2 algorithm that satisfies the best-known first-order complexity bound—and second-order bound when the first- and second-order criticality tolerances ϵ_1 and ϵ_2 are suitably coupled. The method introduces a small regularization term in the trust-region subproblem and the step calculation. The subproblem approximately solved on each iteration of TRϵ uses the model

$$\begin{aligned} m_k(s) &\stackrel{\text{def}}{=} f(x_k) + \langle \nabla_x^1 f(x_k), s\rangle + \tfrac{1}{2}\langle s, \nabla_x^2 f(x_k)s\rangle + \tfrac{1}{2}\epsilon_*\|s\|^2 \\ &= T_{f,2}(x_k, s) + \tfrac{1}{2}\epsilon_*\|s\|^2, \end{aligned} \tag{3.4.1}$$

where $\epsilon_* > 0$ is a fixed regularization weight, reminiscent of the lower bound $\sigma_{\min}$ in the AR2 algorithm. The model (3.4.2) is then approximately minimized over a trust-region ball:

$$\min_{s\in\mathbb{R}^n} m_k(s) \text{ subject to } x_k + s \in R_k = \{s : \|s\| \le \Delta_k\}. \tag{3.4.2}$$

We formally state the TRϵ algorithm as follows.

Algorithm 3.4.1: Trust-region algorithm with a regularized second-order model (TRϵ)

Step 0: Initialization. An initial point x_0, accuracy thresholds $\epsilon_1 \in (0,1]$, $\epsilon_2 \in (0,1]$, a regularization threshold $\epsilon_* \in (0,1]$, and an initial trust-region radius $\Delta_0 > \epsilon_*$ are given. The constants η_1, η_2, γ_1, γ_2, and θ_1 are also given and satisfy

$$0 < \eta_1 \le \eta_2 < 1, \quad \theta_1 > 0, \quad \text{and} \quad 0 < \gamma_1 \le \gamma_2 < 1 < \gamma_3.$$

Compute $f(x_0)$, $\nabla_x^1 f(x_0)$, and $\nabla_x^2 f(x_0)$, and set $k = 0$.

Step 1: Test for termination. If $\epsilon_2 = +\infty$ and $\epsilon_* = \sqrt{\epsilon_1}$, then if

$$\|\nabla_x^1 f(x_k)\| \le \epsilon_1, \tag{3.4.3}$$

terminate with the ϵ_1-approximate first-order minimizer $x_\epsilon = x_k$.

If $\epsilon_2 < \infty$ and $\epsilon_* = \epsilon_2$, then if

$$\|\nabla_x^1 f(x_k)\| \le \epsilon_1 \quad \text{and} \quad \lambda_{\min}[\nabla_x^2 f(x_k)] \ge -\epsilon_2, \tag{3.4.4}$$

terminate with the (ϵ_1, ϵ_2)-approximate second-order minimizer $x_\epsilon = x_k$.

Step 2: Step computation. Compute a step $s_k \in \mathcal{B}_{\Delta_k}$ such that $m_k(s_k) < m_k(0)$, and whenever $\|s_k\| < \Delta_k$ ensure also that

$$\|\nabla_s^1 m_k(s_k)\| \le \tfrac{1}{2}\theta_1\|s_k\|^2. \tag{3.4.5}$$

If $\epsilon_2 < +\infty$ and $\lambda_{\min}[\nabla_x^2 f(x_k)] < -\epsilon_2$, calculate s_k such that $\|s_k\| = \Delta_k$.

Step 3: Acceptance of the trial point. Compute $f(x_k + s_k)$ and define

$$\rho_k = \frac{f(x_k) - f(x_k + s_k)}{f(x_k) - T_{f,2}(x_k, s_k)}. \tag{3.4.6}$$

If $\rho_k \ge \eta_1$, then define $x_{k+1} = x_k + s_k$ and compute $\nabla_x^1 f(x_{k+1})$ and $\nabla_x^2 f(x_{k+1})$. Otherwise define $x_{k+1} = x_k$.

Step 4: Trust-region radius update. Set

$$\Delta_{k+1} \in \begin{cases} [\Delta_k, \gamma_3 \Delta_k] & \text{if } \rho_k \geq \eta_2, \\ [\gamma_2 \Delta_k, \Delta_k] & \text{if } \rho_k \in [\eta_1, \eta_2), \\ [\gamma_1 \Delta_k, \gamma_2 \Delta_k] & \text{if } \rho_k < \eta_1. \end{cases}$$

Increment k by one and go to Step 1.

Clearly the global minimizer of (3.4.2), or even a nonzero local minimizer, satisfies the step conditions in Step 2 of the TRϵ algorithm. Furthermore, in the case when $\epsilon_2 < +\infty$, $\epsilon_* = \epsilon_2$, and $\lambda_{\min}[\nabla_x^2 f(x_k)] > -\epsilon_2$ in Step 2 of TRϵ, the requirement that $\|s_k\| = \Delta_k$ can be achieved by a variant of the eigenpoint condition of the TR2 algorithm, namely, (3.2.5) with $T_{f,2}$ replaced by the regularized model m_k of the TRϵ algorithm.[20]

As for the TR2 and AR2 algorithms, we define successful and unsuccessful iterations by (2.3.9)–(2.3.10) and note that Lemma 2.3.1 still applies to the TRϵ algorithm because it uses the same updating rules for Δ_k as did the TR2 algorithm.

We now proceed to analyze the evaluation complexity of the TRϵ algorithm, and we will see that we will follow common paths to both the TR2 and AR2 algorithms. Firstly, a familiar bound on the model decrease, similar to that for the AR2 algorithm, is derived.

Lemma 3.4.1: Decrease in the Taylor-series model

At every iteration k of the TRϵ algorithm, one has that

$$T_{f,2}(x_k, 0) - T_{f,2}(x_k, s_k) \geq \tfrac{1}{2}\epsilon_* \|s_k\|^2. \tag{3.4.7}$$

Proof. From (3.4.1), we have that

$$T_{f,2}(x_k, 0) - T_{f,2}(x_k, s_k) = m_k(0) - m_k(s_k) + \tfrac{1}{2}\epsilon_* \|s_k\|^2,$$

and (3.4.7) follows from the requirement (in Step 2) that $m_k(s_k) < m_k(0)$. □

Next, as is typical for trust-region methods, we provide a lower bound on the trust-region radius.

Lemma 3.4.2: Lower bound on the trust-region radius

Suppose that **f.HL** *holds, as well as that, in the* TRϵ *algorithm,*

$$\Delta_k < \frac{3(1-\eta_2)}{L_{f,2}}\epsilon_*. \tag{3.4.8}$$

Then $\rho_k \geq \eta_2$, iteration k is successful, and $\Delta_{k+1} \geq \Delta_k$. Furthermore, for all iterations of the TRϵ *algorithm, we have*

$$\Delta_k \geq \frac{(1-\eta_2)\gamma_1}{L_{f,2}}\epsilon_*. \tag{3.4.9}$$

[20] Note that in this case, $\|s_k\| = \|t_k^E u_k\| = t_k^E$, where u_k is normalized. Also, $m_k(tu_k)'' = \lambda_{\min}[\nabla_x^2 f(x_k)] + \epsilon_* < 0$ and so the model along $t > 0$ is concave and achieves its minimum on the boundary of the trust-region constraint.

Proof. Combining the definition of ρ_k in (3.4.6), the bound (A.19) in Corollary A.8.4 (p. 437) with $p = 2$, and the model decrease (3.4.7), we deduce that

$$|\rho_k - 1| \le \frac{L_{f,2}\|s_k\|^3}{3\epsilon_*\|s_k\|^2} \le \frac{L_{f,2}\Delta_k}{3\epsilon_*} < 1 - \eta_2,$$

where we used (3.4.8) to deduce the last inequality; the first desired conclusion follows. The second part follows by arguments similar to those in Lemma 3.2.5 for the TR2 algorithm; namely, assume that $k+1$ is the first iteration when Δ_{k+1} goes below the right-hand side of (3.4.9). This implies that iteration k must have been unsuccessful and so $\Delta_{k+1} = \gamma_1\Delta_k$. Thus Δ_k must satisfy (3.4.8), which means that iteration k must be successful; we have reached a contradiction. Recalling that TRϵ assumes $\Delta_0 > \epsilon_*$ and that $(1-\eta_2)\gamma_1/L_{f,2} \in (0,1]$, we obtain the second desired conclusion of the lemma. □

We now provide a lower bound on the step in the case when it is an interior approximate solution to the trust-region subproblem; the proof follows in part the now familiar path of the bound on the step in the AR2 algorithm.

Lemma 3.4.3: Lower bound on the length of the strictly feasible step

Suppose that **f.HL** *holds. Then, if iteration k of the* TRϵ *algorithm is successful and $\|s_k\| < \Delta_k$, we have that*

$$\|s_k\| \ge \frac{1}{\sqrt{1+2(L_{f,2}+\theta_1)}} \min\left[\epsilon_*, \|\nabla_x^1 f(x_{k+1})\|\epsilon_*^{-1}\right]. \tag{3.4.10}$$

Proof. This follows by the now familiar add and subtract argument. Using the triangle inequality yields that

$$\begin{aligned}
\|\nabla_x^1 f(x_{k+1})\| &\le \left\|\nabla_x^1 f(x_{k+1}) - \nabla_s^1 m_k(s_k) + \nabla_s^1 m_k(s_k)\right\| \\
&\le \|\nabla_x^1 f(x_{k+1}) - \nabla_s^1 T_{f,2}(x_k, s_k)\| + \left\|\nabla_s^1 m_k(s_k)\right\| + \tfrac{1}{2}\epsilon_*\|s_k\| \\
&\le \tfrac{1}{2}L_{f,2}\|s_k\|^2 + \tfrac{1}{2}\theta_1\|s_k\|^2 + \epsilon_*\|s_k\|,
\end{aligned} \tag{3.4.11}$$

where to obtain the first term in the last inequality, we used **f.HL** and (A.20) for $j = 1$ and $p = 2$, while for the second term, we used (3.4.5).

We may rewrite (3.4.11) as a quadratic inequality in $\|s_k\|$,

$$\tfrac{1}{2}L_{f,2} + \theta_1\|s_k\|^2 + \epsilon_*\|s_k\| - \|\nabla_x^1 f(x_{k+1})\| \ge 0,$$

which is satisfied if

$$\begin{aligned}
\|s_k\| &\ge \frac{1}{L_{f,2}}\left[-\epsilon_* + \sqrt{\epsilon_*^2 + 2(L_{f,2}+\theta_1)\|\nabla_x^1 f(x_{k+1})\|}\right] \\
&= \frac{\epsilon_*}{L_{f,2}}\left[-1 + \sqrt{1 + 2(L_{f,2}+\theta_1)\|\nabla_x^1 f(x_{k+1})\|\epsilon_*^{-2}}\right].
\end{aligned} \tag{3.4.12}$$

A simple algebraic property (see Lemma B.3.1 (p. 455) in Appendix B) can be used to show that (3.4.12) implies

$$\|s_k\| \ge \frac{\epsilon_*}{L_{f,2}}\left[-1 + \sqrt{1+2(L_{f,2}+\theta_1)}\right] \min\left[1, \|\nabla_x^1 f(x_{k+1})\|\epsilon_*^{-2}\right]$$

and, furthermore,

$$\begin{aligned}\|s_k\| &\geq \frac{\epsilon_*}{L_{f,2}}\left[-1+\sqrt{1+2(L_{f,2}+\theta_1)}\right]\min\left[1, \|\nabla_x^1 f(x_{k+1})\|\epsilon_*^{-2}\right]\\ &\geq \frac{2}{1+\sqrt{1+2(L_{f,2}+\theta_1)}}\min\left[\epsilon_*, \|\nabla_x^1 f(x_{k+1})\|\epsilon_*^{-1}\right],\end{aligned}$$

which immediately gives the bound (3.4.10). We note that the latter bound is valid in the case $L_{f,2} = 0$. □

Bounds on the function decrease on successful iterations now follow easily by combining (3.4.7) and (3.4.10) in the case when the step is strictly interior to the trust-region constraint, and (3.4.7) and $\|s_k\| = \Delta_k$ otherwise.

Lemma 3.4.4: Lower bound on the function decrease

*Suppose that **f.HL** holds. Then, on each successful iteration k of the* TRϵ *algorithm, we have the sufficient function decrease*

$$f(x_k) - f(x_{k+1}) \geq \kappa_{\Delta f}\min\left[\epsilon_*^3, \|\nabla_x^1 f(x_{k+1})\|^2\epsilon_*^{-1}\right] \tag{3.4.13}$$

whenever x_{k+1} is in the interior of the trust-region ball and

$$f(x_k) - f(x_{k+1}) \geq \kappa_{\Delta f}\epsilon_*^3 \tag{3.4.14}$$

when x_{k+1} is on the trust-region boundary, where

$$\kappa_{\Delta f} \stackrel{\text{def}}{=} \frac{\eta_1}{2}\min\left[\frac{\gamma_1^2(1-\eta_2)^2}{L_{f,2}^2}, \frac{1}{1+2(L_{f,2}+\theta_1)}\right]. \tag{3.4.15}$$

Proof. We have two cases to consider, namely, when s_k is on the boundary of the trust-region constraint or in the interior.
Case 1: $\|s_k\| = \Delta_k$. Then (3.4.6) and (3.4.7) imply that on each such iteration that is also successful, we have

$$f(x_k) - f(x_{k+1}) \geq \frac{\eta_1}{2}\epsilon_*\Delta_k^2 \geq \frac{\eta_1\gamma_1^2(1-\eta_2)^2}{2L_{f,2}^2}\epsilon_*^3,$$

where the second inequality follows from (3.4.9). This clearly implies (3.4.13).
Case 2: $\|s_k\| < \Delta_k$. Then (3.4.6), (3.4.7), and (3.4.10) give

$$f(x_k) - f(x_{k+1}) \geq \frac{\eta_1}{2}\epsilon_*\|s_k\|^2 \geq \frac{\eta_1}{2(1+2(L_{f,2}+\theta_1))}\min\left[\epsilon_*^3, \|\nabla_x^1 f(x_{k+1})\|^2\epsilon_*^{-1}\right],$$

which yields (3.4.14). □

If we are only interested in achieving first-order criticality, then setting $\epsilon_* = \sqrt{\epsilon_1}$ in the TRϵ algorithm will ensure that it has the better (optimal) complexity bound typical of the AR2 algorithm.

Theorem 3.4.5: First-order complexity of the TRϵ algorithm

Suppose that **f.Bb** *and* **f.HL** *hold. Then there exist positive constants* $\kappa^{S}_{\mathsf{TR}\epsilon}$, $\kappa^{A}_{\mathsf{TR}\epsilon}$, $\kappa^{B}_{\mathsf{TR}\epsilon}$, *and* $\kappa^{C}_{\mathsf{TR}\epsilon}$ *such that, for any* $\epsilon_1 \in (0,1]$, *the* TRϵ *algorithm (p. 76) with* $\epsilon_* = \sqrt{\epsilon_1}$ *and* $\epsilon_2 = +\infty$ *requires at most*

$$\#^{F}_{\mathsf{TR}\epsilon,1} \stackrel{\rm def}{=} \kappa^{A}_{\mathsf{TR}\epsilon}\frac{f(x_0)-f_{\rm low}}{\epsilon_1^{3/2}} + \kappa^{B}_{\mathsf{TR}\epsilon}|\log\epsilon_1| + \kappa^{C}_{\mathsf{TR}\epsilon} = \mathcal{O}\left(\epsilon_1^{-3/2}\right) \tag{3.4.16}$$

evaluations of f *and at most*

$$\#^{D}_{\mathsf{TR}\epsilon,1} \stackrel{\rm def}{=} \kappa^{S}_{\mathsf{TR}\epsilon}\frac{f(x_0)-f_{\rm low}}{\epsilon_1^{3/2}} + 2 = \mathcal{O}\left(\epsilon_1^{-3/2}\right) \tag{3.4.17}$$

evaluations of $\nabla^1_x f$ *and* $\nabla^2_x f$ *to produce an iterate* x_ϵ *such that* $\|\nabla^1_x f(x_\epsilon)\| \le \epsilon_1$ *(i.e., (3.4.3) is satisfied).*

Proof. For any $k \ge 0$ such that the algorithm does not terminate at or before iteration k, we have that

$$f(x_0) - f_{\rm low} \ge f(x_0) - f(x_{k+1}) = \sum_{j\in\mathcal{S}_k}[f(x_j) - f(x_{j+1})], \tag{3.4.18}$$

where $\mathcal{S}_k$ is defined as in (2.3.10). We now use Lemma 3.4.4 and combine the two bounds (3.4.13) and (3.4.14) on the function decrease to deduce that for each $j \in \mathcal{S}_{k-1}$,

$$\begin{aligned} f(x_j) - f(x_{j+1}) &\ge \kappa_{\Delta f}\min\left[\epsilon_*^3, \|\nabla^1_x f(x_{j+1})\|^2\epsilon_*^{-1}\right] \\ &= \kappa_{\Delta f}\min\left[\epsilon_1^{3/2}, \|\nabla^1_x f(x_{j+1})\|^2\epsilon_1^{-1/2}\right], \end{aligned} \tag{3.4.19}$$

where we have used $\epsilon_* = \sqrt{\epsilon_1}$. Since the algorithm has not terminated at or before iteration k, we have that $\|\nabla^1_x f(x_{j+1})\| \ge \epsilon_1$ for all $j \in \mathcal{S}_{k-1}$, and hence (3.4.19) becomes

$$f(x_j) - f(x_{j+1}) \ge \kappa_{\Delta f}\epsilon_1^{3/2},$$

where $\kappa_{\Delta f}$ is defined as in (3.4.15). Substituting this in (3.4.18), we obtain

$$f(x_0) - f_{\rm low} \ge \sum_{j\in\mathcal{S}_{k-1}}[f(x_j) - f(x_{j+1})] \ge \kappa_{\Delta f}\epsilon_1^{3/2}\,|\mathcal{S}_{k-1}|,$$

which, after taking into account that $|\mathcal{S}_k| \le |\mathcal{S}_{k-1}| + 1$ and adding 1 for the final iteration, gives (3.4.17) with

$$\kappa^{\rm S}_{\mathsf{TR}\epsilon} \stackrel{\rm def}{=} \kappa_{\Delta f}^{-1} = \frac{2}{\eta_1}\max\left[\frac{L_{f,2}^2}{\gamma_1^2(1-\eta_2)^2}, 1 + 2(L_{f,2}+\theta_1)\right], \tag{3.4.20}$$

recalling also that $\nabla^1_x f$ and $\nabla^2_x f$ are only evaluated at successful iterations. The remaining bound (3.4.16) follows from combining the bound (3.4.17) on $|\mathcal{S}_k|$ with Lemma 2.3.1 and noting that the objective function f is evaluated at every iteration.

The constants derived from Lemma 2.3.1 are

$$\kappa_{\mathsf{TR}\epsilon}^{\mathrm{A}} \stackrel{\text{def}}{=} \kappa_{\mathsf{TR}\epsilon}^{\mathrm{S}}\left(1+\frac{\log\gamma_3}{|\log\gamma_2|}\right), \qquad \kappa_{\mathsf{TR}\epsilon}^{\mathrm{B}} \stackrel{\text{def}}{=} \frac{1}{2|\log\gamma_2|}, \tag{3.4.21}$$

and

$$\kappa_{\mathsf{TR}\epsilon}^{\mathrm{C}} \stackrel{\text{def}}{=} \frac{1}{|\log\gamma_2|}\left\{\log\gamma_3+\left|\log\frac{(1-\eta_2)\gamma_1}{\Delta_0 L_{f,2}}\right|\right\}+2, \tag{3.4.22}$$

where we also used (3.4.9) to derive the expressions of $\kappa_{\mathsf{TR}\epsilon}^{\mathrm{B}}$ and $\kappa_{\mathsf{TR}\epsilon}^{\mathrm{C}}$, and we added 1 in the expression of $\kappa_{\mathsf{TR}\epsilon}^{\mathrm{C}}$ to take the final iteration into account. □

Setting $\epsilon_* = \epsilon_2$ in the TRϵ algorithm allows us to achieve second-order critical points in a quantifiable way, as we show below.

Theorem 3.4.6: Second-order complexity of the TRϵ algorithm

*Suppose that **f.Bb** and **f.HL** hold. Then there exist positive constants $\kappa_{\mathsf{TR}\epsilon}^{S}$, $\kappa_{\mathsf{TR}\epsilon}^{A}$, $\kappa_{\mathsf{TR}\epsilon}^{C}$, and $\kappa_{\mathsf{TR}\epsilon}^{D}$ such that, for any $(\epsilon_1,\epsilon_2)\in(0,1]^2$, the TR$\epsilon$ algorithm (p. 76) with $\epsilon_* = \epsilon_2$ requires at most*

$$\#_{\mathsf{TR}\epsilon,2}^{F} \stackrel{\text{def}}{=} \kappa_{\mathsf{TR}\epsilon}^{A}\frac{f(x_0)-f_{\text{low}}}{\min[\epsilon_1^2\epsilon_2^{-1},\epsilon_2^3]} + \kappa_{\mathsf{TR}\epsilon}^{D}|\log\epsilon_2| + \kappa_{\mathsf{TR}\epsilon}^{C} = \mathcal{O}\left(\max[\epsilon_1^{-2}\epsilon_2,\epsilon_2^{-3}]\right) \tag{3.4.23}$$

evaluations of f and at most

$$\#_{\mathsf{TR}\epsilon,2}^{D} \stackrel{\text{def}}{=} \kappa_{\mathsf{TR}\epsilon}^{S}\frac{f(x_0)-f_{\text{low}}}{\min[\epsilon_1^2\epsilon_2^{-1},\epsilon_2^3]} + 1 = \mathcal{O}\left(\max[\epsilon_1^{-2}\epsilon_2,\epsilon_2^{-3}]\right) \tag{3.4.24}$$

evaluations of $\nabla_x^1 f$ and $\nabla_x^2 f$ to produce an iterate x_ϵ such that $\|\nabla_x^1 f(x_\epsilon)\| \le \epsilon_1$ and $\lambda_{\min}[\nabla_x^2 f(x_\epsilon)] \ge -\epsilon_2$ (i.e., (3.4.4) is satisfied).

Proof. Clearly, (3.4.18) still holds and we return again to Lemma 3.4.4. Combining (3.4.13) and (3.4.14) with $\epsilon_* = \epsilon_2$, we obtain that for each $j \in \mathcal{S}_k$,

$$f(x_j) - f(x_{j+1}) \ge \kappa_{\Delta f}\min\left[\epsilon_2^3, \|\nabla_x^1 f(x_{j+1})\|^2\epsilon_2^{-1}\right]. \tag{3.4.25}$$

If x_{j+1} is on the boundary of the trust-region ball, we obtain from (3.4.14) that

$$f(x_j) - f(x_{j+1}) \ge \kappa_{\Delta f}\epsilon_2^3. \tag{3.4.26}$$

Since the algorithm has not terminated at iteration k, we have that, for each $j \in \mathcal{S}_{k-1}$, either $\|\nabla_x^1 f(x_{j+1})\| \ge \epsilon_1$ or $\lambda_{\min}[\nabla_x^2 f(x_{j+1})] < -\epsilon_2$. In the former case, (3.4.25) implies that for each $j \in \mathcal{S}_{k-1}$ with $\|\nabla_x^1 f(x_{j+1})\| \ge \epsilon_1$,

$$f(x_j) - f(x_{j+1}) \ge \kappa_{\Delta f}\min\left[\epsilon_2^3, \epsilon_1^2\epsilon_2^{-1}\right].$$

In the latter case, when $\lambda_{\min}[\nabla_x^2 f(x_{j+1})] < -\epsilon_2$, x_{j+1} is on the boundary of the trust-region constraint and the bound (3.4.26) holds. Thus (3.4.18) implies

$$f(x_0) - f_{\text{low}} \ge \kappa_{\Delta f}\min\left[\epsilon_2^3, \epsilon_1^2\epsilon_2^{-1}\right]|\mathcal{S}_{k-1}|,$$

which, after once more taking into account that $|\mathcal{S}_k| \le |\mathcal{S}_{k-1}| + 1$ and adding 1 for the final iteration, gives (3.4.24), with $\kappa^{\mathrm{S}}_{\mathrm{TR}\epsilon}$ defined as in (3.4.20). The remaining bound (3.4.23) follows from combining the bound (3.4.24) on $|\mathcal{S}_k|$ with Lemma 2.3.1. The constants derived from Lemma 2.3.1 are $\kappa^{\mathrm{A}}_{\mathrm{TR}\epsilon}$ and $\kappa^{\mathrm{C}}_{\mathrm{TR}\epsilon}$ as given in (3.4.21) and (3.4.22), respectively, and $\kappa^{\mathrm{D}}_{\mathrm{TR}\epsilon} \stackrel{\text{def}}{=} \frac{1}{|\log \gamma_2|}$. □

A few remarks are in order at this point. Firstly, if $\epsilon_* = \epsilon_2 = \sqrt{\epsilon_1}$, then Theorem 3.4.6 implies that the TRϵ algorithm achieves the best-known complexity for finding second-order critical points, namely that of $\mathcal{O}(\epsilon_1^{-3/2})$, at the expense of a lower accuracy requirement on the second-order criticality condition.

Secondly, in the case when $\epsilon_2 < +\infty$ and $\lambda_{\min}[\nabla_x^2 f(x_k)] > -\epsilon_2$, we calculate a step s_k for which $\|s_k\| = \Delta_k$ in Step 2 of TRϵ. Such a requirement may be replaced by a less demanding one. For instance we might simply stipulate that $\|s_k\| \ge \alpha\Delta_k$ for some $\alpha \in (0, 1]$ chosen at the start of the algorithm, and in this case, the complexity bound in Theorem 3.4.6 can be shown to continue to hold, but with slight changes to the constants to include the α constant.

Finally, under the same conditions, we can envisage other choices for the step-calculation conditions in Step 2 of TRϵ, such as the eigencondition used in TR2. That is to say, we might require that (3.2.10) hold with s_k^{E} as defined in (3.2.5), and where we replace $T_{f,2}$ by the regularized model m_k (see also our comments after the summary of TRϵ). Similar complexity bounds to those in Theorem 3.4.6 may be shown to hold then.

The TRϵ algorithm does not allow pure Newton steps to be taken, which is a primary distinction between trust-region and regularization frameworks. Also, the algorithm needs coupled first- and second-order tolerances to achieve the best-known complexity bound for second-order criticality. In the next subsection, we consider another trust-region variant that permits pure Newton steps and that does not depend on optimality tolerances in its construction.

Notes and references for Subsection 3.4.1

The material in this section differs from the approach in [136] by presenting a more general algorithm formulation through the choice of inexact conditions on the step, and by improving the complexity bound by a log factor. In [136], the focus is on exact and inexact Newton-CG subproblem solutions, while here we require generic conditions that are also in keeping with earlier presentations of the TR2 and AR2 algorithms.

3.4.2 Adaptive trust-region and quadratic-regularization variants

The second variation on the promise of AR2-like worst-case behaviour is more far-reaching. It concerns a general algorithmic framework proposed by Curtis, Robinson, and Samadi [139] that essentially incorporates and generalizes earlier work [47, 85, 137]. The main intuition behind the algorithmic design of these variants has been the observation that the key ingredients used in the construction and analysis of the AR2 algorithm are a specific minimal model decrease (proportional to the cubic power of the length of the step) on successful iterations and a lower bound on the length of the step (proportional to the square root of the norm of the gradient).[21] Trust-region and quadratic-regularization methods are then redesigned with the intention of enforcing these two key elements.

[21] To guarantee approximate nonnegative curvature, the complexity bounds for AR2 and trust-region algorithms agree in the order of the accuracy tolerance ϵ_2.

3.4.2.1 The iRN algorithm

At iteration k of the *Inexact Regularized Newton* (iRN) framework, a step s_k and a multiplier estimate λ_k are calculated as an approximate solution of the subproblem

$$\begin{array}{ll} \min\limits_{(s,\lambda)\in\mathbb{R}^n\times\mathbb{R}_+} & m_k(s) \stackrel{\rm def}{=} f(x_k) + \langle \nabla_x^1 f(x_k), s\rangle + \frac{1}{2}\langle s, (\nabla_x^2 f(x_k) + \lambda I)s\rangle \\ \text{subject to} & \sigma_k^{\rm L}\|s\| \leq \lambda \leq \sigma_k^{\rm U}\|s\| \end{array} \tag{3.4.27}$$

for appropriately chosen and updated nonnegative lower and upper bounds $\sigma_k^{\rm L}$ and σ_k^{U}, respectively. The constraint in (3.4.27) indirectly imposes an upper bound on the length of the step, which is reminiscent of trust-region methods, but also a lower bound, which is one of the key ingredients of good complexity. Alternatively, it may be viewed as controlling the size of the Lagrange multiplier for the usual trust-region constraint, something that is not monitored carefully in traditional trust-region methods. The presence of the regularization term in (3.4.27) also encapsulates quadratic-regularization methods, starting with approaches suggested by Goldfield, Quandt, and Trotter [183] and Levenberg, Morrison, and Marquardt [265, 282, 293, 294].

Problem (3.4.27) need not be solved exactly. Instead, the pair (s_k, λ_k) is constructed to satisfy a set of reasonable conditions. Firstly, s_k and λ_k need to be feasible for problem (3.4.27); that is,

$$\sigma_k^{\rm L}\|s_k\| \leq \lambda_k \leq \sigma_k^{\rm U}\|s_k\|. \tag{3.4.28}$$

They also need to ensure a nontrivial slope,

$$\langle s_k, \nabla_s^1 m_k(s_k)\rangle \leq \min\left[\kappa_1\|s_k\|^2, \tfrac{1}{2}\langle s_k, (\nabla_x^2 f(x_k) + \lambda_k I)s_k\rangle + \tfrac{1}{2}\kappa_2\|s_k\|^3\right], \tag{3.4.29}$$

and a bound on the gradient of the model,

$$\|\nabla_s^1 m_k(s_k)\| \leq \lambda_k\|s_k\| + \kappa_3\|s_k\|^2, \tag{3.4.30}$$

where we have

$$\nabla_s^1 m_k(s_k) = \nabla_x^1 f(x_k) + (\nabla_x^2 f(x_k) + \lambda_k I)s_k \tag{3.4.31}$$

from (3.4.27), and κ_1, κ_2, and κ_3 are positive constants. If an approximate second-order critical point is sought, then additionally it is required that

$$\lambda_{\min}\left[Q_k^T \nabla_s^2 m(s_k) Q_k\right] \geq -\kappa_4\|s_k\|, \tag{3.4.32}$$

where

$$\nabla_s^2 m_k(s_k) = \nabla_x^2 f(x_k) + \lambda_k I, \tag{3.4.33}$$

$\kappa_4 > 0$, and the columns of a given suitable Q_k are orthogonal. In particular, if $Q_k = I$, the left-hand side of (3.4.32) is the leftmost eigenvalue of $\nabla_s^2 m(s_k)$, which, using (3.4.33), is connected to the leftmost eigenvalue of $\nabla_x^2 f(x_k)$, and so the requirement (3.4.32) clearly relates to a $\|s_k\|$-approximate second-order criticality as expressed in (3.2.6). The extra generality gained by not insisting that Q_k be square, and thus negative curvature need only be small over a subspace, has practical benefits, particularly in methods that attempt to solve (3.4.27) over an evolving (say Krylov) subspace.

Conditions (3.4.29), (3.4.30), and (3.4.32) are evocative of the approximate model-minimization conditions (3.3.5) and (3.3.6) in AR2. Indeed, conditions (3.4.28)–(3.4.32) can be satisfied by AR2 by choosing $\sigma_k \in [\sigma_k^{\rm L}, \sigma_k^{\rm U}]$ and $\lambda_k = \sigma_k\|s_k\|$ and minimizing (3.3.1) over increasing (say Krylov) subspaces (see Chapter 10). An alternative is to choose

$\lambda_k \geq 0$ and compute s_k by minimizing $m_k(s)$ in (3.4.27) with $\lambda = \lambda_k$ over increasing subspaces. It can be shown that the ensuing (s_k, λ_k) satisfies conditions (3.4.29)–(3.4.30) provided that this pair is feasible; this is the core of the quadratic-regularization approach of Birgin and Martínez [47] and is partly used in the trust-region algorithm TRACE of Curtis, Robinson, and Samadi [137].

The precise iRN algorithm that we shall analyze may be stated as follows.

Algorithm 3.4.2: Inexact regularized Newton algorithm (iRN).

Step 0: Initialization. An initial point x_0 and accuracy thresholds $\epsilon_1 \in (0,1]$ and $\epsilon_2 > 0$ are given. The constants η, γ_1, and γ_2 and the lower and upper regularization bounds $\sigma_{\min}$ and $\sigma_{\max}$, respectively, are also given and satisfy

$$0 < \sigma_{\min} \leq \sigma_{\max}, \quad 0 < \eta < 1, \quad \text{and} \quad 1 < \gamma_1 \leq \gamma_2.$$

Compute $f(x_0)$, $\nabla_x^1 f(x_0)$, and $\nabla_x^2 f(x_0)$, and set $k = 0$, $\sigma_0^{\text{L}} = 0$, and $\sigma_0^{\text{U}} \in [\sigma_{\min}, \sigma_{\max}]$.

Step 1: First-order test for termination. If $\epsilon_2 = +\infty$, then if

$$\|\nabla_x^1 f(x_k)\| \leq \epsilon_1,$$

terminate with the ϵ_1-approximate first-order minimizer $x_\epsilon = x_k$.

Step 2: Step computation. Compute a step and multiplier pair (s_k, λ_k) for which the conditions (3.4.28)–(3.4.30) are satisfied. If $\epsilon_2 < +\infty$, also require that s_k satisfy (3.4.32).

Step 3: Second-order test for termination. If $\epsilon_2 < \infty$ and

$$\|\nabla_x^1 f(x_k)\| \leq \epsilon_1 \quad \text{and} \quad \lambda_{\min}[Q_k^T \nabla_x^2 f(x_k) Q_k] > \epsilon_2,$$

where Q_k is the same as in (3.4.32), terminate with the (ϵ_1, ϵ_2)-approximate strict local minimizer $x_\epsilon = x_k$. Otherwise, if

$$\|\nabla_x^1 f(x_k)\| \leq \epsilon_1 \quad \text{and} \quad \lambda_{\min}[Q_k^T \nabla_x^2 f(x_k) Q_k] \geq -\epsilon_2, \tag{3.4.34}$$

terminate with the (ϵ_1, ϵ_2)-approximate second-order minimizer $x_\epsilon = x_k$.

Step 4: Acceptance of the trial point. Compute $f(x_k + s_k)$, and define

$$\rho_k = \frac{f(x_k) - f(x_k + s_k)}{\|s_k\|^3}. \tag{3.4.35}$$

If $\rho_k \geq \eta$, then define $x_{k+1} = x_k + s_k$ and compute $\nabla_x^1 f(x_{k+1})$ and $\nabla_x^2 f(x_{k+1})$. Otherwise define $x_{k+1} = x_k$.

Step 5: Regularization parameter update. If $\rho_k \geq \eta$, set

$$\sigma_{k+1}^{\text{L}} = 0 \quad \text{and} \quad \sigma_{k+1}^{\text{U}} = \sigma_k^{\text{U}}.$$

Otherwise, if $\lambda_k < \sigma_{\min}\|s_k\|$, set

$$\sigma_{k+1}^{\text{L}} \in [\sigma_{\min}, \sigma_{\max}] \quad \text{and} \quad \sigma_{k+1}^{\text{U}} \in [\sigma_{k+1}^{\text{L}}, \sigma_{\max}].$$

Otherwise, set

$$\sigma_{k+1}^{\text{L}} = \gamma_1 \frac{\lambda_k}{\|s_k\|} \quad \text{and} \quad \sigma_{k+1}^{\text{U}} = \gamma_2 \frac{\lambda_k}{\|s_k\|}. \tag{3.4.36}$$

Increment k by one and go to Step 1.

One ingredient that is atypical of traditional trust-region methods is the measure of progress (3.4.35). This attempts to ensure a decrease in the model that is typical of cubic-regularization methods such as AR2, and which, as we have already mentioned, is another crucial ingredient for achieving good first-order complexity.[22] Just as in our earlier algorithms, we denote successful iterations as $\mathcal{S} = \{k \mid \rho_k \geq \eta\}$, and the remaining (unsuccessful) ones belong to the set $\mathcal{U}$.

3.4.2.2 Complexity analysis of the iRN algorithm

We now provide a complexity analysis of the iRN algorithm. Our first step is to show that the iteration will be successful provided that the multiplier is sufficiently large.

Lemma 3.4.7: Lower bound on the multiplier

Suppose that **f.HL** *holds. If*

$$\lambda_k \geq \left(2\eta + \kappa_2 + L_{f,2}\right) \|s_k\| \tag{3.4.37}$$

at iteration k of the iRN *algorithm, then $\rho_k \geq \eta$ and the iteration is successful.*

Proof. We have, from (A.20) in Corollary A.8.4 (p. 437) for $j = 1$ and $p = 2$, (3.4.31), and (3.4.29), that

$$\begin{aligned}
f(x_k) - f(x_k + s_k) &= f(x_k) - T_{f,2}(x_k, s_k) + T_{f,2}(x_k, s_k) - f(x_k + s_k) \\
&\geq -\langle s_k, \nabla_x^1 f(x_k)\rangle - \tfrac{1}{2}\langle s_k, \nabla_x^2 f(x_k) s_k\rangle - \tfrac{1}{2} L_{f,2}\|s_k\|^3 \\
&= -\langle s_k, \nabla_x^1 f(x_k) + (\nabla_x^2 f(x_k) + \lambda_k I) s_k\rangle \\
&\quad + \tfrac{1}{2}\langle s_k, (\nabla_x^2 f(x_k) + \lambda_k I) s_k\rangle + \tfrac{1}{2}\lambda_k \|s_k\|^2 - \tfrac{1}{2} L_{f,2}\|s_k\|^3 \\
&= -\langle s_k, \nabla_s^1 m_k(s_k)\rangle + \tfrac{1}{2}\langle s_k, (\nabla_x^2 f(x_k) + \lambda_k I) s_k\rangle \\
&\quad + \tfrac{1}{2}\lambda_k \|s_k\|^2 - \tfrac{1}{2} L_{f,2}\|s_k\|^3 \\
&\geq -\tfrac{1}{2}\kappa_2 \|s_k\|^3 + \tfrac{1}{2}\lambda_k\|s_k\|^2 - \tfrac{1}{2} L_{f,2}\|s_k\|^3.
\end{aligned}$$

Given the definition of ρ_k in (3.4.35), it then follows that $\rho_k \geq \eta$ if (3.4.37) holds. □

The next step is to provide an upper bound on the upper regularization parameter.

Lemma 3.4.8: Upper bound on the upper regularization parameter

Suppose that **f.HL** *holds. Then, for all $k \geq 0$, the* iRN *algorithm ensures that*

$$\sigma_k^{\text{U}} \leq \sigma_{\max}^{\text{U}} \stackrel{\text{def}}{=} \max\left[\sigma_{\max}, \gamma_2(2\eta + \kappa_2 + L_{f,2})\right]. \tag{3.4.38}$$

Proof. If iteration k is successful, then σ_k^{U} is unchanged. If the iteration is unsuccessful, then the mechanics of Step 5 in the iRN algorithm imply that $\sigma_k^{\text{U}} \leq \max\left[\sigma_{\max}, \gamma_2\lambda_k/\|s_k\|\right]$. Furthermore, since the iteration is unsuccessful, Lemma 3.4.7 implies that the converse to (3.4.37) must hold, namely that $\lambda_k/|s_k\| \leq \left(2\eta + \kappa_2 + L_{f,2}\right)$. The upper bound (3.4.38) on σ_k^{U} then follows directly. □

[22] At variance with the standard progress measures (3.3.7), (3.3.10), and (3.2.11), the new formula (3.4.35) has a numerator and denominator belonging to different spaces, and so it may scale differently. Thus care must be taken with implementations based on this measure.

We also require a familiar-looking property (cf. Lemma 3.3.3 on p. 68) concerning the minimal length of the step.

Lemma 3.4.9: Lower bound on the length of the step

Suppose that **f.HL** *holds. Then, if iteration k of the* iRN *algorithm is successful,*

$$\|s_k\|^2 \geq \frac{2}{L_{f,2} + 4\sigma_{\max}^{U} + 2\kappa_3} \|\nabla_x^1 f(x_{k+1})\|, \tag{3.4.39}$$

where $\sigma_{\max}^{U}$ is defined as in (3.4.38). If (3.4.32) holds,

$$\|s_k\| \geq -\frac{1}{\kappa_4 + \sigma_{\max}^{U}} \lambda_{\min}\left[Q_k^T \nabla_x^2 f(x_k) Q_k\right]. \tag{3.4.40}$$

Proof. This is similar to the corresponding proof of Lemma 3.3.3 for the AR2 algorithm. It also combines a simple add and subtract argument with the triangle inequality, which gives

$$\begin{aligned}\|\nabla_x^1 f(x_{k+1})\| &\leq \left\|\nabla_x^1 f(x_k + s_k) - \nabla_s^1 m_k(s_k) + \nabla_s^1 m_k(s_k)\right\| \\ &\leq \|\nabla_x^1 f(x_k + s_k) - \nabla_s^1 T_{f,2}(x_k, s_k)\| + \lambda_k \|s_k\| + \left\|\nabla_s^1 m_k(s_k)\right\|,\end{aligned} \tag{3.4.41}$$

on recalling that iteration k is successful and that (3.4.31) holds. Again we may invoke **f.HL** and (A.20) in Corollary A.8.4 (p. 437) for $j = 1$ and $p = 2$ to deduce (3.3.17). This and (3.4.30) can then be used in (3.4.41) to deduce

$$\begin{aligned}\|\nabla_x^1 f(x_{k+1})\| &\leq \tfrac{1}{2} L_{f,2} \|s_k\|^2 + 2\lambda_k \|s_k\| + \kappa_3 \|s_k\|^2 \\ &\leq \left(\tfrac{1}{2} L_{f,2} + 2\sigma_{\max}^{U} + \kappa_3\right) \|s_k\|^2,\end{aligned}$$

where in the last inequality we used (3.4.28) and (3.4.38). This then leads directly to (3.4.39).

To show the remaining second-order property, we use (3.4.32), (3.4.33), (3.4.28), and (3.4.38), respectively, to deduce that

$$\lambda_{\min}\left[Q_k^T \nabla_x^2 f(x_k) Q_k\right] \geq -\kappa_4 \|s_k\| - \lambda_k \geq -\kappa_4 \|s_k\| - \sigma_k^{U} \|s_k\| \geq -(\kappa_4 + \sigma_{\max}^{U}) \|s_k\|.$$

This immediately gives (3.4.40). □

Next we bound the number of unsuccessful steps that may occur between a pair of successive successful ones; the reader may notice that this result has similarities with Lemmas 2.3.1 and 2.4.1, which bound the number of unsuccessful iterations that may have occurred up to a given iteration of the TR2 and AR2 algorithms.

Lemma 3.4.10: Bound on unsuccessful iterations

Suppose that algorithm iRN *is used. Then, between any two consecutive successful iterations, $k_i, k_{i+1} \in \mathcal{S}$, for any $i \geq 0$, there are at most*

$$k_{i+1} - k_i - 1 \leq \frac{1}{\log \gamma_1} \log\left(\frac{\sigma_{\max}^{U}}{\sigma_{\min}}\right) + 1 \tag{3.4.42}$$

unsuccessful iterations.

Proof. If there are no unsuccessful iterations between k_i and k_{i+1}, the result follows trivially. Otherwise, since iteration $k_i + 1$ is unsuccessful, in both cases in Step 5 of the algorithm, we have $\sigma^L_{k_i+2} \geq \sigma_{\min}$. This and the requirement (3.4.28) on the step at the next iteration give that $\lambda_{k_i+2}/\|s_{k_i+2}\| \geq \sigma_{\min}$. Thus, on each iteration $k \in [k_i+2 : k_{i+1}-1]$, $\sigma^{\rm L}_k$ will be updated according to (3.4.36), and so, using (3.4.28), inductively we have

$$\frac{\lambda_{k+1}}{\|s_{k+1}\|} \geq \sigma^{\rm L}_{k+1} = \gamma_1 \frac{\lambda_k}{\|s_k\|} \geq \gamma_1 \sigma_{\min} \geq \sigma_{\min}.$$

We conclude that $\lambda_{k_{i+1}} \geq \gamma_1^{k_{i+1}-k_i-2} \sigma_{\min} \|s_{k_{i+1}}\|$. Using this, (3.4.28), and (3.4.38), we find that

$$\sigma^{\rm U}_{\max} \geq \sigma^{\rm U}_{k_{i+1}} \geq \frac{\lambda_{k_{i+1}}}{\|s_{k_{i+1}}\|} \geq \gamma_1^{k_{i+1}-k_i-2} \sigma_{\min}.$$

The bound (3.4.42) then follows from this by taking logarithms. □

We are now ready to give our evaluation-complexity bounds for the iRN algorithm to achieve approximate first-order criticality.

Theorem 3.4.11: First-order complexity of the iRN algorithm

[Appendix to Chapter 3, p. 456]: Suppose that **f.Bb** *and* **f.HL** *hold. Then there exist positive constants $\kappa^S_{\rm iRN}$, $\kappa^A_{\rm iRN}$, $\kappa^B_{\rm iRN}$, and $\kappa^C_{\rm iRN}$ such that, for any $\epsilon_1 \in (0,1]$, the* iRN *algorithm (p. 84) with $\epsilon_2 = +\infty$ requires at most*

$$\#^F_{\rm iRN,1} \stackrel{\rm def}{=} \kappa^A_{\rm iRN} \frac{f(x_0) - f_{\rm low}}{\epsilon_1^{3/2}} + \kappa^B_{\rm iRN} = \mathcal{O}\left(\epsilon_1^{-3/2}\right) \tag{3.4.43}$$

evaluations of f and at most

$$\#^D_{\rm iRN,1} \stackrel{\rm def}{=} \kappa^S_{\rm iRN} \frac{f(x_0) - f_{\rm low}}{\epsilon_1^{3/2}} + 2 = \mathcal{O}\left(\epsilon_1^{-3/2}\right) \tag{3.4.44}$$

evaluations of $\nabla^1_x f$ and $\nabla^2_x f$ to produce an iterate x_ϵ such that $\|\nabla^1_x f(x_\epsilon)\| \leq \epsilon_1$.

We can do much the same for second-order criticality.

Theorem 3.4.12: Second-order complexity of the iRN algorithm

[Appendix to Chapter 3, p. 457]: Suppose that **f.Bb** *and* **f.HL** *hold. Then there exist positive constants $\kappa^T_{\rm iRN}$, $\kappa^C_{\rm iRN}$, and $\kappa^D_{\rm iRN}$ such that, for any $(\epsilon_1, \epsilon_2) \in (0,1]^2$, the* iRN *algorithm (p. 84) requires at most*

$$\#^F_{\rm iRN,2} \stackrel{\rm def}{=} \kappa^C_{\rm iRN} \frac{f(x_0) - f_{\rm low}}{\min\left[\epsilon_1^{3/2}, \epsilon_2^3\right]} + \kappa^D_{\rm iRN} = \mathcal{O}\left(\max\left[\epsilon_1^{-3/2}, \epsilon_2^{-3}\right]\right) \tag{3.4.45}$$

evaluations of f and at most

$$\#^D_{\rm iRN,2} \stackrel{\rm def}{=} \kappa^T_{\rm iRN} \frac{f(x_0) - f_{\rm low}}{\min\left[\epsilon_1^{3/2}, \epsilon_2^3\right]} + 2 = \mathcal{O}\left(\max\left[\epsilon_1^{-3/2}, \epsilon_2^{-3}\right]\right) \tag{3.4.46}$$

evaluations of $\nabla_x^1 f$ and $\nabla_x^2 f$ to produce an iterate x_ϵ such that $\|\nabla_x^1 f(x_\epsilon)\| \le \epsilon_1$ and $\lambda_{\min}[Q_\epsilon^T \nabla_x^2 f(x_\epsilon) Q_\epsilon] \ge -\epsilon_2$, where $Q_\epsilon = Q_k$ is as defined in (3.4.32) at the terminating iteration k.

Having derived satisfactory iteration-complexity bounds, a few remarks concerning the theoretical aspects of the above development are now in order.

Firstly, conditions (3.4.28)–(3.4.32) can be shown to hold when minimizing problem (3.4.27) with respect to s over increasing (Krylov) subspaces containing $\nabla_x^1 f(x_k)$ [139, Appendix B].

Secondly, the matrix Q_k in (3.4.32) may be a basis for the subspace in which model problem (3.4.27) is considered, and the second-order criticality condition can then only be achieved for the objective within this subspace. If the intent is to guarantee approximate local convexity for f, one needs to ensure model minimization over the whole of $\mathbb{R}^n$, asymptotically or at least during the final iteration.

Finally, the iRN algorithm allows $\sigma_k^{\mathrm{L}} = 0$, for example, by setting it as such on successful iterations. This allows $\lambda_k = 0$, which corresponds to allowing pure Newton steps s_k to be calculated (provided that Q_k is a basis for a full-dimensional subspace).

3.4.2.3 Algorithms that fit into the iRN framework

While the iRN algorithm provides a flexible framework that guarantees a best-known $\mathcal{O}(\max[\epsilon_1^{-3/2}, \epsilon_2^{-3}])$ worst-case evaluation-complexity bound, it remains for us to comment on methods that fit assumptions/conditions (3.4.28)–(3.4.32).

First up, the AR2 algorithm falls within the iRN framework if $\eta_1 = \eta_2 = \eta$, $\sigma_k \in [\sigma_k^{\mathrm{L}}, \sigma_k^{\mathrm{U}}]$, and $\sigma_{\min} \le \sigma_k^{\mathrm{L}}$; together the latter two conditions ensure that (3.4.28) holds. Let s_k be obtained by minimizing the cubic model (3.3.1) over a (Krylov) subspace containing $\nabla_x^1 f(x_k)$ and also satisfying the model-minimization conditions (3.3.5) and (3.3.6) over the subspace (see Chapter 10); for simplicity in what follows, assume here that $Q_k = I$. Letting $\lambda_k = \sigma_k \|s_k\|$ in the iRN algorithm, and assuming that **f.gL** holds, one can easily verify that conditions (3.4.29)–(3.4.32) are satisfied provided that $\kappa_1 \ge \frac{1}{2} L_{f,1}$, $\kappa_2 \ge \theta_1$, and $\kappa_4 \ge \theta_2$. The conditions for step acceptance are also satisfied if $\eta \in (0, \frac{1}{3}\eta_1 \sigma_{\min}]$.

The TRACE (Trust Region Algorithm with Contraction and Expansion) algorithm proposed by Curtis, Robinson, and Samadi [137] is the earliest trust-region variant which achieves the best-known evaluation-complexity bound while also permitting pure Newton steps. The trust-region update is defined as

$$\Delta_k \stackrel{\text{def}}{=} \begin{cases} \|s_k\| & \text{if } \lambda_k = 0, \\ \|\nabla_x^1 f(x_k)\| \dfrac{\|s_k\|}{6\lambda_k} & \text{if } \lambda_k > 0. \end{cases}$$

As is common in trust-region methods, a Cauchy decrease is required on the quadratic model:

$$f(x_k) - T_{f,2}(x_k, s_k) \ge \frac{\|\nabla_x^1 f(x_k)\|}{6\sqrt{2}} \min\left[\frac{\|\nabla_x^1 f(x_k)\|}{1 + \|\nabla_x^2 f(x_k)\|}, \Delta_k\right]. \tag{3.4.47}$$

It can be shown that TRACE belongs to the iRN framework [139, Sect. 4.2] under the additional assumption **f.gL**. The latter assumption is needed to ensure compatibility of (3.4.47) with the choice of ρ_k in (3.4.35)—for instance, we need the model decrease in

(3.4.47) to be a multiple of the product $\|\nabla_x^1 f(x_k)\|\|s_k\|$, and for this we need $\|\nabla_x^2 f(x_k)\|$ to be uniformly bounded above, which is guaranteed by **f.gL**.

Likewise, the quadratic-regularization algorithm with cubic descent given by Birgin and Martínez [47] essentially fits into the iRN framework since it enforces the cubic-descent condition (3.4.35) on successful steps. A practical advantage, which occurs because of the quadratic-only regularization term, is that only a linear-equation subproblem need be solved on the majority of iterations.

A hybrid approach that also fits into the iRN framework was proposed by Curtis, Robinson, and Samadi [139, Sect. 4.3]. If $\sigma_k^{\text{L}} > 0$, the cubic model (3.3.1) (with $\sigma_k = \sigma_k^{\text{L}}$) is minimized approximately over increasing Krylov subspaces containing $\nabla_x^1 f(x_k)$. By contrast, if $\sigma_k^{\text{L}} = 0$, the method attempts to calculate an approximate Newton step using, say, a conjugate gradient method—which is the same as minimizing $T_{f,2}$ over increasing Krylov subspaces containing $\nabla_x^1 f(x_k)$—until the conditions (3.4.28)–(3.4.32) are satisfied with $\lambda_k = 0$. This may not always be possible, for instance, when a direction of negative curvature is encountered. Under these circumstances, σ_k^{L} is reset and the method reverts to approximately minimizing (3.3.1) with $\sigma_k = \sigma_k^{\text{L}}$.

As we have seen, in general, algorithms that permit Newton steps but aim for best-known complexity come at the cost of intricate design, needing to carefully monitor step lengths and the resulting model decreases. Once one is willing to give up the obsession with pure Newton steps, simpler frameworks with best-known complexity ensue, as we have seen for the AR2 and TRϵ algorithms. An important open question remains as to which of these methods offers the most practical advantages in terms of typical/average computational efficiency.

Notes and references for Subsection 3.4.2

In terms of linesearch methods that have best-known complexity, the first proposal is due to Royer and Wright [335]. Their method enforces a cubic-descent condition in the linesearch and carefully alternates between first- and second-order steps, the latter using negative curvature, as well as using the accuracy tolerances in the construction of the search directions, in the same vein as that of Curtis, Robinson, Royer, and Wright [136].

A recent trend in complexity analysis of second-order methods has been to include the cost of solving the subproblem inexactly, so that the ensuing complexity bound counts function, gradient, and Hessian-vector product evaluations; this is a problem-dependent concept because it hinges on the relative costs of evaluations and internal computations. In particular, the steepest-descent method has been proposed and analyzed for solving the cubic-regularization subproblem [72, 74] and the trust-region one [74]. It has been shown that, despite the potential nonconvexity of these subproblems, with positive probability, the steepest-descent method can find the global minimum of either subproblem approximately in a quantifiable way in terms of the required accuracy (ϵ_1, ϵ_2). More efficient and practical approaches for the subproblem solution, such as conjugate gradient and other Krylov methods, have also been analyzed as part of linesearch [334], trust-region [73, 74, 136], or cubic-regularization [73, 74] approaches. The ensuing complexity bounds usually multiply the function-evaluation bound by the cost of solving the subproblem, which is typically, in order, the maximum of $n\epsilon_1^{-3/2}$ and $\epsilon_1^{-7/4}\log\epsilon$ (and $\epsilon_2 = \epsilon_1^{1/2}$). Some of the results also have a probabilistic aspect since the subproblem solver is assumed to be randomly initialized. Furthermore, extensions of Nesterov's accelerated-gradient scheme from the convex to the nonconvex regime have been proposed [75, 76], with similar Hessian-free evaluation-complexity results, including an entirely deterministic result. It is also possible to use cutting-plane methods instead of acceleration to achieve comparable results [235]. As mentioned in the introduction (see Section 1.2.2, p. 8), the total computational complexity of methods, rather than the evaluation complexity, is beyond the scope of this book. We hope that the above pointers will help the interested reader start their journey into this emerging and evolving literature.

3.5 Optimal complexity for methods using second derivatives

Having considered a number of methods that use second-order derivatives and shown that the bounds on their worst-case evaluation complexity can differ materially, it is now time to take a step back and ask ourselves two general questions. The first is, "What exactly do we mean by a second-order method?" In other words, is there an underlying structure which is common to the methods we have considered so far, and possibly to others that we have not considered here? The second is, "Which second-order methods have the best evaluation complexity?" We have already seen in Sections 3.1 and 3.2 that, sadly, Newton's method does not qualify for the second accolade, and neither does the traditional trust-region method, but are the other algorithms we have considered in the succeeding Sections 3.3 and 3.4 the absolutely best possible given the information available? Would the underlying structure of the best second-order methods we have studied so far provide a clue? This section aims at providing some answers to these questions.

3.5.1 The $\mathcal{M}$ class of second-order methods

We start by addressing the first question, describing a class of methods we believe is a reasonable attempt to unify second-order methods. Our intention is, as we have done throughout this chapter, to apply these methods to the minimization of twice continuously differentiable functions f that are bounded below, with Lipschitz-continuous gradient and Hessian, and thus satisfy **f.Bb**, **f.gL**, and **f.HL**.

We say that *the $\mathcal{M}$ class of second-order methods consists of all methods which satisfy the following generic conditions.* Given some initial $x_0 \in \mathbb{R}^n$, a method in the $\mathcal{M}$ class generates a sequence of iterates $\{x_k\}$ such that

$$x_{k+1} = x_k + s_k, \quad k \geq 0, \tag{3.5.1}$$

where the step s_k satisfies

$$(\nabla_x^2 f(x_k) + \lambda_k I)s_k = -\nabla_x^1 f(x_k) + r_k \text{ with } \|r_k\| \leq \min\left[\kappa_{rg}\|\nabla_x^1 f(x_k)\|, \kappa_{rs}\|s_k\|^2\right] \tag{3.5.2}$$

for some residual r_k and constants $\kappa_{rg} \in [0, 1)$ and $\kappa_{rs} > 0$ and for some λ_k such that

$$\lambda_k \geq 0, \;\; \nabla_x^2 f(x_k) + \lambda_k I \text{ is positive semidefinite} \tag{3.5.3}$$

and

$$\lambda_{\min}(\nabla_x^2 f(x_k)) + \lambda_k \leq \kappa_\lambda \max\left\{|\lambda_{\min}(\nabla_x^2 f(x_k))|, \|\nabla_x^1 f(x_k)\|^{\frac{1}{2}}\right\} \tag{3.5.4}$$

for some $\kappa_\lambda > 1$ independent of k. Without loss of generality, we assume that $s_k \neq 0$.

Does this framework cover existing methods? Do conditions (3.5.2)–(3.5.4) relate to local and/or global minimization of the local models/subproblems we have encountered until now? We defer this discussion to Section 3.5.1.2 (p. 92), and consider properties of methods in the class $\mathcal{M}$ here.

3.5.1.1 Properties of methods in class $\mathcal{M}$

The algorithms we have explored in this chapter allow successful and unsuccessful iterations, where the latter are used to adjust algorithm parameters so as to facilitate progress. In order to simplify exposition, in the $\mathcal{M}$-class framework we focus only on successful iterations, when the step is accepted. There is no loss of generality in doing so since although local refinement of the model may be required to ensure function decrease, as we have seen in Chapters 2 and 3, the number of function evaluations to do so does not increase the overall worst-case evaluation complexity by more than a constant multiple and thus does not affect quantitatively the worst-case bounds derived; see, for example, Lemmas 2.3.1 and 2.4.1. Furthermore, the examples of slow convergence we have presented in this book, and those that we propose here, are constructed in such a way that each iteration of the method is successful and provides sufficient decrease of f in a local second-order model of the form

$$m_{k,\beta}(s) = f(x_k) + \nabla_x^1 f(x_k)^T s + \tfrac{1}{2} s^T (\nabla_x^2 f(x_k) + \beta_k \lambda_k I) s \tag{3.5.5}$$

of $f(x_k + s)$, where $0 \leq \beta_k \leq 1$ and its choice is method dependent.

Since the requirement (3.5.4) may at first glance seem rather obscure, we start by giving a sufficient, more concise, condition on the algorithm-generated λ_k that implies this awkward-looking bound. All missing proofs can be found in Appendix B.

Lemma 3.5.1: Bounded multiplier

[Appendix to Chapter 3, p. 458]: Let (3.5.2) and (3.5.3) hold. Assume also that the algorithm-generated λ_k satisfies

$$\lambda_k \leq \overline{\kappa}_\lambda \|s_k\| \quad \text{for some } \overline{\kappa}_\lambda > 1 \text{ independent of } k. \tag{3.5.6}$$

Then (3.5.4) holds with $\kappa_\lambda \stackrel{\text{def}}{=} 2\sqrt{\overline{\kappa}_\lambda(1+\kappa_{rg})}$.

Thus a method satisfying (3.5.1)–(3.5.4) and (3.5.6) belongs to class $\mathcal{M}$, but not every method in $\mathcal{M}$ needs to satisfy (3.5.6). This latter requirement implies the following property regarding the length of the step generated by methods in $\mathcal{M}$ satisfying (3.5.6), a hallmark that is undoubtedly now familiar as a key ingredient of methods possessing good complexity.

Lemma 3.5.2: Lower bound on the step

[Appendix to Chapter 3, p. 458]: Suppose that **f.Bb**, **f.gL**, *and* **f.HL** *hold. Assume also that a method satisfying (3.5.1), (3.5.2), and (3.5.6) is applied to minimize f. Then*

$$\|s_k\| \geq \overline{\kappa}_s \|\nabla_x^1 f(x_{k+1})\|^{\frac{1}{2}}, \quad k \geq 0, \tag{3.5.7}$$

where $\overline{\kappa}_s \stackrel{\text{def}}{=} \left[\tfrac{1}{2} L_{f,2} + \kappa_{rs} + \overline{\kappa}_\lambda\right]^{-1}$.

We next show that (3.5.7) is a necessary condition for fast local convergence of methods of type (3.5.2), under reasonable assumptions; fast local rate of convergence in a neighbourhood of well-behaved minimizers is a "trademark" of what are commonly regarded as second-order methods.

Lemma 3.5.3: A necessary condition for fast rates of convergence

[Appendix to Chapter 3, p. 459]: Suppose that **f.Bb**, **f.gL**, *and* **f.HL** *hold and that we apply an algorithm to minimize f that satisfies (3.5.1) and (3.5.2) and for which*

$$\lambda_k \leq \overline{\kappa}_\lambda, \quad k \geq 0, \quad \text{for some } \overline{\kappa}_\lambda > 0. \tag{3.5.8}$$

Assume also that convergence at a quadratic rate occurs, namely,

$$\|\nabla_x^1 f(x_{k+1})\| \leq \kappa_c \|\nabla_x^1 f(x_k)\|^2, \quad k \geq 0, \tag{3.5.9}$$

for some constant $\kappa_c > 0$. Then (3.5.7) holds.

In Lemma 3.5.3, the requirement that (3.5.9) hold on each iteration may be relaxed to merely asking for it asymptotically. Summarizing, we have shown that (3.5.7) holds for a method in $\mathcal{M}$ if (3.5.6) holds, or if the multiplier λ_k is bounded and quadratic asymptotic convergence takes place for unit steps.

3.5.1.2 Some methods belonging to the class $\mathcal{M}$

It is now time to review if, and under what conditions, familiar algorithms belong to the class $\mathcal{M}$.

Newton's method. Letting $r_k = 0$ and $\beta_k = 0$ in (3.5.2) and (3.5.5), respectively, yields Newton's method (Algorithm 3.1.1). Provided additionally that both $\nabla_x^1 f(x_k) \in \text{Range}(\nabla_x^2 f(x_k))$ and $\nabla_x^2 f(x_k)$ is positive semidefinite, s_k is the nonascent direction, and (3.5.3) holds, while (3.5.4) is trivially satisfied with $\lambda_k = 0$ in this case. Note that allowing $\|r_k\| > 0$ subject to the second part of (3.5.2) covers inexact variants of Newton's method. Thus Newton's method belongs to $\mathcal{M}$ provided that the local Hessians $\nabla_x^2 f(x_k)$ are positive semidefinite. Alternatively, the class $\mathcal{M}$ accepts perturbed Newton variants, when $\lambda_k > 0$ is chosen to satisfy (3.5.4) (which also ensures that the perturbed Hessian $\nabla_x^2 f(x_k) + \lambda_k I$ is positive semidefinite).

Cubic-regularization methods. The AR2 algorithm (p. 65) belongs to the $\mathcal{M}$ class provided that the cubic model (3.3.1) is minimized sufficiently accurately, if necessary going beyond the local optimality conditions required in (3.3.5) and (3.3.6). In particular, if $r_k = 0$ and $\lambda_k = \sigma_k \|s_k\|$, then (3.5.2) and (3.5.3) coincide with the characterization of the global minimizer(s) of the cubic model (3.3.1) (see Theorem 8.2.8 on p. 226). More importantly, an inexact calculation of the global minimizer of the cubic model can be performed so that (3.5.2) and (3.5.3) hold with $r_k \neq 0$ by varying λ_k within a bounded multiple of $\sigma_k \|s_k\|$—see Part IC, and particularly Section 9.3 on p. 238, for full details. We also set $\beta_k = 0$ in (3.5.5) because progress in the AR2 algorithm is measured in (3.3.7) with respect to $T_{f,2}$. The expression of λ_k and Lemma 3.3.2 imply that (3.5.6) holds for the function class of interest. Thus the AR2 algorithm, with sufficiently accurate subproblem solutions, is in class $\mathcal{M}$.

The TR2 algorithm (p. 55). The TR2 algorithm satisfies (3.5.2)–(3.5.3) provided that the trust-region subproblem (2.3.1) is solved exactly (see Theorem 8.2.7 on p. 226), and in this case $r_k = 0$ and λ_k is the exact Lagrange multiplier of the trust-region constraint; in addition, $\beta_k = 0$ in (3.5.5). In the inexact case, if $\lambda_k + \delta\lambda_k$ is chosen to satisfy (3.5.3), then it can be shown that

$$(\nabla_x^2 f(x_k) + \lambda_k I)s_k = -\nabla_x^1 f(x_k) + r_k, \quad \text{where } \|r_k\| = |\delta\lambda|\|s_k\|,$$

and thus if (3.5.6) holds for λ_k, it also holds for $\lambda_k + \delta\lambda$. It is unclear, however, whether condition (3.5.6) or (3.5.4) can be guaranteed in general for the class of functions of interest. However, the worst-case example that we shall shortly reveal does apply to the TR2 algorithm, by choosing a sufficiently large initial trust-region radius, when $\nabla_x^2 f(x_k)$ is positive definite for all $k \geq 0$, because then the TR2 algorithm becomes Newton's method, which belongs to the class $\mathcal{M}$ when $\nabla_x^2 f(x_k)$ is positive semidefinite for all k.

The TRϵ algorithm (p. 76). For the TRϵ algorithm to be included in the class $\mathcal{M}$, it is necessary that the subproblem (3.4.1) be solved sufficiently accurately and that $\epsilon_* = \sqrt{\epsilon_1}$. Furthermore, $\beta_k = 0$ because of (3.3.7). A natural-seeming choice for λ_k is $\epsilon_* + \mu_k$, where μ_k is the Lagrange multiplier of the trust-region constraint for (3.4.1), but in this case, just as for the TR2 algorithm, the size of μ_k is difficult to control. However, choosing λ_k as a positive multiple of ϵ_* is promising. Indeed, for the function class of interest, Lemmas 3.4.2 (p. 77) and 3.4.3 (p. 78) (with $\|\nabla_x^1 f(x_{k+1})\| \geq \epsilon_1 = \epsilon_*^2$) imply that (3.5.6) holds with $\lambda_k = \mathcal{O}(\epsilon_*)$, both for the case when s_k is interior to the trust-region constraint and when $\|s_k\| = \Delta_k$. Thus the TRϵ algorithm is in class $\mathcal{M}$ provided that $\nabla_x^2 f(x_k) + \lambda_k I$ is positive semidefinite. The latter can be guaranteed when $\|s_k\| < \Delta_k$ by a sufficiently accurate (unconstrained) minimization of the model in (3.4.1). It can also be ensured when $\|s_k\| = \Delta_k$ by choosing λ_k as a sufficiently large multiple of ϵ_* (and, as a consequence, possibly slightly adjusting the constants in (3.5.6)). Thus the TRϵ algorithm, with sufficiently accurate subproblem solutions, is in class $\mathcal{M}$.

Inexact regularized Newton (iRN) algorithm (p. 84). iRN algorithms will be in the class $\mathcal{M}$ if the subproblem (3.4.27) is solved sufficiently accurately, possibly more accurately than prescribed by (3.4.29)–(3.4.32). Note that, for the functions of interest, the latter conditions are just inexact requirements on the solution of (3.4.27). The λ_k parameter is chosen as prescribed by the (exact or approximate) solution of (3.4.27).

When applied to functions of interest, Lemma 3.4.8 and (3.4.28) show that (3.5.6) holds, and, furthermore, (3.5.2) also holds as a consequence of (3.4.30) and by ensuring sufficient subproblem accuracy. A distinct aspect of iRN algorithms is the choice of progress measure in (3.4.35), whose denominator $\|s_k\|^3$ can be written as a multiple of the difference $f(x_k) - m_{k,\beta}(s_k)$ for functions of interest, but note that this is not a requirement of the $\mathcal{M}$-class conditions. Thus iRN variants, with appropriately chosen subproblem solutions, belong to $\mathcal{M}$.

Quadratic regularization–type methods. The pioneering Goldfeld–Quandt–Trotter (GQT) method [183] adjusts

$$\lambda_k = \begin{cases} 0 \text{ when } \lambda_{\min}(\nabla_x^2 f(x_k)) \geq \sigma_k \|\nabla_x^1 f(x_k)\|^{1/2} \text{ or} \\ -\lambda_{\min}(\nabla_x^2 f(x_k)) + \omega_k \|\nabla_x^1 f(x_k)\|^{1/2} \text{ otherwise} \end{cases} \tag{3.5.10}$$

to satisfy (3.5.2) with $r_k = 0$, where $\sigma_k > 0$ is a parameter that is updated so as to ensure sufficient objective-function decrease. Related schemes that allow other powers of $\|\nabla_x^1 f(x_k)\|$ in (3.5.10) with similar convergence and complexity analyses are also possible [355]. The choice $\beta_k = 0$ in (3.5.5) is natural for these. It is straightforward to check that (3.5.3) holds for the choice (3.5.10). Thus the GQT approach takes the pure Newton step whenever the Hessian is locally sufficiently positive definite, and a suitable regularization of this step otherwise. The parameter σ_k and x_{k+1} updates are adjusted similarly to AR2. The choice (3.5.10) implies that (3.5.4) holds, provided that σ_k is uniformly bounded above, which is true for the functions of interest here [106, Lemma 2.6]. Thus GQT belongs to class $\mathcal{M}$.

Linesearch methods. Linesearch methods can also be viewed as being in class $\mathcal{M}$ if λ_k in (3.5.2) is chosen so that $\nabla_x^2 f(x_k) + \lambda_k I$ is "sufficiently" positive definite.[23] The residual is selected to be $r_k = (1-\mu_k)\nabla_x^1 f(x_k)$ for some stepsize $\mu_k \in [1-\kappa_{rg}, 1]$, calculated by the *linesearch* to decrease f; this is always possible for sufficiently small μ_k (and hence sufficiently small κ_{rg}). Furthermore, $\beta_k = 1$ is appropriate. Finally, we may assume here that $\mu_k = 1$ is acceptable whenever $m_k(s_k) \geq f(x_k + s_k)$, which is allowed, for instance, by standard Armijo and Goldstein linesearches, with suitable parameters. Thus suitable linesearch variants are in class $\mathcal{M}$.

3.5.2 Complexity and optimality of methods in class $\mathcal{M}$

We now turn our attention to answering our second question, regarding the complexity of methods in $\mathcal{M}$. As in our earlier sharpness constructions, our approach here is to derive a lower bound on their evaluation complexity by producing a suitable example of slow convergence.

Theorem 3.5.4: A complexity lower bound for methods in $\mathcal{M}$

*For every $\epsilon_1 \in (0,1)$ and every method in $\mathcal{M}$, there is a function $f_\epsilon^{\mathcal{M}} : \mathbb{R} \to \mathbb{R}$ that satisfies **f.Bb**, **f.gL**, and **f.HL** with values in a bounded interval independent of ϵ_1, for which the method under consideration terminates exactly at iteration*

$$k_\epsilon = \left\lceil \epsilon_1^{-\frac{3}{2}} \right\rceil,$$

where x_{k_ϵ} is the first iterate that satisfies $\|\nabla_x f_\epsilon^{\mathcal{M}}(x_{k_\epsilon})\| \leq \epsilon_1$.

Proof. Let $\epsilon_1 \in (0,1)$ be given, and consider an arbitrary method in the $\mathcal{M}$ class. Our intent is to construct a univariate function $f_\epsilon^{\mathcal{M}}(x)$ satisfying **f.Bb**, **f.gL**, and **f.HL** such that

$$f_\epsilon^{\mathcal{M}}(0) = 1, \quad f_\epsilon^{\mathcal{M}}(x) \in [a,b] \text{ for } x \geq 0,$$

for some constants $a \leq b$ independent of ϵ_1, and such that the method will terminate in exactly

$$k_\epsilon = \left\lceil \epsilon_1^{-\frac{3}{2}} \right\rceil \tag{3.5.11}$$

iterations (and evaluations of f, $\nabla_x^1 f$, and $\nabla_x^2 f$).

We start by defining the sequences f_k, g_k, and H_k for $k = 0, \ldots, k_\epsilon$ by

$$f_k = 1 - \tfrac{1}{2} k \epsilon_1^{\frac{3}{2}}, \quad g_k = -2\,\epsilon_1\, f_k, \quad \text{and} \quad H_k = 4\,\epsilon_1^{\frac{1}{2}}\, f_k^2. \tag{3.5.12}$$

They are intended to specify the objective function, gradient, and Hessian values at successive iterates generated by the chosen method in $\mathcal{M}$, according to (3.5.1)

[23] Nondiagonal perturbations are most common in linesearch methods, and they are covered in the class defined in [106].

and (3.5.2) for some choice of multipliers $\{\lambda_k\}$ satisfying (3.5.3) and (3.5.4) and residuals r_k satisfying (3.5.2). In other words, we require that $f_k = f_\epsilon^{\mathcal{M}}(x_k)$, $g_k = \nabla_x^1 f_\epsilon^{\mathcal{M}}(x_k)$, and $H_k = \nabla_x^2 f_\epsilon^{\mathcal{M}}(x_k)$ for $k \in \mathcal{K} \stackrel{\rm def}{=} [0 : k_\epsilon]$. Note that f_k, $|g_k|$, and H_k are monotonically decreasing and that, using (3.5.11),

$$f_k \in \left[\tfrac{1}{2}, 1\right] \quad \text{for} \quad k \in \mathcal{K}. \tag{3.5.13}$$

In addition, (3.5.3) and (3.5.4) impose that $\lambda_k \geq 0$ and

$$0 < \lambda_k + 4\epsilon_1^{\frac{1}{2}} f_k^2 \leq \kappa_\lambda \max[4\epsilon_1^{\frac{1}{2}} f_k^2, (2\epsilon_1 f_k)^{\frac{1}{2}}] = 4\kappa_\lambda \epsilon_1^{\frac{1}{2}} f_k^2$$

for $k \in \mathcal{K}$, yielding that

$$\lambda_k \in \left[0, 4(\kappa_\lambda - 1)\epsilon_1^{\frac{1}{2}} f_k^2\right]. \tag{3.5.14}$$

For ease of notation, we let

$$\lambda_k + H_k = \mu_k H_k, \quad \text{where } \mu_k \in [1, \kappa_\lambda]. \tag{3.5.15}$$

In view of (3.5.2), we let $r_k \stackrel{\rm def}{=} \alpha_k g_k$, where

$$|\alpha_k| \leq \min\left[\kappa_{rg}, \kappa_{rs}\frac{s_k^2}{|g_k|}\right]. \tag{3.5.16}$$

As a consequence, we obtain from (3.5.2) that

$$s_k = -\frac{(1-\alpha_k)}{\mu_k} \cdot \frac{g_k}{H_k} = -\theta_k \frac{\epsilon_1^{\frac{1}{2}}}{2 f_k}, \tag{3.5.17}$$

where $\theta_k \stackrel{\rm def}{=} \dfrac{1-\alpha_k}{\mu_k}$, and that

$$\frac{s_k^2}{|g_k|} = \theta_k^2 \frac{|g_k|}{H_k^2} = \frac{\theta_k^2}{8 f_k^3} = \frac{(1-\alpha_k)^2}{8\mu_k^2 f_k^3} \in \left[\frac{(1-\alpha_k)^2}{8\kappa_\lambda^2}, (1-\alpha_k)^2\right].$$

This and (3.5.16) imply $|\alpha_k| \leq \kappa_{rg}$ and

$$|\alpha_k| \leq \kappa\kappa_{rs}(1-\alpha_k)^2, \quad \text{where } \kappa \in [(8\kappa_\lambda^2)^{-1}, 1].$$

The last display leads to quadratic inequalities in α_k,

$$\alpha_k^2 - \left(2 \pm \frac{1}{\kappa\kappa_{rs}}\right)\alpha_k + 1 \geq 0,$$

which hold when $\alpha_k = 0$ and thus for all $|\alpha_k|$ sufficiently small. Thus, there exists $0 \leq \kappa_r \stackrel{\rm def}{=} \kappa_r(\kappa_{rg}, \kappa_{rs}, \kappa_\lambda) \leq \kappa_{rg}$ and independent of k and ϵ_1 such that

$$|\alpha_k| \leq \kappa_r, \quad k \in \mathcal{K}. \tag{3.5.18}$$

We deduce[24] from (3.5.17), (3.5.15), and (3.5.18) that

$$s_k = \theta_k \frac{\epsilon_1^{\frac{1}{2}}}{2 f_k} \quad \text{for some} \quad \theta_k \in \left[\frac{1-\kappa_r}{\kappa_\lambda}, 1+\kappa_r\right] \tag{3.5.19}$$

for $k \in \mathcal{K}$.

[24]We recognize that throughout this example/proof, we could have set $\kappa_{rg} = \kappa_{rs} = 0$ and used exact model minimization to obtain the ideal step. Our construction here holds in the more general setting of a nonzero residual.

Given (3.5.13), we deduce from (3.5.19) that $|s_k| \le 1$ for $k \in \mathcal{K}$. It also follows from (3.5.1) and (3.5.19) that, if $x_0 = 0$,

$$s_k > 0 \;\text{ and }\; x_k = \sum_{i=0}^{k-1} s_i, \quad k = 0, \dots, k_\epsilon.$$

We therefore conclude that the sequences $\{f_k\}_{k=0}^{k_\epsilon}$, $\{g_k\}_{k=0}^{k_\epsilon}$, $\{H_k\}_{k=0}^{k_\epsilon}$, $\{\lambda_k\}_{k=0}^{k_\epsilon-1}$, and $\{s_k\}_{k=0}^{k_\epsilon-1}$ can be viewed as produced by our chosen method in $\mathcal{M}$, and, from (3.5.12), that termination occurs precisely for $k = k_\epsilon$, as desired.

Unsurprisingly, we now assemble the function $f_\epsilon^{\mathcal{M}}(x)$ for $x \in [0, x_{k_\epsilon}]$ using Hermite interpolation, setting

$$f_\epsilon^{\mathcal{M}}(x) = p_k(x - x_k) \;\text{ for } x \in [x_k, x_{k+1}] \text{ and } k = 0, \dots, k_\epsilon - 1, \tag{3.5.20}$$

where p_k is a fifth-degree polynomial satisfying the interpolation conditions

$$\begin{aligned} p_k(0) &= f_k, \quad p_k(s_k) = f_{k+1}, \\ p_k'(0) &= g_k, \quad p_k'(s_k) = g_{k+1}, \\ p_k''(0) &= H_k, \;\text{ and }\; p_k''(s_k) = H_{k+1} \end{aligned} \tag{3.5.21}$$

for $s_k = x_{k+1} - x_k$ defined as in (3.5.19). The construction of a sufficiently smooth $f_\epsilon^{\mathcal{M}}(x)$ (meaning that **f.Bb**, **f.gL**, and **f.HL** hold) is similar to that of the functions used in the examples of sharpness of complexity bounds throughout the book, and we delegate the technical details to the appendix to Chapter 3 on p. 459, where we also show that $f_\epsilon^{\mathcal{M}}(x)$ can be extended to the whole of $\mathbb{R}$ and has an isolated finite global minimizer.

Though not part of the requirements in class $\mathcal{M}$, we end by reassuring the reader that our constructed function does decrease sufficiently on each iteration of the method compared to the decrease in the model (3.5.5). We have

$$\begin{aligned} m_k(s_k) &= f_k + g_k s_k + \tfrac{1}{2} s_k (H_k + \beta_k \lambda_k) s_k \\ &= f_k + g_k s_k + \tfrac{1}{2} s_k [-g_k + r_k + (\beta_k - 1)\lambda_k s_k] \\ &\ge f_k - \tfrac{1}{2}|g_k| s_k - \tfrac{1}{2}\kappa_r |g_k| s_k + \tfrac{1}{2}\theta_k^2 (\kappa_\lambda - 1)(\beta_k - 1)\epsilon_1^{\frac{3}{2}} \\ &\ge f_k - \tfrac{1}{2}\theta_k \epsilon_1^{\frac{3}{2}} [1 + \kappa_r + \theta_k (1 - \beta_k)(\kappa_\lambda - 1)] \\ &\ge f_k - \tfrac{1}{2}\epsilon_1^{\frac{3}{2}} (1 + \kappa_r)^2 [1 + (1 - \beta_k)(\kappa_\lambda - 1)] \\ &\ge f_k - \tfrac{1}{2}\epsilon_1^{\frac{3}{2}} (1 + \kappa_r)^2 \kappa_\lambda, \end{aligned}$$

where we have used (3.5.2), (3.5.12), (3.5.19), (3.5.14), and $\beta_k \le 1$. Hence, again taking (3.5.12) into account,

$$\frac{f_k - f_{k+1}}{f_k - m_k(s_k)} \ge \frac{\frac{1}{2}\epsilon_1^{\frac{3}{2}}}{\frac{1}{2}\epsilon_1^{\frac{3}{2}} \kappa_\lambda (1 + \kappa_r)^2} = \frac{1}{(1 + \kappa_r)^2 \kappa_\lambda} \in (0, 1),$$

and sufficient decrease of the objective function automatically follows. □

The next corollary allows us to answer our second question for methods in $\mathcal{M}$.

Corollary 3.5.5: Optimal complexity

When applied to a function satisfying **f.Bb**, **f.gL**, *and* **f.HL**, *the* AR2, TRϵ, *and* iRN *variants have optimal evaluation complexity within the $\mathcal{M}$ class for generating approximate first-order critical points.*

Proof. Let $\epsilon_2 = \infty$ in the methods discussed here. For the AR2 algorithm, due to (3.5.2) and (3.5.3), we have $m_k^\beta(s_k) = T_{f,2}(x_k, s_k) = f_\epsilon^{\mathcal{M}}(x_k + s_k) = f_{k+1}$, for $k \in \mathcal{K}$. Thus every AR2 iteration is successful and the choice $\sigma_k = \sigma > 0$ for all k is allowed by the method, and thus $\lambda_k = \sigma\|s_k\|$ satisfies (3.5.3) and (3.5.4). The above example in Theorem 3.5.4—and in fact, Theorem 3.3.6 (p. 71)—shows that the complexity bound in Theorem 3.3.5 (p. 70) of the AR2 algorithm is sharp.

Similarly, for TRϵ with $\epsilon_* = \sqrt{\epsilon_1}$, we can choose a sufficiently large initial trust-region radius which remains unchanged during the iteration, so that the subproblem (approximately global) solution is strictly interior to the trust-region constraint. The example above applies with $\beta_k = 0$ and $\lambda_k = \epsilon_*$, showing that the respective complexity bound in ϵ_1 is sharp.

Finally, for iRN, we can fix $\lambda_k = \sigma\|s_k\|$, where $0 \le \sigma_L \le \sigma \le \sigma_U$. This case is also addressed in more detail in [106, Sect. 5].

Recall the upper complexity bounds in Theorems 3.3.5, 3.4.5, and 3.4.11 on the complexity of AR2, TRϵ, and iRN, respectively. Since we have that $f(x_0) - f_{\rm low}$ is a fixed number independent of ϵ_1 for the example, this shows that the ratio

$$\rho_{\rm comp} \stackrel{\rm def}{=} \frac{\text{upper bound on the worst-case evaluation complexity}}{\text{lower bound on the worst-case evaluation complexity}}$$

for the AR2 algorithm is bounded independently of ϵ_1. Given that the above-mentioned upper bounds involve an unspecified constant, this is the best that can be obtained as far as the order in ϵ_1 is concerned. The optimal complexity of the AR2, TRϵ, and iRN methods within $\mathcal{M}$ results from the observation that the example in this section implies that no method in $\mathcal{M}$ can have a worst-case evaluation complexity of a better order than AR2, TRϵ, and iRN. □

By contrast, *Newton's method is not optimal in $\mathcal{M}$ as far as worst-case evaluation complexity is concerned.* While the example in Theorem 3.5.4 provides a complexity lower bound that Newton's method may require $\epsilon_1^{-3/2}$ evaluations when applied to $f_\epsilon^{\mathcal{M}}$ to generate $|\nabla_x^1 f(x_k)| \le \epsilon_1$, Theorem 3.1.1 provides the stronger result that it may in fact require at least ϵ_1^{-2} evaluations for the same task.

Although it is unclear whether the TR2 algorithm sits precisely in $\mathcal{M}$, the example in Theorem 3.5.4 for Newton's method applies to this algorithm since the latter will take only Newton steps if the initial trust-region radius is sufficiently large. In particular, $\Delta_0 > 1$ will be suitable since s_k in the example satisfies $\|s_k\| \le 1$.

With regards to the GQT and similar methods [183, 355], an upper complexity bound $\mathcal{O}(\epsilon_1^{-2})$ is shown for finding first-order stationary points of functions of interest in [355, Thm. 5]. The example in Theorem 3.5.4 provides a lower complexity bound of order $\epsilon_1^{-3/2}$ on the class $\mathcal{M}$. Thus, based on existing analysis, the GQT algorithm does not have optimal complexity in the $\mathcal{M}$ class.

Traditional linesearch variants in $\mathcal{M}$ can be shown to satisfy a worst-case first-order complexity bound of $\mathcal{O}(\epsilon_1^{-2})$ for the class of functions of interest; see, for example, [353,

Thm. 4.2]. Thus (some of) these variants are (un-improvably) suboptimal within $\mathcal{M}$. Nonetheless, modern linesearch methods with optimal complexity can be designed [106, Sect. 6], [335].

Note that the list of methods in the class $\mathcal{M}$ is not exhaustive, nor are our descriptions of existing methods with optimal complexity. In Section 4.2, we discuss a different, and more general, way of constructing lower complexity bounds, using an algorithm-independent function [77] of growing dimension on which all methods of a certain type are guaranteed to be slow.

Notes and references for Section 3.5

This section is a simplified version of a more general framework [106] that improved upon earlier work [87], with some modifications to bring it in line with developments presented in this book. The broader framework [87, 106] describes a nested family of methods $\mathcal{M}.\alpha$ that allow a more expansive set of conditions of the form (3.5.4) and (3.5.6) and that apply to functions that have Hölder rather than Lipschitz-continuous second derivatives. Regularization methods with varying powers $(2+\alpha)$ of the regularization term are found to be optimal for corresponding functions with Hölder Hessians with Hölder exponents $\alpha \in [0,1]$. In addition, [106] allows nondiagonal perturbations of the Hessian in (3.5.2) when defining method classes.

Clearly, we have not covered all existing complexity frameworks for second-order methods. For instance, Grapiglia, Yuan, and Yuan [205, 206] consider the complexity of finding approximate first- and second-order minimizers for a class of "nonlinear stepsize control" methods that generalize ideas presented in [349], themselves based on the similarities between adaptive-regularization and trust-region methodologies.

We also note that while the above univariate example in Theorem 3.5.4 changes with the method, in [249], a multidimensional variant of Rosenbrock's function is considered, which has exponential dimension-dependence, but for which all first- and second-order descent methods are inefficient when finding exact minimizers. Superficially, this seems at odds with the finite, implicitly dimension dependent bounds we have presented here when finding approximate critical points; an explanation of this phenomenon can be found in [95].

Chapter 4

Methods Using Third Derivatives and Beyond

Chapter summary

In the previous chapter, we discussed the improvements in worst-case performance that are achievable if second derivatives of the objective function are used. Here we move a step further by showing that even better worst-case performance is possible for functions that possess derivatives of degree p using regularization methods with access to oracles up to degree p. More precisely, we will demonstrate that convergence to an ϵ-approximate first-order minimizer can be ensured in at most $\mathcal{O}(\epsilon^{-(p+1)/p})$ evaluations, and that at most $\mathcal{O}(\epsilon^{-(p+1)/(p-1)})$ and $\mathcal{O}(\epsilon^{-(p+1)/(p-2)})$ evaluations are required to find ϵ-approximate second-order and third-order minimizers, respectively.

We round off the chapter with a far-reaching result that such a first-order guarantee is actually as good as possible. In particular, in the worst case, no pth-order algorithm can do better.

In the last chapter, we saw that using second derivatives to build a model of the objective function may—for example, using the AR2 algorithm (p. 65)—lead to better worst-case[25] performance than methods that use only first derivatives. A natural question is therefore to ask if even better evaluation-complexity bounds are achievable if one is ready to incorporate higher-order derivatives in the model of the objective function. This presumes, of course, that we are able to solve the related step-computation subproblem.

The purpose of this chapter is to show that the answer is affirmative. We first consider AR2, our best algorithm thus far, and describe how its worst-case evaluation complexity for convergence to first- and second-order approximate minimizers—the primary objective of Chapter 3—can be improved using models of higher degree. An increase in the degree of derivatives available also gives us the hope of computing approximate minimizers of higher order. Indeed, as we will see, it is straightforward to identify third-order approximate minimizers if required.

[25] And also practical!

4.1 The ARp algorithm and its complexity

4.1.1 The ARp algorithm

Using third derivatives and beyond is conceptually simple: instead of basing the model in the neighbourhood of the iterate x_k on the Taylor's expansion of degree two, $T_{f,2}(x_k,s)$, we may simply use the general pth-order Taylor-series expansion $T_{f,p}(x_k,s)$, where

$$T_{f,p}(x,s) \stackrel{\text{def}}{=} f(x) + \sum_{\ell=1}^{p} \frac{1}{\ell!} \nabla_x^\ell f(x)[s]^\ell, \tag{4.1.1}$$

instead, increasing the degree of the regularization term accordingly, giving

$$m_k(s) = T_{f,p}(x_k,s) + \frac{\sigma_k}{(p+1)!}\|s\|^{p+1}. \tag{4.1.2}$$

As was the case for $p=2$ in the AR2 algorithm, this model is bounded below because the high-order regularization term dominates for a large enough step. Note that, since we are computing higher-order derivatives, at least in principle we might check for higher-order critical points at the same time. There is a difficulty, though, that we shall describe and address in Section 12.1 and beyond, but we can at least contemplate an approximate third-order case. Bearing in mind Theorem 1.1.1 (p. 3), we say that x_ϵ is an *ϵ-approximate third-order minimizer* of f if (3.2.6) holds and additionally

$$\left|\nabla_x^3 f(x_\epsilon)[d]^3\right| \le \epsilon_3 \ \text{ for all } \ d \in \mathcal{Z}_\epsilon \tag{4.1.3}$$

whenever

$$\mathcal{Z}_\epsilon \stackrel{\text{def}}{=} \left\{d \mid \|d\| = 1 \ \text{ and } \ |\nabla_x^2 f(x_\epsilon)[d]^2| \le \epsilon_2\right\} \neq \emptyset \tag{4.1.4}$$

for given $\epsilon_1, \epsilon_2, \epsilon_3 > 0$. If $\mathcal{Z}_\epsilon = \emptyset$ but (3.2.6) holds, then $\lambda_{\min}[\nabla_x^2 f(x_\epsilon)] > \epsilon_2$ and, thus, just as in (3.2.7), Theorem 1.1.1 suggests that x_ϵ is instead an ϵ-approximate strict local minimizer.

The resulting algorithm, which we refer to as ARp, is stated below. Notice that we implicitly assume that derivatives of the required orders are computed when needed and that these only need to be recomputed when the iterate has changed, i.e., following a successful iteration.

Algorithm 4.1.1: Adaptive-regularization algorithm with pth-order models (ARp)

Step 0: Initialization. An initial point x_0, accuracy thresholds $\epsilon_1 > 0$, $\epsilon_2 > 0$, and $\epsilon_3 > 0$, and an initial regularization parameter $\sigma_0 > 0$ are given. The constants η_1, η_2, γ_1, γ_2, γ_3, θ_1, θ_2, θ_3, and $\sigma_{\min}$ are also given and satisfy

$$\sigma_{\min} \in (0,\sigma_0], \ \ \theta_1, \ \theta_2, \ \theta_3 \in (0,1], \ \ 0 < \eta_1 \le \eta_2 < 1, \\ \text{and } \ 0 < \gamma_1 < 1 < \gamma_2 < \gamma_3.$$

Compute $f(x_0)$; $\nabla_x^j f(x_0)$ for all $j \in [1:p]$; and, if $\epsilon_3 < \infty$,

$$\mathcal{M}_0 \stackrel{\text{def}}{=} \{d \mid \|d\| = 1 \ \text{ and } \ |\nabla_x^2 f(x_0)[d]^2| \le \epsilon_2\}.$$

Set $k = 0$.

Step 1: Test for termination. If $\epsilon_2 = \epsilon_3 = \infty$ and

$$\|\nabla_x^1 f(x_k)\| \le \epsilon_1, \tag{4.1.5}$$

terminate with the ϵ_1-approximate first-order minimizer $x_\epsilon = x_k$.
If $\epsilon_2 < \infty$, (4.1.5) holds, and

$$\lambda_{\min}[\nabla_x^2 f(x_k)] > \epsilon_2, \tag{4.1.6}$$

terminate with the (ϵ_1, ϵ_2)-approximate strict local minimizer $x_\epsilon = x_k$.
If $\epsilon_2 < \infty$ and $\epsilon_3 = \infty$, (4.1.5) holds, and (4.1.6) fails but

$$\lambda_{\min}[\nabla_x^2 f(x_k)] \ge -\epsilon_2, \tag{4.1.7}$$

terminate with the (ϵ_1, ϵ_2)-approximate second-order minimizer $x_\epsilon = x_k$.
If $\epsilon_3 < \infty$, (4.1.5), (4.1.7), and $\mathcal{M}_k \ne \emptyset$ hold, and

$$\left|\nabla_x^3 f(x_k)[d]^3\right| \le \epsilon_3 \tag{4.1.8}$$

and

$$|\nabla_x^2 f(x_k)[d]^2| \le \epsilon_2 \tag{4.1.9}$$

for all $d \in \mathcal{M}_k$, terminate with the $(\epsilon_1, \epsilon_2, \epsilon_3)$-approximate third-order minimizer $x_\epsilon = x_k$.

Step 2: Step computation. Compute a step s_k such that

$$m_k(s_k) < m_k(0), \tag{4.1.10}$$

where $m_k(s)$ is given by (4.1.2),

$$\|\nabla_s^1 m_k(s_k)\| \le \theta_1 \|s_k\|^p, \tag{4.1.11}$$

and, if $\epsilon_2 < \infty$,

$$\lambda_{\min}[\nabla_s^2 m_k(s_k)] \ge -\theta_2 \|s_k\|^{p-1}, \tag{4.1.12}$$

and, if also $\epsilon_3 < \infty$,

$$\max_{d \in \mathcal{M}_{k+1}} \left|\nabla_s^3 m_k(s_k)[d]^3\right| \le \theta_3 \|s_k\|^{p-2}, \tag{4.1.13}$$

whenever

$$\mathcal{M}_{k+1} \stackrel{\text{def}}{=} \left\{d \mid \|d\| = 1 \text{ and } |\nabla_s^2 m_k(s_k)[d]^2| \le \theta_2 \|s_k\|^{p-1}\right\} \ne \emptyset. \tag{4.1.14}$$

Step 3: Acceptance of the trial point. Compute $f(x_k + s_k)$, and define

$$\rho_k = \frac{f(x_k) - f(x_k + s_k)}{T_{f,p}(x_k, 0) - T_{f,p}(x_k, s_k)}. \tag{4.1.15}$$

If $\rho_k \ge \eta_1$, then define $x_{k+1} = x_k + s_k$ and compute $\nabla_x^j f(x_{k+1})$ for all $j \in [1 : p]$. Otherwise define $x_{k+1} = x_k$.

Step 4: Regularization parameter update. Set

$$\sigma_{k+1} \in \left\{ \begin{array}{ll} [\max(\sigma_{\min}, \gamma_1 \sigma_k), \sigma_k] & \text{if } \rho_k \ge \eta_2, \\ [\sigma_k, \gamma_2 \sigma_k] & \text{if } \rho_k \in [\eta_1, \eta_2), \\ [\gamma_2 \sigma_k, \gamma_3 \sigma_k] & \text{if } \rho_k < \eta_1. \end{array} \right. \tag{4.1.16}$$

Increment k by one and go to Step 1.

The first- and second-order termination conditions in Step 1 of the ARp algorithm reflect well the exact requirements in Theorem 1.1.1 (p. 3). However, the third-order conditions (4.1.8) and (4.1.9) differ from the requirement (4.1.3) and (4.1.4), in that (4.1.3) can only be guaranteed along directions of small curvature for the model, namely, $d \in \mathcal{M}_k$. We will show that at termination, $\mathcal{M}_k \subseteq \mathcal{Z}_\epsilon$, and so the model directions of small curvature also lead to small curvature for the objective, but there may be directions in $\mathcal{Z}_\epsilon$ that the model does not "see" or along which the third-order model derivative is not small. A way to overcome this deficiency can be found in Section 6.2.1 (p. 179) by letting $\mathcal{C} = \mathbb{R}^n$.

A cursory glance may not have revealed the significant difference between the AR2 and ARp algorithms. The fundamental distinction is of course that the model (4.1.2) now involves the pth-order Taylor's expansion (A.10), that is,

$$T_{f,p}(x_k, s) = f(x_k) + \sum_{j=1}^{p} \frac{1}{j!} \nabla_x^j f(x_k)[s]^j,$$

which involves derivative tensors (see Section A.6 for a very brief primer) of order up to $p > 2$ rather than simply gradients and Hessians. This has two consequences. The first is that we may no longer rely on familiar linear algebra for vectors and matrices but have to resort to multilinear algebra—tensor norms in particular; see (A.7) in Appendix A—to derive properties of the algorithm. The second is that approximately minimizing $m_k(s)$ is now more complicated since the model is now a polynomial of degree $p+1$. While this is not impossible—general low-order optimization methods such as those given in Chapters 2 and 3 may be applied to $m_k(x)$ to satisfy (4.1.11) and (4.1.12)—this is clearly an added difficulty because the specialized techniques for regularized quadratic models described in Chapters 9 and 10 may no longer be employed. However, it is crucial to note that *the kth model minimization does not involve any additional evaluations of f or of its derivatives* other than those at x_k. As a consequence, *the effort in computing the step has no theoretical impact on the evaluation complexity of the* ARp *algorithm.* The word "theoretical" in the last sentence is obviously important, because the same cannot be said in general from the computational-complexity and practical points of view.

On the positive side, the fact that Step 4 of the ARp algorithm is identical to the corresponding step in the AR1 and AR2 algorithms means that Lemma 2.4.1 (p. 44) continues to apply.

Notes and references for Subsection 4.1.1

We should inject a note of caution at this stage. While it is realistic to be able to check for approximate first- and second-order minimizers—either by eigenvalue calculations or when solving the subproblems involved, as we shall discuss in Chapters 9 and 10—this is not so for third-order minimizers. In essence, one needs to be able to minimize a symmetric cubic form $\nabla_x^3 f(x)[d]^3$ over the intersection of the Euclidean unit ball $\|d\| = 1$ and a subspace $\nabla_x^2 f(x) d = 0$ of the eigenvectors corresponding to the zero eigenvalues of $\nabla_x^2 f(x)$, although in practice—and in ARp—a perturbation of this problem is involved. Unfortunately, such a problem, like so many in multilinear algebra [234], is NP-hard [234, Thm. 1.10], and so is trying to find a useful approximation [234, Thm. 1.11]. While this does not preclude our solving this global-optimization problem, or doing so approximately using useful heuristics such as those in [324] and [80, Thms. 2.3, 2.4], we have to recognize the associated computational hurdles. Special structures, such as low-rank third-order tensors when the Hessian has a large invariant subspace, may alleviate some of these difficulties, as the growing interest in and importance of this topic may show. No such challenges are present if we simply restrict ourselves to using the ARp algorithm to find first- and second-order approximate minimizers ($\epsilon_3 = \infty$).

4.1.2 Complexity of finding approximate first- to third-order minimizers

We now claim that the complexity analysis for the ARp algorithm is a direct extension of that already covered for AR2, the only differences being that proofs and results use a general p rather than specifically $p = 2$ and that assumption **f.HL** must now be replaced by an assumption on the pth derivative.

f.DpL $f(x)$ is p-times continuously differentiable in $\mathbb{R}^n$, and its pth-order derivative tensor $\nabla_x^p f(x)$ is Lipschitz continuous on $\mathbb{R}^n$; that is, there exists a constant $L_{f,p} \geq 1$ such that

$$\|\nabla_x^p f(x) - \nabla_x^p f(y)\| \leq L_{f,p}\|x - y\| \tag{4.1.17}$$

for all $x, y \in \mathbb{R}^n$, where the norm on the left-hand side of (4.1.17) is the tensor norm (A.7).

The first easy generalization is a lemma giving a lower bound on the model decrease that directly mimics Lemma 3.3.1 (p. 68).

Lemma 4.1.1: Decrease in the pth-degree Taylor-series model

Suppose that $m_k(s)$ is defined by (4.1.2) and that $m_k(s) \leq m_k(0)$. Then

$$T_{f,p}(x_k, 0) - T_{f,p}(x_k, s_k) \geq \frac{\sigma_k}{(p+1)!}\|s_k\|^{p+1}. \tag{4.1.18}$$

Proof. From (4.1.2), we obtain that

$$T_{f,p}(x_k, 0) - T_{f,p}(x_k, s_k) = m_k(0) - m_k(s_k) + \frac{\sigma_k}{(p+1)!}\|s_k\|^{p+1},$$

and (4.1.18) follows from the requirement (4.1.10). □

Note that the assumptions of this lemma are satisfied at every iteration of the ARp algorithm and thus that (4.1.18) holds at each of its iterations.

The upper bound on the regularization parameter similarly follows the arguments of Lemmas 2.4.2 (p. 44) and 3.3.2 (p. 68).

Lemma 4.1.2: Upper bound on the regularization parameter

*Suppose that **f.DpL** holds. Then, for all $k \geq 0$, the ARp algorithm ensures that*

$$\sigma_k \leq \sigma_{\max} \stackrel{\text{def}}{=} \max\left[\sigma_0, \frac{\gamma_3 L_{f,p}}{1 - \eta_2}\right]. \tag{4.1.19}$$

Proof. Successively using (4.1.15), (A.19) in Corollary A.8.4 (p. 437), and (4.1.18), we obtain that

$$\begin{aligned} |\rho_k - 1| &= \frac{|f(x_k) - f(x_k + s_k) - T_{f,p}(x_k, 0) + T_{f,p}(x_k, s_k)|}{T_{f,p}(x_k, 0) - T_{f,p}(x_k, s_k)} \\ &= \frac{|f(x_k + s_k) - T_{f,p}(x_k, s_k)|}{T_{f,p}(x_k, 0) - T_{f,p}(x_k, s_k)} \\ &\leq \frac{L_{f,p}}{\sigma_k}. \end{aligned}$$

Thus, if $\sigma_k \geq L_{f,p}/(1-\eta_2)$, then $\rho_k \geq \eta_2$, iteration k is successful, and (4.1.16) implies that $\sigma_{k+1} \leq \sigma_k$. The mechanism of the algorithm then implies that (4.1.19) must hold. □

The next step along a now well-trodden path is to derive lower bounds on the stepsize using the conditions for step acceptance in Step 3 of the algorithm.

Lemma 4.1.3: Lower bounds on the length of the step for ARp

Suppose that **f.DpL** *holds. Then on every successful iteration k of the* ARp *algorithm, the step s_k calculated in Step 2 satisfies*

$$\|s_k\|^p \geq \frac{p!}{L_{f,p} + \sigma_{\max} + \theta_1 p!} \|\nabla_x^1 f(x_{k+1})\|, \tag{4.1.20}$$

where $\sigma_{\max}$ is given by (4.1.19). If additionally $\epsilon_2 < \infty$, then

$$\|s_k\|^{p-1} \geq \frac{(p-1)!}{L_{f,p} + \sigma_{\max} + \theta_2 (p-1)!} \max\left(0, -\lambda_{\min}[\nabla_x^2 f(x_{k+1})]\right). \tag{4.1.21}$$

Furthermore, if $\epsilon_3 < \infty$, then

$$\|s_k\|^{p-2} \geq \frac{(p-2)!}{L_{f,p} + \sigma_{\max} + \theta_3 (p-2)!} \left|\nabla_x^3 f(x_{k+1})[d]^3\right| \tag{4.1.22}$$

for all d in the set $\mathcal{M}_{k+1}$ given by (4.1.14). Moreover, for any such d,

$$\left|\nabla_x^2 f(x_{k+1})[d]^2\right| \leq \frac{1}{\kappa_{s2}} \|s_k\|^{p-1}, \quad \text{where } \kappa_{s2} \stackrel{\text{def}}{=} \frac{(p-1)!}{L_{f,p} + \sigma_{\max} + \theta_2 (p-1)!}. \tag{4.1.23}$$

Proof. (Cf. Lemmas 3.3.3 and 3.3.8.) Since iteration k is successful, $x_{k+1} = x_k + s_k$. Adding and subtracting terms and using the triangle inequality gives

$$\begin{aligned} &\|\nabla_x^1 f(x_{k+1})\| \\ &= \left\|\nabla_x^1 f(x_{k+1}) - \nabla_s^1 T_{f,p}(x_k, s_k) + \nabla_s^1 T_{f,p}(x_k, s_k) + \frac{\sigma_k}{p!}\|s_k\|^{p-1} s_k - \frac{\sigma_k}{p!}\|s_k\|^{p-1} s_k\right\| \\ &\leq \|\nabla_x^1 f(x_{k+1}) - \nabla_s^1 T_{f,p}(x_k, s_k)\| + \left\|\nabla_s^1 T_{f,p}(x_k, s_k) + \frac{\sigma_k}{p!}\|s_k\|^{p-1} s_k\right\| + \frac{\sigma_k}{p!}\|s_k\|^p. \end{aligned} \tag{4.1.24}$$

Assumption **f.DpL** then allows us to use (A.20) in Corollary A.8.4 (p. 437) for $j = 1$ to deduce that

$$\|\nabla_x^1 f(x_{k+1}) - \nabla_s^1 T_{f,p}(x_k, s_k)\| \leq \frac{L_{f,p}}{p!}\|s_k\|^p. \tag{4.1.25}$$

We also deduce from (4.1.2) and (4.1.11) that

$$\left\|\nabla_s^1 T_{f,p}(x_k,s_k) + \frac{\sigma_k}{p!}\|s_k\|^{p-1}s_k\right\| = \|\nabla_s^1 m_k(s_k)\| \le \theta_1\|s_k\|^p. \tag{4.1.26}$$

Substituting (4.1.25) and (4.1.26) into (4.1.24), and using (4.1.19), immediately gives (4.1.20).

Now suppose that $\epsilon_2 < \infty$. Because $\min_z[a(z)+b(z)] \ge \min_z[a(z)] + \min_z[b(z)]$, it follows that

$$\begin{aligned}
&\lambda_{\min}[\nabla_x^2 f(x_{k+1})]\\
&\quad = \min_{\|d\|=1}\Big[\Big\langle d, [\nabla_x^2 f(x_{k+1}) - \nabla_s^2 T_{f,p}(x_k,s_k)]d\Big\rangle\\
&\qquad\qquad -\frac{\sigma_k}{(p+1)!}\Big\langle d, \nabla_s^2(\|s_k\|^{p+1})d\Big\rangle + \langle d, \nabla_s^2 m_k(s_k)d\rangle\Big]\\
&\quad \ge \min_{\|d\|=1}\Big\langle d, [\nabla_x^2 f(x_{k+1}) - \nabla_s^2 T_{f,p}(x_k,s_k)]d\Big\rangle\\
&\qquad\qquad +\frac{\sigma_k}{(p+1)!}\min_{\|d\|=1}\Big\langle -d, \nabla_s^2(\|s_k\|^{p+1})d\Big\rangle + \lambda_{\min}[\nabla_s^2 m_k(s_k)].
\end{aligned} \tag{4.1.27}$$

Using the relationship (A.4) for $\nabla_x^2 f(x_{k+1}) - \nabla_s^2 T_{f,p}(x_k,s_k)$, **f.DpL**, and (A.20) from Corollary A.8.4 (p. 437) with $j=2$ yields that

$$\begin{aligned}
\min_{\|d\|=1}\Big\langle d, [\nabla_x^2 f(x_{k+1}) - \nabla_s^2 T_{f,p}(x_k,s_k)]d\Big\rangle &\ge -\Big\|\nabla_x^2 f(x_{k+1}) - \nabla_s^2 T_{f,p}(x_k,s_k)\Big\|\\
&\ge -\frac{L_{f,p}}{(p-1)!}\|s_k\|^{p-1}.
\end{aligned} \tag{4.1.28}$$

Furthermore, simple differentiation shows that the Hessian

$$\nabla_s^2(\|s_k\|^{p+1}) = (p+1)\|s_k\|^{p-1}\left(I + (p-1)\frac{s_k}{\|s_k\|}\frac{s_k^T}{\|s_k\|}\right),$$

and thus it is positive definite because $p \ge 1$. But this and Lemma B.4.1 (p. 461) with $\beta = 1$ and $j = 2$ then give that

$$\begin{aligned}
\min_{\|d\|=1}\Big\langle -d, \nabla_s^2(\|s_k\|^{p+1})d\Big\rangle &= -\max_{\|d\|=1}\Big\langle d, \nabla_s^2(\|s_k\|^{p+1})d\Big\rangle\\
&= -\Big\|\nabla_s^2(\|s_k\|^{p+1})\Big\| = -\frac{(p+1)!}{(p-1)!}\|s_k\|^{p-1}.
\end{aligned} \tag{4.1.29}$$

Combining (4.1.27), (4.1.28), (4.1.29), and (4.1.12), we find that

$$\lambda_{\min}[\nabla_x^2 f(x_{k+1})] \ge -\left[\frac{L_{f,p}+\sigma_k}{(p-1)!} + \theta_2\right]\|s_k\|^{p-1},$$

which then implies (4.1.21).

Finally, assume that we also have $\epsilon_3 < \infty$. Let $d \in \mathcal{M}_{k+1}$ and $j \in \{2,3\}$. Then

$$\begin{aligned}
\Big|\nabla_x^j f(x_{k+1})[d]^j\Big| &\le \Big|\nabla_x^j f(x_{k+1})[d]^j - \nabla_s^j m_k(s_k)[d]^j\Big| + \Big|\nabla_s^j m_k(s_k)[d]^j\Big|\\
&\le \Big\|\nabla_x^j f(x_{k+1}) - \nabla_s^j T_p(x_k,s_k)\Big\|\\
&\qquad + \frac{\sigma_k}{(p+1)!}\Big|\nabla_s^j\left(\|s_k\|^{p+1}\right)[d]^j\Big| + \Big|\nabla_s^j m_k(s_k)[d]^j\Big|\\
&\le \frac{L_{f,p}+\sigma_{\max}}{(p-j+1)!}\|s_k\|^{p-j+1} + \Big|\nabla_s^j m_k(s_k)[d]^j\Big|,
\end{aligned} \tag{4.1.30}$$

where the triangle inequality, $\|d\| = 1$, and the definition of m_k give the first two inequalities, and we use **f.DpL** and (A.20) from Corollary A.8.4 (p. 437), Lemma B.4.1 (p. 461) with $\beta = 1$, and Lemma 4.1.2 (p. 103) for the last. Letting $j = 3$ in (4.1.30) and using the model-termination condition (4.1.13), we obtain the required bound (4.1.22). By contrast, letting $j = 2$ in (4.1.30) and using the bound $|\nabla_s^2 m_k(s_k)[d]^2| \leq \theta_2 \|s_k\|^{p-1}$ from (4.1.14) for any d in the nonempty $\mathcal{M}_{k+1}$, we obtain (4.1.23). □

This then leads directly to a minimum guaranteed decrease in the objective function following a successful step.

Lemma 4.1.4: Bounds on the successful function decrease for ARp

*Suppose that **f.DpL** holds and that iteration k of the ARp algorithm is successful. Then*

$$f(x_k) - f(x_{k+1}) \geq \kappa_1 \|\nabla_x^1 f(x_{k+1})\|^{\frac{p+1}{p}}, \tag{4.1.31}$$

where

$$\kappa_1 \stackrel{\text{def}}{=} \frac{\eta_1 \sigma_{\min}}{(p+1)!} \left(\frac{p!}{L_{f,p} + \sigma_{\max} + \theta_1 p!} \right)^{\frac{p+1}{p}}. \tag{4.1.32}$$

If additionally $\epsilon_2 < \infty$, then

$$f(x_k) - f(x_{k+1}) \geq \kappa_2 \max\left(0, -\lambda_{\min}[\nabla_x^2 f(x_{k+1})]\right)^{\frac{p+1}{p-1}}, \tag{4.1.33}$$

where

$$\kappa_2 \stackrel{\text{def}}{=} \frac{\eta_1 \sigma_{\min}}{(p+1)!} \left(\frac{(p-1)!}{L_{f,p} + \sigma_{\max} + \theta_2 (p-1)!} \right)^{\frac{p+1}{p-1}}, \tag{4.1.34}$$

and, if $\epsilon_3 < \infty$, then

$$f(x_k) - f(x_{k+1}) \geq \kappa_3 \left| \nabla_x^3 f(x_{k+1})[d]^3 \right|^{\frac{p+1}{p-2}}, \tag{4.1.35}$$

where

$$\kappa_3 \stackrel{\text{def}}{=} \frac{\eta_1 \sigma_{\min}}{(p+1)!} \left(\frac{(p-2)!}{L_{f,p} + \sigma_{\max} + \theta_3 (p-2)!} \right)^{\frac{p+1}{p-2}} \tag{4.1.36}$$

for all d in the set $\mathcal{M}_{k+1}$ given by (4.1.14).

Proof. For successful iterations, it follows from (4.1.15), Lemma 4.1.1, and the requirement $\sigma_k \geq \sigma_{\min}$ in (4.1.16) that

$$f(x_k) - f(x_{k+1}) \geq \eta_1 \left[T_{f,p}(x_k, 0) - T_{f,p}(x_k, s_k) \right] \geq \frac{\eta_1 \sigma_{\min}}{(p+1)!} \|s_k\|^{p+1}. \tag{4.1.37}$$

The required bounds (4.1.31), (4.1.33), and (4.1.35) then follow from the corresponding bounds on $\|s_k\|$ given in Lemma 4.1.3. □

We now have all the ingredients necessary to establish upper bounds on the number of successful iterations of the ARp algorithm, and thus on the associated number of evaluations of the derivatives of order one to p. This in turn provides an upper bound on the total number of iterations and evaluations of the objective function, because of

Lemma 2.3.1. At variance with Theorems 3.3.5 and 3.3.9, though, we combine the results for first- to third-order approximate minimization into a single theorem: the results for a version of the algorithm that only searches for an approximate first-order or second-order minimizer may be retrieved by setting $\epsilon_2 = \epsilon_3 = \infty$ and $\epsilon_3 = \infty$ in the complexity bounds, respectively.

Theorem 4.1.5: Complexity of finding first- to third-order critical points by the ARp algorithm

Suppose that **f.Bb** *and* **f.DpL** *hold. Then there exist positive constants κ^S_{ARp}, κ^A_{ARp}, and κ^C_{ARp} such that, for any $(\epsilon_1, \epsilon_2, \epsilon_3) \in (0,1]^3$, the* AR$p$ *algorithm (p. 100) requires at most*

$$\#^F_{\text{ARp},3} \stackrel{\text{def}}{=} \kappa^A_{\text{ARp}} \frac{f(x_0) - f_{\text{low}}}{\min\left[\epsilon_1^{\frac{p+1}{p}}, \epsilon_2^{\frac{p+1}{p-1}}, \epsilon_3^{\frac{p+1}{p-2}}\right]} + \kappa^C_{\text{ARp}} = \mathcal{O}\left(\max\left[\epsilon_1^{-\frac{p+1}{p}}, \epsilon_2^{-\frac{p+1}{p-1}}, \epsilon_3^{-\frac{p+1}{p-2}}\right]\right)$$

evaluations of f and at most

$$\#^D_{\text{ARp},3} \stackrel{\text{def}}{=} \kappa^S_{\text{ARp}} \frac{f(x_0) - f_{\text{low}}}{\min\left[\epsilon_1^{\frac{p+1}{p}}, \epsilon_2^{\frac{p+1}{p-1}}, \epsilon_3^{\frac{p+1}{p-2}}\right]} + 2 = \mathcal{O}\left(\max\left[\epsilon_1^{-\frac{p+1}{p}}, \epsilon_2^{-\frac{p+1}{p-1}}, \epsilon_3^{-\frac{p+1}{p-2}}\right]\right) \tag{4.1.38}$$

evaluations of derivatives of order one to p of f to produce an iterate x_ϵ such that $\|\nabla^1_x f(x_\epsilon)\| \le \epsilon_1$, $\lambda_{\min}[\nabla^2_x f(x_\epsilon)] \ge -\epsilon_2$, and $|\nabla^3_x f(x_\epsilon)[d]^3| \le \epsilon_3$ for all $d \in \mathcal{M}_\epsilon$, where $\mathcal{M}_\epsilon \subseteq \mathcal{Z}_\epsilon = \{d : \|d\| = 1 \text{ and } |\nabla^2_x f(x_\epsilon)[d]^2| \le \epsilon_2\}$.

Proof. The proof of Theorem 4.1.5 is a slight variation on a by now familiar pattern. As usual, we have that, for any iteration k of the ARp algorithm,

$$f(x_0) - f_{\text{low}} \ge f(x_0) - f(x_{k+1}) = \sum_{j \in \mathcal{S}_k} [f(x_j) - f(x_{j+1})]. \tag{4.1.39}$$

Our goal is to bound the number of successful iterations required to find a step s_k and resulting point x_{k+1} for which

$$\|\nabla^1_x f(x_{k+1})\| \le \epsilon_1 \quad \text{and} \quad \max\Big(0, -\lambda_{\min}[\nabla^2_x f(x_{k+1})]\Big) \le \epsilon_2 \tag{4.1.40}$$

as well as

$$\left|\nabla^3_x f(x_{k+1})[d]^3\right| \le \epsilon_3 \quad \text{for all } d \in \mathcal{M}_{k+1} \neq \emptyset \tag{4.1.41}$$

and

$$\|s_k\|^{p-1} \le \kappa_{s2}\epsilon_2, \tag{4.1.42}$$

where κ_{s2} is defined as in (4.1.23). Crucially, if $d \in \mathcal{M}_\epsilon \stackrel{\text{def}}{=} \mathcal{M}_{k+1}$ and (4.1.42) holds, (4.1.23) implies that

$$\left|\nabla^2_x f(x_{k+1})[d]^2\right| \le \epsilon_2$$

and consequently $d \in \mathcal{Z}_\epsilon$, where $\mathcal{Z}_\epsilon$ is defined as in (4.1.4); if $\mathcal{Z}_\epsilon = \emptyset$ and (4.1.40) holds, then the termination test (4.1.6) is satisfied. Thus (4.1.40)–(4.1.42) together imply that (4.1.5)–(4.1.8) hold, and algorithm ARp will terminate.

If (4.1.40)–(4.1.42) fail to hold up to iteration $k+1$, it must be that at least one of

$$\|\nabla_x^1 f(x_{j+1})\| > \epsilon_1 \;\text{ and }\; \max\left(0, -\lambda_{\min}[\nabla_x^2 f(x_{j+1})]\right) > \epsilon_2 \;\text{ and }\; \left|\nabla_x^3 f(x_{j+1})[d_j]^3\right| > \epsilon_3$$

holds for some $d_j \in \mathcal{M}_{j+1}$ as defined in (4.1.14), or

$$\|s_j\|^{p+1} > \kappa_{s2}^{\frac{p+1}{p-1}} \epsilon_2^{\frac{p+1}{p-1}} \tag{4.1.43}$$

for $j \in \mathcal{S}_{k-1}$. As a consequence, we may call upon Lemma 4.1.4, or (4.1.37) and (4.1.43), to deduce that

$$\begin{aligned} f(x_j) - f(x_{j+1}) &\geq \min\left[\kappa_1 \epsilon_1^{\frac{p+1}{p}}, \kappa_2 \epsilon_2^{\frac{p+1}{p-1}}, \kappa_3 \epsilon_3^{\frac{p+1}{p-2}}\right] \\ &\geq \min[\kappa_1, \kappa_2, \kappa_3] \min\left[\epsilon_1^{\frac{p+1}{p}}, \epsilon_2^{\frac{p+1}{p-1}}, \epsilon_3^{\frac{p+1}{p-2}}\right] \\ &\stackrel{\text{def}}{=} \kappa_{123} \min\left[\epsilon_1^{\frac{p+1}{p}}, \epsilon_2^{\frac{p+1}{p-1}}, \epsilon_3^{\frac{p+1}{p-2}}\right] \end{aligned} \tag{4.1.44}$$

for all $j \in \mathcal{S}_{k-1}$; note that when (4.1.43) holds, (4.1.37) implies that the function decrease in this case is lower bounded by the same decrease as in the second-order case since $\kappa_2 = \frac{\eta_1 \sigma_{\min}}{(p+1)!} \kappa_{s2}^{\frac{p+1}{p-1}}$. Hence (4.1.39) and (4.1.44) give that

$$f(x_0) - f_{\text{low}} \geq \kappa_{123} |\mathcal{S}_{k-1}| \min\left[\epsilon_1^{\frac{p+1}{p}}, \epsilon_2^{\frac{p+1}{p-1}}, \epsilon_3^{\frac{p+1}{p-2}}\right]$$

and, therefore, using $|\mathcal{S}_k| \leq |\mathcal{S}_{k-1}| + 1$, that

$$|\mathcal{S}_k| \leq \kappa_{\text{ARp}}^{\text{S}} \frac{f(x_0) - f_{\text{low}}}{\min\left[\epsilon_1^{\frac{p+1}{p}}, \epsilon_2^{\frac{p+1}{p-1}}, \epsilon_3^{\frac{p+1}{p-2}}\right]} + 1 \stackrel{\text{def}}{=} N_S,$$

where

$$\begin{aligned} \kappa_{\text{ARp}}^{\text{S}} &\stackrel{\text{def}}{=} \frac{1}{\kappa_{123}} \\ &= \frac{(p+1)!}{\eta_1 \sigma_{\min}} \max\left[\left(\frac{L_{f,p} + \sigma_{\max} + \theta_1 p!}{p!}\right)^{\frac{p+1}{p}}, \left(\frac{L_{f,p} + \sigma_{\max} + \theta_2 (p-1)!}{(p-1)!}\right)^{\frac{p+1}{p-1}}, \right. \\ &\qquad \left. \left(\frac{L_{f,p} + \sigma_{\max} + \theta_3 (p-2)!}{(p-2)!}\right)^{\frac{p+1}{p-2}}\right], \end{aligned} \tag{4.1.45}$$

and, thus, we have an upper bound on the number of successful iterations required to satisfy (4.1.40)–(4.1.42).

Since the derivatives of f are evaluated once at each of these iterations, we deduce the bound (4.1.38) after adding 1 for the final iteration. We may now apply Lemma 2.4.1 (p. 44) to infer an upper bound on the total number of iterations and evaluations of f given by (3.3.29), with

$$\kappa_{\text{ARp}}^{\text{A}} \stackrel{\text{def}}{=} \kappa_{\text{ARp}}^{\text{S}} \left(1 + \frac{|\log \gamma_1|}{\log \gamma_2}\right)$$

and

$$\kappa^{\text{C}}_{\text{ARp}} \stackrel{\text{def}}{=} \frac{1}{\log \gamma_2} \left\{ |\log \gamma_1| + \log\left(\frac{\sigma_{\max}}{\sigma_0}\right)\right\} + 2, \qquad (4.1.46)$$

where $\sigma_{\max}$ is defined as in (4.1.19) and where 1 was added to $\kappa^{\text{C}}_{\text{ARp}}$ to account for the last iteration. □

This streamlined result subsumes Theorems 2.4.4, 3.3.5, and 3.3.9 into a single, concise statement. It also shows that the evaluation complexity to guarantee approximate first- to third-order minimizers decreases, as a function of ϵ_1, ϵ_2, and ϵ_3, with the degree p of the derivatives incorporated in the model. As might be expected from the discussions following our earlier results, the evaluation-complexity bounds provided by Theorem 4.1.5 are also sharp.

Theorem 4.1.6: Slow convergence of the ARp algorithm when finding first- to third-order critical points

The ARp algorithm (p. 100) applied to minimize a function f satisfying **f.Bb** *and* **f.DpL** *may require*

$$\epsilon_1^{-\frac{p+1}{p}}$$

iterations and evaluations of f and its derivatives of order 1 to p to produce a point x_ϵ such that $\|\nabla^1_x f(x_\epsilon)\| \le \epsilon_1$. It may also require

$$\max\left[\epsilon_1^{-\frac{p+1}{p}}, \epsilon_2^{-\frac{p+1}{p-1}}\right]$$

iterations and evaluations of f and its derivatives of order 1 to p to produce a point x_ϵ such that $\|\nabla^1_x f(x_\epsilon)\| \le \epsilon_1$ and $\lambda_{\min}[\nabla^2_x f(x_\epsilon)] \ge -\epsilon_2$. Finally, it may require

$$\max\left[\epsilon_1^{-\frac{p+1}{p}}, \epsilon_2^{-\frac{p+1}{p-1}}, \epsilon_3^{-\frac{p+1}{p-2}}\right]$$

iterations and evaluations of f and its derivatives of order 1 to p to produce a point x_ϵ such that $\|\nabla^1_x f(x_\epsilon)\| \le \epsilon_1$, $\lambda_{\min}[\nabla^2_x f(x_\epsilon)] \ge -\epsilon_2$, and $|\nabla^3_x f(x_\epsilon)[d]^3| \le \epsilon_3$ for all $d \in \mathcal{Z}_\epsilon = \{d : \|d\| = 1$ and $|\nabla^2_x f(x_\epsilon)[d]^2| \le \epsilon_2\}$.

Proof. We again postpone the proof to Section 12.2.2.4 (p. 314); see the paragraph following Corollary 12.2.16. □

Notes and references for Subsection 4.1.2

The material of this section follows the proposals of [45] and [107]. Once again, **f.DpL** is stronger than necessary and may be relaxed so that (4.1.17) simply holds for all x and y in an open convex set containing the iterates and trial steps generated by the algorithm. While computing high-order derivatives may sound computationally expensive for problems of moderate and large size, we note that this might not be the case for the ubiquitous "partially separable" problems, as discussed in Section 5.2.

An early second-order method [7] for computing third-order approximate minimizers of functions with Lipschitz-continuous third derivatives used random sampling of a subspace closely related to $\mathcal{M}_{k+1}$ to escape from second-order saddle points. This results in an evaluation-complexity bound of $\mathcal{O}(\max[\epsilon_1^{-3/2}, \epsilon_2^{-3}, \epsilon_3^{-9/2}])$ for "strict third-order saddle" functions. The practical aspects of using third-order models within fourth-order regularization are discussed in [43]. It is also possible to extend multilevel methods [212] to use high-order regularization [71], and these indicate promising practical performance using an AR2 variant.

Significantly, pth-degree models may also be used by adaptive-regularization methods to find first-order approximate minimizers in real infinite-dimensional uniformly q-smooth Banach spaces[26] (which includes Hilbert spaces) [208]. The motivation for this generalization is a matter of coherence when optimization algorithms are applied to large-scale discretized problems, because then it is important to show that adaptive-regularization methods continue to make sense in the limit as the discretization mesh converges to zero. This coherence, sometimes called "mesh independence," has long been considered an important feature of numerical-optimization methods [5,149,230,253,356]. This has been investigated for trust-region methods in the Hilbert-space context [348] and subsequently developed for Hilbert and Banach spaces [131, Sect. 8.3]. The main difficulty in handling infinite-dimensional spaces is that the existence of a suitable step at a given iteration of the method typically relies on approaching a minimizer of the regularized model, which may no longer exist. The ARp-BS algorithm and its analysis [208] circumvent this problem by proposing a specialized optimization technique for regularized polynomials that guarantees an acceptable step. The proof hinges on properties of the norm in smooth Banach spaces (see [208] for details).

4.1.3 A TRp algorithm?

At this point, the reader may well anticipate the definition of a rival trust-region method, involving a pth-degree Taylor-series model of the form (4.1.1)—using standard constructs from, say TR2—and its subsequent analysis. Unfortunately, there is nothing[27] to gain by doing so. A hint that this might be the case is already present in the details of Theorem 3.2.2 (p. 58), in which we showed that using a second-order Taylor-series model, $T_{f,2}(x,s)$, does not improve on the first-order one, $T_{f,1}(x,s)$, with the worst-case bound in both cases being $\mathcal{O}(\epsilon_1^{-2})$ evaluations when searching for a first-order approximate minimizer. Could it be that considering higher-degree models ($p > 2$) improves the situation? The answer is unsurprisingly no.[28] This can be verified by returning to the example presented in the proof of Theorem 2.2.3 (p. 27), where, following the suggestion made before Theorem 2.3.8 (p. 41), one can choose $\nabla_x^2 f(x_k) = 2t_{\rm ini}^{-1}$ for all iterations $k \geq 0$, but also $\nabla_x^\ell f(x_k) = 0$ for $\ell \in [3:p]$. The model minimizer is of course unchanged, as is the conclusion that $\Delta_k = 1$ is acceptable for all k. Moreover, we have that (2.2.21) and (2.2.22) still hold and

$$\|\nabla_x^\ell f(x_{k+1}) - \nabla_s^\ell T_{f,p}(x_k, s_k)\| = 0 \leq \kappa_f \|s_k\|^{p-\ell+1}$$

for all $\ell \geq 2$ and all $k \geq 0$. Thus the conditions for Hermite interpolation of Theorem A.9.2 (p. 441) are fulfilled and a p-times continuously differentiable function with Lipschitz-continuous pth derivative can be built on which a trust-region algorithm using p exact derivatives would still require $\mathcal{O}(\epsilon_1^{-2})$ evaluations to find an iterate x_k for

[26] Broadly speaking, a uniformly q-smooth Banach space is a Banach space $\mathcal{V}$ (see Appendix A.4) in which the modulus of smoothness $\rho_{\mathcal{V}}(t)$ satisfies

$$\rho_{\mathcal{V}}(t) \stackrel{\rm def}{=} \sup_{x,y\in\mathcal{V},\ \|x\|=1,\ \|y\|=t} \left\{ \frac{\|x+y\| + \|x-y\|}{2} - 1 \right\} \leq \kappa_{\mathcal{V}} t^q$$

for some $q \in (1,2]$ and some $\kappa_{\mathcal{V}} > 0$.

[27] At least from the evaluation-complexity point of view.

[28] But recall that second-order trust-region variants with optimal complexity exist; see, for example, [137] or the earlier Section 3.4 (p. 75). Moreover, in Section 12.2.1 (p. 292), we devise a trust-region variant that uses a high-degree Taylor model and can certify high-order criticality, albeit using more involved and more expensive optimality concepts.

which $\|\nabla_x^1 f(x_k)\| \le \epsilon_1$. Thus we conclude that extending the model degree p beyond the sought optimality order q does not improve the evaluation complexity for finding first-order points, which is why we will not consider this option any further in this book.

4.2 A lower complexity bound for finding first-order stationary points

By this stage of our book, the reader might rightly have started to be suspicious that perhaps we have been overly selective in the algorithms that we have chosen to describe and analyze. While we have exhibited methods with what we have termed "good behaviour," might it be possible that we have only focused on methods we know, and that actually there are hidden gems that we have overlooked and that actually outperform our self-proclaimed champions?

In this section, we review a far-reaching discovery by Carmon, Duchi, Hinder, and Sidford [77], who identify a problem instance that is difficult to optimize for a large class of algorithms. The construction they provide follows the tradition of lower complexity bound constructions for first-order algorithms for minimizing convex functions [296, 299], in which a "worst function in the world" is exhibited whose dimension grows as a polynomial in the inverse of the optimality tolerance, and for which *all* algorithms with access to specified problem oracles (say function and gradient evaluations) will be slow in terms of global iteration and evaluation counts. The example of bad behaviour they suggest is algorithm independent and applies to a larger class of less prescriptive algorithms than simply those that we have encountered thus far.

Unlike for the one- or two-dimensional bad-case examples we have met in Section 3.5 (p. 90) and in Theorems 3.2.2 (p. 58) and 4.1.6 (p. 109), this generality is made possible by constructing a high-dimensional example—whether there are similar examples that do not require large dimensions is an open question. In particular, here we show that for any $\epsilon_1 > 0$, there exists a dimension n such that, for any $\bar{n} \ge n$, there is a function $f : \mathbb{R}^{\bar{n}} \to \mathbb{R}$ such that any algorithm in the class of interest requires at least k_ϵ evaluations of f and its derivatives up to order p to generate $\|\nabla_x^1 f(x_{k_\epsilon})\| \le \epsilon_1$, where $n = 2k_\epsilon$. Because $k_\epsilon = \mathcal{O}(\epsilon_1^{-(p+1)/p})$, the results here imply that ARp has optimal complexity within a general class of algorithms using derivatives up to pth order.

4.2.1 Problems and methods of interest

Just as in Section 3.5 (p. 90), we start by describing the classes of problems and algorithms that will be our focus.

4.2.1.1 The function class $\mathcal{F}(\Delta, p)$ of interest

Let $p \ge 1$. Our intention is, as we have done throughout this chapter, to apply the algorithms that we permit to the minimization of sufficiently smooth functions f that are bounded from below by a given $f_{\rm low} = \inf f(x)$ and have Lipschitz-continuous pth derivatives. For a given $\Delta > 0$, we say *f is in the class of functions $\mathcal{F}(\Delta, p)$* if $f : \mathbb{R}^n \to \mathbb{R}$ is infinitely continuously differentiable[29] and satisfies **f.DpL** and **f.Bb**, where

$$f(0) - f_{\rm low} \le \Delta.$$

[29]This is needed so that the dimension n of f is allowed to grow.

Notice that functions in the class of interest have an important orthogonal invariance property, namely, for any function f in the class, and any $U \in O(\bar{n}, n)$, where

$$O(\bar{n}, n) \stackrel{\text{def}}{=} \{O \mid O \text{ is an } \bar{n} \text{ by } n \text{ real, orthonormal matrix}\},$$

the function $f_{\text{U}} : \mathbb{R}^{\bar{n}} \to \mathbb{R}$ defined by $f_{\text{U}}(z) \stackrel{\text{def}}{=} f(U^T z)$ for all $z \in \mathbb{R}^{\bar{n}}$ also belongs to this class.[30] Thus $\mathcal{F}(\Delta, p)$ contains functions of a variable number of unknowns, and hence n is not fixed.

4.2.1.2 The class $\mathcal{M}(p)$ of methods and its zero-respecting subclass $\mathcal{M}_{\text{ZR}}(p)$

For any dimension n; permitted derivative degree p; and, without loss of generality, $x_0 \stackrel{\text{def}}{=} 0 \in \mathbb{R}^n$, we view a *method in the class* $\mathcal{M}(p)$ as a mapping that associates any given function $f : \mathbb{R}^n \to \mathbb{R}$ with a corresponding sequence of iterates $\{x_k\} \subset \mathbb{R}^n$, where

$$x_{k+1} = M^{(k)}(f(x_0), \nabla_x^1 f(x_0), \ldots, \nabla_x^p f(x_0), \ldots, f(x_k), \ldots, \nabla_x^p f(x_k))$$

and where $M^{(k)}$ is a mapping[31] from $\mathbb{R}^n$ to $\mathbb{R}^n$. This general definition implies that all algorithms considered in this book that use exact derivatives belong to $\mathcal{M}(p)$ for an appropriate choice of p. Furthermore, this general class of algorithms also includes methods that do not explicitly use all available derivatives, such as those we will encounter in Chapter 13 (p. 325). Thus the definition is understood to imply that the generated iterates depend upon derivatives of degree *at most* o, with $o \le p$.

A useful subclass of $\mathcal{M}(p)$ is the class of *zero-respecting methods* $\mathcal{M}_{ZR}(p)$. Specifically, a method in $\mathcal{M}(p)$ that generates iterates $\{x_k\}$ is *zero respecting* with respect to any function $f : \mathbb{R}^n \to \mathbb{R}$ if for each $k \ge 0$,

$$\left\{ \bigcap_{q \in [1:p]} \bigcap_{i \le k} \{j \mid \nabla_x^q f(x_i) e_j = 0\} \right\} \subseteq \{j \mid [x_{k+1}]_j = 0\}; \tag{4.2.1}$$

in other words, if all partial derivatives of f up to order p containing the jth coordinate are zero at the current and all previous iterates (including $x_0 = 0$), then the jth coordinate of the next iterate remains zero. This definition says that zero components of iterates are preserved if corresponding derivative components of f (up to order p) are zero, so such an algorithm does not explore coordinates that do not appear to affect the function.

It is clear that first-order ($p = 1$) algorithms belong to $\mathcal{M}_{\text{ZR}}(p)$, but it is not difficult to see that Newton's method (Algorithm 3.1.1 (p. 50)) can also be viewed as zero respecting (with $p = 2$) if we employ a suitable convention.[32] Under similar conventions, algorithms AR2 and TR2 and other variants of Newton's method, as well as algorithms ARp and the forthcoming TRq and ARqp (which we shall encounter in Chapter 12), can be shown to be members of $\mathcal{M}_{\text{ZR}}(p)$ for $p = 2$ and $p \ge 1$, respectively. We note though that it is not necessary for these algorithms to be zero respecting in order for the lower-bound results to apply to them.

[30] This is because the Lipschitz constants of f and f_{U} and their respective derivatives coincide for all orders, and $f(0) - f_{\text{low}} = f_{\text{U}}(0) - f_{U,\text{low}}$.

[31] We note that in [77], $M^{(k)}$ is assumed to be measurable. However, this assumption does not appear to be used by the authors in the deterministic case that we are considering here.

[32] When the jth row and column of the Hessian matrix of f at x_k are zero, and the corresponding component of the gradient is also zero, then the step component $[s_k]_j$ is arbitrary. Newton's method is zero respecting if, in this case, we choose $[s_k]_j = 0$.

4.2.1.3 The complexity of a class of methods

Thus armed with sensible sets of functions and suitably related minimization algorithms, it is now reasonable to talk of the complexity of approximating a minimizer, or at least of finding an approximate stationary point. Let $k = k_\epsilon(M, f)$ be the first iteration for which $\|\nabla_x^1 f(x_k)\| \leq \epsilon_1$, where x_k is generated by an algorithm M applied to minimize f. The *complexity of the class* $\mathcal{M}$ *on the function class* $\mathcal{F}$ is defined as the best worst-case complexity bound of an algorithm in $\mathcal{M}$ for functions in $\mathcal{F}$, namely,

$$\#_\epsilon(\mathcal{M}, \mathcal{F}) \stackrel{\text{def}}{=} \inf_{M \in \mathcal{M}} \sup_{f \in \mathcal{F}} k_\epsilon(M, f). \tag{4.2.2}$$

In particular, when $\mathcal{F} = \mathcal{F}(\Delta, p)$, Theorem 4.1.5 with $\epsilon_2 = \epsilon_3 = \infty$ shows that

$$\#_\epsilon(\mathcal{M}, \mathcal{F}(\Delta, p)) \leq \sup_{f \in \mathcal{F}(\Delta,p)} k_\epsilon(\mathsf{AR}p, f) = \mathcal{O}\left(\Delta L_{f,p}^{\frac{p+1}{p}} \epsilon_1^{-\frac{p+1}{p}}\right), \tag{4.2.3}$$

where $L_{f,p}$ is the Lipschitz constant of the pth-order derivative $\nabla_x^p f$. We show that these bounds are sharp within p-dependent constant factors.

4.2.2 Constructing a lower complexity bound for the methods in $\mathcal{M}(p)$ when applied to functions in $\mathcal{F}(\Delta, p)$

In order to derive a lower complexity bound, we need to exhibit a "difficult" function, that is, one that requires a significant number of evaluations for all methods in our class. The key here is the idea of a zero-chain function.

A function $f \in C^p$ is a *pth-order zero chain* if for each $x \in \mathbb{R}^n$ and any $j \in [1 : n]$,

$$[x]_j = \cdots = [x]_n = 0 \text{ implies } \nabla_x^q f(x) e_{j+1} = \cdots = \nabla_x^q f(x) e_n = 0 \text{ for all } q \in [1 : p]. \tag{4.2.4}$$

In other words, if the last $n - j$ components of a point x are zero, a zero-chain function looks constant in the last $n - j - 1$ components. See [299, Sect. 2.1.2] and [77, Def. 3] for more details and examples.

This has the following immediate, trivial consequence.

Lemma 4.2.1: Scale-invariance of zero chains

Let μ_1 and $\mu_2 \neq 0$, and suppose that f is a pth-order zero chain. Then the function $\mu_1 f(x/\mu_2)$ for all $x \in \mathbb{R}^n$ is also a pth-order zero chain.

More significantly, if we apply a zero-respecting algorithm to a zero-chain objective function, the following holds.

Lemma 4.2.2: Zero-respecting methods applied to zero chains

[77, Obs. 1] Consider a method $M \in \mathcal{M}_{\text{ZR}}(p)$ and an objective function $f : \mathbb{R}^n \to \mathbb{R} \in \mathcal{F}(\Delta, p)$ that is also a pth-order zero chain. Then the iterates $\{x_k\}_{k\geq 0}$ generated by M satisfy $[x_k]_j = 0$ for all $j \in [k + 1 : n]$ and all $k \leq n - 1$.

Proof. The proof follows by induction on k. If $k = 0$, then $j \geq 1$, and the conclusion clearly holds because $x_0 = 0$. Assume the required property holds for all $k \leq k_0$, in which case $[x_k]_j = 0$ for all $j \in [k+1 : n]$ and all $k \leq k_0$. Since f is a pth-order zero chain, it follows from (4.2.4) that $\nabla_x^q f(x_k) e_j = 0$ for all $j \in [k+2 : n]$, $k \leq k_0$, and $q \in [1 : p]$. But then, since $M \in \mathcal{M}_{\rm ZR}(p)$, (4.2.1) implies that $[k+2 : n] \subseteq \{j : [x_{k+1}]_j = 0\}$, $k \leq k_0$, and thus the required property holds for $k \leq k_0 + 1$, completing the inductive step because the induction stops when $k_0 = n - 1$. □

This lemma indicates that zero-respecting algorithms take many iterations to find the correct values for all the components of a zero-chain objective. The slow behaviour of such algorithms impacts their complexity, as we now show.

Lemma 4.2.3: Zero-respecting methods are slow on zero chains

[77, Obs. 2] Let $\epsilon_1 > 0$ and let $f \in \mathcal{F}(\Delta, p)$ be a pth-order zero chain for which

$$\|\nabla_x f(x)\| > \epsilon_1 \quad \text{for each } x \text{ such that } [x]_n = 0. \tag{4.2.5}$$

Then

$$\#_\epsilon(\mathcal{M}_{\rm ZR}(p), \mathcal{F}(\Delta, p)) \geq \#_\epsilon(\mathcal{M}_{\rm ZR}(p), \{f\}) \geq n,$$

where the best worst-case complexity bound $\#_\epsilon$ is defined by (4.2.3).

Proof. Let $M \in \mathcal{M}_{\rm ZR}(p)$ generate iterates $\{x_k\}$ when applied to f. Then Lemma 4.2.2 implies that $[x_k]_n = 0$ for all $k+1 \leq n$. Due to (4.2.5), method M does not terminate at iteration $n-1$ and so $k_\epsilon(M, f) \geq n$. Since the choice of M was arbitrary, we have from (4.2.2) that

$$\#_\epsilon(\mathcal{M}_{\rm ZR}(p), \mathcal{F}(\Delta, p)) \geq \sup_{f \in \mathcal{F}(\Delta,p)} \inf_{M \in \mathcal{M}_{\rm ZR}(p)} k_\epsilon(M, f) \geq \inf_{M \in \mathcal{M}_{\rm ZR}(p)} k_\epsilon(M, f) \geq n.$$

□

We now connect the complexity of the algorithm class $\mathcal{M}(p)$ to that of the class $\mathcal{M}_{\rm ZR}(p)$ when both are applied to functions in $\mathcal{F}(\Delta, p)$.

Theorem 4.2.4: Connecting the complexity of $\mathcal{M}(p)$ and $\mathcal{M}_{\rm ZR}(p)$

For any $\epsilon_1 > 0$, we have

$$\#_\epsilon(\mathcal{M}(p), \mathcal{F}(\Delta, p)) \geq \#_\epsilon(\mathcal{M}_{\rm ZR}(p), \mathcal{F}(\Delta, p)).$$

Furthermore, for any function $f : \mathbb{R}^n \to \mathbb{R} \in \mathcal{F}(\Delta, p)$ for which $\#_\epsilon(\mathcal{M}_{\rm ZR}(p), \{f\}) \geq \bar{k}$ for some $\bar{k}$, we have

$$\#_\epsilon(\mathcal{M}(p), \mathcal{F}(\Delta, p)) \geq \#_\epsilon(\mathcal{M}(p), \{f_{\rm U}\}) \geq \bar{k},$$

where $\{f_{\rm U}\}$ is the set of all functions $f_{\rm U} : \mathbb{R}^{n+\bar{k}} \to \mathbb{R}$ for which $f_{\rm U}(z) = f(U^T z)$ with $U \in O(n + \bar{k}, n)$.

Proof. See [77, Props. 1 and 2] and their proofs in [77, App. A]. □

The proof of Theorem 4.2.4 relies on the concept of a *resisting oracle* [296, 299]. Namely, for an algorithm $M \in \mathcal{M}(p)$ and objective $f \in \mathcal{F}(\Delta, p)$, a matrix $U \in O(\bar{n}, n)$ is adversarially and sequentially constructed so that for the objective function $f_{\rm U}(z) =$

$f(U^T z) \in \mathcal{F}(\Delta, p)$ the algorithm M behaves in an equivalent way to a zero-respecting algorithm. Thus a zero-respecting algorithm M_{ZR} is constructed such that the iterates $\{x_k^{zr}\}$ it generates when applied to f satisfy $x_k^{zr} = U^T z_k \in \mathbb{R}^n$ for all k, where $\{z_k\} \subseteq \mathbb{R}^{\bar{n}}$ is generated by applying M to f_{U}. Thus $k_\epsilon(M, f_{\text{U}}) = k_\epsilon(M_{\text{ZR}}, f)$. This and (4.2.2) give

$$\begin{aligned} \#_\epsilon(\mathcal{M}, \mathcal{F}(\Delta, p)) &= \inf_{M \in \mathcal{M}} \sup_{f_{\text{U}} \in \mathcal{F}(\Delta,p)} k_\epsilon(M, f_{\text{U}}) = \inf_{M \in \mathcal{M}} \sup_{f \in \mathcal{F}(\Delta,p)} k_\epsilon(M_{\text{ZR}}, f) \\ &\geq \inf_{\bar{M} \in \mathcal{M}_{\text{ZR}}(p)} \sup_{f \in \mathcal{F}(\Delta,p)} k_\epsilon(\bar{M}, f) = \#_\epsilon(\mathcal{M}_{\text{ZR}}(p), \mathcal{F}(\Delta, p)), \end{aligned}$$

where we use that $M_{\text{ZR}} \in \mathcal{M}_{\text{ZR}}(p)$ to obtain the first inequality. Thus the first part of the theorem follows, while the second part ensues by setting $\bar{n} = n + \bar{k}$.

After these general considerations, we are poised to present the main complexity result of this section. But first we need to illuminate a "difficult" function that all algorithms in $\mathcal{M}_{\text{ZR}}(p)$ will find hard. For fixed $\bar{k} > 0$, let $x \in \mathbb{R}^{\bar{k}}$, and let $f^{\text{D}} : \mathbb{R}^{\bar{k}} \to \mathbb{R}$ be

$$f^{\text{D}}(x) \stackrel{\text{def}}{=} -\Psi([x]_1)\Phi([x]_1) + \sum_{i=2}^{\bar{k}} [\Psi(-[x]_{i-1})\Phi(-[x]_i) - \Psi([x]_{i-1})\Phi([x]_i)], \tag{4.2.6}$$

where

$$\Psi(y) = \begin{cases} 0 & \text{if } y \leq \frac{1}{2} \\ \exp\left(1 - (2y-1)^{-2}\right) & \text{otherwise} \end{cases}$$

and

$$\Phi(y) = \exp\left(\tfrac{1}{2}\right) \int_{-\infty}^{y} \exp\left(-\tfrac{1}{2}t^2\right) dt$$

for all $y \in \mathbb{R}$. The function f^{D} has a number of attractive properties.

Lemma 4.2.5: Properties of the difficult example function

Let $f^{\text{D}}(x)$ be as defined in (4.2.6). Then

(i) *$f^{\text{D}}(x)$ is a pth-order zero chain, namely, it satisfies (4.2.4) for any $p \geq 1$;*

(ii) *if $x \in \mathbb{R}^{\bar{k}}$ with $|[x]_i| < 1$ for any $i \leq \bar{k}$, then there exists $j \leq i$ such that $|[x]_j| < 1$ and*

$$\|\nabla_x^1 f^{\text{D}}(x)\| \geq \left| \frac{\partial f^{\text{D}}}{\partial [x]_j}(x) \right| > 1;$$

(iii) *$f^{\text{D}}(0) - f^{\text{D}}_{\text{low}} \leq 12\bar{k}$; and*

(iv) *for every $p \geq 1$, $\nabla_x^p f^{\text{D}}(x)$ satisfies **f.DpL** with Lipschitz constant $L_{f^{\text{D}}_{,p}} \leq \exp\left(\frac{5}{2} p \log p + c_1 p\right)$ for some constant $c_1 > 0$.*

Proof. In order to establish (i), it suffices to show that for any $j > 1$, if $|[x]_{j-1}| < 1/2$ and $|[x]_j| < 1/2$, then $f^{\text{D}}(y) = f^{\text{D}}\left([y]_1, \ldots, [y]_{j-1}, 0, [y]_{j+1}, \ldots, [y]_{\bar{k}}\right)$ for all y is a neighbourhood of x. See [77, Lem. 1 and Obs. 3] and their proofs in [77, App. B].

The large gradient property in (ii) is [77, Lem. 2], and a proof of this is given after the statement of that lemma.

The properties in (iii) and (iv) are given in [77, Lem. 3] and proved in [77, App. B]. □

We now come to the main result. This provides a lower iteration bound for finding approximate stationary points of smooth functions in $\mathcal{F}(\Delta, p)$ using methods in $\mathcal{M}_{\text{ZR}}(p)$, and hence also those in $\mathcal{M}(p)$ in light of Theorem 4.2.4.

Theorem 4.2.6: Lower complexity bound

Let $p \geq 1$, $\Delta > 0$, and $\epsilon_1 > 0$. Then there exist positive constants c_0 and c_1 such that

$$\#_\epsilon(\mathcal{M}(p), \mathcal{F}(\Delta, p)) \geq \#_\epsilon(\mathcal{M}_{\text{ZR}}(p), \mathcal{F}(\Delta, p)) \geq c_0 \Delta \left(\frac{L_{f,p}}{L_{f^{\text{D}}_{,}p}} \right)^{1/p} \epsilon_1^{-\frac{p+1}{p}}, \quad (4.2.7)$$

where $L_{f,p} > 0$ is the Lipschitz constant associated with functions in $\mathcal{F}(\Delta, p)$ and $L_{f^{\text{D}}_{,}p} \leq \exp\left(\frac{5}{2} p \log p + c_1 p\right)$. Furthermore, the bound (4.2.3) is tight, and the ARp *algorithm has the optimal order of evaluation complexity for methods in $\mathcal{M}(p)$.*

Proof. Define $f : \mathbb{R}^{\bar{k}} \to \mathbb{R}$ using f^{D} via (4.2.6) as

$$f(x) \stackrel{\text{def}}{=} \mu^{p+1} \frac{L_{f,p}}{L_{f^{\text{D}}_{,}p}} f^{\text{D}}(x/\mu) \quad (4.2.8)$$

for all $x \in \mathbb{R}^{\bar{k}}$, where the scale parameter $\mu > 0$ is yet to be specified and $L_{f^{\text{D}}_{,}p}$ is the Lipschitz constant of $\nabla_x^p f^{\text{D}}$ as described in Lemma 4.2.5 (iv).

We show that the conditions of Lemma 4.2.3 are satisfied for f. Lemma 4.2.1 and Lemma 4.2.5 (i) imply that f is a pth-order zero chain. To demonstrate that (4.2.5) holds, let $x \in \mathbb{R}^{\bar{k}}$ be such that $[x]_{\bar{k}} = 0$. Recalling Lemma 4.2.5 (ii), we deduce that $\|\nabla_x^1 f^{\text{D}}(x/\mu)\| > 1$, which leads to

$$\|\nabla_x^1 f(x)\| \geq \mu^p \frac{L_{f,p}}{L_{f^{\text{D}}_{,}p}} \|\nabla_x^1 f^{\text{D}}(x/\mu)\| > \mu^p \frac{L_{f,p}}{L_{f^{\text{D}}_{,}p}}$$

using (4.2.8). Thus, to guarantee that $\|\nabla_x^1 f(x)\| > \epsilon_1$, it suffices to choose $\mu \stackrel{\text{def}}{=} (L_{f^{\text{D}}_{,}p} \epsilon_1 / L_{f,p})^{1/p}$. It remains to show that $f \in \mathcal{F}(\Delta, p)$. Lemma 4.2.5 (iv) implies that $\nabla_x^p f^{\text{D}}(x)$ is Lipschitz continuous with constant $L_{f^{\text{D}}_{,}p}$, so $\|\nabla_x^{p+1} f^{\text{D}}(x)\|$ is uniformly bounded above by $L_{f^{\text{D}}_{,}p}$. Thus, since

$$\nabla_x^{p+1} f(x) = (L_{f,p}/L_{f^{\text{D}}_{,}p}) \nabla_x^{p+1} f^{\text{D}}(x/\mu),$$

it follows that $\|\nabla_x^{p+1} f(x)\| \leq L_{f,p}$ and hence $\nabla_x^p f$ is Lipschitz continuous with constant $L_{f,p}$. Finally, using Lemma 4.2.5 (iii), we find

$$f(0) - f_{\text{low}} = \mu^{p+1} \frac{L_{f,p}}{L_{f^{\text{D}}_{,}p}} (f^{\text{D}}(0) - f^{\text{D}}_{\text{low}}) \leq 12 \bar{k} \mu^{p+1} \frac{L_{f,p}}{L_{f^{\text{D}}_{,}p}} = 12 \bar{k} \left(\frac{L_{f^{\text{D}}_{,}p}}{L_{f,p}} \right)^{1/p} \epsilon_1^{\frac{p+1}{p}}.$$

Thus $f(0) - f_{\text{low}} \leq \Delta$, and so $f \in \mathcal{F}(\Delta, p)$, provided

$$\bar{k} = \left\lfloor \frac{\Delta}{12} \left(\frac{L_{f,p}}{L_{f^{\text{D}}_{,}p}} \right)^{1/p} \epsilon_1^{-\frac{p+1}{p}} \right\rfloor.$$

Lemma 4.2.3 implies $\#_\epsilon(\mathcal{M}_{\rm ZR}(p), \mathcal{F}(\Delta, p)) \geq \#_\epsilon(\mathcal{M}_{\rm ZR}(p), \{f\}) \geq \bar{k}$, and then the second part of Theorem 4.2.4 with $n = \bar{k}$ implies $\#_\epsilon(\mathcal{M}(p), \mathcal{F}(\Delta, p)) \geq 2\bar{k}$. The remaining claims follow directly from, for example, Theorem 4.1.5 because the ARp algorithm requires precisely one function evaluation per iteration. □

We conclude this section with two observations. Firstly, Theorem 4.2.6, together with the definition (4.2.2) of $\#_\epsilon(\mathcal{M}(p), \mathcal{F}(\Delta, p))$, says that the lower bound (4.2.7) applies to every method in class $\mathcal{M}(p)$. Furthermore, the lower bound (4.2.7) is valid for any method in $\mathcal{M}(o)$ with $0 < o \leq p$ when applied to functions in $\mathcal{F}(\Delta, p)$. This is because the construction of the counterexample does not depend on whether the algorithm from the respective class actually uses the available derivatives, although it is important that these derivatives exist and are sufficiently smooth. In particular, (4.2.7) holds for first- and second-order methods in previous chapters[33] and elsewhere [75], and for the derivative-free variants in Chapter 13 (p. 325) when applied to functions in $\mathcal{F}(\Delta, p)$. The lower bound also applies, typically loosely, to some unusual methods—such as grid-search [359] and cutting planes [235]—that do not use local Taylor models and hence do not belong to the class in Section 3.5 (p. 90).

Secondly, the difficult function derived to obtain the conclusions of Theorem 4.2.6 for the class $\mathcal{M}(p)$ has dimension $2\bar{k} = \mathcal{O}(\epsilon_1^{-(p+1)/p})$. In other words, the function dimension grows polynomially with ϵ_1 and exponentially with p. This is in contrast to the one-dimensional examples of optimal second-order or sharp pth-order complexity given in Section 3.5 (p. 90) and throughout the book.

Notes and references for Section 4.2

The main source of the material in this section is the paper by Carmon et al. [77], which also features randomized algorithms and their corresponding lower evaluation-complexity bounds. The central lower bound given here may be tightened further for accelerated first-order methods for nonconvex optimization when applied to functions that have Lipschitz-continuous derivatives of order up to three [78]. Indeed, the authors provide an example that has Lipschitz-continuous second derivatives for which all first-order methods must require at least $\mathcal{O}(\epsilon_1^{-12/7})$ function and gradient evaluations. This is significant since we will see later (in Section 13.2) that there is such a method that requires at most $\mathcal{O}(\epsilon_1^{-3/2})$ evaluations. While this would appear to be a contradiction, in fact it isn't, because the constant hidden by the $\mathcal{O}$ notation for the method in Section 13.2 has an explicit dimension dependence, while such a dependence is implicit in the purported worst-case example in [78] for which ϵ_1 and n are linked, and n must be sufficiently large. Arjevani et al. [10] extend the techniques in this section to providing lower bounds on the complexity of finding a second-order stationary point of suitably smooth functions. We note that zero-chain constructions, as in (4.2.6), have previously been used when deriving lower bounds for convex optimization algorithms [296,299] and are summarized in [77, Sect. 3.1]. In the context of nonconvex optimization, Jarre [249] showed that the so-called Chebyshev–Rosenbrock function is difficult to optimize, taking an exponential number of iterations in terms of the problem dimension for any descent method, despite having a single (exact) stationary point. However, this does not contradict the typical ϵ_1-dependent complexity bounds for descent methods, since the function has other approximate stationary points at which such methods terminate [95].

[33] However, note that this lower bound is typically not tight for first- and second-order methods applied to $f \in \mathcal{F}(\Delta, p)$ with $p > 2$.

Chapter 5

Structured Problems

Chapter summary

The results we have discussed so far for unconstrained minimization may be extended in various ways. In this chapter, we consider how problem structure affects our complexity results, and how it may be exploited.

We first consider perhaps the most ubiquitous unconstrained optimization problem encountered in practice, nonlinear least-squares, or more generally the minimization of the Euclidean norm of a vector-valued residual function. We devise a way to exploit the specific nature of the norm, so as to improve the worst-case performance of adaptive-regularization methods. This will come in handy later when we deal with constraints. In this development, we show that evaluation complexity varies when considering equivalent formulations of the original problem that differ according to the exponent to which the norm of the residual is raised. We also indicate how to choose this exponent to obtain the best complexity, given the degree of the available residual derivatives.

Because of their considerable advantage when high-degree derivatives are used, we next very briefly discuss partially separable problems.

Our next topic involves generic classes of "measure-dominated" problems for which considerably better complexity results may be derived. Both trust-region and regularization algorithms are considered, with the latter proving to provide better worst-case behaviour. We also briefly consider the consequences of iterates blundering into regions of strong local convexity.

Finally, as indicated in the introduction, the smoothness assumptions on the objective function may be relaxed in various ways. We discuss the possibility that the highest derivative considered is merely Hölder continuous rather than Lipschitz continuous. We propose modifications to our favourite regularization algorithm and indicate that its worst-case performance varies naturally and continuously as a function of the assumed smoothness.

5.1 Nonlinear least-squares and least-Euclidean norms

5.1.1 Introduction

In our experience, a large proportion of practical unconstrained optimization problems arise from fitting models to data. More specifically, given a vector $r(x)$ of parameterized residual functions $r_i : \mathbb{R}^n \to \mathbb{R}$, $i \in [1 : m]$, the parameters x are chosen to make an appropriate norm of $r(x)$ as small as possible—each residual typically measures the discrepancy between a modeled quantity (using the model parameters x) and a physical observation of the same quantity. Although other norms may arise, the majority prefer the Euclidean norm, or a weighted version thereof [164]. Thus the problem of concern is to[34]

$$\underset{x \in \mathbb{R}^n}{\text{minimize}} \quad \Phi_1(x) \stackrel{\text{def}}{=} \|r(x)\|. \tag{5.1.1}$$

However, because the Euclidean norm is not everywhere smooth, while its square is, it is usually considered convenient to solve the equivalent *least-squares* problem, that is,

$$\underset{x \in \mathbb{R}^n}{\text{minimize}} \quad \Phi_2(x) \stackrel{\text{def}}{=} \tfrac{1}{2}\|r(x)\|^2 = \tfrac{1}{2}\sum_{i=1}^{m} r_i^2(x), \tag{5.1.2}$$

instead. Such a justification is somewhat spurious since, for smooth r, the only point of derivative discontinuity of $\Phi_1(x)$ occurs if by chance $r(x) = 0$, and, as we will shortly see, it is possible to check separately for this possibility.

Least-squares problems are classified as either linear—in which the parameters may simply be found by solving a suitable system of linear equations—or nonlinear—in the case when at least one of the residuals is a nonlinear function. A least-squares fit is often justified by the assumption that modeling errors are normally distributed, an assumption that is partially justified by the central-limit theorem. For linear models, normally distributed errors lead to the least-squares formulation via maximum likelihood estimation. The formulation above assumes that variances are equal, but this is not a restriction since unequal variances may be included either implicitly (via a weighted Euclidean norm) or explicitly (by weighting the residuals).

Examples of problems of this kind abound [226, Chap. 12]. These range throughout experimental science (computational physics, chemistry, biology, and earth sciences), engineering (in disciplines such as transportation, aerospace, and power systems), and finance (pricing and hedging of financial derivatives). Specific applications arise in exploration seismology, NMR spectroscopy, and piezoelectric crystal identification (used in ultrasound for imaging) in medicine; selection of satellite image sensor characteristics, powering internet mapping services; data assimilation for weather forecasting and climate modeling; and building neural networks for artificial intelligence, to mention but a few.

The reader may well have noticed that there is nothing sacred about the least-squares reformulation of (5.1.1); indeed the problem is also equivalent to minimizing

$$\Phi_\ell(x) \stackrel{\text{def}}{=} \tfrac{1}{\ell}\|r(x)\|^\ell \tag{5.1.3}$$

[34]Please be aware that we are using the Euclidean norm $\|\cdot\| = \|\cdot\|_2$ throughout, not the one-norm $\|\cdot\|_1$!

for any positive integer ℓ chosen so that Φ_ℓ inherits the smoothness of r. The following lemma indicates when this is the case, allowing us to focus on (5.1.3) rather than $\Phi_2(x)$ when it is advantageous to do so.

Lemma 5.1.1: Smoothness of Φ_ℓ

Suppose $r(x) \in C^p$. Then $\Phi_\ell \in C^p$ if ℓ is either an even integer or a real $\ell > 2p$.

Proof. We may write $\Phi_\ell = \phi(\pi(x))$, where $\phi(\alpha) = \frac{1}{\ell}\alpha^{\ell/2}$, $\pi(x) = 2\Phi_2(x)$, and the latter is as smooth as $r(x)$. When ℓ is an even integer, ϕ is infinitely differentiable, and thus by the chain rule, $\Phi_\ell \in C^p$. For general ℓ, it follows from the chain rule that $\nabla_x^p \Phi_\ell$ comprises a sum of products of derivatives of the univariate ϕ up to order p evaluated at $\alpha = \pi(x)$ and derivative tensors of $\pi(x) \in C^p$ up to the same order; this is a multivariate extension [133, 245] of the famous Faá di Bruno formula [162] for the derivatives of composite univariate functions. The kth derivative $\phi^k(\alpha) = [\frac{1}{2}\prod_{j=1}^{k-1}(\frac{\ell}{2} - j)]\alpha^{\ell/2-k}$, which is continuous for all $k \leq p$ so long as $\ell/2 - p \geq 0$, and thus $\nabla_x^p \Phi_\ell(x)$ is continuous in this case. □

The requirement stated in the lemma may be reduced to $\ell > p$ (and no further since the pth derivative of the univariate example function $|\alpha|^\ell$ is discontinuous at the origin when $\ell \leq p$) because the potential nondifferentiability at the origin of $\alpha^{\ell/2}$ may be countered by the vanishing of the gradient of $\pi(x)$ when $r(x) = 0$. Since the proof of this is very complicated (due to the combinatorial aspects of the products in Faá Di Bruno's formula), we do not give this result, since we shall not rely on it. All we actually require is that we can pick a large enough ℓ for Φ_ℓ to be p-times continuously differentiable, and actually we will only consider even integers ℓ.

Superficially, we already know a great deal about the evaluation complexity of finding an approximate local solution of (5.1.2) since we may apply the ARp algorithm (p. 100) to obtain an ϵ-approximate qth-order minimizer of $\Phi_\ell(x)$ for $q \leq 3$ so long as Φ_ℓ and its derivatives satisfy the required smoothness assumptions. Indeed, Theorem 4.1.5 implies that the ARp algorithm may require $\mathcal{O}(\max_{j\in[1:q]} \epsilon_j^{-(p+1)/(p-j+1)})$ evaluations of f and its first p derivatives to find an ϵ-approximate qth-order minimizer; actually, we shall see later (Chapter 12) that similar bounds may be derived if we seek higher-order approximate minimizers.

This, though, is far from the end of the story. Consider $\Phi_1(x) = \|r(x)\|$, and suppose that $\|r(x)\| > 0$ and thus that $\Phi_1(x)$ is locally differentiable. Then, since $\Phi_1(x) = (\ell\Phi_\ell(x))^{1/\ell}$, simple differentiation using the chain rule yields

$$\begin{aligned}
\nabla_x^1 \Phi_1(x) &= (\ell\Phi_\ell(x))^{-(\ell-1)/\ell}\nabla_x^1\Phi_\ell(x) \\
&= \frac{\nabla_x^1\Phi_\ell(x)}{\|r(x)\|^{\ell-1}}, \qquad (5.1.4)\\
\nabla_x^2 \Phi_1(x) &= -(\ell-1)(\ell\Phi_\ell(x))^{-(2\ell-1)/\ell}\nabla_x^1\Phi_\ell(x)\otimes\nabla_x^1\Phi_\ell(x) \\
&\quad + (\ell\Phi_\ell(x))^{-(\ell-1)/\ell}\nabla_x^2\Phi_\ell(x) \\
&= -\frac{(\ell-1)}{\|r(x)\|}\left(\frac{\nabla_x^1\Phi_\ell(x)}{\|r(x)\|^{\ell-1}}\right)\otimes\left(\frac{\nabla_x^1\Phi_\ell(x)}{\|r(x)\|^{\ell-1}}\right) + \frac{\nabla_x^2\Phi_\ell(x)}{\|r(x)\|^{\ell-1}}, \quad \text{and}
\end{aligned}$$

$$\begin{aligned}
\nabla_x^3 \Phi_1(x) &= (\ell-1)(2\ell-1)(\ell\Phi_\ell(x))^{-(3\ell-1)/\ell} \nabla_x^1 \Phi_\ell(x) \otimes \nabla_x^1 \Phi_\ell(x) \otimes \nabla_x^1 \Phi_\ell(x) \\
&\quad - 3(\ell-1)(\ell\Phi_\ell(x))^{-(2\ell-1)/\ell} \nabla_x^2 \Phi_\ell(x) \otimes \nabla_x^1 \Phi_\ell(x) \\
&\quad + (\ell\Phi_\ell(x))^{-(\ell-1)/\ell} \nabla_x^3 \Phi_\ell(x) \\
&= \frac{(\ell-1)(2\ell-1)}{\|r(x)\|^2} \left(\frac{\nabla_x^1 \Phi_\ell(x)}{\|r(x)\|^{\ell-1}} \right) \otimes \left(\frac{\nabla_x^1 \Phi_\ell(x)}{\|r(x)\|^{\ell-1}} \right) \otimes \left(\frac{\nabla_x^1 \Phi_\ell(x)}{\|r(x)\|^{\ell-1}} \right) \\
&\quad - \frac{3(\ell-1)}{\|r(x)\|} \left(\frac{\nabla_x^2 \Phi_\ell(x)}{\|r(x)\|^{\ell-1}} \right) \otimes \left(\frac{\nabla_x^1 \Phi_\ell(x)}{\|r(x)\|^{\ell-1}} \right) + \frac{\nabla_x^3 \Phi_\ell(x)}{\|r(x)\|^{\ell-1}},
\end{aligned}$$

where $\otimes$ is the tensor product. Now suppose that x_* is a first-order critical point of $\Phi_\ell(x)$ for which $\|r(x)\| > 0$. Then[35]

$$\nabla_x^1 \Phi_1(x_*) = \frac{\nabla_x^1 \Phi_\ell(x_*)}{\|r(x_*)\|^{\ell-1}} = 0, \quad \nabla_x^2 \Phi_1(x_*) = \frac{\nabla_x^2 \Phi_\ell(x_*)}{\|r(x_*)\|^{\ell-1}}, \quad \text{and } \nabla_x^3 \Phi_1(x_*) = \frac{\nabla_x^3 \Phi_\ell(x_*)}{\|r(x_*)\|^{\ell-1}}, \tag{5.1.5}$$

and thus x_* is first-, second-, or third-order critical for Φ_1 whenever it is for Φ_ℓ. More importantly, (5.1.5) suggests that if we wish to define low-order *approximate* criticality measures for Φ_1 using those for Φ_ℓ, we need to take into account the $\|r(x)\|^{\ell-1}$ term in the denominator of each. All of this is predicated on the assumption that $\Phi_\ell(x_*) \neq 0$. But if $\Phi_\ell(x_*) = 0$, x_* is a global minimizer of $\Phi_1(x)$, and there is no need to check for higher-order criticality. Thus we say that x_ϵ is an *ϵ-approximate global minimizer* of $\Phi_1(x)$ if

$$\|r(x_\epsilon)\| \leq \epsilon_0,$$

while it is an *ϵ-approximate first-order minimizer* of Φ_1 if instead

$$\psi_{\ell,1}(x_\epsilon) \leq \epsilon_1, \text{ where } \psi_{\ell,1}(x) \stackrel{\text{def}}{=} \frac{\|\nabla_x^1 \Phi_\ell(x)\|}{\|r(x)\|^{\ell-1}},$$

for prespecified $\epsilon_0, \epsilon_1 > 0$. An ϵ-approximate first-order minimizer of Φ_1 is an *ϵ-approximate second-order* one if

$$\psi_{\ell,2}(x_\epsilon) \geq -\epsilon_2, \text{ where } \psi_{\ell,2}(x) \stackrel{\text{def}}{=} \frac{\lambda_{\min}[\nabla_x^2 \Phi_\ell(x)]}{\|r(x)\|^{\ell-1}},$$

and an *ϵ-approximate strict local minimizer* if

$$\psi_{\ell,2}(x_\epsilon) > \epsilon_2$$

for a given ϵ_2. Finally, any ϵ-approximate second-order minimizer of Φ_1 is an *ϵ-approximate third-order one* if

$$\psi_{\ell,3}(x_\epsilon, d) \leq \epsilon_3 \text{ for all } d \in \mathcal{Z}_\epsilon, \text{ where } \psi_{\ell,3}(x,d) \stackrel{\text{def}}{=} \left| \frac{\nabla_x^3 \Phi_\ell(x)[d]^3}{\|r(x)\|^{\ell-1}} \right|, \tag{5.1.6}$$

whenever

$$\mathcal{Z}_\epsilon \stackrel{\text{def}}{=} \{d : \|d\| = 1 \text{ and } \psi_{\ell,2}(x_\epsilon, d) \leq \epsilon_2\} \neq \emptyset, \text{ where } \psi_{\ell,2}(x,d) \stackrel{\text{def}}{=} \frac{|\nabla_x^2 \Phi_\ell(x)[d]^2|}{\|r(x)\|^{\ell-1}}, \tag{5.1.7}$$

[35] Interestingly, $\nabla_x^4 \Phi_1(x)$ involves a term $\nabla_x^2 \Phi_\ell(x) \otimes \nabla_x^2 \Phi_\ell(x)$ as well as $\nabla_x^4 \Phi_\ell(x)$ and those containing products with $\nabla_x^1 \Phi_\ell(x)$, and thus $\nabla_x^4 \Phi_1(x_*)$ is not $\nabla_x^4 \Phi_\ell(x_*)/\|r(x_*)\|^{\ell-1}$.

for given $\epsilon_2, \epsilon_3 > 0$; we have introduced the notation $\psi_{\ell,1}$, $\psi_{\ell,2}$, and $\psi_{\ell,3}$ for brevity in what follows.

Armed with these approximate minimization conditions, we now focus on an algorithm that aims to satisfy them.

Algorithm 5.1.1: Adaptive-regularization algorithm with pth-order models for least Euclidean norm problems (ARpLE)

Apply Algorithm 4.1.1 (ARp, on p. 100) to $f(x) = \Phi_\ell(x)$ for a given $\ell \geq 2$ using the model

$$m_k(s) = T_{\Phi_\ell,p}(x_k, s) + \frac{\sigma_k}{(p+1)!}\|s\|^{p+1}, \tag{5.1.8}$$

but replace the termination tests (Step 1) with the following:

Step 1: Test for termination. If

$$\|r(x_k)\| \leq \epsilon_0, \tag{5.1.9}$$

terminate with the ϵ_0-approximate global minimizer $x_\epsilon = x_k$.

If $\epsilon_2 = \epsilon_3 = \infty$ and

$$\psi_{\ell,1}(x_k) \leq \epsilon_1, \tag{5.1.10}$$

terminate with the ϵ_1-approximate first-order minimizer $x_\epsilon = x_k$.

If $\epsilon_2 < \infty$, (5.1.10) holds, and

$$\psi_{\ell,2}(x_k) > \epsilon_2, \tag{5.1.11}$$

terminate with the (ϵ_1, ϵ_2)-approximate strict local minimizer $x_\epsilon = x_k$.

If $\epsilon_2 < \infty$ and $\epsilon_3 = \infty$, (5.1.10) holds, and (5.1.11) fails but

$$\psi_{\ell,2}(x_k) \geq -\epsilon_2, \tag{5.1.12}$$

terminate with the (ϵ_1, ϵ_2)-approximate second-order minimizer $x_\epsilon = x_k$.

If $\epsilon_3 < \infty$, (5.1.10), (5.1.12), and $\mathcal{M}_k \neq \emptyset$ hold, where $\mathcal{M}_k$ is defined as in (4.1.14), and

$$\psi_{\ell,3}(x_k, d) \leq \epsilon_3 \tag{5.1.13}$$

and

$$\psi_{\ell,2}(x_k, d) \leq \epsilon_2 \tag{5.1.14}$$

for all $d \in \mathcal{M}_k$, terminate with the $(\epsilon_1, \epsilon_2, \epsilon_3)$-approximate third-order minimizer $x_\epsilon = x_k$.

Just as when we applied ARp to general objectives, the third-order conditions (5.1.13) and (5.1.14) differ from the requirement (5.1.6) in that the latter can only be guaranteed along directions of small curvature for the model, namely, those in $\mathcal{M}_k$. Again, we will be able to show that at termination, $\mathcal{M}_k \subseteq \mathcal{Z}_\epsilon$, where $\mathcal{Z}_\epsilon$ is defined as in (5.1.7), and so model directions of small curvature also lead to small curvature for the true objective.

Given p, we will consider the choice of ℓ in some detail in Section 5.1.3. For the time being, it suffices to say that if one wishes to obtain a first-, second-, or third-order approximate minimizer, it is clear that p should be at least 1, 2, or 3 (respectively). If p

is a multiple of 1, 2, or 6, then $\ell = p$, $\frac{1}{2}p$, and $\frac{1}{3}p$ (respectively) is reasonable. The larger the p, the better the (order of) worst-case evaluation complexity, but at the cost of more expensive subproblem solves.

5.1.2 Bounds on the residual decrease

Many of the pieces are already in place; indeed, had we been aiming to satisfy the original termination tests (Step 1 of Algorithm 4.1.1), we would have nothing further to say.

In order to proceed, we need an appropriate version of **f.DpL** for the case when $f(x) = \Phi_\ell(x)$.

Φ_ℓ.DpL

$\Phi_\ell(x)$ is p-times continuously differentiable in $\mathbb{R}^n$ and its pth-order derivative tensor $\nabla_x^p \Phi_\ell(x)$ is Lipschitz continuous on $\mathbb{R}^n$; that is, there exists a constant $L_{\Phi_\ell,p} \geq 0$ such that, for all $x, y \in \mathbb{R}^n$,

$$\|\nabla_x^p \Phi_\ell(x) - \nabla_x^p \Phi_\ell(y)\| \leq L_{\Phi_\ell,p}\|x - y\|,$$

where the norm on the left-hand side of this inequality is the tensor norm (A.7).

At this point, the reader might reasonably object that $\Phi_\ell(x)$ is actually composed of the action of one function, $\phi_\ell(r) = \|r\|^\ell/\ell$, on another, $r(x)$, and in this case one might more naturally make assumptions about $r(x)$ and use knowledge of (the fixed form of) $\phi_\ell(r)$ to deduce properties of the composition. Our view is that although this is possible, it can easily become ferociously complicated[36] and actually adds no significant extra insight since derived bounds often become weaker. For example, more generally if $f(x) = h(r(x)$, where $h : \mathbb{R}^m \to \mathbb{R}$ is smooth, the chain rule gives that $\nabla_x f(x) = h^{(1)}(r(x))\nabla_x r(x)$. Thus the gradient of f is sure to be Lipschitz if $|h^{(1)}(r(x))|$ is bounded and $r(x)$ is Lipschitz, since

$$\|h^{(1)}(r(x))\nabla_x r(x) - h^{(1)}(r(y))\nabla_x r(y)\| \leq \max_x |h^{(1)}(r(x))| \|\nabla_x r(x) - \nabla_x r(y)\|.$$

But this is only a necessary condition,[37] and the inequality above may well be weak.

With this in place, we have an immediate corollary of Lemma 4.1.4.

Lemma 5.1.2: Bounds on the successful function decrease for ARpLE

*Suppose that **Φ_ℓ.DpL** holds and that iteration k of the ARpLE algorithm is successful. Then there are constants κ_1, κ_2, $\kappa_3 > 0$ such that*

$$\|r(x_k)\|^\ell - \|r(x_{k+1})\|^\ell \geq \kappa_1 \ell \left[\psi_{\ell,1}(x_{k+1})\right]^{\frac{p+1}{p}} \|r(x_{k+1})\|^{\frac{(\ell-1)(p+1)}{p}}, \tag{5.1.15}$$

together with

$$\|r(x_k)\|^\ell - \|r(x_{k+1})\|^\ell \geq \kappa_2 \ell \max\left[0, -\psi_{\ell,2}(x_{k+1})\right]^{\frac{p+1}{p-1}} \|r(x_{k+1})\|^{\frac{(\ell-1)(p+1)}{p-1}} \tag{5.1.16}$$

[36] We do provide details for the $\ell = 2$ case in Lemma 14.3.6 (p. 416), but, as predicted, even here the details are cumbersome.

[37] See also Lemma 14.3.6 (p. 416) in a slightly more general context.

when $\epsilon_2 < \infty$, *and additionally*

$$\|r(x_k)\|^\ell - \|r(x_{k+1})\|^\ell \geq \kappa_3 \ell \left[\psi_{\ell,3}(x_{k+1},d)\right]^{\frac{p+1}{p-2}} \|r(x_{k+1})\|^{\frac{(\ell-1)(p+1)}{p-2}} \tag{5.1.17}$$

for all d *in the set* $\mathcal{M}_{k+1}$ *given by (4.1.14) when* $\epsilon_3 < \infty$. *Moreover, for any such* d,

$$\psi_{\ell,2}(x_{k+1},d)\,\|r(x_{k+1})\|^{\ell-1} \leq \frac{1}{\kappa_{s2}}\|s_k\|^{p-1}, \text{ where } \kappa_{s2} \stackrel{\text{def}}{=} \frac{(p-1)!}{L_{\Phi_\ell,p}+\sigma_{\max}+\theta_2(p-1)!}. \tag{5.1.18}$$

Proof. The inequalities (5.1.15)–(5.1.17) follow immediately by applying Lemma 4.1.4 when $f(x) = \Phi_\ell(x) \equiv \|r(x)\|^\ell/\ell$. The constants κ_1, κ_2, and κ_3 are those given in (4.1.32), (4.1.34), and (4.1.36), using $\sigma_{\max}$ from (4.1.19), but with the Lipschitz constant $L_{\Phi_\ell,p}$ in place of $L_{f,p}$ throughout. Inequality (5.1.18) follows directly from (4.1.23). □

Our analysis hinges on a trivial bound on cascading powers of differences of a pair of ordered, positive numbers.

Lemma 5.1.3: Cascading differences

[Appendix to Chapter 5, p. 463]: Suppose that $a \geq b \geq 0$. *Then*

$$a^{1/2^i} - b^{1/2^i} \geq \frac{a^\ell - b^\ell}{2^i \ell a^{\frac{2^i-1}{2^i}+\ell-1}}$$

for all integers $i \geq 0$ *and* $\ell \geq 1$.

Next, we combine the two previous results to obtain more useful bounds on the guaranteed residual decrease following successful iterations.

Lemma 5.1.4: Improved bounds on the function decrease for ARpLE

Suppose that $\mathbf{\Phi_\ell.DpL}$ *holds, that iteration* k *of the* ARpLE *algorithm is successful, and that*

$$\|r(x_{k+1})\| > \beta\|r(x_k)\| \tag{5.1.19}$$

for some constant $\beta \in (0,1)$. *Then, for any integer* $i \geq 0$, *there are constants* κ_{r1}, κ_{r2}, $\kappa_{r3} > 0$ *such that*

$$\|r(x_k)\|^{1/2^i} - \|r(x_{k+1})\|^{1/2^i} \geq \kappa_{r1}\frac{\left[\psi_{\ell,1}(x_{k+1})\right]^{\frac{p+1}{p}}}{\|r(x_k)\|^{\iota_1(p,\ell,i)}}, \tag{5.1.20}$$

where

$$\iota_1(p,\ell,i) \stackrel{\text{def}}{=} \frac{2^i-1}{2^i} - \frac{\ell-1}{p} \text{ and } \kappa_{r1} \stackrel{\text{def}}{=} \frac{\kappa_1\beta^{\frac{(\ell-1)(p+1)}{p}}}{2^i}, \tag{5.1.21}$$

together with

$$\|r(x_k)\|^{1/2^i} - \|r(x_{k+1})\|^{1/2^i} \geq \kappa_{r2}\frac{\max\left[0,-\psi_{\ell,2}(x_{k+1})\right]^{\frac{p+1}{p-1}}}{\|r(x_k)\|^{\iota_2(p,\ell,i)}}, \tag{5.1.22}$$

where

$$\iota_2(p,\ell,i) \stackrel{\rm def}{=} \frac{2^i-1}{2^i} - 2\frac{\ell-1}{p-1} \text{ and } \kappa_{r2} \stackrel{\rm def}{=} \frac{\kappa_2\beta^{\frac{(\ell-1)(p+1)}{p-1}}}{2^i} \tag{5.1.23}$$

when $\epsilon_2 < \infty$, *and additionally*

$$\|r(x_k)\|^{1/2^i} - \|r(x_{k+1})\|^{1/2^i} \geq \kappa_{r3}\frac{\left[\psi_{\ell,3}(x_{k+1},d)\right]^{\frac{p+1}{p-2}}}{\|r(x_k)\|^{\iota_3(p,\ell,i)}} \tag{5.1.24}$$

for all d *in the set* $\mathcal{M}_{k+1}$ *given by (4.1.14), where*

$$\iota_3(p,\ell,i) \stackrel{\rm def}{=} \frac{2^i-1}{2^i} - 3\frac{\ell-1}{p-2} \text{ and } \kappa_{r3} \stackrel{\rm def}{=} \frac{\kappa_3\beta^{\frac{(\ell-1)(p+1)}{p-2}}}{2^i} \tag{5.1.25}$$

when $\epsilon_3 < \infty$. *By contrast, if*

$$\|r(x_{k+1})\| \leq \beta\|r(x_k)\|, \tag{5.1.26}$$

then

$$\|r(x_k)\|^{1/2^i} - \|r(x_{k+1})\|^{1/2^i} \geq \kappa_{r0}\|r(x_k)\|^{1/2^i} \geq \kappa_{r0}\|r(x_{k+1})\|^{1/2^i}, \tag{5.1.27}$$

where

$$\kappa_{r0} \stackrel{\rm def}{=} 1-\beta^{1/2^i}. \tag{5.1.28}$$

Proof. Letting $a = \|r(x_k)\|$ and $b = \|r(x_{k+1})\|$ and recalling that the ARpLE algorithm guarantees that $\|r(x_k)\| \geq \|r(x_{k+1})\|$, Lemma 5.1.3 implies that

$$\|r(x_k)\|^{1/2^i} - \|r(x_{k+1})\|^{1/2^i} \geq \frac{\|r(x_k)\|^\ell - \|r(x_{k+1})\|^\ell}{2^i\ell\|r(x_k)\|^{\frac{2^i-1}{2^i}+\ell-1}}. \tag{5.1.29}$$

Combining this with (5.1.15) and using (5.1.19), we deduce the bound (5.1.20). If $\epsilon_2 < \infty$, it follows from (5.1.16), (5.1.19), and (5.1.29) that (5.1.22) holds, while if $\epsilon_3 < \infty$, the inequalities (5.1.17), (5.1.19), and (5.1.29) together imply (5.1.24) for all $d \in \mathcal{M}_{k+1}$.

Conversely, if (5.1.26) holds, then immediately

$$\|r(x_{k+1})\|^{1/2^i} \leq \beta^{1/2^i}\|r(x_k)\|^{1/2^i}.$$

It then follows directly that

$$\|r(x_k)\|^{1/2^i} - \|r(x_{k+1})\|^{1/2^i} \geq (1-\beta^{1/2^i})\|r(x_k)\|^{1/2^i} \geq (1-\beta^{1/2^i})\|r(x_{k+1})\|^{1/2^i},$$

which is (5.1.27), since $\|r(x_k)\| \geq \|r(x_{k+1})\|$. □

5.1.3 The choice of critical parameters

Given p, we now turn to the vital choice of ℓ and consider first the case in which we are simply seeking an ϵ_1-approximate first-order minimizer of $\Phi_1(x)$, that is, $\epsilon_2 = \epsilon_3 = \infty$. Thus we focus on the bound (5.1.20). As we have seen many times in our development of evaluation-complexity bounds, a common ploy is to ensure that a reduction in the objective function limits the criticality measure in question. In (5.1.20), we provide such a relationship, but we must account for the unfortunate extra term involving a power of $\|r(x_k)\|$ in the denominator of the right-hand side.

Suppose that the ARpLE algorithm has not terminated with the iterate x_k. In particular, we must have that

$$\epsilon_0 < \|r(x_k)\| \leq \|r(x_0)\|$$

from (5.1.9) and the fact that the algorithm produces successively improving iterates. Thus superficially the $\|r(x_k)\|$ term in the bound (5.1.20) might be replaced by one involving either ϵ_0 if its exponent $\iota_1(p,\ell,i)$, defined by (5.1.21), is negative or $\|r(x_0)\|$ if $\iota_1(p,\ell,i)$ is positive. However, our aspiration is to avoid an ϵ_0 dependence if at all possible, because otherwise such a term would appear in any complexity bound based on (5.1.20). Thus our goal must be to pick ℓ and i so that $\iota_1(p,\ell,i)$ is nonnegative.

Arguably the simplest way to deal with this is to remove the $\|r(x_k)\|$ term entirely, and we can do this by choosing p, ℓ, and i so that the exponent $\iota_1(p,\ell,i)$ is zero, which requires that

$$\ell = 1 + (2^i - 1)\frac{p}{2^i} \tag{5.1.30}$$

be an even positive integer. We use the following elementary result.

Lemma 5.1.5: Odd factorization of a positive integer

[Appendix to Chapter 5, p. 464]: Let $p \in \mathbb{N}$. Then p may be expressed uniquely as the product of an odd natural number and a power of 2, i.e.,

$$p = o \cdot 2^i \tag{5.1.31}$$

for some odd positive integer o and $i \in \mathbb{N}^0$.

Note that to find o and i from p, one simply successively divides p by increasing powers of 2 until the result is no longer an integer. Given o and i, we now choose

$$\ell = 1 + o \cdot (2^i - 1). \tag{5.1.32}$$

Observe that if p is odd, $i = 0$ and $\ell = 1$, while if p is even, $i > 0$ and ℓ is even. Since we require Φ_ℓ to be smooth, we cannot allow ℓ to be odd (unless $\ell > p$), and thus henceforth we restrict our attention to even p. We illustrate the relationship between a given p and the implied o and i from (5.1.31) and ℓ from (5.1.32) for the first 16 positive even integers in Table 5.1.1. It is easy to show that $1 + p/2 \leq \ell \leq p$, and clearly $\ell = p$ if and only if $p = 2^i$.

We turn now to the case in which we seek an (ϵ_1, ϵ_2)-approximate second-order minimizer of $\Phi_1(x)$, that is, $\epsilon_2 < \infty$ but $\epsilon_3 = \infty$. In this case we will make use of (5.1.20) and

p	2	4	6	8	10	12	14	16	18	20	22	24	26	28	30	32
o	1	1	3	1	5	3	7	1	9	5	11	3	13	7	15	1
i	1	2	1	3	1	2	1	4	1	2	1	3	1	2	1	5
ℓ	2	4	4	8	6	10	8	16	10	16	12	22	14	22	16	32

Table 5.1.1. The selected o, i, and ℓ suggested by (5.1.31) and (5.1.32) to make the exponent $\iota_1(p,\ell,i) = 0$ (first-order case), as a function of the Taylor degree p for the first 16 acceptable positive integers. This and subsequent tables should be read from top to bottom.

(5.1.22), both of which mix an appropriate (respectively first- and second-order) criticality measure with an unwanted power of $\|r(x_k)\|$. Suppose that we focus on (5.1.22) rather than (5.1.20) and pick p, ℓ, and i so that the exponent $\iota_2(p,\ell,i)$, defined as in (5.1.23), of $\|r(x_{k+1})\|$ in (5.1.22) is zero, which requires that

$$\ell = 1 + (2^i - 1)\frac{(p-1)}{2^{i+1}}$$

be an even positive integer. Unlike (5.1.30), this limits p to be odd. Lemma 5.1.5 shows that

$$p - 1 = o_2 \cdot 2^{i+1}$$

for some odd positive integer o_2, and from this we may choose

$$\ell = 1 + o_2 \cdot (2^i - 1). \tag{5.1.33}$$

Note now that, as we illustrate in Table 5.1.2, every p for which $p-1$ is divisible by four (with a dash indicating an unsuitable $i=0$) leads to a suitable $\ell > 1$, and that $\ell = 2^i$ if (and only if) $p = 2^{i+1} + 1$.

p	3	5	7	9	11	13	15	17	19	21	23	25	27	29	31	33
o_2	1	1	3	1	5	3	7	1	9	5	11	3	13	7	15	1
i	0	1	0	2	0	1	0	3	0	1	0	2	0	1	0	4
ℓ	-	2	-	4	-	4	-	8	-	6	-	10	-	8	-	16

Table 5.1.2. The selected o_2, i, and ℓ suggested by (5.1.31) and (5.1.33) to make the exponent $\iota_2(p,\ell,i) = 0$ (second-order case), as a function of the Taylor degree for the first 16 acceptable positive integers.

Of course an ℓ chosen to make $\iota_2(p,\ell,i)$ zero will not do the same for $\iota_1(p,\ell,i)$, but fortunately this is immaterial since trivially

$$\iota_1(p,\ell,i) = \iota_2(p,\ell,i) + \frac{(\ell-1)(p+1)}{p(p-1)} \tag{5.1.34}$$

and thus $\iota_1(p,\ell,i) > 0$ whenever $\iota_2(p,\ell,i) = 0$ and $\ell > 1$. Our forthcoming analysis will rely on this.

It remains to consider the most general case in which we seek an $(\epsilon_1,\epsilon_2,\epsilon_3)$-approximate third-order minimizer of $\Phi_1(x)$, that is, when both ϵ_2 and $\epsilon_3 < \infty$. We focus first on the bound (5.1.24), rather than (5.1.20) or (5.1.22), and once again we aim to remove the dependence of the right-hand side of this on $\|r(x_k)\|$ by ensuring that its exponent $\iota_3(p,\ell,i)$, as defined in (5.1.25), is zero, which requires that

$$\ell = 1 + \frac{(2^i-1)}{3}\frac{(p-2)}{2^i}$$

be an even positive integer. For this to hold, we need $2^i - 1$ to be divisible by 3, which fortunately occurs for all even $i = 2j$, since then

$$2^i - 1 = 2^{2j} - 1 = (2^2-1)(2^{2(j-1)} + \cdots + 2^2 + 1) = 3\sum_{k=0}^{j-1} 2^{2k}.$$

We also require that $p-2$ be divisible by 2^{2j}. Again using Lemma 5.1.5, we have that

$$p - 2 = o_3 \cdot 2^i$$

for some odd positive integer o_3, and we are interested in those for which $i \geq 2$ is even, for which we use

$$\ell = 1 + o_3 \cdot \sum_{k=0}^{j-1} 2^{2k}, \quad \text{where } j = \lfloor i/2 \rfloor. \tag{5.1.35}$$

As before, we illustrate small examples in Table 5.1.3 as follows, with a dash indicating that i is not divisible by 2 and is thus unsuitable. Indeed, in view of the preceding discussion, p will result in a suitable ℓ if and only if the power of two in the prime factorization of $p-2$ is positive and even. Although there are clearly now fewer candidates, we have specifically $\ell = 1 + \sum_{k=0}^{j-1} 2^{2k}$ when $p = 2 + s^{2j}$ for all $j > 0$.

p	4	6	8	10	12	14	16	18	20	22	24	26	28	30	32	34
o_3	1	1	3	1	5	3	7	1	9	5	11	3	13	7	15	1
i	1	2	1	3	1	2	1	4	1	2	1	3	1	2	1	5
j	-	1	-	-	-	1	-	2	-	1	-	-	-	1	-	-
ℓ	-	2	-	-	-	4	-	6	-	6	-	-	-	8	-	-

Table 5.1.3. The selected o_3, i, and ℓ suggested by (5.1.31) and (5.1.35) to make the exponent $\iota_3(p, \ell, i) = 0$ (third-order case), as a function of the Taylor degree p for the first 16 acceptable positive integers.

Once more, the reader might be concerned that an ℓ chosen to make $\iota_3(p, \ell, i)$ zero will not do the same for either $\iota_1(p, \ell, i)$ or $\iota_2(p, \ell, i)$. As before, though, such a worry may be alleviated because trivially the exponents

$$\iota_1(p, \ell, i) = \iota_3(p, \ell, i) + 2\frac{(\ell-1)(p+1)}{p(p-2)} \tag{5.1.36}$$

and

$$\iota_2(p, \ell, i) = \iota_3(p, \ell, i) + \frac{(\ell-1)(p+1)}{(p-1)(p-2)} \tag{5.1.37}$$

and thus $\iota_1(p, \ell, i) > 0$ and $\iota_2(p, \ell, i) > 0$ whenever $\iota_3(p, \ell, i) = 0$ and $\ell > 1$.

There are other ways of ensuring that there are no negative powers of $\|r(x_k)\|$ on the right-hand sides of (5.1.20), (5.1.22), and (5.1.24). Suppose as before that we are just seeking an ϵ_1-approximate first-order minimizer of $\Phi_1(x)$, and consider the exponent $\iota_1(p, \ell, i)$ given by (5.1.21) of $\|r(x_k)\|$ in (5.1.20). Clearly as i increases to infinity, $\iota_1(p, \ell, i)$ approaches $1 - (\ell-1)/p$, and thus if $\ell = p$, $\iota_1(p, \ell, i)$ will be positive for all sufficiently large i. Specifically, if $\ell = p$ and $i > \log_2 p$, it is easy to show (see Lemma B.5.1, p. 463) that $\iota_1(p, \ell, i)$ is positive. Similarly for an (ϵ_1, ϵ_2)-approximate second-order minimizer, the exponent $\iota_2(p, \ell, i)$ from (5.1.23), and hence $\iota_1(p, \ell, i)$ because of (5.1.34), are positive if p is even, $\ell = \frac{1}{2}p$, and $i > \log_2(p-1)$, while if we seek an $(\epsilon_1, \epsilon_2, \epsilon_3)$-approximate third-order minimizer, the exponents $\iota_3(p, \ell, i)$ from (5.1.25), and hence both $\iota_1(p, \ell, i)$ and $\iota_2(p, \ell, i)$ because of (5.1.36) and (5.1.37), are all positive if p is a multiple of three, $\ell = \frac{1}{3}p$, and $i > \log_2(p-2)$ (again, see Lemma B.5.1). So, for example, if

$$\ell = p \quad \text{and} \quad i = 1 + \lfloor \log_2 p \rfloor,$$

the exponent of $\|r(x_k)\|$ in (5.1.20) is

$$\iota_1(p, \ell, i) = \frac{1}{2p}$$

when p is a positive power of two.

We summarize our discussions above as follows.

Lemma 5.1.6: Good exponents of $\|r(x_k)\|$ in residual-decrease bounds

Let $p \in \mathbb{N}$. If

$$\ell = p \text{ and } i > \log_2 p, \tag{5.1.38}$$

then the exponent (5.1.21) of $\|r(x_k)\|$ in (5.1.20) is strictly positive. If p is even and

$$\ell = 1 + o \cdot (2^i - 1), \text{ where } p = o \cdot 2^i, \tag{5.1.39}$$

then the same exponent is zero and $\ell > 1$. If p is even and

$$\ell = \tfrac{1}{2}p \text{ and } i > \log_2(p-1), \tag{5.1.40}$$

then (5.1.21) and the exponent (5.1.23) of $\|r(x_k)\|$ in (5.1.22) are both strictly positive. If $p-1$ is divisible by four and

$$\ell = 1 + o_2 \cdot (2^i - 1), \text{ where } p - 1 = o_2 \cdot 2^{i+1}, \tag{5.1.41}$$

then (5.1.23) is zero, (5.1.21) is strictly positive, and $\ell > 1$. If p is a multiple of three and

$$\ell = \tfrac{1}{3}p \text{ and } i > \log_2(p-2), \tag{5.1.42}$$

then (5.1.21), (5.1.23), and the exponent (5.1.25) of $\|r(x_k)\|$ in (5.1.24) are all strictly positive. If the power of two in the prime factorization of $p-2$ is positive and even and

$$\ell = 1 + o_3 \cdot \sum_{k=0}^{j-1} 2^{2k}, \text{ where } p - 2 = o_3 \cdot 2^i \text{ and } j = \lfloor i/2 \rfloor, \tag{5.1.43}$$

then (5.1.25) is zero, (5.1.21) and (5.1.23) are strictly positive, and $\ell > 1$.

5.1.4 Evaluation-complexity bounds

We now provide our main evaluation-complexity result for problem (5.1.1).

Theorem 5.1.7: Complexity of finding first- to third-order critical points of $\|r(x)\|$ by the ARpLE algorithm

Suppose that $i \geq 0$ and $\ell > 1$ satisfy either (5.1.42) or (5.1.43) for an appropriate p, that $\mathbf{\Phi_\ell.DpL}$ holds, and that $\beta \in (0,1)$ is a fixed, problem-independent constant. Then there exist positive constants $\kappa^S_{\text{ARpLE,3}}$, $\kappa^A_{\text{ARpLE,3}}$, and κ^C_{ARpLE} such that, for any $(\epsilon_0, \epsilon_1, \epsilon_2, \epsilon_3) \in (0,1]^4$, the ARpLE algorithm (p. 123) requires at most

$$\begin{aligned}\#^F_{\text{ARpLE,3}} &\stackrel{\text{def}}{=} \kappa^C_{\text{ARpLE,3}} \frac{\max\left[\|r(x_0)\|^{\frac{1}{2^i}}, \|r(x_0)\|^{1-\frac{\ell-1}{p}}, \|r(x_0)\|^{1-2\frac{\ell-1}{p-1}}, \|r(x_0)\|^{1-3\frac{\ell-1}{p-2}}\right]}{\min\left[\epsilon_0^{\frac{1}{2^i}}, \epsilon_1^{\frac{p+1}{p}}, \epsilon_2^{\frac{p+1}{p-1}}, \epsilon_3^{\frac{p+1}{p-2}}\right]} + \kappa^C_{\text{ARpLE}} \\ &= \mathcal{O}\left(\max\left[\epsilon_0^{-\frac{1}{2^i}}, \epsilon_1^{-\frac{p+1}{p}}, \epsilon_2^{-\frac{p+1}{p-1}}, \epsilon_3^{-\frac{p+1}{p-2}}\right]\right)\end{aligned} \tag{5.1.44}$$

evaluations of r and at most

$$\#^D_{\text{ARpLE,3}} \stackrel{\text{def}}{=} \kappa^S_{\text{ARpLE,3}} \frac{\max\left[\|r(x_0)\|^{\frac{1}{2^i}}, \|r(x_0)\|^{1-\frac{\ell-1}{p}}, \|r(x_0)\|^{1-2\frac{\ell-1}{p-1}}, \|r(x_0)\|^{1-3\frac{\ell-1}{p-2}}\right]}{\min\left[\epsilon_0^{\frac{1}{2^i}}, \epsilon_1^{\frac{p+1}{p}}, \epsilon_2^{\frac{p+1}{p-1}}, \epsilon_3^{\frac{p+1}{p-2}}\right]} + 2$$
$$= \mathcal{O}\left(\max\left[\epsilon_0^{-\frac{1}{2^i}}, \epsilon_1^{-\frac{p+1}{p}}, \epsilon_2^{-\frac{p+1}{p-1}}, \epsilon_3^{-\frac{p+1}{p-2}}\right]\right) \tag{5.1.45}$$

evaluations of derivatives of order one to p of r to produce an iterate x_ϵ such that $\|r(x_\epsilon)\| \le \epsilon_0$ or (together) $\psi_{\ell,1}(x_\epsilon) \le \epsilon_1$, $\psi_{\ell,2}(x_\epsilon) \ge -\epsilon_2$, and $\psi_{\ell,3}(x_\epsilon, d) \le \epsilon_3$ for all $d \in \mathcal{M}_\epsilon$, where $\mathcal{M}_\epsilon \subseteq \mathcal{Z}_\epsilon = \{d : \|d\| = 1 \text{ and } \psi_{\ell,2}(x_\epsilon, d) \le \epsilon_2\}$.

Proof. In the by now familiar way, we have that, for any iteration k of the ARpLE algorithm,

$$\|r(x_0)\|^{1/2^i} \ge \|r(x_0)\|^{1/2^i} - \|r(x_{k+1})\|^{1/2^i} = \sum_{j\in\mathcal{S}_k} \|r(x_j)\|^{1/2^i} - \|r(x_{j+1})\|^{1/2^i}. \tag{5.1.46}$$

Our goal is to bound the number of successful iterations required to find a step s_k and resulting point x_{k+1} for which either

$$\|r(x_{k+1})\| \le \epsilon_0 \tag{5.1.47}$$

or (together)

$$\psi_{\ell,1}(x_{k+1}) \le \epsilon_1 \text{ and } \max\left(0, -\psi_{\ell,2}(x_{k+1})\right) \le \epsilon_2 \tag{5.1.48}$$

and

$$\psi_{\ell,3}(x_{k+1}, d) \le \epsilon_3 \quad \text{for all} \quad d \in \mathcal{M}_{k+1} \ne \emptyset$$

and

$$\|s_k\|^{p-1} \le \kappa_{s2}\|r(x_{k+1})\|^{\ell-1}\epsilon_2. \tag{5.1.49}$$

Crucially, if $d \in \mathcal{M}_\epsilon \stackrel{\text{def}}{=} \mathcal{M}_{k+1}$ and (5.1.49) holds, (5.1.18) implies that

$$\left|\nabla_x^2 \Phi_\ell(x_k + s_k)[d]^2\right| \le \|r(x_{k+1})\|^{\ell-1}\epsilon_2$$

and consequently $d \in \mathcal{Z}_\epsilon$, where $\mathcal{Z}_\epsilon$ is defined as in (5.1.7); if $\mathcal{Z}_\epsilon = \emptyset$ and (5.1.48) holds, then the termination test (5.1.11) is satisfied. Thus (5.1.47)–(5.1.49) together imply that either (5.1.10) or (5.1.10), (5.1.12), and (5.1.13) together hold, and ARpLE algorithm will terminate.

If (5.1.48)–(5.1.49) fail to hold up to iteration $k+1$, it must be that

$$\|r(x_{j+1})\| > \epsilon_0 \tag{5.1.50}$$

and at least one of

$$\psi_{\ell,1}(x_{j+1}) > \epsilon_1 \text{ and } \max\left[0, -\psi_{\ell,2}(x_{j+1})\right] > \epsilon_2 \text{ and } \psi_{\ell,3}(x_{j+1}, d_j) > \epsilon_3 \tag{5.1.51}$$

for some $d_j \in \mathcal{M}_{j+1}$ as defined in (4.1.9) or

$$\|s_j\|^{p-1} > \kappa_{s2}\|r(x_{j+1})\|^{\ell-1}\epsilon_2 \tag{5.1.52}$$

holds for $j \in \mathcal{S}_{k-1}$.

Let

$$\mathcal{S}_{k-1}^{\beta} \stackrel{\text{def}}{=} \{j \in \mathcal{S}_{k-1} \mid \|r(x_{j+1})\| > \beta\|r(x_j)\|\}. \tag{5.1.53}$$

If one of (5.1.51) holds and $j \in \mathcal{S}_{k-1}^{\beta}$, it follow from Lemma 4.1.4 that

$$\|r(x_j)\|^{1/2^i} - \|r(x_{j+1})\|^{1/2^i} > \kappa_{r1} \frac{\epsilon_1^{\frac{p+1}{p}}}{\|r(x_j)\|^{\iota_1(p,\ell,i)}} \tag{5.1.54}$$

or

$$\|r(x_j)\|^{1/2^i} - \|r(x_{j+1})\|^{1/2^i} > \kappa_{r2} \frac{\epsilon_2^{\frac{p+1}{p-1}}}{\|r(x_j)\|^{\iota_2(p,\ell,i)}} \tag{5.1.55}$$

or

$$\|r(x_j)\|^{1/2^i} - \|r(x_{j+1})\|^{1/2^i} > \kappa_{r3} \frac{\epsilon_3^{\frac{p+1}{p-2}}}{\|r(x_j)\|^{\iota_3(p,\ell,i)}}. \tag{5.1.56}$$

If (5.1.52) holds, we may deduce from (4.1.37) (when $f(x) = \Phi_\ell(x)$) that

$$\begin{aligned}
\|r(x_j)\|^{\ell} - \|r(x_{j+1})\|^{\ell} &\geq \frac{\eta_1 \sigma_{\min} \ell}{(p+1)!} \|s_j\|^{p+1} \\
&> \frac{\eta_1 \sigma_{\min} \ell}{(p+1)!} \kappa_{s2}^{\frac{p+1}{p-1}} \|r(x_{j+1})\|^{\frac{(\ell-1)(p+1)}{p-1}} \epsilon_2^{\frac{p+1}{p-1}} \\
&= \ell \kappa_2 \|r(x_{j+1})\|^{\frac{(\ell-1)(p+1)}{p-1}} \epsilon_2^{\frac{p+1}{p-1}} \\
&> \ell \kappa_2 \beta^{\frac{(\ell-1)(p+1)}{p-1}} \|r(x_j)\|^{\frac{(\ell-1)(p+1)}{p-1}} \epsilon_2^{\frac{p+1}{p-1}}
\end{aligned} \tag{5.1.57}$$

for all $j \in \mathcal{S}_{k-1}^{\beta}$, where we recall the definitions of κ_{s2} and κ_2 in (4.1.23) and (4.1.34). As before, invoking Lemma 5.1.3 with $a = \|r(x_j)\|$ and $b = \|r(x_{j+1})\|$, and using (5.1.57), it follows that

$$\|r(x_j)\|^{1/2^i} - \|r(x_{j+1})\|^{1/2^i} \geq \frac{\|r(x_j)\|^{\ell} - \|r(x_{j+1})\|^{\ell}}{2^i \ell \|r(x_j)\|^{\frac{2^i-1}{2^i}+\ell-1}} > \kappa_{r2} \frac{\epsilon_2^{\frac{p+1}{p-1}}}{\|r(x_j)\|^{\iota_2(p,\ell,i)}},$$

where we recall the definition (5.1.23) of $\iota_2(p, \ell, i)$ and κ_{r2}; notice that this is identical to (5.1.55). Thus, so long as the algorithm has not terminated, at least one of (5.1.54)–(5.1.56) holds for all $j \in \mathcal{S}_{k-1}^{\beta}$. Furthermore, the choice of ℓ and i ensures that the exponents $\iota_1(p, \ell, i)$, $\iota_2(p, \ell, i)$, and $\iota_3(p, \ell, i)$ are all nonnegative, and since $\|r(x_j)\| \leq \|r(x_0)\|$, we conclude that

$$\begin{aligned}
&\|r(x_j)\|^{1/2^i} - \|r(x_{j+1})\|^{1/2^i} \\
&\quad > \min\left[\kappa_{r1} \frac{\epsilon_1^{\frac{p+1}{p}}}{\|r(x_0)\|^{\iota_1(p,\ell,i)}}, \kappa_{r2} \frac{\epsilon_2^{\frac{p+1}{p-1}}}{\|r(x_0)\|^{\iota_2(p,\ell,i)}}, \kappa_{r3} \frac{\epsilon_3^{\frac{p+1}{p-2}}}{\|r(x_0)\|^{\iota_3(p,\ell,i)}}\right]
\end{aligned} \tag{5.1.58}$$

holds for all $j \in \mathcal{S}_k^{\beta}$.

Conversely, if $j \in \mathcal{S}_{k-1} \setminus \mathcal{S}_{k-1}^{\beta}$, it follows directly from (5.1.27) and (5.1.50) that

$$\|r(x_j)\|^{1/2^i} - \|r(x_{j+1})\|^{1/2^i} \geq \kappa_{r0} \|r(x_{j+1})\|^{1/2^i} > \kappa_{r0} \epsilon_0^{1/2^i}. \tag{5.1.59}$$

Thus, (5.1.58) and (5.1.59) together ensure that

$$\|r(x_j)\|^{1/2^i} - \|r(x_{j+1})\|^{1/2^i} > \frac{\min\left[\kappa_{r0}, \kappa_{r1}, \kappa_{r2}, \kappa_{r3}\right] \cdot \min\left[\epsilon_0^{\frac{1}{2^i}}, \epsilon_1^{\frac{p+1}{p}}, \epsilon_2^{\frac{p+1}{p-1}}, \epsilon_3^{\frac{p+1}{p-2}}\right]}{\max\left[1, \|r(x_0)\|^{\iota_1(p,\ell,i)}, \|r(x_0)\|^{\iota_2(p,\ell,i)}, \|r(x_0)\|^{\iota_3(p,\ell,i)}\right]}$$

holds for all $j \in \mathcal{S}_{k-1}$. Combining this with (5.1.46), we deduce that

$$\|r(x_0)\|^{1/2^i} > |\mathcal{S}_{k-1}| \frac{\min\left[\kappa_{r0}, \kappa_{r1}, \kappa_{r2}, \kappa_{r3}\right] \cdot \min\left[\epsilon_0^{\frac{1}{2^i}}, \epsilon_1^{\frac{p+1}{p}}, \epsilon_2^{\frac{p+1}{p-1}}, \epsilon_3^{\frac{p+1}{p-2}}\right]}{\max\left[1, \|r(x_0)\|^{\iota_1(p,\ell,i)}, \|r(x_0)\|^{\iota_2(p,\ell,i)}, \|r(x_0)\|^{\iota_3(p,\ell,i)}\right]}$$

and hence that the total number of successful iterations is bounded by

$$\|r(x_0)\|^{1/2^i} \frac{\max\left[1, \|r(x_0)\|^{\iota_1(p,\ell,i)}, \|r(x_0)\|^{\iota_2(p,\ell,i)}, \|r(x_0)\|^{\iota_3(p,\ell,i)}\right]}{\min\left[\kappa_{r0}, \kappa_{r1}, \kappa_{r2}, \kappa_{r3}\right] \cdot \min\left[\epsilon_0^{\frac{1}{2^i}}, \epsilon_1^{\frac{p+1}{p}}, \epsilon_2^{\frac{p+1}{p-1}}, \epsilon_3^{\frac{p+1}{p-2}}\right]} + 1,$$

where we have used that $|\mathcal{S}_k| \leq |\mathcal{S}_{k-1}| + 1$. This gives (5.1.45) with

$$\begin{aligned}\kappa^{\mathrm{S}}_{\mathrm{ARpLE,3}} &\stackrel{\mathrm{def}}{=} \max\left(\frac{1}{\kappa_{r0}}, \frac{1}{\kappa_{r1}}, \frac{1}{\kappa_{r2}}, \frac{1}{\kappa_{r3}},\right) \\ &= \max\left(\begin{array}{l} \dfrac{1}{1-\beta^{1/2^i}}, \\ \dfrac{2^i}{\beta^{\frac{(\ell-1)(p+1)}{p}}} \dfrac{(p+1)!}{\eta_1\sigma_{\min}} \left(\dfrac{L_{\Phi_\ell,p} + \sigma_{\max} + \theta_1 p!}{p!}\right)^{\frac{p+1}{p}}, \\ \dfrac{2^i}{\beta^{\frac{(\ell-1)(p+1)}{p-1}}} \dfrac{(p+1)!}{\eta_1\sigma_{\min}} \left(\dfrac{L_{\Phi_\ell,p} + \sigma_{\max} + \theta_2 (p-1)!}{(p-1)!}\right)^{\frac{p+1}{p-1}}, \\ \dfrac{2^i}{\beta^{\frac{(\ell-1)(p+1)}{p-2}}} \dfrac{(p+1)!}{\eta_1\sigma_{\min}} \left(\dfrac{L_{\Phi_\ell,p} + \sigma_{\max} + \theta_3 (p-2)!}{(p-2)!}\right)^{\frac{p+1}{p-2}} \end{array}\right),\end{aligned}$$

where

$$\sigma_{\max} = \max\left[\sigma_0, \frac{\gamma_3 L_{\Phi_\ell,p}}{1-\eta_2}\right], \tag{5.1.60}$$

since the derivatives of r are evaluated once at each of these iterations.

We may now apply Lemma 2.4.1 (p. 44) to infer an upper bound on the total number of iterations and evaluations of f given by (5.1.44), with

$$\kappa^{\mathrm{A}}_{\mathrm{ARpLE,3}} \stackrel{\mathrm{def}}{=} \kappa^{\mathrm{S}}_{\mathrm{ARpLE,3}} \left(1 + \frac{|\log\gamma_1|}{\log\gamma_2}\right)$$

and

$$\kappa^{\mathrm{C}}_{\mathrm{ARpLE}} \stackrel{\mathrm{def}}{=} \frac{1}{\log\gamma_2}\left\{|\log\gamma_1| + \log\left(\frac{\sigma_{\max}}{\sigma_0}\right)\right\} + 2, \tag{5.1.61}$$

with $\sigma_{\max}$ as above. □

The results for a version of the algorithm that only searches for an approximate first-order or second-order minimizer may be retrieved by setting $\epsilon_2 = \epsilon_3 = \infty$ and picking ℓ and i from (5.1.38) or (5.1.39) or by setting $\epsilon_3 = \infty$ and picking ℓ and i from (5.1.40) or (5.1.41) in the complexity bounds, respectively; the constants $\kappa^{\mathrm{S}}_{\mathrm{ARpLE},3}$ (etc.) need to be adjusted accordingly.

It is worth briefly comparing the result above with that from Theorem 4.1.5 on page 107 for the case when $f(x) = \Phi_2(x)$. Superficially there is little difference; indeed one might argue that little has been gained in complexity terms by treating the structure separately. However, Theorem 5.1.7 applies directly to the nonsmooth objective $\Phi_1(x)$, and the possibility that the residual may be small is exposed in the ϵ_0 term.

For the reader's benefit, Table 5.1.4 summarizes some achievable evaluation-complexity orders as a function of model degree ($p \leq 18$) and required optimality order ($q \leq 3$) under the requirements (5.1.39) ($q = 1$), (5.1.41) ($q = 2$), and (5.1.43) ($q = 3$) for the simplified[38] case where $\epsilon_1 = \epsilon_2 = \epsilon_3 = \epsilon_d$. We also report the corresponding value of ℓ in each case. We note that if (5.1.38) ($q = 1$), (5.1.40) ($q = 2$), and (5.1.42) ($q = 3$) are imposed instead, the dependence on ϵ_0 may be reduced further, but since this is unlikely to be the dominant term in the order bound, we provide no further details here.

	$q=1$		$q=2$		$q=3$	
p	ℓ	order bound	ℓ	order bound	ℓ	order bound
1	-	-	-	-	-	-
2	2	$\max[\epsilon_0^{-1/2}, \epsilon_d^{-3/2}]$	-	-	-	-
3	-	-	-	-	-	-
4	4	$\max[\epsilon_0^{-1/4}, \epsilon_d^{-5/4}]$	-	-	-	-
5	-	-	2	$\max[\epsilon_0^{-1/2}, \epsilon_d^{-3/2}]$	-	-
6	4	$\max[\epsilon_0^{-1/2}, \epsilon_d^{-7/6}]$	-	-	2	$\max[\epsilon_0^{-1/4}, \epsilon_d^{-7/4}]$
7	-	-	-	-	-	-
8	8	$\max[\epsilon_0^{-1/8}, \epsilon_d^{-9/8}]$	-	-	-	-
9	-	-	4	$\max[\epsilon_0^{-1/4}, \epsilon_d^{-5/4}]$	-	-
10	6	$\max[\epsilon_0^{-1/2}, \epsilon_d^{-11/10}]$	-	-	-	-
11	-	-	-	-	-	-
12	10	$\max[\epsilon_0^{-1/4}, \epsilon_d^{-13/12}]$	-	-	-	-
13	-	-	4	$\max[\epsilon_0^{-1/2}, \epsilon_d^{-7/6}]$	-	-
14	8	$\max[\epsilon_0^{-1/2}, \epsilon_d^{-15/14}]$	-	-	4	$\max[\epsilon_0^{-1/4}, \epsilon_d^{-5/4}]$
15	-	-	-	-	-	-
16	16	$\max[\epsilon_0^{-1/16}, \epsilon_d^{-17/16}]$	-	-	-	-
17	-	-	8	$\max[\epsilon_0^{-1/8}, \epsilon_d^{-9/8}]$	-	-
18	10	$\max[\epsilon_0^{-1/2}, \epsilon_d^{-19/18}]$	-	-	6	$\max[\epsilon_0^{-1/8}, \epsilon_d^{-19/16}]$

Table 5.1.4. Residual exponent (ℓ) and order of the achievable evaluation-complexity bound for least-Euclidean-norm problems when $\epsilon_1 = \epsilon_2 = \epsilon_3 = \epsilon_d$. These are those for ℓ and i given by (5.1.39) for $q = 1$, (5.1.41) for $q = 2$, and (5.1.43) for $q = 3$.

[38]For the general case, the bound in $\epsilon_d^{-\pi}$ should be replaced by $\max[\epsilon_1^{-(p+1)/p}, \epsilon_2^{-(p+1)/(p-1)}, \epsilon_3^{-(p+1)/(p-2)}]$.

The reader may well be surprised to see that the table has no entries for odd p when $q = 1$, or even p when $q = 2$, but this simply reflects the stated goal to provide bounds that separate the dependence of ϵ_0 from the remaining tolerances (see the comments at the start of Section 5.1.3). We could choose otherwise by allowing negative exponents (5.1.21), (5.1.23), and (5.1.25), and if we do there is no reason to focus on anything but the $\ell = 2$ case, i.e., on the objective $\Phi_2(x)$. We go into more details, and more generality, in Section 14.3.2.1. As a foretaste, in Table 5.1.5 (p. 135) we illustrate the order of the worst-case evaluation bound when $\ell = 2$ as given by Theorem 14.3.4 (on p. 408) under suitable assumptions. As the reader will see, the bounds illustrated in Table 5.1.4 are always better than those in Table 5.1.5, although the improvement diminishes as p grows.

	$q = 1$	$q = 2$	$q = 3$
p	order bound	order bound	order bound
1	$\epsilon_0^{-1}\epsilon_d^{-2}$	-	-
2	$\epsilon_0^{-1/2}\epsilon_d^{-3/2}$	$\epsilon_0^{-2}\epsilon_d^{-3}$	-
3	$\epsilon_0^{-1/3}\epsilon_d^{-4/3}$	$\epsilon_0^{-1}\epsilon_d^{-2}$	$\epsilon_0^{-3}\epsilon_d^{-4}$
4	$\epsilon_0^{-1/4}\epsilon_d^{-5/4}$	$\epsilon_0^{-2/3}\epsilon_d^{-5/3}$	$\epsilon_0^{-3/2}\epsilon_d^{-5/2}$
5	$\epsilon_0^{-1/5}\epsilon_d^{-6/5}$	$\epsilon_0^{-1/2}\epsilon_d^{-3/2}$	$\epsilon_0^{-1}\epsilon_d^{-6/3}$
6	$\epsilon_0^{-1/6}\epsilon_d^{-7/6}$	$\epsilon_0^{-2/5}\epsilon_d^{-7/5}$	$\epsilon_0^{-3/4}\epsilon_d^{-7/4}$
7	$\epsilon_0^{-1/7}\epsilon_d^{-8/7}$	$\epsilon_0^{-1/3}\epsilon_d^{-4/3}$	$\epsilon_0^{-3/5}\epsilon_d^{-8/5}$
8	$\epsilon_0^{-1/8}\epsilon_d^{-9/8}$	$\epsilon_0^{-2/7}\epsilon_d^{-9/7}$	$\epsilon_0^{-1/2}\epsilon_d^{-9/6}$
9	$\epsilon_0^{-1/9}\epsilon_d^{-10/9}$	$\epsilon_0^{-1/4}\epsilon_d^{-5/4}$	$\epsilon_0^{-3/7}\epsilon_d^{-10/7}$
10	$\epsilon_0^{-1/10}\epsilon_d^{-11/10}$	$\epsilon_0^{-2/9}\epsilon_d^{-11/9}$	$\epsilon_0^{-3/8}\epsilon_d^{-11/8}$
11	$\epsilon_0^{-1/11}\epsilon_d^{-12/11}$	$\epsilon_0^{-1/5}\epsilon_d^{-6/5}$	$\epsilon_0^{-1/3}\epsilon_d^{-12/9}$
12	$\epsilon_0^{-1/12}\epsilon_d^{-13/12}$	$\epsilon_0^{-2/11}\epsilon_d^{-13/11}$	$\epsilon_0^{-3/10}\epsilon_d^{-13/10}$
13	$\epsilon_0^{-1/13}\epsilon_d^{-14/13}$	$\epsilon_0^{-1/6}\epsilon_d^{-7/6}$	$\epsilon_0^{-3/11}\epsilon_d^{-14/11}$
14	$\epsilon_0^{-1/14}\epsilon_d^{-15/14}$	$\epsilon_0^{-2/13}\epsilon_d^{-15/13}$	$\epsilon_0^{-1/4}\epsilon_d^{-15/12}$
15	$\epsilon_0^{-1/15}\epsilon_d^{-16/15}$	$\epsilon_0^{-1/7}\epsilon_d^{-8/7}$	$\epsilon_0^{-3/13}\epsilon_d^{-16/13}$
16	$\epsilon_0^{-1/16}\epsilon_d^{-17/16}$	$\epsilon_0^{-2/15}\epsilon_d^{-17/15}$	$\epsilon_0^{-3/14}\epsilon_d^{-17/14}$
17	$\epsilon_0^{-1/17}\epsilon_d^{-18/17}$	$\epsilon_0^{-1/8}\epsilon_d^{-9/8}$	$\epsilon_0^{-1/5}\epsilon_d^{-18/15}$
18	$\epsilon_0^{-1/18}\epsilon_d^{-19/18}$	$\epsilon_0^{-2/17}\epsilon_d^{-19/17}$	$\epsilon_0^{-3/16}\epsilon_d^{-19/16}$

Table 5.1.5. Order of the achievable evaluation-complexity bound for least-Euclidean-norm problems when $\ell = 2$ and $\epsilon_1 = \epsilon_2 = \epsilon_3 = \epsilon_d$. These are from Theorem 14.3.4.

5.1.5 Improved complexity under a full-rank assumption

A much stronger result is possible if a full-rank condition is satisfied by the iterates as the ARpLE algorithm proceeds. To be specific, suppose that the iterates x_k generated by the ARpLE algorithm satisfy the bound

$$\psi_{\ell,1}(x_k) \geq \psi_{\min} \tag{5.1.62}$$

for all k until termination, for some $\psi_{\min} > 0$. The condition (5.1.62) might, at first sight, seem strange, but it is easy to show that

$$\psi_{\ell,1}(x) = \frac{\|\nabla_x^1 \Phi_\ell(x)\|}{\|r(x)\|^{\ell-1}} = \frac{\|\nabla_x^1 \Phi_2(x)\|}{\|r(x)\|} = \frac{\|(\nabla_x^1 r(x))^T r(x)\|}{\|r(x)\|}. \tag{5.1.63}$$

Since the smallest singular value, $\sigma_{\min}(\nabla_x^1 r(x))$, of $\nabla_x^1 r(x)$ satisfies

$$\frac{\|(\nabla_x^1 r(x))^T v\|}{\|v\|} \geq \sigma_{\min}\left(\nabla_x^1 r(x)\right) \tag{5.1.64}$$

for all (nonzero) v, (5.1.62) may then simply be seen as a requirement that the smallest singular value of the Jacobian of $r(x)$ be bounded away from zero at all points x encountered by the algorithm.

Although it is possible to derive an improved complexity bound based on (5.1.62) using arguments akin to those in the proof of Theorem 5.1.7, the result also follows as a consequence of a more general algorithmic framework that we shall encounter in Section 5.3. Thus we postpone further discussions on this until Section 5.3.4.

Notes and references for Section 5.1

The use of pth-order models to improve the evaluation-complexity bound to find first-order approximate minimizers of least-squares problems was first given in [97], and these built on analyses for second-order models given by [98] and [305]. Moreover, practical methods that use second-order models of the residuals $r(x)$ have been proposed and their evaluation complexity investigated [194]. These are related to earlier Euclidean-residual algorithms [26] and generalize known bounds for first-order Levenberg–Morrison–Marquardt [265, 282, 293, 294] models [300, 354]. The extension given here to find second- and third-order approximate minimizers is new.

Theorem 14.1.10 on page 384 establishes an $\mathcal{O}(\epsilon_1^{-(p+1)/p})$ evaluation bound for Algorithm 14.1.1 (ARqpC) on p. 376 to find an ϵ-approximate first-order minimizer of Φ_1 (since $\|\cdot\|$ is convex) but a less optimistic $\mathcal{O}(\max_{j\in[1:q]}\epsilon_j^{-(j+1)})$ bound to find an ϵ-approximate qth-order minimizer for $q \geq 2$.

There are a large number of software packages for solving nonlinear least-squares problems, many of them Gauss–Newton/Levenberg–Morrison–Marquardt variants based only on first derivatives, but some that also use second derivatives. The first that tries to use second-order models of the residual is the recent `NLS` package from the GALAHAD open-source software library [192].

5.2 Partially separable problems

Another ubiquitous structure in optimization problems is *partial separability*. We say that a function f is partially separable if it can be expressed as the sum of (possibly many) *element functions*, each depending only on a low-rank linear transformation of the variables. Formally, f is partially separable when, for all $x \in \mathbb{R}^n$, it can be written as

$$f(x) = \sum_{i=1}^{m} f_i(U_i x), \tag{5.2.1}$$

where each U_i is an $n_i \times n$ matrix. This structure is most useful when $n_i \ll n$, i.e., when each U_i has low rank. This definition is also equivalent to saying that each element function f_i is invariant along an (ideally large) subspace $\{v \mid U_i v = 0\}$. Partially separable functions occur in many different contexts. Originally introduced to cover the finite-element applications—hence the element-function terminology—they also occur in some least-squares applications (notice the similarity between (5.1.2) and (5.2.1)), in image processing, and, more generally, in all sufficiently smooth problems with sparse Hessian matrices.

Exploiting the partial separability of an optimization problem has a long history and essentially hinges on mimicking the problem structure in any model of f used in the calculation: one then typically replaces a single potentially (very) large model by a collection of m much smaller ones, noting that

$$T_{f,p}(x,s) = \sum_{i=1}^{m} T_{f_i,p}(U_i x, U_i s). \tag{5.2.2}$$

This technique usually brings substantial gains both in computing time and storage, because the derivatives of $T_{f_i,p}$ are now n_i-dimensional symmetric tensors instead of n-dimensional ones. While the gain is often important for gradients and Hessians, it becomes *crucial if derivatives of higher order are considered* (as in Chapters 4 and 12). We note that partially separable models of the form (5.2.2) can be used in trust-region or adaptive-regularization methods *without any modification to the corresponding algorithms*, thereby ensuring that associated worst-case *evaluation-complexity bounds are unchanged.*

Partial separability may also be a crucial ingredient in a parallel-computing environment, since evaluations of the objective function and its derivative can then easily be split into parallel tasks. In this context, the evaluation complexity may be very significantly reduced.

Notes and references for Section 5.2

The notion of a partially separable function was introduced by Griewank and Toint in [219] in the framework of quasi-Newton methods. Its "group-partially-separable" generalization is at the core of the `LANCELOT` optimization package [130] and the `FILTRANE` routine for nonlinear least-squares [200], both now part of the GALAHAD software library [192]. The special "coordinate-partially-separable" case (for which the rows of the U_i matrices are selected rows from the $n \times n$ identity matrix) has been used to great advantage in pattern-search derivative-free methods [319, 323].

While partially separable models of the form (5.2.2) can be used without modification of trust-region or adaptive-regularization methods, one may also consider using the structure in the associated globalization technique, especially if the different element functions have very different nonlinear behaviour. In trust-region methods, this might be done at iteration k by using a structured trust region of the form

$$\prod_{i=1}^{m} \mathcal{R}_{i,k} \quad \text{with} \quad \mathcal{R}_{i,k} = \{s_{i,k} \in \mathbb{R}_i^n \mid \|s_{i,k}\| \le \Delta_{i,k} i\},$$

where $x_{i,k} \stackrel{\text{def}}{=} eU_i x_k$, $s_{i,k} = U_i s_k$, and $\Delta_{i,k}$ is an element-dependent trust-region radius. Convergence properties of such methods have been considered at length in [128] and in Section 10.2 of [131]. In [116, 117], a partially separable adaptive-regularization algorithm is proposed in which the regularized model is defined as

$$m_k(s) = \sum_{i=1}^{m} m_{i,k}(s_i) \stackrel{\text{def}}{=} \sum_{i=1}^{m} \left[T_{f_i,p}(x_{i,k}, s_i) + \frac{\sigma_{i,k}}{(p+1)!} \|s_i\|^{p+1} \right],$$

where each element function's model is regularized independently using an element-specific regularization parameter $\sigma_{i,k}$. The resulting algorithmic framework is simpler than that proposed for structured trust regions in [128] and [131]. When weak approximate minimizers (in the sense of (12.1.40)) are sought, it is shown in [116, 117] that the use of the structure in the regularization does not affect the evaluation-complexity bounds.

5.3 Measure-dominated problems

There is another class of structured problems where complexity results can be improved substantially: the class of "measure-dominated" problems. While, as we will see, this class is somewhat restrictive, its associated complexity analysis provides significantly better bounds than what we have seen so far. Moreover, by introducing an additional assumption on the global behaviour of the objective function, it also provides a stepping stone to the well-mannered world of convex problems. This, we feel, is enough to justify our interest.

Consider an unconstrained optimization problem with global minimum value f_* and some nonnegative optimality measure $\psi(x)$; typical examples of the latter when searching for first- or second-order minimizers are $\|\nabla_x^1 f(x)\|$ and $\min\left[0, -\lambda_{\min}[\nabla_x^2 f(x)]\right]$. We say that f is *measure dominated* of level[39] μ for a real $\mu \geq 1$ if there is a constant $\kappa_\psi > 0$ for which

$$f(x) - f_* \leq \kappa_\psi \, [\psi(x)]^\mu \tag{5.3.1}$$

for all $x \in \mathbb{R}^n$. When $\psi(x) = \|g(x)\|$ and $\mu \in [1, 2]$, the requirement (5.3.1) is sometimes known as the *Polyak–Łojasiewicz property* [275, 317] or *gradient domination* [305]. The restriction $\mu \in [1, 2]$ in this case is motivated by considering the univariate situation (see [160] for details).

Because it is simple to do so, we discuss domination for more general optimality measures $\psi(x)$. A related but different approach is taken in [135], which we briefly discuss in the notes of this section.

5.3.1 Some examples

As we intimated above, many problems are not measure dominated. For instance, functions having saddle points or local but nonglobal minimizers are not measure dominated of level μ for any μ when $\psi(x) = \|\nabla_x^1 f(x)\|$. However, there exist interesting classes of functions which are measure dominated. Perhaps the most frequently occurring are nonlinear least-squares problems with uniform full rank. Indeed, more generally, if, as in Section 5.1, we consider minimizing

$$\Phi_\ell(x) = \frac{1}{\ell}\|r(x)\|^\ell, \quad \text{where } r : \mathbb{R}^n \to \mathbb{R}^m \text{ is continuously differentiable,}$$

and assume, as in Section 5.1.5, that condition (5.1.62) holds,[40] then it follows from (5.1.63) and (5.1.64) that $\min \Phi_\ell(x) = 0$ and

$$\|\nabla_x^1 \Phi_\ell(x)\| \geq \psi_{\min}\|r(x)\|^{\ell-1}. \tag{5.3.2}$$

Hence $\Phi_\ell(x)$ is measure dominated of level $\mu = \ell/(\ell-1)$ for $\psi(x) = \|\nabla_x^1 \Phi_\ell(x)\|$, since

$$\Phi_\ell(x) - \min \Phi_\ell(x) = \Phi_\ell(x) = \frac{1}{\ell}\|r(x)\|^\ell \leq \frac{1}{\ell}\left(\frac{1}{\psi_{\min}}\right)^{\ell/(\ell-1)} \|\nabla_x^1 \Phi_\ell(x)\|^{\ell/(\ell-1)}. \tag{5.3.3}$$

[39] Other authors use the term "degree" instead of "level," but we prefer the latter here in order to avoid confusion with the notion of a model's degree.

[40] That is, the smallest singular value of the Jacobian of $r(x)$ is bounded away from zero at all points x encountered by the algorithm; see the discussion on p. 136.

Another class of measure-dominated problems includes those involving convex continuously differentiable functions on the bounded domain $\{x \in \mathbb{R}^n \mid \|x\| \leq \kappa_u\}$ for some $\kappa_u > 0$. For then

$$f(x) - f_* \leq \langle \nabla_x^1 f(x), x - x_* \rangle \leq 2\kappa_u \|\nabla_x^1 f(x)\|$$

from (1.2.1) (p. 11), and f is measure dominated of level one with $\psi(x) = \|\nabla_x^1 f(x)\|$. In the same vein, if f is strongly convex, then one can verify from the definition (1.2.2) (p. 11) that

$$\begin{aligned} f(z) - f(x) &\geq \langle z - x, \nabla_x^1 f(x) \rangle + \tfrac{1}{2}\lambda_{\rm L} \|z - x\|^2 \\ &\geq \min_z \; \langle z - x, \nabla_x^1 f(x) \rangle + \tfrac{1}{2}\lambda_{\rm L} \|z - x\|^2 \\ &= -\frac{1}{2\lambda_{\rm L}} \|\nabla_x^1 f(x)\|^2 \end{aligned}$$

and thus, if z is a global minimizer of f, that

$$f(x) - f_* \leq \frac{1}{2\lambda_{\rm L}} \|\nabla_x^1 f(x)\|^2,$$

where $\lambda_{\rm L} > 0$ is the constant defining the strong convexity of f. The function f is now measure dominated of level two with $\psi(x) = \|\nabla_x^1 f(x)\|$.

Finally, we note that measure-dominated functions for measures differing from the gradient also exist. For instance, the function $f(x) = \|x\|^{2i}$ $(i \geq 2)$ is measure dominated of level one for the optimality measure $\psi(x) = \phi_{f,2}^1(x)$, where, as we shall see in Section 12.1.3 (p. 278) of Chapter 12,

$$\phi_{f,2}^1(x) \stackrel{\rm def}{=} f(x) - \min_{\|d\| \leq 1} T_{f,2}(x, d)$$

and $T_{f,2}(x, d) = f(x) + \langle \nabla_x^1 f(x), d \rangle + \frac{1}{2}\langle d, \nabla_x^2 f(x) d \rangle$ is the second-degree Taylor expansion (4.1.1) of f around x within the domain $\mathcal{D} = \{x \mid \|x\| \leq 1 + 2(i-1)\}$. Indeed,

$$\nabla_x^1 f(x) = 2i\|x\|^{2i-2} x \;\text{ and }\; \nabla_x^2 f(x) = 2i\|x\|^{2i-2} I + 4i(i-1)\|x\|^{2i-4} x x^T$$

so that

$$T_{f,2}(x, d) = \|x\|^{2i} + 2i\|x\|^{2i-2} x^T d + i\|x\|^{2i-2}\|d\|^2 + 2i(i-1)\|x\|^{2i-4}(x^T d)^2,$$

and it is easy to verify that $T_{f,2}(x, d)$ is convex and attains its minimum value

$$T_{f,2}(x, d_*) = \left(1 - \frac{i}{1 + 2(i-1)}\right)\|x\|^{2i} \;\text{ for }\; d_* = -\frac{x}{1 + 2(i-1)}.$$

We therefore obtain that

$$\phi_{f,2}^1(x) = T_{f,2}(x, 0) - T_{f,2}(x, d_*) = \frac{i}{1 + 2(i-1)} f(x) = \frac{i}{1 + 2(i-1)}(f(x) - f_*)$$

if $x \in \mathcal{D}$, and (5.3.1) holds for $\psi(x) = \phi_{f,2}^1(x)$, $\mu = 1$, and $\kappa_\psi = \frac{1 + 2(i-1)}{i}$ within $\mathcal{D}$.

5.3.2 Applying adaptive regularization–like methods

We now examine what can be said of the rate at which the objective-function gap

$$\Gamma_k = f(x_k) - f_* \tag{5.3.4}$$

converges to zero for iterates $\{x_k\}$ generated by our two favourite classes of optimization methods, those broadly based on adaptive regularization and those on trust-region globalization.

Consider first the case of our archetypal adaptive-regularization methods. These derive a bound on the minimal decrease at successful iterations from the inequalities

$$f(x_k) - f(x_{k+1}) \geq \eta_1[T_{f,p}(x_k, 0) - T_{f,p}(x_k, s_k)] \geq \eta_1 \sigma_{\min} \|s_k\|^{p+1}$$

and then compute a lower bound on $\|s_k\|$ that depends on the value of the chosen optimality measure at x_{k+1}. The proofs of Lemmas 2.4.3 (p. 45) and 3.3.8 (p. 73) illustrate this fact for AR1 and AR2. It is even more obvious for the ARp algorithm (p. 100), for which a minimal decrease per successful iteration of the form

$$f(x_k) - f(x_{k+1}) \geq \kappa_\nu [\psi(x_{k+1})]^\nu \tag{5.3.5}$$

is guaranteed by Lemma 4.1.4 (p. 106). In particular, we have the following three cases:

1. Condition (4.1.31) ensures that (5.3.5) holds for

$$\psi_1(x) = \|\nabla_x^1 f(x)\| \quad \text{with} \quad \nu = \frac{p+1}{p} \quad \text{and} \quad \kappa_\nu = \kappa_1. \tag{5.3.6}$$

2. Condition (4.1.33) ensures that (5.3.5) holds for

$$\psi_2(x) = \max\left(0, -\lambda_{\min}[\nabla_x^2 f(x_{k+1})]\right) \quad \text{with} \quad \nu = \frac{p+1}{p-1} \quad \text{and} \quad \kappa_\nu = \kappa_2. \tag{5.3.7}$$

3. Condition (4.1.35) ensures that (5.3.5) holds for

$$\psi_3(x) = \max_{d \in \mathcal{M}_{k+1}} \left|\nabla_x^3 f(x_{k+1})[d]^3\right| \quad \text{with} \quad \nu = \frac{p+1}{p-2} \quad \text{and} \quad \kappa_\nu = \kappa_3. \tag{5.3.8}$$

For full generality, we let ARψ,ν be the class of monotonic minimization algorithms for which (a) the decrease (5.3.5) is guaranteed for each k in an infinite subset $\mathcal{S}$ of the iterations, (b) $x_{k+1} = x_k$ for all $k \notin \mathcal{S}$, and (c) $\mathcal{S}_k = \mathcal{S} \cap [0:k]$ is dense enough in $[0:k]$ in the sense that

$$k \leq \kappa_{\rm R}^{\mathcal{S}} |\mathcal{S}_k| + \kappa_{\rm A}^{\mathcal{S}} \tag{5.3.9}$$

for some $\kappa_{\rm R}^{\mathcal{S}} \geq 1$ and $\kappa_{\rm A}^{\mathcal{S}} \geq 0$. These properties are obviously satisfied by the ARp algorithm, but also by other variants discussed in Section 3.4. In particular, (5.3.9) is a generic version of the crucial bound (2.4.8) from Lemma 2.4.1 (p. 44).

As the next lemma shows, the combination of these decrease properties and corresponding ψ-domination for ARψ,ν algorithms has profound consequences for the rate of decrease of the objective-function gap.

Lemma 5.3.1: Gap-decrease rates for measure-dominated problems (adaptive regularization–like methods)

Suppose that the objective function f is ψ-dominated of level μ for some $\mu \geq 1$ and that an ARψ,ν *algorithm for which $\nu \geq 1$ is applied to minimize f. Then we have that*

$$\Gamma_{k+1} \leq \Gamma_k \left(1 + \frac{\kappa_\nu}{\kappa_\psi^{\nu/\mu}} \Gamma_{k+1}^{\frac{\nu-\mu}{\mu}}\right)^{-1} \tag{5.3.10}$$

for every $k \in \mathcal{S}$, and[a]

(i) *if $\nu < \mu$, the gaps $\{\Gamma_k\}_{k \in \mathcal{S}}$ converge superlinearly to zero;*

(ii) *if $\nu = \mu$, the gaps $\{\Gamma_k\}_{k \in \mathcal{S}}$ converge linearly to zero with factor $\left(1 + \kappa_\nu/\kappa_\psi\right)^{-1}$; and*

(iii) *if $\nu > \mu$, the gaps $\{\Gamma_k\}_{k \in \mathcal{S}}$ converge sublinearly to zero.*

[a]Convergence rates are discussed in Section A.3.

Proof. Combining (5.3.1) and (5.3.5), we deduce that

$$f(x_k) - f(x_{k+1}) \geq \frac{\kappa_\nu}{\kappa_\psi^{\nu/\mu}} \Gamma_{k+1}^{\frac{\nu}{\mu}}$$

for every $k \in \mathcal{S}$. Adding and subtracting f_* on the left-hand side, we thus deduce that

$$\Gamma_k - \Gamma_{k+1} \geq \frac{\kappa_\nu}{\kappa_\psi^{\nu/\mu}} \Gamma_{k+1}^{\frac{\nu}{\mu}}, \tag{5.3.11}$$

yielding the inequality

$$\Gamma_{k+1} \left(1 + \frac{\kappa_\nu}{\kappa_\psi^{\nu/\mu}} \Gamma_{k+1}^{\frac{\nu-\mu}{\mu}}\right) \leq \Gamma_k.$$

This new inequality is the same as (5.3.10). The other conclusions of the theorem directly follow from (5.3.10). □

Observe that we have not used (5.3.1) at points other than the iterates $\{x_k\}$, and it therefore suffices to assume that (5.3.1) holds at all points encountered by the algorithm. Moreover, the inequality (5.3.10) does not depend on what happened in the algorithm before iteration k. This is important because it shows that the conclusions of the lemma still apply if (5.3.1) is satisfied at iterates for all k sufficiently large.

Suppose now that our (monotonic) algorithm produces a sequence of iterates converging to a local minimizer x_* and that (5.3.1) holds in the neighbourhood of x_* with $f_* = f(x_*)$. Then it is easy to see that the conclusions of the above lemma remain valid (we will return to such a "regional" approach in the notes for this section). This in turn has a remarkable consequence. Namely, since Łojasiewicz's theorem [160] ensures that every critical point of a real analytic function has a neighbourhood in which (5.3.1) holds for $\psi(x) = \|\nabla_x^1 f(x)\|$ for some $\mu \in [1, 2]$ and some associated κ_ψ, this means that any monotonic algorithm converging to a local minimizer of a real analytic function will be restricted to a neighbourhood of this local minimizer for which f is gradient dominated for sufficiently large k and hence that the conclusions of the above lemma must apply asymptotically.

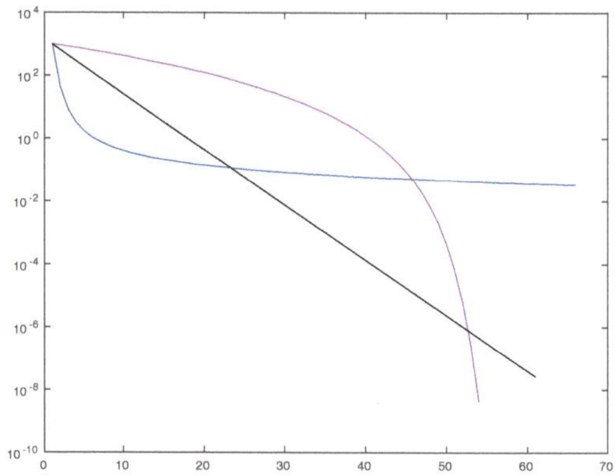

Figure 5.3.1. The bound on the decrease of Γ_k as predicted by Lemma 5.3.1 for $\mu = 2$ and $\nu = \frac{3}{2}$ (magenta), $\nu = 2$ (black), and $\nu = 4$ (blue), using $\kappa_\psi = 1$ and $\kappa_\nu = \frac{1}{2}$.

Finally, observe that the identifications (5.3.6)–(5.3.8) imply that ν tends to one as the degree of the model p increases. Thus the likelihood of the first two (favourable) cases of Lemma 5.3.1 also increases with p for the ARp algorithm.

We illustrate the error decreases for the three cases of the above theorem in Figure 5.3.1. The magenta curve corresponds to case 1 above (first-order point with $p = 2$) when using the ARp algorithm, the black curve to case 2 (second-order point with $p = 3$), and the blue curve to case 3 (third-order point with $p = 3$).

We now define the sequence $\{t_k\}$ as the positive solution of the recurrence relation

$$\kappa\, t_{n+1}^{\tau} + t_{n+1} - t_n = 0 \tag{5.3.12}$$

for given $\tau > 0$ and $\kappa > 0$, starting from $t_0 > 0$. The form of this equation clearly implies that $\{t_k\}$ is monotonically decreasing and converges to zero. Indeed, since $\kappa\, t_{n+1}^{\tau} > 0$, we immediately deduce that $t_{n+1} < t_n$. Moreover, any fixed point of the relation must be zero. On defining

$$\Upsilon_{\kappa,\tau}(t_0, \epsilon) \stackrel{\text{def}}{=} \text{the smallest } k \geq 0 \text{ such that } t_k \leq \epsilon \tag{5.3.13}$$

for given $\epsilon > 0$, we derive the following bound.

Theorem 5.3.2: Complexity of reducing the gap (implicit)

Suppose that the objective function f is ψ-dominated of level μ for some $\mu \geq 1$ and that an ARψ,ν algorithm for which $\nu \geq 1$ is applied to minimize f. Then the algorithm needs at most

$$\#^{SI}_{\mathsf{ARp,gap,0}} \stackrel{\text{def}}{=} \Upsilon_{\frac{\kappa_\nu}{\kappa_\psi^{\nu/\mu}}, \frac{\nu}{\mu}}(\Gamma_0, \epsilon) \tag{5.3.14}$$

successful iterations (and derivative evaluations) and at most

$$\#^{F}_{\mathsf{ARp,gap,0}} \stackrel{\text{def}}{=} \kappa_R^{\mathcal{S}}\, \Upsilon_{\frac{\kappa_\nu}{\kappa_\psi^{\nu/\mu}}, \frac{\nu}{\mu}}(\Gamma_0, \epsilon) + \kappa_A^{\mathcal{S}} \tag{5.3.15}$$

iterations and evaluations of f to compute an iterate x_ϵ such that $\Gamma_\epsilon = f(x_\epsilon) - f_* \leq \epsilon$.

Proof. The bound (5.3.14) immediately results from the definition of Υ in (5.3.13) and the identification of (5.3.12) with (5.3.11), itself equivalent to (5.3.10), where $\kappa = \kappa_\nu / \kappa_\psi^{\nu/\mu}$ and $\tau = \nu/\mu$. We then obtain (5.3.15) from (5.3.9) and (5.3.14). □

Of course, Lemma 5.3.1 and Theorem 5.3.2 do not give complexity bounds in the sense of the other results in this book, since these depend on a termination rule based on $\psi(x_k)$ rather than on Γ_k. Additionally, while the values of (5.3.14) can easily be computed numerically (as we did to produce Figure 5.3.1), such a bound is not explicit. It turns out, however, that more explicit evaluation-complexity bounds may be obtained.

We now state general evaluation-complexity results for ARψ,ν algorithms when applied to ψ-dominated functions. For simplicity, we distinguish our results according to each of the three possibilities (i)–(iii) identified by Lemma 5.3.1 in turn, starting with the first case, for which $\nu < \mu$. Because the proofs of these theorems are lengthy and technical, we do not present them here, but they can be found in Appendix B.

Theorem 5.3.3: Complexity for measure-dominated problems (adaptive regularization–like methods, $\nu < \mu$)

[Appendix to Chapter 5, p. 464]: Suppose that the objective function f is ψ-dominated of level μ for some $\mu \geq 1$ and that an ARψ,ν algorithm for which $\nu \geq 1$ is applied to minimize f. In addition, suppose that $1 \leq \nu < \mu$. Then the gaps $\{\Gamma_k\}_{k\in\mathcal{S}}$ converge superlinearly to zero, and there exist positive constants $\kappa^B_{\text{AR,gap,1}}$, $\kappa^C_{\text{AR,gap,1}}$, $\kappa^E_{\text{AR,gap,1}}$, and $\kappa^F_{\text{AR,gap,1}}$ such that the algorithm needs at most

$$\#^F_{\text{AR,gap,1}} \stackrel{\text{def}}{=} \kappa^B_{\text{AR,gap,1}} \log(|\log(\epsilon)|) + \kappa^C_{\text{AR,gap,1}} = \mathcal{O}\left(\log(|\log(\epsilon)|)\right) \tag{5.3.16}$$

iterations and evaluations of f and at most

$$\#^D_{\text{AR,gap,1}} \stackrel{\text{def}}{=} \kappa^E_{\text{AR,gap,1}} \log(|\log(\epsilon)|) + \kappa^F_{\text{AR,gap,1}} = \mathcal{O}\left(\log(|\log(\epsilon)|)\right) \tag{5.3.17}$$

successful iterations (and derivative evaluations) to compute an iterate x_ϵ such that $\Gamma_\epsilon = f(x_\epsilon) - f_* \leq \epsilon$. Moreover, there exist constants $\kappa^B_{\text{AR,dom,1}}$, $\kappa^C_{\text{AR,dom,1}}$, $\kappa^E_{\text{AR,dom,1}}$, and $\kappa^F_{\text{AR,dom,1}}$ such that the algorithm needs at most

$$\#^F_{\text{AR,dom,1}} \stackrel{\text{def}}{=} \kappa^B_{\text{AR,dom,1}} \log(|\log(\kappa_\nu) + \nu\log(\epsilon)|) + \kappa^C_{\text{AR,dom,1}} = \mathcal{O}\left(\log(|\log(\epsilon)|)\right) \tag{5.3.18}$$

iterations and evaluations of f and at most

$$\#^D_{\text{AR,dom,1}} \stackrel{\text{def}}{=} \kappa^E_{\text{AR,dom,1}} \log(|\log(\kappa_\nu) + \nu\log(\epsilon)|) + \kappa^F_{\text{AR,dom,1}} = \mathcal{O}\left(\log(|\log(\epsilon)|)\right) \tag{5.3.19}$$

successful iterations (and derivative evaluations) to compute an iterate x_ϵ such that $\psi(x_\epsilon) \leq \epsilon$.

We remark that (5.3.17)—and the other bounds derived from it—may well be rather pessimistic, since the proof is based on the decomposition of the convergence into two successive phases: the first, linear, to bring the gap from its initial value to a value below one, and the second, superlinear, to reduce it to ϵ. This view is more an artifact of the proof technique than caused by the true nature of the convergence. We can see this, for instance, by examining the magenta curve of gap values in Figure 5.3.1, which displays no particular sign of a two-phase process. Furthermore, since the rate of linear convergence for the first of the two phases in the proof is determined by Γ_0, this is the worst linear rate possible (the tangent to the magenta curve at iteration zero) and may result in a large overestimate of $\kappa^{\mathrm{C}}_{\mathrm{AR,gap,1}}$ and $\kappa^{\mathrm{F}}_{\mathrm{AR,gap,1}}$.

We next turn to the possibility that $\nu = \mu$.

Theorem 5.3.4: Complexity for measure-dominated problems (adaptive regularization–like methods, $\nu = \mu$)

[Appendix to Chapter 5, p. 467]: Suppose that the objective function f is ψ-dominated of level μ for some $\mu \geq 1$ and that an ARψ,ν algorithm for which $\nu \geq 1$ is applied to minimize f. In addition, suppose that $\nu = \mu$. Then the gaps $\{\Gamma_k\}_{k\in\mathcal{S}}$ converge linearly to zero with factor $(1+(\kappa_\nu/\kappa_\psi))^{-1}$, and there exist positive constants $\kappa^{B}_{\mathrm{AR,gap,2}}$, $\kappa^{C}_{\mathrm{AR,gap,2}}$, $\kappa^{E}_{\mathrm{AR,gap,2}}$, and $\kappa^{F}_{\mathrm{AR,gap,2}}$ such that the algorithm needs at most

$$\#^{F}_{\mathrm{AR,gap,2}} \stackrel{\mathrm{def}}{=} \kappa^{B}_{\mathrm{AR,gap,2}}|\log(\epsilon)| + \kappa^{C}_{\mathrm{AR,gap,2}} = \mathcal{O}\left(|\log(\epsilon)|\right) \tag{5.3.20}$$

iterations and evaluations of f and at most

$$\#^{D}_{\mathrm{AR,gap,2}} \stackrel{\mathrm{def}}{=} \kappa^{E}_{\mathrm{AR,gap,2}}|\log(\epsilon)| + \kappa^{F}_{\mathrm{AR,gap,2}} = \mathcal{O}\left(|\log(\epsilon)|\right) \tag{5.3.21}$$

successful iterations (and derivative evaluations) to compute an iterate x_ϵ such that $\Gamma_\epsilon = f(x_\epsilon) - f_* \leq \epsilon$. Moreover, there exist positive constants $\kappa^{B}_{\mathrm{AR,dom,2}}$, $\kappa^{C}_{\mathrm{AR,dom,2}}$, $\kappa^{E}_{\mathrm{AR,dom,2}}$, and $\kappa^{F}_{\mathrm{AR,dom,2}}$ such that the algorithm needs at most

$$\#^{F}_{\mathrm{AR,dom,2}} \stackrel{\mathrm{def}}{=} \kappa^{B}_{\mathrm{AR,dom,2}}|\log(\epsilon)| + \kappa^{C}_{\mathrm{AR,dom,2}} = \mathcal{O}\left(|\log(\epsilon)|\right) \tag{5.3.22}$$

iterations and evaluations of f and at most

$$\#^{D}_{\mathrm{AR,dom,2}} \stackrel{\mathrm{def}}{=} \kappa^{E}_{\mathrm{AR,dom,2}}|\log(\epsilon)| + \kappa^{F}_{\mathrm{AR,dom,2}} = \mathcal{O}\left(|\log(\epsilon)|\right) \tag{5.3.23}$$

successful iterations (and derivative evaluations) to compute an iterate x_ϵ such that $\psi(x_\epsilon) \leq \epsilon$.

We finally come to the case where $\nu > \mu$.

Theorem 5.3.5: Complexity for measure-dominated problems (adaptive regularization–like methods, $\nu > \mu$)

[Appendix to Chapter 5, p. 467]: Suppose that the objective function f is ψ-dominated of level μ for some $\mu \geq 1$ and that an ARψ,ν algorithm for which $\nu \geq 1$ is applied to minimize f. In addition, suppose that $\nu > \mu$. Then the gaps $\{\Gamma_k\}_{k\in\mathcal{S}}$ converge sublinearly to zero, and there exist positive constants $\kappa^{B}_{\mathrm{AR,gap,3}}$, $\kappa^{C}_{\mathrm{AR,gap,3}}$, $\kappa^{E}_{\mathrm{AR,gap,3}}$, and $\kappa^{F}_{\mathrm{AR,gap,3}}$ such that the algorithm needs at most

$$\#^{F}_{\mathrm{AR,gap,3}} \stackrel{\mathrm{def}}{=} \kappa^{B}_{\mathrm{AR,gap,3}}\,\epsilon^{-\frac{\nu-\mu}{\mu}} + \kappa^{C}_{\mathrm{AR,gap,3}} = \mathcal{O}\left(\epsilon^{-\frac{\nu-\mu}{\mu}}\right) \tag{5.3.24}$$

iterations and evaluations of f and at most

$$\#^{D}_{\mathrm{AR,gap,3}} \stackrel{\mathrm{def}}{=} \kappa^{E}_{\mathrm{AR,gap,3}}\,\epsilon^{-\frac{\nu-\mu}{\mu}} + \kappa^{F}_{\mathrm{AR,gap,3}} = \mathcal{O}\left(\epsilon^{-\frac{\nu-\mu}{\mu}}\right) \tag{5.3.25}$$

successful iterations (and derivative evaluations) to compute an iterate x_ϵ such that $\Gamma_\epsilon = f(x_\epsilon) - f_* \leq \epsilon$. Moreover, the optimality measures $\{\psi(x_k)\}_{k\in\mathcal{S}}$ also *converge to zero sublinearly, and there exist positive constants* $\kappa^{B}_{\text{AR,dom},3}$, $\kappa^{C}_{\text{AR,dom},3}$, $\kappa^{E}_{\text{AR,dom},3}$, *and* $\kappa^{F}_{\text{AR,dom},3}$ *such that the algorithm needs at most*

$$\#^{F}_{\text{AR,dom},3} \stackrel{\text{def}}{=} \kappa^{B}_{\text{AR,dom},3}\,\epsilon^{-(\nu-\mu)} + \kappa^{C}_{\text{AR,dom},3} = \mathcal{O}\left(\epsilon^{-(\nu-\mu)}\right) \tag{5.3.26}$$

iterations and evaluations of f *and at most*

$$\#^{D}_{\text{AR,dom},3} \stackrel{\text{def}}{=} \kappa^{E}_{\text{AR,dom},3}\,\epsilon^{-(\nu-\mu)} + \kappa^{F}_{\text{AR,dom},3} = \mathcal{O}\left(\epsilon^{-(\nu-\mu)}\right) \tag{5.3.27}$$

successful iterations (and derivative evaluations) to compute an iterate x_ϵ *such that* $\psi(x_\epsilon) \le \epsilon$.

It is also interesting to consider the effect of the dependence of ν on the degree of the model p on the evaluation complexity. Table 5.3.1 gives the relevant orders for a few low-degree cases when using the ARp algorithm and recalling (5.3.6)–(5.3.8). We see that these dominated complexity bounds are always better—and often substantially so—than the traditional $\mathcal{O}(\epsilon^{-\nu})$ one predicted by Theorem 4.1.5 (p. 107). The case of full-rank nonlinear least-squares is emblematic, since Theorem 5.3.3 implies that the ARp algorithm will converge superlinearly to zero on such problems using at most $\mathcal{O}(\log(|\log(\epsilon)|))$ function and derivative evaluations.

μ	q	ν	$p=1$	$p=2$	$p=3$	$p=4$
1	1	$\frac{p+1}{p}$	$\mathcal{O}\left(\epsilon^{-1}\right)$	$\mathcal{O}\left(\epsilon^{-1/2}\right)$	$\mathcal{O}\left(\epsilon^{-1/3}\right)$	$\mathcal{O}\left(\epsilon^{-1/4}\right)$
	2	$\frac{p+1}{p-1}$	-	$\mathcal{O}\left(\epsilon^{-2}\right)$	$\mathcal{O}\left(\epsilon^{-1}\right)$	$\mathcal{O}\left(\epsilon^{-2/3}\right)$
	3	$\frac{p+1}{p-2}$	-	-	$\mathcal{O}\left(\epsilon^{-3}\right)$	$\mathcal{O}\left(\epsilon^{-3/2}\right)$
2	1	$\frac{p+1}{p}$	$\mathcal{O}\left(\lvert\log(\epsilon)\rvert\right)$	$\mathcal{O}\left(\lvert\log(\lvert\log(\epsilon)\rvert)\rvert\right)$	$\mathcal{O}\left(\lvert\log(\lvert\log(\epsilon)\rvert)\rvert\right)$	$\mathcal{O}\left(\lvert\log(\lvert\log(\epsilon)\rvert)\rvert\right)$
	2	$\frac{p+1}{p-1}$	-	$\mathcal{O}\left(\epsilon^{-1}\right)$	$\mathcal{O}\left(\lvert\log(\epsilon)\rvert\right)$	$\mathcal{O}\left(\lvert\log(\lvert\log(\epsilon)\rvert)\rvert\right)$
	3	$\frac{p+1}{p-1}$	-	-	$\mathcal{O}\left(\epsilon^{-2}\right)$	$\mathcal{O}\left(\epsilon^{-1/2}\right)$

Table 5.3.1. Evaluation-complexity bounds for reducing the gap below ϵ (Theorems 5.3.3, 5.3.4, and 5.3.5) when using the ARp algorithm and the measure ϕ_q given by (5.3.6)–(5.3.8) for $q \le 3$ to approximately minimize a μ-dominated function.

5.3.3 Applying trust-region methods

Trust-region methods also provide a framework in which a decrease condition holds that can be coupled with ψ-domination to improve evaluation-complexity bounds. As for adaptive-regularization methods, the analysis hinges on the fact that the guaranteed function decrease at successful iterations can be bounded below by some power of a well-chosen optimality measure. In particular, Lemma 2.3.5 (p. 39) for the TR1 and first-order TR2 algorithms provides the key, but, as we shall see in Section 12.2.1 (p. 292) of Chapter 12, this also applies to far more general trust-region methods, those that aim for higher-order criticality.

Given a criticality measure $\psi(x)$, we shall be interested in algorithms that guarantee an objective-function decrease

$$f(x_k) - f(x_{k+1}) \geq \kappa_\nu \left[\min_{i\in[0:k]} \psi(x_i)\right]^\nu \tag{5.3.28}$$

for some $\kappa_\nu > 0$ and $\nu \geq 1$ on all successful iterations k. This is the case for the TR1 and TR2 algorithms for the measure $\psi(x) = \psi_1(x) = \|\nabla_x f(x)\|$ with $\kappa_\nu = \eta_1\kappa_{\rm decr}$ and $\nu = 2$ because of (2.3.19) in Lemma 2.3.5.[41] Since our aim is to reduce $\psi(x_k)$ to zero, this is equivalent to forcing $\min_{i\in[0:k]}\psi(x_i)$ to zero since $\psi(x_k) = \min_{i\in[0:k]}\psi(x_i)$ when the former is small enough.

Just as in the previous section, we let TRψ,ν be the class of trust region–like monotonic minimization algorithms for which (a) the decrease (5.3.28) is guaranteed for each k in an infinite subset $\mathcal{S}$ of the iterations, (b) $x_{k+1} = x_k$ for all $k \notin \mathcal{S}$, and (c) $\mathcal{S}_k = \mathcal{S} \cap [0:k]$ is dense enough in $[0:k]$ in the sense of (5.3.9) for some $\kappa_{\rm R}^{\mathcal{S}} \geq 1$ and $\kappa_{\rm A}^{\mathcal{S}} \geq 0$. The bound (5.3.9) is now a generalization of (2.3.11) from Lemma 2.3.1 (p. 36). In particular, it implies that $\mathcal{S} \neq \emptyset$. We also assume that algorithms in the TRψ,ν class are terminated if $\psi(x_k) = 0$, which must happen if $f(x_k) = f_*$.

To remind the reader, the principal difference between algorithms in TRψ,ν and those in the previous ARψ,ν class is that the required improvement (5.3.10) in the former depends on the measure at the new point x_{k+1}, while that in (5.3.28) hinges on the measure at previous points x_i, $i \in [0:k]$. As we shall see, this has a profound impact on worst-case evaluation complexity for ψ-dominated functions.

We now derive a relative of Lemma 5.3.1 for our new class of algorithms.

Lemma 5.3.6: Gap-decrease rates for measure-dominated problems (trust region–like methods)

Suppose that the objective function f is ψ-dominated of level μ for some $\mu \geq 1$ and that a TRψ,ν algorithm for which $\nu \geq 1$ is applied to minimize f. In addition, suppose that $\Gamma_0 > 0$, and let

$$\Gamma_\square \stackrel{\rm def}{=} \left(\frac{\kappa_\psi^{\nu/\mu}}{\kappa_\nu}\right)^{\frac{\mu}{\nu-\mu}}.$$

Then the following conclusions hold.

(i) *If $\nu = \mu$, then $\kappa_\nu \leq \kappa_\psi$ and the gaps $\{\Gamma_k\}_{k\in\mathcal{S}}$ converge linearly to zero with factor $1 - \kappa_\nu/\kappa_\psi$ when $\kappa_\nu < \kappa_\psi$. If $\kappa_\nu = \kappa_\psi$, then $f(x_{k_{\mathcal{S},1}+1}) = f_*$, where $k_{\mathcal{S},1}$ is the index of the first successful iteration.*

---

[41] Similarly, as we shall see, (12.2.20) in the proof of Lemma 12.2.4 (p. 297) for the TRq algorithm described in Chapter 12 does the same. In particular, (12.2.20) implies that, at successful iterations of the TRq algorithm,

$$f(x_k) - f(x_{k+1}) \geq \frac{\eta_1}{q!}\min\left[1, \kappa_\delta \min_{i\in[0:k]} \widehat{\phi}_{f,i}\right]^{q+1} \geq \frac{\eta_1\kappa_\delta^q}{q!}\min_{i\in[0:k]}\min\left[1, \widehat{\phi}_{f,i}\right]^{q+1}$$

for some constants η_1 and κ_Δ in $(0,1)$, where $\widehat{\phi}_{f,k}$ is an optimality measure defined in (12.2.8) (p. 294) (note that, for $q=1$, $\widehat{\phi}_{f,k} = \|\nabla_x^1 f(x_k)\|$), and we may define

$$\psi(x_k) = \min\left[1, \widehat{\phi}_{f,k}\right]$$

and obtain that successful iterations of the TRq algorithm guarantee an objective function decrease of the form (5.3.28), where $\kappa_\nu = \eta_1\kappa_\delta^q/q! \in (0,1)$ and $\nu = q+1$.

(ii) *If $\nu < \mu$, then*

$$\Gamma_0 \geq \Gamma_\square. \tag{5.3.29}$$

If, in addition, inequality (5.3.29) holds as an equality, then $f(x_{k_{\mathcal{S},1}+1}) = f_*$. If the inequality is strict, then

$$\Gamma_{k+1} \leq \Gamma_k \left[1 - \frac{\kappa_\nu}{\kappa_\psi^{\nu/\mu}} \Gamma_k^{\frac{\nu-\mu}{\mu}}\right] \tag{5.3.30}$$

for all $k \in \mathcal{S}$ such that

$$\Gamma_k \geq \Gamma_\square, \tag{5.3.31}$$

and the gaps $\{\Gamma_k\}_{k \in \mathcal{S}}$ start converging superlinearly to zero until a final value $\Gamma_{k_\square} \geq 1$ is reached, after which $f(x_{k_\square+1}) = f_*$.

(iii) *If $\nu > \mu$, then*

$$\Gamma_0 \leq \Gamma_\square. \tag{5.3.32}$$

If, in addition, inequality (5.3.32) holds as an equality, then $f(x_{k_{\mathcal{S},1}+1}) = f_*$. If the inequality is strict, then (5.3.30) holds for all $k \in \mathcal{S}$ and the gaps $\{\Gamma_k\}_{k \in \mathcal{S}}$ converge sublinearly to zero.

Proof. The decrease condition (5.3.28) implies that, for $k \in \mathcal{S}$,

$$f(x_k) - f(x_{k+1}) \geq \kappa_\nu \psi(x_{i_k})^\nu$$

for some $i_k \in [0 : k]$. Applying (5.3.1) at x_{i_k}, we obtain that

$$f(x_k) - f(x_{k+1}) \geq \frac{\kappa_\nu}{\kappa_\psi^{\nu/\mu}} \Gamma_{i_k}^{\frac{\nu}{\mu}} \geq \frac{\kappa_\nu}{\kappa_\psi^{\nu/\mu}} \Gamma_k^{\frac{\nu}{\mu}},$$

where we used the monotonically decreasing nature of the sequence $\{\Gamma_k\}$ to deduce the last inequality. As in the proof of Lemma 5.3.1, we now add f_* to and subtract it from the left-hand side of this inequality and deduce that

$$\Gamma_k - \Gamma_{k+1} \geq \frac{\kappa_\nu}{\kappa_\psi^{\nu/\mu}} \Gamma_k^{\frac{\nu}{\mu}},$$

that is,

$$0 \leq \Gamma_{k+1} \leq \Gamma_k - \frac{\kappa_\nu}{\kappa_\psi^{\nu/\mu}} \Gamma_k^{\frac{\nu}{\mu}} = \Gamma_k \left[1 - \frac{\kappa_\nu}{\kappa_\psi^{\nu/\mu}} \Gamma_k^{\frac{\nu-\mu}{\mu}}\right]. \tag{5.3.33}$$

If $\nu = \mu$, this gives that

$$0 \leq \Gamma_{k+1} \leq \left(1 - \frac{\kappa_\nu}{\kappa_\psi}\right) \Gamma_k, \tag{5.3.34}$$

and $\kappa_\nu > \kappa_\psi$ is thus impossible. Thus $\kappa_\nu \leq \kappa_\psi$, which, with (5.3.34), ensures the conclusions of outcome (i). Namely, if $\kappa_\nu < \kappa_\psi$, the gaps $\{\Gamma_k\}_{k \in \mathcal{S}}$ converge linearly to zero with rate $1 - \kappa_\nu/\kappa_\psi$, while the gap is reduced to zero in a single successful iteration if $\kappa_\nu = \kappa_\psi$. Otherwise $\nu \neq \mu$. If equality holds in (5.3.29) or (5.3.32), then (5.3.33) implies that $\Gamma_{k+1} = \Gamma_{k_{\mathcal{S},1}+1} = 0$, as stated. Suppose therefore that this is not the case and note that (5.3.33) then implies (5.3.30). If $\nu < \mu$, (5.3.33) is impossible for $\omega_k \in (0, 1)$, yielding (5.3.31). If (5.3.29) holds, then (5.3.33) ensures that the gaps start converging superlinearly to zero. However, since a value of $\Gamma_k \in (0,1)$ is impossible, there must be a finite last successful iteration of index $k_\square$ such that (5.3.31) holds, after which (5.3.33) ensures that termination occurs with $f(x_{k_\square}) = f_*$. This is outcome (ii). Finally, if $\nu > \mu$, (5.3.33) implies (5.3.32) and the sublinear convergence of the gaps to zero, yielding outcome (iii). □

The case $\mu > \nu$ does not occur for first-order criticality measures, because we know from Theorem 3.2.1 (p. 57) that $\nu = 2$ for the TR2 algorithm and we mentioned at the start of this section that $\mu \leq 2$ when $\psi(x) = \|\nabla_x f(x)\|$. Hence $\mu = \nu$ in this case.

The worst-case behaviour of the objective-function gap Γ_k in the context of this lemma is illustrated by the two curves in Figure 5.3.2. For algorithms such that (5.3.28) holds, the curves show, as one might expect, that a higher ν (meaning a smaller guaranteed function decrease when close to an approximate minimizer) results in slower convergence. Observe that, since the situation $\nu < \mu$ cannot arise except in exceptional cases, superlinear convergence of the gaps does not occur.

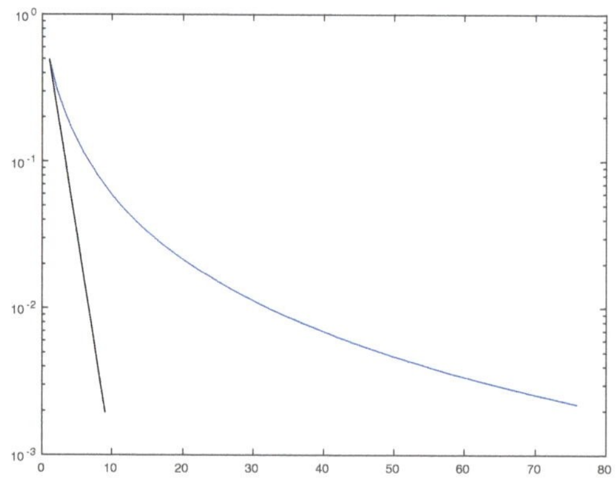

Figure 5.3.2. The decrease of Γ_k as predicted by Lemma 5.3.6 for $\mu = 2$ and $\nu = 2$ (black) and $\nu = 3$ (blue).

We may contemplate complexity bounds for the approximate minimization of ψ-dominated functions using TRψ,ν algorithms (with proofs again in Appendix B).

Theorem 5.3.7: Complexity for measure-dominated problems (trust region–like methods, $\nu < \mu$)

[Appendix to Chapter 5, p. 470]: Suppose that the objective function f is ψ-dominated of level μ for some $\mu \geq 1$ and that a TRψ,ν algorithm for which $\nu \geq 1$ is applied to minimize f. In addition, suppose that $\nu < \mu$. Then either the gap is zero after the first successful iteration or the gaps $\{\Gamma_k\}_{k\in\mathcal{S}}$ start converging superlinearly to zero until a final iteration $k_\square$ such that (5.3.29) holds, after which it is identically zero. Thus there exist positive constants $\kappa^C_{\text{TR},1}$ and $\kappa^F_{\text{TR},1}$ such that the algorithm needs at most

$$\#^F_{\text{TR,gap},1} \stackrel{\text{def}}{=} \kappa^C_{\text{TR},1} = \mathcal{O}(1) \tag{5.3.35}$$

iterations and evaluations of f and at most

$$\#^D_{\text{TR,gap},1} \stackrel{\text{def}}{=} \kappa^F_{\text{TR},1} = \mathcal{O}(1) \tag{5.3.36}$$

successful iterations (and derivative evaluations) to compute an iterate x_ϵ *such that* $\Gamma_\epsilon = f(x_\epsilon) - f_* = 0$. *Moreover, the algorithm needs at most*

$$\#^F_{\mathsf{TR,dom,1}} \stackrel{\text{def}}{=} \kappa^C_{\mathsf{TR,1}} = \mathcal{O}(1) \tag{5.3.37}$$

iterations and evaluations of f *and at most*

$$\#^D_{\mathsf{TR,dom,1}} \stackrel{\text{def}}{=} \kappa^F_{\mathsf{TR,1}} = \mathcal{O}(1) \tag{5.3.38}$$

successful iterations (and derivative evaluations) to compute an iterate x_ϵ *such that* $\psi(x_\epsilon) = 0$.

Theorem 5.3.8: Complexity for measure-dominated problems (trust region–like methods, $\nu = \mu$)

[Appendix to Chapter 5, p. 471]: Suppose that the objective function f *is* ψ*-dominated of level* μ *for some* $\mu \geq 1$ *and that a* TRψ,ν *algorithm for which* $\nu \geq 1$ *is applied to minimize* f. *In addition, suppose that* $\nu = \mu$. *Then either the gap is zero after the first successful iteration or the gaps* $\{\Gamma_k\}_{k\in\mathcal{S}}$ *converge linearly to zero with factor* $1 - \kappa_\nu/\kappa_\psi$, *and there exist positive constants* $\kappa^B_{\mathsf{TR,gap,2}}$, $\kappa^C_{\mathsf{TR,gap,2}}$, $\kappa^E_{\mathsf{TR,gap,2}}$, *and* $\kappa^F_{\mathsf{TR,gap,2}}$ *such that the algorithm needs at most*

$$\#^F_{\mathsf{TR,gap,2}} \stackrel{\text{def}}{=} \kappa^B_{\mathsf{TR,gap,2}} |\log(\epsilon)| + \kappa^C_{\mathsf{TR,gap,2}} = \mathcal{O}(|\log(\epsilon)|) \tag{5.3.39}$$

iterations and evaluations of f *and at most*

$$\#^D_{\mathsf{TR,gap,2}} \stackrel{\text{def}}{=} \kappa^E_{\mathsf{TR,gap,2}} |\log(\epsilon)| + \kappa^F_{\mathsf{TR,gap,2}} = \mathcal{O}(|\log(\epsilon)|) \tag{5.3.40}$$

successful iterations (and derivative evaluations) to compute an iterate x_ϵ *such that* $\Gamma_\epsilon = f(x_\epsilon) - f_* \leq \epsilon$. *Moreover, there exist positive constants* $\kappa^B_{\mathsf{TR,dom,2}}$, $\kappa^C_{\mathsf{TR,dom,2}}$, $\kappa^E_{\mathsf{TR,dom,2}}$, *and* $\kappa^F_{\mathsf{TR,dom,2}}$ *such that the algorithm needs at most*

$$\#^F_{\mathsf{TR,dom,2}} \stackrel{\text{def}}{=} \kappa^B_{\mathsf{TR,dom,2}} |\log(\epsilon)| + \kappa^C_{\mathsf{TR,dom,2}} = \mathcal{O}(|\log(\epsilon)|) \tag{5.3.41}$$

iterations and evaluations of f *and at most*

$$\#^D_{\mathsf{TR,dom,2}} \stackrel{\text{def}}{=} \kappa^E_{\mathsf{TR,dom,2}} |\log(\epsilon)| + \kappa^F_{\mathsf{TR,dom,2}} = \mathcal{O}(|\log(\epsilon)|) \tag{5.3.42}$$

successful iterations (and derivative evaluations) to compute an iterate x_ϵ *such that* $\psi(x_\epsilon) \leq \epsilon$.

Theorem 5.3.9: Complexity for measure-dominated problems (trust region–like methods, $\nu > \mu$)

[Appendix to Chapter 5, p. 472]: Suppose that the objective function f *is* ψ*-dominated of level* μ *for some* $\mu \geq 1$ *and that a* TRψ,ν *algorithm for which* $\nu \geq 1$ *is applied to minimize* f. *In addition, suppose that* $\nu > \mu$. *Then the gaps* $\{\Gamma_k\}_{k\in\mathcal{S}}$ *converge sublinearly to zero, and there exist positive constants* $\kappa^B_{\mathsf{TR,gap,3}}$,

$\kappa^C_{\rm TR,gap,3}$, $\kappa^E_{\rm TR,gap,3}$, and $\kappa^F_{\rm TR,gap,3}$ such that the algorithm needs at most

$$\#^F_{\rm TR,gap,3} \stackrel{\rm def}{=} \kappa^B_{\rm TR,gap,3}\,\epsilon^{-\frac{\nu-\mu}{\mu}} + \kappa^C_{\rm TR,gap,3} = \mathcal{O}\left(\epsilon^{-\frac{\nu-\mu}{\mu}}\right)$$

iterations and evaluations of f and at most

$$\#^D_{\rm TR,gap,3} \stackrel{\rm def}{=} \kappa^E_{\rm TR,gap,3}\,\epsilon^{-\frac{\nu-\mu}{\mu}} + \kappa^F_{\rm TR,gap,3} = \mathcal{O}\left(\epsilon^{-\frac{\nu-\mu}{\mu}}\right) \tag{5.3.43}$$

successful iterations (and derivative evaluations) to compute an iterate x_ϵ such that $\Gamma_\epsilon = f(x_\epsilon) - f_* \le \epsilon$. Moreover, the optimality measures $\{\psi(x_k)\}_{k\in\mathcal{S}}$ also converge to zero sublinearly and there exist positive constants $\kappa^B_{\rm TR,dom,3}$, $\kappa^C_{\rm TR,dom,3}$, $\kappa^E_{\rm TR,dom,3}$, and $\kappa^F_{\rm TR,dom,3}$ such that the algorithm needs at most

$$\#^F_{\rm TR,dom,3} \stackrel{\rm def}{=} \kappa^B_{\rm TR,dom,3}\,\epsilon^{-(\nu-\mu)} + \kappa^C_{\rm TR,dom,3} = \mathcal{O}\left(\epsilon^{-(\nu-\mu)}\right)$$

iterations and evaluations of f and at most

$$\#^D_{\rm TR,dom,3} \stackrel{\rm def}{=} \kappa^E_{\rm TR,dom,3}\,\epsilon^{-(\nu-\mu)} + \kappa^F_{\rm TR,dom,3} = \mathcal{O}\left(\epsilon^{-(\nu-\mu)}\right)$$

successful iterations (and derivative evaluations) to compute an iterate x_ϵ such that $\psi(x_\epsilon) \le \epsilon$.

In the linear and sublinear cases, the complexities of adaptive regularization–like and trust region–like methods applied to measure-dominated functions are therefore identical in order, the main difference being that, barring convergence in a finite number of successful iterations, superlinear convergence (and thus much improved complexity) is only guaranteed to occur for the former.

5.3.4 Consequences for uniformly full rank least-squares problems

We argued at the beginning of Section 5.1 that nonlinear least-squares problems are truly important in practice and devoted a full section of this chapter to the analysis of adaptive-regularization methods to solve them. Moreover, we mentioned in Section 5.3.1 that least-squares problems that are uniformly full rank are a prime example of measure-dominated problems, and we showed that

$$\Phi_\ell(x) = \frac{1}{\ell}\|r(x)\|^\ell$$

with $\ell \ge 2$, the objective function of interest in this case, is measure dominated of level $\mu = \ell/(\ell-1)$, so that $\mu \in (1,2]$. We now specialize the results of Sections 5.3.2 and 5.3.3 to this context and show that a much stronger result than Theorem 5.1.7 is possible under the full-rank assumption mentioned in Section 5.1.5.

Thus, for the rest of this section, we assume that condition (5.1.62) holds, that is, that the smallest singular value of the Jacobian of $r(x)$ is bounded away from zero at all points x encountered by the algorithms we are going to consider. Hence the inequalities (5.3.2) and (5.3.3) hold at each iterate x_k. As a consequence, the convergence of $\|\nabla^1_x\Phi_\ell(x_k)\|$ to zero implies the convergence of $\|r(x_k)\|$ to zero, and also $\min_x \phi_\ell(x) = 0$. This in turn makes the consideration of necessary optimality conditions of orders above one unnecessary, and we therefore focus on the convergence of $\|r(x_k)\|$ to zero.

Consider first applying an adaptive-regularization algorithm ARψ,ν for finding a first-order approximate minimizer of $\Phi_\ell(x)$. We saw in case 1, which we discussed on page 140, that the equality $\nu = (p+1)/p > 1$ then holds. As a consequence, we have that it is always possible to ensure that $\mu > \nu$ by choosing $\ell < p+1$, and we may therefore apply Theorem 5.3.3 to deduce the following result.

Theorem 5.3.10: Complexity for full-rank nonlinear least-squares (adaptive regularization–like methods)

Suppose that an ARψ,ν algorithm is applied to minimize Φ_ℓ for an even positive integer $\ell \le p$, where p is the degree of the model used by this algorithm. Suppose also that (5.1.62) holds. Then the sequence $\{\|r(x_k)\|\}_{k\in\mathcal{S}}$ converges superlinearly to zero and there exist positive constants $\kappa^B_{\text{AR,FRLS}}$, $\kappa^C_{\text{AR,FRLS}}$, $\kappa^E_{\text{AR,FRLS}}$, and $\kappa^F_{\text{AR,FRLS}}$ such that the algorithm needs at most

$$\#^F_{\text{AR,FRLS}} \stackrel{\text{def}}{=} \kappa^B_{\text{AR,FRLS}} \log\left(|\log(\epsilon_0)| + \frac{1}{\ell}\log\ell\right) + \kappa^C_{\text{AR,FRLS}} = \mathcal{O}\left(\log|\log(\epsilon_0)|\right) \quad (5.3.44)$$

iterations and evaluations[a] *of Φ_ℓ and at most*

$$\#^D_{\text{AR,FRLS}} \stackrel{\text{def}}{=} \kappa^E_{\text{AR,FRLS}} \log\left(|\log(\epsilon_0)| + \frac{1}{\ell}\log\ell\right) + \kappa^F_{\text{AR,FRLS}} = \mathcal{O}\left(\log|\log(\epsilon_0)|\right) \quad (5.3.45)$$

successful iterations (and derivative evaluations) to compute an iterate x_ϵ such that $\|r(x_\epsilon)\| \le \epsilon_0$.

[a] Strictly, to show the order estimates $\mathcal{O}(\log|\log(\epsilon_0)|)$ in (5.3.44) and (5.3.45) requires a mild additional assumption such as that ϵ_0 is small enough so that $|\log\epsilon_0| \ge 1$.

Proof. A direct application of Theorem 5.3.3 and (5.3.3) yields that

$$\|r(x_k)\|^\ell = o\big(\ell(\Phi_\ell(x_k) - 0)\big),$$

which in turn implies that $\{\|r(x_k)\|\}_{k\in\mathcal{S}}$ also converges superlinearly to zero. The bounds (5.3.44) and (5.3.45) then result from (5.3.16) and (5.3.17) in Theorem 5.3.3 with $\epsilon = \epsilon_0^\ell/\ell$ and noting that

$$\log\left(\left|\log\left(\frac{1}{\ell}\epsilon_0^\ell\right)\right|\right) = \log\left(\ell\left|\log\left(\frac{\epsilon_0}{\ell^{1/\ell}}\right)\right|\right) \le \log\left(|\log(\epsilon_0)| + \frac{1}{\ell}\log\ell\right) + \log(\ell).$$

Thus the constants $\kappa^B_{\text{AR,FRLS}}$, $\kappa^C_{\text{AR,FRLS}}$, $\kappa^E_{\text{AR,FRLS}}$, and $\kappa^F_{\text{AR,FRLS}}$ are inherited from the corresponding constants in Theorem 5.3.3, except that $\kappa^C_{\text{AR,FRLS}}$ and $\kappa^F_{\text{AR,FRLS}}$ are increased by $\kappa^B_{\text{AR,FRLS}}(\log\ell)$ and $\kappa^F_{\text{AR,FRLS}}(\log\ell)$, respectively. □

This is obviously in stark contrast to Theorem 5.1.7, but at the price of the strong assumption (5.1.62).

If we now turn to TRψ,ν methods, we note that $\nu = 2$ for TR1, TR2, and (the unspecified) TRp, which we touched on in Section 4.1.3, for the first-order measure ψ_1, which is what we need here. If $\ell > 2$, then $\mu = \ell/(\ell-1) < \nu$, and therefore the much less favourable Theorem 5.3.9 holds for TRψ,ν methods.

Theorem 5.3.11: Complexity for full-rank nonlinear least-squares (trust region–like methods, $\ell > 2$)

Suppose that a TRψ, ν *algorithm is applied to minimize* Φ_ℓ *for an even integer* $\ell > 2$*. Suppose also that (5.1.62) holds. Then the sequence* $\{\|r(x_k)\|\}_{k\in\mathcal{S}}$ *converges sublinearly to zero, and there exist positive constants* $\kappa^{B}_{\rm TR2,FRLS}$, $\kappa^{C}_{\rm TR2,FRLS}$, $\kappa^{E}_{\rm TR2,FRLS}$, *and* $\kappa^{F}_{\rm TR2,FRLS}$ *such that the algorithm needs at most*

$$\#^{F}_{\rm TR2,FRLS} \stackrel{\rm def}{=} \kappa^{B}_{\rm TR2,FRLS}\, \epsilon_0^{-(\ell-2)} + \kappa^{C}_{\rm TR2,FRLS} = \mathcal{O}\left(\epsilon_0^{-(\ell-2)}\right)$$

iterations and evaluations of r *and at most*

$$\#^{D}_{\rm TR2,FRLS} \stackrel{\rm def}{=} \kappa^{E}_{\rm TR2,FRLS}\, \epsilon_0^{-(\ell-2)} + \kappa^{F}_{\rm TR2,FRLS} = \mathcal{O}\left(\epsilon_0^{-(\ell-2)}\right)$$

successful iterations (and derivative evaluations) to compute an iterate x_ϵ *such that* $\|r(x_\epsilon)\| \leq \epsilon_0$.

Proof. We now apply Theorem 5.3.9 with $\epsilon = \epsilon_0^\ell/\ell$ and observe that

$$\frac{\nu-\mu}{\mu} = \frac{2-\frac{\ell}{\ell-1}}{\frac{\ell}{\ell-1}} = \frac{\ell-2}{\ell} \quad \text{and} \quad \epsilon^{-\frac{\nu-\mu}{\mu}} = \left(\frac{1}{\ell}\epsilon_0^\ell\right)^{-\frac{\ell-2}{\ell}} = \ell^{\frac{\ell-2}{\ell}}\epsilon_0^{-(\ell-2)}.$$

The constants $\kappa^{\rm B}_{\rm TR,FRLS}$, $\kappa^{\rm C}_{\rm TR,FRLS}$, $\kappa^{\rm E}_{\rm TR,FRLS}$, and $\kappa^{\rm F}_{\rm TR,FRLS}$ are inherited from the corresponding constants in Theorem 5.3.9, except that $\kappa^{\rm B}_{\rm TR,FRLS}$ and $\kappa^{\rm E}_{\rm TR,FRLS}$ are multiplied by the factor $\ell^{(\ell-2)/\ell}$. □

Notice again that although the bounds depend on p (as well as ℓ), the complexity order does not.

A much better bound is possible when $\ell = 2$, for then $\mu = \nu$, and we may appeal to Theorem 5.3.8. This gives the following result.

Theorem 5.3.12: Complexity for full-rank nonlinear least-squares (trust region–like methods, $\ell = 2$)

Suppose that a TRψ, ν *algorithm is applied to minimize* Φ_2*. Suppose also that (5.1.62) holds. Then either* $r(x_k) = 0$ *after the first successful iteration or the sequence* $\{\|r(x_k)\|\}_{k\in\mathcal{S}}$ *converges linearly to zero, and there exist positive constants* $\kappa^{B}_{\rm TR,FRLS}$, $\kappa^{C}_{\rm TR,FRLS}$, $\kappa^{E}_{\rm TR,FRLS}$, *and* $\kappa^{F}_{\rm TR,FRLS}$ *such that the algorithm needs at most*

$$\#^{F}_{\rm TR,FRLS} \stackrel{\rm def}{=} \kappa^{B}_{\rm TR,FRLS}\, |\log(\epsilon_0)| + \kappa^{C}_{\rm TR1,FRLS} = \mathcal{O}\left(|\log(\epsilon_0)|\right)$$

iterations and evaluations of r *and at most*

$$\#^{D}_{\rm TR,FRLS} \stackrel{\rm def}{=} \kappa^{E}_{\rm TR,FRLS}\, |\log(\epsilon_0)| + \kappa^{F}_{\rm TR,FRLS} = \mathcal{O}\left(|\log(\epsilon_0)|\right)$$

successful iterations (and derivative evaluations) to compute an iterate x_ϵ *such that* $\|r(x_\epsilon)\| \leq \epsilon_0$.

Proof. Since $\ell = 2$, we have that $\nu = \mu = 2$ and Theorem 5.3.8 applies. The constants are again directly inherited from its proof. □

On this basis, there seems little reason to favour $\ell > 2$ in the trust-region case. Moreover, the result of Theorem 5.3.12 applies equally to steepest-descent ($p = 1$), Gauss–Newton (the TR1 algorithm on page 35 with $B_k = (\nabla_x^1 r(x_k))^T \nabla_x^1 r(x_k)$), and Newton ($p = 2$) models.

In summary, the take-away message of this section is that a much better evaluation-complexity bound may be derived when a nonlinear least-squares problem is uniformly of full rank, although in practice this assumption may be hard to verify.

Notes and references for Section 5.3

The rates of decrease for "gradient-dominated" problems were analyzed by Nesterov and Polyak [305], in conjunction with the adaptive-regularization algorithm using exact global minimization of a regularized second-order model. This paper also mentions an earlier reference by Polyak [317]. The first of these references also inspired the work by Curtis and Robinson [135], who extended the idea of domination to second order and beyond. For order ℓ, [135] proposes the specific measure

$$\psi(x) = f(x) - \min_{d \in \mathbb{R}^n} \left(\frac{1}{\ell!} \nabla_x^\ell(x)[d]^\ell + \frac{\|d\|^{\ell+1}}{\ell+1} \right),$$

which is a weaker optimality condition than $\phi_{f,\ell}^\delta(x)$, which we shall encounter in Chapter 12 (see Example 12.1.2, p. 277). Curtis and Robinson also introduce the interesting concept of "regional convergence," in which ψ-domination is measured not with respect to the global minimum value f_* but with respect to a local "reference value" $f_{\rm ref}$ for which the relevant variant of (5.3.1) holds in a local neighbourhood, called the "region"—of course, we have already seen this in the special case where the only reference value is $f_{\rm ref} = f_*$. A more general choice of $f_{\rm ref}$ has the advantage of flexibility—the inequality (5.3.10) is only needed locally in relation to a reference value, possibly for some finite subsequence of iterates—and may help us understand specific convergence histories a posteriori, even if determining regions and associated reference values may be difficult.

Note that, at a given successful iteration of the ARp algorithm, the decrease $f(x_k) - f(x_{k+1})$ must exceed all the lower bounds (4.1.31)–(4.1.35) discussed above; that is,

$$\begin{aligned} f(x_k) - f(x_{k+1}) &\geq \max\left[\kappa_1 \psi_1(x_{k+1})^{\frac{p+1}{p}}, \kappa_2 \psi_2(x_{k+1})^{\frac{p+1}{p-1}}, \kappa_3 \psi_3(x_{k+1})^{\frac{p+1}{p-3}}\right] \\ &\geq \min[\kappa_1, \kappa_2, \kappa_3] \max\left[\psi_1(x_{k+1})^{\frac{p+1}{p}}, \psi_2(x_{k+1})^{\frac{p+1}{p-1}}, \psi_3(x_{k+1})^{\frac{p+1}{p-3}}\right]. \end{aligned} \tag{5.3.46}$$

It is then possible, as in [135], to relax the definition (5.3.1) to say that f is $[\psi_1, \ldots, \psi_t]$-dominated of level $(\mu_1, \ldots, \mu_t)$ for real $\mu_i > 0$ ($i \in [1 : t]$) if there is a constant $\kappa_\psi > 0$ such that

$$f(x) - f_* \leq \kappa_\psi \left[\max_{i \in [1:t]} \psi_i(x)^{\mu_i} \right] \tag{5.3.47}$$

for all $x \in \mathbb{R}^n$. It is then clear from (5.3.46) and (5.3.47) that (5.3.10) remains valid, where now the level μ at iteration k of the generic algorithm is determined by which of the terms in the maximum in (5.3.47) is largest. We have not explored this possibility because it involves making quite strong and complicated assumptions on which μ_i is "active" and in what order.

Also note that the weakening of the decrease condition proposed in (5.3.28), in which only the smallest of the past optimality measures appears in the lower bound, can also be applied to the adaptive-regularization case, with a straightforward adaptation of the argument at the beginning of the proof of Lemma 5.3.6.

There is a substantial literature on the interaction between gradient-based methods, gradient flows, and the Polyak–Łojasiewicz property (see, for instance, Bolte et al. [53]; Noll and Rondepierre [308]; or Karimi, Nutini, and Schmidt [252] and the references therein), but this goes beyond the scope of this book. We also note interesting statements[42] by Zhou, Wang,

[42]No proof is provided.

and Liang [380] on the impact of the Kurdyka–Łojasiewicz inequality [261] (a generalization of the Polyak–Łojasiewicz property) on the convergence rate of the ARC algorithm to second-order minimizers. The explicit computations of the Łojasiewicz exponent μ for classes of optimization problems is also the subject of current investigation (see Li and Pong [270]). For simplicity of exposition, we have also ignored an additional feature of the discussions in [135, 305], where a further decomposition of the convergence of errors to zero implied by (5.3.10) in successive phases is presented (as in the proof of Theorem 5.3.3). Finally, our choice to limit the presentation above to low orders is also deliberate. While an extension of the notion to higher levels is technically possible, it would then use a $\psi(x)$-measure based on $\phi_{f,j}^{\delta_j}(x)$ (see Chapter 12), for which one must then specify an optimality radius δ_j in some uniform way. This then complicates the notion beyond what we currently feel is useful.

We finally note that Theorem 5.3.12 is reminiscent of the results of Gratton, Lawless, and Nichols [209], where the *local* linear convergence of the Gauss–Newton method is established under the condition (5.3.1). As we hinted in Section 5.1.5, similar results may be obtained using the proof techniques used in Theorem 5.1.7 [97, Thm. 3.6] and, for the case $p = 2$, by other means [305, Thm. 7].

5.4 Phase complexity

5.4.1 Introduction

We have seen that when nonconvex f has a Lipschitz-continuous Hessian, the AR2 algorithm will find an ϵ-approximate first-order minimizer using at most $\mathcal{O}(\epsilon^{-3/2})$ function and derivative iterations, but that there are fiendish, carefully constructed examples for which this pessimistic bound may be attained. This is particularly unfortunate since, in practice, many years of computational experience with real-world problems have not revealed such bad examples, and evaluations in the tens to hundreds are far more typical.

The question we wish to consider in this section is whether the general bound can be improved at least for certain (nonconvex) problem classes, such as those whose local minimizers are strict. In particular, using known results for strongly convex problems [90] and the results we have already presented for general nonconvex functions, we may characterize such (nonconvex) problem classes by their *complexity phases*, so that the overall algorithm complexity is obtained by *adding* together the complexities of the phases. A notable analysis of such additive phase complexity was given for Nesterov and Polyak's cubic-regularization algorithm on classes that include problems with *gradient-dominated* objectives—which we considered in more general terms in the previous section—and those derived from *nonlinear transformations of convex functions* [305, Sects. 4.2, 4.3]. More recently, Curtis and Robinson have investigated a general formulation that they refer to as regional complexity [135]. Our interest here, aside from employing and analyzing a practical cubic-regularization algorithm that is suitable for large-scale calculations, is to investigate "head-on" nonconvex problems and subclasses.

5.4.2 Phase complexity under strict local convexity

The results in this section establish that strong convexity of f in a neighbourhood of a strict local minimizer implies local Q-quadratic convergence of the AR2 algorithm (p. 65) in this neighbourhood and hence provides an associated evaluation complexity of order $\log_2(\log_2(\epsilon^{-1}))$.

We first show that there is a neighbourhood around any strong local minimizer of f for which f is locally strongly convex.

Lemma 5.4.1: Region of local strong convexity

Let x_* be a local minimizer of $f(x)$, and suppose that **f.HL** *holds and that $\lambda_{\min,*} \stackrel{\text{def}}{=} \lambda_{\min}[\nabla_x^2 f(x_*)] > 0$. Then there is a neighbourhood of x_*,*

$$\mathcal{Q}_* \stackrel{\text{def}}{=} \{x : \|x - x_*\| \leq q_*\}, \quad \text{where } q_* \stackrel{\text{def}}{=} \frac{\lambda_{\min,*}}{4L_{f,2}}, \tag{5.4.1}$$

for which $\lambda_{\min}[\nabla_x^2 f(x)] \geq \frac{1}{2}\lambda_{\min,*}$ for all $x \in \mathcal{Q}_*$.

Proof. Consider $x \in \mathcal{Q}_*$. It then follows that

$$\left|\lambda_{\min}[\nabla_x^2 f(x)] - \lambda_{\min,*}\right| \leq \|\nabla_x^2 f(x) - \nabla_x^2 f(x_*)\| \leq L_{f,2}\|x - x_*\| \leq \tfrac{1}{2}\lambda_{\min,*}, \tag{5.4.2}$$

using, in order, the eigenvalue bound (A.5), **f.HL**, the weaker bound $\|x - x_*\| \leq 2q_*$ and the definition (5.4.1) of q_*. Rearranging (5.4.2) gives the required result. □

Next, we show that tightening the model-termination requirement in the AR2 algorithm leads to a quadratic reduction in the gradient of f if an iterate enters a neighbourhood of a strong local minimizer of f, and that this behaviour persists.

Lemma 5.4.2: Region of quadratic convergence

Suppose that **f.HL** *holds, that x_* is a local minimizer of $f(x)$ for which $\lambda_{\min,*} = \lambda_{\min}[\nabla_x^2 f(x_*)] > 0$, and that the AR2 algorithm (p. 65), with the more stringent model-termination test*

$$\|\nabla_s^1 m_k(s_k)\| \leq \theta_1 \min\left(\|\nabla_x^1 f(x_k)\|, \tfrac{1}{2}\|s_k\|^2\right) \tag{5.4.3}$$

replacing (3.3.5), is applied to minimize f. Furthermore, suppose that $x_k \in \mathcal{Q}_*$, where the set $\mathcal{Q}_*$ is as given in (5.4.1), and that iteration k is successful. Then

$$\|\nabla_x^1 f(x_{k+1})\| \leq \frac{1}{\zeta_*}\|\nabla_x^1 f(x_k)\|^2, \tag{5.4.4}$$

where

$$\begin{aligned}\zeta_* \stackrel{\text{def}}{=} \min\Bigg[& \frac{2}{L_{f,2} + \theta_1 + \sigma_{\max}}\left(\frac{\lambda_{\min,*} + \frac{1}{2}\sigma_{\min}}{1 + \theta_1}\right)^2, \\ & \frac{\lambda_{\min,*}}{4L_{f,2}}\left(\frac{\lambda_{\min,*} + \frac{1}{2}\sigma_{\min}}{1 + \theta_1}\right), \frac{\lambda_{\min,*}^2}{16L_{f,2}}, \\ & \frac{3(1 - \eta_2)\lambda_{\min,*}}{4L_{f,2}}\left(\frac{\lambda_{\min,*} + \frac{1}{2}\sigma_{\min}}{1 + \theta_1}\right), \\ & \frac{\lambda_{\min,*}}{4\max(\theta_1 - \sigma_{\min}, 0)}\left(\frac{\lambda_{\min,*} + \frac{1}{2}\sigma_{\min}}{1 + \theta_1}\right)\Bigg],\end{aligned} \tag{5.4.5}$$

$\sigma_{\min}$ and $\theta_1 > 0$ are as specified for the algorithm, and $\sigma_{\max}$ is given by (3.3.14). Furthermore, let

$$\mathcal{Q}_*^1 \stackrel{\text{def}}{=} \mathcal{Q}_* \cap \{x : \|\nabla_x^1 f(x)\| \leq \zeta_*\}, \tag{5.4.6}$$

and suppose that $x_{k_0} \in \mathcal{Q}_*^1$ for some $k_0 \geq 0$. Then $x_k \in \mathcal{Q}_*^1$ for all iterations $k \geq k_0$, each of these iterations is successful, and (5.4.4) holds for all.

Proof. It follows from Lemma 5.4.1 that

$$\lambda_{\min}[\nabla_x^2 f(x)] \geq \tfrac{1}{2}\lambda_{\min,*} > 0 \tag{5.4.7}$$

whenever $x = x_k \in \mathcal{Q}_*$. The stopping rule (5.4.3) implies (3.3.5), and thus the convergence theory given in Section 3.3 carries over to the modified algorithm. Since Step 4 of the AR2 algorithm ensures that $\sigma_{\min} \leq \sigma_k$, the Rayleigh-quotient inequality, (5.4.7) at x_k, the Cauchy–Schwarz inequality, the definition of the model gradient (3.3.2), and the inner-iteration stopping rule (5.4.3) in turn imply

$$\begin{aligned}
(\lambda_{\min,*} + \tfrac{1}{2}\sigma_{\min})\|s_k\|^2 &\leq \langle s_k, (\nabla_x^2 f(x_k) + \tfrac{1}{2}\sigma_k I)s_k\rangle \\
&\leq \|s_k\|\|(\nabla_x^2 f(x_k) + \tfrac{1}{2}\sigma_k I)s_k\| \\
&= \|s_k\|\left(\|(\nabla_x^2 f(x_k) + \tfrac{1}{2}\sigma_k I)s_k + \nabla_x^1 f(x_k) - \nabla_x^1 f(x_k)\|\right) \\
&\leq \|s_k\|\left(\|(\nabla_x^2 f(x_k) + \tfrac{1}{2}\sigma_k I)s_k + \nabla_x^1 f(x_k)\| + \|\nabla_x^1 f(x_k)\|\right) \\
&= \|s_k\|\left(\|\nabla_s^1 m_k(s_k)\| + \|\nabla_x^1 f(x_k)\|\right) \\
&\leq \|s_k\|(1+\theta_1)\|\nabla_x^1 f(x_k)\|
\end{aligned}$$

and thus, dividing by $\|s_k\|$ and rearranging terms, that

$$\|s_k\| \leq \kappa_s \|\nabla_x^1 f(x_k)\|, \quad \text{where } \kappa_s \stackrel{\text{def}}{=} \frac{1+\theta_1}{\lambda_{\min,*} + \tfrac{1}{2}\sigma_{\min}}. \tag{5.4.8}$$

Combining this with (3.3.15) leads directly to (5.4.4).

Now suppose that $x_k \in \mathcal{Q}_*^1$ and $k \in \mathcal{S}$. Then (5.4.4) implies $\|\nabla_x^1 f(x_{k+1})\| \leq \zeta_*$, and it remains to show that $x_{k+1} \in \mathcal{Q}_*$. Applying again (A.5), for any ξ_{k+1} on the line segment $[x_{k+1}, x_*]$,

$$\begin{aligned}
\left|\lambda_{\min}[\nabla_x^2 f(\xi_{k+1})] - \lambda_{\min,*}\right| &\leq \|\nabla_x^2 f(\xi_{k+1}) - \nabla_x^2 f(x_*)\| \\
&\leq L_{f,2}\|x_{k+1} - x_*\| \\
&\leq L_{f,2}\|s_k\| + L_{f,2}\|x_k - x_*\| \\
&\leq L_{f,2}\kappa_s\|\nabla_x^1 f(x_k)\| + \tfrac{1}{4}\lambda_{\min,*} \\
&\leq \tfrac{1}{2}\lambda_{\min,*},
\end{aligned}$$

where in the remaining inequalities we used **f.HL**, the triangle inequality, (5.4.8), (5.4.1), (5.4.6), and (5.4.5), respectively. Thus (5.4.7) also holds for $x = \xi_{k+1}$, and hence f is strongly convex on the line segment $[x_{k+1}, x_*]$. A Taylor expansion gives

$$f(x_*) = f(x_{k+1}) + \langle x_* - x_{k+1}, \nabla_x^1 f(x_{k+1})\rangle + \tfrac{1}{2}\langle x_* - x_{k+1}, \nabla_x^2 f(\xi_{k+1})(x_* - x_{k+1})\rangle$$

for some ξ_{k+1} on the line segment $[x_{k+1}, x_*]$, which together with (5.4.7) for $x = \xi_{k+1}$ implies that

$$f(x_*) \geq f(x_{k+1}) + \langle x_* - x_{k+1}, \nabla_x^1 f(x_{k+1})\rangle + \tfrac{1}{4}\lambda_{\min,*}\|x_{k+1} - x_*\|^2.$$

Moreover, since f is convex on the line segment $[x_{k+1}, x_*]$ and $\nabla_x^1 f(x_*) = 0$, it follows from (1.2.1) that $f(x_{k+1}) \geq f(x_*)$, and we deduce

$$\langle x_{k+1} - x_*, \nabla_x^1 f(x_{k+1})\rangle \geq \tfrac{1}{4}\lambda_{\min,*}\|x_{k+1} - x_*\|^2.$$

Applying the Cauchy–Schwarz inequality to the left-hand side of this expression reveals that

$$\|x_{k+1} - x_*\| \le \frac{4}{\lambda_{\min,*}} \|\nabla_x^1 f(x_{k+1})\|,$$

which together with $\|\nabla_x^1 f(x_{k+1})\| \le \zeta_*$ and the expression (5.4.5) gives $\|x_{k+1} - x_*\| \le q_*$ and so $x_{k+1} \in \mathcal{Q}_*$.

Finally, using Taylor's theorem and **f.hL**, we have

$$|f(x_k + s_k) - T_{f,2}(x_k, s_k)| \le \tfrac{1}{6} L_{f,2} \|s_k\|^3 \tag{5.4.9}$$

directly from (A.19). Because

$$\nabla_s^1 m_k(s_k) = \nabla_x^1 f(x_k) + \nabla_x^2 f(x_k) s_k + \tfrac{1}{2}\sigma_k \|s_k\| s_k$$

from (3.3.2), we have immediately that

$$\langle \nabla_x^1 f(x_k), s_k \rangle = -\langle s_k, \nabla_x^2 f(x_k) s_k \rangle - \tfrac{1}{2}\sigma_k \|s_k\|^3 + \langle \nabla_s^1 m_k(s_k), s_k \rangle$$

and hence

$$\begin{aligned} f(x_k) - T_{f,2}(x_k, s_k) &= -\langle \nabla_x^1 f(x_k), s_k \rangle - \tfrac{1}{2}\langle s_k, \nabla_x^2 f(x_k) s_k, \rangle \\ &= \tfrac{1}{2}\langle s_k, \nabla_x^2 f(x_k) s_k \rangle + \tfrac{1}{2}\sigma_k \|s_k\|^3 - \langle \nabla_s^1 m_k(s_k), s_k \rangle \\ &\ge \tfrac{1}{4}\lambda_{\min,*} \|s_k\|^2 + \tfrac{1}{2}\sigma_{\min} \|s_k\|^3 - \tfrac{1}{2}\theta_1 \|s_k\|^3, \end{aligned} \tag{5.4.10}$$

using (5.4.7), $\sigma_k \ge \sigma_{\min}$, and (5.4.3). Notice that (5.4.4), (5.4.5) (note the last term in the minimum), and (5.4.8) together imply that

$$\|s_k\| \le \min\left(\frac{\lambda_{\min,*}}{4\max(\theta_1 - \sigma_{\min}, 0)}, \frac{3(1-\eta_2)\lambda_{\min,*}}{4L_{f,2}} \right). \tag{5.4.11}$$

Therefore, we deduce that

$$f(x_k) - T_{f,2}(x_k, s_k) \ge \tfrac{1}{8}\lambda_{\min,*} \|s_k\|^2, \tag{5.4.12}$$

using (5.4.10) and (5.4.11). It then follows from (3.2.11), (5.4.9), (5.4.12), and again (5.4.11) that

$$|\rho_k - 1| = \left| \frac{T_{f,2}(x_k, s_k) - f(x_k + s_k)}{f(x_k) - T_{f,2}(x_k, s_k)} \right| \le \frac{4L_{f,2}}{3\lambda_{\min,*}} \|s_k\| \le 1 - \eta_2$$

and therefore that iteration k is successful since $\rho_k \ge \eta_2$. □

An immediate consequence of the recursion (5.4.4) is a bound on the iteration count required to reach approximate optimality once inside a neighbourhood of quadratic convergence for the AR2 algorithm.

Lemma 5.4.3: Asymptotic evaluation count

*Suppose that **f.HL** holds, that x_* is a local minimizer of $f(x)$ for which $\lambda_{\min}[\nabla_x^2 f(x_*)] > 0$, and that the* AR2 *algorithm (p. 65) with the model-termination test (5.4.3) is applied to minimize f. If there exists k_0 such that*

$$\|x_{k_0} - x_*\| \le q_* \text{ and } \|\nabla_x^1 f(x_{k_0})\| \le \zeta, \tag{5.4.13}$$

where q_* and ζ_* are defined as in (5.4.1) and (5.4.5) and $\zeta \le \zeta_*/2$, then the AR2 *algorithm takes at most*

$$\#^{SI}_{\text{AR2,A}} \stackrel{\text{def}}{=} \log_2\left(\log_2\left(\frac{\zeta}{\epsilon}\right)\right) \tag{5.4.14}$$

(successful) iterations from iteration k_0 onward to generate $\|\nabla_x^1 f(x_j)\| \le \epsilon$ for given $\epsilon > 0$.

Proof. Since $x_{k_0} \in \mathcal{Q}_*^1$, we may apply Lemma 5.4.2 to deduce that $x_k \in \mathcal{Q}_*^1$ for all successful iterations $k \ge k_0$. In this case, we have that $\|\nabla_x^1 f(x_{k_0})\| \le \zeta_*/2$ from (5.4.13), and employing (5.4.4) ℓ times, we deduce that

$$\|\nabla_x^1 f(x_j)\| \le \zeta_* 2^{-2^\ell} \tag{5.4.15}$$

for any $j \ge k_0$, where $\ell = |\{k_0, k_0+1, \ldots, j\} \cap \mathcal{S}|$ denotes the number of successful iterations from k_0 up to j. Then we deduce from (5.4.15) that $\|\nabla_x^1 f(x_j)\| \le \epsilon$ provided that $2^{-2^\ell}\zeta_* \le \epsilon$, which gives the bound (5.4.14). □

But now we simply recall our earlier general result, Theorem 3.3.5, for nonconvex objectives, which provides a bound of order $\mathcal{O}(\epsilon^{-3/2})$ on the total number of evaluations of the AR2 algorithm that may arise for which $\|\nabla_x^1 f(x_k)\| > \epsilon$, for any given $\epsilon > 0$. Thus the algorithm will require at most $\mathcal{O}(\zeta_*^{-3/2})$ evaluations to enter the region $\mathcal{Q}_*^1$, defined by (5.4.6), of quadratic convergence provided that there is no gradient value whose norm is smaller than ζ_* but greater than ϵ outside of $\mathcal{Q}_*^1$.

Theorem 5.4.4: Global complexity bound for locally convex functions

Suppose that **f.Bb** *and* **f.HL** *hold, that x_* is a local minimizer of $f(x)$ for which $\lambda_{\min}[\nabla_x^2 f(x_*)] > 0$, and that the* AR2 *algorithm (p. 65) with the model-termination test (5.4.3) is applied to minimize f. Then there are at most*

$$\#^{SI}_{\text{AR2,G}} \stackrel{\text{def}}{=} \kappa^{S}_{\text{AR2}} \frac{f(x_0) - f_{\text{low}}}{\zeta^{3/2}} + 1 \tag{5.4.16}$$

successful iterations k of the AR2 *algorithm for which $\|\nabla_x^1 f(x_k)\| > \zeta$ for some positive constant κ^{S}_{AR2} defined as in Lemma 3.3.4, and where $\zeta \le \zeta_*/2$ with ζ_* defined as in (5.4.5). Furthermore, the* AR2 *algorithm takes at most*

$$\#^{SI}_{\text{AR2,L}} \stackrel{\text{def}}{=} \kappa^{S}_{\text{AR2}} \frac{f(x_0) - f_{\text{low}}}{\zeta^{3/2}} + \log_2\left(\log_2\left(\frac{\zeta}{\epsilon}\right)\right) + 1 \tag{5.4.17}$$

successful iterations to generate $\|\nabla_x^1 f(x_j)\| < \epsilon$, for given $\epsilon > 0$, so long as there are no x_k such that

$$\|x_k - x_*\| > q_* \text{ and } \epsilon < \|\nabla_x^1 f(x_k)\| \le \zeta, \tag{5.4.18}$$

with q_* given in (5.4.1).

Proof. We invoke Theorem 3.3.5 to deduce the bound (5.4.16), where ζ is viewed as a positive tolerance. Further, the conditions of Lemma 5.4.3 clearly hold, and so (5.4.14)—which is the same as the second term in (5.4.17)—bounds the number of successful iterations if the AR2 algorithm enters the region of quadratic convergence defined by (5.4.13); the latter must happen due to assumption (5.4.18). □

Notes and references for Subsection 5.4.2

Although the requirement (5.4.18) is both strong and unverifiable in general, it may be a clue to reasonable practical performance of algorithms like AR2. For example, consider the minimization of the well-known Rosenbrock function [333] of two variables

$$f(x) = 100\left([x]_2 - [x]_1^2\right)^2 + ([x]_1 - 1)^2,$$

which has the unique stationary point and global minimizer $x_* = (1,1)$. We determined the radius of the neighbourhood of quadratic convergence (5.4.6) for $\nabla_x^1 f(x_k)$ by computing the relevant constants that occur in the expression (5.4.5) of ζ_*. We then set ϵ to a lower value than our ζ_* estimate and found (numerically[43]) that there is no gradient value whose size is less than ζ_* but greater than ϵ outside the neighbourhood of quadratic convergence, so that condition (5.4.13) holds. Thus Theorem 5.4.4 applies.

Lemmas 5.4.2 and 5.4.3 show the asymptotic Q-quadratic convergence behaviour of the AR2 algorithm once an iterate enters a neighbourhood of a nondegenerate minimizer. Note that this local quadratic rate occurs even though the regularization parameter σ_k in AR2 might not converge to zero, and thus the Newton and cubic-regularization steps may differ slightly—the difference between the two steps will normally vanish, though, since the actual and regularized Hessians vary precisely by the term $\sigma_k \|s_k\| I$.

One may wonder about potential extensions of the phase analysis here when the ARp algorithm—rather than the AR2 one—is applied to a locally convex function. For this, one would have to investigate the local rate of convergence of ARp variants. Recent work [150] considers this for globally strongly convex objectives using pth-degree tensor models, and a pth-order Q-rate of convergence is established for such functions in a local neighbourhood of the nondegenerate (global) minimizer. We would expect such results to generalize to ARp variants, with similar conclusions of a pth-order local rate.

5.5 Universal regularization methods: Varying the power and smoothness

5.5.1 Introduction

The aim of this section is twofold. Firstly, we are interested in seeing if decoupling, and varying, the power of the regularization term in the model (4.1.2) from its underlying Taylor series might improve the worst-case complexity bound of the ensuing variant of the ARp algorithm. That is, instead of (4.1.2) we shall calculate a step s_k from the current iterate x_k by approximately minimizing the local model

$$m_k(s) = T_p(x_k, s) + \frac{\sigma_k}{r!}\|s\|^r, \tag{5.5.1}$$

[43]It turns out that $\lambda_{\min}[\nabla_x^2 f(x_*)] \approx 0.3994$, the Lipschitz constant of the Hessian on $[-1.5, 1.5]^2$ (a box containing all iterates from the standard starting point $(-1.2, 1)$) is bounded above by 5000, and, if the algorithm's constant are chosen as $\sigma_{\min} = 0.001$, $\eta_2 = 0.95$, $\gamma_3 = 2$, $\sigma_0 = 1$, and $\theta_1 = 0.1$, then $\sigma_{\max} \approx 2 \times 10^5$, $q_* \approx 1.9968 \times 10^{-5}$, $\frac{1}{2}\zeta_* \approx 6.4619 \times 10^{-7}$, and the maximum distance from x_* such that $\|\nabla_x^1 f(x)\| \leq \frac{1}{2}\zeta_*$ is approximately 7.1440×10^{-7}, which is clearly smaller than q_*. The condition (5.4.18) is thus verified whenever $\epsilon < \frac{1}{2}\zeta_*$.

where $T_p(x_k, s)$ is the pth-order Taylor polynomial (4.1.1) of the objective function f at x_k, the power $r > p \geq 1$, and[44]

$$r! \stackrel{\text{def}}{=} \prod_{\ell=1}^{\lfloor r \rfloor} (\ell + r - \lfloor r \rfloor). \tag{5.5.2}$$

Just as for the ARp algorithm, the parameter $\sigma_k > 0$ may need to be adjusted to ensure sufficient decrease in f occurs when the model value is decreased.

Secondly, we shall not assume that the level of smoothness of the objective is known, or indeed that derivatives above a certain order are smooth at all. We aim to capture this by making, throughout this section, the following assumption.

f.DpH

$f(x)$ is p-times continuously differentiable in $\mathbb{R}^n$ and its pth-order derivative tensor $\nabla_x^p f(x)$ is *Hölder continuous* on $\mathbb{R}^n$; that is, there exists a constant $H_{f,p} \geq 0$ and an *exponent* $\beta_p \in [0, 1]$ such that

$$\|\nabla_x^p f(x) - \nabla_x^p f(y)\| \leq H_{f,p} \|x - y\|^{\beta_p} \tag{5.5.3}$$

for all $x, y \in \mathbb{R}^n$, where the norm on the left-hand side of (5.5.3) is the tensor norm (A.7).

If $\beta_p = 1$, **f.DpH** coincides with **f.DpL**. If $p = 1$, we only allow $\beta_1 \in (0, 1]$ in **f.DpH**, for reasons that we shall explain later. If $\beta_p > 1$ in (5.5.3), the resulting class of objectives is restricted to multivariate polynomials of degree p, and consequently, for simplicity, we exclude this special case.

Our interest in loosening the requirement from Lipschitz to Hölder continuity of derivatives arises both as a result of the needs of some engineering applications—such as flows in gas pipelines [175, Sect. 17] and properties of nonlinear PDE problems [32]—and in its own right in optimization theory, as a bridging case between smooth and nonsmooth classes of problems [296, 302]. In particular, a zero Hölder exponent for a Hölder-continuous derivative corresponds to a bounded derivative, an exponent in $(0, 1)$ corresponds to a continuous but not necessarily differentiable derivative, while an exponent of 1 coincides with a Lipschitz-continuous derivative that is also absolutely continuous and thus may be differentiated again almost everywhere. For functions with Hölder-continuous gradients, methods have already been devised—and their complexity analyzed—both from the perspective of a weaker set of assumptions and as an attempt to have a "smooth" transition between smooth and nonsmooth (albeit convex) problem classes, without knowing a priori the level of smoothness of the gradient (i.e., the Hölder exponent) [302]; in addition, lower bounds on the worst-case complexity are known [296].

5.5.2 The ARp-H algorithm and its complexity

We now introduce the ARp-H algorithm, a modification of the ARp regularization algorithm (p. 100) that ensures the computed steps are "sufficiently long" and that makes the algorithm robust to the value of the regularization power and adaptable to changes in smoothness of the objective. The algorithm is designed to find both a first- and, if required, second-order approximate minimizer, and in view of this we clearly require

[44]The definition (5.5.2) coincides with the traditional factorial when r is an integer. Moreover, the standard identity $r! = r(r-1)!$ holds for all $r \geq 1$.

$p \geq 1$ in the first case and $p \geq 2$ for the second. It is intended for functions f that satisfy **f.DpH**, but crucially no a priori knowledge of the Hölder exponent β_p is required.

Algorithm 5.5.1: Universal adaptive-regularization algorithm with pth-order models (ARp-H)

Step 0: Initialization. An initial point x_0, accuracy thresholds $\epsilon_1 \in (0,1]$, $\epsilon_2 \in (0,1] \cup \{\infty\}$, and an initial regularization parameter $\sigma_0 > 0$ are given. An integer degree $p \geq 1$ ($p \geq 2$ if $\epsilon_2 < \infty$) and a real regularization power $r > p$ are specified. The constants η_1, η_2, γ_1, γ_2, γ_3, θ_1, θ_2, α_1, and α_2 are also given and satisfy

$$\alpha_1 \in \left(0, \tfrac{1}{3}\min[1,(r-1)!]\right], \; \alpha_2 \in \left(0, \tfrac{1}{3}\min[1,(r-2)!]\right],$$
$$\theta_1 \in (0,1], \;\; \theta_2 \in (0,1], \;\; 0 < \eta_1 \leq \eta_2 < 1, \;\; \text{and} \;\; 0 < \gamma_1 < 1 < \gamma_2 < \gamma_3.$$

Compute $f(x_0)$ and set $k = 0$. Check for termination as in Step 2 below with $s_0 = 0$. If termination fails, compute all unavailable derivatives $\nabla_x^j f(x_0)$ for orders $j \leq p$.

Step 1: Step computation. Compute a step s_k such that

$$m_k(s_k) < m_k(0), \tag{5.5.4}$$

where $m_k(s)$ is given by (5.5.1);

$$\|\nabla_s^1 m_k(s_k)\| \leq \theta_1 \|s_k\|^{r-1}; \tag{5.5.5}$$

and, if $\epsilon_2 < \infty$,

$$\lambda_{\min}[\nabla_s^2 m_k(s_k)] \geq -\theta_2 \|s_k\|^{r-2}. \tag{5.5.6}$$

Step 2: Test for termination. If $\epsilon_2 = \infty$, compute $\nabla_x^1 f(x_k + s_k)$, and if

$$\|\nabla_x^1 f(x_k + s_k)\| \leq \epsilon_1, \tag{5.5.7}$$

terminate with the ϵ_1-approximate first-order minimizer $x_\epsilon = x_k + s_k$. If $\epsilon_2 < \infty$, compute both $\nabla_x^1 f(x_k + s_k)$ and $\nabla_x^2 f(x_k + s_k)$, and if (5.5.7) holds and

$$\lambda_{\min}[\nabla_x^2 f(x_k + s_k)] > \epsilon_2, \tag{5.5.8}$$

terminate with the (ϵ_1, ϵ_2)-approximate strict local minimizer $x_\epsilon = x_k + s_k$. If (5.5.7) holds and (5.5.8) fails but

$$\lambda_{\min}[\nabla_x^2 f(x_k + s_k)] \geq -\epsilon_2,$$

terminate with the (ϵ_1, ϵ_2)-approximate second-order minimizer $x_\epsilon = x_k + s_k$.

Step 3: Acceptance of the trial point. Compute $f(x_k + s_k)$ and define

$$\rho_k = \frac{f(x_k) - f(x_k + s_k)}{T_{f,p}(x_k, 0) - T_{f,p}(x_k, s_k)}. \tag{5.5.9}$$

If $\rho_k \geq \eta_1$, then check whether

$$\sigma_k \|s_k\|^{r-1} \geq \alpha_1 \|\nabla_x^1 f(x_k + s_k)\|, \tag{5.5.10}$$

and if $\epsilon_2 < \infty$, check also whether

$$\sigma_k \|s_k\|^{r-2} \geq -\alpha_2 \lambda_{\min}[\nabla_x^2 f(x_k + s_k)]. \tag{5.5.11}$$

If $\rho_k \geq \eta_1$, (5.5.10), and (5.5.11) hold, define $x_{k+1} = x_k + s_k$, and compute all unavailable derivatives $\nabla_x^j f(x_{k+1})$ for orders $j \leq p$. Otherwise, define $x_{k+1} = x_k$.

Step 4: Regularization parameter update. Set

$$\sigma_{k+1} \in \begin{cases} [\gamma_1 \sigma_k, \sigma_k] & \text{if } \rho_k \geq \eta_2, \text{ (5.5.10), and (5.5.11) hold,} \\ [\sigma_k, \gamma_2 \sigma_k] & \text{if } \rho_k \in [\eta_1, \eta_2), \text{ (5.5.10), and (5.5.11) hold,} \\ [\gamma_2 \sigma_k, \gamma_3 \sigma_k] & \text{if } \rho_k < \eta_1 \text{ and (5.5.10) or (5.5.11) fails.} \end{cases}$$

Increment k by one and go to Step 1.

The design of the ARp-H algorithm clearly exhibits many similarities with its predecessor, ARp, when $r = p + 1$ in (5.5.1). However, there are three main differences in addition to the flexible regularization power. Firstly, no explicit lower bound $\sigma_{\min} > 0$ on σ_k is imposed, as we required for the ARp and earlier algorithms. In principle, this might allow the algorithm to take steps that are asymptotically closer to those from a pure Newton method, but actually conditions (5.5.10) and (5.5.11) may implicitly provide a lower bound on σ_k. Secondly, we check whether termination occurs at each trial point, not just on successful iterations. Finally, even when the candidate step s_k provides sufficient decrease according to (5.5.9), we only allow the algorithm to accept the step if it is appropriately long according to the recipes (5.5.10) and possibly (5.5.11). If either (5.5.10) or (5.5.11) fails or $\rho_k \leq \eta_1$, the parameter σ_k is increased. These two additional ingredients—the gradient and Hessian calculation at each trial point and the steplength condition—are directly related to trying to achieve universality of the ARp-H algorithm. Here, *universality* [302] refers to algorithms that are invariant to the regularization power (in our case, so long as $r \geq p + 1$) and applicable to different levels of smoothness in the objective (that are unknown a priori) while maintaining good or even optimal worst-case evaluation complexity.

We turn as always to a worst-case evaluation-complexity analysis of the algorithm. The first ingredient is a by now familiar sufficient-decrease result.

Lemma 5.5.1: Sufficient decrease

Each iteration of the ARp-H *algorithm guarantees the decrease*

$$f(x_k) - T_p(x_k, s_k) \geq \frac{\sigma_k}{r!} \|s_k\|^r. \tag{5.5.12}$$

Proof. Requirement (5.5.4) and the definition of $m_k(s)$ in (5.5.1) immediately give (5.5.12). □

Recall that we require that $r > p \geq 1$ for $r \in \mathbb{R}$ and $p \in \mathbb{N}$. The results that we derive demand that $p + \beta_p > 1$, which can be ensured by assuming that either $p \geq 1$ and $\beta_p \in (0, 1]$ or $p \geq 2$ and $\beta_p \in [0, 1]$. If we desire a second-order approximate minimizer (i.e., when $\epsilon_2 < \infty$), our results also demand that $p + \beta_p > 2$. Again this can be ensured by insisting that either $p > 2$ and $\beta_p \in [0, 1]$ or $p \geq 2$ and $\beta_p \in (0, 1]$.

In what follows, we shall address the case where the regularization power $r \geq p+\beta_p$. Since typically the smoothness exponent $\beta_p \in [0,1]$ is unknown a priori, allowing a sufficiently large regularization power (say $r = p+1$) that covers all useful β_p values is of most interest. Note, however, that the other case in which $r \in (p, p+\beta_p)$ may also be addressed [105], but for simplicity we ignore this possibility here because it is less crucial given the commonly unknown value of β_p.

A useful property involving the length of the step is next—it depends on the definition

$$(p+\beta_p)! \stackrel{\text{def}}{=} \prod_{\ell=1}^{p}(\ell+\beta_p)$$

for any $\beta_p \in [0,1]$ (cf. (5.5.2)), which coincides with the standard factorial if β_p is 0 or 1.

Lemma 5.5.2: Lower bound on the step

Suppose that **f.DpH** *holds. Then on each iteration of the* ARp-H *algorithm, the step s_k calculated in Step 1 satisfies*

$$\|\nabla_x^1 f(x_k+s_k)\| \leq \frac{H_{f,p}}{(p+\beta_p-1)!}\|s_k\|^{p+\beta_p-1} + \frac{\sigma_k}{(r-1)!}\|s_k\|^{r-1} + \theta_1\|s_k\|^{r-1}. \tag{5.5.13}$$

If additionally $\epsilon_2 < \infty$, then

$$-\lambda_{\min}[\nabla_x^2 f(x_k+s_k)] \leq \frac{H_{f,p}}{(p+\beta_p-2)!}\|s_k\|^{p+\beta_p-2} + \frac{\sigma_k}{(r-2)!}\|s_k\|^{r-2} + \theta_2\|s_k\|^{r-2}. \tag{5.5.14}$$

Proof. Using the triangle inequality and adding and subtracting terms, we obtain

$$\begin{aligned}\|\nabla_x^1 f(x_k+s_k)\| &= \|\nabla_x^1 f(x_k+s_k) - \nabla_s^1 m_k(x_k+s_k)\| + \|\nabla_s^1 m_k(x_k+s_k)\| \\ &\leq \|\nabla_x^1 f(x_k+s_k) - \nabla_s^1 T_p(x_k,s_k)\| + \frac{\sigma_k}{(r-1)!}\|s\|^{r-1} + \theta_1\|s\|^{r-1},\end{aligned}$$

where we used $\nabla_s^1 m_k(x_k+s) = \nabla_s^1 T_p(x_k,s) + \sigma_k\|s\|^{r-2}s/(r-1)!$ and (5.5.5) in the last inequality. To bound the error in the Taylor approximation, we use **f.DpH** and apply (A.15) from Theorem A.8.3 (p. 436) with $j=1$ to deduce that

$$\|\nabla_x^1 f(x_k+s_k) - \nabla_s^1 T_p(x_k,s_k)\| \leq \frac{H_{f,p}}{(p+\beta_p-1)!}\|s_k\|^{p+\beta_p-1},$$

and the bound (5.5.13) then follows from this and the previous inequality.

Now suppose that $\epsilon_2 < \infty$. Because $\min_z[a(z)+b(z)] \geq \min_z a(z) + \min_z b(z)$, it follows that

$$\begin{aligned}&\lambda_{\min}[\nabla_x^2 f(x_k+s_k)] \\ &\quad= \min_{\|d\|=1}\Big[\Big\langle d, \Big[\nabla_x^2 f(x_k+s_k) - \nabla_s^2 T_{f,p}(x_k,s_k)\Big]\, d\Big\rangle \\ &\qquad\qquad -\frac{\sigma_k}{r!}\Big\langle d, \nabla_s^2(\|s_k\|^r)d\Big\rangle + \langle d, \nabla_s^2 m_k(s_k)d\rangle\Big] \\ &\quad\geq \min_{\|d\|=1}\Big\langle d, \Big[\nabla_x^2 f(x_k+s_k) - \nabla_s^2 T_{f,p}(x_k,s_k)\Big]\, d\Big\rangle \\ &\qquad\qquad +\frac{\sigma_k}{r!}\min_{\|d\|=1}\Big\langle -d, \nabla_s^2(\|s_k\|^r)d\Big\rangle + \lambda_{\min}[\nabla_s^2 m_k(s_k)].\end{aligned} \tag{5.5.15}$$

Using the relationship (A.4) for $\nabla_x^2 f(x_k+s_k) - \nabla_s^2 T_{f,p}(x_k,s_k)$, **f.DpH**, and (A.15) from Theorem A.8.3 (p. 436) with $j=2$ yields that

$$\begin{aligned}\min_{\|d\|=1} \left\langle d, [\nabla_x^2 f(x_k+s_k) - \nabla_s^2 T_{f,p}(x_k,s_k)]d \right\rangle &\geq - \left\| \nabla_x^2 f(x_k+s_k) - \nabla_s^2 T_{f,p}(x_k,s_k) \right\| \\ &\geq -\frac{H_{f,p}}{(p+\beta_p-2)!}\|s_k\|^{p+\beta_p-2}.\end{aligned} \tag{5.5.16}$$

Furthermore, simple differentiation shows that the Hessian

$$\nabla_s^2(\|s_k\|^r) = r\|s_k\|^{r-2}\left(I + (r-2)\frac{s_k}{\|s_k\|}\frac{s_k^T}{\|s_k\|} \right),$$

and thus it is positive definite since $r > p \geq 2$ is required when $\epsilon_2 < \infty$. But this and Lemma B.4.1 (p. 461) with $r = p+\beta$ and $j=2$ then give that

$$\begin{aligned}\min_{\|d\|=1} \left\langle -d, \nabla_s^2(\|s_k\|^r)d \right\rangle &= -\max_{\|d\|=1} \left\langle d, \nabla_s^2(\|s_k\|^r)d \right\rangle \\ &= -\left\|\nabla_s^2(\|s_k\|^r)\right\| = -\frac{r!}{(r-2)!}\|s_k\|^{r-2}.\end{aligned} \tag{5.5.17}$$

Combining (5.5.15), (5.5.16), (5.5.17), and (5.5.6), we obtain (5.5.14). □

Lemma 5.5.3: Successful iterations

*Suppose that $r \geq p+\beta_p$ and **f.DpH** hold.* (i) *If $\epsilon_2 = \infty$ and*

$$\sigma_k \geq \max\left[\theta_1(r-1)!, \kappa_\beta\|s_k\|^{p+\beta_p-r}\right], \tag{5.5.18}$$

where

$$\kappa_\beta \stackrel{\text{def}}{=} \frac{r!H_{f,p}}{(p+\beta_p-1)!(1-\eta_2)}, \tag{5.5.19}$$

then $\rho_k \geq \eta_2$ and (5.5.10) hold, and so iteration k is very successful.
(ii) *If $\epsilon_2 < \infty$ and*

$$\sigma_k \geq \max\left[\theta_1(r-1)!, \theta_2(r-2)!, \kappa_\beta\|s_k\|^{p+\beta_p-r}\right], \tag{5.5.20}$$

then additionally (5.5.11) holds, and again iteration k is very successful.

Proof. If (5.5.18) or (5.5.20) as appropriate holds, then

$$\sigma_k \geq \kappa_\beta\|s_k\|^{p+\beta_p-r}. \tag{5.5.21}$$

The definition of ρ_k in (5.5.9) gives

$$|\rho_k - 1| = \frac{|f(x_k+s_k) - T_p(x_k,s_k)|}{f(x_k) - T_p(x_k,s_k)},$$

whose numerator we bound from above by (A.14) in Theorem A.8.3 (p. 436) and whose denominator we bound from below by (5.5.12), from which we deduce that

$$|\rho_k - 1| \leq \frac{\dfrac{H_{f,p}}{(p+\beta_p)!}\|s_k\|^{p+\beta_p}}{\dfrac{\sigma_k}{r!}\|s_k\|^r} = \frac{r!H_{f,p}}{(p+\beta_p)!\sigma_k}\|s_k\|^{p+\beta_p-r} \leq \frac{r!H_{f,p}}{(p+\beta_p-1)!\sigma_k}\|s_k\|^{p+\beta_p-r}, \tag{5.5.22}$$

since $p+\beta_p \geq 1$. Substituting (5.5.21) and the definition of κ_β from (5.5.19) into (5.5.22) shows that $|1-\rho_k| \leq 1-\eta_2$, which ensures that $\rho_k \geq \eta_2$.

To show that (5.5.18) also implies (5.5.10), we have that $\sigma_k \geq \theta_1(r-1)!$ from (5.5.18), which together with (5.5.13) gives

$$\|\nabla_x^1 f(x_k+s_k)\| \leq \|s_k\|^{p+\beta_p-1}\left(\frac{H_{f,p}}{(p+\beta_p-1)!} + 2\frac{\sigma_k}{(r-1)!}\|s_k\|^{r-p-\beta_p}\right). \tag{5.5.23}$$

The definition (5.5.19) and the requirements $r \geq 1$ and $\eta_2 \in (0,1)$ imply that

$$\frac{(r-1)!H_{f,p}}{(p+\beta_p-1)!} \leq \frac{r!H_{f,p}}{(p+\beta_p-1)!} \leq \kappa_\beta. \tag{5.5.24}$$

This and (5.5.23) give

$$\|\nabla_x^1 f(x_k+s_k)\| \leq \frac{\|s_k\|^{p+\beta_p-1}}{(r-1)!}\left(\kappa_\beta + 2\sigma_k\|s_k\|^{r-p-\beta_p}\right). \tag{5.5.25}$$

But (5.5.21) is equivalently $\kappa_\beta \leq \sigma_k\|s_k\|^{r-p-\beta_p}$, and inserting this into (5.5.25) reveals that

$$\|\nabla_x^1 f(x_k+s_k)\| \leq \frac{\|s_k\|^{p+\beta_p-1}}{(r-1)!}\left(3\sigma_k\|s_k\|^{r-p-\beta_p}\right) = \frac{3\sigma_k}{(r-1)!}\|s_k\|^{r-1}.$$

Thus $\sigma_k\|s_k\|^{r-1} \geq \frac{1}{3}(r-1)!\|\nabla_x^1 f(x_k+s_k)\|$, which implies (5.5.10) since $\alpha_1 \leq \frac{1}{3}\min[1,(r-1)!]$.

To show (5.5.11) when $\epsilon_2 < \infty$, and consequently $r \geq p+\beta_p \geq 2$, we proceed similarly. Since $\sigma_k \geq \theta_2(r-2)!$, because of (5.5.20), and

$$\begin{aligned}\frac{(r-2)!H_{f,p}}{(p+\beta_p-2)!} &= \frac{(r-1)!H_{f,p}}{(p+\beta_p-1)!}\,\frac{(p+\beta_p-1)}{(r-1)} \\ &\leq \frac{(r-1)!H_{f,p}}{(p+\beta_p-1)!} < \frac{r!H_{f,p}}{(p+\beta_p-1)!} \\ &< \kappa_\beta,\end{aligned} \tag{5.5.26}$$

this and the bound (5.5.14) imply

$$\begin{aligned}-\lambda_{\min}[\nabla_x^2 f(x_k+s_k)] &\leq \|s_k\|^{p+\beta_p-2}\left[\frac{H_{f,p}}{(p+\beta_p-2)!} + 2\frac{\sigma_k}{(r-2)!}\|s_k\|^{r-p-\beta_p}\right] \\ &\leq \frac{\|s_k\|^{p+\beta_p-2}}{(r-2)!}\left[\frac{r!H_{f,p}}{(p+\beta_p-1)!} + 2\sigma_k\|s_k\|^{r-p-\beta_p}\right] \\ &\leq \frac{\|s_k\|^{p+\beta_p-2}}{(r-2)!}\left[\kappa_\beta + 2\sigma_k\|s_k\|^{r-p-\beta_p}\right] \\ &\leq \frac{\|s_k\|^{p+\beta_p-2}}{(r-2)!}\left[3\sigma_k\|s_k\|^{r-p-\beta_p}\right] = 3\sigma_k\frac{\|s_k\|^{p+\beta_p-2}}{(r-2)!},\end{aligned}$$

where we employed (5.5.21) in the last inequality. The result then follows from the requirement that $\alpha_2 \leq \frac{1}{3}\min[1,(r-2)!]$. □

We note that the terms $\theta_1(r-1)!$ and $\theta_2(r-2)!$ in (5.5.18) and (5.5.20) simplify to θ_1 and θ_2 provided that the right-hand side of the model-minimization conditions (5.5.5) and (5.5.6) are scaled by $1/(r-1)!$ and $1/(r-2)!$, respectively.

Lemma 5.5.4: A lower bound on σ_k to ensure success

*Suppose that $r \geq p+\beta_p$, that **f.DpH** holds, and that the* ARp-H *algorithm has not terminated.* (i) *If $\epsilon_2 = \infty$ and*

$$\sigma_k \geq \max\left[\theta_1(r-1)!, \kappa_{\beta,1}\epsilon_1^{\frac{p+\beta_p-r}{p+\beta_p-1}}\right], \tag{5.5.27}$$

where

$$\kappa_{\beta,1} \stackrel{\text{def}}{=} \left(\frac{(r-1)!}{3}\right)^{\frac{p+\beta_p-r}{p+\beta_p-1}} \kappa_\beta^{\frac{r-1}{p+\beta_p-1}} \tag{5.5.28}$$

and κ_β is defined by (5.5.19), then (5.5.18) holds, and consequently iteration k is very successful.
(ii) *If $\epsilon_2 < \infty$ and*

$$\sigma_k \geq \max\left[\theta_1(r-1)!, \theta_2(r-2)!, \kappa_{\beta,1}\epsilon_1^{\frac{p+\beta_p-r}{p+\beta_p-1}}, \kappa_{\beta,2}\epsilon_2^{\frac{p+\beta_p-r}{p+\beta_p-2}}\right], \tag{5.5.29}$$

where

$$\kappa_{\beta,2} \stackrel{\text{def}}{=} \left(\frac{(r-2)!}{3}\right)^{\frac{p+\beta_p-r}{p+\beta_p-2}} \kappa_\beta^{\frac{r-2}{p+\beta_p-2}}, \tag{5.5.30}$$

then (5.5.20) holds, and once again iteration k is very successful.

Proof. Suppose first that $r = p+\beta_p$. In this case, Lemma 5.5.3 implies that if $\epsilon_2 = \infty$ and

$$\sigma_k \geq \max[\theta_1(r-1)!, \kappa_\beta], \tag{5.5.31}$$

or $\epsilon_2 < \infty$ and

$$\sigma_k \geq \max[\theta_1(r-1)!, \theta_2(r-2)!, \kappa_\beta], \tag{5.5.32}$$

then iteration k is very successful. But the right-hand sides of (5.5.31) and (5.5.27) or (5.5.32) and (5.5.29) coincide, since $\kappa_{\beta,1} = \kappa_{\beta,2} = \kappa_\beta$ and the terms in ϵ_1 and ϵ_2 are both equal to one when $r = p+\beta_p$, and thus the conclusion of Lemma 5.5.4 holds in this case.

By comparison, suppose that $r > p+\beta_p$. We then establish the desired result by contradiction. Suppose that (5.5.18) does not hold at iteration k, and since (5.5.27) ensures that $\sigma_k \geq \theta_1(r-1)!$, it must be that

$$\sigma_k \|s_k\|^{r-p-\beta_p} < \kappa_\beta. \tag{5.5.33}$$

The same is true if (5.5.20) does not hold because then (5.5.29) implies that $\sigma_k \geq \max[\theta_1(r-1)!, \theta_2(r-2)!]$. Since ARp-H has not terminated, we have either (a) $\|\nabla_x^1 f(x_k+s_k)\| > \epsilon_1$ or (b) $\lambda_{\min}[\nabla_x^2 f(x_k+s_k)] < -\epsilon_2$ when $\epsilon_2 < \infty$ (or both).

Firstly, consider possibility (a). Then $\sigma_k \geq \theta_1(r-1)!$ because of (5.5.27) and (5.5.29), and so (5.5.13) implies

$$\begin{aligned}\epsilon_1 &\leq \frac{H_{f,p}}{(p+\beta_p-1)!}\|s_k\|^{p+\beta_p-1} + 2\frac{\sigma_k}{(r-1)!}\|s_k\|^{r-1} \\ &= \left(\frac{H_{f,p}}{(p+\beta_p-1)!} + 2\frac{\sigma_k}{(r-1)!}\|s_k\|^{r-p-\beta_p}\right)\|s_k\|^{p+\beta_p-1}.\end{aligned} \tag{5.5.34}$$

Using (5.5.33) to bound the second term in (5.5.34), we find that

$$\epsilon_1 < \left(\frac{H_{f,p}}{(p+\beta_p-1)!} + 2\frac{\kappa_\beta}{(r-1)!}\right)\|s_k\|^{p+\beta_p-1},$$

and combining this with (5.5.24), it follows that

$$\epsilon_1 < \frac{3\kappa_\beta}{(r-1)!}\|s_k\|^{p+\beta_p-1}. \tag{5.5.35}$$

Moreover, using (5.5.33) once again, we have that

$$\|s_k\| \leq \left(\frac{\kappa_\beta}{\sigma_k}\right)^{\frac{1}{r-p-\beta_p}} \tag{5.5.36}$$

since $r > p+\beta_p > 1$, and substituting this bound into (5.5.35), we obtain

$$\epsilon_1 < \frac{3\kappa_\beta}{(r-1)!}\left(\frac{\kappa_\beta}{\sigma_k}\right)^{\frac{p+\beta_p-1}{r-p-\beta_p}}.$$

Coupling this with the definition of $\kappa_{\beta,1}$ in (5.5.28) implies that $\sigma_k < \kappa_{\beta,1}\epsilon_1^{\frac{p+\beta_p-r}{p+\beta_p-1}}$, which contradicts (5.5.27). Thus possibility (a) cannot occur if (5.5.33) does.

By contrast, consider possibility (b), for which we must have $p \geq 2$ and thus $r > p+\beta \geq 2$. Then $\sigma_k \geq \theta_2(r-2)!$ because of (5.5.29), and thus (5.5.14) implies

$$\begin{aligned}\epsilon_2 &\leq \frac{H_{f,p}}{(p+\beta_p-2)!}\|s_k\|^{p+\beta_p-2} + 2\frac{\sigma_k}{(r-2)!}\|s_k\|^{r-2} \\ &= \left(\frac{H_{f,p}}{(p+\beta_p-2)!} + 2\frac{\sigma_k}{(r-2)!}\|s_k\|^{r-p-\beta_p}\right)\|s_k\|^{p+\beta_p-2}.\end{aligned} \tag{5.5.37}$$

Once again, using (5.5.33) to bound the second term in (5.5.37), we have that

$$\epsilon_2 < \left(\frac{H_{f,p}}{(p+\beta_p-2)!} + 2\frac{\kappa_\beta}{(r-2)!}\right)\|s_k\|^{p+\beta_p-2}.$$

Combining this with (5.5.26) and (5.5.36), we find that

$$\epsilon_2 < \frac{3\kappa_\beta}{(r-2)!}\|s_k\|^{p+\beta_p-2} < \frac{3\kappa_\beta}{(r-2)!}\left(\frac{\kappa_\beta}{\sigma_k}\right)^{\frac{p+\beta_p-2}{r-p-\beta_p}}. \tag{5.5.38}$$

It follows from (5.5.38) and the definition of $\kappa_{\beta,2}$ in (5.5.30) that $\sigma_k < \kappa_{\beta,2}\epsilon_1^{\frac{p+\beta_p-r}{p+\beta_p-2}}$, which contradicts (5.5.29), and thus possibility (b) cannot occur if (5.5.33) does.

Thus in both (a) and (b), (5.5.18) must hold and Lemma 5.5.3 provides the desired conclusion. □

The next lemma shows that if σ_k is sufficiently large, a successful iteration ensues. This then enables us to provide an upper bound on the value of the regularization parameter.

Lemma 5.5.5: An upper bound on σ_k

Suppose that **f.DpH** *holds,* $r \geq p+\beta_p$, *and the* ARp-H *algorithm has not terminated. Then*

$$\sigma_k \leq \max\left[\sigma_0, \gamma_3\theta_1(r-1)!, \gamma_3\kappa_{\beta,1}\epsilon_1^{\frac{p+\beta_p-r}{p+\beta_p-1}}\right] \tag{5.5.39}$$

if $\epsilon_2 = \infty$, *and*

$$\sigma_k \leq \max\left[\sigma_0, \gamma_3\theta_1(r-1)!, \gamma_3\theta_2(r-2)!, \gamma_3\kappa_{\beta,1}\epsilon_1^{\frac{p+\beta_p-r}{p+\beta_p-1}}, \gamma_3\kappa_{\beta,2}\epsilon_2^{\frac{p+\beta_p-r}{p+\beta_p-2}}\right] \tag{5.5.40}$$

if $\epsilon_2 < \infty$, *and where* $\kappa_{\beta,1}$ *and* $\kappa_{\beta,2}$ *are defined as in (5.5.28) and (5.5.30), respectively. Furthermore, if* $\epsilon_2 < \infty$, *then*

$$\sigma_k \leq \sigma_H \max\left[\epsilon_1^{\frac{p+\beta_p-r}{p+\beta_p-1}}, \epsilon_2^{\frac{p+\beta_p-r}{p+\beta_p-2}}\right], \tag{5.5.41}$$

where

$$\sigma_H \stackrel{\text{def}}{=} \max\left[\sigma_0, \gamma_3\theta_1(r-1)!, \gamma_3\theta_2(r-2)!, \gamma_3\kappa_{\beta,1}, \gamma_3\kappa_{\beta,2}\right]. \tag{5.5.42}$$

Proof. Suppose that $\epsilon_2 = \infty$, and denote the right-hand side of (5.5.27) by $\overline{\sigma}$. It then follows from Lemma 5.5.4 and the mechanism of the algorithm that

$$\sigma_k \geq \overline{\sigma} \text{ implies that } \sigma_{k+1} \leq \sigma_k. \tag{5.5.43}$$

Thus, when $\sigma_0 \leq \gamma_3\overline{\sigma}$, it follows that $\sigma_k \leq \gamma_3\overline{\sigma}$, where the factor γ_3 is introduced for the case when σ_k is less than $\overline{\sigma}$ and the iteration k is not very successful. Letting $k = 0$ in (5.5.43) gives (5.5.39) when $\sigma_0 \geq \gamma_3\overline{\sigma}$ since $\gamma_3 > 1$. Identical reasoning covers the case $\epsilon_2 < \infty$, but where now $\overline{\sigma}$ is the right-hand side of (5.5.29), and leads to (5.5.40).

The bound (5.5.41) follows immediately from (5.5.40) because

$$\max\left[\epsilon_1^{\frac{p+\beta_p-r}{p+\beta_p-1}}, \epsilon_2^{\frac{p+\beta_p-r}{p+\beta_p-2}}\right] \geq 1$$

since $\epsilon_1, \epsilon_2 \in (0, 1]$ and $r \geq p + \beta_p$. □

We are now in a position to establish a lower bound on the objective-function decrease that occurs on successful iterations.

Lemma 5.5.6: Lower bound on successful iterations

Suppose that **f.DpH** *holds. Then, on each successful iteration k of the* ARp-H *algorithm up to termination, we have the following:*

(i) *If $\epsilon_2 = \infty$ and $r \geq p + \beta_p > 1$, then*

$$f(x_k) - f(x_{k+1}) \geq \kappa_{H,1} \epsilon_1^{\frac{p+\beta_p}{p+\beta_p-1}},$$

where

$$\kappa_{H,1} \stackrel{\text{def}}{=} \frac{\eta_1}{r!} \left(\frac{\alpha_1^r}{\sigma_{H,1}} \right)^{\frac{1}{r-1}} \quad \text{and} \quad \sigma_{H,1} \stackrel{\text{def}}{=} \max\left[\sigma_0, \gamma_3 \theta_1 (r-1)!, \gamma_3 \kappa_{\beta,1}\right],$$

with $\kappa_{\beta,1}$ defined as in (5.5.28).

(ii) *If $\epsilon_2 < \infty$ and $r \geq p + \beta_p > 2$, then*

$$f(x_k) - f(x_{k+1}) \geq \kappa_{H,2} \min\left[\epsilon_1^{\frac{p+\beta_p}{p+\beta_p-1}}, \epsilon_2^{\frac{p+\beta_p}{p+\beta_p-2}} \right], \tag{5.5.44}$$

where

$$\kappa_{H,2} \stackrel{\text{def}}{=} \frac{\eta_1}{r!} \min\left[\left(\frac{\alpha_1^r}{\sigma_H} \right)^{\frac{1}{r-1}}, \left(\frac{\alpha_2^r}{\sigma_H^2} \right)^{\frac{1}{r-2}} \right] \tag{5.5.45}$$

and σ_H is defined as in (5.5.42).

Proof. We establish case (ii) first. On every successful iteration k, we have $\rho_k \geq \eta_1$; this and Lemma 5.5.1 imply that

$$f(x_k) - f(x_{k+1}) \geq \eta_1 (f(x_k) - T_p(x_k, s_k)) \geq \eta_1 \frac{\sigma_k}{r!} \|s_k\|^r. \tag{5.5.46}$$

In this case we also have that (5.5.10) and (5.5.11) hold. Thus, while the algorithm has not terminated, $\|\nabla_x^1 f(x_k + s_k)\| > \epsilon_1$ or $\lambda_{\min}[\nabla_x^2 f(x_k + s_k)] < -\epsilon_2$ or both, and we have

$$\|s_k\| \geq \left(\frac{\alpha_1 \epsilon_1}{\sigma_k} \right)^{\frac{1}{r-1}} \quad \text{or} \quad \|s_k\| \geq \left(\frac{\alpha_2 \epsilon_2}{\sigma_k} \right)^{\frac{1}{r-2}}. \tag{5.5.47}$$

Substituting (5.5.47) into the right-hand side of (5.5.46), we deduce

$$\begin{aligned} f(x_k) - f(x_{k+1}) &\geq \frac{\eta_1}{r!} \sigma_k \min\left[\left(\frac{\alpha_1 \epsilon_1}{\sigma_k} \right)^{\frac{r}{r-1}}, \left(\frac{\alpha_2 \epsilon_2}{\sigma_k} \right)^{\frac{r}{r-2}} \right] \\ &\geq \frac{\eta_1}{r!} \min\left[\left(\frac{\alpha_1^r \epsilon_1^r}{\sigma_k} \right)^{\frac{1}{r-1}}, \left(\frac{\alpha_2^r \epsilon_2^r}{\sigma_k^2} \right)^{\frac{1}{r-2}} \right]. \end{aligned}$$

Since $\max[a, b] = \min[1/a, 1/b]$ for $a, b \neq 0$, we may introduce the bound (5.5.41) into the above to deduce that

$$f(x_k) - f(x_{k+1}) \geq \frac{\eta_1}{r!} \min\left[\left(\frac{\alpha_1^r}{\sigma_H}\right)^{\frac{1}{r-1}}, \left(\frac{\alpha_2^r}{\sigma_H^2}\right)^{\frac{1}{r-2}}\right]$$
$$\times \min\left[\left(\epsilon_1^r \min\left[\epsilon_1^{\frac{r-p-\beta_p}{p+\beta_p-1}}, \epsilon_2^{\frac{r-p-\beta_p}{p+\beta_p-2}}\right]\right)^{\frac{1}{r-1}}, \left(\epsilon_2^r \min\left[\epsilon_1^{\frac{2(r-p-\beta_p)}{p+\beta_p-1}}, \epsilon_2^{\frac{2(r-p-\beta_p)}{p+\beta_p-2}}\right]\right)^{\frac{1}{r-2}}\right],$$

which further simplifies to

$$f(x_k) - f(x_{k+1})$$
$$\geq \kappa_{H,2} \min\left[\epsilon_1^{\frac{p+\beta_p}{p+\beta_p-1}}, \left(\epsilon_1^r \epsilon_2^{\frac{r-p-\beta_p}{p+\beta_p-2}}\right)^{\frac{1}{r-1}}, \left(\epsilon_2^r \epsilon_1^{\frac{2(r-p-\beta_p)}{p+\beta_p-1}}\right)^{\frac{1}{r-2}}, \epsilon_2^{\frac{p+\beta_p}{p+\beta_p-2}}\right], \quad (5.5.48)$$

using the definition (5.5.45). It remains to assess which are the dominant terms in the minimum on the right-hand side of (5.5.48).

If $r = p + \beta_p$, then the second and third terms in (5.5.48), those containing both ϵ_1 and ϵ_2, reduce to $\epsilon_1^{\frac{p+\beta_p}{p+\beta_p-1}}$ and $\epsilon_2^{\frac{p+\beta_p}{p+\beta_p-2}}$, respectively, and thus match the first and last terms in the expression. This establishes (5.5.44) in this special case. By contrast, if $r > p+\beta_p > 2$, we consider two separate possibilities, depending on which of the two terms on the right-hand side of (5.5.41) is the larger.

Specifically, if

$$\epsilon_1^{\frac{1}{p+\beta_p-1}} \geq \epsilon_2^{\frac{1}{p+\beta_p-2}}, \quad (5.5.49)$$

then

$$\left(\epsilon_1^r \epsilon_2^{\frac{r-p-\beta_p}{p+\beta_p-2}}\right)^{\frac{1}{r-1}} \geq \left(\epsilon_2^{r\frac{p+\beta_p-1}{p+\beta_p-2}} \epsilon_2^{\frac{r-p-\beta_p}{p+\beta_p-2}}\right)^{\frac{1}{r-1}} = \epsilon_2^{\frac{p+\beta_p}{p+\beta_p-2}}$$

and

$$\left(\epsilon_2^r \epsilon_1^{\frac{2(r-p-\beta_p)}{p+\beta_p-1}}\right)^{\frac{1}{r-2}} \geq \left(\epsilon_2^r \epsilon_2^{\frac{2(r-p-\beta_p)}{p+\beta_p-1}\frac{p+\beta_p-1}{p+\beta_p-2}}\right)^{\frac{1}{r-2}} = \epsilon_2^{\frac{p+\beta_p}{p+\beta_p-2}}.$$

Thus the smallest term on the right-hand side in the decrease (5.5.48) is the last, and thus

$$f(x_k) - f(x_{k+1}) \geq \kappa_{H,2} \epsilon_2^{\frac{p+\beta_p}{p+\beta_p-2}}. \quad (5.5.50)$$

Otherwise, when (5.5.49) does not hold, essentially identical calculations reveal that

$$\left(\epsilon_1^r \epsilon_2^{\frac{r-p-\beta_p}{p+\beta_p-2}}\right)^{\frac{1}{r-1}} \geq \epsilon_1^{\frac{p+\beta_p}{p+\beta_p-1}} \quad \text{and} \quad \left(\epsilon_2^r \epsilon_1^{\frac{2(r-p-\beta_p)}{p+\beta_p-1}}\right)^{\frac{1}{r-2}} \geq \epsilon_1^{\frac{p+\beta_p}{p+\beta_p-1}},$$

and thus the smallest term on the right-hand side in the decrease (5.5.48) is now the first, and so

$$f(x_k) - f(x_{k+1}) \geq \kappa_{H,2} \epsilon_1^{\frac{p+\beta_p}{p+\beta_p-1}}. \quad (5.5.51)$$

Combining (5.5.50) and (5.5.51) then shows that (5.5.48) holds.

It remains to consider case (i). But this follows immediately from the analysis in case (ii) by letting $\epsilon_2 = \infty$, because the terms involving α_2 are absent. □

We finally arrive at our main complexity result.

Theorem 5.5.7: Complexity of finding first- and second-order critical points by the ARp-H algorithm

Suppose that **f.Bb** *and* **f.DpH** *hold and that* $r \geq p+\beta_p$. *Then there exist positive constants* $\kappa^{S}_{\text{ARp-H}}$, $\kappa^{A}_{\text{ARp-H}}$, $\kappa^{B}_{\text{ARp-H},1}$, $\kappa^{B}_{\text{ARp-H},2}$, *and* $\kappa^{C}_{\text{ARp-H}}$ *for which the* ARp-H *algorithm (p. 161) requires at most*

$$\begin{aligned}\#^{F}_{\text{ARp-H},2} &\stackrel{\text{def}}{=} \kappa^{A}_{\text{ARp-H}} \frac{f(x_0)-f_{\text{low}}}{\min\left[\epsilon_1^{\frac{p+\beta_p}{p+\beta_p-1}}, \epsilon_2^{\frac{p+\beta_p}{p+\beta_p-2}}\right]} + \kappa^{B}_{\text{ARp-H},1}|\log\epsilon_1| + \kappa^{B}_{\text{ARp-H},2}|\log\epsilon_2| + \kappa^{C}_{\text{ARp-H}} \\ &= \mathcal{O}\left(\max\left[\epsilon_1^{-\frac{p+\beta_p}{p+\beta_p-1}}, \epsilon_2^{-\frac{p+\beta_p}{p+\beta_p-2}}\right]\right)\end{aligned} \tag{5.5.52}$$

evaluations of f *and its derivatives of order one and (when* $\epsilon_2 < \infty$*) two and at most*

$$\#^{D}_{\text{ARp-H},2} \stackrel{\text{def}}{=} \kappa^{S}_{\text{ARp-H}} \frac{f(x_0)-f_{\text{low}}}{\min\left[\epsilon_1^{\frac{p+\beta_p}{p+\beta_p-1}}, \epsilon_2^{\frac{p+\beta_p}{p+\beta_p-2}}\right]} + 1 = \mathcal{O}\left(\max\left[\epsilon_1^{-\frac{p+\beta_p}{p+\beta_p-1}}, \epsilon_2^{-\frac{p+\beta_p}{p+\beta_p-2}}\right]\right)$$

evaluations of derivatives of the remaining orders up to p *of* f *to produce an iterate* x_ϵ *such that* $\|\nabla^1_x f(x_\epsilon)\| \leq \epsilon_1$ *and* $\lambda_{\min}[\nabla^2_x f(x_\epsilon)] \geq -\epsilon_2$.

Proof. As usual, we have that, for any iteration k of the ARp-H algorithm,

$$f(x_0) - f_{\text{low}} \geq f(x_0) - f(x_{k+1}) = \sum_{j\in\mathcal{S}_k} [f(x_j) - f(x_{j+1})]. \tag{5.5.53}$$

To bound the number of successful iterations until termination, we may call upon Lemma 5.5.6, in particular (5.5.44), to deduce that for all $j \in \mathcal{S}_k$,

$$f(x_j) - f(x_{j+1}) \geq \kappa_{H,2} \min\left[\epsilon_1^{\frac{p+\beta_p}{p+\beta_p-1}}, \epsilon_2^{\frac{p+\beta_p}{p+\beta_p-2}}\right], \tag{5.5.54}$$

where $\kappa_{H,2}$ is defined as in (5.5.45). Combining (5.5.53) and (5.5.54), we have that

$$f(x_0) - f_{\text{low}} \geq \kappa_{H,2}|\mathcal{S}_k| \min\left[\epsilon_1^{\frac{p+\beta_p}{p+\beta_p-1}}, \epsilon_2^{\frac{p+\beta_p}{p+\beta_p-2}}\right],$$

and therefore

$$|\mathcal{S}_k| \leq \kappa^{S}_{\text{ARp-H}} \frac{f(x_0)-f_{\text{low}}}{\min\left[\epsilon_1^{\frac{p+\beta_p}{p+\beta_p-1}}, \epsilon_2^{\frac{p+\beta_p}{p+\beta_p-2}}\right]} + 1 \stackrel{\text{def}}{=} N_S,$$

where

$$\kappa^{S}_{\text{ARp-H}} \stackrel{\text{def}}{=} \frac{1}{\kappa_{H,2}} \equiv \frac{r!}{\eta_1} \max\left[\left(\frac{\sigma_H}{\alpha_1^r}\right)^{\frac{1}{r-1}}, \left(\frac{\sigma_H^2}{\alpha_2^r}\right)^{\frac{1}{r-2}}\right]$$

and

$$\sigma_H \stackrel{\text{def}}{=} \max \left[\begin{array}{l} \sigma_0,\ \gamma_3\theta_1(r-1)!,\ \gamma_3\theta_2(r-2)!, \\ \quad \gamma_3 \left(\frac{(r-1)!}{3}\right)^{\frac{p+\beta_p-r}{p+\beta_p-1}} \left(\frac{r! H_{f,p}}{(p+\beta_p-1)!(1-\eta_2)}\right)^{\frac{r-1}{p+\beta_p-1}}, \\ \quad \gamma_3 \left(\frac{(r-2)!}{3}\right)^{\frac{p+\beta_p-r}{p+\beta_p-2}} \left(\frac{r! H_{f,p}}{(p+\beta_p-1)!(1-\eta_2)}\right)^{\frac{r-2}{p+\beta_p-2}} \end{array} \right], \quad (5.5.55)$$

and thus we have an upper bound on the number of successful iterations required until termination occurs in Step 2 of ARp-H. By construction this is also a bound on the total number of evaluations of derivatives of f that are only needed on such iterations.

We may now apply Lemma 2.4.1 (p. 44), with $\sigma_{\max}$ defined as the right-hand side of (5.5.41), to infer an upper bound on the total number of iterations and evaluations of f, namely,

$$\kappa^{\text{A}}_{\text{ARp-H}} N_S + \frac{1}{\log \gamma_2}\left[|\log \gamma_1| + \log\left(\frac{\sigma_H}{\sigma_0}\right) + \frac{r-p-\beta_p}{p+\beta_p-1}|\log \epsilon_1| + \frac{r-p-\beta_p}{p+\beta_p-2}|\log \epsilon_2|\right],$$

where σ_H is defined as in (5.5.55) and

$$\kappa^{\text{A}}_{\text{ARp-H}} \stackrel{\text{def}}{=} \kappa^{\text{S}}_{\text{ARp-H}}\left(1 + \frac{|\log \gamma_1|}{\log \gamma_2}\right).$$

Thus

$$\kappa^{\text{B}}_{\text{ARp-H,1}} \stackrel{\text{def}}{=} \frac{1}{\log \gamma_2} \cdot \frac{r-p-\beta_p}{p+\beta_p-1}, \quad \kappa^{\text{B}}_{\text{ARp-H,2}} \stackrel{\text{def}}{=} \frac{1}{\log \gamma_2} \cdot \frac{r-p-\beta_p}{p+\beta_p-2},$$

and

$$\kappa^{\text{C}}_{\text{ARp-H}} \stackrel{\text{def}}{=} \frac{1}{\log \gamma_2}\left[|\log \gamma_1| + \log\left(\frac{\sigma_H}{\sigma_0}\right)\right].$$

Since the value of f, as well as those of the (first and second) derivatives that are needed to check conditions (5.5.10) and (5.5.11), occurs at every iteration, this provides the bound (5.5.52). □

A few comments are in order. As a function of the accuracy tolerances, the bounds in Theorem 5.5.7 (p. 171) do not depend on the regularization power r. However, even though the smoothness parameter β_p is (usually) unknown, its value is captured accurately in the complexity, even for the case when $\beta_p = 0$, which corresponds to the case when the objective has the highest level of nonsmoothness (compared to $\beta_p \in (0,1]$). This property is referred to as *universality* in the works of Nemirovskii and Yudin [296] and Nesterov [299]. The values of the complexity bounds as a function of the accuracies indicate that one should choose $r \geq p+1$ to achieve the best complexity when β_p is unknown, and there seems to be little reason, from an evaluation-complexity point of view, to pick anything other than $r = p+1$. Then, if $\beta_p = 1$, that is, the pth-order derivative is Lipschitz continuous, we recover the second-order complexity bounds of ARp in Theorem 4.1.5 (p. 107). Furthermore, the results in Theorem 5.5.7 (p. 171) show that for the ARp-H variant, this same complexity bound continues to hold for any regularization power $r \geq p+1$.

The bounds in Theorem 5.5.7 vary continuously with p and $\beta_p \in [0,1]$. More significantly, they indicate increasingly improving complexity as p grows, i.e., the more derivatives that are available and used, and as β_p increases in $[0,1]$, i.e., the smoother these derivatives are. In particular, purely as a function of ϵ_1 and ϵ_2, as β_p varies, we observe the following ranges of complexity powers: for $p = 1$, the complexity orders lie between $[\epsilon_1^{-2}, \infty)$; for $p = 2$, they lie in $[\epsilon_1^{-\frac{3}{2}}, \epsilon_1^{-2}]$ and $[\epsilon_2^{-3}, \infty]$; for $p = 3$, in $[\epsilon_1^{-\frac{4}{3}}, \epsilon_1^{-\frac{3}{2}}]$ and $[\epsilon_2^{-2}, \epsilon_2^{-3}]$; for $p = 4$, in $[\epsilon_1^{-\frac{5}{4}}, \epsilon_1^{-\frac{4}{3}}]$ and $[\epsilon_2^{-\frac{5}{3}}, \epsilon_2^{-2}]$; and so on.

Notes and references for Section 5.5

The material in this section is based on, and extends, [105]. In particular, the inclusion of the second-order criticality algorithm and guarantees is novel. We note that [105] also includes an extensive discussion and analysis of the constants involved in the complexity bounds and shows that these remain bounded above as p grows, provided that some user-chosen algorithm parameters are suitably scaled and that $r = \mathcal{O}(p)$. Earlier complexity analysis of second-order regularization variants of ARp-H can be found in [85], where the case $p = 2$ and $r = p + \beta_2$ is addressed. A sharp—indeed optimal among second-order methods applied to Hölder-continuous functions in a similar vein to the approach taken in Section 3.5 (p. 90)—complexity result can be found in [106].

Evaluation-complexity analysis of Hölder-continuous functions dates back at least ten years. Early studies centered on regularized first- [101] and second- [85] order methods, in which the highest-available derivative is Hölder rather than Lipschitz continuous. In the earlier of these, the Hölder exponent is assumed to be known, and the regularization power fixed accordingly, while in [101] the consequences of general rth-order regularization are investigated. The latter also shows that the (first-order) bound in Theorem 5.5.7 (p. 171) for the case $p = 1$ and $r \geq 1 + \beta_1$, $\beta_1 \in (0,1]$, is sharp. The first universal algorithm [203], in the sense we have discussed, was for problems with Hölder-continuous Hessians and using cubic regularization and corresponds to the case $p = 2$, $r = 3$, and $\beta_2 \in [0,1]$ in our framework. The method imposes a more stringent sufficient decrease condition of the objective function than is traditional in order to promote universality. The ARp-H algorithm described in this section includes a modification in a similar, but not identical, vein, generalizes that in [203] to higher-order models and second-order criticality, and achieves complexity bounds with similar universal properties. In particular, we recall that the approach that we have presented monitors the length of the step on each iteration on which the objective function is sufficiently decreased and requires that this step not be too small. It is also possible to restrict the search to satisfy convex constraints [105], but thus far complexity bounds have only been given when finding approximate first-order critical points; the generalization of this to second-order points using the techniques we will encounter in Chapter 6 is straightforward.

Other approaches [38, 39, 40, 116, 117] that relax the Lipschitz-continuity assumption on the objective and its derivatives are designed specifically for classes of problems with discontinuous and possibly infinite derivatives (such as when cusps are present).

Part IB

Evaluation Complexity for Constrained Problems

We now extend our findings and analysis to the case where constraints are present. A priori, constrained optimization problems appear more complicated than unconstrained ones. Does this intuition translate to worst-case complexity bounds? The answer to that question depends on how expensive it is to compute and enforce the constraints (compared to evaluating the objective function and its derivatives).

Previous page: *Vision* by Livia Cartis, 1998

Chapter 6

Convexly Constrained Problems

Chapter summary

If the minimization variables are required to lie within a convex domain, and if the cost of ensuring that iterates remain feasible is reasonable, we now show that many of the complexity results that we obtained for the unconstrained case in our previous chapters remain valid in the presence of such constraints. In particular, we derive an equivalent complexity result for a simple modification of the ARp *algorithm from Chapter 4 when searching for approximate constrained minimizers of order one to three. This analysis is then refined for a variant that has significance for later, more generally constrained problems.*

Finally, we return to the least-Euclidean-norm problem from the previous chapter and show that the algorithm there can also be simply modified to maintain the fast complexity bound when the constraints are convex.

6.1 Problems with convex constraints

We next consider how we might modify the regularization methods introduced in Chapters 2, 3, and 4 to cope with the additional restriction that the variables x might be constrained. We start with the simple case in which the constraints are convex, and in particular $x \in \mathcal{C}$, where we assume the following throughout.

C.CNe	The constraint set $\mathcal{C}$ is a closed, convex, nonempty subset of $\mathbb{R}^n$.

This ensures that the optimization problem in question,

$$\underset{x \in \mathcal{C}}{\text{minimize}}\ f(x), \tag{6.1.1}$$

has a solution, and we remind the reader that Theorem 1.1.2 provides both necessary and sufficient optimality conditions in this case.

6.1.1 Approximate optimality conditions

Before we develop a suitable algorithm for (6.1.1), we need to formalize our aims. We consider

$$\mathcal{C}(x) \stackrel{\text{def}}{=} \{d \in \mathbb{R}^n \mid x + d \in \mathcal{C}\}, \tag{6.1.2}$$

the set of feasible steps from a given $x \in \mathcal{C}$ (as we defined in (1.1.7) on p. 4), as well as the Euclidean unit ball

$$\mathcal{B} \stackrel{\text{def}}{=} \{d \in \mathbb{R}^n \mid \|d\| \leq 1\}. \tag{6.1.3}$$

In view of (1.1.10), we say that x_ϵ is an ϵ_1-*approximate first-order minimizer* of f within $\mathcal{C}$ if

$$\omega_{f,\mathcal{C},1}(x_\epsilon) \leq \epsilon_1 \tag{6.1.4}$$

holds for given $\epsilon_1 > 0$, where the first-order criticality measure [131, Thm. 12.1.6] is

$$\omega_{f,\mathcal{C},1}(x) \stackrel{\text{def}}{=} - \min_{d \in \mathcal{C}(x) \cap \mathcal{B}} \langle d, \nabla_x^1 f(x) \rangle \geq 0. \tag{6.1.5}$$

By analogy, and in view of (1.1.11), x_ϵ is an (ϵ_1, ϵ_2)-*approximate second-order minimizer* if (6.1.4) and

$$\omega_{f,\mathcal{C},2}(x_\epsilon, \epsilon_1) \leq \epsilon_2 \tag{6.1.6}$$

hold for given $(\epsilon_1, \epsilon_2) > 0$, where the second-order measure is

$$\omega_{f,\mathcal{C},2}(x, \epsilon_1) \stackrel{\text{def}}{=} - \min_{d \in \mathcal{C}^1(x, \epsilon_1) \cap \mathcal{B}} \langle d, \nabla_x^2 f(x) d \rangle \geq 0 \tag{6.1.7}$$

and the set of first-order α-restricted feasible steps is

$$\mathcal{C}^1(x, \alpha) \stackrel{\text{def}}{=} \left\{ d \in \mathcal{C}(x) \;\middle|\; |\langle d, \nabla_x^1 f(x) \rangle| \leq \tau_1 \alpha \right\} \tag{6.1.8}$$

for a prespecified $\tau_1 \in (0, 1)$ and for some $\alpha > 0$ that may depend on ϵ_1. Strictly, we must be cautious as it is by no means obvious that $\omega_{f,\mathcal{C},2}(x, \epsilon_1)$ is a criticality measure. Indeed, a simple example [309, Rem. 2.4] illustrates that the related measure $\omega_{f,\mathcal{C},2}(x, 0)$ might not necessarily be continuous. For this example, there is a path to the (exact) first-order minimizer that does not reveal negative curvature in $\mathcal{C}^1(x, 0)$ until the limit is reached. However the enlarged feasibility set $\mathcal{C}^1(x, \epsilon_1)$ for $\epsilon_1 > 0$ prevents such unfortunate behaviour, since feasible negative curvature is always exposed to some extent.

More generally, and bearing in mind (1.1.12), x_ϵ is an $(\epsilon_1, \epsilon_2, \epsilon_3)$-*approximate third-order minimizer* if (6.1.4), (6.1.6), and

$$\omega_{f,\mathcal{C},3}(x_\epsilon, \epsilon_1, \epsilon_2) \leq \epsilon_3$$

hold for given $(\epsilon_1, \epsilon_2, \epsilon_3) > 0$, where the third-order measure is

$$\omega_{f,\mathcal{C},3}(x, \epsilon_1, \epsilon_2) \stackrel{\text{def}}{=} - \min_{d \in \mathcal{C}^2(x, \epsilon_1, \epsilon_2) \cap \mathcal{B}} \nabla_x^3 f(x)[d]^3 \geq 0 \tag{6.1.9}$$

and the set of second-order (α_1, α_2)-restricted feasible steps is

$$\mathcal{C}^2(x, \alpha_1, \alpha_2) \stackrel{\text{def}}{=} \left\{ d \in \mathcal{C}^1(x, \alpha_1) \;\middle|\; |\langle d, \nabla_x^2 f(x) d \rangle| \leq \tau_2 \alpha_2 \right\} \tag{6.1.10}$$

for a prespecified $\tau_2 \in (0, 1)$ and for some $(\alpha_1, \alpha_2) > 0$ that may depend on (ϵ_1, ϵ_2). Letting $\tau_1 \alpha = \tau_1 \alpha_1 = \epsilon_1$ and $\tau_2 \alpha_2 = \epsilon_2$ in these conditions is equivalent to our previously stated conditions in the unconstrained case, namely, (3.2.6), (4.1.3), and (4.1.4).

It is worth remarking at this point that computing the first-order measure $\omega_{f,\mathcal{C},1}(x)$ requires the solution of a convex optimization problem and is thus achievable in polynomial time [299, Sect. 2.2]. By contrast, the problems that define $\omega_{f,\mathcal{C},2}(x, \epsilon_1)$ and

$\omega_{f,\mathcal{C},3}(x, \epsilon_1, \epsilon_2)$ are nonconvex, and certainly the former may be NP-hard[45] even for simple convex sets $\mathcal{C}$ [295, 309]. Additionally, although we defined the unit ball (6.1.3) that we use to assess criticality in terms of the Euclidean norm, this is by no means necessary[46] since all norms are equivalent in finite dimensions (see Appendix A.2). In particular, it is sometimes convenient to pick a norm that aligns with $\mathcal{C}$—for example, the infinity norm $\|\cdot\|_\infty$ is often preferred in practice for box constraints/simple bounds $\mathcal{C} = \{x \mid x^{\rm L} \leq x \leq x^{\rm U}\}$.

6.2 The ARp-CC algorithm and its complexity

6.2.1 The ARp-CC algorithm

The resulting ARp-CC algorithm is the obvious generalization of the ARp algorithm (p. 100) to cope with convex constraints. Progress at the kth iteration is ensured by computing a step s_k to approximately minimize the model $m_k(s)$ defined in (4.1.2) within the region $x_k + s \in \mathcal{C}$. To be precise, consider the model counterpart of the special feasible sets (6.1.2), (6.1.8), and (6.1.10), namely,

$$\begin{aligned} \mathcal{M}_k^0(s) &\stackrel{\rm def}{=} \mathcal{C}(x_k + s), \\ \mathcal{M}_k^1(s) &\stackrel{\rm def}{=} \left\{d \in \mathcal{M}_k^0(s) \mid |\langle d, \nabla_s^1 m_k(s)\rangle| \leq \theta_1 \epsilon_1\right\}, \end{aligned} \tag{6.2.1}$$

and

$$\mathcal{M}_k^2(s) \stackrel{\rm def}{=} \left\{d \in \mathcal{M}_k^1(s) \mid |\langle d, \nabla_s^2 m_k(s) d\rangle| \leq \theta_2 \epsilon_2\right\} \tag{6.2.2}$$

for given $\theta_1, \theta_2 > 0$. We stop as soon as all of the approximate first- to third-order criticality measures,

$$\omega_{m_k,\mathcal{M},1}(s) \stackrel{\rm def}{=} - \min_{d \in \mathcal{M}_k^0(s) \cap \mathcal{B}} \langle d, \nabla_s^1 m_k(s)\rangle \geq 0, \tag{6.2.3}$$

$$\omega_{m_k,\mathcal{M},2}(s) \stackrel{\rm def}{=} - \min_{d \in \mathcal{M}_k^1(s) \cap \mathcal{B}} \langle d, \nabla_s^2 m_k(s) d\rangle \geq 0, \tag{6.2.4}$$

and

$$\omega_{m_k,\mathcal{M},3}(s) \stackrel{\rm def}{=} - \min_{d \in \mathcal{M}_k^2(s) \cap \mathcal{B}} \nabla_s^3 m_k(s)[d]^3 \geq 0, \tag{6.2.5}$$

are sufficiently small. Notice that these are appropriate variants of (6.1.5), (6.1.7), and (6.1.9) for the model, and in particular, these will all be small in the neighbourhood of any third-order critical point or global minimizer of $m_k(s)$.

We state the algorithm formally on the following page. But before we do so, it is important to recognize that we are making the tacit assumption that *computing the criticality measures* for the objective function f in the previous section and those for the model m_k here *is easy*. In particular, it requires that ensuring feasibility does not constitute a significant cost. The measures themselves only depend on f and its derivatives at the current iterate (a cost that we shall consider), but they also require that we maintain feasibility with respect to $\mathcal{C}$ (a cost that we shall discount). For some convex regions (e.g., when the minimization variables have lower and upper bounds or must stay inside a hypersphere), this is a reasonable assumption, but for others it might not be. We will

[45]This need not always be the case; some sets $\mathcal{C}$ are special [288].

[46]Although other norms may change constants in the stated complexity bounds.

make this distinction more precise in Section 14.2, where we refer to such constraints as *simple*. The abstract setting "$x \in \mathcal{C}$" is usually realized either by consulting an inclusion oracle (to answer the question "is x in $\mathcal{C}$ or not?") or by requiring a means to find the projection $P_{\mathcal{C}}[x]$ of any x into $\mathcal{C}$, that is, to compute

$$P_{\mathcal{C}}[x] = \arg\min_{y \in \mathcal{C}} \|x - y\|; \tag{6.2.6}$$

the set $\mathcal{C}$ may have either a geometric or an algebraic description. If convex constraints are so expensive to evaluate that we cannot discount the cost of the inclusion oracle or projection, then we may be forced instead to consider them as general constraints, a topic that we shall not embrace until our next chapter.

Algorithm 6.2.1: Adaptive-regularization algorithm with *p*th-order models and convex constraints (AR*p*-CC)

Step 0: Initialization. An initial point $x_0 \in \mathcal{C}$; accuracy thresholds $\epsilon_1 > 0$, $\epsilon_2 > 0$, $\epsilon_3 > 0$, $\tau_1 \in (0,1)$, $\tau_2 \in (0,1)$; and an initial regularization parameter $\sigma_0 > 0$ are given. The constants η_1, η_2, γ_1, γ_2, γ_3, θ_1, θ_2, θ_3, and $\sigma_{\min}$ are also given and satisfy

$$\begin{gathered} \sigma_{\min} \in (0, \sigma_0], \quad \theta_1, \theta_2, \theta_3 \in (0,1), \quad 0 < \eta_1 \le \eta_2 < 1, \\ 0 < \gamma_1 < 1 < \gamma_2 < \gamma_3, \\ \tfrac{1}{2}[1 + \tau_1] \le \theta_1 < 1, \quad \text{and} \quad \tfrac{1}{2}[1 + \tau_2] \le \theta_2 < 1. \end{gathered} \tag{6.2.7}$$

Compute $f(x_0)$, and set $k = 0$.

Step 1: Test for termination. If $\epsilon_2 = \epsilon_3 = \infty$ and

$$\omega_{f,\mathcal{C},1}(x_k) \le \epsilon_1, \tag{6.2.8}$$

terminate with the ϵ_1-approximate first-order minimizer $x_\epsilon = x_k$.

If $\epsilon_2 < \infty$ and $\epsilon_3 = \infty$, (6.2.8) holds, and

$$\omega_{f,\mathcal{C},2}(x_k, \epsilon_1) \le \epsilon_2, \tag{6.2.9}$$

terminate with the (ϵ_1, ϵ_2)-approximate second-order minimizer $x_\epsilon = x_k$.

If $\epsilon_3 < \infty$, (6.2.8) and (6.2.9) hold, and

$$\omega_{f,\mathcal{C},3}(x_k, \epsilon_1, \epsilon_2) \le \epsilon_3,$$

terminate with the $(\epsilon_1, \epsilon_2, \epsilon_3)$-approximate third-order minimizer $x_\epsilon = x_k$.

Step 2: Step computation. Compute a step $s_k \in \mathcal{C}(x_k)$ such that

$$m_k(s_k) < m_k(0),$$

where $m_k(s)$ is given by (4.1.2), and

$$\omega_{m_k,\mathcal{M},1}(s_k) \le \theta_1 \epsilon_1, \tag{6.2.10}$$

and, if $\epsilon_2 < \infty$,

$$\omega_{m_k,\mathcal{M},2}(s_k) \le \theta_2 \epsilon_2, \tag{6.2.11}$$

and, if also $\epsilon_3 < \infty$,

$$\omega_{m_k,\mathcal{M},3}(s_k) \le \theta_3 \epsilon_3. \tag{6.2.12}$$

Step 3: Acceptance of the trial point. Compute $f(x_k + s_k)$, and define

$$\rho_k = \frac{f(x_k) - f(x_k + s_k)}{T_{f,p}(x_k, 0) - T_{f,p}(x_k, s_k)}.$$

If $\rho_k \geq \eta_1$, then define $x_{k+1} = x_k + s_k$. Otherwise define $x_{k+1} = x_k$.

Step 4: Regularization parameter update. Set

$$\sigma_{k+1} \in \begin{cases} [\max(\sigma_{\min}, \gamma_1 \sigma_k), \sigma_k] & \text{if } \rho_k \geq \eta_2, \\ [\sigma_k, \gamma_2 \sigma_k] & \text{if } \rho_k \in [\eta_1, \eta_2), \\ [\gamma_2 \sigma_k, \gamma_3 \sigma_k] & \text{if } \rho_k < \eta_1. \end{cases}$$

Increment k by one and go to Step 1.

6.2.2 Complexity of finding approximate first- to third-order minimizers over convex sets

Much of the analysis follows the same pattern as that given in Section 4.1.2 for the unconstrained case. We do need, however, to adjust the objective boundedness and Lipschitz assumptions that we rely on.

f.bB.C

$f(x)$ is bounded below on $\mathcal{C}$; that is, there exists a constant $f_{\rm low}$ such that, for all $x \in \mathcal{C}$,

$$f(x) \geq f_{\rm low}.$$

f.DpL.C

$f(x)$ is p-times continuously differentiable on an open set[a] $\mathcal{D}$ containing $\mathcal{C}$, and its pth-order derivative tensor $\nabla_x^p f(x)$ is Lipschitz continuous on $\mathcal{C}$; that is, there exists a constant $L_{f,p} \geq 0$ such that

$$\|\nabla_x^p f(x) - \nabla_x^p f(y)\| \leq L_{f,p} \|x - y\|$$

for all $x, y \in \mathcal{C}$.

[a] The set $\mathcal{D}$ is a convenience; we could instead have relied on properties in $\mathcal{C}$ had we used Jordan's definition of continuity rather than Weierstrass's.

Lemmas 4.1.1 and (under **f.DpL.C**) 4.1.2 continue to hold, so we turn to deriving useful lower bounds on the step generated by the algorithm.

Lemma 6.2.1: Lower bounds on the length of the step for ARp-CC

Suppose that **C.CNe** *and* **f.DpL.C** *hold. Then on every successful iteration k of the ARp-CC algorithm (p. 180), if $\omega_{f,\mathcal{C},1}(x_{k+1}) > \epsilon_1$, the step s_k calculated in Step 2 satisfies*

$$\|s_k\|^p > \frac{p!(1-\theta_1)}{L_{f,p} + \sigma_{\max}} \epsilon_1, \tag{6.2.13}$$

where $\sigma_{\max}$ *is given by (4.1.19). If additionally* $\epsilon_2 < \infty$ *and* $\omega_{k+1,2} > \epsilon_2$*, where*

$$\omega_{k+1,2} \stackrel{\rm def}{=} -\min_{d \in \mathcal{M}_k^1(s_k) \cap \mathcal{B}} \nabla_x^2 f(x_{k+1})[d]^2 \geq 0, \tag{6.2.14}$$

then

$$\|s_k\|^{p-1} > \frac{(p-1)!(1-\theta_2)}{L_{f,p} + \sigma_{\max}} \epsilon_2. \tag{6.2.15}$$

Furthermore, if $\epsilon_3 < \infty$ *and if* $\omega_{k+1,3} > \epsilon_3$*, where*

$$\omega_{k+1,3} \stackrel{\rm def}{=} -\min_{d \in \mathcal{M}_k^2(s_k) \cap \mathcal{B}} \nabla_x^3 f(x_{k+1})[d]^3 \geq 0, \tag{6.2.16}$$

then

$$\|s_k\|^{p-2} > \frac{(p-2)!(1-\theta_3)}{L_{f,p} + \sigma_{\max}} \epsilon_3. \tag{6.2.17}$$

Moreover, for any $d \in \mathcal{C}^1(x_{k+1}, \epsilon_1)$,

$$|\nabla_s^1 m_k(s_k)[d]^1| \leq \frac{L_{f,p} + \sigma_{\max}}{p!} \|s_k\|^p + \tau_1 \epsilon_1, \tag{6.2.18}$$

while for any $d \in \mathcal{C}^2(x_{k+1}, \epsilon_1, \epsilon_2)$,

$$|\nabla_s^2 m_k(s_k)[d]^2| \leq \frac{L_{f,p} + \sigma_{\max}}{(p-1)!} \|s_k\|^{p-1} + \tau_2 \epsilon_2. \tag{6.2.19}$$

We note, for clarity, that the objective functions in the definitions of $\omega_{k+1,2}$ and $\omega_{k+1,3}$ in the above statement involve f, while the constraint sets are based on the model m_k.

Proof. The fact that $k \in \mathcal{S}$ and the definition of the trial point give that $x_{k+1} = x_k + s_k$. Clearly,

$$\begin{aligned} \left| \nabla_x^j f(x_{k+1})[d]^j - \nabla_s^j m_k(s_k)[d]^j \right| &= \left| [\nabla_x^j f(x_{k+1}) - \nabla_s^j T_{f,p}(x_k, s_k)][d]^j \right. \\ &\quad \left. + \frac{\sigma_k}{(p+1)!} \nabla_s^j (-\|s_k\|^{p+1})[d]^j \right| \\ &\leq \left| [\nabla_x^j f(x_{k+1}) - \nabla_s^j T_{f,p}(x_k, s_k)][d]^j \right| \\ &\quad + \frac{\sigma_k}{(p+1)!} \left| \nabla_s^j (-\|s_k\|^{p+1})[d]^j \right| \end{aligned} \tag{6.2.20}$$

for all $d \in \mathcal{B}$, where here and hereafter $j \in [1:q]$ with $q = 1$ if $\epsilon_2 = \epsilon_3 = \infty$, $q = 2$ if $\epsilon_2 < \infty$ and $\epsilon_3 = \infty$, and $q = 3$ otherwise. Furthermore, for all $d \in \mathcal{B}$, we have that both

$$\begin{aligned} \left| [\nabla_x^j f(x_{k+1}) - \nabla_s^j T_{f,p}(x_k, s_k)][d]^j \right| &\leq \|\nabla_x^j f(x_{k+1}) - \nabla_s^j T_{f,p}(x_k, s_k)\| \\ &\leq \frac{L_{f,p}}{(p-j+1)!} \|s_k\|^{p-j+1} \end{aligned} \tag{6.2.21}$$

and

$$\begin{aligned} \left| \nabla_s^j (-\|s_k\|^{p+1})[d]^j \right| &\leq \|\nabla_s^j (-\|s_k\|^{p+1})\| \\ &\leq \frac{(p+1)!}{(p-j+1)!} \|s_k\|^{p-j+1}, \end{aligned} \tag{6.2.22}$$

where we used the definition of the norm (A.7); the Taylor-series error bound (A.20), which relies on **f.DpL.C**; and Lemma B.4.1 (p. 461) with $\beta = 1$. Combining (6.2.20)–(6.2.22) and using the guaranteed upper bound (4.1.19) on σ_k, it follows that, for all $d \in \mathcal{B}$,

$$\left|\nabla_x^j f(x_{k+1})[d]^j - \nabla_s^j m_k(s_k)[d]^j\right| \leq \frac{L_{f,p} + \sigma_{\max}}{(p-j+1)!}\|s_k\|^{p-j+1}. \tag{6.2.23}$$

When $j = 1$, we deduce from (6.2.23) that

$$\begin{aligned}
\omega_{f,\mathcal{C},1}(x_{k+1}) &= -\min_{d\in\mathcal{C}(x_{k+1})\cap\mathcal{B}} \nabla_x^1 f(x_{k+1})[d]^1 \\
&\leq \frac{L_{f,p} + \sigma_{\max}}{p!}\|s_k\|^p - \min_{d\in\mathcal{C}(x_{k+1})\cap\mathcal{B}} \nabla_s^1 m_k(s_k)[d]^1 \\
&= \frac{L_{f,p} + \sigma_{\max}}{p!}\|s_k\|^p - \min_{d\in\mathcal{M}_k^0(s_k)\cap\mathcal{B}} \langle d, \nabla_s^1 m_k(s_k)\rangle \\
&= \frac{L_{f,p} + \sigma_{\max}}{p!}\|s_k\|^p + \omega_{m_k,\mathcal{M},1}(s_k),
\end{aligned}$$

since $\langle d, \nabla_x^1 f(x_{k+1})\rangle = \nabla_x^1 f(x_{k+1})[d]^1$, $\langle d, \nabla_s^1 m_k(s_k)\rangle = \nabla_s^1 m_k(s_k)[d]^1$, and $\mathcal{M}_k^0(s_k) \equiv \mathcal{C}(x_{k+1})$. The desired bound (6.2.13) then follows by combining this inequality with the requirement (6.2.10) and the assumption $\omega_{f,\mathcal{C},1}(x_{k+1}) > \epsilon_1$ and rearranging terms. When $j = 2$ and 3, (6.2.23) together with (6.2.11) and (6.2.12) gives that

$$\begin{aligned}
-\min_{d\in\mathcal{M}_k^{j-1}(s_k)\cap\mathcal{B}} \nabla_x^j f(x_{k+1})[d]^j &\leq \frac{L_{f,p} + \sigma_{\max}}{(p-j+1)!}\|s_k\|^{p-j+1} - \min_{d\in\mathcal{M}_k^{j-1}(s_k)\cap\mathcal{B}} \nabla_s^j m_k(s_k)[d]^j \\
&= \frac{L_{f,p} + \sigma_{\max}}{(p-j+1)!}\|s_k\|^{p-j+1} + \omega_{m_k,\mathcal{M},j}(s_k) \\
&\leq \frac{L_{f,p} + \sigma_{\max}}{(p-j+1)!}\|s_k\|^{p-j+1} + \theta_j\epsilon_j,
\end{aligned}$$

which, together with the assumptions $\omega_{k+1,2}(x_{k+1}) > \epsilon_2$ and $\omega_{k+1,3}(x_{k+1}) > \epsilon_3$, gives (6.2.15) and (6.2.17).

For any $d \in \mathcal{C}^1(x_{k+1}, \epsilon_1)$, (6.2.23) with $j = 1$, using (6.1.8), leads to

$$\begin{aligned}
|\nabla_s^j m_k(s_k)[d]^1| &\leq \frac{L_{f,p} + \sigma_{\max}}{p!}\|s_k\|^p + |\nabla_x^1 f(x_{k+1})[d]^1| \\
&\leq \frac{L_{f,p} + \sigma_{\max}}{p!}\|s_k\|^p + \tau_1\epsilon_1,
\end{aligned}$$

which is (6.2.18).

Similarly, if $d \in \mathcal{C}^2(x_{k+1}, \epsilon_1, \epsilon_2)$, (6.2.23) with $j = 2$, using (6.1.10), gives

$$\begin{aligned}
|\nabla_s^j m_k(s_k)[d]^2| &\leq \frac{L_{f,p} + \sigma_{\max}}{(p-1)!}\|s_k\|^{p-1} + |\nabla_x^2 f(x_{k+1})[d]^2| \\
&\leq \frac{L_{f,p} + \sigma_{\max}}{(p-1)!}\|s_k\|^{p-1} + \tau_2\epsilon_2,
\end{aligned}$$

which is (6.2.19). □

Lemma 6.2.1 then leads directly to the following guaranteed reduction in the objective function following a successful step of the ARp-CC algorithm; the proof is identical to that of Lemma 4.1.4.

Lemma 6.2.2: Bounds on the successful function decrease for ARp-CC

*Suppose that **C.CNe** and **f.DpL.C** hold and that iteration k of the ARp-CC algorithm is successful. Then if $\omega_{f,\mathcal{C},1}(x_{k+1}) > \epsilon_1$, then*

$$f(x_k) - f(x_{k+1}) > \kappa_1^{CC} \epsilon_1^{\frac{p+1}{p}},$$

where

$$\kappa_1^{CC} \stackrel{\text{def}}{=} \frac{\eta_1 \sigma_{\min}}{p!} \left(\frac{p!(1-\theta_1)}{L_{f,p} + \sigma_{\max}} \right)^{\frac{p+1}{p}}.$$

If additionally $\epsilon_2 < \infty$ and $\omega_{k+1,2} > \epsilon_2$, then

$$f(x_k) - f(x_{k+1}) > \kappa_2^{CC} \epsilon_2^{\frac{p+1}{p-1}},$$

where

$$\kappa_2^{CC} \stackrel{\text{def}}{=} \frac{\eta_1 \sigma_{\min}}{p!} \left(\frac{(p-1)!(1-\theta_2)}{L_{f,p} + \sigma_{\max}} \right)^{\frac{p+1}{p-1}}$$

and $\omega_{k+1,2}$ is defined in (6.2.14), and if $\epsilon_3 < \infty$ and $\omega_{k+1,3} > \epsilon_3$, then

$$f(x_k) - f(x_{k+1}) > \kappa_3^{CC} \epsilon_3^{\frac{p+1}{p-2}},$$

where

$$\kappa_3^{CC} \stackrel{\text{def}}{=} \frac{\eta_1 \sigma_{\min}}{p!} \left(\frac{(p-2)!(1-\theta_3)}{L_{f,p} + \sigma_{\max}} \right)^{\frac{p+1}{p-2}}$$

and $\omega_{k+1,3}$ is defined by (6.2.16).

We are now ready to establish an evaluation-complexity bound for the ARp-CC algorithm.

Theorem 6.2.3: Complexity of finding first- to third-order critical points by the ARp-CC algorithm

*Suppose that **C.CNe**, **f.DpL.C**, and **f.bB.C** hold. Then there exist positive constants κ^S_{ARpCC}, κ^A_{ARpCC}, and κ^C_{ARpCC} such that, for any $(\epsilon_1, \epsilon_2, \epsilon_3) \in (0,1]^3$, the ARp-CC algorithm (p. 180) requires at most*

$$\#^F_{\text{ARpCC},3} \stackrel{\text{def}}{=} \kappa^A_{\text{ARpCC}} \frac{f(x_0) - f_{\text{low}}}{\min\left[\epsilon_1^{\frac{p+1}{p}}, \epsilon_2^{\frac{p+1}{p-1}}, \epsilon_3^{\frac{p+1}{p-2}}\right]} + \kappa^C_{\text{ARpCC}} = \mathcal{O}\left(\max\left[\epsilon_1^{-\frac{p+1}{p}}, \epsilon_2^{-\frac{p+1}{p-1}}, \epsilon_3^{-\frac{p+1}{p-2}} \right] \right)$$

evaluations of f and at most

$$\#^D_{\text{ARpCC},3} \stackrel{\text{def}}{=} \kappa^S_{\text{ARpCC}} \frac{f(x_0) - f_{\text{low}}}{\min\left[\epsilon_1^{\frac{p+1}{p}}, \epsilon_2^{\frac{p+1}{p-1}}, \epsilon_3^{\frac{p+1}{p-2}}\right]} + 2 = \mathcal{O}\left(\max\left[\epsilon_1^{-\frac{p+1}{p}}, \epsilon_2^{-\frac{p+1}{p-1}}, \epsilon_3^{-\frac{p+1}{p-2}} \right] \right)$$

evaluations of derivatives of order one to p of f to produce an iterate x_ϵ such that $\omega_{f,\mathcal{C},1}(x_\epsilon) \le \epsilon_1$, $\omega_{f,\mathcal{C},2}(x_\epsilon, \epsilon_1) \le \epsilon_2$, and $\omega_{f,\mathcal{C},3}(x_\epsilon, \epsilon_1, \epsilon_2) \le \epsilon_3$.

Proof. The proof follows along lines similar to that of Theorem 4.1.5 (p. 107). In particular, (4.1.39) holds for every iteration of the ARp-CC algorithm, and our goal is now to bound the number of successful iterations required to find a step s_k and resulting point x_{k+1} for which

$$\omega_{f,\mathcal{C},1}(x_{k+1}) \leq \epsilon_1, \tag{6.2.24}$$

$$\omega_{k+1,2} \leq \epsilon_2, \tag{6.2.25}$$

and

$$\omega_{k+1,3} \leq \epsilon_3, \tag{6.2.26}$$

as well as

$$\|s_k\|^p \leq \frac{p!(1-\theta_1)}{L_{f,p}+\sigma_{\max}}\epsilon_1 \tag{6.2.27}$$

and

$$\|s_k\|^{p-1} \leq \frac{(p-1)!(1-\theta_2)}{L_{f,p}+\sigma_{\max}}\epsilon_2. \tag{6.2.28}$$

If at least one of (6.2.24)–(6.2.26) fails to hold, we may call upon Lemma 6.2.2 to deduce that

$$\begin{aligned} f(x_j) - f(x_{j+1}) &\geq \min\left[\kappa_1^{\text{CC}}\epsilon_1^{\frac{p+1}{p}}, \kappa_2^{\text{CC}}\epsilon_2^{\frac{p+1}{p-1}}, \kappa_3^{\text{CC}}\epsilon_3^{\frac{p+1}{p-2}}\right] \\ &\geq \min\left[\kappa_1^{\text{CC}}, \kappa_2^{\text{CC}}, \kappa_3^{\text{CC}}\right] \min\left[\epsilon_1^{\frac{p+1}{p}}, \epsilon_2^{\frac{p+1}{p-1}}, \epsilon_3^{\frac{p+1}{p-2}}\right] \\ &\stackrel{\text{def}}{=} \kappa_{123}^{\text{CC}} \min\left[\epsilon_1^{\frac{p+1}{p}}, \epsilon_2^{\frac{p+1}{p-1}}, \epsilon_3^{\frac{p+1}{p-2}}\right] \end{aligned}$$

for all $j \in \mathcal{S}_{k-1}$, which is (4.1.44). Note that if (6.2.27) or (6.2.28) fails, $f(x_j)-f(x_{j+1})$ is lower bounded by the same amount as in the first-order or second-order decrease, respectively; thus (4.1.44) still holds. The remainder of the proof for the calculation of the bound to satisfy (6.2.24)–(6.2.28) is identical to that of Theorem 4.1.5 (p. 107).[47] For completeness, the constants involved are

$$\begin{aligned} \kappa_{\text{ARpCC}}^{\text{S}} &\stackrel{\text{def}}{=} \frac{1}{\kappa_{123}^{\text{CC}}} \\ &\equiv \frac{(p+1)!}{\eta_1\sigma_{\min}} \max\left[\left(\frac{L_{f,p}+\sigma_{\max}}{p!(1-\theta_1)}\right)^{\frac{p+1}{p}}, \left(\frac{L_{f,p}+\sigma_{\max}}{(p-1)!(1-\theta_2)}\right)^{\frac{p+1}{p-1}},\right. \\ &\qquad\qquad\left.\left(\frac{L_{f,p}+\sigma_{\max}}{(p-2)!(1-\theta_3)}\right)^{\frac{p+1}{p-2}}\right], \end{aligned}$$

$$\kappa_{\text{ARpCC}}^{\text{A}} \stackrel{\text{def}}{=} \kappa_{\text{ARpCC}}^{\text{T}}\left(1+\frac{|\log\gamma_1|}{\log\gamma_2}\right), \quad \text{and} \quad \kappa_{\text{ARpCC}}^{\text{C}} \stackrel{\text{def}}{=} \frac{1}{\log\gamma_2}\left\{|\log\gamma_1| + \log\left(\frac{\sigma_{\max}}{\sigma_0}\right)\right\}+2,$$

where $\sigma_{\max}$ is defined as in (4.1.19) (p. 103) and where 1 was added to $\kappa_{\text{ARpCC}}^{\text{C}}$ to account for the last iteration.

[47] We note that this bound accounts for all iterations that can occur for which one of (6.2.24)–(6.2.28) fails.

When (6.2.24)–(6.2.26) hold, as well as (6.2.27) and (6.2.28), we wish now to relate these measures to approximate criticality measures for f. Clearly, (6.1.4) is satisfied due to (6.2.24) and $\mathcal{C}(x_{k+1}) = \mathcal{M}^0(s_k)$.

Let $d \in \mathcal{C}^1(x_{k+1}, \epsilon_1)$. Then (6.2.18) and (6.2.27) give

$$|\nabla_s^1 m_k(s_k)[d]^1| \leq \frac{L_{f,p} + \sigma_{\max}}{p!} \cdot \frac{p!(1-\theta_1)}{L_{f,p} + \sigma_{\max}} \epsilon_1 + \tau_1 \epsilon_1 = [1 - \theta_1 + \tau_1]\epsilon_1 \leq \theta_1 \epsilon_1,$$

where in the last inequality we also used (6.2.7), and so $d \in \mathcal{M}_k^1(s_k)$. Thus $\mathcal{C}^1(x_{k+1}, \epsilon_1) \subseteq \mathcal{M}_k^1(s_k)$ and so $\omega_{f,\mathcal{C},2}(x_{k+1}, \epsilon_1) \leq \omega_{k+1,2} \leq \epsilon_2$.

Similarly, now letting $d \in \mathcal{C}^2(x_{k+1}, \epsilon_1, \epsilon_2)$, (6.2.19) and (6.2.28) give

$$|\nabla_s^2 m_k(s_k)[d]^2| \leq \frac{L_{f,p} + \sigma_{\max}}{(p-1)!} \|s_k\|^{p-1} + \tau_2 \epsilon_2 \leq [1 - \theta_2 + \tau_2]\epsilon_2 \leq \theta_2 \epsilon_2,$$

where in the last inequality we also used (6.2.7), and so $d \in \mathcal{M}_k^2(s_k)$. Thus, $\mathcal{C}^2(x_{k+1}, \epsilon_1, \epsilon_2) \subseteq \mathcal{M}_k^2(s_k)$. Furthermore, by definition, $\mathcal{C}^2(x_{k+1}, \epsilon_1, \epsilon_2) \subseteq \mathcal{C}^1(x_{k+1}, \epsilon_1) \subseteq \mathcal{M}_k^1(s_k)$. We then conclude that $\omega_{f,\mathcal{C},3}(x_{k+1}, \epsilon_1, \epsilon_2) \leq \omega_{k+1,3} \leq \epsilon_3$. □

In the case when $\mathcal{C} = \mathbb{R}^n$, the bounds in Theorem 6.2.3 coincide with those in Theorem 4.1.5 (p. 107) in the order of ϵ but differ in their constants due mainly to the different model-minimization conditions and the second- and third-order optimality conditions that they secure for the objective function. In particular, the ARp-CC algorithm guarantees approximately nonnegative objective curvature on a smaller (by a constant fraction) set of small objective gradients but is able to secure the third-order optimality condition on a quantifiable set of small objective-curvature directions. We may contrast the latter with the requirements in ARp, where approximate third-order criticality can be guaranteed only on the subset of small model-curvature directions. However, changing the right-hand sides of the second- and third-order model-termination conditions (4.1.12)–(4.1.14) in the ARp algorithm to the same right-hand sides as the ARp-CC ones (namely, (6.2.11), (6.2.12), and (6.2.2)) leads to an ARp variant that can satisfy a similarly improved (but unconstrained) third-order criticality condition as ARp-CC, while still guaranteeing the same first- and second-order criticality conditions as ARp.

6.2.3 A variant for finding approximate first-order minimizers

The ARp-CC algorithm happily generalizes the evaluation-complexity bounds derived when solving unconstrained optimization problems using the ARp algorithm to the convexly constrained case. One unfortunate effect, though, is that the inequalities derived in Lemma 4.1.4 that relate the objective decrease to the first- to third-order approximate criticality measures are no longer present in the corresponding Lemma 6.2.2 in the previous subsection. This has ramifications when we turn to the least-Euclidean-norm problem considered in Section 5.1, since Lemma 4.1.4 was crucial when deriving the satisfactory evaluation-complexity bounds given in Theorem 5.1.7. In our next section, we shall examine the least-Euclidean-norm problem when there are additional convex constraints on the variables, and to achieve this we now briefly consider how we might develop an appropriate version of Lemma 4.1.4 for such a case.

To do so, we lower our expectations and examine a variant of the ARp-CC algorithm that merely aims for an ϵ_1-approximate first-order minimizer of (6.1.1). Consider the following algorithm.

Algorithm 6.2.2: Adaptive-regularization algorithm with pth-order models and convex constraints for a first-order approximate minimizer (ARp1-CC)

This algorithm is the same as Algorithm 6.2.1 with $\epsilon_2 = \epsilon_3 = \infty$, but replacing the model-minimization stopping rule (6.2.10) in Step 2 by

$$\omega_{m_k,\mathcal{M},1}(s_k) \le \theta_1 \|s_k\|^p. \tag{6.2.29}$$

Notice that (6.2.29) is the obvious extension of (4.1.11) to the constrained model. The importance of this simple change is obvious in the following result.

Lemma 6.2.4: Bounds on the successful function decrease for ARp1-CC

*Suppose that **C.CNe** and **f.DpL.C** hold and that iteration k of the* ARp1-CC *algorithm is successful. Then*

$$f(x_k) - f(x_{k+1}) \ge \kappa_1 [\omega_{f,\mathcal{C},1}(x_{k+1})]^{\frac{p+1}{p}}, \tag{6.2.30}$$

where κ_1 is given by (4.1.32).

Proof. The first part of the proof of Lemma 6.2.1 remains true for the revised algorithm. When $j = 1$, we deduce from (6.2.23) that

$$\begin{aligned}
\omega_{f,\mathcal{C},1}(x_{k+1}) &= - \min_{d \in \mathcal{C}(x_{k+1}) \cap \mathcal{B}} \nabla_x^1 f(x_{k+1})[d]^1 \\
&\le \frac{L_{f,p} + \sigma_{\max}}{p!} \|s_k\|^p - \min_{d \in \mathcal{C}(x_{k+1}) \cap \mathcal{B}} \nabla_s^1 m_k(s_k)[d]^1 \\
&= \frac{L_{f,p} + \sigma_{\max}}{p!} \|s_k\|^p - \min_{d \in \mathcal{M}_k^0(s_k) \cap \mathcal{B}} \langle d, \nabla_s^1 m_k(s_k) \rangle \\
&= \frac{L_{f,p} + \sigma_{\max}}{p!} \|s_k\|^p + \omega_{m_k,\mathcal{M},1}(s_k),
\end{aligned}$$

since $\langle d, \nabla_x^1 f(x_{k+1}) \rangle = \nabla_x^1 f(x_{k+1})[d]^1$, $\langle d, \nabla_s^1 m_k(s_k) \rangle = \nabla_s^1 m_k(s_k)[d]^1$, and $\mathcal{M}_k^0(s_k) \equiv \mathcal{C}(x_{k+1})$. The bound

$$\|s_k\|^p \ge \frac{p!}{L_{f,p} + \sigma_{\max} + \theta_1 p!} \omega_{f,\mathcal{C},1}(x_{k+1})$$

then follows by combining this inequality with the requirement (6.2.29) and rearranging terms. The resulting bound (6.2.30) may be derived from this exactly as in Lemma 4.1.4. □

We may then use this result, just as in the proof of Theorem 4.1.5, and remembering that $\epsilon_2 = \epsilon_3 = \infty$, to deduce the following bound. Notice that, aside from constants, this is equivalent to the result in Theorem 6.2.3.

Theorem 6.2.5: Complexity of finding first-order critical points by the ARp1-CC algorithm

Suppose that **C.CNe**, **f.DpL.C**, *and* **f.bB.C** *hold. Then there exist positive constants* $\kappa^{S}_{\text{ARp1CC}}$, $\kappa^{A}_{\text{ARp1CC}}$, *and* κ^{C}_{ARp} *such that, for any* $\epsilon_1 \in (0,1]$, *the* ARp1-CC *algorithm (p. 180) requires at most*

$$\#^{F}_{\text{ARp1CC}} \stackrel{\text{def}}{=} \kappa^{A}_{\text{ARp1CC}} \frac{f(x_0) - f_{\text{low}}}{\epsilon_1^{\frac{p+1}{p}}} + \kappa^{C}_{\text{ARp}} = \mathcal{O}\left(\epsilon_1^{-\frac{p+1}{p}}\right)$$

evaluations of f *and at most*

$$\#^{D}_{\text{ARp1CC}} \stackrel{\text{def}}{=} \kappa^{S}_{\text{ARp1CC}} \frac{f(x_0) - f_{\text{low}}}{\epsilon_1^{\frac{p+1}{p}}} + 2 = \mathcal{O}\left(\epsilon_1^{-\frac{p+1}{p}}\right)$$

evaluations of derivatives of order one to p *of* f *to produce an iterate* x_ϵ *such that* $\omega_{f,\mathcal{C},1}(x_\epsilon) \le \epsilon_1$.

For completeness, in the preceding statement,

$$\kappa^{S}_{\text{ARp1CC}} \stackrel{\text{def}}{=} \frac{(p+1)!}{\eta_1 \sigma_{\min}} \left(\frac{L_{f,p} + \sigma_{\max} + \theta_1 p!}{p!}\right)^{\frac{p+1}{p}},$$

where $\sigma_{\max}$ is given by (4.1.19);

$$\kappa^{A}_{\text{ARp1CC}} \stackrel{\text{def}}{=} \kappa^{S}_{\text{ARp1CC}} \left(1 + \frac{|\log \gamma_1|}{\log \gamma_2}\right);$$

and κ^{C}_{ARp} is given by (4.1.46).

The importance of this result is not so much in itself, but in the impact it has on the convexly constrained least-squares problem, as we shall now see.

6.3 Convexly constrained least-Euclidean-norm problems

The *convexly constrained least-Euclidean-norm problem* is to

$$\underset{x \in \mathcal{C}}{\text{minimize}} \;\; \Phi_1(x) \stackrel{\text{def}}{=} \|r(x)\|,$$

where as before $r(x)$ is a smooth vector-valued residual function and $\mathcal{C}$ is a closed, convex, nonempty set. Just as in Section 5.1, it is often more convenient to minimize $\Phi_\ell(x)$ from (5.1.3) rather than $\Phi_1(x)$ over the same set, for some suitable ℓ chosen so that $\Phi_\ell(x)$ is sufficiently smooth for our purposes. We argued in Section 5.1 that if we do so, using a first-order approximate minimizer of $\Phi_\ell(x)$ as a surrogate for one of $\Phi_1(x)$ should account for the power ℓ, and that this may be achieved by dividing the criticality measure by $\|r(x)\|^{\ell-1}$. Thus, as before, we say that $x_\epsilon \in \mathcal{C}$ is an *ϵ-approximate global minimizer* of $\Phi_1(x)$ within $\mathcal{C}$ if $\|r(x_\epsilon)\| \le \epsilon_0$, but now it is an *$\epsilon$-approximate first-order minimizer* of Φ_1 over the same set if instead

$$\frac{\omega_{\Phi_\ell,\mathcal{C},1}(x_\epsilon)}{\|r(x_\epsilon)\|^{\ell-1}} \le \epsilon_1,$$

where

$$\omega_{\Phi_\ell,\mathcal{C},1}(x) \stackrel{\text{def}}{=} -\min_{d\in\mathcal{C}(x)\cap\mathcal{B}} \langle d, \nabla_x^1 \Phi_\ell(x)\rangle \geq 0,$$

just as in (6.1.5) and (6.1.2), for prespecified $\epsilon_0, \epsilon_1 > 0$.

To find such an x_ϵ, we consider the following algorithm.

Algorithm 6.3.1: Adaptive-regularization algorithm with *p*th-order models for first-order minimization of convexly constrained Euclidean-norm problems (AR*p*LE-CC)

Apply Algorithm 6.2.2 (AR*p*1-CC, on p. 187) to $f(x) = \Phi_\ell(x)$ for a given $\ell \geq 2$ using the model (5.1.8), but replace the termination tests (Step 1) with the following:

Step 1: Test for termination. If

$$\|r(x_k)\| \leq \epsilon_0, \tag{6.3.1}$$

terminate with the ϵ_0-approximate global minimizer $x_\epsilon = x_k$.

Otherwise, if

$$\frac{\omega_{\Phi_\ell,\mathcal{C},1}(x_k)}{\|r(x_k)\|^{\ell-1}} \leq \epsilon_1, \tag{6.3.2}$$

terminate with the ϵ_1-approximate first-order minimizer $x_\epsilon = x_k$.

To analyze this algorithm, we require the following generalization of $\mathbf{\Phi_\ell}$**.DpL**.

Φ_ℓ.DpL.C $\Phi_\ell(x)$ is p-times continuously differentiable on an open set $\mathcal{D}$ containing $\mathcal{C}$, and its pth-order derivative tensor $\nabla_x^p \Phi_\ell(x)$ is Lipschitz continuous in $\mathcal{C}$; that is, there exists a constant $L_{\Phi_\ell,p} \geq 0$ such that

$$\|\nabla_x^p \Phi_\ell(x) - \nabla_x^p \Phi_\ell(y)\| \leq L_{\Phi_\ell,p}\|x-y\|$$

for all $x, y \in \mathcal{C}$.

We may then follow the proof of Lemma 5.1.2, using Lemma 6.2.4 instead of 4.1.4, to deduce that

$$\|r(x_k)\|^\ell - \|r(x_{k+1})\|^\ell \geq \kappa_1 \ell \left[\omega_{\Phi_\ell,\mathcal{C},1}(x_{k+1})\right]^{\frac{p+1}{p}} \|r(x_{k+1})\|^{\frac{(\ell-1)(p+1)}{p}}$$

for successful iteration k, and this bound leads directly to

$$\|r(x_k)\|^{1/2^i} - \|r(x_{k+1})\|^{1/2^i} \geq \kappa_{r1} \frac{\left[\omega_{\Phi_\ell,\mathcal{C},1}(x_{k+1})\right]^{\frac{p+1}{p}}}{\|r(x_k)\|^{\iota_1(p,\ell,i)}} \tag{6.3.3}$$

when (5.1.19) holds, and (5.1.27) otherwise, just as in the proof of Lemma 5.1.4. It is then simply a matter of repeating the argument in the proof of Theorem 5.1.7 and taking into account the stopping rules (6.3.1) and (6.3.2) to deduce our main evaluation-complexity result.

Theorem 6.3.1: Complexity of finding a first-order critical point of $\|r(x)\|$ within $\mathcal{C}$ by the ARpLE-CC algorithm

Suppose that $i \geq 0$ and $\ell > 1$ satisfy either (5.1.38) or (5.1.39) for an appropriate p; that **C.CNe** *and* **Φ_ℓ.DpL.C** *hold; and that $\beta \in (0,1)$ is a fixed, problem-independent constant. Then there exist positive constants $\kappa^S_{\text{ARpLE,1}}$, $\kappa^A_{\text{ARpLE,1}}$, and κ^C_{ARpLE} such that, for any $(\epsilon_0, \epsilon_1) \in (0,1]^2$, the ARpLE-CC algorithm (p. 189) requires at most*

$$\#^F_{\text{ARpLE-CC}} \stackrel{\text{def}}{=} \kappa^A_{\text{ARpLE,1}} \frac{\max\left[\|r(x_0)\|^{\frac{1}{2^i}}, \|r(x_0)\|^{1-\frac{\ell-1}{p}}\right]}{\min\left[\epsilon_0^{\frac{1}{2^i}}, \epsilon_1^{\frac{p+1}{p}}\right]} + \kappa^C_{\text{ARpLE}} = \mathcal{O}\left(\max\left[\epsilon_0^{-\frac{1}{2^i}}, \epsilon_1^{-\frac{p+1}{p}}\right]\right)$$

evaluations of r and at most

$$\#^D_{\text{ARpLE-CC}} \stackrel{\text{def}}{=} \kappa^S_{\text{ARpLE,1}} \frac{\max\left[\|r(x_0)\|^{\frac{1}{2^i}}, \|r(x_0)\|^{1-\frac{\ell-1}{p}}\right]}{\min\left[\epsilon_0^{\frac{1}{2^i}}, \epsilon_1^{\frac{p+1}{p}}\right]} + 2 = \mathcal{O}\left(\max\left[\epsilon_0^{-\frac{1}{2^i}}, \epsilon_1^{-\frac{p+1}{p}}\right]\right)$$

evaluations of derivatives of order one to p of r to produce an iterate x_ϵ such that $\|r(x_\epsilon)\| \leq \epsilon_0$ or $\omega_{\Phi_\ell,\mathcal{C},1}(x_\epsilon)/\|r(x_\epsilon)\|^{\ell-1} \leq \epsilon_1$.

The constants mentioned in the statement of this theorem are

$$\kappa^S_{\text{ARpLE,1}} \stackrel{\text{def}}{=} \max\left(\frac{1}{1-\beta^{1/2^i}}, \frac{2^i}{\beta^{\frac{(\ell-1)(p+1)}{p}}} \frac{(p+1)!}{\eta_1 \sigma_{\min}} \left(\frac{L_{\Phi_\ell,p} + \sigma_{\max} + \theta_1 p!}{p!}\right)^{\frac{p+1}{p}}\right),$$

where $\sigma_{\max}$ is given by (5.1.60),

$$\kappa^A_{\text{ARpLE,1}} \stackrel{\text{def}}{=} \kappa^S_{\text{ARpLE,1}} \left(1 + \frac{|\log \gamma_1|}{\log \gamma_2}\right),$$

and κ^C_{ARpLE} is given by (5.1.61).

Notes and references for Section 6.3

The complexity analysis of the ARp-CC algorithm for optimality orders from one to three is new, but a precursor can be found in [88], where the complexity of using a variant of AR2 adapted to convex constraints for finding approximate first-order minimizers is investigated. The resulting bound is (in order) the same as that obtained in Theorem 6.2.3. The variant considered in Section 6.2.3 and its application to the least-Euclidean-norm problem in Section 6.3 are a slight improvement on what is presented in [97]. It is also possible to generalize the significantly improved (linear rate of) convergence under a full-rank assumption to the convexly constrained case [97, Thm. 3.6]; this requires that the measure $\omega_{\Phi_\ell,\mathcal{C},1}(x_k)/\|r(x_k)\|^{\ell-1}$ be uniformly bounded away from zero, which is clearly an extension of (5.1.62). Although we have concentrated our presentation on the first-order criticality measure $\omega_{f,\mathcal{C},1}(x)$ from (6.1.5), it is straightforward to consider the alternative measure

$$\|P_{\mathcal{C}}[x - \nabla^1_x f(x)] - x\|$$

instead, where here $P_{\mathcal{C}}[x]$ is the projection (6.2.6) [48,103]. It is also easy to extend the analysis of methods that use second-order models of the residuals [194] to cope with convex constraints [195].

Finally, we note that although we have focused on regularization methods, these are not the only possibilities. In particular, it is possible to design trust region–like methods, akin to those we encountered in Section 3.4.2, that are capable of finding an (ϵ_1, ϵ_2)-approximate second-order minimizer of f within $\mathcal{C}$ in $\mathcal{O}(\max[\epsilon_1^{-3/2}, \epsilon_2^{-3}])$ evaluations [310].

In [117], a variant of AR2 is also considered for finding approximate first-order minimizers in the context of a more general problem. Although the constrained set is restricted to be "kernel centered," the class of allowed objective functions is extended to cover non-Lipschitzian singularities of the type induced by fractional norms, that is, $\|\cdot\|^\alpha$ for $\alpha \in (0,1]$. The analysis is extended to weak approximate minimizers of general optimality orders in [116]. Both references make extensive use of the inherent rotational symmetry of the norms to construct a nonsmooth but Lipschitz-continuous model of the objective function, which is then used in a partially separable version of the adaptive-regularization algorithm. The results indicate that the complexity of this method is identical (in order) to that of the corresponding method for unconstrained smooth problems.

Chapter 7

Problems with Nonconvex Constraints

Chapter summary

Finally, we come to consider problems that involve explicit nonconvex constraints. Broadly, these are constraints whose defining functions are both available and nonconvex.[48] *If the problem also involves convex constraints, they will generally be dealt with separately and explicitly, but in theory there is nothing to stop us from including some of them in the nonconvex set if there is good reason to do so.*[49] *The nonconvex constraints may be equations or inequalities, but the latter will often be converted to equations by adding slack variables that are treated explicitly as extra "simple-bound" convex constraints.*

We construct a "penalty-like" function for the reformulated problem, and this leads to a parameter-dependent Euclidean-norm minimization problem subject to convex constraints. We then appeal to the method we designed in the previous chapter, together with an intricate mechanism to refine the parameter, to derive an $\mathcal{O}(\epsilon_0^{-1/p}\epsilon_1^{-(p+1)/p})$ *evaluation-complexity bound to find a point that is within* ϵ_0 *of feasibility (if possible) and* ϵ_1 *of a scaled KKT point.*

We conclude by comparing this result favourably to other known bounds for existing methods.

7.1 The general constrained problem

At last, we near the end of our journey of exploration into algorithms for smooth, nonconvex optimization. Thus far, we hope that we have persuaded the reader that there are methods with provable worst-case evaluation complexity for the approximate qth-order ($q \leq 3$) solution of many classes of optimization problems as well as demonstrating that the bounds we provide cannot be improved by more than constants. Moreover, as we have hinted, while the methods we describe may behave badly in rare extreme cases—those that give rise to the complexity bounds—in general they are typical of the

[48] We refine this in Chapter 14 to distinguish nonconvex constraints whose evaluation is expensive from others.

[49] See the comment just before Algorithm 6.2.1 (p. 180).

best-performing currently known methods for their problem classes. In this sense, we maintain that the fields addressed thus far are relatively mature, and we do not expect large improvements in the foreseeable future.

This comforting prospect changes as we address our most general problem. Indeed, we will consider algorithms that are unlikely to be implemented by the optimization-software community, nor to be efficient, general-purpose methods, but which nonetheless exhibit reasonable worst-case evaluation complexity. Regrettably, the state of the art in generally constrained nonlinear optimization is less mature than for the unconstrained and convexly constrained cases; even the best methods sometimes fail to converge, for a variety of technical reasons, when exposed to large sets of test examples. Although there are well-known classes of algorithms designed to cope with general constraints—names such as interior-point, augmented-Lagrangian, and SQP methods may be familiar to the reader—they universally rely on relatively strong, and often difficult to verify, assumptions. We wish that this were not the case, and perhaps this is simply a reflection of a rapidly evolving field.[50]

The problem we address can be expressed in many forms, but we focus on one in which we aim to

$$\underset{x \in \mathcal{C}}{\text{minimize}} \;\; f(x) \;\text{ subject to }\; c(x) = 0, \tag{7.1.1}$$

where $f : \mathbb{R}^n \to \mathbb{R}$ and $c : \mathbb{R}^n \to \mathbb{R}^m$ are suitably smooth and $\mathcal{C} \subseteq \mathbb{R}^n$ is a closed, bounded, nonempty convex set. In particular, such a formulation covers the general problem involving both equality constraints $c_{\mathrm{E}}(x) = 0$ and inequalities $c_{\mathrm{I}}(x) \geq 0$, by converting the latter into equations $c_{\mathrm{I}}(x) - s_{\mathrm{I}} = 0$ using slack variables s_{I} and including the associated simple bounds $s_{\mathrm{I}} \geq 0$ in the convex set $\mathcal{C}$.

Right from the outset, the reader may see a potential complication. Just because $\mathcal{C}$ is nonempty, how do we know that the set of feasible points

$$\mathcal{F}(\mathcal{C}, c) \stackrel{\text{def}}{=} \{x \in \mathcal{C} \mid c(x) = 0\}$$

is nonempty, that is, that there is any $x \in \mathcal{C}$ for which $c(x) = 0$? Unfortunately, checking this is itself a global optimization problem, and the best we can do is try and see. As we shall discover, this is an essential ingredient of the algorithm that we shall develop.

Let us be optimistic. Suppose that $\mathcal{F}(\mathcal{C}, c) \neq \emptyset$, and consider the Lagrangian

$$\mathcal{L}(x, y) \stackrel{\text{def}}{=} f(x) - \langle y, c(x) \rangle$$

for the explicit constraints, as we defined in (1.1.1). This enables us to characterize, to some extent, local minimizers of (7.1.1).

Theorem 7.1.1: Local sufficiency of the Lagrangian

Let $\mathcal{S}$ be a closed subset of $\mathcal{C}$, and suppose that $x_* \in \mathcal{S}$ and $y_* \in \mathbb{R}^m$ are such that

$$\mathcal{L}(x_*, y_*) = \min_{x \in \mathcal{S}} \; \mathcal{L}(x, y_*)$$

and $c(x_*) = 0$. Then

$$x_* = \underset{x \,\in\, \mathcal{F}(\mathcal{S}, c)}{\arg\min} \; f(x).$$

[50] Maybe we can persuade the reader to return in a few years' time for our second edition?

Proof. We have that

$$\begin{aligned}\min_{x\in\mathcal{F}(\mathcal{S},c)} f(x) &= \min_{x\in\mathcal{F}(\mathcal{S},c)} \mathcal{L}(x,y_*) \geq \min_{x\in\mathcal{S}} \mathcal{L}(x,y_*)\\ &= f(x_*) - \langle y_*, c(x_*)\rangle = f(x_*),\end{aligned}$$

where the first identity holds because $c(x) = 0$ when $x \in \mathcal{F}(\mathcal{S},c)$, the inequality follows because the minimum is considered over a larger set, the second identity holds because x_* minimizes $\mathcal{L}$, and the final identity is true since $c(x_*) = 0$. □

In particular, if x_* in the above does not lie on the "artificial" boundary $\partial\mathcal{S}\setminus\partial\mathcal{C}$ of $\mathcal{S}$, x_* provides a local solution to (7.1.1); recall that we denote the boundary of any set $\mathcal{S}$ by $\partial\mathcal{S}$. While this is perhaps not the perfect justification[51] for the use of the Lagrangian, we shall follow tradition by replacing the local minimization of $f(x)$ over $\mathcal{F}(\mathcal{C},c)$ with that of minimizing $\mathcal{L}(x,y_*)$ over $\mathcal{C}$, where y_* is chosen by some as yet unspecified means to make $c(x) = 0$ at the minimizer.

As in the previous chapter, given $h \in C^1(\mathcal{C})$, we let

$$\omega_{h,\mathcal{C},1}(x) \stackrel{\text{def}}{=} -\min_{d\in\mathcal{C}(x)\cap\mathcal{B}} \langle d, \nabla_x^1 h(x)\rangle, \tag{7.1.2}$$

where the set of feasible directions $\mathcal{C}(x)$ of $\mathcal{C}$ at x and the unit ball $\mathcal{B}$ are as in (1.1.7) (p. 4) and (6.1.3) (p. 178), respectively. We shall say that x_* is a *first-order minimizer of (7.1.1)* if

$$\omega_{\mathcal{L},\mathcal{C},1}(x_*,y_*) = -\min_{d\in\mathcal{C}(x)\cap\mathcal{B}} \langle d, \nabla_x^1 \mathcal{L}(x,y_*)\rangle = 0 \tag{7.1.3}$$

for some y_* chosen so that $c(x_*) = 0$. Likewise, x_ϵ is an *(ϵ_0,ϵ_1)-approximate first-order minimizer* of f within $\mathcal{F}(\mathcal{C},c)$ if

$$\omega_{\mathcal{L},\mathcal{C},1}(x_\epsilon,y_\epsilon) = -\min_{d\in\mathcal{C}(x)\cap\mathcal{B}} \langle d, \nabla_x^1 \mathcal{L}(x,y_\epsilon)\rangle \leq \epsilon_1\|(y_\epsilon,1)\| \tag{7.1.4}$$

for some y_ϵ chosen so that $\|c(x_\epsilon)\| \leq \epsilon_0$.

The reader might well wonder why we include the scaling term $\|(y_\epsilon,1)\|$ on the right-hand side of (7.1.4). The reasons are twofold and simple. Firstly, the explicit equality constraints for problem (7.1.1) may each be multiplied by a nonzero constant without changing the problem in the slightest, and the scaling we suggest endeavours to address this. Secondly, and best illustrated when $\mathcal{C} = \mathbb{R}^n$, it follows in this case from (7.1.3) that we seek x_* and y_* for which

$$\nabla_x^1 f(x_*) - [\nabla_x^1 c(x_*)]^T y_* = 0 \text{ and } c(x_*) = 0. \tag{7.1.5}$$

Now, for small $\delta = (\delta_x,\delta_y) \in \mathbb{R}^{n+m}$, consider the effect the perturbations $x_* + \delta_x$ and $y_* + \delta_y$ have on the residuals. Using a Taylor expansion and (7.1.5), we find that[52]

$$\begin{aligned}&\nabla_x^1 f(x_*+\delta_x) - [\nabla_x^1 c(x_*+\delta_x)]^T(y_*+\delta_y)\\ &= \left[\nabla_x^2 f(x_*) - \sum_{i=1}^m \nabla_x^2 c_i(x_*)[y_*]_i\right]\delta_x - [\nabla_x^1 c(x_*)]^T\delta_y + \mathcal{O}\left(\|\delta\|^2\right)\\ &= \mathcal{O}\left(\|(y_*,1)\|\cdot\|\delta\|\right),\end{aligned}$$

[51]Some would argue that the Lagrangian is a convex-optimization construct.

[52]The order bound follows when the derivatives of f and c are bounded since

$$\begin{aligned}&\|[\nabla_x^2 f(x_*) - \sum_{i=1}^m \nabla_x^2 c_i(x_*)[y_*]_i]\delta_x - [\nabla_x^1 c(x_*)]^T\delta_y + \mathcal{O}(\|\delta\|^2)\\ &\leq \left[\|\nabla_x^2 f(x_*)\| + m\max\|\nabla_x^2 c_i(x_*)\|\|[y\| + \|\nabla_x^1 c(x_*)\|\right]\delta + \mathcal{O}(\|\delta\|^2).\end{aligned}$$

and thus the residuals may proportionately be magnified by $\|(y_*, 1)\|$ for well-scaled problem functions. It is, of course, certainly possible to choose not to adopt such a scaling, but in this case the appropriate version of (7.1.4) may be more costly to achieve.

7.2 Finding first-order approximate minimizers

Since generally constrained optimization is more complicated than the topics we have addressed thus far in this book, both in theory and in practice, we shall take things at a gentle pace. In this chapter, we consider the simplest case, in which we aim to compute an (ϵ_0, ϵ_1)-approximate first-order minimizer of f within $\mathcal{F}(\mathcal{C}, c)$; that is, we aim to satisfy (7.1.4). It is fair to say that almost all the current research on evaluation complexity for generally constrained optimization focuses on achieving first-order optimality, but as we shall see later, in Chapter 14, there are hints of generalizations to higher-order conditions.

7.2.1 The basic approach

Most, but not all, methods for generally-constrained optimization recognize that there is an unstated conflict between reducing the objective on one hand and satisfying the constraints on the other. A common ploy is to build an artificial function that balances these two conflicting goals; penalty-function and augmented-Lagrangian methods are cases in point, while interior-point and SQP methods depend on suitable merit functions of this kind. The method that will form the focus of this chapter is also of this flavour but differs in one key aspect. This is that it first tries to satisfy the constraints to a reasonable degree and, if it achieves this, thereafter reduces the objective as much as possible while keeping a tight rein on subsequent constraint infeasibility. This very cautious approach is the key to its reasonable worst-case behaviour but mitigates against the bravado that the best practical methods exhibit.

The essential idea is straightforward and dates back to a class of methods for nonlinear convex optimization from the 1960s [260, 294]. Consider, for simplicity, the equality-constrained problem of minimizing $f(x)$ subject to the equality constraints $c(x) = 0$, and let $f_{\min}$ be the smallest value of f allowed by the constraints. Then, if $t = f_{\min}$, the roots of the nonlinear system

$$c(x) = 0 \quad \text{and} \quad f(x) - t = 0 \tag{7.2.1}$$

are the global minimizers of the problem. If $t < f_{\min}$, (7.2.1) has no solution, while if $t > f_{\min}$, (7.2.1) may have many. In particular, if $x = x_{\text{F}}$ is any vector that satisfies

$$c(x) = 0, \tag{7.2.2}$$

(7.2.1) has a solution when $t = f(x_{\text{F}})$.

Thus, our basic strategy is to find a solution to (7.2.2) if that is possible, to record the corresponding $t = f(x_{\text{F}})$, and then to gradually reduce t so long as (7.2.1) continues to hold. But rather than solving the nonlinear equations (7.2.2) and (7.2.1), we consider instead the equivalent optimization problems

$$\min_{x \in \mathbb{R}^n} \|c(x)\| \tag{7.2.3}$$

and, if the former (approximately) achieves (7.2.2),

$$\min_{x \in \mathbb{R}^n} \|r(x,t)\|, \quad \text{where } r(x,t) \stackrel{\text{def}}{=} \begin{pmatrix} c(x) \\ f(x) - t \end{pmatrix}, \tag{7.2.4}$$

where the latter is solved for a well-chosen sequence of decreasing t. There is no difficulty, at least in principle, in adapting such a scheme to cope with additional convex constraints by restricting the minimizations in (7.2.3) and (7.2.4) to $x \in \mathcal{C}$. Fortunately, as the reader will recall, we have already studied good methods for the problems (7.2.3) and (7.2.4) both with and without additional constraints in Sections 5.1 and 6.3.

7.2.2 The TPp-GC algorithm and its complexity

7.2.2.1 The TPp-GC algorithm

We now expand on the method we sketched in the previous section. To say anything more, we need to define more thoroughly the class of constrained optimization problems (7.1.1) that will be our focus in this chapter. We consider first the constraints on the problem and aim at solving (7.2.3) by solving the equivalent convexly constrained least-Euclidean-norm problem

$$\min_{x \in \mathcal{C}} \Phi_1(x) \stackrel{\text{def}}{=} \|c(x)\|$$

using the ARpLE-CC algorithm (p. 189). This requires that we choose a particular ℓ satisfying either (5.1.38) or (5.1.39). Moreover, we must ensure the necessary smoothness conditions to be able to use the complexity result of Theorem 6.3.1, that is, **Φ_ℓ.DpL.C** (for the chosen ℓ). As we argued after the statement of assumption **Φ_ℓ.DpL** on p. 124, a sufficient condition for **Φ_ℓ.DpL.C** to hold is that $\|c(x)\|$ remain bounded in $\mathcal{C}$ and that the following hold.

c.D1pL.C	The constraint functions $c_i(x)$, $i \in [1 : m]$, are all p-times continuously differentiable in an open set $\mathcal{D}$ containing $\mathcal{C}$, and their derivative tensors $\nabla_x^j c_i(x)$, $j \in [1 : p]$, are Lipschitz continuous and bounded in $\mathcal{C}$.

As in previous chapters, we prefer to use the weaker **Φ_ℓ.DpL.C**. This is then sufficient to try to find an approximately feasible point x_1 for which $\|c(x_1)\| \le \epsilon_0$ for a predefined, small $\epsilon_0 > 0$. If we fail, there is little more that we can do; indeed, we shall link failure to finding instead an infeasible approximately critical point of $\|c(x)\|$. But if we succeed, our focus then necessarily changes to the objective function.

Armed with the approximately feasible point x_1, we may then consider using the ARp-CC or ARp1-CC algorithm once again to approximately solve the problem

$$\min_{x \in \mathcal{C}} \mu_2(x, t_k) \stackrel{\text{def}}{=} \tfrac{1}{2}\|r(x, t_k)\|^2 \tag{7.2.5}$$

or ARpLE-CC to tackle the equivalent

$$\min_{x \in \mathcal{C}} \mu_1(x, t_k) \stackrel{\text{def}}{=} \|r(x, t_k)\| \tag{7.2.6}$$

for some monotonically decreasing sequence of *targets* $\{t_k\}$, $k \ge 1$. Again this requires that we specify the assumptions on the objective function that make this possible. Let $\beta > \epsilon_0 > 0$, and consider the region of near feasibility

$$\mathcal{F}_\beta = \{x \in \mathcal{C} \mid \|c(x)\| \le \beta\}. \tag{7.2.7}$$

We first require that the objective function satisfy an assumption on this region.

f.B.Fb If $\mathcal{F}_\beta \neq \emptyset$, $f(x)$ is bounded above and below on $\mathcal{F}_\beta$; that is, there are constants $f_{\rm low}$ and $f_{\rm up}$ for which

$$f_{\rm low} \leq f(x) \leq f_{\rm up}$$

for all $x \in \mathcal{F}_\beta$.

In line with our approach for ensuring feasibility, we also require that μ_2.**DpL.C** hold,[53] where μ_2 is considered as a function of its first argument. We note once again, in the light of our comment on p. 124, that a sufficient condition to ensure μ_2.**DpL.C** is that $|f(x) - t|$ is bounded[54] in $\mathcal{C}$ and that the following holds.

f.D1pL.C The objective function $f(x)$ is p-times continuously differentiable in an open set $\mathcal{D}$ containing $\mathcal{C}$, and its derivative tensors $\nabla_x^j f_i(x)$, $j \in [1:p]$, are Lipschitz continuous and bounded in $\mathcal{C}$.

However, we prefer to focus on μ_2.**DpL.C**. Of course, Φ_ℓ.**DpL.C** and μ_2.**DpL.C** ensure that f and the components of c are p-times continuously differentiable in $\mathcal{C}$.

Of course, this is simply a sketch of an algorithm, and we now need to fill in crucial details, most particularly of how we pick the targets t_k, how accurately we need to solve the crucial subproblems (7.2.5)–(7.2.6), and when to stop. Although, as we mention above, it is possible to apply a variety of algorithms during the feasibility and optimality phases, in the description below we have selected those that lead to the currently best complexity in each phase.

Algorithm 7.2.1: Two-phase adaptive-regularization algorithm using pth-order models for general constraints (TPp-GC)

Initialization: An initial point $x_0 \in \mathcal{C}$, an infeasibility bound $\beta > 0$ defining $\mathcal{F}_\beta$, and a constant $\theta \in (0,1)$ are given. Infeasibility and optimality tolerances

$$0 < \epsilon_0 \leq \min(1, \beta) \quad \text{and} \quad 0 < \epsilon_1 \leq 1 \tag{7.2.8}$$

are also given.

[53] By μ_2.**D1pL.C** we mean $\boldsymbol{\Phi_\ell}$.**DpL.C** on page 189 in the case where $\Phi_\ell = \mu_2$. Assumption μ_2.**DpL.C** imposes smoothness requirements on $\mu_2(\cdot, t)$ throughout $\mathcal{C}$. This is actually stronger than needed and made purely for simplicity. A more reasonable assumption might be that the required smoothness hold instead only for x in $\mathcal{F}_\beta$. To do so is delicate and needs the parameter β to be sufficiently large to ensure that iterates and trial points of the ARp-CC algorithm in Phase 2 of the forthcoming TPp-GC algorithm remain in this restricted subset. The construction of the ARp-CC algorithm ensures that this is trivially satisfied for successful iterates. At unsuccessful trial points, one needs to show that the length of the trial steps $\|s_{k,i}\|$ remains uniformly bounded above, and this is so provided that the derivatives of the generated iterates are bounded [105, Lem. 3.2]. Thus a suitable β ultimately depends on Lipschitz constants and other algorithmic parameters.

[54] Note that boundedness of $|f(x)|$ is ensured in $\mathcal{F}_\beta$ by **f.B.Fb** and will ensure in what follows that t remains bounded.

Phase 1: Feasibility phase: Pick a Taylor degree p along with suitable values $i \geq 0$ and $\ell > 1$ satisfying either (5.1.38) or (5.1.39), and apply the ARpLE-CC algorithm (p. 189), starting from x_0, to find a point $x_1 \in \mathcal{C}$ for which

$$\|c(x_1)\| \leq \theta\epsilon_0 \quad \text{or} \quad \omega_{\Phi_\ell,\mathcal{C},1}(x_1) \leq \epsilon_1 \|c(x_1)\|^{\ell-1}. \tag{7.2.9}$$

If $\|c(x_1)\| > \theta\epsilon_0$, terminate with $x_\epsilon = x_1$.

Phase 2: Optimality phase:

Step 0: Initialization. Set $t_1 = f(x_1) - \sqrt{\epsilon_0^2 - \|c(x_1)\|^2}$ and $k = 1$.

Step 1: Inner iteration to compute an improved objective value. Starting from x_k, apply the ARp-CC algorithm (p. 180) to minimize $\mu_2(x, t_k)$ as a function of $x \in \mathcal{C}$ until an iterate $x_{k+1} \in \mathcal{C}$ is found such that

$$\|r(x_{k+1}, t_k)\| \leq \theta\epsilon_0 \quad \text{or} \quad f(x_{k+1}) < t_k \quad \text{or} \quad \omega_{\mu_2,\mathcal{C},1}(x_{k+1}, t_k) \leq \epsilon_1 \|r(x_{k+1}, t_k)\|. \tag{7.2.10}$$

Step 2: Decrease the target and check for termination.

(a) If $\|r(x_{k+1}, t_k)\| < \theta\epsilon_0$, set

$$t_{k+1} = f(x_{k+1}) - \sqrt{\epsilon_0^2 - \|c(x_{k+1})\|^2}, \tag{7.2.11}$$

and terminate with $(x_\epsilon, t_\epsilon) = (x_{k+1}, t_{k+1})$ if

$$\omega_{\mu_2,\mathcal{C},1}(x_{k+1}, t_{k+1}) \leq \epsilon_1 \|r(x_{k+1}, t_{k+1})\|. \tag{7.2.12}$$

(b) Otherwise, if $\|r(x_{k+1}, t_k)\| \geq \theta\epsilon_0$ and $f(x_{k+1}) < t_k$, set

$$t_{k+1} = 2f(x_{k+1}) - t_k, \tag{7.2.13}$$

and terminate with $(x_\epsilon, t_\epsilon) = (x_{k+1}, t_{k+1})$ if (7.2.12) holds.

(c) Otherwise, if $\|r(x_{k+1}, t_k)\| \geq \theta\epsilon_0$ and $f(x_{k+1}) \geq t_k$, terminate with $(x_\epsilon, t_\epsilon) = (x_{k+1}, t_k)$.

Step 3: Move to the next iteration. Set $k = k + 1$, and go to Step 1.

Just to remind the reader, the function Φ_ℓ used by the ARpLE-CC algorithm and in the stopping test (7.2.9) is that defined by (5.1.3) (p. 120). Observe that the computation of $\omega_{\mu_2,\mathcal{C},1}(x_{k+1}, t_{k+1})$ in Step 2 does not require that $f(x_{k+1})$, $c(x_{k+1})$, or any of their derivatives be reevaluated.

The mechanism devised to adjust the targets may seem at a first glance to be rather complicated. To reassure the reader that there is some purpose, it is relatively straightforward to show that if the algorithm terminates, it does so at either an ϵ-approximate infeasible local minimizer of $\|c(x)\|$ or an ϵ-approximate KKT point.

Lemma 7.2.1: Consequences of the termination of the TPp-GC algorithm

Suppose that **C.CNe**, Φ_ℓ**.DpL.C**, *and* μ_2**.DpL.C** *hold. Then if the* TPp-GC *algorithm stops in Phase 1,*

$$\omega_{\Phi_\ell,\mathcal{C},1}(x_\epsilon) \leq \epsilon_1 \|c(x_\epsilon)\|^{\ell-1} \text{ and } \|c(x_\epsilon)\| > \theta\epsilon_0. \tag{7.2.14}$$

Otherwise, if it stops in Phase 2, either

$$\omega_{\Phi_2,\mathcal{C},1}(x_\epsilon) \leq \epsilon_1\|c(x_\epsilon)\| \text{ and } \|c(x_\epsilon)\| > \theta\epsilon_0 \tag{7.2.15}$$

or

$$\omega_{\mathcal{L},\mathcal{C},1}(x_\epsilon, y_\epsilon) \leq \epsilon_1\|(y_\epsilon, 1)\| \text{ and } \|c(x_\epsilon)\| \leq \epsilon_0, \tag{7.2.16}$$

where $\mathcal{L}(x, y)$ is the Lagrangian (1.1.1) with respect to the equality constraints and y_ϵ is a vector of Lagrange multipliers associated with the equality constraints.

Proof. If $\|c(x_1)\| > \theta\epsilon_0$ at the end of Phase 1, we may deduce (7.2.14) directly from the termination requirement (7.2.9).

Turning to Phase 2, we first show that on termination,

$$\|r(x_\epsilon, t_\epsilon)\| \geq \theta\epsilon_0, \quad f(x_\epsilon) \geq t_\epsilon, \quad \text{and} \quad \omega_{\mu_2,\mathcal{C},1}(x_\epsilon, t_\epsilon) \leq \epsilon_1\|r(x_\epsilon, t_\epsilon)\|. \tag{7.2.17}$$

In order to establish this result, we need to consider, in turn, each of the three possible terminating cases in Step 2. In the first, in Step 2(a), (7.2.11) holds, and thus $\|r(x_{k+1}, t_{k+1})\| = \epsilon_0 > \theta\epsilon_0$ is automatically satisfied. Moreover, since $\|r(x_{k+1}, t_k)\| < \theta\epsilon_0$,

$$\|c(x_{k+1})\|^2 \leq \|r(x_{k+1}, t_k)\|^2 \leq \theta^2\epsilon_0^2 < \epsilon_0^2,$$

and thus $f(x_{k+1}) > t_{k+1}$ from (7.2.11). If termination ensues, (7.2.12) occurs, and then clearly (7.2.17) holds at $(x_\epsilon, t_\epsilon) = (x_{k+1}, t_{k+1})$.

In the second case, in Step 2(b), it follows from (7.2.13) that

$$t_{k+1} - f(x_{k+1}) = f(x_{k+1}) - t_k, \tag{7.2.18}$$

and thus

$$\|r(x_{k+1}, t_{k+1})\| = \|r(x_{k+1}, t_k)\|. \tag{7.2.19}$$

But the Step 2(b) qualifying rules along with (7.2.18) and (7.2.19) then ensure that $\|r(x_{k+1}, t_{k+1})\| \geq \theta\epsilon_0$ and $f(x_{k+1}) - t_{k+1} > 0$. As before, if termination occurs then so does (7.2.12), and once again (7.2.17) holds at $(x_\epsilon, t_\epsilon) = (x_{k+1}, t_{k+1})$.

The final case, in Step 2(c), is when $\|r(x_{k+1}, t_k)\| \geq \theta\epsilon_0$ and $f(x_{k+1}) \geq t_k$. But then the third possibility in (7.2.10) must hold, and the inequalities (7.2.17) are again satisfied, this time at $(x_\epsilon, t_\epsilon) = (x_{k+1}, t_k)$.

It remains to consider the consequences of (7.2.17). In particular, we have that

$$\epsilon_0 \geq \|r(x_\epsilon, t_\epsilon)\| \geq \theta\epsilon_0, \tag{7.2.20}$$

where the upper bound follows because $\|r(x_k, t_k)\|$ decreases during Steps 1 and 2(b) of the algorithm and is reset to ϵ_0 in Step 2(a), and

$$\omega_{\mu_2,\mathcal{C},1}(x_\epsilon, t_\epsilon) \leq \epsilon_1\|r(x_\epsilon, t_\epsilon)\|. \tag{7.2.21}$$

Suppose first that $f(x_\epsilon) = t_\epsilon$. Then, using (7.1.2) and the definition of $r(x, t)$ in (7.2.4), we deduce that

$$\omega_{\Phi_2,\mathcal{C},1}(x_\epsilon) = \omega_{\mu_2,\mathcal{C},1}(x_\epsilon, t_\epsilon) \leq \epsilon_1\|c(x_\epsilon)\|,$$

and (7.2.15) is satisfied since (7.2.20) gives that $\|c(x_\epsilon)\| = \|r(x_\epsilon, t_\epsilon)\| \geq \theta\epsilon_0$. Since the possibility that $f(x_\epsilon) < t_\epsilon$ is excluded by (7.2.17), suppose otherwise that $f(x_\epsilon) > t_\epsilon$ and let

$$y_\epsilon \stackrel{\text{def}}{=} -\frac{c(x_\epsilon)}{f(x_\epsilon) - t_\epsilon}. \tag{7.2.22}$$

Then, since

$$\nabla_x^1 \mu_2(x_\epsilon, t_\epsilon) = (f(x_\epsilon) - t_\epsilon)\nabla_x^1 f(x_\epsilon) + [\nabla_x^1 c(x_\epsilon)]^T c(x_\epsilon),$$

we deduce that

$$\frac{\nabla_x^1 \mu_2(x_\epsilon, t_\epsilon)}{f(x_\epsilon) - t_\epsilon} = \nabla_x^1 f(x_\epsilon) - [\nabla_x^1 c(x_\epsilon)]^T y_\epsilon = \nabla_x^1 \mathcal{L}(x_\epsilon, y_\epsilon).$$

It then follows from the definition (7.1.4), the inequality $f(x_\epsilon) \geq t_\epsilon$, the termination condition (7.2.21), and the definition (7.2.22) of y_ϵ that

$$\omega_{\mathcal{L},\mathcal{C},1}(x_\epsilon, y_\epsilon) = \frac{\omega_{\mu_2,\mathcal{C},1}(x_\epsilon, t_\epsilon)}{f(x_\epsilon) - t_\epsilon} \leq \epsilon_1 \frac{\|r(x_\epsilon, t_\epsilon)\|}{f(x_\epsilon) - t_\epsilon} = \epsilon_1 \|(y_\epsilon, 1)\|.$$

This then implies (7.2.16) since $\|c(x_\epsilon)\| \leq \|r(x_\epsilon, t_\epsilon)\| \leq \epsilon_0$. □

The reader may recall the discussions in Sections 5.1.1 and 6.3 that both $\omega_{\Phi_2,\mathcal{C},1}(x_\epsilon)/\|c(x_\epsilon)\|$ and $\omega_{\Phi_\ell,\mathcal{C},1}(x_\epsilon)/\|c(x_\epsilon)\|^{\ell-1}$ are surrogates for $\omega_{\Phi_1,\mathcal{C},1}(x_\epsilon)$ because of (5.1.4), and thus the termination conditions (7.2.14) and (7.2.15) are effectively the same.

7.2.2.2 Evaluation complexity of the TPp-GC algorithm

We start our evaluation-complexity analysis of the TPp-GC algorithm by examining that of its initial feasibility phase.

Lemma 7.2.2: Evaluation complexity of Phase 1 of Algorithm 7.2.1

Suppose that ***C.CNe*** *and* Φ_ℓ***.DpL.C*** *hold. Then Phase 1 of Algorithm 7.2.1 (*TPp-GC*) terminates at a point x_ϵ for which either*

$$\|c(x_\epsilon)\| \leq \theta\epsilon_0 \text{ or } \omega_{\Phi_\ell,\mathcal{C},1}(x_\epsilon) \leq \epsilon_1 \|c(x_\epsilon)\|^{\ell-1}$$

after at most

$$\#^F_{\text{ARpLE},1} \stackrel{\text{def}}{=} \kappa^A_{\text{ARpLE},1} \frac{\max\left[\|c(x_0)\|^{\frac{1}{2^i}}, \|c(x_0)\|^{1-\frac{\ell-1}{p}}\right]}{\min\left[\theta^{\frac{1}{2^i}}\epsilon_0^{\frac{1}{2^i}}, \epsilon_1^{\frac{p+1}{p}}\right]} + \kappa^C_{\text{ARpLE}} = \mathcal{O}\left(\max\left[\epsilon_0^{-\frac{1}{2^i}}, \epsilon_1^{-\frac{p+1}{p}}\right]\right) \tag{7.2.23}$$

evaluations of c and

$$\#^D_{\text{ARpLE},1} \stackrel{\text{def}}{=} \kappa^S_{\text{ARpLE},1} \frac{\max\left[\|c(x_0)\|^{\frac{1}{2^i}}, \|c(x_0)\|^{1-\frac{\ell-1}{p}}\right]}{\min\left[\theta^{\frac{1}{2^i}}\epsilon_0^{\frac{1}{2^i}}, \epsilon_1^{\frac{p+1}{p}}\right]} = \mathcal{O}\left(\max\left[\epsilon_0^{-\frac{1}{2^i}}, \epsilon_1^{-\frac{p+1}{p}}\right]\right) \tag{7.2.24}$$

evaluations of its derivatives, where the constants $\kappa^S_{\text{ARpLE},1}$, $\kappa^A_{\text{ARpLE},1}$, and κ^C_{ARpLE} are as for Theorem 6.3.1.

Proof. It follows from Theorem 6.3.1 that the ARpLE-CC algorithm requires at most (7.2.23) and (7.2.24) evaluations of c and its derivatives to find an x_1 satisfying the Phase-1 stopping test (7.2.9). □

Next, we turn to the optimality phase. We refer to Steps 1 to 3 of Phase 2 of the ARp-CC algorithm as the kth *outer* iteration, and we know that, for each $k \in \mathcal{K} \stackrel{\text{def}}{=} [1:\infty)$, the *inner* iteration performed by the ARp-CC algorithm in Step 1 terminates after e_k iterations, where we have a bound on e_k using Theorem 6.2.3. Thus, we may label the complete sequence of iterates generated in Phase 2 as $\{x_{k,i}\}$, $i \in [0:e_k]$, $k \in \mathcal{K}$, where $x_{k,0} = x_k$ and $x_{k+1} = x_{k,e_k}$. Moreover, since the final point x_{k,e_k} from the kth inner iteration is the same as the first $x_{k+1,0}$ for the $(k+1)$st by construction in Step 1, we actually need only consider iterates $\{x_{k,i}\}$, $i \in [0:e_k)$, $k \in \mathcal{K}$.

We partition the outer iterations into two[55] subsets, whose indices are given by

$$\mathcal{K}_a \stackrel{\text{def}}{=} \{k \geq 1 \mid \text{Step 2(a) occurs}\} \text{ and } \mathcal{K}_b \stackrel{\text{def}}{=} \{k \geq 1 \mid \text{Step 2(b) occurs}\}. \tag{7.2.25}$$

This partition allows us to prove the following technical results.

Lemma 7.2.3: Properties of the Phase 2 iterates in Algorithm 7.2.1

*The sequence of targets $\{t_k\}$ generated in Phase 2 of Algorithm 7.2.1 (*TPp-GC*) is monotonically decreasing. Moreover, for all $k \geq 1$, we have that*

$$f(x_k) - t_k \geq 0, \tag{7.2.26}$$

$$\|r(x_{k+1}, t_{k+1})\| = \epsilon_0 \text{ if } k \in \mathcal{K}_a, \tag{7.2.27}$$

$$\|r(x_{k+1}, t_{k+1})\| = \|r(x_{k+1}, t_k)\| \leq \epsilon_0 \text{ if } k \in \mathcal{K}_b, \tag{7.2.28}$$

$$\|c(x_k)\| \leq \epsilon_0 \text{ and } f(x_k) - t_k \leq \epsilon_0, \text{ and} \tag{7.2.29}$$

$$t_k - t_{k+1} \geq (1-\theta)\epsilon_0 \text{ if } k \in \mathcal{K}_a. \tag{7.2.30}$$

Furthermore, if **f.B.Fb** *holds, then, for $k \geq 1$,*

$$t_k \in [f_{\text{low}} - 1, f_{\text{up}}]. \tag{7.2.31}$$

Proof. By construction, t_{k+1} is given by (7.2.11) when $k \in \mathcal{K}_a$ and by (7.2.13) when $k \in \mathcal{K}_b$. The inequality (7.2.26) follows from (7.2.11) for $k-1 \in \mathcal{K}_a$ and from (7.2.13) for $k-1 \in \mathcal{K}_b$. The identity (7.2.27) follows directly from (7.2.11). The equality in (7.2.28) is a direct result of (7.2.13), and since the inner-iteration algorithm ARp-CC in Step 2 generates monotonically decreasing $\mu_2(x, t_k)$, it follows that $\|r(x_{k+1}, t_k)\| \leq \|r(x_k, t_k)\|$. Thus, since $\|r(x_{k+1}, t_{k+1})\| \leq \|r(x_k, t_k)\|$ for all $k \in \mathcal{K}_b$, $\|r(x_{k+1}, t_{k+1})\| = \epsilon_0$ for all $k \in \mathcal{K}_a$, and $\|r(x_1, t_1)\| = \epsilon_0$ by construction in Step 1, the inequality in (7.2.28) holds. The bounds (7.2.29) then follow from (7.2.26)–(7.2.28) and the definition of $\|r(x,t)\|$ in (7.2.4).

Turning to (7.2.30), if $k \in \mathcal{K}_a$, it must be that $\|r(x_{k+1}, t_k)\| \leq \theta\epsilon_0$, that is,

$$(f(x_{k+1}) - t_k)^2 + \|c(x_{k+1})\|^2 \leq \theta^2 \epsilon_0^2. \tag{7.2.32}$$

From (7.2.11), we then have that

$$t_k - t_{k+1} = -(f(x_{k+1}) - t_k) + \sqrt{\epsilon_0^2 - \|c(x_{k+1})\|^2}. \tag{7.2.33}$$

[55] The other outcome of Step 2, Step 2(c), leads immediately to termination.

It is straightforward to show that the global minimum of the problem

$$\min_{(f,c)\in\mathbb{R}^2} -f + \sqrt{\epsilon_0^2 - c^2} \text{ subject to } f^2 + c^2 \le \omega^2 \tag{7.2.34}$$

for $\omega \in [0, \epsilon_0]$ has the value $\epsilon_0 - \omega$ (see Lemma B.7.1, p. 472). Thus, setting $f = f(x_{k+1}) - t_k$, $c = \|c(x_{k+1})\|$, and $\omega = \theta\epsilon_0$ in (7.2.34), we deduce from (7.2.32) and (7.2.33) that

$$t_k - t_{k+1} \ge \epsilon_0 - \omega = (1-\theta)\epsilon_0$$

for $k \in \mathcal{K}_a$, which is (7.2.30). Finally, by definition, $t_k > f(x_{k+1})$ when $k \in \mathcal{K}_b$, and thus (7.2.13) ensures that $t_{k+1} < t_k$. This observation and (7.2.30) then allow us to conclude that the sequence $\{t_k\}$ is monotonically decreasing.

The upper bound in (7.2.31) follows from **f.B.Fb**, the definition of the initial target $t_1 < f(x_1) \le f_{\rm up}$ in Step 0 of Phase 2 of the TPp-GC algorithm, the decreasing nature of the sequence $\{t_k\}$, and (7.2.26). The lower bound follows from (7.2.29), the restriction $\epsilon_0 < 1$, and, once again, **f.B.Fb**. □

The final conclusion in the lemma above leads directly to a bound on the number of outer iterations whose indices lie in $\mathcal{K}_a$.

Lemma 7.2.4: Bound on the size of $\mathcal{K}_a$

Suppose that **f.B.Fb** *holds. Then*

$$|\mathcal{K}_a| \le \frac{[f_{\rm up} - f_{\rm low} + 1]}{1-\theta}\,\epsilon_0^{-1}.$$

Proof. We know from Lemma 7.2.3 that t_k decreases monotonically, with a decrease of at least $(1-\theta)\epsilon_0$ for each $k \in \mathcal{K}_a$ because of (7.2.30). The desired conclusion then follows directly from (7.2.31). □

It is crucial, in what follows, that certain quantities associated with the sequence of inner iterations in Step 1 of Phase 2 of the algorithm be uniformly bounded.

Lemma 7.2.5: Uniform Lipschitz and regularization bounds

*Suppose that ϕ_ℓ.***DpL.C**, *μ_2.***DpL.C**, *and* **f.B.Fb** *hold. Then there exists a constant $L_{\mu,p} \ge 0$ such that the pth derivative of $\mu_2(x, t_k)$ with respect to x is Lipschitz continuous with Lipschitz constant $L_{\mu_2,p}$ for all values of t_k computed in Phase 2 of the* TPp-GC *algorithm. Furthermore, there exists a constant $\sigma_{\mu_2,\max} > \sigma_{\min}$ such that all regularization parameters arising in the* ARp-CC *algorithm used within Step 1 of algorithm* TPp-GC *are bounded above by $\sigma_{\mu_2,\max}$.*

Proof. It follows from ϕ_ℓ.**D1pL.C** and μ_2.**D1pL.C** that $\mu_2(x,t)$ is p-times continuously differentiable in $\mathcal{F}_\beta$ and its pth derivative is Lipschitz continuous for all $t \in [f_{\rm low} - 1, f_{\rm up}]$. Moreover, since $\mu_2(x,t)$ and its derivatives depend continuously on $t \in [f_{\rm low} - 1, f_{\rm up}]$, we may deduce the existence of $L_{\mu,p}$, which is an upper bound on the pth-derivative Lipschitz constant associated with each t; a precise but not especially informative upper bound on $L_{\mu_2,p}$ is known [103, Lem. 5]. The bound on the regularization parameter follows by introducing $L_{\mu,p}$ into (4.1.19) as specified by Lemma 4.1.2 to obtain $\sigma_{\mu,\max}$. □

The main consequence of this result is that we may apply the ARp-CC algorithm to the minimization of $\mu_2(x, t_k)$ in Step 1 of algorithm TPp-GC and use all the properties of the former, as derived in Section 6.1, using problem constants valid for every possible t_k.

Consider the sequence of outer iterates in Phase 2, and, in particular, for any iteration k, define $\ell(k)$ so that iteration $k + \ell(k)$ is the first iteration after k that lies in $\mathcal{K}_a$. If $\ell(k) > 1$, necessarily iterations $[k + 1 : k + \ell(k) - 1]$ lie in $\mathcal{K}_b$. Now let

$$\mathcal{I}_k \stackrel{\text{def}}{=} \{(j, i) \mid i \in [0 : e_j), j \in [k + 1 : k + \ell(k) - 1]\}, \tag{7.2.35}$$

the index set of all inner iterations necessary to find $x_{k+\ell(k)}$ from x_k, as well as

$$\mathcal{I}_k^{\text{S}} \stackrel{\text{def}}{=} \{(j, i) \in \mathcal{I}_k \mid i = 0 \text{ or iteration } i \text{ is successful}\}. \tag{7.2.36}$$

Observe that, by the definitions (7.2.25) and (7.2.35), the index set of all[56] inner iterations before termination is given by $\bigcup_{k\in\mathcal{K}_a} \mathcal{I}_k$, and therefore that the number of problem-function evaluations required to terminate Phase 2 is bounded above by

$$\left|\bigcup_{k\in\mathcal{K}_a} \mathcal{I}_k\right| + 1 \leq \max_{k\in\mathcal{K}_a} |\mathcal{I}_k| \frac{(f_{\text{up}} - f_{\text{low}} + 1)}{1 - \theta} \epsilon_0^{-1} + 1, \tag{7.2.37}$$

where we added 1 to take the final evaluation into account, and we used Lemma 7.2.4 to deduce the inequality. Likewise, the total number of derivative evaluations, which is one plus the total number of successful iterations, is

$$\left|\bigcup_{k\in\mathcal{K}_a} \mathcal{I}_k^{\text{S}}\right| + 1 \leq \max_{k\in\mathcal{K}_a} |\mathcal{I}_k^{\text{S}}| \frac{(f_{\text{up}} - f_{\text{low}} + 1)}{1 - \theta} \epsilon_0^{-1} + 1. \tag{7.2.38}$$

We now invoke the complexity properties of the ARp-CC algorithm applied to problem (7.2.5) to obtain an upper bound on the cardinality of each $\mathcal{I}_k$ and $\mathcal{I}_k^{\text{S}}$.

Lemma 7.2.6: Bound on the size of $\mathcal{I}_k$

Suppose that **C.CNe**, Φ_ℓ.**DpL.C**, μ_2.**DpL.C**, *and* **f.B.Fb** *hold. Then there are constants* $\kappa_{\text{ARpGC}}^{S}$, $\kappa_{\text{ARpGC}}^{A}$, *and* $\kappa_{\text{ARpGC}}^{C}$ *such that*

$$|\mathcal{I}_k^{\text{S}}| \leq \kappa_{\text{ARpGC}}^{S} (1 - \theta) \epsilon_0^{\frac{p-1}{p}} \epsilon_1^{-\frac{p+1}{p}} \tag{7.2.39}$$

and

$$|\mathcal{I}_k| \leq \kappa_{\text{ARpGC}}^{A} (1 - \theta) \epsilon_0^{\frac{p-1}{p}} \epsilon_1^{-\frac{p+1}{p}} + \kappa_{\text{ARpGC}}^{C} \tag{7.2.40}$$

for each $k \in \mathcal{K}_a$ *before termination.*

[56] The terminal iterate from subproblem j is the same as the initial one for subproblem $j + 1$ by construction, so it is not counted twice.

Proof. Let $k \in \mathcal{K}_a$. Then, for each $j \in [k+1 : k+\ell(k)]$, the residuals of the inner iterates $\|r(x_{j,i}, t_j)\|$, $i \in [0 : e_j]$, generated by the ARp-CC algorithm in Step 1 decrease monotonically. Moreover, if $j \in [k+1 : k+\ell(k))$, j lies in $\mathcal{K}_b$, by construction in Step 1 we have $x_{j,e_j} = x_{j+1} = x_{j+1,0}$, and hence from (7.2.28) we have

$$\|r(x_{j,e_j}, t_j)\| = \|r(x_{j+1}, t_j)\| = \|r(x_{j+1}, t_{j+1})\| = \|r(x_{j+1,0}, t_{j+1})\|. \tag{7.2.41}$$

Thus the entire sequence $\{\|r(x_{j,i}, t_j)\|\}_{(j,i)\in\mathcal{I}_k}$ decreases monotonically if $k \in \mathcal{K}_a$. In addition, since $k \in \mathcal{K}_a$, (7.2.27) ensures that $\|r(x_{k+1,0}, t_{k+1})\| = \|r(x_{k+1}, t_{k+1})\| = \epsilon_0$, while $\|r(x_{k+\ell(k),e_{k+\ell(k)}-1}, t_{k+\ell(k)})\| > \theta\epsilon_0$ since otherwise the $[k+\ell(k)]$th inner iteration would have stopped at or before its $(e_{k+\ell(k)} - 1)$st iteration. Thus each member of the monotonic sequence of residuals lies in the range

$$\theta\epsilon_0 \le \|r(x_{j,i}, t_j)\| \le \epsilon_0, \tag{7.2.42}$$

and the entire sequence of iterates $x_{j,i}$ lies in $\mathcal{F}_\beta$.

Let $\mathcal{I}_k^{\rm S}$ be the subset of indices of successful iterations in $\mathcal{I}_k$. If $(j,s) \in \mathcal{I}_k^{\rm S}$, $k \in \mathcal{K}_a$, is the index of a successful inner iteration, then, so long as the third termination test in (7.2.10) has not been satisfied, we may then invoke Lemma 7.2.5, apply Lemma 6.2.4 (p. 187) to $f(x) = \mu_2(x, t_k)$, and use the monotonicity $\|r(x_{j,s}, t_j)\| \ge \|r(x_{j,s+1}, t_j)\|$ to deduce that

$$\begin{aligned}
&\|r(x_{j,s}, t_j)\| \left(\|r(x_{j,s}, t_j)\| - \|r(x_{j,s+1}, t_j)\|\right)\\
&\quad\ge \tfrac{1}{2}\left(\|r(x_{j,s}, t_j)\| + \|r(x_{j,s+1}, t_j)\|\right)\left(\|r(x_{j,s}, t_j)\| - \|r(x_{j,s+1}, t_j)\|\right)\\
&\quad= \tfrac{1}{2}\|r(x_{j,s}, t_j)\|^2 - \tfrac{1}{2}\|r(x_{j,s+1}, t_j)\|^2\\
&\quad= \mu_2(x_{j,s}, t_j) - \mu_2(x_{j,s+1}, t_j)\\
&\quad\ge \kappa_1 \left[\omega_{\mu_2,\mathcal{C},1}(x_{j,s+1}, t_j)\right]^{\frac{p+1}{p}}\\
&\quad= \kappa_1 \left[\left(\frac{\omega_{\mu_2,\mathcal{C},1}(x_{j,s+1}, t_j)}{\|r(x_{j,s+1}, t_j)\|}\right)\|r(x_{j,s+1}, t_j)\|\right]^{\frac{p+1}{p}}\\
&\quad\ge \kappa_1 \epsilon_1^{\frac{p+1}{p}} \|r(x_{j,s+1}, t_j)\|^{\frac{p+1}{p}}
\end{aligned} \tag{7.2.43}$$

for $0 \le s < e_j$ and for some constant

$$\kappa_1 \stackrel{\rm def}{=} \frac{\eta_1\sigma_{\min}}{(p+1)!}\left(\frac{p!}{L_{\mu_2,p} + \sigma_{\mu_2,\max} + \theta_1 p!}\right)^{\frac{p+1}{p}}$$

independent of ϵ_0, ϵ_1, s, and j; when $s = e_j$, we have instead

$$\|r(x_{j,e_j}, t_j)\| - \|r(x_{j+1,0}, t_{j+1})\| = 0 \tag{7.2.44}$$

from (7.2.41).

Now consider two possibilities. Firstly, suppose that for some index $(j,s) \in \mathcal{I}_k^{\rm S}$,

$$\|r(x_{j,s+1}, t_j)\| < \theta\|r(x_{j,s}, t_j)\|. \tag{7.2.45}$$

Then we know from (7.2.27) and the monotonicity of the sequence of residuals that

$$\|r(x_{j,s}, t_j)\| \le \|r(x_{k+1,0}, t_{k+1})\| = \epsilon_0, \tag{7.2.46}$$

and thus $\|r(x_{j,s+1},t_j)\| < \theta\epsilon_0$. But, in this case, the first inner-iteration termination test in (7.2.10) holds, (j,s) is not in $\mathcal{I}_k$, and $j+1 \in \mathcal{K}_a$. Thus, for all $(j,s) \in \mathcal{I}_k^{\text{S}}$, it must be that

$$\|r(x_{j,s+1},t_j)\| \geq \theta\|r(x_{j,s},t_j)\|$$

for all $(j,s) \in \mathcal{I}_k^{\text{S}}$. In this case (7.2.43) and the lower bound $\theta\epsilon_0$ on the sequence of residuals from (7.2.42) give that

$$\|r(x_{j,s},t_j)\| - \|r(x_{j,s+1},t_j)\| \geq \kappa_1 \epsilon_1^{\frac{p+1}{p}} \theta^{\frac{p+1}{p}} \|r(x_{j,s+1},t_j)\|^{\frac{1}{p}} \geq \kappa_1 \theta^{\frac{p+2}{p}} \epsilon_1^{\frac{p+1}{p}} \epsilon_0^{\frac{1}{p}}$$

for all $(j,s) \in \mathcal{I}_k^{\text{S}}$. Then, using the standard telescoping-sum argument and the bounds (7.2.42), we have that

$$\begin{aligned}(1-\theta)\epsilon_0 &\geq \|r(x_{k+1,0},t_{k+1})\| - \|r(x_{j_k,s_k+1},t_{j_k})\| \\ &\geq \sum_{j,s\in\mathcal{I}_k^{\text{S}}} \left(\|r(x_{j,s},t_j)\| - \|r(x_{j,s+1},t_j)\|\right) \\ &\geq \kappa_1 \theta^{\frac{p+2}{p}} \epsilon_0^{\frac{1}{p}} \epsilon_1^{\frac{p+1}{p}} |\mathcal{I}_k^{\text{S}}|\end{aligned}$$

and hence that

$$|\mathcal{I}_k^{\text{S}}| \leq \kappa_{\text{ARpGC}}^{\text{S}} (1-\theta) \epsilon_0^{\frac{p-1}{p}} \epsilon_1^{-\frac{p+1}{p}}, \quad \text{where } \kappa_{\text{ARpGC}}^{\text{S}} \stackrel{\text{def}}{=} \kappa_1^{-1} \theta^{-\frac{p+2}{p}}, \tag{7.2.47}$$

which is (7.2.39). Using Lemma 7.2.5 once again, and invoking Lemma 2.4.1 to account for possible unsuccessful inner iterations, we obtain (7.2.40) with

$$\kappa_{\text{ARpGC}}^{\text{A}} \stackrel{\text{def}}{=} \kappa_{\text{ARpGC}}^{\text{S}} \left(1 + \frac{|\log \gamma_1|}{\log \gamma_2}\right) \tag{7.2.48}$$

and

$$\kappa_{\text{ARpGGC}}^{\text{C}} \stackrel{\text{def}}{=} \frac{1}{\log \gamma_2} \log\left(\frac{\sigma_{\mu_2,\max}}{\sigma_0}\right). \tag{7.2.49}$$

□

We combine our results in a final theorem stating our evaluation-complexity bound for algorithm TPp-GC applied on the general smooth nonlinear optimization problem.

Theorem 7.2.7: Complexity of finding first-order critical points by the TPp-GC algorithm

Suppose that ***C.CNe***, ***Φ_ℓ.DpL.C***, ***μ_2.DpL.C***, *and* ***f.B.Fb*** *hold. Then there are constants $\kappa_{\text{ARpGC}}^{F}$ and $\kappa_{\text{ARpGC}}^{D}$ such that, for any $(\epsilon_0,\epsilon_1) \in (0,1]^2$, the* TP$p$-GC *algorithm (p. 198) requires at most*

$$\#_{\text{ARpGC}}^{F} \stackrel{\text{def}}{=} \#_{\text{ARpLE,1}}^{F} + \#_{\text{ARpGC,2}}^{F} = \mathcal{O}\left(\max\left[\epsilon_0^{-1}, \epsilon_0^{-\frac{1}{p}} \epsilon_1^{-\frac{p+1}{p}}\right]\right) \tag{7.2.50}$$

evaluations of c and f and

$$\#_{\text{ARpGC}}^{D} \stackrel{\text{def}}{=} \#_{\text{ARpLE,1}}^{D} + \#_{\text{ARpGC,2}}^{D} = \mathcal{O}\left(\max\left[\epsilon_0^{-\frac{1}{2^i}}, \epsilon_0^{-\frac{1}{p}} \epsilon_1^{-\frac{p+1}{p}}\right]\right) \tag{7.2.51}$$

evaluations of their derivatives up to order p, where

$$\#^{F}_{\mathsf{ARpGC,2}} \stackrel{\rm def}{=} (f_{\rm up} - f_{\rm low} + 1)\left(\kappa^{A}_{\mathsf{ARpGC}}\epsilon_0^{-\frac{1}{p}}\epsilon_1^{-\frac{p+1}{p}} + \frac{\kappa^{C}_{\mathsf{ARpGC}}}{1-\theta}\epsilon_0^{-1}\right) + 1 \tag{7.2.52}$$

and

$$\#^{D}_{\mathsf{ARpGC,2}} \stackrel{\rm def}{=} (f_{\rm up} - f_{\rm low} + 1)\kappa^{S}_{\mathsf{ARpGC}}\epsilon_0^{-\frac{1}{p}}\epsilon_1^{-\frac{p+1}{p}} + 1; \tag{7.2.53}$$

$\#^{F}_{\mathsf{ARpLE,1}}$ and $\#^{D}_{\mathsf{ARpLE,1}}$ are given by (7.2.23) and (7.2.24); and $\kappa^{A}_{\mathsf{ARpGC}}$, $\kappa^{C}_{\mathsf{ARpGC}}$, and $\kappa^{S}_{\mathsf{ARpGC}}$ are given by (7.2.47)–(7.2.49), to compute a point x_ϵ at which either (7.2.14) or (7.2.15) or (7.2.16) holds.

Proof. If algorithm TPp-GC terminates in Phase 1, the evaluation bounds $\#^{F}_{\mathsf{ARpLE,1}}$ and $\#^{D}_{\mathsf{ARpLE,1}}$ follow immediately from Lemma 7.2.2. Otherwise, the algorithm enters Phase 2, and we then deduce from (7.2.38) and (7.2.39) that the total number of derivative evaluations in this phase is bounded above by $\#^{D}_{\mathsf{ARpGC,2}}$ as given in (7.2.53), while (7.2.37) and (7.2.40) bound the total number of function evaluations by $\#^{F}_{\mathsf{ARpGC,2}}$ in (7.2.52). □

Notice that the bounds (7.2.50) and (7.2.51) are both $\mathcal{O}(\epsilon^{-(p+2)/p})$ if we choose $\epsilon_0 = \epsilon_1 = \epsilon$. Furthermore, if $\epsilon_1 = \epsilon_0^{p/(p+1)}$, these same bounds become $\mathcal{O}(\epsilon_0^{-(p+1)/p})$, matching the complexity order of the ARp algorithm in the accuracy ϵ_0. If, as is likely, the $\mathcal{O}(\epsilon_0^{-1/p}\epsilon_1^{-(p+1)/p})$ term dominates, the complexity of solving the generally constrained problem is marginally—by a factor $\mathcal{O}(\epsilon_0^{-1/p})$—worse than in the unconstrained and convexly constrained cases. One must, of course, anticipate there being some cost associated with satisfying the general constraints, but whether this is the best possible bound for algorithms for this class of problems is, as it stands, unknown.

Notes and references for Subsection 7.2.2

The perceptive reader may wonder why we use different methods—ARpLE-CC and ARp-CC—to solve the subproblems that arise in the two phases of the algorithm since superficially both involve the solution of minimum-norm problems. The choice of the ARpLE-CC algorithm in Phase 1 is clear; of the methods we have considered in this book, it gives the best complexity guarantees by judicious choice of the parameter ℓ. Superficially, one might hope to use the same algorithm in Step 1 of Phase 2, and indeed one could. Unfortunately, it appears that $\ell = 2$ is the best choice, and this leads to essentially the same bounds as we obtain from using ARp-CC. Let us explain.

The key is the proof of Lemma 7.2.6, and especially the bound (7.2.43) and what follows. Let us suppose that instead, we had applied the ARpLE-CC algorithm for some suitable ℓ. Then, rather than (7.2.43), we could invoke (6.3.3) along with (7.2.46) and the violation of the third termination test in (7.2.10) to deduce that

$$\|r(x_{j,s},t_j)\|^{1/2^i} - \|r(x_{j,s+1},t_j)\|^{1/2^i} \geq \kappa_{r1}\frac{[\omega_{\Phi_\ell,\mathcal{C},1}(x_{j,s+1},t_j)]^{\frac{p+1}{p}}}{\|r(x_{j,s},t_j)\|^{\iota_1(p,\ell,i)}} \geq \kappa_{r1}\epsilon_1^{\frac{p+1}{p}}\epsilon_0^{-\iota_1(p,\ell,i)}$$

for $0 \leq s < e_j$, where κ_{r1} is given by (5.1.21) when $\beta = \theta$, as well as (7.2.44) when $s = e_j$. Thus a telescoping-sum argument and the bounds (7.2.42) give that

$$\begin{aligned}(1-\theta^{1/2^i})\epsilon_0^{1/2^i} &\geq \|r(x_{k+1,0},t_{k+1})\|^{1/2^i} - \|r(x_{j_k,s_k+1},t_{j_k})\|^{1/2^i}\\ &\geq \sum_{j,s\in\mathcal{I}_k^{\rm S}}\left(\|r(x_{j,s},t_j)\|^{1/2^i} - \|r(x_{j,s+1},t_j)\|^{1/2^i}\right)\\ &\geq \kappa_{r1}\epsilon_1^{\frac{p+1}{p}}\epsilon_0^{-\iota_1(p,\ell,i)}|\mathcal{I}_k^{\rm S}|\end{aligned}$$

and hence that

$$|\mathcal{I}_k^{\mathrm{S}}| \leq \kappa_{r1}^{-1}(1-\theta^{1/2^i})\epsilon_0^{1/2^i+\iota_1(p,\ell,i)}\epsilon_1^{-\frac{p+1}{p}} = \kappa_{r1}^{-1}(1-\theta^{1/2^i})\epsilon_0^{\frac{p-\ell+1}{p}}\epsilon_1^{-\frac{p+1}{p}}, \tag{7.2.54}$$

using the definition (5.1.21) of $\iota_1(p,\ell,i)$. But if we compare (7.2.54) with (7.2.39), we see that (aside from constants) the only difference is the exponent of ϵ_0, and that (7.2.39) will be asymptotically smaller if $\ell > 2$. Since for Φ_ℓ to be everywhere differentiable, we require $\ell \geq 2$, this then leads us to conclude that $\ell = 2$ produces the smallest evaluation-complexity bound from this line of analysis.

The genesis of the method proposed and analyzed in this and the previous section was for the equality-constrained ($\mathcal{C} = \mathbb{R}^n$) case and used first-order ($p = 1$) models to derive an $\mathcal{O}(\epsilon_1^{-2})$ evaluation-complexity bound whenever $\epsilon_1 \leq \epsilon_0$ [85, 99]. Such an approach was soon extended to second-order ($p = 2$) methods, yielding an $\mathcal{O}(\epsilon_0^{-1/2}\epsilon_1^{-3/2})$ bound [91, 96], then to cover the general ($\mathcal{C} \subseteq \mathbb{R}^n$) problem [98] and, more recently, employing essentially the method described in this section, to use higher-order ($p \geq 2$) models [100, 103]. All of these bounds rely on variants of the scaled termination test (7.1.4). It is also possible to derive a similar algorithm that provides an $\mathcal{O}(\epsilon_0\epsilon_1^{-(p+1)/p}\min[\epsilon_0,\epsilon_1]^{-(p+1)/p})$ bound to find a point for which (7.1.4) is replaced by the unscaled test

$$-\min_{d\in\mathcal{C}(x)\cap\mathcal{B}} \langle d, \nabla_x^1\mathcal{L}(x, y_\epsilon)\rangle \leq \epsilon_1$$

for some y_ϵ chosen so that $\|c(x_\epsilon)\| \leq \epsilon_0$; additional precautions must be taken to detect infeasible critical points [44]. Notice that such a complexity bound is of the same order as (7.2.50) so long as $\epsilon_0 \leq \epsilon_1$.

As we have said from the outset, we do not regard the TPpGC algorithm as being a practical method. An obvious objection is that the objective function plays no role in Phase 1, which may lead to $f(x_1)$ at the start of Phase 2 being far from its optimal value. This defect may be tempered if Phase 1 is altered so that it takes some account of f. Such an idea is central to a proposal by Curtis, Robinson, and Samadi [138], in which Phase 1 uses a trust-funnel approach [201] that aims simultaneously to reduce $f(x)$ and $\|c(x)\|$, with the latter taking precedence—strictly, the analysis provided only uses a second-degree model and is for the equality-constrained ($\mathcal{C} = \mathbb{R}^n$) case, but there is no reason to suspect that it might not be generalized. Somewhat unfortunately, while such an approach might promise better practical behaviour, there appears to be no gain in worst-case complexity in Phase 1, nor does an advanced starting point for Phase 2 lead to complexity improvement overall.

7.2.3 Other methods to find first-order approximate minimizers

7.2.3.1 Nondifferentiable exact-penalty methods

There have been other attempts to provide efficient algorithms for solving the generally constrained problem. The first that we are aware of [86] is a first-order method that applies a nondifferentiable exact-penalty-function approach to the equality-constrained problem and achieves a rather pessimistic $\mathcal{O}(\epsilon^{-5})$ evaluation-complexity bound for $\epsilon_0 = \epsilon_1 = \epsilon$. The bound can be shown to be $\mathcal{O}(\epsilon^{-2})$ in all but degenerate cases, for which a suitable constraint qualification fails and the penalty parameter may grow without bound.

7.2.3.2 Quadratic-penalty methods

More recently, attention has moved to smooth penalty-type approaches. Naively, the idea is to recast the generally constrained optimization problem as one or more of a sequence of convexly constrained ones. It is then possible to appeal to the body of theory that we have already developed for the latter to derive evaluation-complexity bounds for the former. There are, nonetheless, obstacles to overcome.

For simplicity, suppose we wish to solve the equality-constrained problem

$$\underset{x\in\mathbb{R}^n}{\text{minimize}}\ \ f(x)\ \text{ subject to }\ c(x)=0. \tag{7.2.55}$$

Our aim, as before, is to satisfy (7.1.4), and in the absence of $\mathcal{C}$, this requires that we find x_ϵ and y_ϵ so that

$$\|\nabla_x^1\mathcal{L}(x_\epsilon,y_\epsilon)\|\le\epsilon_1\|(y_\epsilon,1)\|\ \text{ and }\ \|c(x_\epsilon)\|\le\epsilon_0. \tag{7.2.56}$$

An early method for solving the equality problem (7.2.55) is to define the *quadratic-penalty function* [167]

$$\phi^{\mathrm{Q}}(x,\mu)\stackrel{\text{def}}{=}f(x)+\|c(x)\|^2/2\mu$$

and to minimize this for a sequence of decreasing *penalty* parameters μ that approach zero. Clearly the gradient of ϕ^{Q} satisfies

$$\nabla_x^1\phi^{\mathrm{Q}}(x,\mu)=\nabla_x^1f(x)+[\nabla_x^1c(x)]^Tc(x)/\mu=\nabla_x^1\mathcal{L}(x,y),\ \text{ where }\ y=-c(x)/\mu,$$

and thus any critical point of ϕ^{Q} satisfies the perturbed first-order optimality conditions

$$\nabla_x^1\mathcal{L}(x,y)=0\ \text{ and }\ c(x)=-\mu y \tag{7.2.57}$$

for (7.2.55); cf. (1.1.13) as an indication of why $\mu=0$ is needed in the limit. Given $\mu_k>0$, now suppose that we find x_k, along with a corresponding $y_k=-c(x_k)/\mu_k$, for which

$$\|\nabla_x^1\phi^{\mathrm{Q}}(x_k,\mu_k)\|\le\mu_k \tag{7.2.58}$$

and hence

$$\|\nabla_x^1\mathcal{L}(x_k,y_k)\|\le\mu_k,\ \text{ where }\ \|c(x_k)\|=\mu_k\|y_k\|. \tag{7.2.59}$$

So long as a suitable constraint qualification holds—that $\nabla_x^1c(x)$ is uniformly of full rank—it is straightforward to show that if $\{\mu_k\}$ decreases to zero, there is a subsequence for which y_k converges to some vector of Lagrange multipliers y_* [186, Thm. 4.1]. Indeed, for k sufficiently large in this subsequence,

$$\tfrac{1}{2}\|y_*\|\le\|y_k\|\le\|(y_k,1)\|\le1+\|y_k\|\le1+2\|y_*\|. \tag{7.2.60}$$

In this case, if

$$\mu_k\le\mu_*\stackrel{\text{def}}{=}\max\left(\frac{\epsilon_0}{1+2\|y_*\|},\tfrac{1}{2}\|y_*\|\epsilon_1\right), \tag{7.2.61}$$

it follows from (7.2.59) and (7.2.60) that (7.2.56) holds for $(x_\epsilon,y_\epsilon)=(x_k,y_k)$.

Thus, to achieve (7.2.56), we might aim to find a μ_k-approximate first-order minimizer (7.2.58) of $\phi^{\mathrm{Q}}(x,\mu_k)$, and once μ_k satisfies (7.2.61) we will have achieved our goal. To do so, we may apply any one of the many appropriate methods we have considered in earlier chapters. For instance, we might apply the ARp algorithm (p. 100), and at a first glance this might suggest that the algorithm will require $\mathcal{O}(\mu_*^{-(p+1)/p})=\mathcal{O}(\max(\epsilon_0,\epsilon_1)^{-(p+1)/p})$ evaluations. However, this is misleading on at least two fronts. Firstly, there is no reason $\phi^{\mathrm{Q}}(x,\mu)$ should be bounded from below. Indeed, even for a relatively benign (quadratic-programming) example such as that of minimizing $-x^2$ subject to $x=1$ leads to $\phi^{\mathrm{Q}}(x,\mu)=-x^2+(x-1)^2/2\mu$, which is unbounded from below if $\mu>1$, while changing the objective function to $-x^3$ leads to an unbounded penalty function for all

μ; such a defect is not restricted to the quadratic-penalty function; there are similar examples for all penalty (and augmented-penalty) functions that we are aware of. More fundamentally, though, such an evaluation-count estimate ignores the constants hidden by the $\mathcal{O}$ notation. A more careful consideration of the constant (4.1.45) in Theorem 4.1.5 (p. 107) applied to the problem at hand reveals the reliance on the Lipschitz constant for the pth derivative of $\phi^{\mathrm{Q}}(x,\mu)$. Unfortunately, even if $f(x)$ and $\|c(x)\|^2$ are well behaved, we have

$$\nabla_x^p \phi^{\mathrm{Q}}(x,\mu) = \nabla_x^p f(x) + \nabla_x^p \|c(x)\|^2/2\mu.$$

Thus the Lipschitz constant is itself likely of $\mathcal{O}(\mu_*^{-\frac{p+1}{p}})$, and this leads to an $\mathcal{O}(\mu_*^{-2(p+1)/p})$ $= \mathcal{O}(\max(\epsilon_0,\epsilon_1)^{-2(p+1)/p})$ evaluation-complexity bound. Such a bound is significantly worse than that given in Theorem 7.2.7 for the TPp-GC algorithm.

Of course, we should be careful not to condemn the quadratic-penalty-function method outright, since the line of analysis outlined above ignores the fact that in practice a μ_{k-1}-approximate first-order minimizer usually provides an excellent starting point for the minimization of $\phi^{\mathrm{Q}}(x,\mu_k)$, particularly for small μ_k. Indeed, under stronger (second-order) assumptions, at most two Newton-like steps suffice to find x_k from x_{k-1} for sufficiently large k [186, Cor. 5.12]. At this stage, however, we are unaware of any attempt to quantify what "sufficiently large k" means and thus to improve on the bound stated above.

7.2.3.3 Augmented-Lagrangian methods

The main impediment to a satisfactory evaluation-complexity bound exposed above relates to the need for the penalty parameter μ in the quadratic-penalty approach to converge to zero, since this gives rise to large (and ultimately unbounded) derivative Lipschitz constants. This has long been recognized as undesirable, and a cure from the 1970s is to use an augmented-Lagrangian approach instead [232, 320]. The *augmented-Lagrangian function* for the problem (7.2.55),

$$\phi^{\mathrm{A}}(x,u,\mu) \stackrel{\text{def}}{=} f(x) - \langle u, c(x)\rangle + \|c(x)\|^2/2\mu,$$

depends on a vector of auxiliary *Lagrange-multiplier estimates* u and can be viewed as either a shifted version of the quadratic-penalty function or, as its name suggests, an augmentation (or attempted convexification) of the Lagrangian function. Once again, a sequence of minimizations of $\phi^{\mathrm{A}}(x,u,\mu)$ is performed, each with a particular $(u,\mu) = (u_k,\mu_k)$. The intention is that by adjusting u appropriately, it will not be necessary to force μ to zero. To see this, observe that the gradient of ϕ^{A} satisfies

$$\nabla_x^1 \phi^{\mathrm{A}}(x,u,\mu) = \nabla_x^1 f(x) - [\nabla_x^1 c(x)]^T(u - c(x)/\mu) = \nabla_x^1 \mathcal{L}(x,y),$$

where now

$$y = u - c(x)/\mu, \tag{7.2.62}$$

and thus any critical point of ϕ^{A} satisfies the perturbed first-order optimality conditions

$$\nabla_x^1 \mathcal{L}(x,y) = 0 \quad \text{and} \quad c(x) = \mu(u-y)$$

for (7.2.55). It is then clear that—at variance with (7.2.57)—we may achieve (7.2.55) if either $\mu = 0$ or, crucially, $y = u$ in the limit.

Suitable rules to encourage convergence of y to u while preventing small μ are known [35, 46, 129] and have resulted in generally effective software [8, 130]. Typically, given x_k,

y_k, and μ_k, a new estimate x_{k+1} is computed by approximately minimizing $\phi^{\mathrm{A}}(x, y_k, \mu_k)$. Then, so long as $\|c(x_{k+1})\| \leq \tau_k \|c(x_k)\|$ for some $0 < \tau_k \leq \tau < 1$, new estimates $y_{k+1} = y_k - c(x_{k+1})/\mu_k$ are computed according to (7.2.62) and the penalty parameter is unaltered, that is, $\mu_{k+1} = \mu_k$. Conversely, if x_{k+1} does not provide the required decrease in infeasibility, the Lagrange-multiplier estimates are left untouched,[57] i.e., $y_{k+1} = y_k$, but the penalty parameter is reduced, i.e., $\mu_{k+1} \leq \gamma_k \mu_k$ for some $0 < \gamma_k < \gamma < 1$. It is easy to show that so long as the penalty parameter is bounded away from zero (i.e., $\mu_k \geq \mu_{\min}$ for some $\mu_{\min} > 0$)—or equivalently, given the above, that μ is reduced a finite number of times—and $c(x_k)$ remains bounded (i.e., $\|c(x_k)\| \leq c_{\max}$ for some $c_{\max} > 0$), then only a logarithmic number (in μ_0, $\mu_{\min}$, $c_{\max}$, τ, γ, and ϵ_0) of minimizations are needed to ensure that $\|c(x_{k+1})\| \leq \epsilon_0$ [49, Thm. 3.1], [207, Thm. 2.5]. Thus if each minimization of $\phi^{\mathrm{A}}(x, y_k, \mu_k)$ guarantees that $\|\nabla_x^1 \phi^{\mathrm{A}}(x_{k+1}, y_k, \mu_k)\| \leq \epsilon_1 \|(y_k, 1)\|$, we will achieve the stopping criteria (7.2.56) in at most the product of this logarithmic bound and the cost of such a minimization.[58] While this appears to be a favourable outcome, as always worst-case complexity has to cope with all eventualities, including the possibility that μ might shrink to zero, indeed that it must if the problem is infeasible. Without going into details, it is this unfortunate possibility that prevents, to date, existing analyses and bounds, e.g., [49, Thm. 3.3], [207, Thm. 2.6], from competing favourably with that we gave in our Section 7.2.2.2, there often being an additional multiplying factor of $\mathcal{O}(\epsilon_0^{\nu})$ for some ν ($\nu = (p+1)/p$ in the best methods we have discussed), as well as restricting relationships between ϵ_0 and ϵ_1. Thus, although the augmented-Lagrangian approach raises the prospect of better evaluation complexity than the quadratic-penalty method in many cases, the worst-case bound does not appear to be any better.

There is another popular class of augmented-Lagrangian methods, those based on proximal regularization [330]. The idea is essentially to control the size of the step taken by choosing x_{k+1} to approximately minimize $\phi^{\mathrm{A}}(x, y_k, \mu_k) + \beta \|x - x_k\|^2$ for some fixed $\beta > 0$. Although potentially promising complexity results have appeared in the literature [241, 273], [368, Thm. 4], these make uniform full-rank assumptions on $\nabla_x^1 c(x)$ (which ensures that there is a feasible solution), require that the starting point be feasible, and expect that μ_k is below an a priori unknown value. It will be interesting to see if similar results can be established under less stringent conditions, akin to those required in Section 7.2.2.2.

7.2.3.4 Interior-point methods

Although the discussion in the two preceding subsections focused on the equality-constrained problem (7.2.55), much of what we said carries over to the inequality case

$$\underset{x \in \mathbb{R}^n}{\text{minimize}} \ f(x) \ \text{subject to} \ c(x) \geq 0 \tag{7.2.63}$$

by suitably adapting the quadratic-penalty and augmented-Lagrangian functions [329]. A more obvious approach, though, is based on the *logarithmic-barrier function*

$$\phi^{\mathrm{B}}(x, \mu) = f(x) - \mu \langle \log c(x), e \rangle$$

that is defined on the set $\mathcal{F} \stackrel{\text{def}}{=} \{x \mid c(x) > 0\}$ of strictly feasible (*interior*) points for (7.2.63); here log denotes the vector of componentwise logarithms, e is the vector of ones,

[57] Some methods, e.g., [207], update the multipliers every iteration.

[58] But of course we need to be aware that $\mu_{\min}$ will likely appear in the bound on the cost of the approximate minimization of $\phi^{\mathrm{A}}(x, y_k, \mu_k)$, just as it did for the quadratic-penalty method.

and $\mu > 0$ is the *barrier parameter.* Clearly

$$\nabla_x^1 \phi^{\mathrm{B}}(x,\mu) = \nabla_x^1 f(x) - [\nabla_x^1 c(x)]^T \mu C^{-1}(x)e = \nabla_x^1 \mathcal{L}(x,y), \quad \text{where } y = \mu C^{-1}(x)e$$

and $C(x) = \mathrm{diag}(c(x))$, the diagonal matrix whose diagonal entries are the components of $c(x)$. Thus any strictly feasible critical point of $\phi^{\mathrm{B}}(x,\mu)$ satisfies the perturbed optimality conditions

$$\nabla_x^1 \mathcal{L}(x,y) = 0, \; c(x) > 0, \; y > 0, \; \text{ and } \; C(x)y = \mu e$$

for (7.2.63), and (1.1.16) suggests that μ needs to approach zero. More importantly, if we pick $\mu = \epsilon_{\mathrm{C}}/\|e\|$ for a given $\epsilon_{\mathrm{C}} > 0$, find an ϵ_1-approximate first-order minimizer x_ϵ of $\phi^{\mathrm{B}}(x,\mu)$, and set $y_\epsilon = \mu C^{-1}(x)e$, we have that

$$\|\nabla_x^1 \mathcal{L}(x_\epsilon, y_\epsilon)\| \le \epsilon_1, \; c(x_\epsilon) > 0, \; y_\epsilon > 0, \; \text{ and } \; \|c(x_\epsilon).y_\epsilon\| \le \epsilon_{\mathrm{C}}, \tag{7.2.64}$$

where $c(x_\epsilon)\cdot y_\epsilon$ is the componentwise product of the pair of vectors. It is of course easy to view (7.2.64) as ϵ-approximate first-order criticality criteria for (7.2.63), and consequently to consider the approximate unconstrained minimization of $\phi^{\mathrm{B}}(x,\mu)$ as a good way to achieve them. But, just as for the penalty methods before, there are difficulties lurking.

The first issue is, of course, the tacit assumption that $\mathcal{F} \neq \emptyset$. Any algorithm must verify this; indeed it needs to exhibit a point $x_0 \in \mathcal{F}$ from which to start the minimization of $\phi^{\mathrm{B}}(x,\mu)$, if at all possible. While this presumption appears to rule out infeasible regions, or regions without an interior, a pragmatic response is to ask instead for a point in the interior of the region

$$\mathcal{F}_\beta = \{x \mid c(x) \ge -\beta e\}$$

(cf. (7.2.7)) for some small $\beta > 0$ and thereafter widen the minimization to this slightly enlarged region. So long as $\mathcal{F}_\beta$ is nonempty, we shall focus on the shifted barrier function

$$\phi^{\mathrm{S}}(x,\mu,\beta) = f(x) - \mu\langle \log(c(x) + \beta e), e\rangle.$$

Clearly we can apply the ARpLE-CC algorithm (p. 189) or any other suitable method to try to find an $x_0 \in \mathcal{F}_{\gamma\beta}$ for some $\gamma \in (0,1)$—or to report if (locally) there is no such point—just as we did in Phase 1 of the TPp-GC algorithm. Should we succeed, such a point will then be interior to $\mathcal{F}_\beta$.

If we apply a suitable minimization method to $\phi^{\mathrm{S}}(x,\mu,\beta)$, we may hope to find a point x_ϵ for which

$$\|\nabla_x^1 \mathcal{L}(x_\epsilon, y_\epsilon)\| \le \epsilon_1, \; c(x_\epsilon) + \beta e > 0, \; y_\epsilon > 0, \; \text{ and } \; \|(c(x_\epsilon) + \beta e).y_\epsilon\| = \mu\|e\|,$$

where $y_\epsilon = \mu[\mathrm{diag}(c(x_\epsilon) + \beta e)]^{-1}e$ just as for (7.2.64). This then implies that

$$\|\nabla_x^1 \mathcal{L}(x_\epsilon, y_\epsilon)\| \le \epsilon_1, \; c(x_\epsilon) \ge -\epsilon_0 e, \; y_\epsilon > 0, \; \text{ and } \; \|c(x_\epsilon).y_\epsilon\| \le \epsilon_{\mathrm{C}}(\|y_\epsilon\| + 1)$$

if we select $\mu = \epsilon_{\mathrm{C}}/\|e\|$ and $\epsilon_0 \le \beta \le \epsilon_{\mathrm{C}}$, and once again this may be considered as a reasonable definition of an ϵ-approximate first-order minimizer of (7.2.63).

This, however, is not the end of the story. All of the methods we have analyzed for unconstrained minimization require that problem function values and derivatives be Lipschitz continuous (with bounded Lipschitz constants). Unfortunately, the log function violates this assumption whenever its argument is too close to zero (and of course log is undefined for nonpositive arguments). Thus it is questionable whether one can use off-the-shelf evaluation-complexity bounds from our previous chapters. A common strategy when assessing the global convergence of descent-based methods that use the logarithmic barrier function is to argue that the region

$$\mathcal{F}_k = \{x \mid \phi^{\mathrm{B}}(x, \mu) \leq \phi^{\mathrm{B}}(x_k, \mu)\}$$

defined at the interior point x_k precludes future iterates from getting too close to the boundary of the feasible region. While this is true, it does not prevent exponentially large (in μ) Lipschitz constants at trial points in $\mathcal{F}_k$. For example [236, p. 5], consider the barrier function $\phi^{\mathrm{B}}(x, \mu) = x - \mu \log(x) - \mu \log(2 - x)$ for the linear program $\min x$ for $x \in [0, 2]$ at $x_0 = 1$. Then the point $x = \exp(-1/\mu) \in \mathcal{F}_0$, and at that point $\nabla_x^2 \phi^{\mathrm{B}}(x, \mu) = \mu[1/x^2 + 1/(2-x)^2] \geq \mu \exp(2/\mu)$. Thus, without further precautions, any promising-sounding complexity bounds suggested by earlier chapters may be ruined by large derivatives. There is a way around this difficulty using very cautious, and small, steps, but to date the proposal with the best bounds [221], [236, Thm. 1, Claim 3] relies on knowing function and derivative Lipschitz constants of f and c in advance.

7.2.3.5 SQP and the future

Among the most efficient methods in practice are the large class of sequential linear or quadratic programming (SLP/SQP) algorithms [52, 199], [307, Chap. 18]. Broadly, this covers most methods that build first-degree-accurate linear or quadratic approximations to the objective function, and first-degree Taylor approximations to the (nonlinear) constraint functions. A step from the current iterate is found by approximately minimizing the surrogate objective function subject to the linearized constraints. Since the reader will recognize this as a Newton-like procedure in the unconstrained case, it is of no surprise that additional tricks are required to ensure convergence to a suitable critical point. These include enlarging the linearized feasible region to ensure that the constraints are feasible; restricting the step either explicitly, using a trust region or regularization, or implicitly via a line search; and measuring progress with respect to a suitable merit function. In addition, and in common with all of the methods in this chapter, precautions are needed to cope with infeasibility.

Given the vast array of possible combinations, it is perhaps somewhat surprising that little has been said about the evaluation complexity of this important class of methods. The exception is a very recent linesearch-like approach [163] that focuses on a basic form of the iteration in which a merit function plays no part (except in the analysis) but convergence is ensured by cautious control of the steplength. Since little is assumed beyond Lipschitz smoothness, the most general evaluation-complexity result is somewhat pessimistic (akin to $\mathcal{O}(\epsilon^{-4})$ evaluations) but can be improved if a feasible point is known ($\mathcal{O}(\epsilon^{-3})$ evaluations) and more so if a suitable constraint qualification holds ($\mathcal{O}(\epsilon^{-2})$ evaluations). The analysis is quite involved, and certainly beyond the space we have here, but it is interesting in setting the pace for future improvements.

Part IC

Computational Aspects

Having considered the evaluation complexity of solving nonconvex optimization problems of various guises with our favourite algorithms, we now focus on the practical aspects of those algorithms and show that, at least for orders one and two, these can be made very efficient. As well as being very important for users of these algorithms, it also makes the evaluation complexity of previous chapters even more significant because the costs of solving the subproblems that occur remain under control.

Previous page: Eryngium Giganteum by Penelope Gould, 2010

Chapter 8

Subproblem Fundamentals

Chapter summary

In this chapter, we investigate the fundamental characteristics of the subproblems occurring when computing a step from the current iterate, or when checking an iterate for approximate criticality. We focus first on the "secular" formulation of the step computation and analyze the optimality conditions associated with these subproblems, highlighting the similarities between the trust-region and adaptive-regularization approaches. We then address the solution of the secular equations in generic terms.

8.1 The context of the step computation

The development thus far in our book has been directed at the cost, in terms of function and derivative evaluations, of solving classes of optimization problems. At the heart of each of the algorithms we have considered has been the need to find a step s_k from the current estimate x_k and to compute function/derivative values at the trial point $x_k + s_k$. Additionally, it has sometimes been necessary to check whether this trial point is an appropriate approximate critical point of the desired order. Neither the mechanics nor the efficiency of finding a suitable step or checking for criticality has been of direct concern.

In this chapter we return to these overlooked but nonetheless crucial components of our algorithms. For the sake of simplicity, our main focus will be on methods for problems without constraints; we will return briefly to see how constraints may be incorporated in due course.

If we are content to use an algorithm that uses only gradients—with no attempt to use or estimate second- or higher-order derivatives—to find a first-order critical point, the step is invariably in the negative gradient direction, or a scaled variant thereof; the precise size of the stepsize results either explicitly, from a trust-region radius, or implicitly, via regularization or a linesearch. Likewise, checking for criticality simply involves finding an appropriate norm of the gradient. Thus these tasks offer no significant challenges.

Things becomes more interesting when a local quadratic model

$$m(s) \stackrel{\text{def}}{=} \langle g, s\rangle + \tfrac{1}{2}\langle s, Hs\rangle, \tag{8.1.1}$$

for which the vector g and symmetric matrix H are (or are suitable approximations of) the gradient and Hessian, $\nabla_x^1 f(x_k)$ and $\nabla_x^2 f(x_k)$, of the objective function $f(x)$ at x_k, is considered. In the trust-region case, the step is determined to approximately

$$\underset{s\in\mathbb{R}^n,\ \|s\|_M\leq\Delta}{\text{minimize}}\ m(s) \tag{8.1.2}$$

involving the radius $\Delta = \Delta_k > 0$, while in the regularization case it is found to approximately

$$\underset{s\in\mathbb{R}^n}{\text{minimize}}\ m^{\mathrm{R}}(s,\sigma,r) \tag{8.1.3}$$

for the given regularization parameter $\sigma = \sigma_k > 0$, where

$$m^{\mathrm{R}}(s,\sigma,r) \stackrel{\text{def}}{=} m(s) + \tfrac{1}{r}\sigma\|s\|_M^r \tag{8.1.4}$$

for a fixed $r \geq 2$ and a suitable norm $\|\cdot\|_M$ defined below.[59]

Although we have used the Euclidean norm to define trust regions and regularizations throughout our theoretical development so far, in practice it is convenient and often desirable to use a norm that better reflects the nonuniform, possibly evolving, scaling of the problem. To this end, we have stated (8.1.2) and (8.1.4) in terms of $\|v\|_M \stackrel{\text{def}}{=} \sqrt{\langle v, Mv\rangle}$, where $M = M_k$ are iteration-dependent, symmetric, uniformly positive definite matrices. Doing so in our previous analyses would merely have introduced additional (annoying) constants that relate M_k to I in the bounds provided, while the generality we mention in this chapter has favourable implementational implications, as we shall see. In practice, we require access to M_k and the ability to solve systems $M_k u = v$ efficiently for given v, and this frequently determines (and limits) the choice of possible M_k.

In reality, the precise solutions of (8.1.2) and (8.1.3) are not required, rather that their models decrease and the norms of their gradients are sufficiently small. Notwithstanding, since the true solutions satisfy these approximating conditions, studying methods for solving (8.1.2) and (8.1.3) is illuminating and forms the basis for later approximations.

8.2 A digression on all things secular

Before we address (8.1.2) and (8.1.3) directly, we focus on a crucial component of what is to follow: the secular function and its related equations.

8.2.1 The secular function and equations

Consider first the trust-region subproblem (8.1.2). There are clearly two possibilities: either the trust-region constraint $\|s\|_M \leq \Delta$ is active (binding) at a solution s_* or it isn't. If the former occurs, (8.1.2) is equivalent to minimizing $m(s)$ subject to the *equality* constraint $c(s) = \frac{1}{2}\langle s, Ms\rangle - \frac{1}{2}\Delta^2 = 0$, and thus, from the theory of Lagrange multipliers, s_* must satisfy $\nabla_s^1 m(s_*) + \lambda_* \nabla_s^1 c(s_*) = 0$, i.e.,

$$g + Hs_* + \lambda M s_* = 0, \quad \text{where } \|s_*\|_M = \Delta, \tag{8.2.1}$$

for some scalar λ_*. There is, of course, the other possibility, namely that the solution lies inside the trust region, and for this $g + Hs_* = 0$ with $\|s_*\|_M < \Delta$; we ignore this for now.

[59]The scale factor $1/r$ is for simplicity of exposition. Other factors, such as the additional $1/2$ that arises in (3.3.1), will be assumed to have been absorbed into σ in our presentation.

Turning to the regularization subproblem (8.1.3), a solution s_* to this necessarily satisfies the first-order criticality condition $\nabla_s^1 m^{\mathrm{R}}(s_*, \sigma, r) = 0$, i.e.,

$$g + Hs_* + \sigma \|s_*\|_M^{r-2} Ms_* = 0$$

or, equivalently,

$$g + Hs_* + \lambda_* Ms_* = 0, \quad \text{where} \quad \lambda_* = \sigma \|s_*\|_M^{r-2}.$$

The common theme here is that we are seeking a vector $s(\lambda)$ for which

$$(H + \lambda M)s(\lambda) = -g, \tag{8.2.2}$$

where λ additionally satisfies a *secular equation*

$$\|s(\lambda)\|_M^2 = \tau(\lambda) \tag{8.2.3}$$

for some monotonically nondecreasing function $\tau(\lambda)$; we have $\tau(\lambda) = \Delta^2$ in the trust region–boundary case and $\tau(\lambda) = (\lambda/\sigma)^{2/(r-2)}$ when using regularization. With this in mind, we now examine properties of the *secular function*

$$\pi(\lambda) \stackrel{\text{def}}{=} s^T(\lambda)Ms(\lambda) \equiv \|s(\lambda)\|_M^2, \tag{8.2.4}$$

where $s(\lambda)$ satisfies (8.2.2).

Let $\lambda_1 \leq \cdots \leq \lambda_n$ be the eigenvalues of the pencil (H, M), and let v_i, $1 \leq i \leq n$, be the corresponding generalized eigenvectors, i.e.,

$$HV = MV\Lambda, \quad \text{where} \quad V^T MV = I; \tag{8.2.5}$$

Λ is the diagonal matrix with entries λ_i, $1 \leq i \leq n$; and V is the matrix whose columns are the v_i, $1 \leq i \leq n$ (see Appendix A.5). Define $g_{\mathrm{V}} \stackrel{\text{def}}{=} V^T g$ and $s_{\mathrm{V}}(\lambda) \stackrel{\text{def}}{=} V^T Ms(\lambda)$, from which we may deduce that $s(\lambda) = Vs_{\mathrm{V}}(\lambda)$ and $g = MVg_{\mathrm{V}}$ using (8.2.5). It then follows from (8.2.2) and (8.2.5) that

$$(H + \lambda M)s(\lambda) = (H + \lambda M)Vs_{\mathrm{V}}(\lambda) = MV(\Lambda + \lambda I)s_{\mathrm{V}}(\lambda) = -g = -MVg_{\mathrm{V}}$$

and thus, since M and V and nonsingular, that

$$s_{\mathrm{V}}(\lambda) = -(\Lambda + \lambda I)^{-1} g_{\mathrm{V}} \tag{8.2.6}$$

so long as λ is not $-\lambda_i$ for any $1 \leq i \leq n$. Since

$$\pi(\lambda) = \langle s(\lambda), Ms(\lambda) \rangle = \left\langle s_{\mathrm{V}}(\lambda), V^T MV s_{\mathrm{V}}(\lambda) \right\rangle = \langle s_{\mathrm{V}}(\lambda), s_{\mathrm{V}}(\lambda) \rangle,$$

it follows immediately from (8.2.6) that

$$\pi(\lambda) = \sum_{i=1}^{n} \left(\frac{\gamma_i}{\lambda + \lambda_i} \right)^2, \tag{8.2.7}$$

where γ_i is the ith component of g_{V}. It is then clear that $\pi(\lambda)$ is positive; has poles at negative the eigenvalues of the pencil (H, M), except those for which the corresponding $\gamma_i = 0$; and decays to zero as λ approaches $\pm$ infinity. We illustrate this in Figure 8.2.1.

Of course the secular equation (8.2.3) or, equivalently,

$$\pi(\lambda) = \tau(\lambda), \tag{8.2.8}$$

may have more than one root, but as we shall shortly see, the one that interests us lies in the interval $(-\lambda_1, \infty)$. It follows immediately by considering (8.2.7) that $\pi(\lambda)$ is continuous and convex and monotonically decreases to zero on this interval. More importantly, we have the following trivial result.

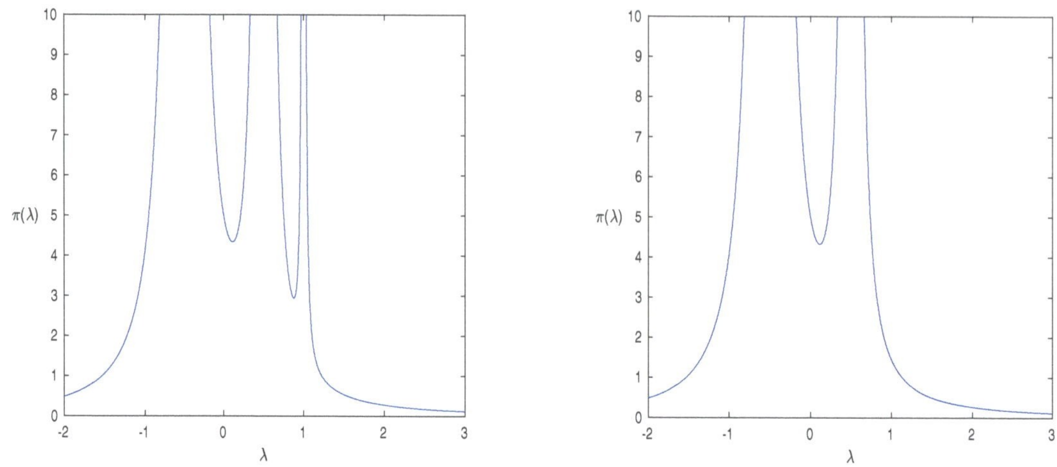

Figure 8.2.1. The secular functions $\pi(\lambda)$ for the quadratic models $m(s) = -\frac{1}{2}s_1^2 - \frac{1}{4}s_2^2 + \frac{1}{4}s_3^2 + \frac{1}{10}s_1 + \frac{1}{2}s_2 + s_3$ (left) and $-\frac{1}{2}s_1^2 - \frac{1}{4}s_2^2 + \frac{1}{4}s_3^2 + \frac{1}{2}s_2 + s_3$ (right) in the Euclidean norm (i.e., $M = I$). Notice how the pole at $\lambda = 1$ corresponding to the eigenvalue -1 vanishes for the second case because the corresponding gradient term is zero.

Lemma 8.2.1: Root of the secular equation

[Appendix to Chapter 8, p. 475]: Let

$$\lambda_s \stackrel{\text{def}}{=} \max(0, -\lambda_1),$$

where λ_1 is the leftmost eigenvalue of the pencil (H, M). Let $\pi(\lambda)$ be the secular function (8.2.4), and suppose that $\tau(\lambda)$ is a nonzero, nonnegative, monotonically nondecreasing, continuous function on (λ_s, ∞) and that $\pi(\lambda_s) > \tau(\lambda_s) \geq 0$. Then there is a unique root, say $\lambda_* > \lambda_s$, of the secular equation (8.2.8).

Both of the functions of interest to us,

$$\tau(\lambda) = \Delta^2 \text{ (trust-region) and } \tau(\lambda) = (\lambda/\sigma)^{2/(r-2)} \text{ (regularization)}, \tag{8.2.9}$$

satisfy the requirements of Lemma 8.2.1.

The nonlinearity of the secular function makes a closed-form solution of (8.2.8) unlikely in all but the simplest cases. Thus we will usually resort to iteration. That is, equipped with an estimate, λ_c, of the desired root of (8.2.8), we aim for a (significantly) improved one, $\lambda_c + \delta$, and then repeat the procedure. Most sophisticated iteration depends on approximation; that is, we replace the true secular equation, written in terms of the correction δ,

$$\pi(\lambda_c + \delta) = \tau(\lambda_c + \delta), \tag{8.2.10}$$

by the surrogate

$$t(\delta) = \tau(\lambda_c + \delta) \tag{8.2.11}$$

for a suitable approximation $t(\delta) \approx \pi(\lambda_c + \delta)$. Of course, we might also approximate $\tau(\lambda_c + \delta)$, but, for the cases we are interested in, τ is often simple enough—for example,

$\tau = \Delta^2$ in the trust-region case or $\tau = \lambda^2/\sigma^2$ for cubic ($r = 3$) and $\tau = \lambda/\sigma$ for quartic ($r = 4$) regularization—that we can treat it directly.[60]

Perhaps the most obvious approximation to $\pi(\lambda_c + \delta)$ is its truncated Taylor series of order k,

$$t_k(\delta) \stackrel{\text{def}}{=} \sum_{j=0}^{k} \frac{\pi^{(j)}(\lambda_C)}{j!}\delta^j, \tag{8.2.12}$$

where $\pi^{(0)}(\lambda) \equiv \pi(\lambda)$ and $\pi^{(j)}(\lambda)$ is the jth derivative of π at λ. Such an approximation has a number of favourable properties.

Theorem 8.2.2: Lower and upper Taylor estimates of $\pi(\lambda)$

[Appendix to Chapter 8, p. 475]: Let $\pi(\lambda)$ be the secular function (8.2.4). Suppose that λ_C is a given value such that $\lambda_C > \lambda_S$, and let $t_k(\delta)$ be the kth-order Taylor-series approximation (8.2.12) to $\pi(\lambda_C + \delta)$. Then

$$\pi(\lambda_C + \delta) \leq t_k(\delta) \text{ for even } k > 0 \text{ and } \pi(\lambda_C + \delta) \geq t_k(\delta) \text{ for odd } k > 0 \tag{8.2.13}$$

when $\delta > 0$, while

$$\pi(\lambda_C + \delta) \geq t_{k+1}(\delta) \geq t_k(\delta) \text{ for all } k > 0 \tag{8.2.14}$$

when $\lambda_s - \lambda_C < \delta < 0$. The inequalities in (8.2.13) and (8.2.14) are strict whenever $g \neq 0$.

We illustrate these properties in Figure 8.2.2. Notice that, since the secular function is not everywhere analytic, the Taylor-series approximations deteriorate as δ increases no matter what degree of approximation is used. Specifically, there is a threshold λ_B—for our example, $\lambda_B \approx 0.98$—above which no approximant t_k is close to π; the actual value depends on the distance of λ_C to the nearest singularity of $\pi(\lambda)$ in the complex plane [9, Thm. 16.20 and what follows]. This implies that the suggested root $\delta_k^T(\lambda_C)$ of

$$t_k(\delta) = \tau(\lambda_c + \delta) \tag{8.2.15}$$

for odd-degree approximations may be a poor estimate of λ_* if $\lambda_S < \lambda_C < \lambda_*$ and $\tau(\lambda_B)$ is significantly smaller than $\pi(\lambda_B)$, and as a consequence many iterations may be required to determine λ_*. Conversely, there appears to be good agreement for negative $\delta_k^T(\lambda_C)$ as the degree of approximation k increases, and thus there is scope for optimism that reasonable-order Taylor approximations will perform well if $\lambda > \lambda_*$. Similar results concerning the monotonic (and rapid) convergence of Taylor-series methods for roots of more general functions, whose derivatives satisfy appropriate sign conditions, are known [350, Thm. 4.2].

Now suppose, just as in Lemma 8.2.1 and our subproblems, that $\tau(\lambda)$ is a nonzero, nonnegative, monotonically nondecreasing, continuous function on (λ_S, ∞), and that

[60] In the cubic-regularization case, it is actually important that we don't use a first-order Taylor approximation $s_1(\delta) = (\lambda^2 + 2\lambda\delta)/\sigma^2$ to $\tau(\lambda) = \lambda^2/\sigma^2$. Such an approximation underestimates τ, and the resulting root of $t_1(\delta) = s_1(\delta)$ might then overestimate λ_*.

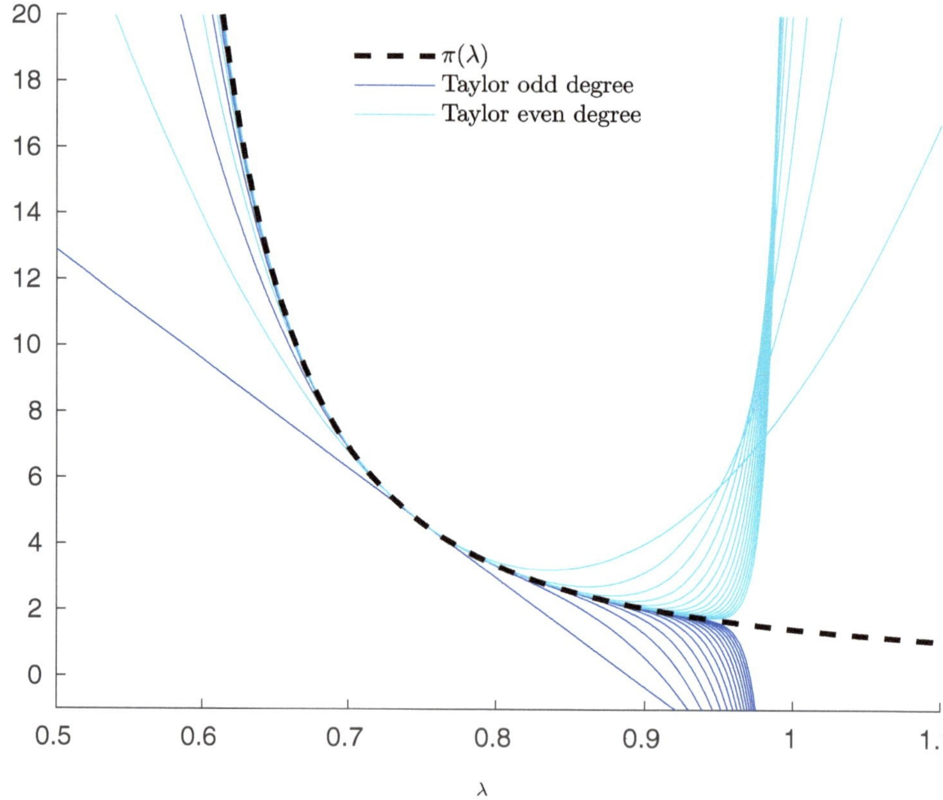

Figure 8.2.2. The secular function $\pi(\lambda)$ for the objective $-\frac{1}{4}x_1^2 + \frac{1}{4}x_2^2 + \frac{1}{2}x_1 + x_2$ (dashed line) along with the Taylor-series approximants $\pi_k(\lambda - \lambda_{\mathrm{C}})$ of degrees $k = 1$ to 100 about $\lambda_{\mathrm{C}} = 0.75$ (solid lines). Observe the threshold $\lambda_{\mathrm{B}} \approx 0.98$ above which no approximant is good.

$\pi(\lambda_{\mathrm{S}}) > \tau(\lambda_{\mathrm{S}})$. In this case, Lemma 8.2.1 implies that there is a unique root, say $\lambda_* > \lambda_{\mathrm{S}}$, of the equation (8.2.8) or, equivalently, a correction δ_* to λ_{C} for (8.2.10).

Consider $\lambda_{\mathrm{C}} < \lambda_*$, in which case $t_k(0) = \pi(\lambda_{\mathrm{C}}) > \tau(\lambda_{\mathrm{C}})$, and let $\delta_* \stackrel{\text{def}}{=} \lambda_* - \lambda_{\mathrm{C}} > 0$. If k is odd, Theorem 8.2.2 implies that $t_k(\delta_*) \leq \pi(\lambda_{\mathrm{C}} + \delta_*) = \tau(\lambda_{\mathrm{C}} + \delta_*)$. Thus the equation (8.2.15) has at least one root in $(0, \delta_*)$ for odd k, and Newton and other odd-degree Taylor-series methods for (8.2.8) based on finding corrections as positive real roots of (8.2.15) will underestimate λ_*. Moreover, since $\tau(\lambda_{\mathrm{C}}+\delta) > \pi(\lambda_{\mathrm{C}}+\delta) \geq t_k(\delta)$ for $\lambda_{\mathrm{C}} + \delta > \lambda_*$, all positive roots of (8.2.15) give underestimators. Conversely, if k is even, $t_k(\delta_*) \geq \pi(\lambda_{\mathrm{C}} + \delta_*) = \tau(\lambda_{\mathrm{C}} + \delta_*)$ and any positive root of (8.2.15) will overestimate λ_*.

By contrast, consider the alternative $\lambda_{\mathrm{C}} > \lambda_*$, in which case $t_k(0) = \pi(\lambda_{\mathrm{C}}) < \tau(\lambda_{\mathrm{C}})$ and $\delta_* < 0$. Theorem 8.2.2 gives that $\tau(\lambda_{\mathrm{C}} + \delta) > \pi(\lambda_{\mathrm{C}} + \delta) \geq t_k(\delta)$ for all $\delta \in (\delta_*, 0]$, and thus the least-negative root (if any) of (8.2.15) will lie to the left of δ_*. Moreover, as k increases, (8.2.14) indicates that the least-negative roots move to the right, and thus the higher the degree of approximation used, the better the lower bound on λ_* provided by the least-negative root of (8.2.15).

So long as $\tau(\lambda)$ is a polynomial—as it is for the trust-region and cubic- and quartic-regularization cases—(8.2.15) is a polynomial root-finding problem, for which there are many fast and reliable solution methods. Of course, when a linear Taylor approximation is used, solving (8.2.11) is trivial in the trust-region case, and we recover Newton's method. Whatever the degree $k \geq 1$, we will need to calculate (or approximate) derivatives of $\pi(\lambda)$, and we will return to this shortly.

8.2.2 The transformed secular function and equations

There is nothing intrinsically special about the equation (8.2.8), and we might equally aim instead to satisfy

$$[\pi(\lambda)]^\beta = \tau(\lambda;\beta) \stackrel{\text{def}}{=} [\tau(\lambda)]^\beta \tag{8.2.16}$$

for arbitrary $\beta \neq 0$. Thus it is of interest to investigate how the *transformed* secular function

$$\pi(\lambda;\beta) \stackrel{\text{def}}{=} [\pi(\lambda)]^\beta \equiv \|s(\lambda)\|_M^{2\beta} \tag{8.2.17}$$

behaves over the interval $(-\lambda_s, \infty)$. To do so, we differentiate with respect to λ and use the chain rule to obtain

$$\pi^{(1)}(\lambda;\beta) = \beta\,[\pi(\lambda)]^{\beta-1}\,\pi^{(1)}(\lambda) \tag{8.2.18}$$

$$\text{and } \pi^{(2)}(\lambda;\beta) = \beta\,[\pi(\lambda)]^{\beta-1}\,\pi^{(2)}(\lambda) + \beta\,(\beta-1)\,[\pi(\lambda)]^{\beta-2}\left[\pi^{(1)}(\lambda)\right]^2. \tag{8.2.19}$$

It is then simple to examine the signs of the derivatives in question.

Lemma 8.2.3: Derivatives of $\pi(\lambda;\beta)$

[Appendix to Chapter 8, p. 476]: Let $\pi(\lambda;\beta)$ be the transformed secular function (8.2.17). Suppose that $\lambda > \lambda_s$. Then

(i) $\pi^{(1)}(\lambda;\beta) \leq 0$ *for all* $\beta > 0$, *while* $\pi^{(1)}(\lambda;\beta) \geq 0$ *for all* $\beta < 0$;

(ii) $\pi^{(2)}(\lambda;\beta) \geq 0$ *for all* $\beta > 0$, *while* $\pi^{(2)}(\lambda;\beta) \leq 0$ *for all* $\beta \in [-1/2, 0)$.

The above inequalities are strict whenever $g \neq 0$.

The signs of the derivatives implied by Lemma 8.2.3 immediately provide crucial monotonicity properties.

Corollary 8.2.4: Monotonicity of transformed secular function

The transformed secular function (8.2.17) is strictly decreasing from $\pi(\lambda_s;\beta)$ to zero and strictly convex for $\lambda \in (\lambda_s, \infty)$ when $\beta > 0$ and strictly increasing from $\pi(\lambda_s;\beta)$ to infinity and concave for $\lambda \in (\lambda_s, \infty)$ when $\beta \in [-1/2, 0)$.

Of course we have already seen these properties in the case $\beta = 1$ because then $\pi(\lambda, 1) \equiv \pi(\lambda)$. We illustrate the inverse secular function $\pi(\lambda, -1/2)$ in Figure 8.2.3.

We now consider the first-order Taylor-series approximation, $t_1(\delta;\beta)$, to $\pi(\lambda_c + \delta;\beta)$; a higher-order analysis is possible [151], but the details are rather complicated.

From this we may deduce the following result.

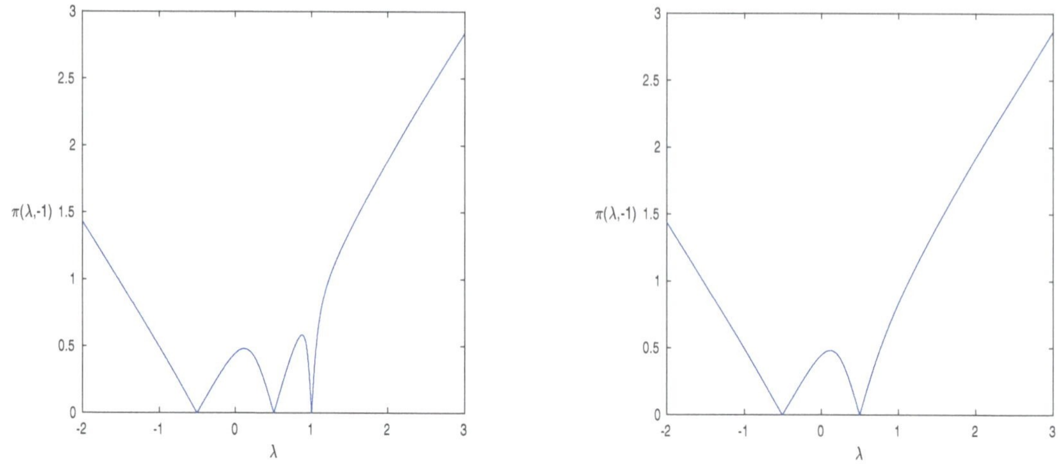

Figure 8.2.3. The inverse secular functions $\pi(\lambda, -1/2)$ for the pair of functions given in Figure 8.2.1 in the Euclidean norm. Now the zero at $\lambda = 1$ corresponding to the eigenvalue -1 vanishes for the second case since the corresponding gradient term is zero.

Theorem 8.2.5: Lower and upper first-order Taylor estimates of $\pi(\lambda; \beta)$

[Appendix to Chapter 8, p. 476]: Let $\pi(\lambda; \beta)$ be the transformed secular function (8.2.17). Suppose that the value λ_C and perturbation δ satisfy $\lambda_C > \lambda_s$ and $\delta > \lambda_s - \lambda_C$. Let $t_1(\delta; \beta) = \pi(\lambda_C; \beta) + \pi^{(1)}(\lambda_C; \beta)\delta$ be the first-order Taylor series approximation to $\pi(\lambda_C + \delta; \beta)$. Then

(i) *for $\beta > 0$,*

$$\pi(\lambda_C + \delta; \beta) \geq t_1(\delta; \beta); \tag{8.2.20}$$

(ii) *otherwise, for $\beta \in [-1/2, 0)$,*

$$\pi(\lambda_C + \delta; \beta) \leq t_1(\delta; \beta). \tag{8.2.21}$$

The inequalities in (8.2.20) and (8.2.21) are strict whenever $g \neq 0$.

We can repeat the discussion following Theorem 8.2.2 concerning $\pi(\lambda)$ for the more general function $\pi(\lambda; \beta)$. For positive β, all we said about solving (8.2.8) remains true for

$$\pi(\lambda; \beta) = \tau(\lambda; \beta). \tag{8.2.22}$$

In particular, if $\lambda_C < \lambda_*$, the largest positive root of

$$t_1(\delta, \beta) = \tau(\lambda_C + \delta; \beta) \tag{8.2.23}$$

leads to underestimators of λ_*, while if $\lambda_* < \lambda_C$, the least-negative root of (8.2.23) gives an estimate to the left of λ_*. If β is negative and $\tau(\lambda; \beta)$ is a given monotonically nonincreasing function on (λ_S, ∞) so that $\pi(\lambda_S; \beta) < \tau(\lambda_S; \beta)$, then the same results are true, but now only so long as β is constrained to be larger than $-1/2$.

Since each permitted β gives rise to a root of (8.2.23) as described, it is natural to ask if there is a particularly advantageous choice of a specific β. In the trust-region case, it is known that $\beta = -1/2$ gives the longest step,

$$\delta^{\mathrm{N}} = \frac{2\pi(\lambda_{\mathrm{C}})}{\pi^{(1)}(\lambda_{\mathrm{C}})}\left(1 - \frac{\|s(\lambda_{\mathrm{C}})\|_M}{\Delta}\right), \tag{8.2.24}$$

in the permitted range $[-1/2, \infty)$ (see Lemma B.8.2) and is thus the natural choice. In the regularization case for which $\tau(\lambda;\beta) = (\lambda/\sigma)^{2\beta/(r-2)}$, it is less obvious. However, it is clear that the function $-\lambda^\gamma$ is convex and decreasing for $\gamma \in (0,1]$ and strictly concave and increasing for $\gamma < 0$ when $\lambda \geq \lambda_{\mathrm{S}}$. Combining this with Corollary 8.2.4, we deduce that $\theta(\lambda;\beta) \stackrel{\mathrm{def}}{=} \pi(\lambda;\beta) - \tau(\lambda;\beta)$ is strictly convex and decreasing for $\beta \in (0,(r-2)/2]$ and strictly concave and increasing for $\beta \in [-1/2, 0)$ when $\lambda \geq \lambda_{\mathrm{S}}$. This then implies that Newton's method, as well as the secant method, applied to $\theta(\lambda;\beta)$ for nonzero $\beta \in [-1/2,(r-2)/2]$ starting from $\lambda_{\mathrm{C}} \in (\lambda_{\mathrm{S}}, \lambda_*)$, converges globally, at a Q-linear rate,[61] and ultimately Q-superlinearly (see Lemma B.8.1) to λ_*. The Newton correction is

$$\begin{aligned}
\delta^{\mathrm{N}}(\lambda_{\mathrm{C}},\beta) &= -\frac{\theta(\lambda_{\mathrm{C}};\beta)}{\theta^{(1)}(\lambda_{\mathrm{C}};\beta)} \\
&= -\frac{[\pi(\lambda_{\mathrm{C}})]^\beta - \left(\frac{\lambda_{\mathrm{C}}}{\sigma}\right)^{2\beta/(r-2)}}{\beta\,[\pi(\lambda_{\mathrm{C}})]^{\beta-1}\,\pi^{(1)}(\lambda_{\mathrm{C}}) - \frac{2\beta}{(r-2)\sigma}\left(\frac{\lambda_{\mathrm{C}}}{\sigma}\right)^{2\beta/(r-2)-1}} \\
&= -\frac{\pi(\lambda_{\mathrm{C}})}{\pi^{(1)}(\lambda_{\mathrm{C}})}\,\frac{1-\mu_{\mathrm{C}}^\beta}{\beta\left(1-\omega_{\mathrm{C}}\mu_{\mathrm{C}}^\beta\right)},
\end{aligned} \tag{8.2.25}$$

$$\text{where } \omega_{\mathrm{C}} = \frac{2\pi(\lambda_{\mathrm{C}})}{(r-2)\lambda_{\mathrm{C}}\pi^{(1)}(\lambda_{\mathrm{C}})} \text{ and } \mu_{\mathrm{C}} = \frac{\left(\frac{\lambda_{\mathrm{C}}}{\sigma}\right)^{2/(r-2)}}{\pi(\lambda_{\mathrm{C}})} < 1.$$

This then offers us the opportunity to pick β to maximize the length of correction within the permitted range $[-1/2,(r-2)/2]$ specifically by solving the problem

$$\underset{\beta\in[-1/2,(r-2)/2]}{\text{maximize}} \quad \eta_{\mathrm{C}}(\beta) \stackrel{\mathrm{def}}{=} \frac{1-\mu_{\mathrm{C}}^\beta}{\beta\left(1-\omega_{\mathrm{C}}\mu_{\mathrm{C}}^\beta\right)} \tag{8.2.26}$$

for the specific data ω_{C} and μ_{C} available at λ_{C}. While there is no obvious closed-form solution to (8.2.26)—indeed the objective may be multimodal—it is possible to solve the problem by iteration. Alternatively, one might simply sample η_{C} at points β within the feasible interval and take the one that gives the largest value of η_C from the sampled points.

8.2.3 Derivatives of the secular function

Finally, if as seems reasonable, we choose to solve the secular equation (8.2.8) or (8.2.16) using a Taylor series or any other derivative-based root-finding method, we need to know how to compute the required derivatives of $\pi(\lambda)$ for $\lambda > \lambda_{\mathrm{S}}$, i.e., when $H + \lambda M$ is positive definite.

---

[61] We refer the reader to Ortega and Rheinboldt [311, Thm. 10.2.2] for the notion of Q-rates of convergence.

Theorem 8.2.6: Derivatives of the secular function

[Appendix to Chapter 8, p. 477]: Suppose that $H + \lambda M$ is positive definite, that $s(\lambda)$ satisfies (8.2.2), and that $s^{(0)}(\lambda) \stackrel{\text{def}}{=} s(\lambda)$ and $\alpha_0 \stackrel{\text{def}}{=} 1$. Then, for $k = 0, 1, \ldots$, the derivatives of $\pi(\lambda) = \langle s(\lambda), M s(\lambda)\rangle$ satisfy

$$\pi^{(2k+1)}(\lambda) = 2\alpha_k \left\langle s^{(k)}(\lambda), M s^{(k+1)}(\lambda)\right\rangle \tag{8.2.27}$$

$$\text{and } \pi^{(2k+2)}(\lambda) = \alpha_{k+1} \left\langle s^{(k+1)}(\lambda), M s^{(k+1)}(\lambda)\right\rangle, \tag{8.2.28}$$

where

$$(H + \lambda M) s^{(k+1)}(\lambda) = -(k+1) M s^{(k)}(\lambda) \tag{8.2.29}$$

and

$$\alpha_{k+1} = 2\frac{(2k+3)}{(k+1)}\alpha_k. \tag{8.2.30}$$

The formulae given rely on being able to solve the system (8.2.29). In practice, we achieve this via a suitable factorization of the positive-definite $H + \lambda M$. If we require the derivatives of the transformed secular function (8.2.17), we simply apply the chain rule, just as we did in (8.2.18) and (8.2.19). We shall return to this in Section 9.1.

8.2.4 Subproblem optimality

We now return to the crucial subproblems (8.1.2) and (8.1.3), and our first step is to characterize their solutions. The trust-region result is well known, but since the regularization case is probably less familiar, especially when $r \neq 3$, we prove it here.

Theorem 8.2.7: Trust-region subproblem global optimality

[178, Thm. 2.1], [292, Lem. 2.1]: Any global minimizer s_* of (8.1.2) satisfies the equation

$$(H + \lambda_* M) s_* = -g, \tag{8.2.31}$$

where $H + \lambda_* M$ is positive semidefinite, $\lambda_* \geq 0$, and

$$\lambda_*(\|s_*\|_M^2 - \Delta^2) = 0. \tag{8.2.32}$$

If $H + \lambda_* M$ is positive definite, then s_* is unique.

Theorem 8.2.8: Regularization subproblem global optimality

[84, Thm. 3.1], [305, Lem. 10] when $r = 3$, [151, Thm. 2.3], [218, Sect. 3] for general $r > 2$]: Any global minimizer s_* of (8.1.3) when $r > 2$ satisfies (8.2.31), where $H + \lambda_* M$ is positive semidefinite and

$$\lambda_* = \sigma \|s_*\|_M^{r-2}. \tag{8.2.33}$$

If $H + \lambda_* M$ is positive definite, then s_* is unique.

Proof. We may express the derivatives of the regularization function $m^{\mathrm{R}}(s,\sigma,r)$ in (8.1.3) as

$$\begin{aligned}\nabla_s^1 m^{\mathrm{R}}(s,\sigma,r) &= g + Hs + \lambda Ms \;\text{ and}\\ \nabla_s^2 m^{\mathrm{R}}(s,\sigma,r) &= H + \lambda M + (r-2)\lambda\left(\frac{Ms}{\|s\|_M}\right)\left(\frac{Ms}{\|s\|_M}\right)^T,\end{aligned} \tag{8.2.34}$$

where $\lambda = \sigma\|s\|_M^{r-2}$.

Let s_* be a global minimizer of $m^{\mathrm{R}}(s,\sigma,r)$. Since $r>2$, $m^{\mathrm{R}}(s,\sigma,r)$ approaches $+\infty$ as s approaches infinity; thus s_* is bounded. It follows from (8.2.34) and the first- and second-order necessary optimality conditions at s_* that

$$g + (H+\lambda_* M)s_* = 0,$$

i.e., that (8.2.31) holds, and that

$$\left\langle w, \left(H + \lambda_* M + (r-2)\lambda_*\left(\frac{Ms_*}{\|s_*\|_M}\right)\left(\frac{Ms_*}{\|s_*\|_M}\right)^T\right) w\right\rangle \geq 0 \tag{8.2.35}$$

for all vectors w.

If $s_* = 0$, then by definition $\lambda_* = 0$, and (8.2.35) is equivalent to H being positive semidefinite, which immediately gives the required result. Thus we need only consider $s_* \neq 0$.

There are two cases to consider. Firstly, suppose that $\langle w, Ms_*\rangle = 0$. In this case, it immediately follows from (8.2.35) that

$$\langle w, (H+\lambda_* M)w\rangle \geq 0 \;\text{ for all }\; w \;\text{ for which }\; \langle w, Ms_*\rangle = 0. \tag{8.2.36}$$

It thus remains to consider vectors w for which $\langle w, Ms_*\rangle \neq 0$. Since Mw and s_* are not orthogonal, the line $s_* + \alpha w$ intersects the ellipsoid $\{s \mid \|s\|_M = \|s_*\|_M\}$ at two points, s_* and

$$u_* \stackrel{\text{def}}{=} s_* - 2\frac{\langle w, Ms_*\rangle}{\langle w, Mw\rangle} w \neq s_*, \tag{8.2.37}$$

and thus

$$\|u_*\|_M = \|s_*\|_M. \tag{8.2.38}$$

We let $w_* = u_* - s_*$ and note that w_* is parallel to w from the definition (8.2.37). Since s_* is a global minimizer, we immediately have that

$$\begin{aligned} 0 &\leq m^{\mathrm{R}}(u_*,\sigma,r) - m^{\mathrm{R}}(s_*,\sigma,r)\\ &= \langle g, u_* - s_*\rangle + \tfrac{1}{2}\langle u_*, Hu_*\rangle - \tfrac{1}{2}\langle s_*, Hs_*\rangle + \tfrac{1}{r}\sigma(\|u_*\|_M^r - \|s_*\|_M^r)\\ &= \langle g, u_* - s_*\rangle + \tfrac{1}{2}\langle u_*, Hu_*\rangle - \tfrac{1}{2}\langle s_*, Hs_*\rangle, \end{aligned} \tag{8.2.39}$$

where the last equality follows from (8.2.38). But (8.2.31) gives that

$$\langle g, u_* - s_*\rangle = \langle s_* - u_*, Hs_*\rangle + \lambda_*\langle s_* - u_*, Ms_*\rangle.$$

In addition, (8.2.37) shows that

$$\langle s_* - u_*, Ms_*\rangle = \tfrac{1}{2}\langle s_*, Ms_*\rangle + \tfrac{1}{2}\langle u_*, Mu_*\rangle - \langle u_*, Ms_*\rangle = \tfrac{1}{2}\langle w_*, Mw_*\rangle. \tag{8.2.40}$$

Thus, combining (8.2.39)–(8.2.40), we find that

$$\begin{aligned} 0 &\leq \tfrac{1}{2}\lambda_*\langle w_*, Mw_*\rangle + \tfrac{1}{2}\langle u_*, Hu_*\rangle - \tfrac{1}{2}\langle s_*, Hs_*\rangle + \langle s_*, Hs_*\rangle - \langle u_*, Hs_*\rangle \\ &= \tfrac{1}{2}\langle w_*, (H + \lambda_* M)w_*\rangle, \end{aligned}$$

from which we deduce that

$$\langle w, (H + \lambda_* M)w\rangle \geq 0 \text{ for all } w \text{ for which } \langle w, Ms_*\rangle \neq 0 \tag{8.2.41}$$

since w_* is parallel to w. Hence (8.2.36) and (8.2.41) together show that $H + \lambda_* M$ is positive semidefinite.

Now suppose that λ_* is a root of the nonlinear equation

$$\|s(\lambda)\|_M^{r-2} - \lambda/\sigma = 0 \tag{8.2.42}$$

for which (8.2.31) holds and that $H + \lambda_* M$ is positive definite and thus $\lambda_* > \lambda_{\rm S}$. But it follows from Corollary 8.2.4 that $\|s(\lambda)\|_M^{r-2}$ strictly decreases from $\|s(\lambda_{\rm S})\|_M^{r-2}$ to zero on $(\lambda_{\rm S}, \infty)$ as $r > 2$, while clearly λ/σ strictly increases from $\lambda_{\rm S}/\sigma$ to infinity on the same interval. Thus (8.2.42) has at most one, and hence precisely one, root λ_* there. The unique solution s_* from (8.2.31) for this λ_* is thus the unique global minimizer of $m^{\rm R}(s, \sigma, r)$. □

8.3 Subproblem solution

8.3.1 Adaptive-regularization methods

Since a whole chapter of *Trust-Region Methods* [131, Chap. 7] has been devoted to methods for solving subproblem (8.1.2), in the interests of variety we shall henceforth concentrate on the less well known regularization problem (8.1.3). In general, essentially identical ideas are used for both, but where appropriate we will comment on any important differences.

Theorem 8.2.8 is effectively constructive. With one notable exception—the analogue, for regularization problems, of the "hard case" of Moré and Sorensen [292] that we will discuss shortly—the multiplier $\lambda_* > \lambda_{\rm S} \stackrel{\rm def}{=} \max(0, -\lambda_1)$ and $\|s_*\|_M = (\lambda_*/\sigma)^{1/(r-2)}$. In this, by contrast, "easy" case, we thus seek the unique root λ_* of the transformed secular equation (8.2.22), involving $\pi(\lambda; \beta)$ from (8.2.17) and $\tau(\lambda; \beta) \stackrel{\rm def}{=} (\lambda/\sigma)^{\beta/(r-2)}$, in the interval $(\lambda_{\rm S}, \infty)$. We are helped in this task by Corollary 8.2.4, which shows that $\pi(\lambda; \beta)$ is convex and strictly decreasing if $\beta > 0$ and concave and strictly increasing for $\beta \in [-1, 0)$, while it is clear that $-\tau(\lambda, \sigma, \beta)$ and hence the sum $\theta(\lambda, \sigma, \beta) \stackrel{\rm def}{=} \theta(\lambda, \sigma, \beta) - \tau(\lambda, \sigma, \beta)$ have the same properties for $\beta \in (0, r-2]$ and $\beta \in [-1, 0)$, respectively. In particular, if we partition the real line into the disjoint sets

$$\begin{aligned} \mathcal{N} &\stackrel{\rm def}{=} \{\lambda \mid \lambda \in (-\infty, \lambda_{\rm S}]\} \equiv \{\lambda \mid H + \lambda M \text{ is not positive definite}\}, \\ \mathcal{L} &\stackrel{\rm def}{=} \{\lambda \mid \lambda \in (\lambda_{\rm S}, \lambda_*]\} \\ &\equiv \{\lambda \mid H + \lambda M \text{ is positive definite and } \|s(\lambda)\|_M \geq (\lambda/\sigma)^{1/(r-2)}\}, \quad \text{and} \\ \mathcal{G} &\stackrel{\rm def}{=} \{\lambda \mid \lambda \in (\lambda_*, \infty)\} \\ &\equiv \{\lambda \mid H + \lambda M \text{ is positive definite and } \|s(\lambda)\|_M < (\lambda/\sigma)^{1/(r-2)}\} \end{aligned}$$

and define $\mathcal{F} \stackrel{\rm def}{=} \mathcal{L} \cup \mathcal{G}$ (see Figure 8.3.1), we have seen in Section 8.2 that a variety

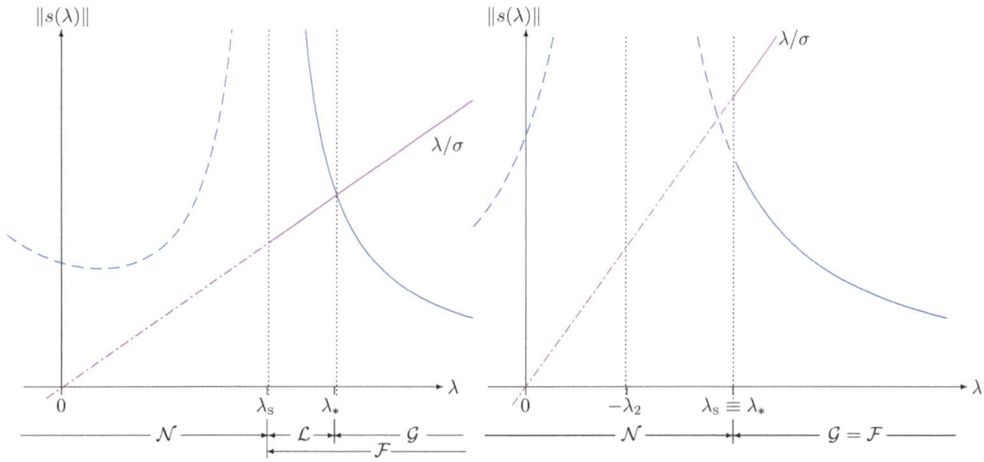

Figure 8.3.1. The sets $\mathcal{N}$, $\mathcal{L}$, $\mathcal{G}$, and $\mathcal{F}$ and $\|s(\lambda)\|$ for the problem of minimizing $-\frac{1}{4}s_1^2 + \frac{1}{4}s_2^2 + \frac{1}{2}s_1 + s_2 + \frac{\sigma}{3}\|x\|^3$ when $\sigma = 0.2$ ("easy" case, left) and those for $-\frac{1}{4}s_1^2 - \frac{1}{8}s_2^2 + s_2 + \frac{\sigma}{3}\|x\|^3$ when $\sigma = 0.1$ ("hard" case, right).

of simple iterations are guaranteed to converge monotonically to λ_* if started in $\mathcal{L}$. Notably, if $\beta = 1$, approximating $\pi(\lambda)$ by any odd-ordered truncated Taylor series t_k and iterating via (8.2.15) will ultimately converge at a Q-kth-order rate (see Section 8.2.1), while for general $\beta \in [-1/2, (r-2)/2]$ the Newton correction (8.2.25), as well as its secant approximation, converges globally Q-linearly and ultimately Q-superlinearly (see Section 8.2.2).[62] Since the dominant cost of any of these iterations is in finding the required derivatives of $\pi(\lambda)$ (see Subsection 8.2.3), one is therefore at liberty to calculate each of the corrections outlined above and to choose the largest. Thus fast ultimate convergence is assured in the easy case, and the main task is to find an initial value in $\mathcal{L}$.

We return to this in Chapter 9, but we summarize this discussion thus far in Algorithm 8.3.1; here we pick integers $i_{\max} \geq 1$ and $j_{\max} \geq 0$, where $k = 2i_{\max} - 1$ is the maximum (odd) degree of the Taylor series we plan to use, and we will provide $j_{\max}$ different values of β in the permitted range $[-1/2, (r-2)/2]$.

Algorithm 8.3.1: Given $\lambda_C \in \mathcal{L}$, estimate the solution λ_* of (8.1.3)

Given initial $\lambda_C \in \mathcal{L}$, $i_{\max} \geq 1$, $j_{\max} \geq 0$, and some small $\epsilon > 0$.
Loop:
Compute $\pi(\lambda_C)$ and its first $2i_{\max}-1$ derivatives as described in Theorem 8.2.6.
For $i = 1, \ldots, i_{\max}$, find

$$\lambda_i^{\mathrm{T}} = \lambda_C + \delta_k^{\mathrm{T}}(\lambda_C),$$

where $\delta_k^{\mathrm{T}}(\lambda_C)$ is the rightmost real root of (8.2.15) for Taylor order $k = 2i-1$.
For $j = 1, \ldots, j_{\max}$, find

$$\lambda_j^{\mathrm{N}} = \lambda_C + \delta^{\mathrm{N}}(\lambda_C, \beta_j),$$

where $\delta^{\mathrm{N}}(\lambda_C, \beta_j)$ is given by (8.2.25) for arbitrary $\beta_j \in [-1/2, (r-2)/2]$.

[62] In addition, finding the required root by linearizing $\pi(\lambda;\beta)$ but not $(\lambda/\sigma)^{\beta/(r-2)}$ is known to outperform Newton's method when $\beta = -1$ and $r \in (2,3]$ [82, Lem. 3.3].

Compute

$$\lambda_+ = \max\left[\max_{1\le i\le i_{\max}} \lambda_i^{\mathrm{T}}, \max_{1\le j\le j_{\max}} \lambda_j^{\mathrm{N}}\right].$$

If $|\lambda_+ - \lambda_{\mathrm{C}}| \le \epsilon$,

exit loop with $\lambda_* \leftarrow \lambda_+$.

Else

set $\lambda_{\mathrm{C}} \leftarrow \lambda_+$.

It is easy to see that Algorithm 8.3.1 is applicable when H is positive semidefinite. There are three possibilities. If $g = 0$, then $s_* = 0$. Otherwise, if g lies in the range of H, which must happen when $\lambda_1 > 0$, but might not if $\lambda_1 = 0$, the equation $Hs = -g$, that is, (8.2.2) with $\lambda = 0$, is consistent and has a minimum-norm solution s_{S}. In this case $\pi(0) = \|s_{\mathrm{S}}\|^2 > 0$, while $\tau(0) = 0$. Thus the required root of (8.2.3) is strictly positive and may be found by the algorithm above. Finally, g does not lie in the range of H, which may be the case if $\lambda_1 = 0$, and $\pi(0) = +\infty$, so that once again the required root of (8.2.3) is strictly positive.

The *hard case* may happen when $\lambda_1 < 0$ and $\langle v_i, g\rangle = 0$ for all i for which $\lambda_i = \lambda_1$, where, as we recall, v_i is the generalized eigenvector of the pencil (H, M) corresponding to the eigenvalue λ_i. Under these circumstances, (8.2.2) will be consistent when $\lambda = \lambda_{\mathrm{S}} = -\lambda_1$—denote the minimum-norm solution in this case by s_{S}, and hence $g+(H+\lambda_{\mathrm{S}}M)s_* = 0$—but if $\|s_{\mathrm{S}}\|_M^{r-2} < \lambda_{\mathrm{S}}/\sigma$, there is no solution to the secular equation in $[\lambda_{\mathrm{S}}, \infty)$ or, equivalently, $\mathcal{L}$ is empty (see Figure 8.3.1). However, since $-\lambda_{\mathrm{S}}$ is a generalized eigenvalue of (H, M), it follows that $(H + \lambda_{\mathrm{S}}M)v_i = 0$ for all eigenvectors v_i corresponding to λ_1, and thus

$$g + (H + \lambda_{\mathrm{S}}M)(s_* + \alpha v_i) = 0$$

for any scalar α. Thus, in order to satisfy the necessary optimality conditions in Theorem 8.2.8, we must additionally choose the scalar $\alpha = \alpha_{\mathrm{S}}$ so that $\|s_{\mathrm{S}} + \alpha_{\mathrm{S}} v_i\|_M = (\lambda_{\mathrm{S}}/\sigma)^{1/(r-2)}$. Notice here that to obtain an exact solution in the hard case requires the eigenvalue λ_1, a corresponding eigenvector v_i of the pencil (H, M), and the "trajectory" vector

$$s_{\mathrm{S}} = \lim_{\lambda\to\lambda_{\mathrm{S}}} s(\lambda) = -(H + \lambda_{\mathrm{S}}M)^+ g,$$

involving the generalized inverse $(H + \lambda_{\mathrm{S}}M)^+$ of $H + \lambda_{\mathrm{S}}M$, from (8.2.2).

We may summarize our discussion as follows:

Corollary 8.3.1: Global minimizers of the regularization subproblem

Suppose that $r > 2$. Then any global minimizer of (8.1.3) may be expressed as

$$s_* = \begin{cases} s_s \text{ (uniquely) if } \lambda_* > -\lambda_1 \text{ or} \\ s_s + \alpha v_1 \text{ if } \lambda_* = -\lambda_1, \end{cases} \tag{8.3.1}$$

where

$$s_s = -(H + \lambda_* M)^+ g, \tag{8.3.2}$$

$\lambda_* = \sigma\|s_*\|_M^{r-2}$, λ_1 is the leftmost eigenvalue of the pencil (H, M), v_1 is any corresponding eigenvector, and α_s is one of the two roots of

$$\|s_s + \alpha v_1\|_M = (\lambda_*/\sigma)^{1/(r-2)}.$$

In all cases, the minimum value of the objective function (8.1.4) is

$$m^{\mathrm{R}}(s_*, \sigma, r) = \tfrac{1}{2}\left(\langle g, s_{\mathrm{S}}\rangle - \frac{(r-2)}{r}\lambda_* \left(\frac{\lambda_*}{\sigma}\right)^{2/(r-2)}\right). \tag{8.3.3}$$

Proof. The identity (8.3.1) follows immediately from the discussion above. It follows from (8.2.31) that

$$\langle s_*, Hs_*\rangle = -\langle g, s_*\rangle - \lambda_*\langle s_*, Ms_*\rangle = -\langle g, s_*\rangle - \lambda_*\|s_*\|_M^2. \tag{8.3.4}$$

Combining this with (8.1.4), (8.2.33), and the fact that $\langle g, s_*\rangle = \langle g, s_{\mathrm{S}}\rangle$ since $\langle g, v_1\rangle = 0$ in the hard case, we find

$$\begin{aligned} m^{\mathrm{R}}(s_*, \sigma, r) &= \langle g, s_*\rangle + \tfrac{1}{2}\langle s_*, Hs_*\rangle + \tfrac{1}{r}\lambda_*\|s_*\|_M^2 \\ &= \tfrac{1}{2}\left(\langle g, s_*\rangle - \frac{(r-2)}{r}\lambda_*\|s_*\|_M^2\right) \\ &= \tfrac{1}{2}\left(\langle g, s_{\mathrm{S}}\rangle - \frac{(r-2)}{r}\lambda_*\left(\frac{\lambda_*}{\sigma}\right)^{2/(r-2)}\right), \end{aligned}$$

which is (8.3.3). □

It is worth noting that the minimum value of the objective function requires the leftmost eigenvalue of the pencil (H, M) in the hard case but does not need a corresponding eigenvector.

8.3.2 Trust-region methods

For the trust-region problem, just as in the regularization case, Theorem 8.2.7 suggests how to find a global minimizer of $m(s)$ within the trust region. Such a minimizer lies strictly within the region only if H is positive semidefinite and $\|H^+g\|_M < \Delta$, while if H is positive semidefinite and $\|H^+g\|_M = \Delta$, the trust-region constraint is active but effectively irrelevant at the minimizer. Otherwise, in all but a similar "hard" case, the multiplier $\lambda_* > \lambda_{\mathrm{S}} \stackrel{\mathrm{def}}{=} \max(0, -\lambda_1)$ and $\|s_*\|_M = \Delta$, and we seek the (unique) root of the transformed secular equation (8.2.22), involving $\pi(\lambda;\beta)$ from (8.2.17) but now $\tau(\lambda;\beta) \stackrel{\mathrm{def}}{=} \Delta^2$ in the interval $(\lambda_{\mathrm{S}}, \infty)$. With $\mathcal{N}$ as before, but redefining

$$\begin{aligned} \mathcal{L} &\stackrel{\mathrm{def}}{=} \{\lambda \mid \lambda \in (\lambda_{\mathrm{S}}, \lambda_*]\} \equiv \{\lambda \mid H + \lambda M \text{ is positive definite and } \|s(\lambda)\|_M \geq \Delta\} \quad \text{and} \\ \mathcal{G} &\stackrel{\mathrm{def}}{=} \{\lambda \mid \lambda \in (\lambda_*, \infty)\} \equiv \{\lambda \mid H + \lambda M \text{ is positive definite and } \|s(\lambda)\|_M < \Delta\}, \end{aligned}$$

we may again partition the real line into disjoint sets, as we illustrate in Figure 8.3.2.

As before, we showed in Section 8.2 that a variety of simple iterations are guaranteed to converge monotonically to λ_* if started in $\mathcal{L}$. In particular, if $\beta = 1$, the iteration based on (8.2.15) will ultimately converge at a Q-kth order rate (see Section 8.2.1), while for general $\beta \in [-1/2, (r-2)/2]$ the Newton correction (8.2.25), as well as its secant approximation, converges globally Q-linearly and ultimately Q-superlinearly, and of these the choice $\beta = -1/2$ is best[63] (see Section 8.2.2). We formalize this in Algorithm 8.3.2.

[63] This formalizes earlier suggestions that it might be wise to consider the secular equation $1/\|s(\lambda)\|_M = 1/\Delta$ with negative β, since this avoids the poles present at $\lambda = -\lambda_i$ when $\beta > 0$ [228, 326].

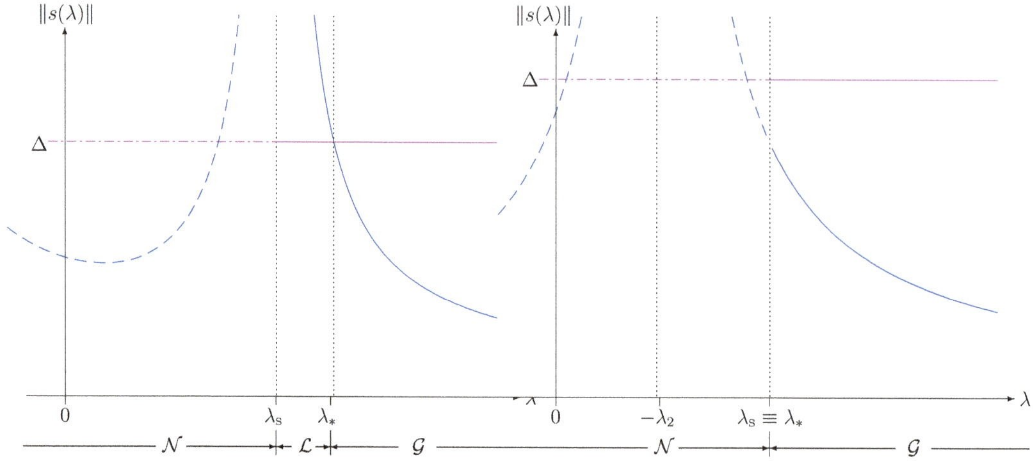

Figure 8.3.2. The sets $\mathcal{N}$, $\mathcal{L}$, and $\mathcal{G}$ and $\|s(\lambda)\|$ for the problem of minimizing $-\frac{1}{4}s_1^2 + \frac{1}{4}s_2^2 + \frac{1}{2}s_1 + s_2$ within an ℓ_2-norm trust region of radius four ("easy" case, left) and those for $-\frac{1}{4}s_1^2 - \frac{1}{8}s_2^2 + s_2$ within a trust region of radius five ("hard" case, right).

With fast ultimate convergence assured in the easy case, the art, as for the regularization subproblem, is in finding an initial $\lambda \in \mathcal{L}$.

Algorithm 8.3.2: Given $\lambda_C \in \mathcal{L}$, estimate the solution λ_* of (8.1.2)

This is identical to Algorithm 8.3.1 except that $j_{\max} = 1$ and $\lambda_1^N = \lambda_C + \delta^N$, where δ^N is given by (8.2.24).

The *hard case* happens under similar circumstances as those that occur for the regularization problem, namely, when $\|s_S\| < \Delta$, there is no solution to the secular equation in $[\lambda_S, \infty)$ when $\lambda_S > 0$ (see Figure 8.3.2). The remedy is the same, namely, that $s_S + \alpha_S v_1$ is a solution to the trust-region subproblem, where the scalar v_i is any eigenvector corresponding to the eigenvalue λ_1 and α_S is chosen so that $\|s_S + \alpha_S v_1\|_M^{r-2} = \Delta$.

Once again, we summarize our findings as follows.

Corollary 8.3.2: Global minimizers of the trust-region subproblem

Any global minimizer of (8.1.2) is given by (8.3.1), where s_S satisfies (8.3.2), $\lambda_* \geq 0$, $\lambda_*(\|s_*\|_M^2 - \Delta^2) = 0$, λ_1 is the leftmost eigenvalue of the pencil (H, M), v_1 is any corresponding eigenvector, and α_S is one of the two roots of

$$\|s_S + \alpha v_1\|_M = \Delta.$$

In all cases, the minimum value of the objective function (8.1.1) is

$$m(s_*) = \tfrac{1}{2}\left(\langle g, s_S\rangle - \lambda_* \Delta^2\right). \tag{8.3.5}$$

Proof. The identity (8.3.1) again follows from the preceding discussion. The identity (8.3.4) holds as before, and combining this with (8.1.1), (8.2.32), and again the fact that $\langle g, s_* \rangle = \langle g, s_\mathrm{S} \rangle$ since $\langle g, v_1 \rangle = 0$ in the hard case, we deduce

$$m(s_*) = \langle g, s_* \rangle + \tfrac{1}{2} \langle s_*, H s_* \rangle = \tfrac{1}{2} \left(\langle g, s_* \rangle - \lambda_* \| s_* \|_M^2 \right) = \tfrac{1}{2} \left(\langle g, s_\mathrm{S} \rangle - \lambda_* \Delta^2 \right),$$

which is (8.3.5). □

Notes and references for Section 8.3

The basis of much of what is summarized in this section first appeared in a far-reaching but unpublished technical report by Griewank [218]. Griewank actually provided a more thorough analysis in which he identified, among other things, bounds on the numbers of local minimizers, saddle points, and maximizers of $m^\mathrm{R}(s, \sigma, r)$. This predates similar results by Martínez [285] and Lucidi et al. [277] for the trust-region subproblem and Hsia et al. [243] in the regularization case.

Chapter 9

Step Computation with Factorizations

Chapter summary

Armed with a better knowledge of the subproblem characteristics, we now build on this knowledge and consider computationally efficient algorithms for the subproblem solution. This chapter focuses on methods using matrix factorizations, which is realistic for problems of moderate size or problems in which sparsity makes this approach feasible.

In this chapter we consider how matrix factorization may be used to help us solve subproblems (8.1.2) and (8.1.3). Our main task will be to design methods that deliver a value $\lambda \in \mathcal{L}$ (if possible) since thereafter Algorithm 8.3.1 or 8.3.2 may be used to find λ_*.

9.1 Matrix factorization

In order to compute $\pi(\lambda)$ and its derivatives, we need to be able to solve the linear systems (8.2.2) and (8.2.29) involving the symmetric matrix $H + \lambda M$. We shall only be concerned with $\lambda \in \mathcal{F}$, since we shall use other techniques to escape from $\mathcal{N}$. Thus $\lambda \geq \lambda_{\mathrm{s}}$, our target matrices are positive semidefinite, and Cholesky (LL^T) or LDL^T factorization (with permutations in the sparse or singular case) is appropriate. That is,

$$H + \lambda M = LL^T \quad \text{or} \quad H + \lambda M = LDL^T, \tag{9.1.1}$$

where L is (row-permuted) lower triangular and D is diagonal with nonnegative entries. However, because a priori we do not know whether $H + \lambda M$ is positive definite, precautions should be in place to report if an indefinite matrix has been encountered (and immediately stop the factorization if this occurs). These features are common to a number of well-known sparse, symmetric linear equation solvers—such methods are generally reliable and effective [187, 197].

Once we have a factorization of the form (9.1.1), it is easy to develop Theorem 8.2.6 into a suitable algorithm for finding the derivatives of $\pi(\lambda)$.

Algorithm 9.1.1: Given $\lambda \in \mathcal{F}$, find $\pi(\lambda)$ and its first $l_{\max}$ derivatives

Given initial $\lambda \in \mathcal{F}$, the factorization $H + \lambda M = LDL^T$, $\alpha_0 \stackrel{\text{def}}{=} 1$, and $l_{\max} > 0$.
Solve $Ly^{(0)}(\lambda) = -g$ to find $y^{(0)}(\lambda)$.
Solve $Dz^{(0)}(\lambda) = y^{(0)}(\lambda)$ to find $z^{(0)}(\lambda)$.
Solve $L^T s^{(0)}(\lambda) = z^{(0)}(\lambda)$ to find $s^{(0)}(\lambda)$.
Set $\pi(\lambda) = \langle s^{(0)}(\lambda), Ms^{(0)}(\lambda)\rangle$.
For $k = 0, \ldots, \lceil l_{\max}/2 \rceil$:
  Set $\beta_k = 2\alpha_k/(k+1)$.
  Solve $Ly^{(k+1)}(\lambda) = (k+1)Ms^{(k)}(\lambda)$ to find $y^{(k+1)}(\lambda)$.
  Solve $Dz^{(k+1)}(\lambda) = y^{(k+1)}(\lambda)$ to find $z^{(k+1)}(\lambda)$.
  Set $\pi^{(2k+1)}(\lambda) = -\beta_k \langle y^{(k+1)}(\lambda), z^{(k+1)}(\lambda)\rangle$.
  If $2k+1 < l_{\max}$:
    Set $\alpha_{k+1} = (2k+3)\beta_k$.
    Solve $L^T s^{(k+1)}(\lambda) = z^{(k+1)}(\lambda)$ to find $s^{(k+1)}(\lambda)$.
    Set $\pi^{(2k+2)}(\lambda) = \alpha_{k+1}\langle s^{(k+1)}(\lambda), Ms^{(k+1)}(\lambda)\rangle$.

This is justified as follows.

Corollary 9.1.1: Derivatives of the secular function via LDL^T factors

[Appendix to Chapter 9, p. 477]: Suppose that $H + \lambda M$ is positive definite with LDL^T factorization $H + \lambda M = LDL^T$. Then Algorithm 9.1.1 provides the first $l_{\max}$ derivatives of $\pi(\lambda)$. The odd-powered derivatives $\pi^{(2k+1)}(\lambda)$ may also be expressed as $2\alpha_k\langle s^{(k)}(\lambda), Ms^{(k+1)}(\lambda)\rangle$.

Notice how each odd-powered derivative of π requires a product with M and solves with L and D, while every even-powered derivative needs a solve with L^T. A slight simplification occurs if a Cholesky rather than an LDL^T factorization is used. In particular, $y^{(k)}(\lambda)$ and $z^{(k)}(\lambda)$ are identical, and the odd-order derivatives become $\pi^{(2k+1)}(\lambda) = -2\beta_k\|y^{(k+1)}(\lambda)\|^2$.

Once we have found LDL^T or Cholesky factors, we can, of course, refine Algorithm 8.3.1 by substituting Algorithm 9.1.1 for the first line in the iteration loop.

9.2 Finding a point in $\mathcal{L}$

The standard method [178, 292] for finding an initial point in $\mathcal{L}$ (if one exists) is to determine bounds $\lambda^{\text{L}} \geq 0$ and λ^{U} for which $\lambda_* \in [\lambda^{\text{L}}, \lambda^{\text{U}}]$. If $\lambda^{\text{U}} > \lambda_*$, then necessarily $\lambda^{\text{U}} \in \mathcal{G}$, while $\lambda^{\text{L}} \in \mathcal{N} \cup \mathcal{L}$. The interval bounds λ^{L} and λ^{U} are adjusted by iteration so that the length of $[\lambda^{\text{L}}, \lambda^{\text{U}}]$ shrinks by a nontrivial amount at each step and thus ultimately will collapse to $\{\lambda_*\}$. If $\mathcal{L}$ is nonempty (the "easy case"), the interval will be adjusted a finite number of times; as soon as a point in $\mathcal{L}$ has been determined, no further adjustments of λ^{L} and λ^{U} are required because subsequent iterates remain in $\mathcal{L}$. The cases when $\mathcal{L}$ is empty (either the "hard case" or when the solution lies interior to the trust region) will be discussed later.

Suppose that we have found an interval $[\lambda^{\text{L}}, \lambda^{\text{U}}]$ surrounding λ_* but that the current estimate $\lambda_{\text{C}} \in [\lambda^{\text{L}}, \lambda^{\text{U}}]$ of λ_* is not in $\mathcal{L}$. If $\lambda_{\text{C}} \in \mathcal{G}$, an improvement λ_+ may be sought by applying one or more iterations of a suitable root finder—Newton's method applied to (8.2.22) for $\beta \geq -1/2$ is guaranteed to overshoot the root (and thus lie in $\mathcal{N} \cup \mathcal{L}$), but other iterations might not; we will return to this later. There are three outcomes. With luck, $\lambda_+ \in \mathcal{L}$ and we are done. Otherwise, we have the opportunity to improve one of the interval bounds: the lower one if $\lambda_+ \in \mathcal{N}$ and the upper if $\lambda_+ \in \mathcal{G}$. If $\lambda_{\text{C}} \in \mathcal{N}$, root-finding is unlikely to be fruitful because we are on the wrong side of the pole of $\|s(\lambda)\|$. In addition—either as a by-product of the root-finding when $\lambda_{\text{C}} \in \mathcal{G}$ or from some auxiliary calculation, for instance, from a suitably chosen Rayleigh quotient, when $\lambda_{\text{C}} \in \mathcal{N}$—we might obtain a new upper bound on λ_1, which may lead to a further improvement in λ^{L}; again we will examine this in detail later. Having refined known lower and upper bounds, it remains to choose a new estimate of λ_* with the goal of ensuring that the bounding interval continues to shrink at an at worst linear rate. Many possibilities have been suggested in the past [131, Sect. 7.3.6], usually involving a convex and/or geometric combination of the current bounds.

We formalize this discussion as Algorithm 9.2.1. The proposed formula for computing the next λ_{C} ensures that the ratio of widths of successive bounding intervals is at most

$$\max\left[1-\theta, \theta, \frac{\gamma\sqrt{\lambda^{\text{U}}}}{\sqrt{\lambda^{\text{L}}}+\sqrt{\lambda^{\text{U}}}}\right]$$

for some $\theta \in (\theta_{\text{L}}, \theta_{\text{U}}) \subset (0,1)$ and $\gamma \in \{0,1\}$ [131, Sect. 7.3.6] and thus that the algorithm has the desired effect of ensuring finite convergence if $\mathcal{L}$ is nonempty.

Algorithm 9.2.1: Find $\lambda \in \mathcal{L}$

Given initial λ^{L} and λ^{U} satisfying

$$\lambda^{\text{L}} \leq \lambda_* \leq \lambda^{\text{U}} \tag{9.2.1}$$

and λ_{C}, set constants $0 < \theta_{\text{L}} \leq \theta_{\text{U}} < 1$.

Loop:

If $\lambda_{\text{C}} \in \mathcal{L}$:

Exit loop with $\lambda \leftarrow \lambda_{\text{C}}$.

Else if $\lambda_{\text{C}} \in \mathcal{N}$:

Set $\lambda^{\text{L}} \leftarrow \max(\lambda^{\text{L}}, \lambda_{\text{C}})$.

Possibly compute an estimate $\lambda_{\text{E}} \geq \lambda_1$, and if so set $\lambda^{\text{L}} \leftarrow \max(\lambda^{\text{L}}, -\lambda_{\text{E}})$.

Select $\theta \in [\theta_{\text{L}}, \theta_{\text{U}}]$ and $\gamma \in \{0,1\}$, and set

$$\lambda_{\text{C}} \leftarrow \max(\gamma\sqrt{\lambda^{\text{L}}\lambda^{\text{U}}}, \lambda^{\text{L}} + \theta(\lambda^{\text{U}} - \lambda^{\text{L}})).$$

Else (i.e., $\lambda_{\text{C}} \in \mathcal{G}$):

Set $\lambda^{\text{U}} \leftarrow \min(\lambda^{\text{U}}, \lambda_{\text{C}})$.

Possibly compute an estimate $\lambda_{\text{E}} \geq \lambda_1$, and if so set $\lambda^{\text{L}} \leftarrow \max(\lambda^{\text{L}}, -\lambda_{\text{E}})$.

Select $\theta_l \leq \theta_u \in [\theta_{\text{L}}, \theta_{\text{U}}]$ and $\gamma_1 \leq \gamma_2 \in \{0,1\}$, and set

$$\lambda_{\text{C}} \in \left[\max\left(\gamma_1\sqrt{\lambda^{\text{L}}\lambda^{\text{U}}}, \lambda^{\text{L}} + \theta_l(\lambda^{\text{U}} - \lambda^{\text{L}})\right), \max\left(\gamma_2\sqrt{\lambda^{\text{L}}\lambda^{\text{U}}}, \lambda^{\text{L}} + \theta_u(\lambda^{\text{U}} - \lambda^{\text{L}})\right)\right]. \tag{9.2.2}$$

We now consider the unspecified aspects of Algorithm 9.2.1 in detail for both the regularization and the trust-region subproblems.

9.3 The regularization subproblem

Once again, we first focus on the missing details for the regularization subproblem.

9.3.1 Initial values for λ^{L} and λ^{U}

To start Algorithm 9.2.1, we require suitable initial values λ^{L} and λ^{U} satisfying (9.2.1). The generalized Rayleigh-quotient inequality (A.3) gives

$$(\lambda_* + \lambda_1)^2 \leq \frac{\langle s_*, (H + \lambda M)M^{-1}(H + \lambda M)s_* \rangle}{\langle s_*, M s_* \rangle} \leq (\lambda_* + \lambda_n)^2, \tag{9.3.1}$$

and it follows from (8.2.31) that

$$\frac{\langle s_*, (H + \lambda M)M^{-1}(H + \lambda M)s_* \rangle}{\langle s_*, M s_* \rangle} = \frac{\sigma^{2/(r-2)} \|g\|^2_{M^{-1}}}{\lambda_*^{2/(r-2)}}. \tag{9.3.2}$$

Thus, combining (9.3.1) and (9.3.2), we deduce that

$$(\lambda_* + \lambda_1)\lambda_*^{1/(r-2)} \leq \sigma^{1/(r-2)} \|g\|_{M^{-1}} \leq (\lambda_* + \lambda_n)\lambda_*^{1/(r-2)}. \tag{9.3.3}$$

For (9.3.3) to be useful, it is necessary to find outer bounds (i.e., a lower bound $\lambda_1^{\text{L}} \leq \lambda_1$ and an upper bound $\lambda_n^{\text{U}} \geq \lambda_n$) on the extreme eigenvalues of the pencil (H, M). Replacing λ_1 and λ_n by outer bounds in (9.3.3) gives simple nonlinear inequalities

$$(\lambda_* + \lambda_1^{\text{L}})\lambda_*^{1/(r-2)} \leq \sigma^{1/(r-2)} \|g\|_{M^{-1}} \leq (\lambda_* + \lambda_n^{\text{U}})\lambda_*^{1/(r-2)} \tag{9.3.4}$$

from which bounds on λ_* may be deduced. Specifically, the function $\eta_\mu(\lambda) = (\lambda + \mu)\lambda^{1/(r-2)}$ monotonically increases when $\lambda \geq \max(0, -\mu)$. Thus if λ^{U} and λ^{L} are, respectively, the rightmost real roots of the nonlinear equations

$$\eta_{\lambda_1^{\text{L}}}(\lambda) = \sigma^{1/(r-2)} \|g\|_{M^{-1}} \quad \text{and} \quad \eta_{\lambda_n^{\text{U}}}(\lambda) = \sigma^{1/(r-2)} \|g\|_{M^{-1}}, \tag{9.3.5}$$

it follows from (9.3.4) that (9.2.1) holds. When $r = 3$, (9.3.5) are quadratic equations, while for other values of r iteration may be required to find the desired roots. We also know that $\lambda_* \geq \lambda_{\text{S}} = \max(0, -\lambda_1)$, so any known upper bound $\lambda_1^{\text{U}} \geq \lambda_1$ may be used.

Usable outer bounds when $M = I$ are normally found from Gershgorin's theorems or computable overestimates of $\|H\|$ such as

$$-\min(\|H\|_\infty, \|H\|_F) \leq -\|H\| \leq \lambda_1 \leq \lambda_n \leq \|H\| \leq \min(\|H\|_\infty, \|H\|_F)$$

involving the infinity norm $\|H\|_\infty$ or the Frobenius norm $\|H\|_F$ [131, Sect. 7.3.8], [178, 292]. For nonunit M, Gershgorin-like methods are also possible so long as M is strictly diagonally dominant. To see this, suppose that $(H - \lambda M)u = 0$ and that k is such that $|u_k| \geq |u_i|$ for $i = 1, \ldots, n$. In this case, $(h_{k,k} - \lambda m_{k,k})u_k = -\sum_{i=1, i\neq k}^n (h_{k,i} - \lambda m_{k,i})u_i$ and thus

$$|h_{k,k} - \lambda m_{k,k}| \leq \left| \sum_{i=1, i\neq k}^{n} (h_{k,i} - \lambda m_{k,i}) \right| \leq o_k^{\text{H}} + |\lambda| o_k^{\text{M}}, \tag{9.3.6}$$

where

$$o_k^{\text{H}} \stackrel{\text{def}}{=} \sum_{i=1, i\neq k}^{n} |h_{k,i}| \quad \text{and} \quad o_k^{\text{M}} \stackrel{\text{def}}{=} \sum_{i=1, i\neq k}^{n} |m_{k,i}|.$$

Hence the eigenvalues of (H, M) lie in the union of the regions defined by

$$|h_{k,k} - \lambda m_{k,k}| \leq o_k^{\text{H}} + |\lambda| o_k^{\text{M}}, \quad k = 1, \ldots, n; \tag{9.3.7}$$

it is easy to see that each region (9.3.7) is a trivially computable closed interval because M is both positive definite and strictly diagonally dominant and thus $m_{k,k} > o_k^{\text{M}}$. Thus outer bounds may easily be found by computing the extrema of each of these interval bounds, giving

$$\lambda_1^{\text{L}} = \min_{1 \leq k \leq n} \left(\frac{h_{k,k} - o_k^{\text{H}}}{m_{k,k} - o_k^{\text{M}}}, \frac{h_{k,k} - o_k^{\text{H}}}{m_{k,k} + o_k^{\text{M}}} \right) \text{ and } \lambda_n^{\text{U}} = \max_{1 \leq k \leq n} \left(\frac{h_{k,k} + o_k^{\text{H}}}{m_{k,k} - o_k^{\text{M}}}, \frac{h_{k,k} + o_k^{\text{H}}}{m_{k,k} + o_k^{\text{M}}} \right).$$

This technique fails if M is not strictly diagonally dominant because then at least one of the sets defined by (9.3.7) may be unbounded; it is not clear to us how to get suitable outer bounds in this case. Note also that the first inequality in (9.3.6) may provide a tighter bound albeit at a slightly higher computational cost. It is also possible to apply optimized diagonal scalings, as suggested by Gay [178], to improve the interval bounds.

In exceptional cases, finding λ_1 may be practicable—for example, if H is tridiagonal and $M = I$, the Lanczos method is a possibility. But usually the cost of computing λ_1 is high, and an upper bound is preferable. Suitable bounds may be deduced from the Rayleigh-quotient inequality (A.3) for specially chosen x. In particular, if H_{S} and M_{S} are symmetric submatrices of H and M (involving the same index sets), if $\lambda_{\text{S}}^{\text{L}}$ is the leftmost eigenvalue of the pencil $(H_{\text{S}}, M_{\text{S}})$ with associated eigenvector u_{S}, then, appropriately padding u_{S} with zeros to obtain a vector $u \in \mathbb{R}^n$, we have that $\lambda_1 \leq \rho_{H,M}(u) = \langle u_{\text{S}}, H_{\text{S}} u_{\text{S}} \rangle / \langle u_{\text{S}}, M_{\text{S}} u_{\text{S}} \rangle = \lambda_{\text{S}}^{\text{L}}$. Thus, for example, considering one-by-one symmetric submatrices gives the bound

$$\lambda_1 \leq \min_{1 \leq i \leq n} h_{i,i} / m_{i,i}.$$

9.3.2 New estimates from $\mathcal{N}$

If $\lambda_{\text{C}} \in \mathcal{N}$, it lies to the left of the rightmost pole of $\pi(\lambda)$ and there is little point in using Taylor-series approximations to try to estimate λ_*. Moreover, since we use the failure of the factorization to identify that $\lambda_{\text{C}} \in \mathcal{N}$, we are unable to use the factors to apply inverse iteration to estimate λ_1—even if we had used an indefinite factorization, there would of course be no guarantee that inverse iteration from an arbitrary $-\lambda_{\text{C}} > \lambda_1$ would converge to the desired eigenvalue. For these reasons alone, and in the absence of any better suggestions, Algorithm 9.2.1 simply aims to reduce the size of the interval $[\lambda^{\text{L}}, \lambda^{\text{U}}]$ as fast as possible.

9.3.3 New estimates from $\mathcal{G}$

Suppose that we have found a $\lambda_{\text{C}} \in \mathcal{G}$ and now wish to find an improvement λ_+. Since $s(\lambda)$ exists and thus the value and derivatives of $\pi(\lambda; \beta)$ may be computed, the obvious idea outlined in Sections 8.2.1 and 8.2.2 is to estimate the required root of (8.2.10) by replacing $\pi(\lambda)$ by its kth-order Taylor approximant $t_k(\delta)$ or, alternatively $\pi(\lambda; \beta)$ in (8.2.16) by its first-order approximant $t_1(\delta, \beta)$. As we mentioned there, every estimate found in this way must lie in $\mathcal{N} \cup \mathcal{L}$, and to encourage the estimate to lie in $\mathcal{L}$, ideally we should pick the largest best prediction $\lambda_+ = \lambda_{\text{I}}$, that is, where

$$\lambda_{\text{I}} \stackrel{\text{def}}{=} \max \left(\lambda_{\text{C}} + \delta_{k_{max}}^{\text{T}}(\lambda_{\text{C}}), \max_{1 \leq j \leq j_{\max}} \lambda_{\text{C}} + \delta^{\text{N}}(\lambda_{\text{C}}, \beta_j) \right) \tag{9.3.8}$$

for given $k_{\max}, j_{\max} \geq 1$ and arbitrary $\beta_k \in [-1/2, (r-2)/2]$, that we can afford; recall that $\delta_k^{\mathrm{T}}(\lambda_{\mathrm{C}})$ is the rightmost real root of (8.2.15) and $\delta^{\mathrm{N}}(\lambda_{\mathrm{C}},\beta)$ is given by (8.2.25). In practice, $k_{\max} = j_{\max} = 3$ with $\beta_1 = -1/2$ and $\beta_2 = (r-2)/2$ and small positive β_3 has proved effective, while larger $k_{\max}$ can be helpful if the cost is reasonable. Note additionally that such a λ_{T} may be used to improve λ^{L}.

Since we cannot be sure that $\lambda_+ \in \mathcal{L}$, we might at the same time try to improve the current lower bound λ^{L} by other means. More specifically, we aim to construct a close upper bound λ_{E} on λ_1. As we already have a factorization of $H + \lambda_{\mathrm{C}} M$, we can reuse this to apply one or more inverse iterations

$$v \leftarrow (H + \lambda_{\mathrm{C}} M)^{-1} M v, \quad v \leftarrow v / \|v\|_M \tag{9.3.9}$$

to estimate the extreme eigenvector of (H, M) and then use the Rayleigh quotient $\lambda_{\mathrm{E}} = \rho_{H,M}(u)$ as our estimate of λ_1. Although the starting vector v used is not critical, it is prudent to use the result from any previous calculation to estimate λ_1 (if any) since this is already an approximation to the desired eigenvector v_1. We consider this next.

9.3.4 Fast convergence in the hard case

We now consider an inverse iteration/Rayleigh quotient–based algorithm for computing approximations to $-\lambda_1$ that are asymptotically greater than $-\lambda_1$. In the easy case, the algorithm generates iterates that will ultimately lie in $\mathcal{L}$, but the main purpose is to cope with the hard case or *near* hard case when $|\mathcal{L}|$ may be quite small.

Before we start, we must mention a small but nonetheless important technicality. Let

$$\mathcal{Z}_1 = \{z \mid \langle z, v_i \rangle \neq 0 \text{ for some } i \text{ such that } \lambda_i = \lambda_1\},$$

where as before (λ_i, v_i) are the eigenpairs of the pencil (H, M). With probability one, any randomly chosen vector lies in $\mathcal{Z}_1$, but of course that offers no deterministic guarantee. The only cast-iron assurance would require that we knew precisely the eigenspace $\mathcal{V}_1 = \{v_i\}$ for all i such that $\lambda_i = \lambda_1$. We might contemplate finding this space using, for instance, the QR algorithm to find the spectral decomposition (8.2.5), when n is small, but it is likely out of the question in the large-dimensional case. Thus, regrettably, here and later, we fall back on randomness. Actually, for the class of methods we shall use, inexact (floating-point) arithmetic may well come to our aid, since even if we are unfortunate enough that our given random starting guess is orthogonal to $\mathcal{V}_1$, small rounding errors are most likely to introduce tiny but growing components in $\mathcal{V}_1$ as the computation proceeds.

We consider the following algorithm.

Algorithm 9.3.1: Potential hard case

Given $\lambda_0^{\mathrm{G}} \in \mathcal{G}$ and $z_0 \in \mathcal{Z}_1$ such that $\|z_0\|_M = 1$, set constants $0 < \omega_{\mathrm{L}} \leq \omega_{\mathrm{U}}$, $0 < \gamma_{\mathrm{U}} \leq 1$, $1 < \gamma_{\mathrm{L}} \leq 2 - \gamma_{\mathrm{U}}$, and $1 \leq n_{\mathrm{U}} \leq \infty$.
For $k = 0$ **until** *converged*
  Choose $1 \leq n_k \leq n_{\mathrm{U}}$. [number of inverse iterations]
  Initialize $w_0 = z_k$.
  For $i = 1 : n_k$: [inverse iteration]
    Set $w_i = (H + \lambda_k^{\mathrm{G}} M)^{-1} M w_{i-1}$ and normalize $w_i \leftarrow w_i / \|w_i\|_M$.
  Set $z_{k+1} = w_{n_k}$ and compute $\rho_{H,M}(z_{k+1}) = \langle z_{k+1}, H z_{k+1} \rangle$. [Rayleigh quotient]
  Choose $\omega_k \in [\omega_{\mathrm{L}}, \omega_{\mathrm{U}}]$ and $\gamma_k \in [\gamma_{\mathrm{L}}, 2n_k - \gamma_{\mathrm{U}}]$.
  Set $\lambda_{k+1}^{\mathrm{G}} = -\rho_{H,M}(z_{k+1}) + \omega_k (\lambda_k^{\mathrm{G}} + \rho_{H,M}(z_{k+1}))^{\gamma_k}$.

Algorithm 9.3.1 assumes that an initial estimate $\lambda_0^{\mathrm{G}} \in \mathcal{G}$ is available and thus that $\lambda_0^{\mathrm{G}} > -\lambda_1$. The algorithm then repeats the following steps until convergence. Firstly, a positive integer n_k is chosen which represents the number of inverse iterations that will be performed. Next, the requested number of inverse iterations are computed using $-\lambda_k^{\mathrm{G}}$ as the fixed estimate of λ_1, resulting in a new best approximation z_{k+1} to the extreme eigenvector v_1—for simplicity, in this discussion we presume that λ_1 has algebraic multiplicity one, although the theoretical justification we provide does not require this. The generalized Rayleigh quotient $\rho_{H,M}(z_{k+1})$ (p. 432) is then computed to (over)estimate λ_1. Values $\omega_k \in [\omega_{\mathrm{L}}, \omega_{\mathrm{U}}]$ and $\gamma_k \in [\gamma_{\mathrm{L}}, 2n_k - \gamma_{\mathrm{U}}]$ are now assigned; the restrictions imposed by ω_{L} and $2n_k - \gamma_{\mathrm{U}}$ are required to guarantee (ultimately) that $\lambda_k^{\mathrm{G}} > -\lambda_1$, while the restrictions $\omega_k \leq \omega_{\mathrm{U}}$ and $\gamma_k \geq \gamma_{\mathrm{L}} > 1$ are needed to ensure that the sequence $\{\lambda_k^{\mathrm{G}}\}$ is monotonically decreasing [196, Lem. 5]. Using these, an improved estimate $\lambda_{k+1}^{\mathrm{G}}$ of $-\lambda_1$ is computed with the aim of being greater than $-\lambda_1$ and thus in $\mathcal{F}$; this is in contrast to the negative of the Rayleigh-quotient estimate $-\rho_{H,M}(z_{k+1})$, which is *always* less than $-\lambda_1$ and thus lies in the $\mathcal{N}$ region. The exact form of the correction is chosen so that it does not interfere with the superlinear convergence of the Rayleigh-quotient iteration but at the same time is ultimately large enough to ensure that that iteration converges from $\mathcal{G}$ in the hard case. This process is then repeated.

Numerous standard results relating to both inverse iteration and the Rayleigh quotient have been established [146, 315, 351]; for simplicity, these results typically assume that the eigenvalue for which convergence occurs is simple and that $M = I$. Fortunately, such guarantees may be extended to the generalized eigenvalue problem and account for the possibility that eigenvalues may not be simple.

Theorem 9.3.1: Fast convergence in the hard case

[196, Thm. 7]:[64] *Let λ_1 be the leftmost eigenvalue of the pencil (H, M). Suppose that there exist positive constants $\bar{\gamma}$ and $\bar{n}$ such that $\gamma_k = \bar{\gamma}$ and $n_k = \bar{n}$ for all k sufficiently large. Then the sequence $\{\lambda_k^{\mathrm{G}}\}$ generated by Algorithm 9.3.1 converges to $-\lambda_1$ with Q-rate equal to $\bar{\gamma}$ provided that λ_0^{G} is chosen sufficiently close to $-\lambda_1$.*

This result essentially says that we can obtain any Q-order convergence we wish at the expense of performing an ever-increasing number of inverse iterations. More precisely, we can obtain the Q-convergence "goal" $\bar{\gamma}$ by setting $\gamma_k = \bar{\gamma}$ and consequently choosing n_k, the number of inverse iterations performed, to satisfy $2n_k - \gamma_{\mathrm{U}} > \bar{\gamma}$ (this should be done for all k sufficiently large). For example, we could obtain superlinear convergence by ultimately setting $\gamma_k = 1.5$ and by performing a single inverse iteration ($n_k = 1$), or we could obtain supercubic convergence by ultimately setting $\gamma_k = 3.5$ and by performing two steps of inverse iteration ($n_k = 2$).

A reasonable implementation would be to use Algorithm 9.3.1 once $\lambda^{\mathrm{U}} - \lambda^{\mathrm{L}}$ is relatively small. In the easy case, Algorithm 9.3.1 will quickly produce an iterate that lands in the $\mathcal{L}$-region, while in the hard case the algorithm converges rapidly to $-\lambda_1 = \lambda_s$. We also note that the algorithm produces iterates z_k that approximate an eigenspace associated with λ_1, which is required for computing a solution to the regularization problem (8.1.3) in the hard case. Initially, using a single step of inverse iteration with $\gamma_k \approx 1.5$ is reasonable. In general, this will force λ_+ into $\mathcal{L}$ rapidly, with subsequent fast convergence via Algorithm 8.3.1, as described in Section 8.3. If the hard case is suspected, it may be wise to increase γ_k from 1.5 to 3 and to perform two steps of inverse iteration, but

[64] The argument in [196] erroneously omits the assumption that $z_0 \in \mathcal{Z}_1$ but is otherwise correct.

only after the value $\gamma_k = 1.5$ has been "successful"; the resultant cubic convergence is sufficient for all practical purposes.

Formally, if $\lambda_{\rm C} \in \mathcal{G}$, we compute a new estimate λ_+ as follows. Let $\lambda_{\rm I}$ be the underestimate (9.3.8), let $\lambda_{\rm E}$ be the closest upper bound of λ_1 found so far, let $\bar{\gamma}$ be the desired order of convergence in the hard case, and let $\theta_{\rm G}$ and $\theta_{\rm H}$ be given constants in $(0,1)$ and $\omega_{\rm L} > 0$—for example, in practice we use $\theta_{\rm G} = 0.5 = \theta_{\rm H}$ and $\omega_{\rm L} = 1$. Then whenever $|\lambda_{\rm C} + \lambda_{\rm E}| \leq \theta_{\rm H}\lambda_{\rm C}$, we suspect we might be in the hard case. In this case, we use the Rayleigh quotient $\rho_{H,M}(z_{k+1})$ to try to improve $\lambda_{\rm E}$ and $\lambda^{\rm L}$ by assigning $\lambda_{\rm E} \leftarrow \min(\lambda_{\rm E}, \rho_{H,M}(z_{k+1}))$ and $\lambda^{\rm L} \leftarrow \max(\lambda^{\rm L}, -\lambda_{\rm E})$ and subsequently set

$$\lambda_+ = \min\left(\lambda^{\rm L} + \theta_{\rm G}(\lambda_{\rm C} - \lambda^{\rm L}), \max\left(\lambda_{\rm I}, \lambda^{\rm L}, -\lambda_{\rm E} + \omega_{\rm L}\left(\frac{\lambda_{\rm C} + \lambda_{\rm E}}{\lambda_{\rm C}}\right)^{\bar{\gamma}}\right)\right) \tag{9.3.10}$$

so as to mimic Algorithm 9.3.1. Otherwise, if $\lambda_{\rm I} \geq \lambda^{\rm L}$, the value (9.3.8) gives the best estimate in $\mathcal{F}$ found so far, and we assign

$$\lambda_+ = \lambda_{\rm I}. \tag{9.3.11}$$

If neither happens, we simply revert to improving the interval of uncertainty by setting

$$\lambda_+ = \lambda^{\rm L} + \theta_{\rm G}(\lambda_{\rm C} - \lambda^{\rm L}). \tag{9.3.12}$$

9.3.5 Initial values and sequences of related problems

In the absence of better initial information, we recommend choosing the initial λ as

$$\lambda_{\rm C} = \max\left(\gamma\sqrt{\lambda^{\rm L}\lambda^{\rm U}}, \lambda^{\rm L} + \theta(\lambda^{\rm U} - \lambda^{\rm L})\right) \tag{9.3.13}$$

for some $\theta \in (0,1)$ and $\gamma \in \{0,1\}$ (cf. Algorithm 9.2.1).

Of course, the main purpose of solving (8.1.3) is to find a step within one of the many regularization methods discussed elsewhere in the book. For these, a sequence of problems of the form (8.1.3) will be solved. There will generally be two possibilities. In the first, the data H and g will be unchanged, but the regularization parameter σ will be increased to σ_+. Consider the usual sets $\mathcal{G}$, $\mathcal{L}$, and $\mathcal{N}$ with respect to σ, and let $\mathcal{G}_+$, $\mathcal{L}_+$, and $\mathcal{N}_+$ be their analogues with respect to σ_+. In this case, it is clear that $\mathcal{N}_+ = \mathcal{N}$ and, more importantly, $\mathcal{L} \subseteq \mathcal{L}_+$. If the previous $\lambda_- \in \mathcal{L} \cup \{\lambda_{\rm S}\}$, then $\lambda_- \in \mathcal{L}_+$ and is likely a good starting point for the new problem. Potentially better, λ may have been sampled at points $\lambda_+ > \lambda_-$ when solving the previous problem, and corresponding values of $\pi(\lambda_+)$ will be known. Thus finding the largest previous λ_+ for which $\pi(\lambda_+) \geq (\lambda_+/\sigma_+)^{2/(r-2)}$ will also give a value in $\mathcal{L}_+$. The other possibility is that H and g might have changed but $\sigma_+ \leq \sigma$. Little useful information is then available, but as a heuristic, starting from λ_- is a possibility; if small changes to H and g have occurred, it is likely that $\lambda_- \in \mathcal{G}_+$.

9.3.6 Early termination

We have focused so far on finding a solution to (8.1.3), but in practice all that is required is that

$$m^{\rm R}(s,\sigma,r) \leq m^{\rm R}(0,\sigma,r) = 0, \tag{9.3.14a}$$

$$\|\nabla_s^1 m^{\rm R}(s,\sigma,r)\|_{M^{-1}} \leq \tfrac{1}{2}\theta_1\|s\|_M^2, \text{ and} \tag{9.3.14b}$$

$$\lambda_{\min}[\nabla_s^2 m^{\rm R}(s,\sigma,r)] \geq -\theta_2\|s\|_M \tag{9.3.14c}$$

for given constants θ_1, $\theta_2 > 0$, where $m^{\mathrm{R}}(s,\sigma,r)$, $\nabla_s^1 m^{\mathrm{R}}(s,\sigma,r)$, and $\nabla_s^2 m^{\mathrm{R}}(s,\sigma,r)$ are as given in (8.1.4) and (8.2.34)—see, for example, requirements (3.3.5) and (3.3.6) in Step 2 of Algorithm 3.3.1, with the appropriate M-scalings of the primal and dual norms. We know from Theorem 8.2.8 that these conditions are all satisfied when $s = s_*$ and thus by continuity within some neighbourhood of s_*. We may quantify this as follows.

Theorem 9.3.2: Constructive early termination test

[Appendix to Chapter 9, p. 477]: Let $s(\lambda)$ satisfy (8.2.2) and $\pi(\lambda) = \|s(\lambda)\|_M^2$. Suppose that $\lambda_{\mathrm{C}} \in \mathcal{F}$ satisfies the conditions

$$\pi(\lambda_{\mathrm{C}}) \leq \left(\frac{r}{2}\right)^{2/(r-2)} \left(\frac{\lambda_{\mathrm{C}}}{\sigma}\right)^{2/(r-2)}, \tag{9.3.15a}$$

$$\left| [\pi(\lambda_{\mathrm{C}})]^{(r-2)/2} - \frac{\lambda_{\mathrm{C}}}{\sigma} \right| \leq \frac{\theta_1}{2\sigma} [\pi(\lambda_{\mathrm{C}})]^{1/2}, \quad \textit{and} \tag{9.3.15b}$$

$$[\pi(\lambda_{\mathrm{C}})]^{(r-2)/2} - \frac{\lambda_{\mathrm{C}}}{\sigma} \geq -\frac{\theta_2}{\sigma} [\pi(\lambda_{\mathrm{C}})]^{1/2}. \tag{9.3.15c}$$

Then $s = s(\lambda_{\mathrm{C}})$ satisfies the stopping rules (9.3.14).

Clearly $\lambda_{\mathrm{C}} = \lambda_* = \sigma\|s(\lambda_*)\|_M^{r-2}$ satisfies (9.3.15), as do a range of λ's to either side of λ_*. For example, when $r = 3$ (cubic regularization), it is easy to show that (9.3.15) is equivalent to

$$\begin{gathered} \eta^{\mathrm{L}}(\sigma,\theta_1,\theta_2)\frac{\lambda_{\mathrm{C}}}{\sigma} \leq \|s(\lambda_{\mathrm{C}})\| \leq \eta^{\mathrm{U}}(\sigma,\theta_1)\frac{\lambda_{\mathrm{C}}}{\sigma}, \quad \text{where} \\ \eta^{\mathrm{L}}(\sigma,\theta_1,\theta_2) \stackrel{\mathrm{def}}{=} \max\left(\frac{1}{(1+\frac{\theta_1}{2\sigma})}, \frac{1}{(1+\frac{\theta_2}{\sigma})}\right) \quad \text{and} \quad \eta^{\mathrm{U}}(\sigma,\theta_1) \stackrel{\mathrm{def}}{=} \min\left(\frac{3}{2}, \frac{1}{(1-\frac{\theta_1}{2\sigma})_+}\right) \end{gathered} \tag{9.3.16}$$

and the subscript + means that the term is absent if it is negative; we illustrate this in Figure 9.3.1. In practice, checking (9.3.15) is trivial, and Algorithms 8.3.1 and 9.2.1 may be terminated as soon as these conditions are satisfied.

9.3.7 Summary

In summary, a complete algorithm may be formulated as follows. First choose λ^{L} and λ^{U} as in Section 9.3.1, and pick the initial estimate $\lambda_{\mathrm{C}} = 0$ via (9.3.13). Now use Algorithm 9.2.1 to try to find a $\lambda \in \mathcal{L}$, choosing the new estimate when $\lambda_{\mathrm{C}} \in \mathcal{G}$ by projecting λ_+ from (9.3.10)–(9.3.12) as appropriate into the interval (9.2.2) and finding approximations to λ_{E} as mentioned in Sections 9.3.3 and 9.3.2. If Algorithm 9.2.1 succeeds in finding a $\lambda_{\mathrm{C}} \in \mathcal{L}$, use this as input to Algorithm 8.3.1 to find an approximation to λ_*, stopping as soon as (9.3.15) holds. Otherwise, if (9.3.15) holds for some $\lambda \in \mathcal{G}$, stop. Finally, if $\lambda^{\mathrm{U}} - \lambda^{\mathrm{L}} < \epsilon$ for some specified small $\epsilon > 0$ in Algorithm 9.2.1, terminate with the approximate hard-case estimate $\lambda_* = \frac{1}{2}(\lambda^{\mathrm{L}} + \lambda^{\mathrm{U}})$.

Notes and references for Subsection 9.3.7

The bulk of this material is based on [196], but with antecedents from [178, 292]. Fortran and MATLAB software that implements this is available as the package RQS as part of GALAHAD [192].

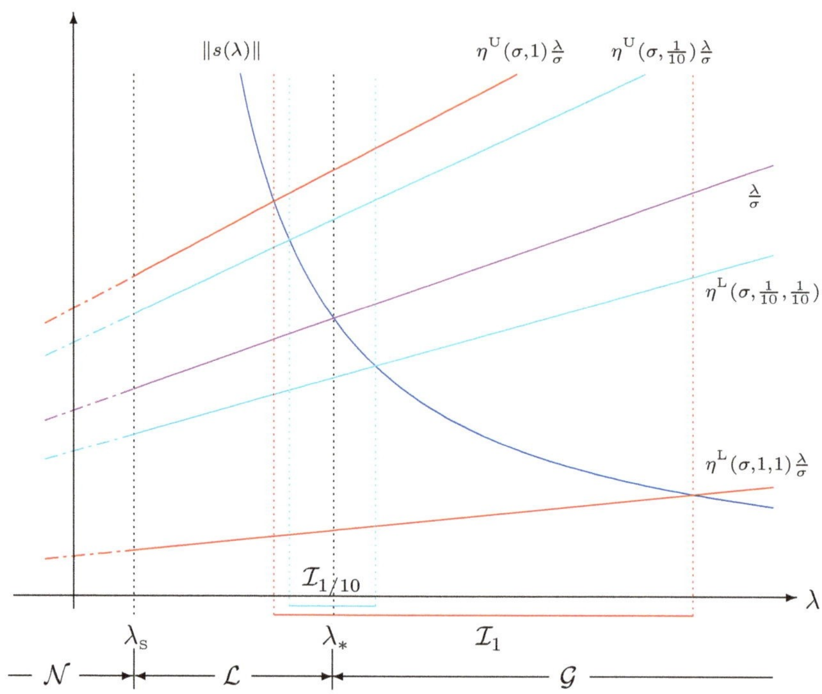

Figure 9.3.1. The intervals of acceptable λ_C that satisfy (9.3.16) for the problem of minimizing $-\frac{1}{4}s_1^2 + \frac{1}{4}s_2^2 + \frac{1}{2}s_1 + s_2 + \frac{\sigma}{3}\|x\|^3$ when $\sigma = 0.2$ for $\theta_1 = \theta_2 = 1/10$ ($\mathcal{I}_{1/10}$, cyan) and $\theta_1 = \theta_2 = 1$ ($\mathcal{I}_1$, red).

An alternative to inverse iteration given by (9.3.9) is to use the so-called LINPACK technique to provide a suitable estimate of v_1; this is a vital component of Moré and Sorensen's [292] method and meshes well with their goal to provide a usable, low-accuracy solution to (8.1.2) but is equally applicable to (8.1.3). The LINPACK technique depends crucially on the ability to intercept and alter components of the solution to linear systems involving the Cholesky or LDL^T factors of $H + \lambda_C M$ and specially crafted right-hand sides as the solution process proceeds [119]. Unfortunately, this limits its utility for modern sparse-factorization packages, where the factors are generally hidden from the user.

Another possibility is to use the method at the heart of the LAPACK condition-number estimator [222,233] to provide a lower bound λ_H, say, on $\|(\overline{H}+\lambda_C I)^{-1}\|_1$, where $\overline{H} \stackrel{\text{def}}{=} M^{-T/2}HM^{-1/2}$. We may then use the inequalities

$$\frac{1}{\lambda + \lambda_C} \geq \|(\overline{H} + \lambda_C I)^{-1}\|_2 \geq \frac{\|(\overline{H} + \lambda_C I)^{-1}\|_1}{\sqrt{n}} \geq \frac{\lambda_H}{\sqrt{n}}$$

to derive the upper bound

$$\lambda_1 \leq \sqrt{n}/\lambda_H - \lambda_C$$

on λ_1. The LAPACK method, like the LINPACK technique, uses multiple applications of the Cholesky or LDL^T factors of $H + \lambda_C M$ to provide the estimate λ_H. This method has been used successfully in the trust-region context [179, Sect. 5].

As Gay [178] points out, when $\lambda_C \in \mathcal{N}$, an LDL^T or Cholesky factorization will continue so long as the leading submatrix of $H + \lambda_C M$ is positive definite, and the factors generated up until this point may be used to improve the upper bound on λ_1. Specifically, if failure first occurs when factorizing the k by k submatrix of $H + \lambda_C M$, this submatrix may be factorized as $L_k D_k L_k^T$, where L_k is k by k unit lower triangular and $D_k = \text{diag}(d_{i,i})$, $1 \leq i \leq k$, with $d_{k,k} \leq 0$. If y_k satisfies $L_k^T y_k = e_k$ and $w_k = (y_k^T \; 0)^T$, it follows that

$$\langle w_k, (H + \lambda_C M) w_k \rangle = \langle y_k, L_k D_k L_k^T y_k \rangle = d_{k,k} \leq 0,$$

and hence

$$\rho_{H,M}(w_k) = \frac{d_{k,k}}{\langle w_k, M w_k \rangle} - \lambda_c$$

provides an upper bound on λ_1 which may be used to improve λ^{L}. In practice, a partial factorization of $H + \lambda_C M$—or more specific the means to find y_k—is not always easy to recover from sophisticated sparse-factorization packages, so this somewhat limits this possibility.

9.4 The trust-region subproblem

Much of what we described in the previous section may easily be adapted to solve the trust-region subproblem (8.1.2). The main differences result from our need to solve the secular equation $\|s(\lambda)\|_M = \Delta$ rather than (8.2.42), but additionally there is the possibility of an "interior solution" to worry about.

9.4.1 The basic algorithm

To find a point in $\mathcal{L}$ for the trust-region subproblem, Algorithm 9.2.1 is still applicable. Finding initial values for λ^{L} and λ^{U} is slightly different. Instead of (9.3.2), we have that

$$\frac{\langle s_*, (H + \lambda M) M^{-1} (H + \lambda M) s_* \rangle}{\langle s_*, M s_* \rangle} = \frac{\|g\|_{M^{-1}}^2}{\Delta^2}$$

from (8.2.31) for solutions on the trust-region boundary, and combining this with the Rayleigh-quotient bound (9.3.1), we deduce that

$$\frac{\|g\|_{M^{-1}}}{\Delta} - \lambda_n \leq \lambda_* \leq \frac{\|g\|_{M^{-1}}}{\Delta} - \lambda_1. \tag{9.4.1}$$

Unlike in the regularization case, (9.4.1) directly provides suitable λ^{L} and λ^{U} so long as outer bounds $\lambda_1^{\text{L}} \leq \lambda_1$ and $\lambda_n^{\text{U}} \geq \lambda_n$ are available because then clearly

$$\frac{\|g\|_{M^{-1}}}{\Delta} - \lambda_n^{\text{U}} \leq \lambda_* \leq \frac{\|g\|_{M^{-1}}}{\Delta} - \lambda_1^{\text{L}}.$$

Outer bounds, such as that provided by the generalization of the Gershgorin theorems described in Section 9.3.1, may be used. In addition, as before, the lower bound $\lambda_* \geq \lambda_{\text{S}} = \max(0, -\lambda_1)$ may be employed, so any known upper bound on λ_1 is useful. Inverse iteration/Rayleigh quotient–based estimates as described in Sections 9.3.3 and 9.3.4 are appropriate.

Algorithm 9.2.1 is just as appropriate in the trust-region case as it was for regularization, and should a $\lambda_{\text{C}} \in \mathcal{L}$ arise, Algorithm 8.3.2 may be applied to find an excellent approximation to λ_*. As we mentioned in Section 9.3.2, there is little useful information available to guide us in the choice of an improved step from a $\lambda_{\text{C}} \in \mathcal{N}$, so once again, we are simply content to construct an improvement with the aim of rapidly reducing the bounding interval $[\lambda^{\text{L}}, \lambda^{\text{U}}]$. The only real issue is how to compute improvements from a given λ_{C} in $\mathcal{G}$.

Just as in Section 9.3.3, we follow the suggestions in Sections 8.2.1 and 8.2.2, namely, to replace $\pi(\lambda)$ in the equation $\pi(\lambda) = \Delta^2$ by its kth-order Taylor approximant $t_k(\delta)$ or, alternatively, $\pi(\lambda;\beta)$ in $\pi(\lambda;\beta) = \Delta^{2\beta}$ by its first-order approximant $t_1(\delta,\beta)$. But now, rather than using the estimate (9.3.8) that was appropriate in the regularization case, we choose $\lambda_+ = \lambda_{\text{I}}$, where

$$\lambda_{\text{I}} \stackrel{\text{def}}{=} \max\left(\lambda_{\text{C}} + \delta^{\text{T}}_{k_{max}}(\lambda_{\text{C}}), \lambda_{\text{C}} + \delta^{\text{N}}\right)$$

for the largest $k_{\max}$ that we can afford and $\delta^{\text{T}}_k(\lambda_{\text{C}})$ is the rightmost real root of $t_k(\delta) = \Delta^2$, since as we saw in Section 8.2.2, δ^{N} given by (8.2.24) gives the best Newton step as a function of $\beta \in [1/2, \infty)$ (see Lemma B.8.2).

The one remaining issue is when, if at all, to test for the possibility that the solution to (8.1.2) lies interior to the trust region, and thus that the required $\lambda_* = 0$. Clearly this is impossible if $\lambda^{\text{L}} > 0$, so any investigation should be delayed until the initial λ^{L} has been computed [292]. In the absence of better initial information, we simply choose the initial λ as 0 if $\lambda^{\text{L}} = 0$ and as (9.3.13) otherwise.

9.4.2 Early termination

Of course, we may not need to solve the subproblem exactly; an approximation may suffice. A particular case in point is the calculation of the criticality measure $\phi^{\delta}_{f,j}(x)$ in (12.1.10), especially in the case when $j = 2$ for which

$$\phi^{\Delta}_{f,2}(x) = f(x) - \min_{\|s\|\leq\Delta} T_{f,2}(x,s) = -\min_{\|s\|\leq\Delta} m(s),$$

where $m(s)$ is the quadratic model (8.1.1) for the particular data $g = \nabla^1_x f(x)$ and $H = \nabla^2_x f(x)$. Such a measure plays two roles in algorithms such as TRq (p. 292), ARqp (p. 300), and their successors. The first is simply to check whether

$$\phi^{\Delta}_{f,2}(x) \leq \phi^{\text{L}} \text{ or } \phi^{\Delta}_{f,2}(x) \geq \phi^{\text{U}} \tag{9.4.2}$$

for given positive ϕ^{L} and ϕ^{U}; for example, this is all that is required in Step 1 of the ARqp algorithm. The second aim is to verify (9.4.2) but at the same time to provide an exemplar s for which $m(s)$ satisfies the required bound; for instance, this is needed in Step 1 of TRq and Step 2 of ARqp.

Fortunately, we can achieve these goals without the need for global minimization. We return to the familiar, but slightly more general, problem (8.1.2), and we write

$$m_\Delta \stackrel{\text{def}}{=} \min_{\|s\|_M\leq\Delta} m(s) \leq 0$$

and, in view of (9.4.2), ask whether

$$m_\Delta \leq m^{\text{L}} \tag{9.4.3}$$

or

$$m_\Delta \geq m^{\text{U}} \tag{9.4.4}$$

for given negative m^{L} and m^{U}. Let

$$s_\Delta \stackrel{\text{def}}{=} \arg\min_{\|s\|_M \le \Delta} m(s).$$

Then clearly

$$m_{\Delta_2} \le m_{\Delta_1} \text{ if and only if } \Delta_2 \ge \Delta_1, \tag{9.4.5}$$

since the feasible region of the subproblem enlarges as Δ increases.

When H is positive definite and $\|H^{-1}g\|_M \le \Delta$, then plainly $s_\Delta = -H^{-1}g$ and $m_\Delta = -\frac{1}{2}\langle g, H^{-1}g\rangle$. Otherwise, we know from Corollary 8.3.2 that $\|s_\Delta\|_M = \Delta$. Let $\Delta^+ > \Delta$, in which case (9.4.5) implies that $m(s_{\Delta^+}) \equiv m_{\Delta^+} \le m_\Delta$. Furthermore, since $\|s_{\Delta^+}\|_M > \Delta$, there is an $\alpha^+ = \Delta/\Delta^+ \in (0,1)$ for which $\|\alpha^+ s_{\Delta^+}\|_M = \Delta$, and thus $m(\alpha^+ s_{\Delta^+}) \ge m(s_\Delta) \equiv m_\Delta$. Hence

$$m_{\Delta^+} \le m_\Delta \le m(\alpha^+ s_{\Delta^+}), \tag{9.4.6}$$

which provides both lower and upper bounds on m_Δ. In particular, (9.4.3) holds if $m(\alpha^+ s_{\Delta^+}) \le m^{\mathrm{L}}$ and (9.4.4) follows when $m_{\Delta^+} \ge m^{\mathrm{U}}$. Similarly, (9.4.5) implies that $m(s_{\Delta^-}) \equiv m_{\Delta^-} \ge m_\Delta$ if $\Delta^- < \Delta$, and it follows that

$$m_\Delta \le \min[m(\alpha^- s_{\Delta^-}), m_{\Delta^-}], \tag{9.4.7}$$

where $\alpha^- = \Delta/\Delta^- > 1$, and thus (9.4.3) results whenever $\min[m(\alpha^- s_{\Delta^-}), m_{\Delta^-}] \le m^{\mathrm{L}}$.

The bounds (9.4.6) and (9.4.7) require that we can find suitable s_{Δ^+} and s_{Δ^-}. To do so, we recall Section 8.3.2 and particularly the sets $\mathcal{L}$ and $\mathcal{G}$ defined there. Let $s(\lambda)$ be the solution of the linear system (8.2.2), and define $\Delta(\lambda) \stackrel{\text{def}}{=} \|s(\lambda)\|_M$. If $\lambda^+ \in \mathcal{L}$, $\Delta^+ = \Delta(\lambda^+) > \Delta$ and is thus suitable. Likewise, if $\lambda^- \in \mathcal{G}$, $\Delta^- = \Delta(\lambda^-) \le \Delta$ is apt. Moreover, we recall from the discussion in Section 9.4.1 that Algorithm 9.2.1 will almost certainly[65] find a $\lambda \in \mathcal{L}$, possibly finding a sequence of decreasing $\lambda \in \mathcal{G}$ along the way, while Algorithm 8.3.2 will monotonically (and ultimately rapidly) improve $\lambda \in \mathcal{L}$ toward λ_* for which $\|s(\lambda_*)\|_M = \Delta$. In addition, the interval (9.4.6) will become ever tighter. In passing, it is worth noting that, given $\lambda \in \mathcal{L} \cup \mathcal{G}$,

$$m_{\Delta(\lambda)} = \tfrac{1}{2}\left(\langle g, s(\lambda)\rangle - \lambda\Delta^2(\lambda)\right),$$

because of (8.3.5).

In summary, we may apply Algorithm 9.2.1 and its enhancements, with the stated aim of finding λ_*, but stop prematurely as soon as the goal (9.4.3) and/or (9.4.4) is achieved.

Notes and references for Section 9.4

Once again, further details are given in [196] and heavily influenced by [178, 292]. As before, Fortran and MATLAB implementations are available in the package **TRS** as part of GALAHAD [192].

[65]The exception is the hard case, in which case $\mathcal{L}$ is empty, and the iterates generated by Algorithm 9.2.1 will ultimately converge to λ_* from $\mathcal{G}$ when augmented by Algorithm 9.3.1 (see Section 9.3.4 for details). The bound $m_{\Delta(\lambda)} \ge \frac{1}{2}(\langle g, s(\lambda)\rangle - \lambda\Delta^2)$ for all $\lambda \in \mathcal{G}$ may be useful.

While the outline given in this section may be used to find a highly accurate solution, there is no reason why a lower-accuracy estimate along the lines of those proposed by Moré and Sorensen [292] or Gertz and Gill [179, Sect. 5] should not be produced. Gertz and Gill [179, Lem. 5.2], based on [292, Lem. 3.4], show that any $s = s(\lambda) + u$ for which $\|s(\lambda) + u\|_M \leq (1+\eta)\Delta$ and

$$\langle u, (H + \lambda M)u \rangle \leq \eta(2 - \eta) \left[\langle x(\lambda), (H + \lambda M)s(\lambda) \rangle + \lambda \Delta^2 \right] \tag{9.4.8}$$

for some (possibly zero) u and (small) $\eta \in (0, 1)$ gives a suitable approximate solution to (8.1.2). These rules may be applied to terminate the algorithm we have sketched: we reduce the interval $[\lambda^{\mathrm{L}}, \lambda^{\mathrm{U}}]$ until either $\|s(\lambda)\|_M \leq (1+\eta)\Delta$ (in which case $u = 0$ suffices) or, ultimately, if the hard case occurs, the eigenvector estimates $u = z_k$ generated by Algorithm 9.3.1 will satisfy (9.4.8).

Chapter 10

Step Computation without Factorizations

Chapter summary

While the general approach we described in the previous chapter is recommended in many circumstances, it relies on the availability of the Hessian matrix, and on the ability to factorize its sum with a suitable multiple of the identity. For large problems, or problems that arise from some "black-box" applications, this may be unreasonable. Fortunately, we show in this chapter that there are alternatives that rely not on having the Hessian but simply on being able to provide the product of this matrix with a given vector, without needing to factorize anything beyond tridiagonal matrices.

10.1 Krylov-subspace methods

10.1.1 Basics

The main idea is very simple and is based on three observations. The first is that frequently a suitable approximation to the solution of the problem, such as that outlined in Subsection 9.3.6, suffices, and full accuracy is not required. Secondly, even if the Hessian is available, unless H has very special structure, the problem dimension will ultimately be the limiting factor in the applicability of the method. Thus, if a limited-accuracy approximation suffices, one might hope to achieve this by solving the problem in a sequence of evolving subspaces $\mathcal{S}_k$ of dimension k for $k = 1, 2, \ldots$ until the required accuracy is achieved, rather than solving over the whole of $\mathbb{R}^n$. To be specific, one might replace the solutions to (8.1.2) or (8.1.3) by the approximations

$$s_k = \underset{s \,\in\, \mathcal{S}_k \,:\, \|s\|_M \,\leq\, \Delta}{\arg\min} \; m(s) \;\text{ or }\; s_k = \underset{s \,\in\, \mathcal{S}_k}{\arg\min}\, m^{\mathrm{R}}(s, \sigma, r). \tag{10.1.1}$$

More particularly, if the columns of S_k are a basis for $\mathcal{S}_k$, the solutions to (10.1.1) may be expressed as $s_k = S_k \bar{s}_k$, where

$$\bar{s}_k = \underset{\bar{s} \,\in\, \mathbb{R}^k \,:\, \|\bar{s}\|_{\bar{M}_k} \,\leq\, \Delta}{\arg\min} \; \bar{m}_k(\bar{s}) \stackrel{\text{def}}{=} \langle \bar{g}_k, \bar{s} \rangle + \tfrac{1}{2} \langle \bar{s}, \bar{H}_k \bar{s} \rangle \tag{10.1.2}$$

and

$$\bar{s}_k = \arg\min_{\bar{s}\,\in\,\mathbb{R}^k} \bar{m}_k^{\mathrm{R}}(\bar{s},\sigma,r) \stackrel{\text{def}}{=} \bar{m}_k(\bar{s}) + \tfrac{1}{r}\sigma\|\bar{s}\|_{\bar{M}_k}^r,$$

as well as $\bar{g}_k \stackrel{\text{def}}{=} S_k^T g$, $\bar{H}_k \stackrel{\text{def}}{=} S_k^T H S_k$, and $\bar{M}_k \stackrel{\text{def}}{=} S_k^T M S_k$. The main point here is that manipulating the k by k matrix pencil $\bar{H}_k + \lambda \bar{M}_k$ is possible for modest k, and thus the methods we advocated in Chapter 9 are viable for the subproblems (10.1.1).

Of course this does not address the explicit need for H and M. Furthermore, as k grows, so do S_k, $\bar{H}_k$, and $\bar{M}_k$, and ultimately it may still prove impossible to solve (10.1.1). Our third and crucial observation is that there is a very special subspace—a Krylov subspace—that deals with both defects at once.

Given $A \in \mathbb{R}^{n\times n}$ and $b \in \mathbb{R}^n$, we define the kth *Krylov subspace* generated by A and b to be the space

$$\mathcal{K}(A,b,k) \stackrel{\text{def}}{=} \text{span}\left\{A^i b\right\}_{i=0}^{k-1},$$

and we let

$$\mathcal{K}(A,b) \stackrel{\text{def}}{=} \lim_{k\longrightarrow\infty} \mathcal{K}(A,b,k) \equiv \mathcal{K}(A,b,n)$$

be the limiting *Krylov space*. Notice that the subspaces are nested, i.e., $\mathcal{K}(A,b,k+1) = \mathcal{K}(A,b,k) \cup \{A^k b\}$; the kth subspace is of dimension at most k; and there is a smallest k, $\text{grade}(A,b) \le n$, the *grade* of A with respect to b, for which $\mathcal{K}(A,b,k) = \mathcal{K}(A,b)$. The particular sequence of Krylov subspaces that we are interested in is $\{\mathcal{K}_k\}_{k>0}$, where

$$\mathcal{K}_k \stackrel{\text{def}}{=} \mathcal{K}(M^{-1}H, M^{-1}g, k).$$

The key, though, is that there is an M-orthogonal basis matrix $Q_k = (q_1,\ldots,q_k)$ for $\mathcal{K}$ for which

$$Q_k^T H Q_k = T_k, \quad Q_k^T M Q_k = I_k, \quad \text{and} \quad Q_k^T g = \gamma_1 e_1 \tag{10.1.3}$$

for some tridiagonal matrix T_k and scalar

$$\gamma_1 \stackrel{\text{def}}{=} \|g\|_{M^{-1}} > 0. \tag{10.1.4}$$

Thus, picking $\mathcal{S}_k$ in (10.1.1) to be $\mathcal{K}_k$, and as a consequence the basis matrix $S_k = Q_k$, it follows immediately from (10.1.2)–(10.1.3) that $s_k = Q_k \bar{s}_k$, where

$$\bar{s}_k = \arg\min_{\bar{s}\,\in\,\mathbb{R}^k\,:\,\|\bar{s}\|\le\Delta} \bar{m}_k(\bar{s}) \stackrel{\text{def}}{=} \gamma_1\langle e_1,\bar{s}\rangle + \tfrac{1}{2}\langle \bar{s}, T_k\bar{s}\rangle \tag{10.1.5}$$

or

$$\bar{s}_k = \arg\min_{\bar{s}\,\in\,\mathbb{R}^k} \bar{m}_k^{\mathrm{R}}(\bar{s},\sigma,r) \stackrel{\text{def}}{=} \bar{m}_k(\bar{s}) + \tfrac{1}{r}\sigma\|\bar{s}\|^r. \tag{10.1.6}$$

The subspace model problems (10.1.5) and (10.1.6) have three important properties: the Euclidean norm is used throughout; the gradient and Hessian are both sparse; and critically the pencil involved, $T_k + \lambda I_k$, is tridiagonal and may thus be factorized in $\mathcal{O}(k)$ operations for a given λ. Moreover, the data for the kth problem is simply related to that for the $(k-1)$st via

$$Q_k = \begin{pmatrix} Q_{k-1} & q_k \end{pmatrix} \text{ and } T_k = \begin{pmatrix} T_{k-1} & \gamma_k e_{k-1} \\ \gamma_k e_{k-1}^T & \rho_k \end{pmatrix}, \tag{10.1.7}$$

where the vector q_k and scalars γ_k and ρ_k may be computed via the preconditioned[66] Lanczos method presented in Algorithm 10.1.1.

[66] "Preconditioned" here refers to the application of M^{-1}.

Algorithm 10.1.1: The preconditioned Lanczos method

Given g, set $t_1 = g$ and $w_0 = 0$, and, for $k = 1, 2, \ldots, n$, perform the iteration

$$\begin{aligned} y_k &= M^{-1} t_k, \\ \gamma_k &= \sqrt{\langle t_k, y_k \rangle} \text{ and stop if } \gamma_k = 0, \\ w_k &= t_k / \gamma_k, \\ q_k &= y_k / \gamma_k, \\ \rho_k &= \langle q_k, H q_k \rangle, \\ t_{k+1} &= H q_k - \rho_k w_k - \gamma_k w_{k-1}. \end{aligned}$$

The terminating iteration, for which $\gamma_k = 0$ or $k = n$, defines grade$(M^{-1}H, M^{-1}g)$, and if this is smaller than n, the Krylov space $\mathcal{K}(M^{-1}H, M^{-1}g)$ revealed is an invariant subspace of $\mathbb{R}^n$. Crucially, the recurrence requires a single matrix-vector product involving H and a single product with M^{-1} (or equivalently the solution of the linear system $Mt_k = y_k$ involving M) per iteration. Thus H itself is not required.

10.1.2 The hard case

While the above appears to paint a rosy picture of the Krylov approach, there is one unavoidable defect, and it relates to the hard case. Recall that the hard case can only occur if $\langle v_i, g \rangle = 0$ for all $v_i \in \mathcal{V}_1 = \{v_j\}_{j=1}^{i_1}$, where v_j is the generalized eigenvector of the pencil (H, M) corresponding to the eigenvalue λ_j and $\lambda_i = \lambda_1 < 0$ for $i = 1, \ldots, i_1$. Since the set of eigenvectors are M-orthogonal and provide a basis for $\mathbb{R}^n$, we may write

$$M^{-1} g = \sum_{j=1}^{n} \alpha_j v_j$$

for some scalars α_j, and thus in the hard case using the M-orthogonality we have that

$$\alpha_i = \alpha_i \langle v_i, M v_i \rangle = \left\langle v_i, \sum_{j=1}^{n} \alpha_j M v_j \right\rangle = \langle v_i, g \rangle = 0$$

for $i = 1, \ldots i_1$. Hence

$$M^{-1} g = \sum_{j=i_1+1}^{n} \alpha_j v_j,$$

and since $M^{-1} H v_j = \lambda_j v_j$, we deduce that

$$(M^{-1} H)^i M^{-1} g = \sum_{j=i_1+1}^{n} \alpha_j \lambda_j^i v_j.$$

Thus $\mathcal{K}_k \subseteq \{v_j\}_{j=i_1+1}^{n}$ for all k. This has two consequences. Firstly, the grade of $M^{-1}H$ with respect to $M^{-1}g$ is at most[67] $n - i_1$, and thus Algorithm 10.1.1 will terminate in fewer than n iterations. But, more importantly, the approximations s_k provided by (10.1.5) or (10.1.6) will never see the subspace of crucial eigenvectors $\mathcal{V}_1$ that are needed in the hard case.

[67] It may be smaller than $n - i_1$ if by chance any other $\alpha_i = 0$ for $i \geq i_1$.

Of course, one can argue—and in practice it is true—that the likelihood that $\langle v_i, g\rangle = 0$ for all $i = 1, \ldots, i_1$ is tiny—of probability zero if g, H, and M are random—and so is of little concern. However, from the perspective of worst-case complexity, it remains an issue. As we saw in Section 8.3, the solution to (8.1.2) or (8.1.3) has an additional component in $\mathcal{V}_1$. There are, in general, two types of methods for finding eigenvalues/vectors: those that compute all of them, such as the Jacobi or QR method, and those that aim to find a subset, such as the power, Lanczos, or Jacobi–Davidson method. The former require access to the matrices themselves, while the latter only require matrix-vector products starting from one or more initial "guesses." Unfortunately, as far as we are aware, the later cannot ensure that such guesses have a component in $\mathcal{V}_1$ and thus offer no more than a probabilistic guarantee. This is regrettable because it would appear that we are trying to deal with a probability-zero event by employing a method that might still fail (with probability zero).

10.2 Computational details for regularization methods

For the remainder of this section, we shall ignore the possibility of the hard case and concentrate instead on solving the subspace model problems (10.1.5)–(10.1.6). As before, we focus on the regularization model, but we will return to the trust-region case in due course.

10.2.1 Solving shifted Lanczos tridiagonal systems

Our first observation is that the quantities computed by the preconditioned Lanczos method (Algorithm 10.1.1) may be written compactly as

$$HQ_k - MQ_kT_k = \gamma_{k+1}w_{k+1}e_k^T. \tag{10.2.1}$$

Since $\bar{s}_k$ is a global minimizer of $\bar{m}_k^{\mathrm{R}}(\bar{s}, \sigma, r)$, we have from Theorem 8.2.8 that

$$(T_k + \lambda_k I_k)\bar{s}_k = -\gamma_1 e_1 \tag{10.2.2}$$

and that $T_k + \lambda_k I_k$ is positive semidefinite, where

$$\lambda_k = \sigma\|s_k\|_M^{r-2} \equiv \sigma\|\bar{s}_k\|^{r-2}. \tag{10.2.3}$$

Moreover, since the tridiagonal Hessian T_k is irreducible (i.e., its off-diagonal entries γ_i are nonzero, because otherwise the Lanczos iteration would have halted), it can be shown that the hard case cannot occur [189, Thm. 5.3] and that $T_k + \lambda_k I_k$ is actually positive definite [189, Thm. 5.5].

Since the hard case does not occur for the subproblems in (10.1.5)–(10.1.6), the dominant cost of applying the methods described in Chapter 9 will be in factorizing $T_k + \lambda I_k$ and solving systems of the form

$$(T_k + \lambda I_k)x(\lambda) = b \tag{10.2.4}$$

for given λ and b. Suppose that λ is large enough that $T_{k-1} + \lambda I_{k-1}$ is positive definite and thus has an LDL^T factorization

$$T_{k-1} + \lambda I_{k-1} = L_{k-1}(\lambda)D_{k-1}(\lambda)L_{k-1}^T(\lambda), \tag{10.2.5}$$

where $L_{k-1}(\lambda)$ is unit lower triangular and $D_{k-1}(\lambda)$ is diagonal with positive diagonal entries $d_i(\lambda)$. It then follows immediately from (10.1.7) and (10.2.5) that

$$T_k + \lambda I_k = L_k(\lambda) D_k(\lambda) L_k^T(\lambda), \tag{10.2.6}$$

where

$$L_k(\lambda) = \begin{pmatrix} L_{k-1}(\lambda) & 0 \\ l_k(\lambda) e_{k-1}^T & 1 \end{pmatrix} \quad \text{and} \quad D_k(\lambda) = \begin{pmatrix} D_{k-1}(\lambda) & 0 \\ 0 & d_k(\lambda) \end{pmatrix}$$

with

$$l_k(\lambda) = \frac{\gamma_k}{d_{k-1}(\lambda)} \quad \text{and} \quad d_k(\lambda) = \rho_k + \lambda - \frac{\gamma_k^2}{d_{k-1}(\lambda)}.$$

Thus we see that L_k has a single off-diagonal entry $l_k(\lambda)$ in row k and column $k-1$ and therefore that it is unit lower bidiagonal. Moreover, $T_k + \lambda I_k$ is positive definite if and only if $\rho_k + \lambda > \gamma_k^2 / d_{k-1}(\lambda)$. To summarize, the factorization (10.2.6) may be found—or $T_k + \lambda I_k$ may be detected not to be positive definite—in $\mathcal{O}(k)$ operations using Algorithm 10.2.1.

Algorithm 10.2.1: Factorize the shifted Lanczos tridiagonal $T_k + \lambda I_k$

Given λ, ρ_i, $i = 1, \ldots, k$, and γ_i, $i = 2, \ldots, k$, if $\rho_1 + \lambda > 0$, set

$$r_1(\lambda) = \frac{1}{\rho_1 + \lambda};$$

otherwise, exit (the matrix is not positive definite).
For $i = 2, \ldots, k$, set

$$l_i(\lambda) = \gamma_i r_{i-1}(\lambda),$$

and if $\rho_i + \lambda > \gamma_i l_i(\lambda)$, set

$$r_i(\lambda) = \frac{1}{\rho_i + \lambda - \gamma_i l_i(\lambda)};$$

otherwise, exit (the matrix is not positive definite).

Notice that we intentionally store and use the reciprocals $r_i(\lambda) = 1/d_i(\lambda)$ of the diagonal entries. This is so that when we use the factors to solve systems, floating-point division operations are performed as more efficient multiplications. The precise cost of computing the factors is k additions and divisions, $k-1$ subtractions, and $2k-2$ multiplications.

Armed with the factors, it is then easy to solve the linear system (10.2.4) using Algorithm 10.2.2.

Algorithm 10.2.2: Solve the shifted Lanczos system $(T_k + \lambda I_k)x(\lambda) = b$

Given b and the LDL^T factors of $T_k + \lambda I_k$ from Algorithm 10.2.1, set $x_1(\lambda) = b_1$.
For $i = 2, \ldots, k$, set $x_i(\lambda) = b_i - l_i(\lambda) x_{i-1}(\lambda)$.
For $i = 1, \ldots, k$, reset $x_i(\lambda) \leftarrow r_i(\lambda) x_i(\lambda)$.
For $i = k, \ldots, 2$, reset $x_{i-1}(\lambda) \leftarrow x_{i-1}(\lambda) - l_i(\lambda) x_i(\lambda)$.

The total cost of this is $3k-1$ multiplications (because we have stored the reciprocals of the $d_i(\lambda)$) and $2k-2$ subtractions. Further savings are obviously possible in the first "For" loop if b is sparse.

10.2.2 Solving the kth subspace subproblem

Having considered the factorizations at the core of the solution of each successive subproblem, our next consideration concerns the subproblem itself, and most particularly how the solutions s_{k-1} and s_k relate as the Krylov space evolves. In many aspects, we have already covered this in Chapter 9, since for subproblem k we have a matrix T_k and means to solve identity-perturbed systems involving it. We may thus apply a suitable rapidly converging iteration to the subproblem's (possibly transformed) secular equation. All that we really need to be concerned about is how to start each iteration. To do so, we need to understand T_k and λ_k in more detail.

To set the scene, for any integer $k \in [2, \text{grade}(M^{-1}H, M^{-1}g)]$, we have that T_k from (10.1.7) is symmetric, tridiagonal, and unreduced (i.e., $\gamma_i > 0$ for $i = 2, \ldots, k$) with distinct real eigenvalues [315, Lem. 7-7], or *Ritz values* as they are often known, $\theta_{i,k}$, $i = 1, \ldots, k$, that we shall write in increasing order of i. The Ritz values for T_k and T_{k+1} interlace, that is,

$$\theta_{i,k+1} \leq \theta_{i,k} \leq \theta_{i+1,k+1}$$

for $i = 1, \ldots, k$ [185, Thm. 8.1.7]. As we have seen, $T_k + \lambda_k I_k$ is positive definite and thus has distinct, positive left- and rightmost eigenvalues

$$0 < \lambda_{\min}[T_k + \lambda_k I_k] \equiv \theta_{1,k} + \lambda_k < \lambda_{\max}[T_k + \lambda_k I_k] \equiv \theta_{k,k} + \lambda_k.$$

We also have the following monotonic relationship between the multipliers λ_k of successive subspace subproblems.

Lemma 10.2.1: Nested subspace solutions increase in norm

[81, Thm. 2.5 and Sect. 3]: Let $s_k = Q_k \bar{s}_k$, where $\bar{s}_k$ satisfies (10.1.6); let λ_k be its multiplier (10.2.3); let s_* be a global minimizer of (8.1.4); and let $\lambda_* = \sigma\|s_*\|_M^{r-2}$. Then $\lambda_k \leq \lambda_\ell \leq \lambda_*$ and $\|s_k\|_M \leq \|s_\ell\|_M \leq \|s_*\|_M$ for $1 \leq k \leq \ell \leq \text{grade}(M^{-1}H, M^{-1}g)$.

Thus λ_k lies to the left of the desired solution, λ_{k+1}, to the secular equation for the $(k+1)$st subspace subproblem, and using the terminology of Section 8.3, $\lambda_k \in \mathcal{N}_k \cup \mathcal{L}_k$ for that subproblem. It is not clear, a priori, if $\lambda_k \in \mathcal{L}_k$ or $\mathcal{N}_k$—since $\theta_{1,k+1} \leq \theta_{1,k}$ it does not follow that $\theta_{1,k+1} + \lambda_k > 0$ just because $\theta_{1,k} + \lambda_k > 0$—but it is easy to check. For, just as in Algorithm 10.2.1, given the reciprocal $r_k(\lambda_k)$ of the kth diagonal of the LDL^T factorization of $T_k + \lambda_k I_k$, the expanded matrix $T_{k+1} + \lambda_k I_{k+1}$ is positive definite if and only if

$$\lambda_k > \gamma_{k+1}^2 r_k(\lambda_k) - \rho_{k+1},$$

where, as we recall, ρ_{k+1} and γ_{k+1} are the diagonal and off-diagonal entries in row $k+1$ of T_{k+1}. If we find that $\lambda_k \in \mathcal{L}_k$, as always Newton or higher-order methods applied to the secular equation for the $(k+1)$st subspace subproblem will converge rapidly to λ_{k+1}. If $\lambda_k \in \mathcal{N}_k$, choosing $\lambda_k^+ \stackrel{\text{def}}{=} \lambda_k + \delta\lambda$ for any $\delta\lambda > \gamma_{k+1}^2 r_k(\lambda_k) - \rho_{k+1} - \lambda_k$ ensures that $T_k + \lambda_k^+ I_k$ is positive definite. Thus either $\lambda_k^+ \in \mathcal{L}_k$ and again provides a suitable starting point for fast monotonic convergence to λ_{k+1}, or $\lambda_k + \delta\lambda \in \mathcal{G}_k$, in which case the interval $[\lambda_k, \lambda_k^+]$ encloses λ_{k+1}, and we proceed just as in Section 9.3.3.

10.2.3 Starting the Lanczos process

Of course, we need to start the process. When $k = 1$, the subproblem (10.1.6) is simply

$$\bar{s}_1 = \arg\min_{\bar{s} \in \mathbb{R}} \gamma_1 \bar{s} + \tfrac{1}{2}\rho_1 \bar{s}^2 + \tfrac{1}{r}\sigma|\bar{s}|^r,$$

and necessarily

$$\gamma_1 + (\rho_1 + \lambda_1)\bar{s}_1 = 0, \quad \text{where} \quad \lambda_1 = \sigma|\bar{s}_1|^{r-2}. \tag{10.2.7}$$

Since additionally $\rho_1 + \lambda_1 > 0$, (10.2.7) implies that λ_1 is the larger of the two roots of the simple rearranged secular equation

$$\lambda_1(\rho_1 + \lambda_1)^{r-2} = \sigma|\gamma_1|^{r-2}$$

and thus is easy to obtain by iteration, or even by formula in special cases; for example, for cubic ($r = 3$) regularization, this is just a quadratic equation.

10.2.4 Stopping the Lanczos process

The final issue relates to stopping the iteration. Recall that in our applications we generally require a step s so that (9.3.14) holds. Since both 0 and s_k lie in $\mathcal{K}_k$, the requirement (9.3.14a) naturally occurs when $s = s_k$. We now investigate (9.3.14b), and the crux is Theorem 10.2.2, which provides a bound on the norm of the gradient (8.2.34) of the regularization model at the subspace minimizer s_k and uses the spectral condition number

$$\kappa_k \stackrel{\text{def}}{=} \kappa(T_k + \lambda_k I_k) = \frac{\theta_{k,k} + \lambda_k}{\theta_{1,k} + \lambda_k}. \tag{10.2.8}$$

Theorem 10.2.2: Bound on the gradient at the subspace minimizer

[Appendix to Chapter 10, p. 479]: The gradient (8.2.34) for the kth iterate $s_k = Q_k\bar{s}_k$, generated by the regularization subproblem (10.1.6), satisfies the bound

$$\|\nabla_s^1 m^{\mathrm{R}}(s_k, \sigma, r)\|_{M^{-1}} \leq 2\|M^{-1}H\|\kappa_k \left(\frac{\lambda_k}{\sigma}\right)^{\frac{1}{r-2}} \left(\frac{\sqrt{\kappa_k} - 1}{\sqrt{\kappa_k} + 1}\right)^{k-1}, \tag{10.2.9}$$

where λ_k satisfies (10.2.3) and κ_k is given by (10.2.8).

We may then use Theorem 10.2.2 to deduce an upper bound on the number of Lanczos iterations required to satisfy the termination test (9.3.14b). To do so, we define the *condition number*

$$\kappa_* \stackrel{\text{def}}{=} \max_{1 \leq k \leq \text{grade}(M^{-1}H, M^{-1}g)} \kappa_k, \tag{10.2.10}$$

using κ_k from (10.2.8).

Corollary 10.2.3: Iteration bound for the subspace approximation

[Appendix to Chapter 10, p. 480]: The iteration defined by $s_k = Q_k \bar{s}_k$ *and the subspace minimization (10.1.6) satisfies (9.3.14b) for some*

$$k \leq \min\left[\left\lceil \frac{\log\left(\frac{4\|M^{-1}H\|\kappa_*}{\theta_1}\left(\frac{\lambda_*}{\sigma}\right)^{\frac{1}{r-2}}\left(\frac{\lambda_{\max}[H,M]+\lambda_*}{\|g\|_{M^{-1}}}\right)^2\right)}{\log\left(\frac{\sqrt{\kappa_*}+1}{\sqrt{\kappa_*}-1}\right)}\right\rceil + 1, \ \operatorname{grade}(M^{-1}H, M^{-1}g)\right],$$

where κ_* *is given by (10.2.10).*

We note that this bound may be very pessimistic due to the large number of inequalities used in its derivation. In practice, the better that M is able to cluster the eigenvalues of $M^{-1}H$, the faster the convergence.

Since there is no way of knowing whether the Krylov space $\mathcal{K}(M^{-1}H, M^{-1}g)$ is an invariant subspace of $\mathbb{R}^n$ rather than $\mathbb{R}^n$ itself, it is not possible to guarantee the second-order termination requirement (9.3.14c). It can be shown, however, that a weaker version of (9.3.14c) in which

$$\min_{v=Q_k\bar{v}, \|v\|=1} \left\langle v, \nabla_s^2 m^{\mathrm{R}}(s,\sigma,r)v \right\rangle \geq -\theta_H \|s\|_M$$

for all $s = Q_k\bar{s}$ in a neighbourhood of s_k may be attained. This is likely the best one may hope for without examining the whole of $\mathbb{R}^n$.

10.3 The trust-region case

There are only minor differences when adapting the approach we have outlined for the regularization case to the trust-region subproblem (10.1.5). The first is that for small k, the solution to the subproblem may lie interior to the trust region. Indeed there is a trust-region analogue of Lemma 10.2.1.

Lemma 10.3.1: Nested subspace trust-region solutions norm increase

[279, Thm. 2.6], [343, Thm. 2.1]: Let $s_k = Q_k\bar{s}_k$*, where* $\bar{s}_k$ *satisfies (10.1.5); let* λ_k *be the corresponding Lagrange multiplier for the trust-region constraint; and let* s_* *be a global solution of (8.1.2) with corresponding multiplier* λ_**. Then* $\lambda_k \leq \lambda_\ell \leq \lambda_*$ *and* $\|s_k\|_M \leq \|s_\ell\|_M \leq \|s_*\|_M$ *for* $1 \leq k \leq \ell \leq \operatorname{grade}(M^{-1}H, M^{-1}g)$*.*

Thus the iterates s_k increase in norm until the trust region is attained (if it ever is), and thereafter the iterates remain on the trust-region boundary. The Lagrange multipliers, by contrast, are zero until the boundary is attained, and thereafter increase in size. So long as T_k is positive definite, it pays to start each subproblem solution with the guess $\lambda = 0$ and to test the norm of $\|s\|_M$, where $T_k s = -\gamma_1 e_1$, to see if $\|s\|_M \leq \Delta$. Thereafter, starting the $(k+1)$st subproblem with λ_k if possible, or λ_k^+ otherwise, just as in Section 10.2.2, is justified. It is worth noting that so long as the subspace solution lies inside the trust-region, the Lanczos iteration is identical to the conjugate-gradient (CG) method. In practice, this is exploited since the CG method is slightly cheaper to execute. Earlier methods [343, 347] actually terminate on the trust region boundary as soon as a CG iterate tries to move outside the trust region or $\rho_k \leqslant 0$.

Knowing when to stop the Krylov iteration is just as important for (10.1.5) as it was for (10.1.6). Since we are necessarily aiming for a critical point s_* of (8.1.2) for which (8.2.1) holds, we might stop the process as soon as the norm of the residual

$$r_k \stackrel{\text{def}}{=} g + (H + \lambda_k M) s_k \tag{10.3.1}$$

satisfies

$$\|r_k\|_{M^{-1}} \leq \epsilon \tag{10.3.2}$$

for some suitably small $\epsilon > 0$.

With this in mind, we have the following analogue of Theorem 10.2.2.

Theorem 10.3.2: Bound on the residual at the subspace minimizer

[Appendix to Chapter 10, p. 481]: The residual (10.3.1) for the kth iterate $s_k = Q_k \bar{s}_k$, generated by the trust-region subproblem (10.1.5), satisfies the bound

$$\|g + (H + \lambda_k M) s_k\|_{M^{-1}} \leq 2\|M^{-1}H\| \kappa_k \Delta \left(\frac{\sqrt{\kappa_k} - 1}{\sqrt{\kappa_k} + 1} \right)^{k-1},$$

where λ_k is the Lagrange multiplier for (10.1.5) and κ_k is given by (10.2.8).

It is then trivial to deduce an upper bound on the number of Lanczos iterations required to satisfy the termination test (10.3.2).

Corollary 10.3.3: Iteration bound for the trust-region subspace approximation

[Appendix to Chapter 10, p. 481]: The iteration defined by $s_k = Q_k \bar{s}_k$ and the subspace trust-region minimization (10.1.5) satisfies (10.3.2) for some

$$k \leq \min \left[\left\lceil \frac{\log \left(\dfrac{2\|M^{-1}H\| \kappa_* \Delta}{\epsilon} \right)}{\log \left(\dfrac{\sqrt{\kappa_*} + 1}{\sqrt{\kappa_*} - 1} \right)} \right\rceil + 1, \;\; \text{grade}(M^{-1}H, M^{-1}g) \right],$$

where κ_* is given by (10.2.10).

Finally, just as in Section 10.2.3, it is easy to start the subspace minimization as

$$\bar{s}_1 = \underset{\bar{s} \in \mathbb{R} \,:\, |\bar{s}| \leq \Delta}{\arg\min} \;\; \gamma_1 \bar{s} + \tfrac{1}{2} \rho_1 \bar{s}^2.$$

If $\rho_1 > 0$ and $|\gamma_1/\rho_1| \leq \Delta$, then $\bar{s}_1 = -\gamma_1/\rho_1$ and $\lambda_1 = 0$. Otherwise $\bar{s}_1 = -(\gamma_1/|\gamma_1|)\Delta$ and $\lambda_1 = -\rho_1 + |\gamma_1|/\Delta$, since we are presuming that $\gamma_1 \neq 0$.

Notes and references for Section 10.3

The basic methods that we have described here are based on [84, 131, 189] and firmly reflect our prejudices that the methods should be effective in practice; packages GLTR and GLRT are available as part of the GALAHAD open-source software library [192]. A completely different eigenvalue-based approach is available in the *LSTRS* package [332]. Although [189] suggests how to extend Krylov methods to cope with the hard case, this has rarely been implemented. An exception is provided by [264]. The elementary convergence bounds provided in Corollaries 10.2.3 and 10.3.3 are given in [198], but more sophisticated estimates are available [376, 377, 378]. It is also possible to exploit the invariance of a particular Krylov space to different λ so as to "precompute" solutions to a set of potentially useful λ_k all at once [159].

There are other possibilities, mostly for the trust-region case but easily adaptable for regularization. Some authors have noted that the optimality conditions may be posed as a parametric eigenvalue problem and solved in this way using existing sophisticated eigenvalue software. Others have examined the subproblems as general convexly constrained optimization problems and applied projection-based approaches. Some of these appear under the "DC" programming banner, while others use conditional gradient ideas. See [131, Sects. 7.5.6 and 7.6] for citations from the previous century and [2, 25, 161, 237, 331], for example, for more recent ones.

Chapter 11

Variable Scaling and Other Practicalities

Chapter summary

This short chapter focuses on an issue which has not been considered so far, in the interests of keeping our exposition free of too many constants: the choice of variable scaling. We will investigate simple scaling strategies and more elaborate ones, potentially exploiting the structure of the problem.

We will also consider another important practical issue, the update of the trust-region radius or regularization parameter.

11.1 Variable scaling

As we mentioned at the start of Chapter 8, it is preferable in practice to measure changes in variables in a way that reflects proper scaling. Thus, while our theoretical development avoids this issue for simplicity—specifically because otherwise unwieldy constants would proliferate in our complexity bounds—in practice it is unwise to do so.

We get a hint of the importance of scaling in Corollaries 10.2.3 and 10.3.3. For example, if H is positive definite and we choose $M = H$, the condition number $\kappa_* = 1$, and the Lanczos process only requires a single iteration. More generally, if M can be found so that the generalized eigenvalues of the pencil (H, M) are clustered around a few distinct values, convergence is generally rapid; this phenomenon is well known for Krylov-based methods for linear systems and in this context is commonly referred to as *preconditioning* rather than scaling; we shall use the terminology interchangeably. Indeed, for many linear systems that arise from the discretization of differential equations, preconditioning is vital and may often be designed to achieve discretization-independent convergence. Unfortunately, such preconditioners often rely heavily on underlying physical properties relating to the specific differential equation in question, and there appears to be no general-purpose recipe for finding universal preconditioners. Thus, we shall be content in this section to explore advantageous heuristic possibilities.

11.1.1 Simple scalings

Suppose that H is available. Then the simplest scaling, and one that is very inexpensive to apply, is to select M as the diagonal matrix formed from the diagonals of H, possibly perturbed to ensure that the result is positive definite, e.g., $M = \text{diag}(\max(h_{i,i}, \delta))$ for some (small) $\delta > 0$. For positive-definite H, such a choice generally improves the condition number κ_* and provides the optimal diagonal scaling in many cases [357]. This may be generalized by picking a band—of specified bandwidth m—from H and perturbing it to guarantee positive definiteness. For example,

$$m_{ij} = \begin{cases} \max\left(h_{i,i}, \sum_k |h_{i,j}| + \delta\right) & \text{if } i = j \text{ and } \max(1, i-m) \leq k \neq i \leq \min(n, i+m), \\ h_{i,j} & \text{if } \max(1, i-m) \leq j \neq i \leq \min(n, i+m), \text{ or} \\ 0 & \text{otherwise} \end{cases}$$

is strictly diagonally dominant and hence positive definite.

Another popular approach, particularly for differential equation–based problems, is to construct M from the Cholesky factors of a modification of H but drop some fill-ins as the factorization proceeds [33, 287]. Since H may not be positive definite, modification may be necessary, either in advance, usually by adding a suitable multiple of I [281], or as part of the factorization process [4, 15, 254, 346]—there is unfortunately a danger of overmodification. Knowing which entries to drop can also be tricky, but good heuristics are available [274, 338]. While such incomplete-Cholesky ideas are aimed at approximating H, the heuristics involved are often less successful in clustering eigenvalues of the pencil (H, M) for general H.

Other popular preconditioners for linear systems, such as matrix splittings and multigrid, multilevel, domain, and hierarchical decomposition [33, 363], are also possible but, as far as we are aware, infrequently used in general optimization applications. When H is only available indirectly through matrix-vector products, it is possible to build limited-memory preconditioners when solving a sequence of problems [290]. Alternatively, if the sparsity pattern of H is known, it is often possible to recreate H from a few "sparse" matrix-vector products [120, 322]. As yet, neither of these ideas has gained widespread acceptance.

11.1.2 The ideal scaling

The remainder of our development is based on what might be considered to be the ideal, albeit often prohibitive, case. Suppose that we can afford to compute the spectral decomposition

$$H = V \Lambda V^T, \tag{11.1.1}$$

where V is an orthonormal matrix whose columns are eigenvectors of H and Λ is a diagonal matrix of corresponding eigenvalues λ_i, $1 \leq i \leq n$. Then a suitable choice of the scaling matrix is

$$M = V \Lambda^{\text{M}} V^T \tag{11.1.2}$$

for some appropriate positive-definite diagonal matrix Λ^{M}. While some authors [50, 182, 286, 369, 375] advocate the simple choice $\Lambda^{\text{M}} = I$, others [190] argue that choosing

$$\Lambda^{\text{M}} = |\Lambda_\delta|, \tag{11.1.3}$$

where the ith diagonal of the *modified absolute-value matrix* $|\Lambda_\delta|$ is defined by

$$|\Lambda_\delta|_{ii} \stackrel{\text{def}}{=} \begin{cases} |\lambda_i| & \text{if } |\lambda_i| \geq \delta \text{ and} \\ \delta & \text{otherwise,} \end{cases} \tag{11.1.4}$$

reflects the proper scaling of the problem, in the sense that directions for which the model changes fastest are restricted more than those for which the curvature is small; the perturbation $0 < \delta \ll 1$ is simply provided to cope with tiny curvature and to ensure that the smallest eigenvalue of the scaling matrix is bounded away from zero. The modified absolute-value matrix was originally proposed in conjunction with linesearch methods [217].

Although the cost of obtaining a spectral factorization may be exorbitant, especially for large problems, a change of variables $s^{\text{D}} = |\Lambda_\delta|^{1/2} V^T s$ in (8.1.2) or (8.1.3) exposes a key feature of (11.1.2)–(11.1.3), namely, that the transformed problems

$$\underset{s^{\text{D}} \in \mathbb{R}^n}{\text{minimize}} \; m^{\text{D}}(s^{\text{D}}) \stackrel{\text{def}}{=} \langle g^{\text{D}}, s^{\text{D}} \rangle + \tfrac{1}{2} \langle s^{\text{D}}, D s^{\text{D}} \rangle \text{ subject to } \|s^{\text{D}}\| \leq \Delta \tag{11.1.5}$$

and

$$\underset{s^{\text{D}} \in \mathbb{R}^n}{\text{minimize}} \; m^{\text{RD}}(s^{\text{D}}, \sigma, r) \stackrel{\text{def}}{=} m^{\text{D}}(s^{\text{D}}) + \tfrac{1}{r} \sigma \|s\|^r \tag{11.1.6}$$

involve only a specially structured diagonal Hessian matrix with diagonals

$$d_{ii} = \begin{cases} 1 & \text{if } \lambda_i > \delta, \\ -1 & \text{if } \lambda_i < -\delta, \text{ and} \\ \lambda_i/\delta \in (-1, 1) & \text{otherwise} \end{cases} \tag{11.1.7}$$

for $i \in \mathbb{N}_n \stackrel{\text{def}}{=} \{1, \ldots, n\}$. Here $g^{\text{D}} = |\Lambda_\delta|^{-1/2} V^T g$, and the solutions to the original problem (8.1.2) or (8.1.3) may be recovered from those, s_*^{D}, of (11.1.5) or (11.1.6) via $s_* = V |\Lambda_\delta|^{-1/2} s_*^{\text{D}}$. Since solving problems of the specific form (11.1.5)–(11.1.7) is central to what will follow, we now consider this aspect in detail—there are other variations on this theme [34].

11.1.3 Solving specially structured regularization subproblems

Let

$$\mathcal{D}^+ \stackrel{\text{def}}{=} \{i \in [1:n] \mid d_{ii} = 1\}, \; \mathcal{D}^- \stackrel{\text{def}}{=} \{i \in [1:n] \mid d_{ii} = -1\},$$

and

$$\mathcal{D}^0 \stackrel{\text{def}}{=} \{i \in [1:n] \mid d_{ii} \in (-1, 1)\}$$

be a partition of $[1:n]$, and define

$$g_+^{\text{D}} \stackrel{\text{def}}{=} \sum_{i \in \mathcal{D}^+} g_i^{\text{D}} e_i \text{ and } g_-^{\text{D}} \stackrel{\text{def}}{=} \sum_{i \in \mathcal{D}^-} g_i^{\text{D}} e_i.$$

According to Theorems 8.2.7 and 8.2.8, the required solutions s_*^{D} of (11.1.5)–(11.1.6) must satisfy

$$(D + \lambda_* I) s_*^{\text{D}} = -g^{\text{D}} \tag{11.1.8}$$

or—except when λ_* is the negative of one of the eigenvalues of D—equivalently

$$s_*^{\text{D}} = -\frac{1}{(\lambda_* + 1)} g_+^{\text{D}} - \frac{1}{(\lambda_* - 1)} g_-^{\text{D}} - \sum_{i \in \mathcal{D}^0} \frac{g_i^{\text{D}}}{(\lambda_* + d_{ii})} e_i, \tag{11.1.9}$$

where the nonnegative Lagrange multiplier λ_* is sufficiently large to ensure that $D+\lambda_* I$ is positive semidefinite and is zero if s^{D} lies within the trust region $\|s^{\text{D}}\| \leq \Delta$ for the transformed problem (11.1.5). This leads to a number of possibilities [190, Sect. 3.3], which we now summarize.

Suppose, as is most likely, since we are assuming that δ is tiny, that $\mathcal{D}^0 = \emptyset$, and thus from (11.1.9)

$$s_*^{\text{D}} = -\frac{1}{(\lambda_*+1)} g_+^{\text{D}} - \frac{1}{(\lambda_*-1)} g_-^{\text{D}} \quad (\lambda_* \neq \pm 1). \tag{11.1.10}$$

Then, if additionally $\mathcal{D}^- = \emptyset$, it follows that $\mathbb{N}_n = \mathcal{D}^+$ and $g_+^{\text{D}} = g^{\text{D}}$, and hence (11.1.10) reduces to

$$s_*^{\text{D}} = -\frac{1}{(\lambda_*+1)} g^{\text{D}}. \tag{11.1.11}$$

For the trust-region subproblem (11.1.5), two alternatives then arise. If $\|g^{\text{D}}\| \leq \Delta$, the solution to (11.1.8) is given by $s^{\text{D}} = -g^{\text{D}}$ with $\lambda_* = 0$. This corresponds to the case where the unconstrained minimizer of m^{D} lies interior to the trust region. By contrast, if $\|g^{\text{D}}\| > \Delta$, the solution to (11.1.8) is obtained by finding the value of $\lambda_* > 0$ for which

$$\frac{1}{(\lambda_*+1)} \|g^{\text{D}}\| = \Delta. \tag{11.1.12}$$

This is a linear equation in λ_* and thus the solution gives trivially

$$s^{\text{D}} = -\frac{\Delta}{\|g^{\text{D}}\|} g^{\text{D}}. \tag{11.1.13}$$

This corresponds to the case where m^{D} is convex, but the trust region excludes the unconstrained minimizer of the model. Notice, also, in this case, a reduction in the trust-region radius following an unsuccessful iteration merely reduces the length of the step in the direction $-g^{\text{D}}$. Such a strategy is identical in its effect (if not in its motivation) to a backtracking linesearch for the original problem along the quasi-Newton direction $-H^{-1}g$, and thus there is a strong similarity between trust-region and linesearch methods with this choice of trust-region scaling.

By contrast, for the regularization subproblem (11.1.6), the required λ_* satisfies $\lambda_* = \sigma\|s_*^{\text{D}}\|^{r-2}$ and thus, using (11.1.11),

$$\lambda_* = \sigma \left(\frac{\|g^{\text{D}}\|}{\lambda_*+1} \right)^{r-2},$$

which gives a quadratic equation for λ_* for cubic ($r = 3$) regularization. Once again, (11.1.11) shows that the step lies in the steepest-descent direction $-s^{\text{D}}$, and increasing the regularization parameter following an unsuccessful iteration simply reduces the length of the step. It is worth noting, though, that while (11.1.12) depends linearly on Δ, the dependence of (11.1.11) on σ is nonlinear.

Whenever $\mathcal{D}^- \neq \emptyset$, the solution to (11.1.5) must lie on the trust-region boundary and $\lambda_* \geq -\lambda_{\min}[D] = 1$. In the special case for which $\mathbb{N}_n = \mathcal{D}^-$, once again (11.1.13) holds, except when $\|g_-^{\text{D}}\| = 0$, in which case *any* vector s^{D} for which $\|s^{\text{D}}\| = \Delta$ is a solution. Otherwise $\mathbb{N}_n = \mathcal{D}^+ \cup \mathcal{D}^-$ but neither $\mathcal{D}^+$ nor $\mathcal{D}^-$ is empty. In this case

$$\|s^{\text{D}}\|^2 = \frac{\|g_+^{\text{D}}\|^2}{(\lambda_*+1)^2} + \frac{\|g_-^{\text{D}}\|^2}{(\lambda_*-1)^2} = \Delta^2 \tag{11.1.14}$$

if $\lambda_* \neq 1$. If $\|g_-^{\text{D}}\| \neq 0$, this is equivalent to a quartic equation (for which either a long-established technique such as Ferrari's method [336] or a safeguarded Newton root finder is appropriate), so we seek the single root $\lambda_* > 1$ and then recover the required solution via (11.1.10). By contrast, if $\|g_-^{\text{D}}\| = 0$, (11.1.14) reduces to (11.1.12), and this only has a root $\lambda \geq 1$ and solution (11.1.13) when $\frac{1}{2}\|g^{\text{D}}\| \geq \Delta$. If not, there are a multitude of solutions

$$s^{\text{D}} = -\tfrac{1}{2} g_+^{\text{D}} + \sum_{i \in \mathcal{D}^-} \alpha_i e_i, \text{ where } \sum_{i \in \mathcal{D}^-} \alpha_i^2 = \Delta^2 - \tfrac{1}{4}\|g^{\text{D}}\|^2,$$

correspond to the hard case. Thus we see that, barring the hard case, the solution we seek is either along $-g^{\text{D}}$ or a linear combination of $-g_+^{\text{D}}$ and $-g_-^{\text{D}}$. Reducing the radius Δ simply rebalances the required combination.

The same is true for the regularization case, except that the analogue of (11.1.14) is that we seek the root $\lambda_* \geq 1$ of

$$\frac{\|g_+^{\text{D}}\|^2}{(\lambda_* + 1)^2} + \frac{\|g_-^{\text{D}}\|^2}{(\lambda_* - 1)^2} = \left(\frac{\lambda_*^2}{\sigma^2}\right)^{2/(r-2)}$$

and that the hard case may occur if $\|g_-^{\text{D}}\| = 0$ and $\frac{1}{2}\|g^{\text{D}}\| \geq 1/\sigma^{1/(r-2)}$, in which case

$$s^{\text{D}} = -\tfrac{1}{2} g_+^{\text{D}} + \sum_{i \in \mathcal{D}^-} \alpha_i e_i, \text{ where } \sum_{i \in \mathcal{D}^-} \alpha_i^2 = \left(\frac{1}{\sigma}\right)^{2/(r-2)} - \tfrac{1}{4}\|g^{\text{D}}\|^2.$$

Increasing the regularization parameter σ once again rebalances the contributions of $-g_+^{\text{D}}$ and $-g_-^{\text{D}}$ to the solution in all but the hard case.

In the very unlikely case that $\mathcal{D}^0 \neq \emptyset$, we lose the simplicity of (11.1.10) and have instead (11.1.9) (aside from the hard case). The required λ_* is best computed via a suitable secular equation as in Section 8.3 but should account for the trivial knowledge of the leftmost eigenvalue(s), -1 (and its/their vector(s)), of D. See [190, Sect. 3.4] or [131, Sect. 7.7.4] for more details.

11.1.4 Scalings using symmetric, indefinite factorizations of *H*

Since finding the factorization (11.1.1) is often out of the question, we now consider alternatives. The best-known [50, 190, 369] are based on factorizations of the form

$$H = PLVD^{\text{B}}V^TL^TP^T, \tag{11.1.15}$$

where P is an orthogonal matrix, L is lower triangular, V is a diagonal orthogonal matrix with blocks of order one and two, and D^{B} is diagonal. Such a factorization may be found, for example, via a Bunch–Fletcher–Kaufman–Parlett [66, 67, 166] *symmetric, indefinite factorization*

$$H = PLBL^TP^T,$$

in which P is simply a permutation matrix, B is block diagonal with blocks of order at most two, and additionally

$$B = VD^{\text{B}}V^T$$

is the spectral (eigenvalue-vector) decomposition of B or a sparse variant thereof [6, 155, 156, 239, 240]. For simplicity, we write

$$F = PLV, \text{ and thus } H = FD^{\text{B}}F^T,$$

and note that solving systems with F and its transpose is straightforward since F is the product of easily invertible matrices. Thus, armed with $D^{\rm B}$, it is natural to contemplate the scaling matrix

$$M = F|D_\delta^{\rm B}|F^T,$$

where $|D_\delta^{\rm B}|$ is the modified absolute value (11.1.4) of $D^{\rm B}$. Making the change of variables $s^{\rm D} = |D_\delta^{\rm B}|^{1/2}F^T s$, (8.1.2) and (8.1.3) become (11.1.5) and (11.1.6), where now $g^{\rm D} = |D_\delta^{\rm B}|^{-1/2}F^{-1}g$ and the solution to the original problems (8.1.2) and(8.1.3) given $s_*^{\rm D}$ is $s_* = F^{-T}|D_\delta^{\rm B}|^{-1/2}s_*^{\rm D}$.

The advantages and disadvantages of this approach are considered in [190]. Some care is required to ensure that the factors in (11.1.15) are bounded relative to H since the standard convergence theory for trust-region (and regularization) methods [84, 131] needs M to be uniformly positive definite, and this can be ensured so long as L in (11.1.15) is itself bounded. Fortunately it can be shown [13] that some of the best-known symmetric-indefinite factorizations [67, 155, 156, 166] produce satisfactory L.

11.2 Trust-region radius and regularization weight updates

By far potentially the most expensive aspects of either of the second-order TR2 (p. 55) and AR2 (p. 65) algorithms that we have considered so far in this book are the computations of the required function and derivative values and of the step. In the preceding sections we have shown that calculating the step may be performed efficiently, while the evaluation of functions lies in the users' rather than the algorithm designers' domain and is generally outside our control. Nonetheless, the pessimistic worst-case iteration/evaluation bounds we described in Chapter 3 hardly encourage us to make the best of either the trust-region or the regularization framework. After all, if these methods behaved as badly as the worst-case bounds indicate, no one would contemplate using them. Fortunately, in practice, both are highly successful methods, and it is fair to say that they reflect the current state of the art when second derivatives are available in some form (either directly via Hessians or indirectly via Hessian-vector products).

If we examine the TR2 and AR2 algorithms in detail, we are immediately struck by how much flexibility is provided through the myriad algorithmic parameters, such as the initial trust-region radius, the regularization parameter, and those given in (3.2.8) and (3.3.3). There have been many attempts [14, 147, 191, 193, 229, 251, 255, 318, 337, 360] over the years to suggest "good" values, but in practice the algorithms have proved on average to be relatively insensitive to the choices made, and thus simplicity dictates obvious defaults.

As we have seen, both trust-region and regularization methods control the size of the steps taken in order to ensure convergence. For trust-region methods, the control is explicit via the size of the trust-region radius. By contrast, for regularization methods the control is implicit—the larger the regularization parameter, the shorter the step—but the precise relationship between σ and $\|s\|$ is generally not known.

Since trust-region methods were developed and employed much earlier than their regularization counterparts, much of the accrued knowledge and prejudices concerns them. In practice the trust-region radius-update strategy (3.2.12) is replaced by

$$\Delta_{k+1} = \begin{cases} \max[\Delta_k, \gamma_3\|s_k\|] & \text{if } \rho_k \geq \eta_2, \\ \Delta_k & \text{if } \rho_k \in [\eta_1, \eta_2), \\ \gamma_1\|s_k\| & \text{if } \rho_k < \eta_1, \end{cases} \tag{11.2.1}$$

with $\gamma_2 = \gamma_1$, since (3.2.12) may be inefficient when the trust region is inactive. A variety of typical values have been proposed, e.g., $\eta_1 = 0.1$, $\eta_2 = 0.75$, $\gamma_1 = 0.5$, and $\gamma_3 = 2$ [147, Sect. 6.4.3]; $\eta_1 = 0.25$, $\eta_2 = 0.75$, $\gamma_1 = 0.25$, and $\gamma_3 = 2$ [307, Chap. 4]; $\eta_1 = 0.05$, $\eta_2 = 0.9$, $\gamma_1 = 0.25$, and $\gamma_3 = 2.5$ [131, Sect. 17.1]; and $\eta_1 = 0.0001$, $\eta_2 = 0.99$, $\gamma_1 = 0.25$, and $\gamma_3 = 3.5$ [191], but there appears to be no strong evidence that any one set is preferable.

For the regularization problem, the analogue of (11.2.1) for the strategy (3.3.8) is to update the regularization parameter using[68]

$$\sigma_{k+1} = \begin{cases} \max(\sigma_{\min}, \gamma_1^{\text{R}}\sigma_k) & \text{if } \rho_k \geq \eta_2^{\text{R}}, \\ \sigma_k & \text{if } \rho_k \in [\eta_1^{\text{R}}, \eta_2^{\text{R}}), \\ \gamma_3^{\text{R}}\sigma_k & \text{if } \rho_k < \eta_1^{\text{R}}, \end{cases} \tag{11.2.2}$$

with $\gamma_2^{\text{R}} = \gamma_3^{\text{R}}$. However, despite the requirement that Δ should shrink to zero—and σ increase to infinity—to guarantee a successful step, and the obvious similarities between (11.2.1) and (11.2.2), to naively view σ as being somehow proportional to the reciprocal of Δ is deceptive. In particular, choosing $\gamma_1^{\text{R}} = 1/\gamma_3$ and $\gamma_3^{\text{R}} = 1/\gamma_1$ for regularization in (11.2.2) of the basis of the values suggested above for the trust-region case may lead to inefficiencies.

To see this, consider the example in Figure 11.2.1. Let us suppose that the data in this example occurs as a trust-region subproblem starting with radius $\Delta = 3$ but that the objective function that led to this subproblem is very nonlinear and that the steps generated for a sequence of halving radii are unsuccessful.[69] The obvious question is then: how would regularization (and for this example, we focus on cubic, $r = 3$, regularization) likely perform if started with compatible data? We see from the figure that the trust-region subproblems with radii 3 and 3/2 that would result from (11.2.1) with the traditional $\gamma_1 = 1/2$, $\gamma_3 = 2$ rule produce the same steps as cubic-regularization subproblems with regularization parameters 0.074837 and 0.28984, respectively. The crucial observation is that the ratio of the two successive regularization parameters, 3.873, is significantly larger than would have resulted from $\gamma_1^{\text{R}} = 2$, $\gamma_3^{\text{R}} = 1/2$ using (11.2.2). Indeed, almost two increases (and hence evaluations of the problem objective function) would have been required to achieve the same step. Thus, for the example parameter values we are contemplating, it might well be that the cubic-regularization algorithm (AR2) will require more (wasted) evaluations than its trust-region (TR2) counterpart unless $\gamma_1^{\text{R}} > 2$.

In Table 11.2.1 on page 267, we repeat this calculation for twenty radius reductions. In the left-hand section of the table, we observe that the ratio of regularization parameters σ corresponding to halving radii Δ varies nonmonotonically from 3.8730 to 4.7722 but appears to settle to 4.0. This is no coincidence.

For, consider the secular function $\pi(\lambda)$ in (8.2.7) and the corresponding secular equations (8.2.8)–(8.2.9) as Δ shrinks to zero, and consequently λ becomes very large. In this case, λ is the largest root of

$$\Delta^2 = \pi(\lambda) = \sum_{i=1}^{n} \left(\frac{\gamma_i}{\lambda + \lambda_i} \right)^2 \approx \sum_{i=1}^{n} \left(\frac{\gamma_i}{\lambda} \right)^2 = \frac{\|g\|_{M^{-1}}^2}{\lambda^2}$$

and hence

$$\lambda \approx \Delta / \|g\|_{M^{-1}}, \tag{11.2.3}$$

[68]Here we have appended the superscript R to distinguish the regularization parameters from their trust-region counterparts.

[69]Of course, so long as the objective has a Lipschitz Hessian and the model is a second-order Taylor one, Lemma 2.3.3 shows that this behaviour cannot continue indefinitely.

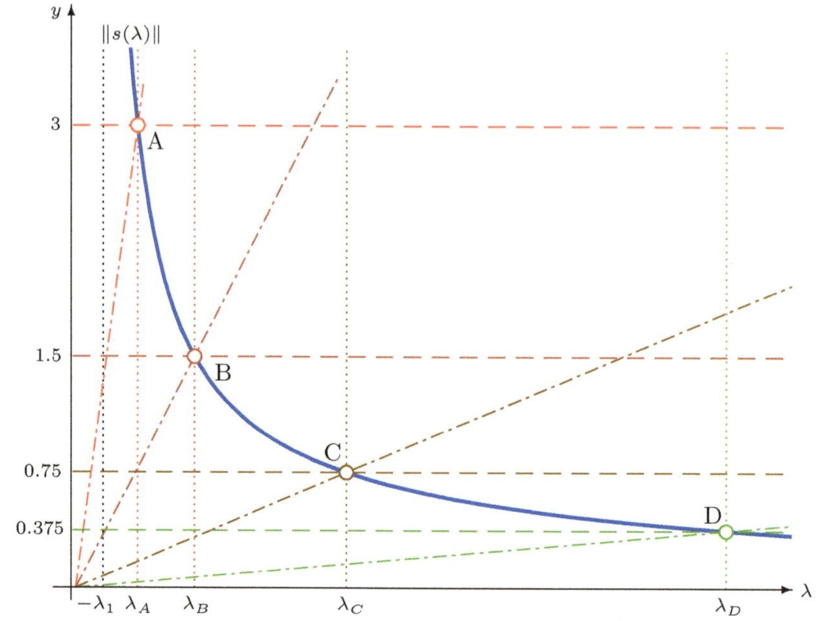

Figure 11.2.1. Trust-region radius versus regularization parameter adjustment for the indefinite quadratic with data $H = \text{diag}(-0.05, 0.9)$ and $g = (0.5, 1)^T$. The points marked A, B, C, and D indicate where the (blue) curve $y = \|s(\lambda)\|$ intersects $y = \Delta$ for decreasing trust-region radii $\Delta = 3$, 1.5, 0.75, and 0.375, respectively; the relevant λ are 0.22451, 0.43476, 0.99042, and 2.3633 to five significant digits. The horizontal dashed (– –) lines are $y = \Delta$ for the given radii, and the slanted dash-dotted (– ·) lines are the corresponding curves $y = \lambda/\sigma$ for regularization parameters $\sigma = 0.074837$, 0.28984, 1.3206, and 6.302.

where $\|g\|_{M^{-1}}^2 = \sum_{i=1}^n \gamma_i^2$ follows from the definition of γ_i and (8.2.5). If instead we consider (8.2.8)–(8.2.9) for large σ and σ, we have

$$\left(\frac{\lambda}{\sigma}\right)^{2/(r-2)} = \pi(\lambda) = \sum_{i=1}^{n} \left(\frac{\gamma_i}{\lambda + \lambda_i}\right)^2 \approx \sum_{i=1}^{n} \left(\frac{\gamma_i}{\lambda}\right)^2 = \frac{\|g\|_{M^{-1}}^2}{\lambda^2},$$

and therefore

$$\lambda \approx \sigma^{1/(r-1)} \|g\|_{M^{-1}}^{(r-2)/(r-1)}. \tag{11.2.4}$$

Thus, we deduce from (11.2.3) and (11.2.4) that the solutions for the trust-region subproblem (8.1.2) and the regularization subproblem (8.1.3) coincide when

$$\sigma \approx \frac{\Delta^{r-1}}{\|g\|_{M^{-1}}}. \tag{11.2.5}$$

Δ	σ	ratio	fidelity	σ	ratio	fidelity
3.0	0.074837	-	0.60242	0.0	-	0.0
1.5	0.28984	3.8730	0.5833	0.14739	∞	0.29662
0.75	1.3206	4.5562	0.6644	1.0792	7.322	0.54296
0.375	6.302	4.7722	0.79266	5.9018	5.4687	0.74232
0.1875	28.235	4.4802	0.88783	27.509	4.6611	0.865
0.09375	119.84	4.2445	0.94209	118.45	4.3059	0.93116
0.046875	493.88	4.1212	0.97063	491.16	4.1465	0.96527
0.023438	2005.2	4.0601	0.98521	1999.8	4.0716	0.98256
0.011719	8080.9	4.0299	0.99258	8070.1	4.0354	0.99126
0.0058594	32444.0	4.0149	0.99629	32423.0	4.0176	0.99563
0.0029297	1.3002^{+5}	4.0075	0.99814	1.2998^{+5}	4.0088	0.99781
0.0014648	5.2056^{+5}	4.0037	0.99907	5.2047^{+5}	4.0044	0.99891
0.00073242	2.0832^{+6}	4.0019	0.99953	2.083^{+6}	4.0022	0.99945
0.00036621	8.3347^{+6}	4.0009	0.99977	8.3344^{+6}	4.0011	0.99973
0.00018311	3.3343^{+7}	4.0005	0.99988	3.3342^{+7}	4.0005	0.99986
9.1553^{-5}	1.3338^{+8}	4.0002	0.99994	1.3338^{+8}	4.0003	0.99993
4.5776^{-5}	5.3353^{+8}	4.0001	0.99997	5.3353^{+8}	4.0001	0.99997
2.2888^{-5}	2.1342^{+9}	4.0001	0.99999	2.1341^{+9}	4.0001	0.99998
1.1444^{-5}	8.5367^{+9}	4.0	0.99999	8.5367^{+9}	4.0	0.99999
5.722^{-6}	3.4147^{+10}	4.0	1.0	3.4147^{+10}	4.0	1.0

Table 11.2.1. Data for the problem in Figure 11.2.1 (left) and that in Figure 11.2.2 (right). A superscript attached to a value indicates its decimal exponent. The column headed "Δ" gives the trust-region radius considered; notice we have chosen each successive radius to be half the value of its predecessor. The data in the columns "σ" gives the values of the regularization parameter that provide the same solution for subproblem (8.1.3) as that for the subproblem (8.1.2) for the corresponding Δ. That in columns headed "ratio" gives the ratio of the current regularization parameter σ to its predecessor. Finally, the values headed "fidelity" are $\sigma\Delta^2/\|g\|_{M^{-1}}$, which will approach 1 as the radius decreases; see (11.2.5).

We can observe this phenomenon for $r = 3$ in the columns headed "fidelity" in Table 11.2.1. Thus for cubic regularization we will expect a reduction in the trust-region radius by a factor of two to correspond to an increase in the regularization parameter by a factor of four, and in general comparing (11.2.3) and (11.2.5) suggests choosing

$$\gamma_1^{\text{R}} = 1/\gamma_3^{r-1} \text{ and } \gamma_3^{\text{R}} = 1/\gamma_1^{r-1}, \tag{11.2.6}$$

at least when σ is large.

Essentially identical behaviour when the model is convex is illustrated in Figure 11.2.2, but of course in this case when the trust-region radius is large enough—for the example, when $\Delta > 3$—the solution to (8.1.2) is inside the trust region. At the critical value at which the trust region becomes active, the corresponding regularization parameter $\sigma = 0$, and for slightly smaller values of Δ, σ changes relatively much more rapidly than when σ is large, and (11.2.6) is unjustified.

It is possible to define more sophisticated rules for adjusting Δ and σ depending on the achieved objective decrease relative to that predicted by the step model, particularly when the prediction turns out to be hopelessly optimistic or inaccurate. However, the details are somewhat involved, and we refer the reader instead to the relevant literature [130, Sect. 3.3.4], [131, Sect. 17.1], [193, 229, 360].

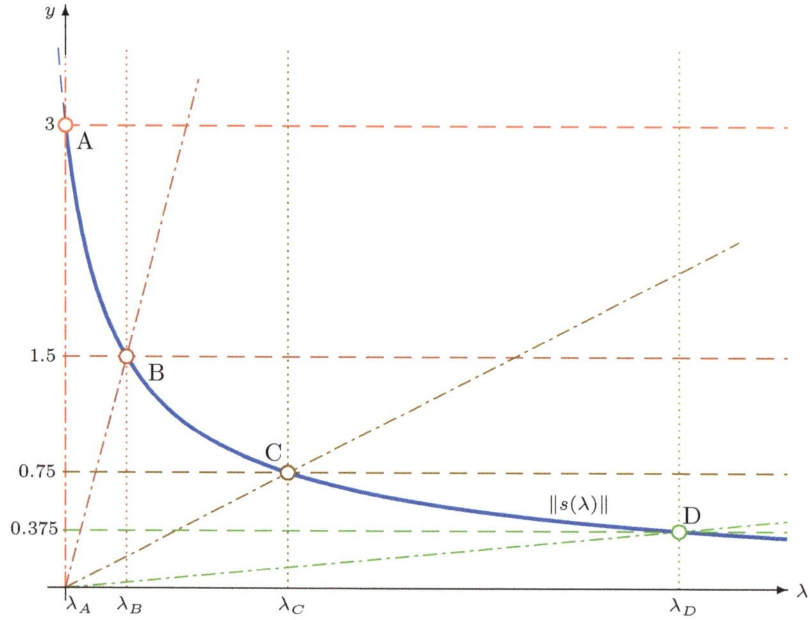

Figure 11.2.2. Radius versus weight adjustment for the convex quadratic with data $H = \text{diag}(1/4\sqrt{2}, 1)$ and $g = (0.5, 1)^T$. The points marked A, B, C, and D indicate where the (blue) curve $y = \|s(\lambda)\|$ intersects $y = \Delta$ for decreasing trust-region radii $\Delta = 3$, 1.5, 0.75, and 0.375, respectively; the relevant λ are 0, 0.22109, 0.8094, and 2.2132 to five significant digits. The horizontal dashed (– –) lines are $y = \Delta$ for the given radii, and the slanted dash-dotted (– ·) lines are the corresponding curves $y = \lambda/\sigma$ for regularization weights $\sigma = 0$, 0.14739, 1.0792, and 5.9018. For any $\Delta > 3$, the solution to the subproblem lies interior to the trust region.

Notes and references for Section 11.2

The choice of the initial trust-region radius and regularization parameter can also have a significant influence on the progress of the relevant algorithm. Indeed, the first step often determines the path of the iterates to critical points, and differing first steps frequently result in convergence to distinct critical points via separate regions of the search space. Thus, it is remarkable that so little effort appears to have been devoted to such an important issue. The exception is the proposal due to Sartenaer [337], in which a sequence of sample objective-function values along a predefined direction—usually the steepest-descent direction—are compared with a quadratic model in that direction. Intricate rules are suggested but with the aim of getting a long trust-region step but at the same time good prediction between the objective function and its model. Of course, such a method adds marginally to the evaluation cost. Similar ideas can clearly be applied in the regularization case.

11.3 Least-squares models

A related and important subproblem occurs when we try to model the least-squares objective function $\Phi_2(x) = \frac{1}{2}\|r(x)\|^2$ for the residuals $r(x)$ that we considered in depth in Section 5.1. The relevant first-order model of $r(x)$ in this case gives rise to the structured quadratic model

$$m(s) \stackrel{\text{def}}{=} \tfrac{1}{2}\|r + Js\|_2^2 = \tfrac{1}{2}\langle r, r\rangle + \langle J^T r, s\rangle + \tfrac{1}{2}\langle s, J^T J s\rangle, \tag{11.3.1}$$

for which the vector r and matrix J are (or are suitable approximations of) the value and Jacobian, $r(x_k)$ and $\nabla_x^1 r(x_k)$, of the residual function $r(x)$ at x_k. The trust-region and regularization subproblems of interest are then (8.1.2) and (8.1.3), respectively, but with (11.3.1) replacing (8.1.1). While we could simply write $g = J^T r$ and $H = J^T J$ and apply the ideas we developed in Chapters 8 to 10, to do so would ignore the special structure involved.

There are a number of important considerations. Firstly, the vital equation (8.2.2) from which we define the secular function (8.2.4) is now

$$(J^T J + \lambda M)s(\lambda) = -J^T r. \tag{11.3.2}$$

The matrix $J^T J + \lambda M$ is positive semidefinite for all $\lambda \geq 0$, and thus λ_{s} in Lemma 8.2.1 is zero. Furthermore, the hard case cannot occur because $\lambda_1 \geq 0$, but in the trust-region case we must be prepared for an interior solution since the vector $J^T r$ lies in the range of $J^T J$. If this doesn't happen, we can restrict the subproblem solution in Section 8.3 to the sets $\mathcal{L}$ and $\mathcal{U}$, $\mathcal{L} \neq \emptyset$, and the secular equation (8.2.3) has a unique root that we may find by Algorithm 8.3.1 or 8.3.2, starting from a tiny positive λ_{c}, as appropriate.

Secondly, on defining auxiliary variables $z \stackrel{\text{def}}{=} Js + r$, we may rewrite (11.3.2) as the equivalent (symmetric) augmented system

$$\begin{pmatrix} \lambda M & J^T \\ J & -I \end{pmatrix} \begin{pmatrix} s \\ z \end{pmatrix} = - \begin{pmatrix} 0 \\ r \end{pmatrix}. \tag{11.3.3}$$

At first sight this might seem to be a retrograde step since the resulting system is order $n + m$ rather than n, but in some cases there may also be a significant advantage. In particular, if n is large and J is sparse, $J^T J$ often has significantly more nonzeros that J. Indeed, in the worst case, if J has a dense row, $J^T J$ will be dense. Thus, if a direct method is to be employed to solve (11.3.2), careful consideration should be given to the alternative (11.3.3) instead; (11.3.3) is symmetric and indefinite, and there is a large choice of appropriate solvers in this case [187].

Lastly, because the hard case does not occur, the Krylov-space methods considered in Chapter 10 converge to the required solution, and Corollaries 10.2.3 and 10.3.3 give some idea of the expected (worst-case) behaviour. Moreover, while the Lanczos process is the key to the iteration in Chapter 10, the fact that $H = J^T J$ allows one to apply the Lanczos bidiagonalization process [184] to J instead. Indeed, it is possible to use the very efficient LSQR method [312] as the basis of effective iterative methods for solving both of the regularized least-squares problems that concern us.

Notes and references for Section 11.3

Iterative methods along the lines mentioned above have been analyzed [82] and implemented (as packages `LSTR` and `LSRT`) as part of GALAHAD [192].

Part II

Evaluation Complexity for Higher-Order Optimality

Our final topic concerns the verification of approximate criticality of order four and above. This is a much less mature area than those that have gone before and illustrates the significant differences and difficulties when we demand more of our algorithms. It is nonetheless the most satisfying in its aspirations.

Previous page: From the combook 4, by Philippe Toint, 2020

Chapter 12

Finding High-Order Approximate Minimizers

Chapter summary

We saw in Chapter 4 how using high-order derivatives benefits the evaluation complexity of finding first-, second-, and third-order ϵ-approximate minimizers. If such derivatives are available, it is natural to ask how they might also be used productively to find ϵ-approximate qth-order minimizers for $q > 3$ ..., provided of course that we are even able to define such minimizers sensibly. With this aim in mind, we start by discussing general optimality conditions and their significant limitations. As a result, we then define reasonable concepts of high-order criticality based on Taylor series and use them to design a trust-region method that requires at most $\mathcal{O}(\epsilon^{-(q+1)})$ evaluations of the objective function and its first q derivatives to compute an ϵ-approximate qth-order minimizer. We then prove that regularization methods require at most $\mathcal{O}(\epsilon^{-(p+1)/(p-q+1)})$ evaluations of the objective function and its first $p \geq q$ derivatives to achieve the same goal for $q = 1$ and 2 and at most $\mathcal{O}(\epsilon^{-q(p+1)/p})$ such evaluations when $q > 2$. We conclude our discussion of evaluation complexity for unconstrained problems by showing that this latter, very general bound is sharp and consider the not-inconsiderable computational issues involved.

12.1 Optimality conditions revisited

12.1.1 The task at hand and its limits

Now that we have exploited high-degree models to speed up convergence to first- and second-order approximate minimizers using the ARp algorithm—and to find third-order minimizers if desired—it is reasonable to wonder whether finding a qth-order approximate minimizer using a model of degree $p \geq q$ is possible in general. The focus of the rest of this chapter is to address this question, and as we shall see this raises a number of interesting, but delicate, issues.

The main point of this section is to explore what we really mean by a qth-order ϵ-approximate minimizer. To do so requires that we return to the basic aims of local optimization (minimization, in this discussion). As is standard, we say that x_* is a local minimizer of f if and only if there exists a (sufficiently small) neighbourhood $\mathcal{B}_*$ of x_* such that

$$f(x) \geq f(x_*) \quad \text{for all} \quad x \in \mathcal{B}_*. \tag{12.1.1}$$

Our aim, as is common in local optimization, is to infer characteristics of such points using derivatives. However, before we start, we must be wary of important intrinsic limitations, as follows.

Example 12.1.1 Consider the smooth two-dimensional problem

$$\min_{x\in\mathbb{R}^2} f(x) = \begin{cases} x_2\left(x_2 - \exp(-1/x_1^2)\right) & \text{if } x_1 \neq 0, \\ x_2^2 & \text{if } x_1 = 0, \end{cases} \tag{12.1.2}$$

which is a simplified version of a problem presented by Hancock more than a century ago [224, p. 36], itself a variation of a famous problem stated even earlier by Peano [316, Nos. 133–136]. The contour lines of its objective function are shown in Figure 12.1.1 below.

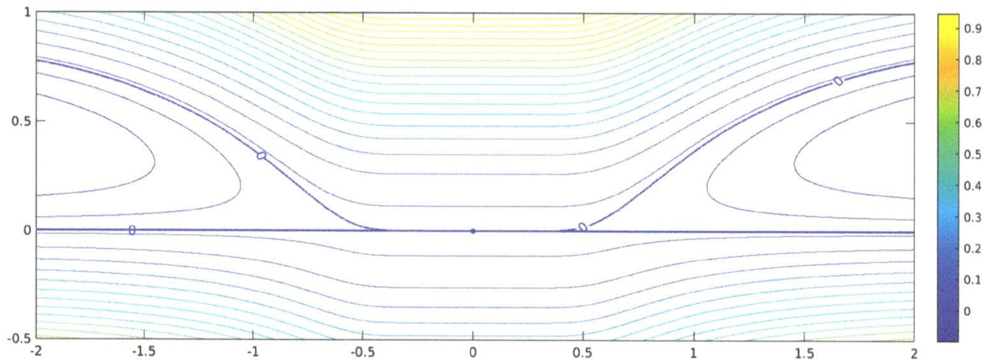

Figure 12.1.1. The contour lines of the objective function in (12.1.2). The bold lines are the contours for $f(x) = 0$. Note that the function is zero at the origin, takes positive values in the upper and lower parts of the picture, and takes negative values in the two aerofoil-shaped areas on the left and right.

The first conclusion that can be drawn by examining this example is that, in general, assessing whether a given point x (the origin in this case) is a local minimizer needs more than simply verifying that no (straight) line from this point is a strict descent direction. Indeed, such a property holds for the example, but the origin is not a local minimizer, but rather a saddle point. This is caused by the fact that objective-function decrease may occur along specific curves starting from the point under consideration, but these curves need not be lines (such as $x(\alpha) = 0 + \alpha e_2 + \frac{1}{2}\exp(-1/2\alpha^2)e_1$ for $\alpha \geq 0$ in the example).

The second deduction that we can make from the example above is that a local minimizer may not always be characterizable via a set of conditions involving only the Taylor expansion of f at x_*. In our example, the difficulty arises because the coefficients of the Taylor expansion of $\exp(-1/x_1^2)$ at x all vanish as x_1 approaches the origin, and therefore the (non)minimizing nature of this point cannot be determined from the values of these coefficients. Thus the gap between necessary and sufficient optimality conditions (of any order) cannot be closed if we restrict our attention to using derivatives of the objective function at a putative solution.

Note that worse situations may also occur, for instance, if we consider the variant

$$\min_{x\in\mathbb{R}^2} f(x) = \begin{cases} x_2\left(x_2 - \sin(1/x_1)\exp(-1/x_1^2)\right) & \text{if } x_1 \neq 0, \\ x_2^2 & \text{if } x_1 = 0 \end{cases} \tag{12.1.3}$$

of Hancock's simplified example (12.1.2), for which no continuous descent curve exists in a neighbourhood of the origin, even though the origin is not a local minimizer.

Notes and references for Subsection 12.1.1

The example (12.1.3) was suggested by Adrian Lewis [267].

12.1.2 Curve-based necessary optimality conditions

The above examples show that fully characterizing a local minimizer in terms of general continuous descent curves is in general impossible. However, the fact that no such curve exists remains a necessary condition for such a point, even if Hancock's example shows that these curves may not always be characterized using derivatives along them. In what follows, we therefore propose derivative-based necessary optimality conditions by focusing on a specific (yet reasonably general) class of descent curves $x_*(\alpha)$ of the form

$$x_*(\alpha) = x_* + \sum_{i=1}^{q} \alpha^i s_i + o(\alpha^q), \tag{12.1.4}$$

where $\alpha > 0$. It is then possible to state conditions that preclude the existence of feasible descent curves of the form (12.1.4), yielding necessary conditions for x_* to be a local minimizer.

Theorem 12.1.1: Curve-based necessary conditions for an unconstrained local minimum

[102, Thm. 3.1]: Suppose that f is q-times continuously differentiable in an open neighbourhood of x_* and that x_* is a local unconstrained minimizer of f. Let

$$\mathcal{P}(j,k) \stackrel{\text{def}}{=} \left\{(\ell_1,\ldots,\ell_k) \in [1:j]^k \mid \sum_{i=1}^{k} \ell_i = j\right\}. \tag{12.1.5}$$

Then, for $j \in [1:q]$, the inequality

$$\sum_{k=1}^{j} \frac{1}{k!}\left(\sum_{(\ell_1,\ldots,\ell_k)\in\mathcal{P}(j,k)} \nabla_x^k f(x_*)[s_{\ell_1},\ldots,s_{\ell_k}]\right) \geq 0 \tag{12.1.6}$$

holds for all $(s_1,\ldots,s_j) \in (\mathbb{R}\setminus\{0\})^j$ such that, for $i \in [1:j-1]$,

$$\sum_{k=1}^{i} \frac{1}{k!}\left(\sum_{(\ell_1,\ldots,\ell_k)\in\mathcal{P}(i,k)} \nabla_x^k f(x_*)[s_{\ell_1},\ldots,s_{\ell_k}]\right) = 0. \tag{12.1.7}$$

Table 12.1.1 gives the values of the index sets $\mathcal{P}(j,k)$ for $k \le j \le 4$.

$j \downarrow$	$k \rightarrow$ 1	2	3	4
1	{(1)}			
2	{(2)}	{(1,1)}		
3	{(3)}	{(1,2),(2,1)}	{(1,1,1)}	
4	{(4)}	{(1,3),(2,2),(3,1))}	{(1,1,2),(1,2,1),(2,1,1)}	{(1,1,1,1)}

Table 12.1.1. The sets $\mathcal{P}(j,k)$ for $k \le j \le 4$.

Following a long tradition, we say that x_* is a *qth-order minimizer* if the conclusions of Theorem 12.1.1 hold for all $j \in [1:q]$. Of course, a qth-order minimizer need not be a local minimizer, but every local minimizer is a qth-order minimizer.

Theorem 12.1.1 can be interpreted as the requirement that, for all $j \in [1:q]$, the jth-order expansion of $f(x_*(\alpha))$ (i.e., the left-hand side of (12.1.6)) not decrease along any curve for which the decrease of the developments of order up to $j-1$ is identically zero (which is expressed by (12.1.7)). Alternatively, (12.1.6) may be interpreted as imposing a condition on s_j (via $\nabla_x^1 f(x_*)[s_j]$), given the directions $\{s_i\}_{i=1}^{j-1}$ satisfying (12.1.7), for increasing order j. In general situations, the fact that conditions (12.1.6) and (12.1.7) not only depend on the behaviour of the objective function in some well-chosen subspace but involve the geometry of all possible curves makes this result difficult to apply (as we discuss below). Because of this, we skip the proof of Theorem 12.1.1, instead referring the reader to [102, Thm. 3.1] for full details (in the more general context of convexly constrained problems).

We now consider some special cases. Because, for $q = 1$, one is free to choose the vectors $\{s_i\}_{i=1}^{q}$ (and their sign) arbitrarily, (12.1.6) implies that, unsurprisingly,

$$\nabla_x^1 f(x_*) = 0.$$

For the second-order condition, we obtain from (12.1.6), again unsurprisingly, that, because $\ker^1[\nabla_x^1 f(x_*)] = \mathbb{R}^n$,

$$\nabla_x^2 f(x_*) \text{ is positive semidefinite on } \mathbb{R}^n,$$

where we recall more generally that $\ker^p[T]$ is the p-kernel of the multilinear p-form $T[v]^p$ defined by (A.8). Hence, if there exists a vector $s_1 \in \ker^2[\nabla_x^2 f(x_*)]$, we have that $\|[\nabla_x^2 f(x_*)]^{\frac{1}{2}} s_1\| = 0$ and therefore that $\nabla_x^2 f(x_*)[s_1, s_2] = 0$ for all $s_2 \in \mathbb{R}^n$. Thus the term for $k = 1$ vanishes from (12.1.6), as well as all terms involving $\nabla_x^2 f(x_*)$ applied to a vector $s_1 \in \ker^2[\nabla_x^2 f(x_*)]$. This implies in particular that the third-order condition may now be written as

$$\nabla_x^3 f(x_*)[s_1]^3 = 0 \quad \text{for all } s_1 \in \ker^2[\nabla_x^2 f(x_*)],$$

where the equality follows since if s_1 lies in $\ker^2[\nabla_x^2 f(x_*)]$ then so does $-s_1$.

Unfortunately, complications arise with fourth-order conditions, even when the objective function is a polynomial.

Example 12.1.2 Consider the following variant of Peano's [316] problem:

$$\min_{x\in\mathbb{R}^2} f(x) = x_2^2 - \kappa_1 x_1^2 x_2 + \kappa_2 x_1^4, \tag{12.1.8}$$

where κ_1 and κ_2 are parameters. Then one can verify that

$$\nabla_x^1 f(0) = 0, \quad \nabla_x^2 f(0) = \begin{pmatrix} 0 & 0 \\ 0 & 2 \end{pmatrix};$$

$$[\nabla_x^3 f(0)]_{ijk} = \begin{cases} -2\kappa_1 & \text{for } \ (i,j,k) \in \{(1,2,1),(1,1,2),(2,1,1)\} = \mathcal{P}(4,3), \\ 0 & \text{otherwise;} \end{cases}$$

and

$$[\nabla_x^4 f(0)]_{ijk\ell} = \begin{cases} 24\kappa_2 & \text{for } \ (i,j,k,\ell) = (1,1,1,1), \\ 0 & \text{otherwise.} \end{cases}$$

Hence

$$\ker^1[\nabla_x^1 f(0)] = \mathbb{R}^2, \quad \ker^2[\nabla_x^2 f(0)] = \operatorname{span}\{e_1\},$$

and

$$\ker^3[\nabla_x^3 f(0)] = \operatorname{span}\{e_1\} \cup \operatorname{span}\{e_2\}.$$

The necessary condition (12.1.6) states that if the origin is a minimizer, then, using the curve defined by $s_1 = e_1$ and $s_2 = \frac{1}{2}\kappa_1 e_2$ and the fact that $\mathcal{P}(4,3)$ contains three elements,

$$\tfrac{1}{2}\nabla_x^2 f(0)[s_2]^2 + \tfrac{1}{2}\nabla_x^3 f(0)[s_1,s_1,s_2] + \tfrac{1}{24}\nabla_x^4 f(0)[s_1]^4 = \tfrac{1}{4}\kappa_1^2 - \tfrac{1}{2}\kappa_1^2 + \kappa_2 = -\tfrac{1}{4}\kappa_1^2 + \kappa_2 \geq 0.$$

This shows that although the condition $\nabla_x^4 f(x_*)[s_1]^4 \geq 0$ on $\cap_{i=1}^3 \ker^i[\nabla_x^i f(x_*)]$ is necessary, it is arbitrarily far away from the better necessary condition

$$\tfrac{1}{2}\nabla_x^2 f(0)[s_2]^2 + \tfrac{1}{2}\nabla_x^3 f(0)[s_1,s_1,s_2] + \tfrac{1}{24}\nabla_x^4 f(0)[s_1]^4 \geq 0 \tag{12.1.9}$$

when κ_1 grows. As we have already seen in the case of problem (12.1.2), choosing, for example, $\kappa_1 = 1$ and $\kappa_2 = 2$ in (12.1.8) demonstrates that a function may admit a saddle point ($x_* = 0$) which is a maximum along a curve ($x_2 = \frac{3}{2}x_1^2$ in this case) while at the same time being minimal along every (straight) line passing through x_*. Figure 12.1.2 indicates the contour lines of the objective function of (12.1.8) for increasing values of κ_2, keeping $\kappa_1 = 3$.

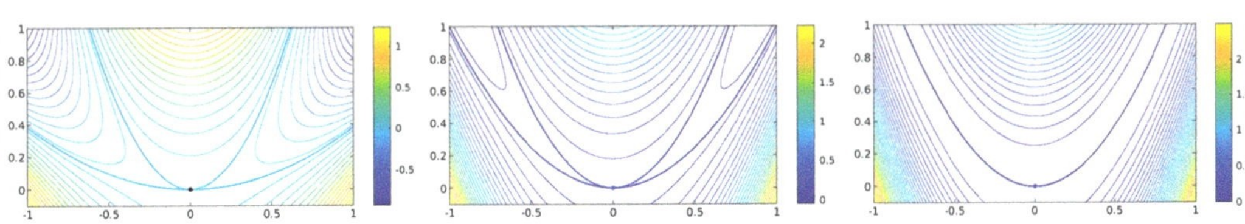

Figure 12.1.2. The contour lines of (12.1.8) for $\kappa_1 = 3$ and $\kappa_2 = 1$ (left), $\kappa_2 = 2$ (center), and $\kappa_2 = 2.25$ (right). The bold lines are the contours for $f(x) = 0$. Notice the negative values of $f(x)$ in the left two pictures, and note how there is no straight line emanating from $x = 0$ into these regions that doesn't initially increase the function.

One may attribute the difficulty that not every term in (12.1.6) vanishes to the fact that switching signs of s_1 or s_2 does not imply that any of the terms in (12.1.9) are zero (as we have verified) because of the terms $\nabla_x^2 f(0)[s_2]^2$ and $\nabla_x^4 f(0)[s_1]^4$. Is this a feature of even orders only? Unfortunately, this is not the case for $q = 7$. Indeed it is not difficult to verify that the terms whose multi-index $(\ell_1, \ldots, \ell_k)$ is a permutation of $(1,2,2,2)$ belong to $\mathcal{P}(7,4)$ and those whose multi-index is a permutation of $(1,1,1,1,1,2)$ belong to $\mathcal{P}(7,6)$. Moreover, the contribution of these terms to the sum (12.1.6) cannot be distinguished by varying s_1 or s_2, for instance, by switching their signs, since this technique yields only one equality in two unknowns.

Thus, in general, we conclude that the requirement (12.1.6) must involve a *mixture of terms with derivative tensors of various degrees* for minimizers of order beyond three. The Peano example above also has another important consequence.

Theorem 12.1.2: Linesearch algorithms are unsuitable for finding high-order approximate local minimizers

Linesearch methods (i.e., methods in which the next iterate is found by approximately minimizing the objective function f along a descent direction) cannot be proved to be locally convergent to approximate local minimizers of orders larger than three.

Proof. Suppose a linesearch method is started from the origin in the above example. Since $f(x)$ is locally minimized at the origin along any line passing through the origin, every search direction is an ascent direction. This then causes the algorithm to terminate prematurely at a point which is not a fourth-order local minimizer. □

This unfortunate conclusion is the reason we shall not consider linesearch methods in this chapter. Note that this result does not apply to "curvilinear linesearches" [57, 126, 291] if the curve along which the search is conducted takes conditions (12.1.6)–(12.1.7) into account.

Notes and references for Subsection 12.1.2

Curve-based approaches for necessary optimality conditions have been discussed by several authors for first- and second-order conditions (see Ben-Tal and Teboulle [31], Cominetti [124], Bonnans, Cominetti, and Shapiro [54], or Gould and Leyffer [188], for example). Much of the material here, and additional details covering problems involving convex constraints, may be found in [102].

12.1.3 A necessary condition using Taylor-series models

As we have already noted, the conditions expressed in Theorem 12.1.1 may, in general, be very difficult to verify by an algorithm because of their dependence on a potentially infinite set of curves. To avoid this difficulty, we shall now explore a different approach: we abandon the complicated curve-based conditions for a neighbourhood-based technique. To this end, we define, for some given *optimality radius* $\delta > 0$ and $j \in [1:p]$, the measure

$$\phi_{f,j}^{\delta}(x) \stackrel{\text{def}}{=} f(x) - \min_{\|d\| \le \delta} T_{f,j}(x,d), \tag{12.1.10}$$

which is the *largest decrease of the jth-order Taylor-series model $T_{f,j}(x,s)$ achievable by a point at distance at most δ from x*. Note that $\phi_{f,j}^{\delta}(x)$ is a continuous function of x and

δ for $f \in C^j$ [238, Thm. 7]. It is also important to observe that $\phi_{f,j}^{\delta}(x_k)$ is *independent of the value of* $f(x_k)$, because the zeroth-degree terms cancel in (12.1.10). Indeed, if we define the *Taylor decrement*

$$\begin{aligned}\Delta T_{f,j}(x,d) &\stackrel{\text{def}}{=} T_{f,j}(x,0) - T_{f,j}(x,d)\\ &= -\sum_{i=1}^{j} \frac{1}{i!}[\nabla_v^i T_{f,j}(x,v)]_{v=0}[d]^i \\ &\equiv -\sum_{i=1}^{j} \frac{1}{i!}\nabla_x^i f(x)[d]^i\end{aligned} \tag{12.1.11}$$

at x and for a step d, then

$$\phi_{f,j}^{\delta}(x) = \max_{\|d\|\leq\delta} \Delta T_{f,j}(x,d).$$

In what follows, we will mostly consider $\delta \leq 1$, but this is not necessary.[70] This new definition avoids the complicated curves used in Theorem 12.1.1 but comes at a price: we must be prepared to perform the global optimization that is central to (12.1.10), at least as we approach optimality.

It is comforting to note that the standard first-, second-, and third-order optimality conditions for unconstrained problems are directly related to our new measure, as we now show.

Theorem 12.1.3: Relating the old and new criticality measures

Suppose that f is continuously differentiable. Then

(i) *for any $\delta > 0$ and any $x \in \mathbb{R}^n$, we have*

$$\|\nabla_x^1 f(x)\| = \frac{\phi_{f,1}^{\delta}(x)}{\delta}, \tag{12.1.12}$$

and so $\phi_{f,1}^{\delta}(x) = 0$ if and only if $\nabla_x^1 f(x) = 0$;

(ii) *if $f \in C^2$, then if*

$$\phi_{f,2}^{\delta}(x) = 0 \text{ for some } \delta > 0 \implies \nabla_x^1 f(x) = 0 \text{ and } \lambda_{\min}[\nabla_x^2 f(x)] \geq 0,$$

and so any such x is a first- and second-order minimizer; and

(iii) *if $f \in C^3$, $\mathcal{M}(x) \stackrel{\text{def}}{=} \ker^1[\nabla_x^1 f(x)] \cap \ker^2[\nabla_x^2 f(x)]$, $P_{\mathcal{M}(x)}$ is the exact orthogonal projection onto that subspace, and x satisfies the first- and second-order conditions exactly (i.e., $\nabla_x^1 f(x) = 0$ and $\nabla_x^2 f(x)$ is positive semidefinite), then*

$$\|P_{\mathcal{M}(x)}[\nabla_x^3 f(x)]\| \leq \frac{\phi_{f,3}^{\delta}(x)}{(\delta^3/3!)}, \tag{12.1.13}$$

and so x is also a third-order minimizer provided that $\phi_{f,3}^{\delta}(x)/\delta^3 \to 0$ as $\delta \to 0$.

---

[70] A more general, scaling-dependent, maximum-value $\delta_{\max}$ on δ is permissible in theory (as can easily be verified by scaling the variables of the problem). We have refrained from such an additional dependence on scaling in our exposition so as to limit the already significant notational burden.

Proof. In the first-order case (i), it is easy to verify that

$$\begin{aligned}\|\nabla_x^1 f(x)\| &= \frac{1}{\delta}\left[-\min_{\|d\|\le\delta} \nabla_x^1 f(x)[d]\right]\\ &= \frac{1}{\delta}\left[f(x) - \min_{\|d\|\le\delta}\left(f(x)+\nabla_x^1 f(x)[d]\right)\right]\\ &= \frac{\phi_{f,1}^{\delta}(x)}{\delta},\end{aligned}$$

where the first equality is justified by the convexity of $\nabla_x^1 f(x)[d]$ as a function of d. Thus if $\phi_{f,1}^{\delta}(x)$ is zero, so is $\nabla_x^1 f(x)$ regardless of the value of $\delta > 0$, and thus x is a first-order minimizer.

For the second-order case (ii), we may appeal to (8.3.5) to see that

$$\begin{aligned}&\frac{1}{\delta^2}\left\langle \nabla_x^1 f(x), \left(\nabla_x^2 f(x) + \lambda I\right)^{+} \nabla_x^1 f(x)\right\rangle + \lambda\\ &\quad= \frac{2}{\delta^2}\left[-\min_{\|d\|\le\delta}\left(\nabla_x^1 f(x)[d] + \tfrac{1}{2}\nabla_x^2 f(x)[d]^2\right)\right]\\ &\quad= \frac{2}{\delta^2}\left[f(x) - \min_{\|d\|\le\delta}\left(f(x) + \nabla_x^1 f(x)[d] + \tfrac{1}{2}\nabla_x^2 f(x)[d]^2\right)\right]\\ &\quad= \frac{\phi_{f,2}^{\delta}(x)}{(\delta^2/2)},\end{aligned}\tag{12.1.14}$$

where $\lambda \ge \max(0, -\lambda_{\min}[\nabla_x^2 f(x)])$ and crucially $\langle \nabla_x^1 f(x), (\nabla_x^2 f(x) + \lambda I)^{+}\nabla_x^1 f(x)\rangle > 0$ unless $\nabla_x^1 f(x) = 0$; the matrix-related notation $^{+}$ denotes the generalized inverse of the matrix in question; see Appendix A.5 (p. 431). In this case, both $\nabla_x^1 f(x) = 0$ and $\lambda = 0$ occur when $\phi_{f,2}^{\delta}(x) = 0$, for any $\delta > 0$, and $\lambda = 0$ ensures that $\lambda_{\min}[\nabla_x^2 f(x)] \ge 0$. Thus such an x is a first- and second-order minimizer.

Finally, for the third-order case (iii), we have

$$\begin{aligned}&\|P_{\mathcal{M}(x)}[\nabla_x^3 f(x)]\|\\ &\quad= \frac{1}{\delta^3}\left[-\min_{\|d\|\le\delta, d\in\mathcal{M}(x)} \nabla_x^3 f(x)[d]^3\right]\\ &\quad= \frac{6}{\delta^3}\left[f(x) - \min_{\|d\|\le\delta, d\in\mathcal{M}(x)}\left(f(x) + \nabla_x^1 f(x)[d] + \tfrac{1}{2}\nabla_x^2 f(x)[d]^2 + \tfrac{1}{6}\nabla_x^3 f(x)[d]^3\right)\right]\\ &\quad\le \frac{\phi_{f,3}^{\delta}(x)}{(\delta^3/3!)},\end{aligned}\tag{12.1.15}$$

where the first equality results from the definition of the tensor norm. Thus, if the right-hand side of (12.1.15) converges to zero with δ, and since the left-hand side is independent of δ, we deduce from Theorem 1.1.1 (p. 3) that x is also a third-order minimizer. □

Theorem 12.1.3 and its proof show that even though global optimization must be performed to calculate the new measure, this is feasible for low criticality orders. In particular, the global optimization problem in (12.1.14) is of trust-region type, and thus it is solvable by any of the methods discussed in Chapters 9 and 10 or [131, Sect. 7.3].

For third-order minimizers, the global optimization on the right-hand side of (12.1.13) still occurs in a subspace ($\mathcal{M}(x)$) as (12.1.15) clearly shows, but we saw at the end of Section 12.1.2 that $q = 3$ is the highest order for which this is possible. And, again, computing $\|P_{\mathcal{M}(x)}[\nabla_x^3 f(x)]\|$ directly involves global optimization in (A.7). Although we justify (12.1.13) by assuming second-order criticality, this is not necessary. Indeed it follows from Lemma B.12.1 (p. 481) that $\phi_{f,2}^{\delta}(x) = 0$ whenever $\phi_{f,3}^{\delta}(x) = 0$.

Reassuringly, the following general result holds for the measure (12.1.10).

Theorem 12.1.4: The ϕ measure and local minimizers

Suppose that f is j-times continuously differentiable and that x is a local minimizer of f. Then

$$\phi_{f,j}^{\delta}(x) = 0 \quad \textit{for} \quad j \in \{1, 2\} \tag{12.1.16}$$

for all $\delta > 0$. Moreover, for all $j \geq 1$,

$$\lim_{\delta \to 0} \frac{\phi_{f,j}^{\delta}(x)}{\delta^j} = 0. \tag{12.1.17}$$

Proof. If x is a local minimizer and $j \in \{1, 2\}$, the standard optimality conditions, (12.1.12) and (12.1.14), imply that $T_{f,1}(x, d)$ and $T_{f,2}(x, d)$ are convex functions of d with a global minimizer at zero. Hence $\phi_{f,j}^{\delta}(x) = 0$ for $j \in \{1, 2\}$ for any $\delta \in [0, 1]$, which proves (12.1.16). We now prove (12.1.17). Since x is a local minimizer, there exists a neighbourhood of x of radius $r > 0$ such that $f(x + d) \geq f(x)$ for every d such that $\|d\| \leq r$. Consider d with $\|d\| = \delta \leq r$ and δ arbitrarily small. Then, using the weak Taylor-series bound (A.12), we have

$$f(x) \leq f(x + d) = T_{f,j}(x, d) + o(\delta^j).$$

Thus

$$0 \leq \phi_{f,j}^{\delta}(x) = f(x) - \min_{\|d\| \leq \delta} T_{f,j}(x, d) \leq o(\delta^j),$$

and the conclusion follows by dividing this inequality by δ^j. □

Unfortunately, (12.1.16) does not necessarily hold for $j > 2$ if δ is chosen too large, as the following example shows.

Example 12.1.3 Consider the function[71]

$$m(x) = -\beta \langle e_1, x \rangle + \sigma \frac{\|x\|^{p+1}}{(p+1)!} \tag{12.1.18}$$

for some $\beta > 0$ and $\sigma > 0$. It is easy to verify that this function has a unique global minimizer at

$$x_* = (\beta p!/\sigma)^{1/p} e_1, \tag{12.1.19}$$

at which

$$m(x_*) = -p\,(p!/\sigma)^{1/p} \beta^{(p+1)/p}/(p+1). \tag{12.1.20}$$

We now investigate the behaviour of the Taylor expansions $T_{m,j}(x_*, d)$ for increasing j. Lemma B.4.1 (p. 461) allows us to deduce that $T_{m,j}(x_*, d)$ is convex in any direction

[71]We call this function m rather than f since we will return to it later for another purpose.

orthogonal to x_* for all $j \in [1:q]$, and thus restricting δ will not affect deductions about approximate minimizers in these directions. Thus we focus on the behaviour of $T_{m,j}(x_*, \alpha x_*)$ for $\alpha \in \mathbb{R}$. Since $\nabla_s^1 m(x_*) = 0$, we obtain that

$$\begin{aligned} T_{m,j}(x_*, \alpha x_*) &= \sum_{\ell=0}^{j} \frac{\nabla_x^\ell m(x_*)[\alpha x_*]^\ell}{\ell!} \\ &= m(x_*) + \frac{\sigma}{(p+1)!} \sum_{\ell=2}^{j} \frac{\nabla_x^\ell (\|x_*\|^{p+1})[\alpha x_*]^\ell}{\ell!} \\ &= m(x_*) + \frac{\sigma}{(p+1)!} \sum_{\ell=2}^{j} \frac{\alpha^\ell \nabla_x^\ell (\|x_*\|^{p+1})[x_*]^\ell}{\ell!} \\ &= m(x_*) + \sigma \sum_{\ell=2}^{j} \frac{\alpha^\ell \|x_*\|^{p+1}}{\ell!(p+1-\ell)!} \end{aligned}$$

from (B.4.2). The shape of these approximations for $j \in [0:q]$ (for the case $\beta = 1 = \|x_*\|$, $p = q = 5$, and $\sigma = p!$) is shown in Figure 12.1.3.

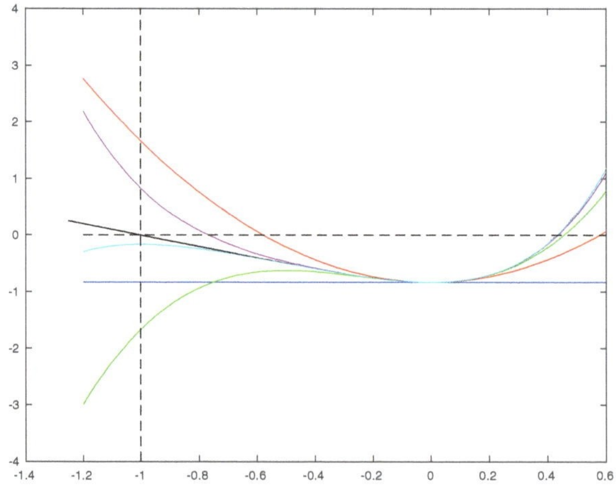

Figure 12.1.3. The function $m(x)$ of Example 12.1.3 (in black) and its first five Taylor approximations computed at the global minimizer $x_* = 1$ (i.e., $\alpha = 0$), as a function of α (first degree in blue, second degree in red, third degree in green, fourth degree in magenta, fifth degree in cyan).

Clearly $T_{m,1}(x_*, \alpha x_*) = T_{m,1}(x_*, 0) = m(x_*)$ for all α because the standard first-order optimality condition at x_* gives that $\nabla_d^1 T_{m,1}(x_*, 0) = 0$. The second-order optimality condition implies that $T_{m,2}(x_*, \alpha x_*)$ is convex (as can be seen in the figure), but it is apparent that *approximations of degree larger than 2 are no longer convex for odd values of j and negative values of α*. For $j = 3$ we have that

$$T_{m,3}(x_*, 0) - T_{m,3}(x_*, \alpha x_*) = -\frac{\sigma\alpha^2 \|x_*\|^{p+1}}{2(p-1)!} - \frac{\sigma\alpha^3 \|x_*\|^{p+1}}{6(p-2)!}, \qquad (12.1.21)$$

and thus, if we choose $\delta_3 = 1$,

$$\begin{aligned}\phi^1_{m,3}(x_*) &= T_{m,3}(x_*,0) - T_{m,3}(x_*,-x_*) = -\frac{\sigma\|x_*\|^{p+1}}{2(p-1)!} + \frac{\sigma\|x_*\|^{p+1}}{6(p-2)!}\\ &= \frac{(p-4)\sigma\|x_*\|^{p+1}}{6(p-1)!} > 0\end{aligned}$$

whenever $p > 4$. However, the figure suggests that we may ensure that $\phi^{\delta_3}_{m,3}(x_*) = 0$ if we are prepared to reduce δ_3. Indeed, for this to hold, we need that

$$T_{m,3}(x_*,0) \le T_{m,3}(x_*,\alpha x_*)$$

for $\|\alpha x_*\| \le \delta$, which, in view of (12.1.21), requires that

$$-\frac{\sigma\alpha^2\|x_*\|^{p+1}}{2(p-1)!} - \frac{\sigma\alpha^3\|x_*\|^{p+1}}{6(p-2)!} \le 0$$

and is true for all

$$\alpha \in \left[-\frac{3}{(p-1)}, \infty\right).$$

Hence $\phi^{\delta_3}_{m,3}(x_*) = 0$ for all $\delta_3 \le \bar\delta_3 \stackrel{\rm def}{=} 3\|x_*\|/(p-1) = 3(\beta p!/\sigma)^{1/p}/(p-1)$. For $p = 5$, $\beta = 1$, and $\sigma = p!$, this requires that $\delta_3 \le \frac{3}{4}$, which is coherent with Figure 12.1.3. Similar reasoning enables us to deduce that $\phi^\delta_{m,j}(x_*) = 0$ for all $\delta \le \bar\delta_3$ and all $3 \le j \le p$. To see this, first suppose that j is odd, that is, $j = 2k+1$ for some $k \ge 1$. Then

$$\begin{aligned}T_{m,2k+1}(x_*,0) - T_{m,2k+1}(x_*,-\alpha x_*) &= -\sigma\|x_*\|^{p+1}\sum_{\ell=2}^{2k+1}\frac{(-\alpha)^\ell}{\ell!(p+1-\ell)!}\\ &= -\sigma\|x_*\|^{p+1}\sum_{i=1}^{k}\alpha^{2i}\left(\frac{1}{(2i)!(p+1-2i)!} - \alpha\frac{1}{(2i+1)!(p-2i)!}\right)\end{aligned}$$

for $\alpha \ge 0$. The ith term in brackets in the last sum is positive when $\alpha \le (2i+1)/(p+1-2i)$, and thus $T_{m,2k+1}(x_*,0) - T_{m,2k+1}(x_*,-\alpha x_*) \le 0$ whenever

$$\|\alpha x_*\| \le \min_{i\in[1:k]}\frac{(2i+1)}{(p+1-2i)}\|x_*\| = \frac{3}{(p-1)}\|x_*\| = \bar\delta_3.$$

As a consequence, $\phi^\delta_{m,2k+1}(x_*) = 0$ for all $\delta \le \bar\delta_3$. For even $j = 2k+2$, $\phi^\delta_{m,2k+2}(x_*) = 0$ for all $\delta \le \bar\delta_3$ because

$$\begin{aligned}&T_{m,2k+2}(x_*,0) - T_{m,2k+2}(x_*,-\alpha x_*)\\ &\quad= T_{m,2k+1}(x_*,0) - T_{m,2k+1}(x_*,-\alpha x_*) - \frac{\sigma\alpha^{2k+2}(2k+2)!\|x_*\|^{p+1}}{(p-2k-1)!}\end{aligned}$$

involves an extra concave term.

In fact, $\phi^\delta_{f,j}(x)$ gives more information than the mere potential proximity of a jth-order minimizer if δ is chosen too large: it is able to see beyond an infinitesimal neighbourhood of x and provides information on possible further descent beyond such a neighbourhood. Rather than a true criticality measure, it can be viewed (for large δ) as an indicator of further possible progress in the minimization, but its use for detecting local minimizers is clearly imperfect. As we have just seen in the example, correcting this requires that we restrict[72] δ or, as we will see later, that we relax the requirement that $\phi^{\delta}_{f,j}(x_*) = 0$.

12.1.4 A sufficient condition using Taylor-series models

Having described a strong necessary condition for a point x to be a local minimizer, we now consider a sufficient condition for x to be an *isolated* local minimizer.

Theorem 12.1.5: A sufficient optimality condition

Let f be $(j+1)$-times continuously differentiable. Suppose that $s = 0$ is an isolated minimizer of $T_{f,j}(x,s)$ and that $\|\nabla_x^{j+1} f(x)\| \leq L$ for some $L \geq 0$. Then $s = 0$ is also an isolated minimizer of $T_{f,j+1}(x,s)$, and x is an isolated local minimizer of f.

Proof. Consider any direction d for which $\|d\| = 1$. We have that $T_{f,j}(x,\alpha d) > T_{f,j}(x,0)$ for $|\alpha|$ sufficiently small. Hence, since $T_{f,j+1}(x,0) = f(x) = T_{f,j}(x,0)$,

$$\begin{aligned}
T_{f,j+1}(x,\alpha d) - T_{f,j+1}(x,0) &= T_{f,j}(x,\alpha d) - T_{f,j+1}(x,0) + \frac{\alpha^{j+1}}{(j+1)!}\nabla_x^{j+1} f(x)[d]^{j+1} \\
&\geq T_{f,j}(x,\alpha d) - T_{f,j}(x,0) - L\frac{|\alpha|^{j+1}}{(j+1)!} \\
&\geq \tfrac{1}{2}\left[T_{f,j}(x,\alpha d) - T_{f,j}(x,0)\right] > 0
\end{aligned}$$

if we further restrict $|\alpha|$. But this implies that $s = 0$ is also an isolated minimizer of $T_{f,j+1}(x,s)$. To obtain the second conclusion, we may repeat the argument using

$$\begin{aligned}
f(x+\alpha d) - f(x) &= T_{f,j}(x,\alpha d) - T_{f,j+1}(x,0) - \frac{|\alpha|^{j+1}}{(j+1)!}\max_{\xi\in[0,\alpha]}\|\nabla_x^{j+1} f(x+\xi d)\| \\
&\geq T_{f,j}(x,\alpha d) - T_{f,j}(x,0) - 2L\frac{|\alpha|^{j+1}}{(j+1)!} \\
&\geq \tfrac{1}{2}\left[T_{f,j}(x,\alpha d) - T_{f,j}(x,0)\right] > 0,
\end{aligned}$$

where we restricted $|\alpha|$ sufficiently to ensure that $\max_{\xi\in[0,\alpha]}\|\nabla_x^{j+1} f(x+\xi d)\| \leq 2L$ (which is possible by continuity) and that the first term on the right-hand side of the first inequality above dominates. □

As a consequence, if f is p-times continuously differentiable and has uniformly bounded derivatives of degree $j+1$ to p, the fact that $s = 0$ is an isolated minimizer of $T_{f,j}(x,s)$ for $j < p$ implies that the origin is also an isolated minimizer of $T_{f,\ell}(x,s)$ for $\ell \in [j:p]$.

12.1.5 Approximate minimizers

Finding a point at which necessary optimality conditions are satisfied exactly is, of course, all but impossible in practice. Consequently, it is imperative instead to seek points that

[72] Observe that we have been able to find a nonzero value for δ_3 because $f(x)$ has an isolated global minimizer (and thus $T_{f,3}(x_*,\alpha x_*)$ is strictly convex in a neighbourhood of x_*). In the general case where x_* is not an isolated minimizer, obtaining a suitable value for δ_j may be far from trivial.

satisfy relaxed versions of the desired conditions. As we have done thus far in this book—recall both Theorem 1.1.1 (p. 3) and the definition (3.2.6)—if a first-order approximate minimizer is sought, we customarily require that

$$\|\nabla_x^1 f(x_\epsilon)\| \leq \epsilon_1, \tag{12.1.22}$$

while, for second-order approximate minimizers, we typically require additionally that

$$\lambda_{\min}[\nabla_x^2 f(x_\epsilon)] \geq -\epsilon_2. \tag{12.1.23}$$

And as we have seen in (4.1.3) in Chapter 4 (p. 99), the relaxation of the requirement that (12.1.13) be zero can be written as

$$\|P_{\mathcal{M}_{\epsilon_2}(x)}[\nabla_x^3 f(x)]\| \leq \epsilon_3, \tag{12.1.24}$$

where $\mathcal{M}_\epsilon(x)$ is now the subspace spanned by all eigenvectors of $\nabla_x^2 f(x)$ associated with eigenvalues that lie in the interval $[-\epsilon, \epsilon]$. However, the latter requirement has a significant drawback, namely that the nullspace of the Hessian may not be continuous in x as a local minimizer is approached—the nullspace may only appear in the limit—and thus the measure itself may not be continuous in x. Unfortunately, in addition, this appears to us to be the brink of what can be done with standard approximate necessary conditions.

It is natural to ask instead whether we might use $\phi_{f,j}^\delta(x)$ to provide a relaxation of local optimality conditions that is generalizable to arbitrary orders. To this end, we introduce a vital concept for what follows.

Key Definition

Given $\epsilon \stackrel{\text{def}}{=} (\epsilon_1, \ldots, \epsilon_q) \in (0,1]^q$, the point x_ϵ is an **(ϵ,δ)-approximate qth-order minimizer** whenever, for some $\delta \stackrel{\text{def}}{=} (\delta_1, \ldots, \delta_q) \in (0,1]^q$,

$$\phi_{f,j}^{\delta_j}(x_\epsilon) \leq \epsilon_j \frac{\delta_j^j}{j!} \quad \text{for all} \quad j \in [1:q]. \tag{12.1.25}$$

The $\{\delta_j\}_{j=1}^q$ are called the *optimality radii* and the vectors $\{d_j\}_{j=1}^q$, the solutions of the global optimization problems defining $\phi_{f,j}^{\delta_j}(x_\epsilon)$ via (12.1.10), are called *optimality displacements*. This condition is a direct relaxation of (12.1.17) and, in view of (12.1.12) and (12.1.14), provides strong coherence with the standard optimality measures, as we show next.

Theorem 12.1.6: Comparing the old and new approximate measures

Suppose that f is continuously differentiable, $\delta > 0$, and $x_\epsilon \in \mathbb{R}^n$. Then

(i) *if $\phi_{f,1}^{\delta_1}(x_\epsilon) \leq \epsilon_1 \delta_1$ (and so (12.1.25) holds with $q = j = 1$), then $\|\nabla_x^1 f(x_\epsilon)\| \leq \epsilon_1$ (and so (12.1.22) is satisfied); and*

(ii) *if $\phi_{f,2}^\delta(x_\epsilon) \leq \epsilon_2 \frac{\delta^2}{2}$, then $\lambda_{\min}[\nabla_x^2 f(x_\epsilon)] \geq -\epsilon_2$ (and so (12.1.23) holds) and $\|\nabla_x^1 f(x_\epsilon)\| \leq \delta\kappa(x_\epsilon)\sqrt{\epsilon_2}$, where $\kappa(x_\epsilon)$ depends on (the eigenvalues of) $\nabla_x^2 f(x_\epsilon)$.*

Proof. For case (i), from (12.1.12), we have that

$$\|\nabla_x^1 f(x_\epsilon)\| = \frac{\phi_{f,1}^\delta(x_\epsilon)}{\delta} \leq \epsilon_1$$

whenever (12.1.25) holds with $q = j = 1$, and thus (12.1.22) is satisfied regardless of the value of $\delta > 0$.

For case (ii), when $j = 2$, (12.1.14) and (12.1.25) together imply that

$$\lambda + \frac{1}{\delta^2}\|\nabla_x^1 f(x_\epsilon)\|^2_{(\nabla_x^2 f(x_\epsilon)+\lambda I)^+} = \frac{\phi_{f,2}^\delta(x_\epsilon)}{(\delta^2/2)} \leq \epsilon_2 \tag{12.1.26}$$

for all $\delta > 0$, where $\lambda \geq \max(0, -\lambda_{\min}[\nabla_x^2 f(x_\epsilon])$ and

$$\|\nabla_x^1 f(x_\epsilon)\|^2_{(\nabla_x^2 f(x_\epsilon)+\lambda I)^+} \stackrel{\text{def}}{=} \left\langle \nabla_x^1 f(x_\epsilon), (\nabla_x^2 f(x_\epsilon) + \lambda I)^+ \nabla_x^1 f(x_\epsilon) \right\rangle > 0$$

unless $\nabla_x^1 f(x_\epsilon) = 0$. In particular, (12.1.26) ensures that (12.1.23) holds. It also guarantees that

$$\|\nabla_x^1 f(x_\epsilon)\|_{(\nabla_x^2 f(x_\epsilon)+\lambda I)^+} \leq \delta\sqrt{\epsilon_2}, \tag{12.1.27}$$

which superficially resembles (12.1.22) (in a weighted norm) when $\epsilon_2 \leq \epsilon_1^2$ and for all $\delta \leq 1$. In particular, if the trust-region bound in (12.1.10) is inactive (and thus $\lambda = 0$ and $\nabla_x^2 f(x_\epsilon)$ is positive definite), we deduce from the Rayleigh-quotient inequality and (the square of) (12.1.27) that $\lambda_{\min}[\nabla_x^2 f(x_\epsilon)^+]\|\nabla_x^1 f(x_\epsilon)\|^2 \leq \delta^2\epsilon_2$. Since $\lambda_{\min}[\nabla_x^2 f(x_\epsilon)^+] = 1/\lambda_{\max}[\nabla_x^2 f(x_\epsilon)]$, we obtain

$$\|\nabla_x^1 f(x_\epsilon)\| \leq \delta\sqrt{\lambda_{\max}[\nabla_x^2 f(x_\epsilon])}\sqrt{\epsilon_2}. \tag{12.1.28}$$

(Note that this is weaker than (12.1.22) unless $\epsilon_2 \leq \epsilon_1^2/\lambda_{\max}[\nabla_x^2 f(x_\epsilon]\delta^2$.) The true character of (12.1.27) is revealed, though, if δ is sufficiently small that the trust-region bound is active at the solution to (12.1.10). If so, it follows from the Cauchy–Schwarz inequality that

$$\delta = \|d\| = \left\|\left(\nabla_x^2 f(x_\epsilon) + \lambda I\right)^+ \nabla_x^1 f(x_\epsilon)\right\| \leq \lambda_{\max}\left[(\nabla_x^2 f(x_\epsilon) + \lambda I)^+\right]\|\nabla_x^1 f(x_\epsilon)\|,$$

and since the Rayleigh-quotient inequality provides the bound

$$\|\nabla_x^1 f(x_\epsilon)\|^2_{(\nabla_x^2 f(x_\epsilon)+\lambda I)^+} \geq \lambda_{\min}\left[(\nabla_x^2 f(x_\epsilon) + \lambda I)^+\right]\|\nabla_x^1 f(x_\epsilon)\|^2,$$

we may combine this pair of bounds with (12.1.27) to see that

$$\|\nabla_x^1 f(x_\epsilon)\| \leq \delta\kappa\left((\nabla_x^2 f(x_\epsilon) + \lambda I)^+\right)\epsilon_2, \tag{12.1.29}$$

where the condition number

$$\kappa\left((\nabla_x^2 f(x_\epsilon) + \lambda I)^+\right) = \frac{\lambda_{\max}\left[(\nabla_x^2 f(x_\epsilon) + \lambda I)^+\right]}{\lambda_{\min}\left[(\nabla_x^2 f(x_\epsilon) + \lambda I)^+\right]}.$$

Combining (12.1.28) and (12.1.29) gives the required bound since $\epsilon_2 \leq 1$. □

In the proof of Theorem 12.1.6 (ii), we show in fact a stronger asymptotic bound of order $\mathcal{O}(\epsilon_2)$, on $\|\nabla_x^1 f(x_\epsilon)\|$, rather than $\mathcal{O}(\sqrt{\epsilon_2})$. To see this, consider (12.1.29) and let δ tend to zero. Then ultimately the trust-region bound is active, the bound (12.1.29) holds, λ approaches infinity, $\kappa((\nabla_x^2 f(x_\epsilon)+\lambda I)^+) \to 1$, and

$$\|\nabla_x^1 f(x_\epsilon)\| \preceq \delta\epsilon_2 \leq \epsilon_1$$

for all $\epsilon_2 \leq \epsilon_1$, where $\preceq$ means "asymptotically less than or equal to." This is essentially the same as (12.1.22), and we see that the stopping rule (12.1.25) for $j=2$ effectively ensures both first- and second-order approximate optimality in the traditional sense for small δ. Of course, this is simply a reinterpretation of Theorem 12.1.4 and directly leads to the following result.

Theorem 12.1.7: Cascading approximate optimality

[Appendix to Chapter 12, p. 482]: Suppose that, for some $j \in [2:q]$,

$$\phi_{f,j}^{\delta_j}(x) \leq \epsilon_j \frac{\delta_j^j}{j!} \tag{12.1.30}$$

for some $\epsilon_j \in (0,1]$, some $\delta_j \in (0,1]$, and some $x \in \mathbb{R}^n$. Suppose also that $\epsilon_j \leq \epsilon_{j-1}$. Then

$$\phi_{f,j-1}^{\delta_j}(x) \leq \epsilon_{j-1}\frac{\delta_j^{j-1}}{(j-1)!} \tag{12.1.31}$$

whenever $\delta_j \leq \min\left[1, \frac{j\epsilon_j}{2\|\nabla_x^j f(x)\|}\right]$.

Thus, if an algorithm were to search for a qth-order approximate minimizer of a function f with bounded derivatives of orders 2 to q by only attempting to satisfy (12.1.30) for $j=q$, lower-order guarantees of the form (12.1.30) would result for $j \in [1:q]$ from the recursive application of the above theorem, provided that

$$\epsilon_1 \leq \epsilon_1 \leq \ldots \leq \epsilon_q \quad \text{and} \quad \delta_q \leq \tfrac{1}{2} \min_{j\in[2:q]} \frac{j\epsilon_j}{\|\nabla_x^j f(x)\|}.$$

These constraints are quite limiting and an important reason for considering the full (12.1.25). This latter definition ensures the *nested-optimality property*, that is, the property that if x is an (ϵ,δ)-approximate qth-order minimizer, it is also an (ϵ,δ)-approximate jth-order one (with its own unconstrained optimality threshold $\epsilon_j \in (0,1]$) for j ranging from 1 to q. Nested optimality with order-dependent thresholds is central in the cumulative definition of the standard measures (12.1.22)–(12.1.24).

We conclude our comments on the approximate criticality measure (12.1.25) by a further, important observation, namely, that if the optimality radius is too small, this may hinder optimality detection during the run of an algorithm. To see this, suppose that the sequence $\{x_k\}$ of iterates generated by a minimization algorithm converges to x_*, a local minimizer of f, but that δ_j, the jth optimality radius at x_* (i.e., associated with $T_{f,j}(x_*,d)$), is allowed to become arbitrarily small. Then there is no hope that the algorithm will deduce that it has produced an approximate minimizer in a finite number of iterations. We illustrate this with the following simple example.

Example 12.1.4 Consider minimizing the univariate function $f(x) = \frac{1}{24}x^4$, whose global minimizer is at the origin. Consider now any iterate $x_k \neq 0$. Because the pth-order Taylor expansion of a polynomial of degree p at any point is the polynomial itself, we obtain that $d_{4,k}$, the fourth optimality displacement at x_k, is equal to $-x_k$ and $\phi_{f,4}^{|x_k|}(x_k) = \frac{1}{24}x_k^4$. Hence, if we ignore the warning above and make the choice $\delta_{4,k} = |x_k|$, condition (12.1.25) requires that $\frac{1}{24}x^4 \leq \frac{1}{24}\epsilon_4 x^4$. Since this always fails for $\epsilon_4 < 1$, we see that, as anticipated, detection of a fourth-order (ϵ, δ)-approximate minimizer is impossible when $\delta_{4,j} = |x_k|$ and $\epsilon_4 < 1$, unless $x_k = x_*$ for some k.

Alternatively, if the optimality radius is not too small, then we should expect the error $x_k - x_*$, and thus the norm of the jth optimality displacement $d_{k,j}$, ultimately to be smaller than $\delta_{j,k}$. Fortunately, because reasonable assumptions on the problem will ensure that this favourable situation always occurs,[73] we conclude that the use of (12.1.25) is suitable for our evaluation-complexity analysis.

12.1.6 Testing for approximate minimizers

In the course of any algorithm, testing for termination is obviously crucial. Thus, given a point x, the derivatives of f at x, the thresholds $\epsilon \in (0,1]^q$, and the radii $\delta \in (0,1]^q$, we need to test whether (12.1.25) holds or not. The most obvious method is to verify this condition by increasing order (i.e., in a loop $j = 1, 2, \ldots, q$ with increasing j) and to declare that the current iterate is not a qth-order (ϵ, δ)-approximate minimizer as soon as (12.1.25) is violated for some j. As a consequence, this may require substantially less work than what would be required if (12.1.25) were tested blindly. We also note that the algorithm *need not call for the full power of global minimization.* For a given j, it is easy to see that the minimization involved in computing $\phi_{f,j}^{\delta_k}(x)$ may be terminated as soon as $T_{f,j}(x,d) < f(x) - \epsilon_j \delta_j^j / j!$ for any iterate d—a few minimization steps often suffice until the iterates get close to optimality. When approaching optimality and the model is linear or quadratic, trust-region technology (see Section 9.4.2 on p. 246 for $p = 2$) allows for early termination of the minimization using a simple bracketing strategy. It is, however, no surprise that verifying (12.1.25) close to optimality remains a potentially daunting task[74] when $p \geq 3$.

Although such a dire prognosis might seem to cast a shadow over our ability to compute $\phi_{f,j}^{\delta}(x)$ and therefore to use it to assess approximate minimality, there is nonetheless some cause for optimism. This is because of an emerging body of work that proposes algorithms that are capable of producing increasingly accurate lower bounds on $\phi_{f,j}^{\delta}(x)$ and therefore enables us to use them as a proxy. To see how this might work, suppose that $\phi_{f,j}^{\delta}(x) = \Delta T_{f,j}(x, d_*)$ for the unknown d_* but that we can compute an approximation $d_{\text{A}} \in \mathcal{B}_\delta \stackrel{\text{def}}{=} \{d \in \mathbb{R}^n \mid \|d\| \leq \delta\}$ for which

$$\varsigma \Delta T_{f,j}(x, d_*) \leq \Delta T_{f,j}(x, d_{\text{A}}) \tag{12.1.32}$$

for some specified $\varsigma \in (0,1]$; crucially we are not assuming here that we have d_*, simply

[73] We will shortly show (in Lemma 12.2.3, taking into account that $\min_{i\in[0:k]} \widehat{\phi}_{f,i} \geq \min_{j\in[1:q]} \epsilon_j$, and in Lemma 12.2.9) that the optimality radius is bounded below by a multiple of $\min_{j\in[1:q]} \epsilon_j$.

[74] It involves the global minimization of a multivariate polynomial function over a convex set, which is an NP-hard problem even when $p = 4$ [297].

that we can ensure (12.1.32) by some means. Suppose furthermore that

$$\Delta T_{f,j}(x, d_{\text{A}}) \le \varsigma \epsilon_j \, \delta_j^j / j!.$$

The preceding pair of inequalities then imply that

$$\phi_{f,j}^{\delta_j}(x) \le \epsilon_j \, \delta_j^j / j!,$$

and if this is true for all $j \in [1 : q]$, it follows from (12.1.25) that x is an (ϵ, δ)-approximate qth-order minimizer of f.

Let $\Delta T_{f,j}(x, d_*) = \Delta T_{f,j}(x, d_{\text{A}}) + e_{\text{A}}$. Then (12.1.32) holds whenever

$$\frac{\varsigma}{1 - \varsigma} e_{\text{A}} \le \Delta T_{f,j}(x, d_{\text{A}}). \tag{12.1.33}$$

But it is possible to construct a sequence of estimates $\{d_{\text{A}_r}\} \in \mathcal{B}_\delta$ for which $\Delta T_{f,j}(x, d_{\text{A}_r})$ increase monotonically to $\Delta T_{f,j}(x, d_*)$ and $e_{\text{A}_r} = \mathcal{O}(r^{-2})$ for increasing r by solving a sequence of semidefinite programs of order r [143, 144, 339].[75] Thus (12.1.33) will hold for sufficiently large r. While there is no denying that this is expensive, it at least opens the door to the possibility of finding computable bounds. Thus the requirement (12.1.32) will be an important ingredient in algorithms that we consider.

We could also think of exploiting the sufficient condition given by Theorem 12.1.5. This might, for instance, be implemented within a loop of increasing j as follows: if $\phi_{f,j}^{\delta_j}(x) = 0$, then globally solve the problem

$$\min_{d \,|\, \|d\| \le \delta_j} u_j(d) \stackrel{\text{def}}{=} T_{f,j}(x, d) - \|d\|^{j+1}$$

for d_j^*. If $d_j^* = 0$, the origin is an isolated global minimizer of $T_{f,j}(x, d)$ and the loop on j may be terminated with the conclusion that a local minimizer has been found. However, the main drawback of the theorem persists, namely that, to activate that test, we need x to be an *exact* isolated local minimizer, a very unlikely situation, and certainly one we would not expect in a worst-case analysis. Because of this shortcoming (and also because it would complicate the presentation of the algorithms to come), we ignore the possibility of such a refinement in what follows.

12.1.7 Stability of (approximate) minimizers

In general, there is no reason to believe that f will be well behaved in a neighbourhood of a point satisfying approximate qth-order optimality conditions—(12.1.22)–(12.1.24) or (12.1.25)—in the absence of conditions that prevent the qth derivative from varying too quickly. We will say that x_ϵ, an approximate qth-order minimizer, is *stable* if there exists a neighbourhood of x_ϵ such that $\nabla_x^q f$ is Lipschitz continuous in that neighbourhood with a Lipschitz constant independent of x_ϵ. This definition is motivated by the following property.

[75] These methods are equally applicable to other norms and actually allow for a variety of additional semialgebraic side constraints, which will be useful in Chapter 14.

Theorem 12.1.8: Stability of approximate unconstrained minimizers

Suppose that f is j-times continuously differentiable and that

$$\phi_{f,j}^{\delta_j}(x_\epsilon) \leq \epsilon_j \frac{\delta_j^j}{j!} \tag{12.1.34}$$

for some $\delta_j \in (0,1]$ and $\epsilon_j \in (0,1]$. Suppose also that $\nabla_x^j f$ is Lipschitz continuous with constant $L_{f,j}$ in an open neighbourhood of a point x_ϵ of radius larger than δ_j. Then, if $L_{f,j} > 0$,

$$f(x_\epsilon + d) \geq f(x_\epsilon) - 2\epsilon_j \frac{\delta^j}{j!} \quad \textit{for all } d \textit{ with } \|d\| \leq \min\left[\delta_j, \frac{(j+1)\,\epsilon_j}{L_{f,j}}\right] \tag{12.1.35}$$

and, otherwise,

$$f(x_\epsilon + d) \geq f(x_\epsilon) - \epsilon_j \frac{\delta_j^j}{j!} \quad \textit{for all } d \textit{ with } \|d\| \leq \delta_j. \tag{12.1.36}$$

Proof. Consider $d \in \mathbb{R}^n$. Using the triangle inequality, we have that

$$\begin{aligned} f(x_\epsilon + d) &= f(x_\epsilon + d) - T_{f,j}(x_\epsilon, d) + T_{f,j}(x_\epsilon, d) \\ &\geq -|f(x_\epsilon + d) - T_{f,j}(x_\epsilon, d)| + T_{f,j}(x_\epsilon, d). \end{aligned} \tag{12.1.37}$$

Now condition (12.1.34) implies that

$$T_{f,j}(x_\epsilon, d) \geq T_{f,j}(x_\epsilon, 0) - \epsilon_j \frac{\delta_j^j}{j!} = f(x_\epsilon) - \epsilon_j \frac{\delta_j^j}{j!} \tag{12.1.38}$$

if $\|d\| \leq \delta_j$. Hence, because $\nabla_x^j f$ is Lipschitz continuous, we may substitute (A.19) of Corollary A.8.4 (p. 437) with $p = j$ and (12.1.38) into (12.1.37) and deduce that

$$f(x_\epsilon + d) \geq f(x_\epsilon) - \frac{L_{f,j}}{(j+1)!}\|d\|\delta_j^j - \epsilon_j \frac{\delta_j^j}{j!}, \tag{12.1.39}$$

and (12.1.35) follows if $L_{f,j} > 0$, while (12.1.36) follows if $L_{f,j} = 0$. □

The obtained accuracy on the true objective function therefore increases with the criticality order j, as one might expect. Interestingly, *this does not indicate any benefit exists in using the cumulative condition (12.1.25) instead of the simpler (12.1.34) with* $j = q$: the best accuracy obtained is $2\epsilon\delta_q^q/q!$ in the special case where $\epsilon_j = \epsilon$ and $\delta_j \leq \delta_q$ for $j \in [1:q]$ in both cases. This indicates that finding a point satisfying (12.1.34) with $j = q$ may also be an interesting goal.

We note that the Lipschitz constants $L_{f,j}$ in Theorem 12.1.8 need not be independent of x_ϵ, in which case the obvious, more general version of this theorem still holds. But, since for simplicity we have opted only to use global Lipschitz constants, along with their ensuing properties, throughout this book, and because this allows us to compare stability neighbourhoods of minimizers (as per our discussion below), we have opted for global properties here as well. If we had chosen to use local Lipschitz constants instead, the corresponding variant of Theorem 12.1.8 would still allow comparisons of stability properties by fixing the optimality radii δ_j.

This discussion allows a further important interpretation of the evaluation-complexity results obtained so far. The results for algorithms using first derivatives alone (covered in Chapter 2) all assume **f.gL**, which means that all approximate first-order minimizers are stable in the sense just defined. The situation is more involved for algorithms using second derivatives or higher (Chapters 3 and 4 and this chapter). For the second-order trust-region TR2 algorithm, Theorems 3.2.1 and 3.2.6 both assume bounded Hessians (**f.HBa**), which implies Lipschitz-continuous gradients (**f.gL**). As a consequence, approximate first-order minimizers are stable. Since **f.HL** is also assumed in Theorem 3.2.6, this is also the case for approximate second-order minimizers.

However, Theorems 3.3.5 and 3.3.9 for the AR2 algorithm only require **f.HL**, and Lipschitz-continuous gradients are not necessary for proving the better (compared with TR2) complexity results. But, without also assuming **f.gL**, there is no guarantee that approximate first-order minimizers are stable, although second-order ones are. Similar conclusions also hold for the ARp algorithm when $p > 2$: the assumptions necessary for obtaining Theorem 4.1.5, i.e., **f.DpL**, do not ensure the stability of (more easily computed) approximate either first- or second-order minimizers.

Stability of first- and second-order approximate minimizers is, however, guaranteed when using the AR2 or ARp algorithm for *p-Lipschitz-stable* functions, that is, functions satisfying **f.DpL** and such that, for every $j \in [1 : p]$, there exist $\Delta_j > 0$ and $L_{f,j} \geq 0$ such that, for every $x, y \in \mathbb{R}^n$ with $\|x - y\| \leq \Delta_j$, one has that $\|\nabla_x^j f(x) - \nabla_x^j f(y)\| \leq L_{f,j}\|x - y\|$. Note that this property is significantly weaker than requiring that all derivatives $\{\nabla_x^j f\}_{j=1}^p$ be Lipschitz continuous.

Notes and references for Subsection 12.1.7

Theorem 12.1.8 is Theorem 2.2 in [108]. This paper also considered a weaker notion of (ϵ, δ)-approximate qth-order minimizers (which we now call weak (ϵ, δ)-approximate qth-order optimality), where x is such a point if

$$\phi_{f,q}^{\delta}(x) \leq \epsilon \chi_q(\delta) \quad \text{with} \quad \chi_q(\delta) = \sum_{\ell=1}^{q} \frac{\delta^\ell}{\ell!}, \tag{12.1.40}$$

where ϵ and δ are now scalars. That this (noncumulative) notion is weaker than (12.1.25) for $q > 1$ results from the observation that $\chi_q(\delta) \in [\delta, 2\delta]$, which, for $\delta \leq 1$, can be substantially larger than $\delta^q/q!$. This limitation motivates our focus on strong approximate minimizers.

The concepts of stable approximate minimizer and of Lipschitz-stable functions are new.

12.2 Finding approximate minimizers of arbitrary qth order

We now turn to the question of designing algorithms that are capable of finding approximate qth-order minimizers as defined in (12.1.25). Because these algorithms will need to find suitable approximations of $\phi_{f,j}^{\delta}(x)$ a number of times and because this computation relates to a complicated global minimization problem, they are better seen, in general, as conceptual tools rather than practical ones. We stress again, however, that once the derivatives of f at x have been computed, *the burden of approximating $\phi_{f,j}^{\delta}(x)$ does not involve any additional evaluations: it therefore has no impact on our evaluation-complexity analysis.*

12.2.1 The TRq algorithm and its complexity

As in our previous chapters, we start our exploration with trust-region methods.

12.2.1.1 The TRq algorithm

The theoretical trust-region algorithm TRq, which we are about to introduce, capitalizes on the observation that computing $\phi^{\delta}_{f,j}(x)$ amounts to solving a trust-region subproblem. The underlying idea is that, unless termination occurs, this computation identifies a step guaranteeing a significant model decrease; any further decrease is obviously welcome. The resulting algorithm is stated as follows.

Algorithm 12.2.1: TRq for (ϵ, δ)-approximate qth-order minimizers

Step 0: Initialization. A criticality order q, accuracy thresholds $\epsilon \in (0,1]^q$, a starting point x_0, and an initial trust-region radius $\Delta_0 > 0$ are given, as well as algorithmic parameters $\Delta_{\max} \geq \Delta_0$, $0 < \gamma_1 \leq \gamma_2 < 1 \leq \gamma_3$, $0 < \eta_1 \leq \eta_2 < 1$, $\varsigma \in (0,1]$, and $\vartheta \in [\min_{j\in[1:q]} \epsilon_j, 1]$. Evaluate $f(x_0)$ and $\nabla_x^j f(x_0)$ for all $j \in [1:q]$, and set $k = 0$.

Step 1: Termination test. Set $\delta_k = \min[\Delta_k, \vartheta]$. For $j = 1, \ldots, q$:

1. Evaluate $\nabla_x^j f(x_k)$ and compute $d_{k,j} \in \mathcal{B}_{\delta_k}$ so that

$$\varsigma \phi^{\delta_k}_{f,j}(x_k) \leq \Delta T_{f,j}(x_k, d_{k,j}). \tag{12.2.1}$$

2. If

$$\Delta T_{f,j}(x_k, d_{k,j}) > \varsigma \epsilon_j \frac{\delta_k^j}{j!}, \tag{12.2.2}$$

go to Step 2 with $j_k = j$ and $d_k = d_{k,j}$.

Terminate with $x_\epsilon = x_k$ and $\delta_\epsilon = \delta_k$.

Step 2: Step computation. Compute a step $s_k \in \mathcal{B}_{\Delta_k}$ for which

$$\Delta T_{f,j_k}(x_k, s_k) \geq \Delta T_{f,j_k}(x_k, d_k). \tag{12.2.3}$$

Step 3: Accept the new iterate. Compute $f(x_k + s_k)$ and

$$\rho_k = \frac{f(x_k) - f(x_k + s_k)}{\Delta T_{f,j_k}(x_k, s_k)}. \tag{12.2.4}$$

If $\rho_k \geq \eta_1$, set $x_{k+1} = x_k + s_k$ and compute $\nabla_x^j f(x_{k+1})$ for all $j \in [1:q]$. Otherwise set $x_{k+1} = x_k$.

Step 4: Update the trust-region radius. Set

$$\Delta_{k+1} \in \begin{cases} [\gamma_1 \Delta_k, \gamma_2 \Delta_k] & \text{if } \rho_k < \eta_1, \\ [\gamma_2 \Delta_k, \Delta_k] & \text{if } \rho_k \in [\eta_1, \eta_2), \\ [\Delta_k, \min(\Delta_{\max}, \gamma_3 \Delta_k)] & \text{if } \rho_k \geq \eta_2. \end{cases} \tag{12.2.5}$$

Increment k by one and go to Step 1.

Notice that, just as in Section 12.1.6, we *do not presume that* $\phi_{f,j}^{\delta_k}(x_k)$ *is known* when computing $d_{k,j}$ to satisfy (12.2.1), merely that such a $d_{k,j}$ may be found. Of course, if the global minimizer is known, it may be used as $d_{k,j}$, and, in addition, if $\varsigma = 1$ and $\Delta_k \leq \vartheta$, then $s_k = d_k$ is acceptable for (12.2.3) in Step 2. Observe that the desired approximate optimality condition (12.1.10) occurs as soon as (12.2.1) holds but (12.2.2) fails for all $j \in [1:q]$ in Step 1. Also note that one could return to Step 2 instead of Step 1 at the end of Step 4 provided that $x_{k+1} = x_k$ and $\Delta_{k+1} \geq \vartheta$, saving the (re)computation of the approximations of $\phi_{f,i}^{\delta_k}(x_k)$. For simplicity, though, we have avoided this possibility. The basic structure of the TRq algorithm is the same as that of the trust-region methods TR1 (p. 35) and TR2 (p. 55). As a consequence, it immediately follows that Lemma 2.3.1 (p. 36) remains valid. Observe also that derivatives are evaluated in Step 1 only *after a successful iteration*, and thus only once per successful iteration, just as in the TR1 and TR2 algorithms.

A distinguishing new feature is the distinct roles played by the two radii δ_k and Δ_k employed. The trust-region radius Δ_k involved in the step calculation may be larger than the optimality radius δ_k, and this allows potentially longer steps s_k to be generated. The role of ϑ is to bring more flexibility to the choice of δ_k, in particular to reflect proper problem scaling.

12.2.1.2 Complexity of finding approximate minimizers

Given the more general context, we have to adapt our assumptions. In Chapter 3, we could prove convergence to first- and second-order approximate minimizers under the assumptions that f is bounded below (**f.Bb**) and has bounded and Lipschitz-continuous Hessian (**f.HBa** and **f.HL**, respectively). Moreover, we already noted that **f.HBa** is equivalent to **f.gL** for twice continuously differentiable functions. Hence our previous theory for the trust-region method required **f.Bb**, **f.gL**, and **f.HL**. We now extend this framework for the case where q is arbitrary. While we still require **f.Bb** to make the problem well defined, **f.gL** and **f.HL** will now be subsumed by the following assumption requiring all derivatives of order one to q to be Lipschitz continuous.

f.D1qL f is q-times continuously differentiable in $\mathbb{R}^n$, and, for all $j \in [1:q]$, the jth derivative of f is Lipschitz continuous with Lipschitz constant $L_{f,j}$; that is, there exist constants $L_{f,j} \geq 1$ such that, for all $x, y \in \mathbb{R}^n$ and all $j \in [1:q]$,

$$\|\nabla_x^j f(x) - \nabla_x^j f(y)\| \leq L_{f,j}\|x - y\|.$$

For simplicity of notation, define

$$L_f \stackrel{\text{def}}{=} \max_{j\in[1:q]} L_{f,j} \geq 1. \tag{12.2.6}$$

Observe that this assumption guarantees that approximate minimizers of orders one to q are stable in the sense of Section 12.1.7.

We now follow the outline we have used thus far to analyze trust-region methods, but first we establish a useful technical result. For the sake of simplifying notation, we define

$$\epsilon_{\min} = \min_{j\in[1:q]} \epsilon_j. \tag{12.2.7}$$

We also introduce the notationally convenient scaled approximate criticality gap

$$\widehat{\phi}_{f,k} \stackrel{\rm def}{=} \frac{j_k!\,\Delta T_{f,j_k}(x_k,d_k)}{\delta_k^{j_k}}, \tag{12.2.8}$$

where j_k is the index j and d_k the direction $d_{k,j}$ for which (12.2.2) holds in Step 1. This allows us to rewrite the requirement (12.2.1) as

$$\widehat{\phi}_{f,k} \geq \zeta \frac{j_k!}{\delta_k^{j_k}} \phi_{f,j_k}^{\delta_k}(x_k).$$

We now establish an important lower bound on $\widehat{\phi}_{f,k}$ and the model decrease $\Delta T_{f,j_k}(x_k,s_k)$.

Lemma 12.2.1: A useful lower bound

At iteration k before termination of the TRq *algorithm, $\widehat{\phi}_{f,k}$ in* (12.2.8) *satisfies*

$$\min_{i\in[0:k]} \widehat{\phi}_{f,i} \geq \varsigma\epsilon_{\min}. \tag{12.2.9}$$

Moreover,

$$\Delta T_{f,j_k}(x_k,s_k) \geq \widehat{\phi}_{f,k}\frac{\delta_k^{j_k}}{j_k!}. \tag{12.2.10}$$

Proof. Let k be the index of an iteration before termination. Then the mechanism (12.2.2) in Step 1 ensures the existence of j_k such that

$$\Delta T_{f,j_k}(x_k,d_k) = \Delta T_{f,j_k}(x_k,d_{k,j_k}) > \varsigma\epsilon_{j_k}\frac{\delta_k^{j_k}}{j_k!} \geq \epsilon_{\min}\varsigma\frac{\delta_k^{j_k}}{j_k!}.$$

The definition of $\widehat{\phi}_{f,k}$ then directly implies that

$$\widehat{\phi}_{f,k} \geq \varsigma\epsilon_{\min}.$$

Since termination has not yet occurred, this inequality must hold for all iterations $i \in [0:k]$, which yields (12.2.9). The bound (12.2.10) results directly from

$$\Delta T_{f,j_k}(x_k,s_k) \geq \Delta T_{f,j_k}(x_k,d_k) = \widehat{\phi}_{f,k}\frac{\delta_k^{j_k}}{j_k!},$$

where the inequality follows from (12.2.3), and the subsequent identity follows from the definition of $\widehat{\phi}_{f,k}$. □

The definition of $\widehat{\phi}_{f,k}$ in (12.2.8) is instrumental in the next lemma, in which we establish a condition under which an iteration must be very successful.

Lemma 12.2.2: Condition for success (TRq)

Suppose that **f.D1qL** *holds, that $\widehat{\phi}_{f,k}$ is defined by (12.2.8), and that L_f is given by (12.2.6). Suppose also that*

$$\Delta_k \le \min\left[\vartheta, \frac{(1-\eta_2)}{L_f}\,\widehat{\phi}_{f,k}\right] \tag{12.2.11}$$

at iteration k. Then $\rho_k \ge \eta_2$, iteration k is very successful, and $\Delta_{k+1} \ge \Delta_k$.

Proof. We first note that (12.2.11) implies that $\delta_k = \min[\vartheta, \Delta_k] = \Delta_k$. Now, we obtain from (12.2.4) and (12.2.10) that

$$\begin{aligned}
|\rho_k - 1| &= \frac{|f(x_k+s_k) - T_{f,j_k}(x_k,s_k)|}{\Delta T_{f,j_k}(x_k,s_k)}\\
&\le \frac{j_k!|f(x_k+s_k) - T_{f,j_k}(x_k,s_k)|}{\widehat{\phi}_{f,k}\delta_k^{j_k}}\\
&\le \frac{L_{f,j_k}\|s_k\|^{j_k+1}}{(j_k+1)\,\widehat{\phi}_{f,k}\delta_k^{j_k}}\\
&\le \frac{L_f\Delta_k}{(j_k+1)\widehat{\phi}_{f,k}}\\
&\le 1-\eta_2,
\end{aligned} \tag{12.2.12}$$

where we used (A.19) in Corollary A.8.4 (p. 437) (implied by **f.D1qL**) to deduce the second inequality, the bound $\|s_k\| \le \Delta_k = \delta_k$ and (12.2.6) to derive the third, and (12.2.11) to deduce the last. Thus $\rho_k \ge \eta_2$ and the mechanism of the TRq algorithm implies that the conclusion of the lemma holds. □

This result is crucial in the proof of the following lower bound on the trust-region radius.

Lemma 12.2.3: Lower bound on the trust-region radius (TRq)

Suppose that **f.D1qL** *holds and L_f is given by (12.2.6). Then*

$$\Delta_k \ge \min\left\{\gamma_1\vartheta, \kappa_\Delta \min_{i\in[0:k]} \widehat{\phi}_{f,i}\right\}, \tag{12.2.13}$$

where $\widehat{\phi}_{f,i}$ is defined as in (12.2.8), and

$$\kappa_\Delta \stackrel{\rm def}{=} \frac{\gamma_1(1-\eta_2)}{L_f}\min\left[\vartheta, \frac{\Delta_0\delta_0^q}{2q!\max_{i\in[1:q]}\|\nabla_x^i f(x_0)\|}\right] \in (0,1). \tag{12.2.14}$$

Proof. We have that

$$\begin{aligned}\widehat{\phi}_{f,0} &\leq \max_{j\in[1:q]} \frac{j!\,\Delta T_{f,j}(x_0,d_{0,j})}{\delta_0^j}\\ &\leq q!\max_{j\in[1:q]} \frac{\max_{i\in[1:j]}\|\nabla_x^i f(x_0)\|}{\delta_0^j}\sum_{i=1}^{j}\frac{\delta_0^i}{i!}\\ &\leq 2q!\,\frac{\max_{i\in[1:q]}\|\nabla_x^i f(x_0)\|}{\delta_0^q}\end{aligned}$$

using (12.2.8), (12.1.11), the sum bound (A.34), and $\delta_0 \leq 1$, and thus

$$\kappa_\Delta \leq \frac{\gamma_1(1-\eta_2)}{L_f}\min\left[\vartheta, \frac{\Delta_0}{\widehat{\phi}_{f,0}}\right] \leq \frac{\Delta_0}{\widehat{\phi}_{f,0}},$$

since $\gamma_1(1-\eta_2) < 1$ and $L_f \geq 1$. As a consequence, (12.2.13) holds for $k = 0$.

Now suppose that $k \geq 1$ is the first iteration such that (12.2.13) is violated and thus that

$$\Delta_k < \frac{\gamma_1(1-\eta_2)}{L_f}\min_{i\in[0:k]}\widehat{\phi}_{f,i} \text{ and } \Delta_k \leq \gamma_1\vartheta. \tag{12.2.15}$$

The updating rule (12.2.5) that implies that $\Delta_k \geq \gamma_1\Delta_{k-1}$ then ensures that

$$\Delta_{k-1} < \frac{(1-\eta_2)}{L_f}\min_{i\in[0:k]}\widehat{\phi}_{f,i} \text{ and } \Delta_{k-1} \leq \vartheta. \tag{12.2.16}$$

Moreover, since

$$\widehat{\phi}_{f,k-1} \geq \min_{i\in[0:k-1]}\widehat{\phi}_{f,i} \geq \min_{i\in[0:k]}\widehat{\phi}_{f,i},$$

we deduce that

$$\frac{\Delta_{k-1}}{\widehat{\phi}_{f,k-1}} \leq \frac{\Delta_{k-1}}{\min\limits_{i\in[0:k-1]}\widehat{\phi}_{f,i}} \leq \frac{\Delta_{k-1}}{\min\limits_{i\in[0:k]}\widehat{\phi}_{f,i}} < \frac{(1-\eta_2)}{L_f}. \tag{12.2.17}$$

Lemma 12.2.2 and the second part of (12.2.16) then ensure that $\Delta_{k-1} \leq \Delta_k$. Using this bound, the second inequality of (12.2.17), and (12.2.15), we obtain that

$$\frac{\Delta_{k-1}}{\min\limits_{i\in[0:k-1]}\widehat{\phi}_{f,i}} \leq \frac{\Delta_k}{\min\limits_{i\in[0:k]}\widehat{\phi}_{f,i}} < \gamma_1\frac{(1-\eta_2)}{L_f} \text{ and } \Delta_{k-1} \leq \gamma_1\vartheta.$$

As a consequence, (12.2.13) is also violated at iteration $k-1$. But this contradicts the assumption that iteration k is the first such that this inequality fails. This latter assumption is thus impossible, and no such iteration can exist. □

Armed with the bound given by this lemma, we may now analyze the guaranteed model decrease.

Lemma 12.2.4: Model and function decreases (TRq)

Suppose that **f.D1qL** *holds. Then*

$$\Delta T_{f,j_k}(x_k, s_k) \geq \frac{(\varsigma\kappa_\Delta)^{q+1}}{q!}\,\epsilon_{\min}^{q+1} \tag{12.2.18}$$

for every iteration of algorithm TRq *before termination, where* κ_Δ *is given by (12.2.14) and* $\epsilon_{\min}$ *by* (12.2.7). *If, in addition,* $k \in \mathcal{S}$, *then*

$$f(x_k) - f(x_{k+1}) \geq \frac{\eta_1(\varsigma\kappa_\Delta)^{q+1}}{q!}\,\epsilon_{\min}^{q+1}. \tag{12.2.19}$$

Proof. Suppose first that $\Delta_k > \vartheta$ and therefore $\delta_k = \vartheta$. Then (12.2.10), (12.2.9), and the bound $\vartheta \geq \epsilon_{\min}$ give that

$$\Delta T_{f,j_k}(x_k, s_k) \geq \frac{\vartheta^{j_k}}{j_k!}\,\widehat{\phi}_{f,k} \geq \frac{\vartheta^q}{q!}\,\widehat{\phi}_{f,k} \geq \varsigma\frac{\epsilon_{\min}^{q+1}}{q!},$$

which yields (12.2.18) since $\varsigma\kappa_\Delta \leq \kappa_\Delta < 1$. By contrast, if $\Delta_k \leq \vartheta$, then $\delta_k = \Delta_k$, and (12.2.13) implies that

$$\delta_k \geq \min\left\{\gamma_1\vartheta, \kappa_\Delta \min_{i\in[0:k]} \widehat{\phi}_{f,i}\right\}.$$

Therefore, using (12.2.10) once again,

$$\begin{aligned}\Delta T_{f,j_k}(x_k, s_k) &\geq \frac{1}{j_k!}\,\widehat{\phi}_{f,k}\min\left\{\gamma_1\vartheta, \kappa_\Delta \min_{i\in[0:k]} \widehat{\phi}_{f,i}\right\}^{j_k} \\ &\geq \frac{1}{q!}\min\left\{\gamma_1\vartheta, \kappa_\Delta \min_{i\in[0:k]} \widehat{\phi}_{f,i}\right\}^{q+1},\end{aligned} \tag{12.2.20}$$

because $j_k \leq q$. Moreover, if termination has not occurred at iteration k, we have that (12.2.9) holds, and, because $\kappa_\Delta < \gamma_1$ and $\vartheta \geq \epsilon_{\min}$, (12.2.20) in turn implies (12.2.18). If additionally $k \in \mathcal{S}$, then $\rho_k \geq \eta_1$ and (12.2.19) follows from (12.2.18) and (12.2.4). □

Our worst-case evaluation-complexity results now follow from a suitable variant of our traditional telescoping-sum argument.

Theorem 12.2.5: qth-order complexity for the TRq **algorithm**

Suppose that **f.Bb** *and* **f.D1qL** *hold. Then there exist positive constants* κ_{TRq}^A, κ_{TRq}^B, *and* κ_{TRq}^C *such that, for any* $\epsilon \in (0,1]^q$, *the* TRq *algorithm (p. 292) requires at most*

$$\begin{aligned}\#_{\text{TRq}}^F &\stackrel{\text{def}}{=} \kappa_{\text{TRq}}^A\frac{f(x_0) - f_{\text{low}}}{\epsilon_{\min}^{q+1}} + \kappa_{\text{TRq}}^B\left|\log\left(\min_{j\in[1:q]}\epsilon_j\right)\right| + \kappa_{\text{TRq}}^C \\ &= \mathcal{O}\left(\max_{j\in[1:q]}\epsilon_j^{-(q+1)}\right)\end{aligned} \tag{12.2.21}$$

evaluations of f and at most

$$\#^{D}_{\mathsf{TRq}} \stackrel{\text{def}}{=} \kappa^{S}_{\mathsf{TRq}} \frac{f(x_0) - f_{\text{low}}}{\epsilon_{\min}^{q+1}} + 1 = \mathcal{O}\left(\max_{j\in[1:q]} \epsilon_j^{-(q+1)} \right) \tag{12.2.22}$$

evaluations of $\{\nabla_x^f\}_{j=1}^q$ to produce an iterate x_ϵ and an optimality radius $\delta_\epsilon > 0$ such that, for all $j \in [1:q]$, $\phi_{f,q}^{\delta_\epsilon}(x_\epsilon) \le \epsilon_j \delta_\epsilon^j / j!$ for some $\delta_\epsilon \in (0,1]$.

Proof. Let k be the index of an arbitrary iteration before termination. Using **f.Bb**, the nature of successful iterations, and (12.2.19), we deduce that

$$f(x_0) - f_{\text{low}} \ge f(x_0) - f(x_{k+1}) = \sum_{i\in\mathcal{S}_k} [f(x_i) - f(x_{i+1})] \ge |\mathcal{S}_k| \left[\kappa^{S}_{\mathsf{TRq}}\right]^{-1} \epsilon_{\min}^{q+1},$$

where

$$\kappa^{S}_{\mathsf{TRq}} = \frac{q!}{\eta_1 (\varsigma \kappa_\Delta)^{q+1}}$$

and κ_Δ is defined as in (12.2.14), and thus that

$$|\mathcal{S}_k| \le \kappa^{S}_{\mathsf{TRq}} \frac{f(x_0) - f_{\text{low}}}{\epsilon_{\min}^{q+1}}.$$

Given that Step 1 of the $\mathsf{TR}q$ algorithm only evaluates derivatives after a successful iteration and adding one for the final evaluation at termination, this last bound gives (12.2.22). Now note that (12.2.13), (12.2.9), (12.2.14), the bounds $\kappa_\Delta < \gamma_1$, $\vartheta \ge \epsilon_{\min}$, and $1 - \varsigma \le 1$, and the fact that termination has not occurred imply that

$$\Delta_k \ge \min\left(\gamma_1 \vartheta, \kappa_\Delta \min_{i\in[0:k]} \widehat{\phi}_{f,i} \right) \ge \varsigma \kappa_\Delta \epsilon_{\min} \stackrel{\text{def}}{=} \Delta_{\min}.$$

Finally, we use this lower bound on Δ_k and call upon Lemma 2.3.1 (p. 36) to compute an upper bound on the total number of iterations before termination (obviously, there must be at least one successful iteration unless termination occurs for $k = 0$) and again add one for the evaluation at termination. This implies (12.2.21) with

$$\kappa^{A}_{\mathsf{TRq}} \stackrel{\text{def}}{=} \kappa^{S}_{\mathsf{TRq}} \left(1 + \frac{\log \gamma_3}{|\log \gamma_2|} \right), \quad \kappa^{B}_{\mathsf{TRq}} \stackrel{\text{def}}{=} \frac{1}{|\log \gamma_2|},$$

and

$$\kappa^{C}_{\mathsf{TRq}} \stackrel{\text{def}}{=} \frac{1}{|\log \gamma_2|} \left| \log\left(\frac{\varsigma \kappa_\Delta}{\Delta_0} \right) \right| + 1.$$

□

Theorem 12.2.5 extends both Theorem 2.3.7 and Theorem 3.2.6 to arbitrary optimality order.

It remains to show again that the bound provided by Theorem 12.2.5 is sharp in the order in ϵ.

Theorem 12.2.6: Slow convergence of the TRq algorithm

*Let $\epsilon_j = j!\,\epsilon$ for given $\epsilon \in (0, 1/q!]$ and all $j \in [1 : q]$. Then the TRq algorithm (p. 292) applied to minimize a function f satisfying **f.Bb** and **f.D1qL** may require $\epsilon^{-(q+1)}$ iterations and evaluations of f and $\{\nabla_x^j f\}_{j=1}^q$ to produce a point x_ϵ such that, for all $j \in [1 : q]$, $\phi_{f,j}^{\delta_\epsilon}(x_\epsilon) \leq \epsilon_j \delta_\epsilon^j / j!$ for some $\delta_\epsilon \in (0, 1]$.*

Proof. The proof is a direct generalization to general optimality order of that presented for Theorem 3.2.7. Details are given in Appendix B (p. 483). □

Notes and references for Subsection 12.2.1

The TRq algorithm and its evaluation-complexity analysis are a generalization of the method discussed in [102] to the case in which different ϵ are used for successive nested-optimality levels. That paper also described an example of slow convergence, showing that the bound in $\mathcal{O}(\epsilon^{-(q+1)})$ is essentially sharp in order. The new example presented in the proof of Theorem 12.2.6 proves that this order is sharp.

12.2.2 The ARqp algorithm and its complexity

As the reader might suspect by now, there is a regularization counterpart to the trust-region framework we explored in the previous subsection. And it will probably be of no surprise that there are theoretical advantages in pursuing such a development.

12.2.2.1 The ARqp algorithm

We shall focus on the ARqp algorithm, whose purpose is to find an (ϵ, δ)-approximate qth-order minimizer by successive approximate minimizations of the model

$$m_k(s) = T_{f,p}(x_k, s) + \frac{\sigma_k}{(p+1)!}\|s\|^{p+1} \tag{12.2.23}$$

of degree p that we first encountered as (4.1.2) in the ARp algorithm. We have already noted that such a model is bounded from below, and it is also easy to verify that the level set $\{s \in \mathbb{R}^n \mid m_k(s) \leq m_k(0)\}$ is bounded. Given $m_k(s)$, we will additionally use the "model decrement"

$$\Delta m_k(s) \stackrel{\text{def}}{=} m_k(0) - m_k(s).$$

As was the case for the TRq algorithm, we assume in what follows that all derivatives involved in the model are Lipschitz continuous.

f.D1pL

f is p-times continuously differentiable in $\mathbb{R}^n$, and, for $j \in [1 : p]$, the jth derivative of f is Lipschitz continuous with Lipschitz constant $L_{f,j}$; that is, there exist constants $L_{f,j} \geq 1$ such that

$$\|\nabla_x^j f(x) - \nabla_x^j f(y)\| \leq L_{f,j}\|x - y\|$$

for all $x, y \in \mathbb{R}^n$ and all $j \in [1 : p]$.

In particular, this assumption and Theorem A.8.5 (p. 438) imply that, for any $x \in \mathbb{R}^n$,

$$\|\nabla_x^j f(x)\| \le L_{j-1} \quad \text{for} \quad j \in [2:p]. \tag{12.2.24}$$

We also (re)define

$$L_f \stackrel{\text{def}}{=} \max_{j\in[1:p]} L_{f,j} \ge 1 \tag{12.2.25}$$

to cover all constants up to p.

The ARqp algorithm is formally stated below. As before, the algorithm relies on our ability to find suitably approximate global minimizers of surrogate multivariate polynomials. But as we noted in Section 12.1.6, while this is undoubtedly a significant task for $q > 2$, it is now within the grasp of currently emerging methods.

Algorithm 12.2.2: ARqp for (ϵ, δ)-approximate qth-order minimizers

Step 0: Initialization. A criticality order q, a degree $p \ge q$, an initial point x_0, and an initial regularization parameter $\sigma_0 > 0$ are given, as well as accuracy levels $\epsilon \in (0,1)^q$ and optimality radii $\delta_0 \in (0,1]^q$. The constants θ, ς, η_1, η_2, γ_1, γ_2, γ_3, and $\sigma_{\min}$ are also given and satisfy

$$\theta \in (0,1), \quad \varsigma \in (0,1], \quad \sigma_{\min} \in (0, \sigma_0], \quad 0 < \eta_1 \le \eta_2 < 1, \\ \text{and} \quad 0 < \gamma_1 < 1 < \gamma_2 < \gamma_3.$$

Compute $f(x_0)$ and set $k = 0$.

Step 1: Test for termination. For $j = 1, \dots, q$:

1. Evaluate $\nabla_x^j f(x_k)$ if it is not yet available.
2. Compute $d_{k,j} \in \mathcal{B}_{\delta_{k,j}}$ such that
$$\varsigma \phi_{f,j}^{\delta_{k,j}}(x_k) \le \Delta T_{f,j}(x_k, d_{k,j}). \tag{12.2.26}$$
3. If
$$\Delta T_{f,j}(x_k, d_{k,j}) \ge \varsigma \epsilon_j \frac{\delta_{k,j}^j}{j!},$$
go to Step 2.

Terminate with the approximate solution $x_\epsilon = x_k$ and $\delta_\epsilon = \delta_k$.

Step 2: Step calculation. Evaluate any unavailable $\{\nabla_x^i f(x_k)\}_{i=j+1}^p$. Compute an approximate minimizer s_k of the model $m_k(s)$ and optimality radii $\delta_{s_k} \in (0,1]^q$ for which

$$m_k(s_k) \le m_k(0) \tag{12.2.27}$$

and either

$$\|s_k\| > 1 \quad \text{with} \quad \delta_{s_k} = \delta_k \tag{12.2.28}$$

or

$$\varsigma \phi_{m_k,j}^{\delta_{s_k,j}}(s_k) \le \Delta T_{m_k,j}(s_k, d_{k,j}^m) \le \varsigma \theta \epsilon_j \frac{\delta_{s_k,j}^j}{j!} \tag{12.2.29}$$

for each $j \in [1:q]$, where $d_{k,j}^m \in \mathcal{B}_{\delta_{s_k,j}}$. If $m_k(s_k) = m_k(0)$, terminate with the approximate solution $x_\epsilon = x_k$ and $\delta_\epsilon = \delta_k$.

Step 3: Acceptance of the trial point. Compute $f(x_k + s_k)$ and define

$$\rho_k = \frac{f(x_k) - f(x_k + s_k)}{\Delta T_{f,p}(x_k, s_k)}. \tag{12.2.30}$$

If $\rho_k \geq \eta_1$, then define $x_{k+1} = x_k + s_k$ and $\delta_{k+1} = \delta_{s_k}$. Otherwise define $x_{k+1} = x_k$ and $\delta_{k+1} = \delta_k$.

Step 4: Regularization parameter update. Set

$$\sigma_{k+1} \in \begin{cases} [\max(\sigma_{\min}, \gamma_1 \sigma_k), \sigma_k] & \text{if } \rho_k \geq \eta_2, \\ [\sigma_k, \gamma_2 \sigma_k] & \text{if } \rho_k \in [\eta_1, \eta_2), \\ [\gamma_2 \sigma_k, \gamma_3 \sigma_k] & \text{if } \rho_k < \eta_1. \end{cases} \tag{12.2.31}$$

Increment k by one and go to Step 1 if $\rho_k \geq \eta_1$ or to Step 2 otherwise.

Take note that, unlike in the TRq algorithm, δ_k and δ_{s_k} are vectors, and here and in what follows we use the shorthand

$$\delta_{k,j} = [\delta_k]_j \quad \text{and} \quad \delta_{s_k,j} = [\delta_{s_k}]_j \tag{12.2.32}$$

for their j components, $j \in [1:q]$. We obviously inherit the notions of successful and unsuccessful iterations that are common to all adaptive-regularization algorithms, and thus the conclusion of Lemma 2.4.1 (p. 44) continues to hold.

Observe that the step computation (Step 2) is now more complicated than was necessary for the ARp algorithm (p. 100). Condition (12.2.29)[76] still aims to ensure that the model is approximately minimized, and thus that the norm of the step is large enough, just as was the case for (4.1.11) and (4.1.12) in the ARp algorithm. But the approximate-minimization property is now expressed using the $\phi^{\delta}_{m_k,j}$ measure instead of the standard approximate first-, second-, and third-order conditions. In the same way, $\phi^{\delta}_{f,j}$ is now used in the algorithm's termination rule in Step 1. This implies that δ_s must be computed along with s_k in Step 2 of the algorithm. In addition, we do not require that we compute the global minima $\phi^{\delta}_{f,j}$ and $\phi^{\delta}_{m_k,j}$, merely that there are vectors $d_{k,j}$ and $d^m_{k,j}$ so that the approximations (12.2.26) and the first part of (12.2.29) hold.

Notice that the efficient procedure of Section 9.4.2 (p. 246) can be used to assess (12.2.26) when $j \in \{1,2\}$. Moreover, the subproblem-termination tests (12.2.29) in Step 2 can be stopped (as in Step 1) as soon as a j is found for which the inequality is violated, and the procedure of Section 9.4.2 (p. 246) can also be used there when $j \in \{1,2\}$. As a consequence, the ARqp algorithm is at least practical for $q \in \{1,2\}$, just like its AR2 and ARp predecessors.

We now need to verify that the algorithm is well defined in the sense that a step s_k and associated δ_{s_k} satisfying (12.2.27)–(12.2.29) can always be found.

While Theorem 12.1.4 applied to the model ensures that

$$\phi^{\delta}_{m_k,j}(s^*_k) = 0 \quad \text{for} \quad j \in \{1,2\}$$

and all $\delta > 0$ (and, in particular, for $\delta = 1$), the situation is more complicated for $j \geq 3$ because a global minimizer of the model (12.2.23) may not be a global minimizer of its jth-order Taylor expansion in a ball of arbitrary radius: we may have to restrict this radius for this property to hold. To improve our understanding of what might happen, we now analyze two simple examples.

[76] It is possible to use distinct θ_j for each $j \in [1:q]$, but we opted not to for simplicity.

Example 12.2.1 Consider an iterate x_k at which $f(x_k) = 0$, $\nabla_x^1 f(x_k) = -\beta e_1$ for some $\beta > 0$, and $\nabla_x^j f(x_k) = 0$ for $j \in [2:p]$. Then the model is given by

$$m_k(s) = \nabla_x^1 f(x_k)[s] + \sigma_k \frac{\|s\|^{p+1}}{(p+1)!},$$

which is (12.1.18) from Example 12.1.3 with a change of independent variable (x becomes s) and specifically $\sigma = \sigma_k$. Reviewing the earlier discussion, we recall that the unique global minimizer of $m_k(s)$ is at $s_k^* = (\beta p!/\sigma_k)^{1/p} e_1$, the jth-degree Taylor expansion $T_{m_k,j}(s_k^*, d)$ of this model at s_k^* is convex in all directions d orthogonal to s_k^*, and therefore no limitation on the size of $\delta_{s_k,j}$ can result from the behaviour of $T_{m_k,j}(s_k^*, d)$ in those directions. It then suffices to focus on the behaviour of $T_{m_k,j}(s_k^*, \alpha s_k^*)$ for $\alpha \in \mathbb{R}$ and to refer to Figure 12.1.3 on page 282 for inspiration. As before, $T_{m_k,1}(s_k^*, \alpha s_k^*) - T_{m_k,1}(s_k^*, 0) = 0$ for all α, while $T_{m_k,2}(s_k^*, \alpha s_k^*)$ is convex, and so our attention turns to $T_{m_k,3}(s_k^*, \alpha s_k^*)$.

We are interested in computing an upper bound on $\delta_{s_k,j}$ so that (12.2.29) holds for $s_k = s_k^*$ and for odd j. Consider the case where $j = 3$, and in this event (12.1.21) and (12.2.29) require that

$$T_{m_k,3}(s_k^*, 0) - T_{m_k,3}(s_k^*, \alpha s_k^*) = -\frac{\sigma_k \alpha^2 \|s_k^*\|^{p+1}}{2(p-1)!} - \frac{\sigma_k \alpha^3 \|s_k^*\|^{p+1}}{6(p-2)!} \leq \theta \epsilon_3 \frac{|\alpha|^3 \|s_k^*\|^3}{6}$$

for all $\|\alpha s_k^*\| \leq \delta_{s_k,3}$ or, equivalently,

$$\tfrac{1}{3}\alpha \left(p(p-1)\|s_k^*\|^{p-2} + \operatorname{sgn}(\alpha) \frac{\theta p! \epsilon_3}{\sigma_k} \right) \geq -p\|s_k^*\|^{p-2}. \tag{12.2.33}$$

For simplicity now suppose that $\sigma_k = p!$ and $\beta = 1$, in which case $\|s_k\| = 1$. Then (12.2.33) reveals that the smallest suitable α is given by

$$\alpha = -3 \left(\frac{p\|s_k^*\|^{p-2}}{p(p-1)\|s_k^*\|^{p-2} - \theta\epsilon_3} \right) = -3 \left(\frac{p}{p(p-1) - \theta\epsilon_3} \right),$$

the largest being infinite. If $p = 5$ and taking into account that $\theta\epsilon_3 \leq 1$, we deduce that

$$\alpha \geq -3 \left(\frac{p}{p(p-1) - 1} \right) = -\frac{15}{19},$$

which is coherent with Figure 12.1.3. As a consequence, $\delta_{s_k,3}$, the diameter of the neighbourhood of $s_k = s_k^*$ in which (12.2.29) holds must be at most $\frac{15}{19}$. Hence, by continuity, $\delta_{s_k,3} < \|s_k\|$ and $\delta_{s_k,3} < 1$ for any alternative s_k close to s_k^*, a situation which will haunt us later (see Lemma 12.2.12). We note that although the relaxation (12.2.29) allows us to use a slightly larger radius than is needed for exact criticality (15/19 rather than the 3/4 as we illustrated in Example 12.1.3), neither is especially close to the ideal $\delta_j = 1$ that is permitted when $j = 1$ or 2.

A related issue can also arise if Step 1 of the algorithm fails to detect an approximate local minimizer. In the following example, we show that potential termination in Step 2 cannot be avoided if the optimality radius inherited from the previous iteration is too large.

Example 12.2.2 Consider $m(s) = s^2(s-\frac{1}{2})^2 = \frac{1}{2}s^2(1-2s)+s^4$. This function has the form (12.2.23) (with $\sigma = 4!$) and has two global minima, at $s=0$ and $s=\frac{1}{2}$. Unfortunately, it is easy to verify that the third-order Taylor expansion at the origin is $T_{f,3}(0,s) = \frac{1}{2}s^2(1-2s)$, whose global minimizer within a ball of radius one centered at zero is at $s=1$ and has the value $-\frac{1}{2}$, even though the origin is a local minimizer of $T_{f,3}(0,s)$. However, the origin is the global minimum of $T_{f,3}(0,s)$ in every ball of radius smaller than $\frac{1}{2}$. We illustrate this in Figure 12.2.1. (Note that the problem disappears as soon as σ_k deviates from the very unlucky choice $\sigma_k = 4!$.)

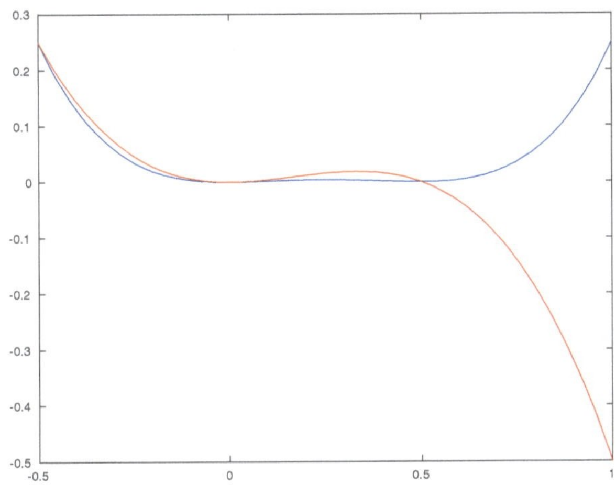

Figure 12.2.1. The model $m(s)$ of Example 12.2.2 (in blue) and its third-order Taylor approximation at the origin (in red)

Fortunately, despite these concerns, we will show that a step can always be found and the optimality radii do not become too small. Specifically, we consider the following technique for computing a suitable pair (s_k, δ_k).

Algorithm 12.2.3: A detailed Step 2 for the ARqp algorithm

Step 2: Step calculation. Suppose that the model $m_k(s)$ from (12.2.23) and the constants $\varsigma \in (0,1]$ and $\theta \in (0,1)$ are given.

Step 2.1: Initialize $\mu \in (0,1]$, and set $\delta_{s_k,j} = 1$ for all $j \in [1:q]$.

Step 2.2: Find s_μ such that

$$\mu(m_k(0) - m_k^*) \leq \Delta m_k(s_\mu), \tag{12.2.34}$$

where m_k^* is the (unknown) globally minimal value of $m_k(s)$.

Step 2.3: If $\|s_\mu\| \geq 1$, return the pair (s_k, δ_{s_k}), where $s_k = s_\mu$.

Step 2.4: For $j \in [1:q]$:

Step 2.4.1: Compute $d_{\mu,j}$ such that $\|d_{\mu,j}\| \leq \delta_{s_k,j}$ and

$$\varsigma\phi_{m_k,j}^{\delta_{s_k,j}}(s_\mu) \leq \Delta T_{m_k,j}(s_\mu, d_{\mu,j}). \tag{12.2.35}$$

Step 2.4.2: If

$$\Delta T_{m_k,j}(s_\mu, d_{\mu,j}) \leq \varsigma\theta\epsilon_j \frac{\delta^j_{s_k,j}}{j!}, \tag{12.2.36}$$

consider the next j (if $j < q$) or go to Step 2.5 (if $j = q$).

Step 2.4.3: Otherwise, if $j \geq 3$ and

$$m_k(s_\mu) - m_k(s_\mu + d_{\mu,j}) < \tfrac{1}{2}\varsigma\theta\epsilon_j \frac{\delta^j_{s_k,j}}{j!}, \tag{12.2.37}$$

set $\delta_{s_k,j} = \frac{1}{2}\delta_{s_k,j}$ and go back to Step 2.4.1.

Step 2.4.4: Otherwise set $\mu = \frac{1}{2}(1+\mu)$ and go back to Step 2.2.

Step 2.5: Return the pair (s_k, δ_{s_k}), where $s_k = s_\mu$.

Notice that each time that Step 2.4.3 is executed, $\delta_{s_k,j}$ is reduced linearly toward zero, while Step 2.4.4 adjusts μ so that it increases linearly to 1. The constant 1 in the test in Step 2.3 is purely arbitrary and chosen for convenience; any positive value is possible, although constants in later-derived bounds would then need to be adjusted accordingly.

In order to verify the validity of Algorithm 12.2.3, we first analyze properties of the model m_k.

Lemma 12.2.7: Properties of the regularized model

Suppose that **f.D1pL** *holds, that $m_k(s)$ is the model (12.2.23), and that L_f is given by (12.2.25). Then m_k is p-times continuously differentiable and has a Lipschitz-continuous pth derivative (with Lipschitz constant σ_k). Moreover, if $\|s\| \leq 1$, then*

$$\|\nabla^\ell_s m_k(s)\| \leq 3L_f + \sigma_k \quad \text{for} \quad \ell \in [2:p]. \tag{12.2.38}$$

Proof. We first note that

$$\nabla^p_s m_k(s) = \nabla^p_s T_{f,p}(x_k, s) + \frac{\sigma_k}{(p+1)!}\nabla^p_s\|s\|^{p+1} = \nabla^p_x f(x_k) + \frac{\sigma_k}{(p+1)!}\nabla^p_s\|s\|^{p+1}$$

from (12.2.23) and thus that

$$\|\nabla^p_s m_k(s_1) - \nabla^p_s m_k(s_2)\| \leq \frac{\sigma_k}{(p+1)!}\left\|\nabla^p_s\|s_1\|^{p+1} - \nabla^p_s\|s_2\|^{p+1}\right\|. \tag{12.2.39}$$

But since (B.4.3) on p. 461 gives that $\|\nabla^{p+1}_s\|s\|^{p+1}\| = (p+1)!$ for all s, we deduce from Theorem A.8.5 on p. 438 that $\nabla^p_s\|s\|^{p+1}$ is Lipschitz continuous with Lipschitz constant $(p+1)!$. Hence (12.2.39) implies that

$$\|\nabla^p_s m_k(s_1) - \nabla^p_s m_k(s_2)\| \leq \sigma_k\|s_1 - s_2\|,$$

and the first conclusion of the lemma follows. From (12.2.23), the triangle inequality, and (B.4.1) (p. 461), we also have that

$$\begin{aligned}\|\nabla^\ell_s m_k(s)\| &\leq \|\nabla^\ell_s T_{f,p}(x_k, s)\| + \frac{\sigma_k}{(p+1)!}\|\nabla^\ell_s\|s\|^{p+1}\| \\ &\leq \|\nabla^\ell_s T_{f,p}(x_k, s)\| + \frac{\sigma_k}{(p-\ell+1)!}\|s\|^{p-\ell+1}\end{aligned} \tag{12.2.40}$$

for $\ell \in [1:p]$. But the definition of the ℓth derivative of the pth-order Taylor-series approximation of f given in (A.11) on p. 435, the triangle inequality, **f.D1pL**, and (12.2.25) ensure that

$$\|\nabla_s^\ell T_{f,p}(x_k, s)\| \le \sum_{i=\ell}^{p} \frac{1}{(i-\ell)!}\|\nabla_x^i f(x_k)\| \, \|s\|^{i-\ell} \le L_f \sum_{i=\ell}^{p} \frac{\|s\|^{i-\ell}}{(i-\ell)!}$$

for $\ell \in [2:p]$. Combining this inequality, (12.2.40), the bound $\|s\| \le 1$, and the series bound (A.33) on p. 442 then gives (12.2.38). □

We next show that the model and its Taylor decrements cannot be very different for sufficiently small displacements.

Lemma 12.2.8: Relationship between model and Taylor decrements

*Suppose that **f.D1pL** holds, that $m_k(s)$ is the model (12.2.23), that $s \in \mathcal{B}_1$, and that L_f is given by (12.2.25). Given $\alpha > 0$ and*

$$\delta \in \left(0, \min\left[1, \frac{\alpha}{4(3L_f + \sigma_k)}\right]\right], \tag{12.2.41}$$

suppose that there is a displacement $d \in \mathcal{B}_\delta$ for which

$$\Delta T_{m_k,j}(s,d) \ge \alpha \frac{\delta^j}{j!} \tag{12.2.42}$$

for some $j \in [1:q]$. Then

$$m_k(s) - m_k(s+d) \ge \tfrac{1}{2}\alpha \frac{\delta^j}{j!}. \tag{12.2.43}$$

Proof. Lemma 12.2.7 ensures that m_k is p-times continuously differentiable and has a Lipschitz-continuous pth derivative with constant σ_k. As a consequence, using the triangle inequality and the fact that $\|d\| \le \delta$, we deduce from **f.D1pL** and Corollary A.8.4 (p. 437) that

$$\begin{aligned} m_k(s) - m_k(s+d) &\ge \Delta T_{m_k,j}(s,d) + \sum_{\ell=j+1}^{p} \frac{1}{\ell!}\nabla_s^\ell m_k(s)[d]^\ell - \frac{\sigma_k}{(p+1)!}\|d\|^{p+1} \\ &\ge \Delta T_{m_k,j}(s,d) - \sum_{\ell=j+1}^{p} \frac{\delta^\ell}{\ell!}\|\nabla_s^\ell m_k(s)\| - \frac{\sigma_k}{(p+1)!}\delta^{p+1}. \end{aligned}$$

This, the fact that $\|s\| \le 1$, (12.2.38), and (12.2.42) give that

$$m_k(s) - m_k(s+d) \ge \alpha\frac{\delta^j}{j!} - (3L_f + \sigma_k)\sum_{\ell=j+1}^{p+1} \frac{\delta^\ell}{\ell!},$$

and thus we deduce that

$$m_k(s) - m_k(s+d) \ge \alpha\frac{\delta^j}{j!} - 2(3L_f + \sigma_k)\frac{\delta^{j+1}}{j!} = \frac{\delta^j}{j!}(\alpha - 2(3L_f + \sigma_k)\delta)$$

using the bounds $\delta \le 1$ and consequently (A.33). Substituting the imposed restriction on δ from (12.2.41) into the above yields (12.2.43). □

Observe that the bound on δ given by (12.2.41) does not depend on j. We now apply this result to show that Algorithm 12.2.3 is well defined and produces a suitable pair (s_k, δ_{s_k}).

Lemma 12.2.9: Existence of a step

Suppose that ***f.D1pL*** *holds and that L_f is given by (12.2.25). Then a pair (s_k, δ_{s_k}) always exists such that (12.2.27) and either $\|s_k\| > 1$ or (12.2.29) hold. Moreover, it is always possible to ensure that*

$$\delta_{s_k,j} = 1 \quad \text{for} \quad j \in [1 : \min(q, 2)] \tag{12.2.44}$$

and, if $q \geq 3$ and $\|s_k\| \leq 1$,

$$\delta_{s_k,j} \geq \kappa_\delta(\sigma_k)\epsilon_j \quad \text{for} \quad j \in [3 : q], \tag{12.2.45}$$

where

$$\kappa_\delta(\sigma) \stackrel{\text{def}}{=} \frac{\varsigma\theta}{8(3L_f + \sigma)}.$$

Proof. Note that (12.2.34) ensures that $m_k(s_\mu) \leq m_k(0)$ for each computed s_μ. Suppose, for the purpose of deriving a contradiction, that Algorithm 12.2.3 does not terminate. Then we have from the mechanism of Step 2.3 that $\|s_\mu\| \leq 1$ and Lemma 12.2.8 is applicable for all values of μ. As a consequence, the situation where both (12.2.36) is violated and (12.2.37) holds can only happen if

$$\delta_{s_k,j} > \frac{\varsigma\theta\epsilon_j}{4(3L_f + \sigma_k)}, \tag{12.2.46}$$

where we also used that the right-hand side of (12.2.46) is less than one. It is thus impossible that the algorithm passes through Step 2.4.3 decreasing $\delta_{s_k,j}$ infinitely often, and we deduce that (12.2.45) holds, with the extra factor $1/2$ arising since $\delta_{s_k,j}$ may be decreased at most one more time. As a consequence, since Algorithm 12.2.3 does not terminate, it must be that (12.2.36) is violated infinitely often, and either (12.2.37) is violated or $j \leq 2$. In all such cases, μ tends to one and $m_k(s_\mu)$ tends to m_k^*. If $j > 2$, the fact that (12.2.37) is violated implies that $m_k^* \leq m(s_\mu + d_{\mu,j}) < m(s_\mu)$, and thus $m_k(s_\mu)$ and $m_k(s_\mu + d_{\mu,j})$ both converge to m_k^*. Thus, given that (12.2.46) holds, (12.2.37) must be satisfied eventually, which is a contradiction. If $j \leq 2$, the fact that (12.2.36) is violated implies that

$$\phi^1_{m_k,j}(s_\mu) > \varsigma\theta\epsilon_j \frac{\delta^j_{s_k,j}}{j!} \geq \tfrac{1}{2}\varsigma\theta\epsilon_j; \tag{12.2.47}$$

recall that $\delta_{s_k,j}$ remains fixed at one when $j \leq 2$. Since the level set $\{s \in \mathbb{R}^n \mid m_k(s) \leq m_k(0)\}$ is bounded, there must be a subsequence $\{s_{\mu_i}\}$ of the successively computed trial steps s_μ for μ converging to one, which itself converges to some global minimizer s_k^*. By virtue of Theorem 12.1.4, we know that $\phi^1_{m_k,j}(s_k^*) = 0$. The continuity of $\phi^1_{m_k,j}(s)$ as a function of s and (12.2.47) therefore imply that

$$0 = \phi^1_{m_k,j}(s_k^*) = \lim_{i\to\infty} \phi^1_{m_k,j}(s_{\mu_i}) > \tfrac{1}{2}\varsigma\theta\epsilon_j, \tag{12.2.48}$$

which is again a contradiction. Hence the algorithm terminates and a pair (s_k, δ_{s_k}) satisfying (12.2.27)–(12.2.29) exists; observe that $d^m_{k,j}$ in (12.2.29) is the final $d_{\mu,j}$. Moreover, since $\delta_{s_k,j}$ can only be adjusted by the algorithm if $j > 2$, it must be that $\delta_{s_k,1}$ and $\delta_{s_k,2}$ remain fixed at their initial values, one, which is (12.2.44). □

Clearly a step to any global minimizer of m_k is always acceptable as it corresponds to starting Algorithm 12.2.3 with $\mu = 1$. On the other hand, the full power of global minimization is not required in this algorithm, because we have seen that the loop between Steps 2.4.4 and 2.2 is finite, which means that, if initialized with a value less than one, μ will remain strictly below one.

We should inject a slight note of caution here. While we have mentioned that recent methods based on semidefinite programming are capable of finding a suitable approximation such as that needed in (12.2.35), these methods are only applicable if the objective in question is a (multivariate) polynomial. This may not be the case for $m_k(s)$, as required in (12.2.34), when p is even, since the regularization term $\sigma_k\|s\|^{p+1}/(p+1)!$ is then not a polynomial; this is not an issue when p is odd. A simple fix is to "overregularize" when p is even, that is, to use the regularizer $\sigma_k\|s\|^{p+2}/(p+2)!$, but we give no further details here.

Notice that we have not excluded the possibility that $m_k^* = m_k(0)$ in Algorithm 12.2.3 and Lemma 12.2.9; this is what happens in Example 12.2.2. In this case, (12.2.34) implies that $\Delta m_k(s_\mu) = 0$ and Algorithm 12.2.3 must return a step s_k such that as $m_k(s_k) = 0$. Step 2 of ARqp then enforces termination. It is important to establish that this is a satisfactory outcome.

Lemma 12.2.10: Termination in Step 2

Termination cannot occur in Step 2 of the ARqp algorithm if $q \leq 2$. If $q \geq 3$ and the ARqp algorithm terminates in Step 2 of iteration k with $x_\epsilon = x_k$, then there exists a $\delta \in (0,1]^q$ such that (12.1.25) holds for $x = x_\epsilon$ and x_ϵ is an (ϵ, δ)-approximate qth-order minimizer.

Proof. Given the mechanism of Algorithm 12.2.3, if termination occurs in Step 2, this must be because every global minimizer s_k^* of $m_k(s)$ is such that $m_k(s_k^*) = 0$. Hence $s_k^* = 0$ is one such global minimizer. Thus

$$T_{m_k,p}(0,d) = T_{f,p}(x_k,d) \text{ for all } d. \tag{12.2.49}$$

Suppose that $q \leq 2$. Then, for $j \in \{1,2\}$, (12.2.49) and Theorem 12.1.4 imply that

$$\phi^{\delta_{k,j}}_{f,j}(x_k) = \phi^{\delta_{k,j}}_{m_k,j}(0) = 0 < \epsilon_j \frac{\delta^j_{k,j}}{j!},$$

which contradicts the fact that termination did not occur in Step 1. Since this is impossible, termination cannot happen in Step 2 when $q \leq 2$.

By contrast, suppose that $q \geq 3$ and that termination occurs in Step 2. Then (12.2.49) and (12.2.29) for $s_k = 0$ ensure (12.1.25), and thus x_k is an (ϵ, δ_k^*)-approximate minimizer. □

Lemmas 12.2.9 and 12.2.10 ensure that the conceptual ARqp algorithm is well defined. Using (12.2.44) and (12.2.45), we may therefore assume henceforth, without loss of generality, that, if termination does not occur in Step 2,

$$\|s_k\| > 1 \text{ or } \delta_{s_k,j} \geq \kappa_\delta(\sigma_k)\,\epsilon_j \tag{12.2.50}$$

for all $k \geq 0$ and $j \in [1:q]$. Furthermore, we choose $\delta_{s_k,j} = 1$ for $j = 1$ and 2.

12.2.2.2 Complexity of finding approximate minimizers

The evaluation-complexity analysis for the ARqp algorithm is once again similar in structure to that of other adaptive-regularization methods that we have examined so far. Because the basis of the algorithm is identical to that of ARp (they only differ in Steps 1 and 2 in the use of different optimality tests) and because **f.D1pL** implies **f.DpL**, it follows that both Lemmas 4.1.1 (p. 103) and 4.1.2 (p. 103) still hold as stated. In particular, (4.1.19) holds and defines $\sigma_{\max}$, and this and the fact that $\delta_{k+1} = \delta_{s_k}$ at successful iterations allow us to deduce from (12.2.50) that

$$\|s_k\| > 1 \quad \text{or} \quad \delta_{k+1,j} \geq \kappa_{\delta,\min}\epsilon_j \tag{12.2.51}$$

for $j \geq 3$ if iteration k is successful, where

$$\kappa_{\delta,\min} \stackrel{\text{def}}{=} \kappa_\delta(\sigma_{\max}) \leq \kappa_\delta(\sigma_k),$$

while $\delta_{k,j} = 1$ for $j = 1$ and 2.

The significant difference in the analysis is, as usual, how one proves a suitable lower bound on the length of the step, given the new termination rule in Step 2. To this end, we first derive a crucial inequality relating δ_{k+1} and the steplength.

Lemma 12.2.11: A useful relationship between the stepsize and radii

Suppose that **f.D1pL** *holds. Suppose also that iteration k is successful and that the* ARqp *algorithm does not terminate at iteration $k+1$. Then there exists $j \in [1:q]$ such that*

$$(1-\theta)\,\epsilon_j\,\frac{\delta_{k+1,j}^j}{j!} \leq (L_{f,p} + \sigma_{\max}) \sum_{\ell=1}^{j} \frac{\delta_{k+1,j}^\ell}{\ell!}\|s_k\|^{p-\ell+1}, \tag{12.2.52}$$

where $\sigma_{\max}$ is given by (4.1.19).

Proof. If the algorithm does not terminate at iteration $k+1$, there must exist a $j \in [1:q]$ such that (12.1.25) fails at order j at iteration $k+1$. Consider such a j and let d be the argument of the minimization in the definition of $\phi_{f,j}^{\delta_{k+1,j}}(x_{k+1})$ for which $\|d\| \leq \delta_{k+1,j} \leq 1$. The definition of $\phi_{f,j}^{\delta_{k+1,j}}(x_{k+1})$ in (12.1.10) then gives that

$$\begin{aligned}
\epsilon_j \frac{\delta_{k+1,j}^j}{j!} < \phi_{f,j}^{\delta_{k+1,j}}(x_{k+1}) &= -\sum_{\ell=1}^{j} \frac{1}{\ell!}\nabla_x^\ell f(x_{k+1})[d]^\ell \\
&= -\sum_{\ell=1}^{j} \frac{1}{\ell!}\nabla_x^\ell f(x_{k+1})[d]^\ell + \sum_{\ell=1}^{j} \frac{1}{\ell!}\nabla_s^\ell T_{f,p}(x_k,s_k)[d]^\ell \\
&\quad - \sum_{\ell=1}^{j} \frac{1}{\ell!}\nabla_s^\ell T_{f,p}(x_k,s_k)[d]^\ell - \frac{\sigma_k}{(p+1)!}\sum_{\ell=1}^{j} \frac{1}{\ell!}\nabla_s^\ell \left(\|s_k\|^{p+1}\right)[d]^\ell \\
&\quad + \frac{\sigma_k}{(p+1)!}\sum_{\ell=1}^{j} \frac{1}{\ell!}\nabla_s^\ell \left(\|s_k\|^{p+1}\right)[d]^\ell.
\end{aligned} \tag{12.2.53}$$

To bound the terms on the right-hand side of (12.2.53), first, using Corollary A.8.4,

$$\begin{aligned} -\sum_{\ell=1}^{j}\frac{1}{\ell!}\nabla_x^\ell f(x_{k+1})[d]^\ell &+ \sum_{\ell=1}^{j}\frac{1}{\ell!}\nabla_s^\ell T_{f,p}(x_k,s_k)[d]^\ell \\ &\le \sum_{\ell=1}^{j}\frac{\delta_{k+1,j}^\ell}{\ell!}\left\|\nabla_x^\ell f(x_{k+1})-\nabla_s^\ell T_{f,p}(x_k,s_k)\right\| \\ &\le L_{f,p}\sum_{\ell=1}^{j}\frac{\delta_{k+1,j}^\ell}{\ell!(p-\ell+1)!}\|s_k\|^{p-\ell+1}, \end{aligned} \tag{12.2.54}$$

where $L_{f,p}$ is defined as in **f.D1pL**. Furthermore, because of (12.2.23) and (12.2.29),

$$\begin{aligned} -\sum_{\ell=1}^{j}\frac{1}{\ell!}\nabla_s^\ell T_{f,p}(x_k,s_k)[d]^\ell &- \frac{\sigma_k}{(p+1)!}\sum_{\ell=1}^{j}\frac{1}{\ell!}\nabla_s^\ell\left(\|s_k\|^{p+1}\right)[d]^\ell \\ &\le \phi_{m_k,j}^{\delta_{s_k,j}}(s_k) \le \theta\epsilon_j\frac{\delta_{s_k,j}^j}{j!} = \theta\epsilon_j\frac{\delta_{k+1,j}^j}{j!}, \end{aligned} \tag{12.2.55}$$

where the last equality follows because $\delta_{s_k,j} = \delta_{k+1,j}$ if iteration k is successful. Moreover, in view of (B.4.1) in Lemma B.4.1, with $\beta = 1$, and Lemma 14.1.6 (p. 381), we also have that

$$\begin{aligned} \frac{\sigma_k}{(p+1)!}\sum_{\ell=1}^{j}\frac{1}{\ell!}\nabla_s^\ell\left(\|s_k\|^{p+1}\right)[d]^\ell &\le \frac{\sigma_k}{(p+1)!}\sum_{\ell=1}^{j}\frac{1}{\ell!}\left\|\nabla_s^\ell\left(\|s_k\|^{p+1}\right)\right\|\|d\|^\ell \\ &= \sigma_k\sum_{\ell=1}^{j}\frac{\|s_k\|^{p-j+1}\|d\|^\ell}{\ell!(p-j+1)!} \\ &\le \sigma_{\max}\sum_{\ell=1}^{j}\frac{\delta_{k+1,j}^\ell}{\ell!(p-j+1)!}\|s_k\|^{p-j+1}. \end{aligned} \tag{12.2.56}$$

We may now substitute (12.2.54)–(12.2.56) into (12.2.53) and use the inequality $(p-\ell+1)! \ge 1$ to obtain (12.2.52). □

Next, we use this inequality to derive the vital required lower bound on the stepsize.

Lemma 12.2.12: Lower bound on the length of step for ARqp

Suppose that iteration k is successful and that the ARqp algorithm does not terminate at iteration $k+1$. Suppose also that (as Lemma 12.2.9 allows) the algorithm ensures that $\delta_{k+1,j} = 1$ for $j \in [1:\min(q,2)]$ and that (12.2.51) holds for $j \in [3:q]$. Then there exists a $j \in [1:q]$ such that

$$\|s_k\| \ge \begin{cases} \left(\dfrac{1-\theta}{2j!(L_{f,p}+\sigma_{\max})}\right)^{\frac{1}{p-j+1}}\epsilon_j^{\frac{1}{p-j+1}} & \text{if } q\in\{1,2\}, \\[2ex] \left(\dfrac{(1-\theta)\kappa_{\delta,\min}^{j-1}}{2j!(L_{f,p}+\sigma_{\max})}\right)^{\frac{1}{p}}\epsilon_j^{\frac{j}{p}} & \text{if } q>2, \end{cases} \tag{12.2.57}$$

where $\kappa_{\delta,\min}$ is defined as in (12.2.51) and $\sigma_{\max}$ as in (4.1.19).

Proof. If $\|s_k\| > 1$, (12.2.57) clearly holds since $\epsilon_j \leq 1$ and $\kappa_{\delta,\min} < 1$. So suppose instead that $\|s_k\| \leq 1$. Because the algorithm does not terminate, Lemma 12.2.11 ensures that (12.2.52) holds for some $j \in [1:q]$. It is easy to verify that this inequality is equivalent to

$$\tau\, \epsilon_j\, \delta_{k+1,j}^j \leq \|s_k\|^{p+1} \chi_j \left(\frac{\delta_{k+1,j}}{\|s_k\|} \right), \tag{12.2.58}$$

where the function χ_j is defined by

$$\chi_j(t) = \sum_{\ell=1}^{j} \frac{t^\ell}{\ell!} \tag{12.2.59}$$

and where we have set

$$\tau = \frac{1-\theta}{j!(L_{f,p} + \sigma_{\max})}.$$

In particular, since $\chi_j(t) \leq 2t^j$ for $t \geq 1$ from the sum bound (A.35), we have that

$$\tau\, \epsilon_j \leq 2\|s_k\|^{p+1} \left(\frac{1}{\|s_k\|} \right)^j = 2\|s_k\|^{p-j+1} \tag{12.2.60}$$

whenever $\|s_k\| \leq \delta_{k+1,j}$.

If j is one or two, by assumption $\delta_{k+1,j} = 1$ and $\|s_k\| \leq 1 = \delta_{k+1,j}$. Thus (12.2.60) yields the first case of (12.2.57). Otherwise, if $j \geq 3$, by assumption (12.2.51) holds. In this case, if $\|s_k\| \leq \delta_{k+1,j}$, we may again deduce from (12.2.60) that the first case of (12.2.57) holds, which implies that the second case also holds since $\kappa_{\delta,\min} < 1$ and[77] $1/(p-j+1) \leq j/p$ for $p \geq j \geq 1$. Consider therefore the alternative for which $\|s_k\| > \delta_{k+1,j}$. Then, using (12.2.58) and the fact that the sum bound (A.34) ensures $\chi_j(t) < 2t$ for $t \in [0,1]$, we deduce that

$$\tau\, \epsilon_j\, \delta_{k+1,j}^j \leq 2\|s_k\|^{p+1} \left(\frac{\delta_{k+1,j}}{\|s_k\|} \right),$$

which, with (12.2.51), implies the second case of (12.2.57). □

We consolidate the above results by stating lower bounds on the minimal model and function decreases at successful iterations.

Lemma 12.2.13: Model and function decreases (ARqp)

Suppose that **f.D1qL** *holds. Then*

$$\Delta T_{f,p}(x_k, s_k) \geq \kappa_{\Delta m} \min_{j \in [1:q]} \epsilon_j^{\pi_j} \tag{12.2.61}$$

[77]This is easily verified by noting that, if $\lambda(t) = t/(j(t-j+1))$, then $\lambda(j) = 1$ and $\lambda^{(1)}(t) \leq 0$ for all $t \geq j$.

for every iteration of algorithm ARqp *before termination, where*

$$\kappa_{\Delta m} \stackrel{\text{def}}{=} \begin{cases} \dfrac{\sigma_{\min}}{(p+1)!}\left(\dfrac{1-\theta}{2q!(L_{f,p}+\sigma_{\max})}\right)^{\frac{p+1}{p-q+1}} & \text{if } q \in \{1,2\}, \\ \dfrac{\sigma_{\min}}{(p+1)!}\left(\dfrac{(1-\theta)\kappa_{\delta,\min}^{q-1}}{2q!(L_{f,p}+\sigma_{\max})}\right)^{\frac{q(p+1)}{p}} & \text{if } q > 2, \end{cases}$$

with $\kappa_{\delta,\min}$ *being given by (12.2.51),* $\sigma_{\max}$ *by (4.1.19) on p. 103, and*

$$\pi_j \stackrel{\text{def}}{=} \begin{cases} \dfrac{p+1}{p-j+1} & \text{if } q \in \{1,2\}, \\ \dfrac{j(p+1)}{p} & \text{if } q > 2. \end{cases}$$

If, in addition, $k \in \mathcal{S}$,

$$f(x_k) - f(x_{k+1}) \geq \eta_1 \kappa_{\Delta m} \min_{j\in[1:q]} \epsilon_j^{\pi_j}. \tag{12.2.62}$$

Proof. Lemma 4.1.1 (p. 103), (12.2.31), and Lemma 12.2.12 together imply that for every $k \in \mathcal{S}$, there exists a $j \in [1:q]$ such that

$$\Delta T_{f,p}(x_k, s_k) \geq \begin{cases} \dfrac{\sigma_{\min}}{(p+1)!}\left(\dfrac{1-\theta}{2j!(L_{f,p}+\sigma_{\max})}\right)^{\frac{p+1}{p-j+1}} \epsilon_j^{\frac{p+1}{p-j+1}} & \text{if } q\in\{1,2\}, \\ \dfrac{\sigma_{\min}}{(p+1)!}\left(\dfrac{(1-\theta)\kappa_{\delta,\min}^{j-1}}{2j!(L_{f,p}+\sigma_{\max})}\right)^{\frac{j(p+1)}{p}} \epsilon_j^{\frac{j(p+1)}{p}} & \text{if } q>2, \end{cases}$$

and (12.2.61) follows. When $k \in \mathcal{S}$, then necessarily $\rho_k \geq \eta_1$, and (12.2.62) results from (12.2.61) and (12.2.30). □

This result naturally leads to the following complexity theorem.

Theorem 12.2.14: qth-order complexity for the ARqp algorithm

Suppose that **f.Bb** *and* **f.D1pL** *hold. Suppose also that the* ARqp *algorithm ensures, for each* k, *either that* $\delta_{k+1,j} = 1$ *for* $j \in [1:q]$ *if* $q \in \{1,2\}$ *(as allowed by Lemma 12.2.9) or that (12.2.51) holds otherwise.*

1. *If* $q \in \{1,2\}$, *then there exist positive constants* $\kappa_{\text{ARqp}}^{s,1}$, $\kappa_{\text{ARqp}}^{a,1}$, *and* κ_{ARqp}^{c} *such that, for any* $\epsilon \in (0,1]^q$, *the* ARqp *algorithm requires at most*

$$\#_{\text{ARqp}}^{F} \stackrel{\text{def}}{=} \kappa_{\text{ARqp}}^{a,1} \frac{f(x_0) - f_{\text{low}}}{\min_{j\in[1:q]} \epsilon_j^{\frac{p+1}{p-j+1}}} + \kappa_{\text{ARqp}}^{c} = \mathcal{O}\left(\max_{j\in[1:q]} \epsilon_j^{-\frac{p+1}{p-j+1}}\right) \tag{12.2.63}$$

evaluations of f *and at most*

$$\#_{\text{ARqp}}^{D} \stackrel{\text{def}}{=} \kappa_{\text{ARqp}}^{s,1} \frac{f(x_0) - f_{\text{low}}}{\min_{j\in[1:q]} \epsilon_j^{\frac{p+1}{p-j+1}}} + 2 = \mathcal{O}\left(\max_{j\in[1:q]} \epsilon_j^{-\frac{p+1}{p-j+1}}\right) \tag{12.2.64}$$

evaluations of the derivatives of f of orders one to p to produce an iterate x_ϵ such that $\phi^1_{f,j}(x_\epsilon) \le \epsilon_j/j!$ for all $j \in [1:q]$.

2. *If $q > 2$, then there exist positive constants $\kappa^{s,2}_{\mathsf{ARqp}}$, $\kappa^{a,2}_{\mathsf{ARqp}}$, and $\kappa^{c}_{\mathsf{ARqp}}$ such that, for any $\epsilon \in (0,1]^q$, the ARqp algorithm requires at most*

$$k_{\mathsf{ARqp}} \stackrel{\rm def}{=} \kappa^{a,2}_{\mathsf{ARqp}} \frac{f(x_0) - f_{\rm low}}{\min\limits_{j\in[1:q]} \epsilon_j^{\frac{j(p+1)}{p}}} + \kappa^{c}_{\mathsf{ARqp}} = \mathcal{O}\left(\max_{j\in[1:q]} \epsilon_j^{-\frac{j(p+1)}{p}} \right) \tag{12.2.65}$$

evaluations of f and at most

$$\kappa^{s,2}_{\mathsf{ARqp}} \frac{f(x_0) - f_{\rm low}}{\min\limits_{j\in[1:q]} \epsilon_j^{\frac{j(p+1)}{p}}} + 2 = \mathcal{O}\left(\max_{j\in[1:q]} \epsilon_j^{-\frac{j(p+1)}{p}} \right) \tag{12.2.66}$$

evaluations of the derivatives of f of orders one to p to produce an iterate x_ϵ such that $\phi^{\delta_\epsilon}_{f,j}(x_\epsilon) \le \epsilon_j\, \delta^j_{\epsilon,j}/j!$ for some $\delta_\epsilon \in (0,1]^q$ and all $j \in [1:q]$.

Proof. By now, the reader has come to expect an argument to bound $f(x_0) - f_{\rm low}$ from below by accumulating a minimum decrease of f per successful iteration, resulting from a decrease in the pth-order Taylor expansion $T_{f,p}(x_k, s_k)$. Specifically, as $\{f(x_k)\}$ decreases monotonically, we find that

$$f(x_0) - f(x_{k+1}) \ge \eta_1 \kappa_{\Delta m} \min_{j\in[1:q]} \epsilon_j^{\pi_j} |\mathcal{S}_{k-1}|,$$

using (12.2.62) in Lemma 12.2.13 (p. 310), where π_j is either $(p+1)/(p-j+1)$ or $j(p+1)/p$, depending on the two cases distinguished in the lemma. This in turn yields an upper bound on the number of successful iterations. That is, using **f.Bb**, we conclude that

$$|\mathcal{S}_k| \le (\eta_1 \kappa_{\Delta m})^{-1} \frac{f(x_0) - f_{\rm low}}{\min\limits_{j\in[1:q]} \epsilon_j^{\omega}} + 1$$

because $|\mathcal{S}_k| \le |\mathcal{S}_{k-1}| + 1$. Lemma 2.4.1 (p. 44) may then be invoked to bound the total number of iterations and, consequently, the total number of evaluations. The details are straightforward and result in constants given by

$$\begin{aligned}
\kappa^{s,1}_{\mathsf{ARqp}} &\stackrel{\rm def}{=} \frac{(p+1)!}{\eta_1 \sigma_{\min}} \left(\frac{2q!(L_{f,p} + \sigma_{\max})}{1-\theta} \right)^{\frac{p+1}{p-q+1}} && \text{if } q \in \{1,2\}, \\
\kappa^{s,2}_{\mathsf{ARqp}} &\stackrel{\rm def}{=} \frac{(p+1)!}{\eta_1 \sigma_{\min}} \left(\frac{2q!(L_{f,p} + \sigma_{\max})}{(1-\theta)\kappa^{q-1}_{\delta,\min}} \right)^{\frac{q(p+1)}{p}} && \text{if } q > 2, \\
\kappa^{a,1}_{\mathsf{ARqp}} &\stackrel{\rm def}{=} \kappa^{s,1}_{\mathsf{ARqp}} \left(1 + \frac{|\log \gamma_1|}{\log \gamma_2} \right), \\
\kappa^{a,2}_{\mathsf{ARqp}} &\stackrel{\rm def}{=} \kappa^{s,2}_{\mathsf{ARqp}} \left(1 + \frac{|\log \gamma_1|}{\log \gamma_2} \right), \quad \text{and} \\
\kappa^{c}_{\mathsf{ARqp}} &\stackrel{\rm def}{=} \frac{1}{\log \gamma_2} \left\{ |\log \gamma_1| + \log\left(\frac{\sigma_{\max}}{\sigma_0} \right) \right\} + 2,
\end{aligned}$$

where

$$\sigma_{\max} = \max\left[\sigma_0, \frac{\gamma_3 L_{f,p}}{1-\eta_2}\right] \quad \text{and} \quad \kappa_{\delta,\min} = \frac{\varsigma\theta}{8(3\max_{j\in[1:p]} L_{f,j} + \sigma_{\max})}.$$

□

We remark that if each $\delta_{s_k,j}$ is bounded below by a constant (as is the case when $q \in \{1,2\}$), the better bounds (12.2.63) and (12.2.64) hold with adapted constants.

12.2.2.3 Comparisons with the TRq algorithm

We note that the bounds (12.2.63) and (12.2.65) are never worse in order than those obtained in Theorem 12.2.5 (p. 297) for the TRq algorithm because

$$\max_{j\in[1:q]} \epsilon_j^{-\frac{p+1}{p-j+1}} \leq \max_{j\in[1:q]} \epsilon_j^{-(q+1)} \quad \text{and} \quad \max_{j\in[1:q]} \epsilon_j^{-\frac{j(p+1)}{p}} \leq \max_{j\in[1:q]} \epsilon_j^{-(q+1)}.$$

However, they can be substantially better, for instance, if the smallest ϵ_j is not ϵ_q. If we compare the complexity bounds of ARqp with those of ARp in Chapter 4, in particular, Theorem 4.1.5 (p. 107) for criticality orders up to three, we see that in the order of the accuracy ϵ, they coincide for $q = 1, 2$ but differ for $q = 3$ since

$$\max_{j\in[1:3]} \epsilon_j^{-\frac{p+1}{p-j+1}} \leq \max_{j\in[1:3]} \epsilon_j^{-\frac{j(p+1)}{p}}.$$

For reference, in Table 12.2.1, we compare and contrast the orders of worst-case bounds given in Theorems 4.1.5, 12.2.5, and 12.2.14 for the ARp, TRq, and ARqp algorithms for the simple but popular case where all the ϵ_j's are equal. Notice, in particular, that the ARqp algorithm only offers an order improvement over TRq for $p > q$ but that as p rises, the improvement may be significant. In particular, were it feasible to pick ever-increasing p, ultimately we would gain an extra power of ϵ using the ARqp algorithm compared to TRq in their worst-case bounds.

Furthermore, the bounds predicted by Theorems 12.2.5 and 12.2.14 for the TRq and ARqp algorithms are reassuringly in line with those obtained earlier for the TR1 and AR1

q		1	2	3	$q > 3$
ARp		$\mathcal{O}\left(\epsilon^{-\frac{(p+1)}{p}}\right)$	$\mathcal{O}\left(\epsilon^{-\frac{(p+1)}{p-1}}\right)$	$\mathcal{O}\left(\epsilon^{-\frac{(p+1)}{p-2}}\right)$	–
TRq		$\mathcal{O}\left(\epsilon^{-2}\right)$	$\mathcal{O}\left(\epsilon^{-3}\right)$	$\mathcal{O}\left(\epsilon^{-4}\right)$	$\mathcal{O}\left(\epsilon^{-(q+1)}\right)$
ARqp	$p \geq q$	$\mathcal{O}\left(\epsilon^{-\frac{(p+1)}{p}}\right)$	$\mathcal{O}\left(\epsilon^{-\frac{(p+1)}{(p-1)}}\right)$	$\mathcal{O}\left(\epsilon^{-3\frac{(p+1)}{p}}\right)$	$\mathcal{O}\left(\epsilon^{-q\frac{(p+1)}{p}}\right)$
	$p = q$	$\mathcal{O}\left(\epsilon^{-2}\right)$	$\mathcal{O}\left(\epsilon^{-3}\right)$	$\mathcal{O}\left(\epsilon^{-4}\right)$	$\mathcal{O}\left(\epsilon^{-(q+1)}\right)$
	$p = q+1$	$\mathcal{O}\left(\epsilon^{-3/2}\right)$	$\mathcal{O}\left(\epsilon^{-4/3}\right)$	$\mathcal{O}\left(\epsilon^{-15/4}\right)$	$\mathcal{O}\left(\epsilon^{-\frac{q(q+2)}{(q+1)}}\right)$
	$p \to \infty$	$\mathcal{O}\left(\epsilon^{-1}\right)$	$\mathcal{O}\left(\epsilon^{-2}\right)$	$\mathcal{O}\left(\epsilon^{-3}\right)$	$\mathcal{O}\left(\epsilon^{-q}\right)$

Table 12.2.1. Order bounds with respect to the accuracy for the worst-case evaluation complexity of ARp to satisfy (12.1.22)–(12.1.24) and of TRq and ARqp to satisfy (12.1.25) when $\epsilon_j = \epsilon$ for all $j \in [1:q]$.

algorithms (Theorem 2.3.7 on p. 40 and Theorem 2.4.4 on p. 46) when $q = 1$ and for the TR2 and AR2 algorithms (Theorem 3.2.1 on p. 57, Theorem 3.2.6 on p. 61, Theorem 3.3.5 on p. 70, and Theorem 3.3.9 on p. 74) when $q = 2$ and $p = 1$ and 2. This is also true for the ARp algorithm (Theorem 4.1.5 on p. 107) when $q = 1$ and $q = 2$. For $q = 3$, Theorem 4.1.5 suggests an advantage when $p > 3$, but note that ARp uses standard approximate criticality measures.

12.2.2.4 Sharp evaluation complexity

Finally (and predictably), the evaluation-complexity bounds of Theorem 12.2.14 can be proved to be sharp in the order of ϵ.

We first consider the case where the choice of $\delta_{s_k,j} = 1$ is allowed in (12.2.29) by the global minimizer of the model $m_k(s)$ at every iteration.

Theorem 12.2.15: Slow qth-order convergence of the ARqp algorithm (good case)

Suppose that the choice $\delta_{s_k,j} = 1$ is possible (and made) for all k and all $j \in [1:q]$ and that $\epsilon_j = \epsilon \le \frac{1}{2}$ for all j. Then the ARqp algorithm applied to minimize a function f satisfying **f.Bb** *and* **F.D1pL** *may require*

$$\epsilon^{-\frac{p+1}{p-q+1}}$$

iterations and evaluations of f and of its derivatives of order one up to p to produce a point x_ϵ such that $\phi^1_{f,j}(x_\epsilon) \le \epsilon/j!$ for all $j \in [1:q]$.

Proof. Our aim is to show that, for each choice of $p \ge 1$ and whenever $\delta_{s_k,j} = 1$, there is an objective function satisfying **f.D1pL** and **f.Bb** for which finding an (ϵ, δ)-approximate qth-order minimizer may require at least $\epsilon^{-(p+1)/(p-q+1)}$ evaluations of the objective function and its derivatives using the ARqp algorithm.

Given a model degree $p \ge 1$ and an optimality order q, we define the sequences $\{f_k^{(j)}\}$ for $j \in [1:p]$ and $k \in [0:k_\epsilon]$ by

$$f_k^{(j)} = 0 \;\text{ for }\; j \in [1:q-1] \cup [q+1:p] \;\text{ and }\; f_k^{(q)} = -(\epsilon + \omega_k) < 0,$$

where

$$\omega_k = \epsilon\,\frac{k_\epsilon - k}{k_\epsilon} \in [0, \epsilon] \tag{12.2.67}$$

and

$$k_\epsilon = \left\lceil \epsilon^{-\frac{p+1}{p-q+1}} \right\rceil. \tag{12.2.68}$$

Thus

$$T_{f,p}(x_k, s) = \sum_{j=0}^{p} \frac{f_k^{(j)}}{j!} s^j = f_k^{(0)} - (\epsilon + \omega_k)\frac{s^q}{q!} \tag{12.2.69}$$

for some as yet undefined sequences $\{x_k\}$ and $\{f_k^{(0)}\}$. We also set $\sigma_k = p!/(q-1)!$ for all $k \in [0:k_\epsilon]$; we verify below that this is acceptable. It is easy to show using (12.2.69) and this definition of σ_k that the model (12.2.23) is globally minimized at

$$s_k = |f_k^{(q)}|^{\frac{1}{p-q+1}} = [\epsilon + \omega_k]^{\frac{1}{p-q+1}} \quad (k \in [0:k_\epsilon]). \tag{12.2.70}$$

We now suppose that Step 2 of the ARqp algorithm produces the step s_k given by (12.2.70) and the optimality radius $\delta_{s_k,j} = 1$ leads to (12.2.29) (as allowed by our assumption) for $j \in [1:q]$ and all $k \in [0:k_\epsilon]$ since $\epsilon \leq \frac{1}{2}$. The step (12.2.70) yields that

$$\begin{aligned} m_k(s_k) &= f_k^{(0)} - \frac{\epsilon+\omega_k}{q!}\left[\epsilon+\omega_k\right]^{\frac{q}{p-q+1}} + \frac{\sigma_k}{(p+1)!}\left[\epsilon+\omega_k\right]^{\frac{p+1}{p-q+1}} \\ &= f_k^{(0)} - \frac{\epsilon+\omega_k}{q!}\left[\epsilon+\omega_k\right]^{\frac{q}{p-q+1}} + \frac{1}{(p+1)(q-1)!}\left[\epsilon+\omega_k\right]^{\frac{p+1}{p-q+1}} \\ &= f_k^{(0)} - \tfrac{p-q+1}{(p+1)q!}\left[\epsilon+\omega_k\right]^{\frac{p+1}{p-q+1}}. \end{aligned}$$

Thus $m_k(s_k) < m_k(0)$ and (12.2.27) holds. We then define

$$f_0^{(0)} = 2^{1+\frac{p+1}{p-q+1}} \quad \text{and} \quad f_{k+1}^{(0)} = f_k^{(0)} - \frac{1}{q!}\left[\epsilon+\omega_k\right]^{\frac{p+1}{p-q+1}}, \tag{12.2.71}$$

which provide the identity

$$T_{f,p}(x_k, s_k) = f_{k+1}^{(0)}. \tag{12.2.72}$$

Thus iteration k is successful because $\rho_k = 1$ in (12.2.30), and therefore our choice of a constant σ_k is acceptable and $x_{k+1} = x_k + s_k$; for simplicity, we set

$$x_0 = 0 \text{ and therefore } x_k = \sum_{i=0}^{k-1} s_i. \tag{12.2.73}$$

Now observe that since (12.2.69) is unbounded from below, the minimizer of $T_{f,q}(x_k, s)$ within the ball $\|s\| \leq 1$ lies at $s = 1$, and thus

$$\phi_{f,q}^{1}(x_k) = (\epsilon+\omega_k)\frac{1}{q!}.$$

Therefore

$$\omega_k \in (0,\epsilon], \quad \phi_{f,j}^{1}(x_k) = 0 \text{ for } j \in [1:q-1], \quad \phi_{f,q}^{1}(x_k) > \epsilon\frac{1}{q!}, \tag{12.2.74}$$

and (12.1.25) fails for $(x_k, 1)$ for $k \in [0:k_\epsilon - 1]$, while

$$\omega_{k_\epsilon} = 0, \quad \phi_{f,j}^{1}(x_{k_\epsilon}) = 0 \text{ for } j \in [1:q-1], \quad \phi_{f,q}^{1}(x_{k_\epsilon}) = \epsilon\frac{1}{q!},$$

and (12.1.25) holds at x_{k_ϵ}.

In addition, using (12.2.71), (12.2.67), and the inequality $k_\epsilon \leq 1+\epsilon^{-\frac{p+1}{p-q+1}}$ that follows from (12.2.68) gives that

$$\begin{aligned} f_0^{(0)} \geq f_k^{(0)} &\geq f_0^{(0)} - k\left[2\epsilon\right]^{\frac{p+1}{p-q+1}} \\ &> f_0^{(0)} - k_\epsilon \epsilon^{\frac{p+1}{p-q+1}} 2^{\frac{p+1}{p-q+1}} \\ &\geq f_0^{(0)} - \left(1+\epsilon^{\frac{p+1}{p-q+1}}\right) 2^{\frac{p+1}{p-q+1}} \\ &\geq f_0^{(0)} - 2^{1+\frac{p+1}{p-q+1}} \end{aligned}$$

for all $k \in [0:k_\epsilon]$, and hence from (12.2.71) that

$$f_k^{(0)} \in \left(0, 2^{1+\frac{p+1}{p-q+1}}\right] \quad \text{for } k \in [0:k_\epsilon]. \tag{12.2.75}$$

Then (12.2.23) and (12.2.72) give that

$$|f_{k+1}^{(0)} - T_{f,p}(x_k, s_k)| = 0 \le |s_k|^{p+1}. \tag{12.2.76}$$

Now note that, using (12.2.69) and the first equality in (12.2.70),

$$T_{f,p}^{(j)}(x_k, s_k) = \frac{f_k^{(q)}}{(q-j)!}\, s_k^{q-j}\, \delta_{[j\le q]} = -\frac{1}{(q-j)!}\, s_k^{p-j+1}\, \delta_{[j\le q]},$$

where $\delta_{[\cdot]}$ is the standard indicator function. We now see that, for $j \in [1 : q-1]$,

$$|f_{k+1}^{(j)} - T_{f,p}^{(j)}(x_k, s_k)| = |0 - T_{f,p}^{(j)}(x_k, s_k)| \le \frac{1}{(q-j)!}\, |s_k|^{p-j+1} \le |s_k|^{p-j+1}, \tag{12.2.77}$$

while, for $j = q$, we have that

$$|f_{k+1}^{(q)} - T_{f,p}^{(q)}(x_k, s_k)| = |-s_k^{p-q+1} + s_k^{p-q+1}| = 0 \tag{12.2.78}$$

and, for $j \in [q+1 : p]$,

$$|f_{k+1}^{(j)} - T_{f,p}^{(j)}(x_k, s_k)| = |0 - 0| = 0. \tag{12.2.79}$$

Combining (12.2.76)–(12.2.79) and observing that $|f_k^{(j)}| \le \epsilon + \omega_k \le 2$ for all k and $j \in [1 : p]$ and $|s_k| \le 2^{\frac{1}{p-q+1}} \le 2$, we may then apply Theorem A.9.2 (p. 441) with $\kappa_f = 2$ and deduce the existence of a p-times continuously differentiable function f_{ARqp} from $\mathbb{R}$ to $\mathbb{R}$ with bounded continuous derivatives of order 0 to p and Lipschitz-continuous derivative of order p (hence satisfying **f.D1pL**) which interpolates $\{f_k^{(j)}\}$ at $\{x_k\}$ for $k \in [0 : k_\epsilon]$ and $j \in [0 : p]$. Moreover, (12.2.75), (12.2.69), (12.2.70), and the same Hermite-interpolation theorem imply that $|f^{(j)}(x)|$ is bounded by a constant depending only on p and q for all $x \in \mathbb{R}$ and $j \in [0 : p]$ and that f_{ARqp} is bounded below (ensuring **f.Bb**) and that its range only depends on p and q. This concludes our proof. □

This immediately provides the following important corollary.

Corollary 12.2.16: Slow qth-order convergence of the ARqp algorithm for $q \in \{1, 2\}$

*Suppose that $q \in \{1, 2\}$ and that $\epsilon_j = \epsilon \le \frac{1}{2}$ for $j \le 2$. Then the ARqp algorithm applied to minimize a function f satisfying **F.Bb** and **f.D1pL** may require*

$$\epsilon^{-\frac{p+1}{p-q+1}}$$

iterations and evaluations of f and of its derivatives of orders one up to p to produce a point x_ϵ such that $\phi_{f,j}^{1}(x_\epsilon) \le \epsilon_j/j!$ for all $j \in [1 : q]$.

Proof. We note that, in both cases covered by our assumptions, Lemma 12.2.9 allows the choice $\delta_{s_k,j} = 1$ for all k and all $j \in [1 : q]$. We conclude by applying Theorem 12.2.15. □

Another useful consequence of Theorem 12.2.15 is that it can be used to prove Theorems 3.3.6 (p. 71), 3.3.10 (p. 75), and 4.1.6 (p. 109), which consider slow convergence of the AR2 and ARp algorithms for orders one, two, or three. The main idea is to observe first that, because of Theorem A.8.5 (p. 438), **f.D1pL** clearly implies **f.HL** (p. 51) for $p = 2$ or **f.D1qL** (p. 293) for $p = 3$. We also note that the step (12.2.70) globally minimizes the regularized model of degree p and is therefore an acceptable step for both the AR2 and ARp algorithms. Finally, we see that, in the example of the proof of Theorem 12.2.15, all derivatives at x_k except the qth one are identically zero (see (12.2.69)). As a consequence, we have that for all $k \in [0 : k_\epsilon]$,

- if $q = 1$ and $p \in \{2, 3\}$,

$$\|\nabla_x^1 f(x_k)\| = |f_k^{(1)}| = (\epsilon + \omega_k);$$

- if $q = 2$ and $p \in \{2, 3\}$,

$$\|\nabla_x^1 f(x_k)\| = |f_k^{(1)}| = 0 \quad \text{and} \quad \lambda_{\min}[\nabla_x^2 f(x_k)] = f_k^{(2)} = (\epsilon + \omega_k); \text{ and}$$

- if $q = 3$ and $p = 3$,

$$\|\nabla_x^1 f(x_k)\| = |f_k^{(1)}| = 0, \quad \lambda_{\min}[\nabla_x^2 f(x_k)] = f_k^{(2)} = 0,$$

$$\mathcal{Z}_k = \mathbb{R}, \quad \text{and} \quad \|\nabla_x^3 f(x_k)\| = |f_k^{(3)}| = (\epsilon + \omega_k).$$

From these observations and the fact that termination occurs at iteration $k_\epsilon = \left\lceil \epsilon^{-\frac{p+1}{p-q+1}} \right\rceil$ when ω_k vanishes (see (12.2.68) and (12.2.67)), we deduce that the bounds stated in Theorems 3.3.6, 3.3.10, and 4.1.6 are achieved by the example discussed in the proof of Theorem 12.2.15.

We now return to the ARqp algorithm and consider the case where the choice $\delta_{s_k,j} = 1$ for all $k \geq 0$ and all $j \in [1 : q]$ is not possible—as we know from Example 12.2.1, this may happen. At variance with Theorem 12.2.15, we consider a class of smoother functions than those given by **f.D1pL**, namely those satisfying the following assumption for a given integer $r \geq p$.

f.D1rL

f is r-times continuously differentiable in $\mathbb{R}^n$ and, for all $j \in [1 : r]$, the jth derivative of f is Lipschitz continuous with Lipschitz constant $L_{f,j}$; that is, there exist constants $L_{f,j} \geq 1$ such that

$$\|\nabla_x^j f(x) - \nabla_x^j f(y)\| \leq L_{f,j} \|x - y\|$$

for all $x, y \in \mathbb{R}^n$ and all $j \in [1 : r]$

In other words, we consider functions with (possibly) more available Lipschitz-continuous derivatives than those used by the ARqp algorithm. This allows a significantly stronger result.

Theorem 12.2.17: Slow qth-order convergence of the ARqp algorithm (general case)

Let $1 \le q \le p \le r$. If $\epsilon_j = \epsilon \in (0,1)$ is sufficiently small for all $j \in [1:q]$, and if the ARqp algorithm is applied to minimize a function f satisfying **f.Bb** *and* **f.D1rL**, *and we can choose an arbitrary $\delta_{k,j} > 0$ satisfying (12.2.51), it may then require*

$$\epsilon^{-\frac{q(p+1)}{p}}$$

iterations and evaluations of f and of its derivatives of orders one up to p to produce a point x_ϵ such that $\phi_{f,j}^{\delta_{\epsilon,j}}(x_\epsilon) \le \epsilon_j \delta_{\epsilon,j}^j / j!$ for all $j \in [1:q]$ and some $\delta_\epsilon \in (0,1]^q$.

Proof. Our aim is to show that, for each choice of $p \ge q \ge 1$, there is an objective function satisfying **f.Bb** and **f.D1rL** for which finding an (ϵ, δ)-approximate qth-order minimizer may require at least $\epsilon^{-q(p+1)/p}$ evaluations of the objective function and its derivatives using the ARqp algorithm. The proof follows broadly the same pattern as that of Theorem 12.2.15 but with significant, subtle differences.

Once again, without loss of generality, we assume that $\epsilon \le \frac{1}{2}$. Given a model degree $p \ge 1$ and an optimality order $q \ge 1$, we set

$$k_\epsilon = \left\lceil \epsilon^{-q(p+1)/p} \right\rceil, \tag{12.2.80}$$

$$\gamma \stackrel{\text{def}}{=} q\left(\frac{r}{p} - 1\right), \tag{12.2.81}$$

and

$$\omega_k = \epsilon^{q+\gamma} \frac{k_\epsilon - k}{k_\epsilon} \in [0, \epsilon^q], \quad (k \in [0:k_\epsilon]). \tag{12.2.82}$$

Moreover, for $j \in [1:p]$ and each $k \in [0:k_\epsilon]$, we define the sequence $\{f_k^{(j)}\}$ by

$$f_k^{(1)} = -\frac{\epsilon^q + \omega_k}{q!} < 0 \text{ and } f_k^{(j)} = 0 \text{ for } j \in [2:p], \tag{12.2.83}$$

and therefore

$$T_{f,p}(x_k, s) = \sum_{j=0}^{p} \frac{f_k^{(j)}}{j!} s^j = f_k^{(0)} - \frac{\epsilon^q + \omega_k}{q!} s \tag{12.2.84}$$

for some sequences $\{x_k\}$ and $\{f_k^{(0)}\}$ to be defined later. We also suppose that

$$\sigma_k = \sigma \stackrel{\text{def}}{=} \begin{cases} \left(\dfrac{3}{p-1}\right)^p \dfrac{p!}{q!} & \text{if } p > 2, \\ p! & \text{if } p \le 2; \end{cases} \tag{12.2.85}$$

we shall verify below that a fixed σ_k is allowed by the algorithm. The model (12.2.23) is then

$$m_k(s) = f_k^{(0)} - \frac{\epsilon^q + \omega}{q!} s + \sigma \frac{|s|^{p+1}}{(p+1)!}, \tag{12.2.86}$$

which is the same as (12.1.18), which we studied in detail in Examples 12.1.3 and 12.2.1, with a change of the single independent variable (x becomes s), the addition of an irrelevant constant term $f_k^{(0)}$, and specifically

$$\beta = \frac{\epsilon^q + \omega_k}{q!}. \tag{12.2.87}$$

We now assume that, for each k, Step 2 returns the global minimizer (12.1.19) of the model

$$s_k = \left(\frac{p!(\epsilon^q + \omega_k)}{\sigma q!}\right)^{1/p} \quad (k \in [0 : k_\epsilon]) \tag{12.2.88}$$

and the optimality radius

$$\delta_{s_k,j} = \epsilon \quad (j \in [1 : q]). \tag{12.2.89}$$

Crucially, the derivation of Example 12.1.3 on which the example in this proof is based shows that any radius

$$\delta_{k,j} \leq \frac{3}{p-1}\|s_k\| = \frac{3}{p-1}\left(\frac{p!}{\sigma}\right)^{1/p}\left(\frac{\epsilon^q + \omega_k}{q!}\right)^{1/p}$$

ensures that $\phi^{\delta}_{m_k,j}(s_*) = 0$ when $p \geq 3$, while any radius at all suffices if p is 1 or 2. Thus (12.2.89) is both possible and acceptable[78] to (12.2.29) for all j and $p \geq 3$ since (12.2.85), $q \leq p$, and $\epsilon < 1$ give

$$\epsilon \leq \frac{3}{p-1}\left(\frac{p!}{\sigma q!}\right)^{1/p}\epsilon^{q/p} = \frac{3}{p-1}\left(\frac{p!}{\sigma}\right)^{1/p}\left(\frac{\epsilon^q}{q!}\right)^{1/p} \leq \frac{3}{p-1}\left(\frac{p!}{\sigma}\right)^{1/p}\left(\frac{\epsilon^q + \omega_k}{q!}\right)^{1/p}.$$

Now (12.1.20) and (12.2.87) give that

$$m_k(s_k) - m_k(0) = m_k(s_k) - f_k^{(0)} = -\frac{p}{p+1}\left(\frac{p!}{\sigma}\right)^{1/p}\left(\frac{\epsilon^q + \omega_k}{q!}\right)^{(p+1)/p}.$$

Thus $m_k(s_k) < m_k(0)$ and (12.2.27) holds. We then let

$$\kappa_{\rm F} \stackrel{\rm def}{=} \frac{1}{q!}\left(\frac{p!}{\sigma q!}\right)^{1/p}$$

and define

$$f_0^{(0)} = 2^{1+q(p+1)/p}\kappa_{\rm F} \quad \text{and} \quad f_{k+1}^{(0)} = f_k^{(0)} - \left(\frac{p!}{\sigma}\right)^{1/p}\left(\frac{\epsilon^q + \omega_k}{q!}\right)^{(p+1)/p}, \tag{12.2.90}$$

providing the identity

$$T_{f,p}(x_k, s_k) = f_{k+1}^{(0)}. \tag{12.2.91}$$

Thus iteration k is successful because $\rho_k = 1$ in (12.2.30), our choice of a constant $\sigma_k = \sigma$ is permissible, and $x_{k+1} = x_k + s_k$; for simplicity, just as in (12.2.73), we set $x_0 = 0$. Moreover, $\delta_{k,j} = \delta_{s_{k-1},j} = \epsilon$ for all $k \in [1 : k_\epsilon]$ and $j \in [1 : q]$.

[78] Note that this example also shows that $\delta_{k,3} = 1$ is not acceptable.

Assume now that $\delta_{0,j} = \epsilon$ for $j \in [1:q]$. Since (12.2.84) is unbounded from below, the minimizer of $T_{f,j}(x_k, s)$ within the ball $\|s\| \le \epsilon$ lies at $s = \epsilon$, and thus

$$\phi_{f,j}^{\delta_{k,j}}(x_k) = (\epsilon^q + \omega_k)\frac{\epsilon}{q!}$$

for $j \in [1:q]$ and $k \in [0:k_\epsilon]$ because of (12.2.89). Using (12.2.82) and (12.2.80), we find that if $q \ge 2$,

$$\frac{\epsilon^q + \omega_k}{q!} \le \frac{2\epsilon^q}{q!} \le \frac{\epsilon^j}{j!} = \frac{\delta_{k,j}^j}{j!}$$

for $j \in [1:q-1]$ when $\epsilon \le \frac{1}{2}$. We may then deduce that

$$\phi_{f,j}^{\delta_{k,j}}(x_k) \le \epsilon\frac{\delta_{k,j}^j}{j!} \text{ for } j \in [1:q-1], \quad \phi_{f,q}^{\delta_{k,q}}(x_k) > \epsilon\frac{\delta_{k,q}^q}{q!},$$

and (12.1.25) fails for $(x_k, \delta_{k,q})$ for $k \in [0:k_\epsilon - 1]$, while

$$\phi_{f,j}^{\delta_{k,j}}(x_{k_\epsilon}) < \epsilon\frac{\delta_{k,j}^j}{j!} \text{ for } j \in [1:q-1], \quad \phi_{f,q}^{\delta_{k,q}}(x_{k_\epsilon}) = \epsilon\frac{\delta_{k,q}^q}{q!},$$

and (12.1.25) holds at x_{k_ϵ}. In addition, (12.2.90) gives that

$$\begin{aligned} f_0^{(0)} \ge f_k^{(0)} &\ge f_0^{(0)} - \kappa_{\rm F} k\,[2\epsilon]^{q(p+1)/p} \\ &\ge f_0^{(0)} - \kappa_{\rm F} k_\epsilon \epsilon^{q(p+1)/p} 2^{q(p+1)/p} \\ &\ge f_0^{(0)} - \kappa_{\rm F}\left(1 + \epsilon^{q(p+1)/p}\right) 2^{q(p+1)/p} \\ &\ge f_0^{(0)} - \kappa_{\rm F} 2^{1+q(p+1)/p} \end{aligned}$$

for $k \in [0:k_\epsilon]$, using (12.2.82) and the inequality $k_\epsilon \le 1 + \epsilon^{-q(p+1)/p}$ that results from (12.2.80). Hence

$$f_k^{(0)} \in \left[0, 2^{1+\frac{q(p+1)}{p}}\kappa_{\rm F}\right] \text{ for } k \in [0:k_\epsilon]$$

by definition of $f_0^{(0)}$ in (12.2.90). Then (12.2.91) gives immediately that

$$|f_{k+1}^{(0)} - T_{f,p}(x_k, s_k)| = 0, \tag{12.2.92}$$

while (12.2.83) and (12.2.84) lead to

$$|f_{k+1}^{(j)} - T_{f,p}^{(j)}(x_k, s_k)| = |0 - 0| = 0 \tag{12.2.93}$$

for $j \in [2:p]$, as well as to

$$|f_{k+1}^{(1)} - T_{f,p}^{(1)}(x_k, s_k)| = \left|-\frac{(\epsilon^q + \omega_{k+1})}{q!} + \frac{(\epsilon^q + \omega_k)}{q!}\right| \le \frac{1}{q!}|\omega_{k+1} - \omega_k|. \tag{12.2.94}$$

But

$$|\omega_{k+1} - \omega_k| \le \frac{\epsilon^{q+\gamma}}{k_\epsilon} \le \epsilon^{q+\gamma} = \epsilon^{qr/p} \le (\epsilon^q + \omega_k)^{r/p}$$

because of (12.2.82), and therefore, combining this with (12.2.94), (12.2.81), and (12.2.88),

$$|f_{k+1}^{(1)} - T_{f,p}^{(1)}(x_k, s_k)| \leq \frac{(q!)^{r/p-1}}{(q!)^{r/p}} (\epsilon^q + \omega_k)^{r/p} = (q!)^{r/p-1} \left(\frac{\sigma}{p!}\right)^{r/p} |s_k|^r. \quad (12.2.95)$$

Thus we obtain from (12.2.92), (12.2.93), and (12.2.95) that

$$|f_{k+1}^{(j)} - T_{f,p}^{(j)}(x_k, s_k)| \leq (q!)^{r/p-1} \left(\frac{\sigma}{p!}\right)^{r/p} |s_k|^{r-j+1} \quad (12.2.96)$$

for $j \in [1 : p]$.

The remainder of the proof then follows along familiar lines, but with a twist. Instead of building a piecewise polynomial interpolating (12.2.83) and (12.2.90), we require that it additionally interpolate the extra conditions

$$f_k^{(j)} = 0 \quad \text{for} \quad j \in [p+1 : r],$$

and note that (12.2.93), and thus (12.2.96), now hold for $j \in [1 : r]$. This and the inequalities $|f_k^{(j)}| \leq (\epsilon^q + \omega_k)/q! \leq 2$ for all k and $j \in [1 : r]$ and $|s_k| \leq [(\epsilon^q + \omega_k)/q!]^{1/p}(p!/\sigma)^{1/p} \leq 2$ then allow us to apply Theorem A.9.2 (p. 441) with $\kappa_f = \max[2, (q!)^{\frac{r}{p}-1}(\sigma/p!)^{r/p}]$ and deduce the existence of an r-times continuously differentiable function f_{ARqp} from $\mathbb{R}$ to $\mathbb{R}$ with bounded continuous derivatives of order 0 to r and Lipschitz derivative of order r (hence satisfying **f.D1rL**) which interpolates $\{f_k^{(j)}\}$ at $\{x_k\}$ for $k \in [0 : k_\epsilon]$ and $j \in [0 : r]$. Moreover, the Hermite theorem, (12.2.83), and (12.2.88) also guarantee that $|f^{(j)}(x)|$ is bounded by a constant depending only on r and q for all $x \in \mathbb{R}$ and $j \in [0 : r]$. As a consequence, **f.Bb** also holds. This concludes the proof. □

The main consequence of this theorem is to show that the complexity bounds given by the second item of Theorem 12.2.14 are sharp in order not only for functions satisfying **f.D1pL** (the bare minimum for applying the ARqp algorithm), but also for arbitrarily smoother functions (i.e., satisfying the stronger **f.D1rL** for any finite[79] $r \geq p$). Thus *the existence of smooth derivatives beyond those used by the algorithm does not improve the lower complexity bound.* If one focuses on the case of first-order approximate minimizers ($q = 1$), the "smoothness-independent" lower complexity bound we have just derived gives that the ARqp algorithm applied for finding such approximate minimizers may require ϵ^{-2} evaluations when using a model of degree one, or $\epsilon^{-3/2}$ when using a model of degree two, irrespective of the objective function's (finite) degree of smoothness. This is of course reminiscent of the example in Carmon et al. [77] of slow convergence to first-order approximate minimizers in $\epsilon^{-(p+1)/p}$ evaluations of derivatives of degree one to p (as is the case for the bound discussed above when $q = 1$), which involves a very large dimensional even smoother (**f.D1∞L**) objective function (see Section 4.2). Whether our result for arbitrary criticality order can be extended to infinitely smooth functions remains open at this stage. The result however extends to the TR1 algorithm and variants of the steepest-descent method because, as we have seen in Chapter 2, linesearch, first-order trust-region, and first-order regularization methods have the same examples of slow convergence.

[79] Our theory does not cover infinitely continuously differentiable functions with Lipschitz derivatives because this would require $\gamma = \infty$ and thus $\omega_k = 0$ in (12.2.81)–(12.2.82).

We conclude this section with a few additional comments. The attentive reader may have noticed that Theorem 12.2.17 is valid for all $q \geq 1$. Is this a contradiction with the first case of Theorem 12.2.14, which implies that an approximate multiplier can be found faster?[80] Fortunately not, but this illustrates the delicate interplay of the various algorithmic ingredients. The difference between the two results is explained by our choice to set $\delta_{k,j} = \epsilon$ for all j in Theorem 12.2.17, in order to allow for high criticality orders (as is shown necessary by Example 12.2.1), but we know from Lemma 12.2.9 that, for $j \leq \min[2, q]$, we could have chosen $\delta_{k,j} = 1$ instead. Had we made that choice, we would then have needed to consider (as in Corollary 12.2.16) the example of Theorem 12.2.15. Instead, our choice of $\delta_{k,j} = \epsilon$ makes it more difficult to satisfy the approximate optimality condition (12.1.25), hence the need for more evaluations.

For simplicity, we have chosen, in the above examples, to minimize the model $m_k(s)$ globally at every iteration, but we might consider other pairs (s, δ_s). Similar examples of slow convergence may in fact be constructed along the lines used above[81] for any sequence of acceptable model-reducing steps and associated optimality radii—in the sense of (12.2.29)—provided that the optimality radii remain bounded away from zero.[82] This means that *our example of slow convergence in Theorem 12.2.17 applies* not only to the ARqp algorithm but also to a much broader class of minimization methods containing *all known methods using Taylor series that attempt to achieve approximate qth-order criticality* as defined here. Indeed, much of the definition of the function in our example is independent of the algorithm, and one could for instance replace regularization with a trust-region mechanism or a linesearch—of course, the complexity would be worse for these if one used standard frameworks. Moreover, it is also possible to weaken the constraints on the step further by relaxing (12.2.91) and only insisting on an acceptable decrease of the objective function value in Step 3 of the algorithm.

Notes and references for Subsection 12.2.2

The discussion in this section is a special case of the results discussed in [109] and strengthens results presented in [108]. A result similar to Theorem 12.1.8 is shown to hold for weak approximate minimizers (see (12.1.40)), namely, that a variant of the ARqp algorithm finds a weak (ϵ, δ)-approximate qth-order minimizer in at most $\mathcal{O}(\epsilon^{-(p+1)/(p-q+1)})$ evaluations of f and its derivatives, provided that the pth-order derivative of f is Lipschitz continuous (this is **f.DpL**, which is significantly weaker than **f.D1pL**). Moreover, as is also the case for the stronger setting, this bound is shown to be sharp.

An interesting feature of the algorithm discussed in [108] is that computing and testing the value of $\phi^{\delta}_{m_k,j}(s_k)$ is unnecessary if the step is long enough. The same feature can easily be introduced into the ARqp algorithm above. Specifically, we may redefine Step 2 to accept a step as soon as (12.2.27) holds and

$$\|s_k\| \geq \begin{cases} \varpi \min\limits_{j\in[1:q]} \epsilon_j^{\frac{1}{p-q+1}} & \text{if } q \in \{1,2\}, \\ \varpi \min\limits_{j\in[1:q]} \epsilon_j^{\frac{q}{p}} & \text{if } q > 2 \end{cases}$$

for some $\varpi \in (\theta, 1]$. If these conditions fail, then one still needs to verify the requirements (12.2.27) and (12.2.29), as we have done previously. Given Lemma 4.1.1 and the proof of Theorem 12.2.14, it is possible to verify that this modification does not affect the conclusions of the complexity theorem, while potentially avoiding significant computations.

[80] Think of $p = 3$ and $q = 2$. Theorem 12.2.17 gives an example of convergence in $\epsilon^{-8/3}$ evaluations, while the first item of Theorem 12.2.14 says that $\mathcal{O}(\epsilon^{-2})$ suffices.

[81] At the price of possibly larger constants.

[82] Remember that $\delta_{s_k} = 1$ is always possible for $q \in \{1, 2\}$.

Theorem 12.2.17 is new. Theorem 12.2.15, which is adapted from [102] for the stronger approximate optimality measure, generalizes the bound obtained by Carmon et al. [77], described in Section 4.2, for the case $q = 1$ to arbitrary orders q.

12.3 Inexact global minimization in the ARqp algorithm

We have seen in both the TRq and ARqp algorithms that approximate global minimization can be accommodated in the computation of an approximate criticality measure by suitably choosing $\varsigma \in (0, 1]$ without worsening their complexity bounds. This is also the case for the computation of the step s_k in the ARqp algorithm, as we noted after Lemma 12.2.9. However, further relaxation is often possible in ARqp, as we now discuss.

Consider the ARqp step computation first, and remember that the ultimate purpose of Step 2 of the ARqp algorithm is to find a step s_k guaranteeing a sufficient decrease of the Taylor series at x_k, in that

$$\Delta T_{f,j}(x_k, s_k) \geq \kappa_{\rm decr} \epsilon_j^{\omega} \tag{12.3.1}$$

for some fixed $\kappa_{\rm decr} > 0$ and $j \in [1 : q]$, where ω is either $(p+1)/(p-q+1)$ or $q(p+1)/p$ (see the proof of Theorem 12.2.14). Of course, if a step s_k that satisfies (12.3.1) for some given $\kappa_{\rm decr}$ can be found simply[83] without resorting to Algorithm 12.2.3, so much the better (and we may then choose $\delta_{k,j} = 1$ for $j \in [1 : q]$). In other cases, the decrease guarantee (12.3.1) may be ensured in one of two possible ways. If $\|s_k\| \geq 1$ and given that $\epsilon_j \in (0, 1]$, sufficient decrease follows from Lemma 4.1.1 (p. 103) with $\kappa_{\rm decr} = \sigma_{\min}/(p + 1)!$. Conversely, if $\|s_k\| \leq 1$, we then have to enforce (12.2.29) for some $\delta_{s,j} \in (0, 1]$ and use the more complicated Lemma 12.2.12 to reach the desired conclusion. In our development, the constant 1 in the inequality $\|s_k\| \geq 1$ was chosen solely for simplicity of exposition, but in reality it may be replaced by any positive constant independent of k. In particular, it can be replaced by $\kappa_{\rm decr}^{1/(p+1)} \epsilon_{\min}^{\omega/(p+1)}$, where $\epsilon_{\min} = \min_{j \in [1:q]} \epsilon_j$, so that sufficient decrease still immediately follows from Lemma 4.1.1 if

$$\|s_k\| \geq \kappa_{\rm decr}^{1/(p+1)} \epsilon_{\min}^{\omega/(p+1)}. \tag{12.3.2}$$

As a consequence, we see that performing (approximate) global optimization in Step 2 is only necessary whenever a descent step cannot be found that satisfies either (12.3.1) or (12.3.2). From a practical point of view, the failure of these two conditions might be viewed as a reasonable termination rule for small enough $\epsilon_{\min}$, even if there is then no guarantee that the iterate x_k at which the algorithm appears to be stuck is an approximate minimizer.

If one insists on true optimality, the details of Algorithm 12.2.3 and of the involved approximate global optimization (in Steps 2.2 and 2.4.1) become relevant. If these approximate calculations cannot be performed exactly, it may happen that Algorithm 12.2.3 is terminated prematurely (for instance, if $\delta_{s,j}$ becomes too small or if a maximum number of inner iterations have been taken), returning a value of $\delta_{s,j}$ which is potentially too large for the computed step, compared to what would have resulted if the algorithm had been run to termination. Thus, given that $\delta_{k+1} = \delta_s$ at successful iterations, we next have to consider what happens in Step 1 of iteration $k + 1$ when one or more of the $\delta_{k+1,j}$ is too large. In this case, the definition of $\phi_{f,j}^{\delta_{k+1,j}}(x_{k+1})$ in (12.1.10) implies that

[83] Say, by applying some trusted local minimization method.

there might be a move $d_{k+1,j}$ with $\|d_{k+1,j}\| \leq \delta_{k+1,j}$ such that

$$\Delta T_{f,j}(x_{k+1}, d_{k+1,j}) > \epsilon_j \frac{\delta_{k+1,j}^j}{j!},$$

preventing termination even if x_{k+1} is a suitable (ϵ, δ)-approximate minimizer. This is obviously a serious problem from the point of view of bounding evaluation complexity, since the algorithm will continue and evaluate further, unnecessary, values of f and its derivatives. Two possibilities may then occur. If iteration $k+1$ is unsuccessful, σ_k will be increased, causing a subsequent stepsize reduction and, should the behaviour persist, forcing convergence to x_k. Otherwise, the iteration is successful,[84] yielding a further objective function reduction and allowing the algorithm to progress toward an alternative approximate minimizer with a lower objective-function value. The complexity bound is maintained if (12.3.1) or (12.3.2) holds, or if an insufficient decrease only occurs at most a number of times independent of $\epsilon_{\min}$. However, even if this is not the case and the complexity bound we have derived evaporates as a consequence, the fact that the algorithm moves on can be viewed as beneficial for the optimization process from a more global perspective.

Notes and references for Section 12.3

Although it would be foolish to suggest that computing full global minimizers for $q, p \geq 3$ is in any sense practical, it is worth briefly assessing the task. All of the global optimization problems mentioned concern the minimization of a multivariate polynomial of degree p or q within a Euclidean ball. At first sight one might be reassured that we are only dealing with a polynomial objective function and nothing more general, but unfortunately even for polynomials of degree three, the problem is NP-hard in general [298]. While semidefinite programming approximation schemes, along the lines we have already mentioned [143, 144, 306, 339], and algorithms for specific cases [242] and alternative norms [180], have been proposed, currently there is little to suggest that the required global optimization will really be practicable in anything but low dimensions—unless the special structure of the problem is exploited—for the foreseeable future.

[84] As suggested by the fact that minimization in Step 2 of iteration $k+1$ may obviously be started from $x_{k+1} + d_{k+1,j}$, a point already providing descent on a good approximation of f.

Chapter 13

Inexact Function Values and Derivatives

Chapter summary

An important practical generalization arises when the value of the objective function and those of its derivatives are computed inexactly, due to noise, or numerical approximations such as embedded iterative processes or sampling, or simply because derivatives are not available. We will show that the worst-case performance of a modified trust-region algorithm remains essentially unchanged compared to that of the same algorithm using exact, fully accurate information: at most $\mathcal{O}(\epsilon^{-(q+1)})$ evaluations of the inexact objective function and its first q derivatives are necessary to compute an exact ϵ-approximate qth-order minimizer. Similarly, a modification of the adaptive-regularization algorithm to handle inexact evaluations provides an evaluation bound of $\mathcal{O}(\epsilon^{-(p+1)/(p-q+1)})$ (for $q \in \{1,2\}$) or $\mathcal{O}(\epsilon^{-q(p+1)/p})$ (for $q > 2$) inexact evaluations of the first p derivatives, which is identical in order of the accuracy demanded to that available for the case where evaluations are exact. These results are obtained in two complementary frameworks. Both consider "dynamic accuracy" (DA), a situation where the accuracy of the function values and derivatives can be improved at the request of the minimization algorithm. The first (IDA) is when this improvement is implicit *in that it is known to exist but cannot be guaranteed algorithmically. The second (EDA) is when the improvement is* explicitly *enforceable. The methods we derive are fully implementable for orders one and two and provide guidelines for designing practical model-based derivative-free algorithms (when derivatives are approximated by combining function values) or methods using variable computer accuracy. Our results also indicate that the distinction between first-, second-, third-, and higher-order methods is not absolute and that (complexitywise) fast algorithms can be designed using derivatives of low degree or no derivative at all.*

13.1 Inexact algorithms using implicit dynamic accuracy

13.1.1 Introduction to dynamic accuracy

We have presumed thus far that evaluations of the objective function, as well as those of its derivatives of relevant orders, are performed exactly. Unfortunately, such a presumption is not always justified in practice, and there are many applications where either the objective-function values or those of its derivatives (or both) are only known approximately. This may arise in several circumstances. The first is when the values in question are computed by some kind of experimental process whose accuracy can be adjusted, with the understanding that more accurate values may be, sometimes substantially, more expensive in terms of computational effort. A second related case is when objective-function or derivative values result from some (one trusts, convergent) iteration: obtaining more accuracy is possible by letting the iteration proceed further, but again at the price of possibly significant additional computing. A third case, currently much in vogue in the context of machine learning, is when the values of the objective function and/or its derivatives are obtained by sampling—say from among the terms of a sum of an enormous number of subfunctions. Again, using a larger sample size results in probabilistically better accuracy, but at a cost.

The aim of this chapter is to consider variants of the optimization methods that we have encountered thus far that can handle such considerations under what we call a "dynamic-accuracy" (DA) requirement. That is, we shall assume that *the required values* (objective function or derivatives) *can always be computed with an accuracy that is specified, either implicitly or explicitly, before the calculation, by the algorithm itself.* It is also understood in what follows that *the algorithm should require high accuracy only if necessary* but nonetheless *guarantee final results to full requested accuracy.* In this situation, it is hoped that many function or derivative evaluations can be carried out with a fairly loose accuracy (we will refer to these as "inexact values"), resulting in a significantly cheaper optimization process.

13.1.2 The implicit dynamic accuracy framework

Both trust-region and regularization methods are based on Taylor-series models, computed using derivatives of the objective function. In addition, both the TRq and ARqp algorithms in the previous chapter use the stopping criterion (12.1.25), which features $\phi_{f,j}^{\delta_j}(x)$, which is itself based on a Taylor-series model. Such models clearly depend on values and derivatives of the objective function at a sequence of points, but in the scenarios we are about to consider, we do not assume that we can calculate them exactly. That is, rather than having true problem-function and derivative values, $f(x)$ and $\nabla_x^i f(x)$ for $i \in [1 : j]$ at x, we are provided with approximations $\overline{f}(x)$ and $\overline{\nabla_x^i f}(x)$—here and hereafter, we denote inexact quantities and approximations with an overbar.

Consequently, while we have been used to dealing with Taylor-series approximations of the form

$$T_{f,j}(x,s) = f(x) + \sum_{i=1}^{j} \frac{1}{i!} \nabla_x^i f(x)[s]^i \equiv T_{f,j}(x,0) + \sum_{i=1}^{j} \frac{1}{i!} [\nabla_v^i T_{f,j}(x,v)]_{v=0}[s]^i$$

of f for perturbations s around x, in our new framework, we have to be content with the inexact equivalent

$$\overline{T}_{f,j}(x,s) = \bar{f}(x) + \sum_{i=1}^{j} \frac{1}{i!}\overline{\nabla_x^i f(x)}[s]^i \equiv \overline{T}_{f,j}(x,0) + \sum_{i=1}^{j} \frac{1}{i!}[\nabla_v^i \overline{T}_{f,j}(x,v)]_{v=0}[s]^i. \quad (13.1.1)$$

It is therefore pertinent to investigate the effect of inexact derivatives on such approximations. More specifically, we will be concerned with the Taylor decrement $\Delta T_{f,j}(x,s)$, which we introduced in (12.1.11), at x and for a step s. While our traditional algorithms depend on this quantity, it is of course out of the question to use it in the present context, since we only have approximate values. But an obvious alternative is instead to consider the inexact Taylor decrement

$$\overline{\Delta T}_{f,j}(x,s) \stackrel{\rm def}{=} \overline{T}_{f,j}(x,0) - \overline{T}_{f,j}(x,s) = -\sum_{i=1}^{j} \frac{1}{i!}[\nabla_v^i \overline{T}_{f,j}(x,v)]_{v=0}[s]^i.$$

In this section, we shall consider the Implicit Dynamic Accuracy (IDA) framework, in which the errors in the derivatives involved in the Taylor decrement are required to satisfy a set of bounds of the form

$$\| \overline{\nabla_x^i f(x_k)} - \nabla_x^i f(x_k)\| \leq \kappa_{\nabla,i}\, h_{k,i} \quad (i \in [1:j]) \quad (13.1.2)$$

for all x_k and for some *unknown* constants $\kappa_{\nabla,i} \geq 0$ for all $i \in [1:j]$ and some *requested accuracy* $h_{k,i}$ that will be specified by an algorithm; for example, this could be a finite-difference stepsize or a discretization meshsize (see Section 13.2 for more details). Thus an algorithm might require an improvement in accuracy by reducing the requested accuracies, but this does *not* imply that a bound on the error $\| \overline{\nabla_x^i f(x_k)} - \nabla_x^i f(x_k)\|$ is known to—not to mention enforceable by—the algorithm. It then follows that the error between the exact and inexact Taylor expansions satisfies the bound

$$\begin{aligned} |\overline{\Delta T}_{f,j}(x_k,s) - \Delta T_{f,j}(x_k,s)| &\leq \sum_{i=1}^{j} \|\overline{\nabla_x^i f(x_k)} - \nabla_x^i f(x_k)\| \frac{\|s\|^i}{i!} \\ &\leq \sum_{i=1}^{j} \kappa_{\nabla,i} h_{k,i} \frac{\|s\|^i}{i!}, \end{aligned} \quad (13.1.3)$$

and

$$\begin{aligned} |\overline{T}_{f,j}(x_k,s) - T_{f,j}(x_k,s)| &\leq |\bar{f}(x_k) - f(x_k)| + |\overline{\Delta T}_{f,j}(x_k,s) - \Delta T_{f,j}(x_k,s)| \\ &\leq |\bar{f}(x_k) - f(x_k)| + \sum_{i=1}^{j} \kappa_{\nabla,i} h_{k,i} \frac{\|s\|^i}{i!}, \end{aligned} \quad (13.1.4)$$

using the triangle inequality and the requirement (13.1.2). As such, the assumption (13.1.2) is insufficient to define the algorithms we have in mind, and an additional requirement on the size of the requested accuracies $h_{k,i}$ (and on the quality of the function evaluations) will be made by each algorithm in order to enforce the necessary coherence between exact and inexact quantities typified by (13.1.4).

For any $j \in [1:q]$, we also have the following crucial relationship between the exact but unknown optimality measure $\phi_{f,j}^{\delta}(x_k)$ in (12.1.10) (p. 278) and its inexact incarnation:

$$\overline{\phi}_{f,j}^{\delta}(x_k) \stackrel{\rm def}{=} \bar{f}(x_k) - \min_{\|d\|\leq\delta} \overline{T}_{f,j}(x_k,d) = \max_{\|d\|\leq\delta} \overline{\Delta T}_{f,j}(x_k,d). \quad (13.1.5)$$

This is that

$$\begin{aligned}\phi^{\delta}_{f,j}(x_k) &\le \max_{\|d\|\le\delta}\left(\overline{\Delta T}_{f,j}(x_k,d) + |\overline{\Delta T}_{f,j}(x_k,d) - \Delta T_{f,j}(x_k,d)|\right)\\ &\le \overline{\phi}^{\delta}_{f,j}(x_k) + \kappa_{\nabla,\max} h_{k,\max}\sum_{i=1}^{j}\frac{\|d\|^i}{i!}\\ &\le \overline{\phi}^{\delta}_{f,j}(x_k) + 2\kappa_{\nabla,\max} h_{k,\max},\end{aligned} \tag{13.1.6}$$

where

$$\kappa_{\nabla,\max} \stackrel{\text{def}}{=} \max_{i\in[1:q]} \kappa_{\nabla,i} \text{ and } h_{k,\max} \stackrel{\text{def}}{=} \max_{i\in[1:q]} h_{k,i},$$

and where we successively used the triangle inequality, (12.1.10), (13.1.3) with $s = d$, the fact that $\|d\| \le \delta \le 1$, and the sum bound (A.34).

In addition to (13.1.2), our IDA framework requires *function values* $f(x)$ *to be computed exactly.*[85] A consequence of this assumption is that when $s = 0$, (13.1.3) implies that

$$T_{f,j}(x_k,0) = \overline{T}_{f,j}(x_k,0) = f(x_k) = \bar{f}(x_k) \tag{13.1.7}$$

and

$$|T_{f,j}(x_k,s) - \overline{T}_{f,j}(x_k,s)| = |\overline{\Delta T}_{f,j}(x_k,s) - \Delta T_{f,j}(x_k,s)|. \tag{13.1.8}$$

13.1.3 The TRqIDA algorithm and its complexity

13.1.3.1 A trust-region algorithm with implicit dynamic accuracy

We shall first consider a trust-region optimization algorithm that extends the TRq framework from Section 12.2.1 (p. 292), which we shall refer to as TRqIDA—the IDA tag here refers to the Implicit Dynamic Accuracy Framework. The goal is to find a vector $x = x_\epsilon$ for which "approximate optimality is achieved" in the sense that (12.1.25) (p. 285) holds for some vector of optimality radii $\delta \in (0,1]^q$. Here all components of the vector δ will be identical (as was the case for TRq) and we will only use inexact derivatives of degree at most q. Thus effectively the index j in Section 13.1.2 will range from one to q.

The key to the IDA trust-region approach is to require that the requested accuracies $h_{k,i}$ in (13.1.2) satisfy the bounds

$$h_{k,i} \le \kappa_s \|s_k\|^{q-i+1} \text{ for all } i \in [1:q] \tag{13.1.9}$$

and for some $\kappa_s \in (0,1]$ before we contemplate taking a step s_k from x_k. Note that κ_s is now an algorithmic parameter independent of the problem whose value is known.

Of course, not all derivatives need be inexact in our framework. If derivatives of order $i \in \mathcal{E} \subseteq [1:q]$ were exact, then the left-hand side of (13.1.2) would vanish for $i \in \mathcal{E}$ and we could have defined κ_s to depend on i in (13.1.9), with $\kappa_{s,i}$ (say) being equal to zero for $i \in \mathcal{E}$. However, we avoid such a generalization in the arguments that follow for the sake of simplicity.

[85] We shall relax this requirement in the upcoming Section 13.3. See also the comment right at the end of the current section on p. 351.

If conditions (13.1.2) are met for $j \leq q$, it is then simple to improve the bound (13.1.3) with $s = s_k$ using (13.1.9) and the sum bound (A.34) to obtain

$$\begin{aligned} |\overline{\Delta T}_{f,j}(x_k, s_k) - \Delta T_{f,j}(x_k, s_k)| &\leq \sum_{i=1}^{j} \kappa_{\nabla,i} \kappa_s \frac{\|s_k\|^{q+1}}{i!} \\ &\leq \kappa_s \kappa_{\nabla,\max} \|s_k\|^{q+1} \sum_{i=1}^{j} \frac{1}{i!} \\ &\leq 2\kappa_s \kappa_{\nabla,\max} \|s_k\|^{q+1} \end{aligned} \tag{13.1.10}$$

and also, from (13.1.8),

$$|T_{f,j}(x_k, s_k) - \overline{T}_{f,j}(x_k, s_k)| \leq 2\kappa_s \kappa_{\nabla,\max} \|s_k\|^{q+1}. \tag{13.1.11}$$

Because of the inequality (13.1.6), we observe that the key optimality test (12.1.25) (p. 285) will be satisfied whenever

$$\overline{\phi}_{f,j}^{\delta_k}(x_k) \leq \tfrac{1}{2}\epsilon_j \frac{\delta_k^j}{j!} \quad \text{and} \quad h_{k,\max} \leq \frac{\epsilon_j}{4\kappa_{\nabla,\max}} \frac{\delta_k^j}{j!}, \tag{13.1.12}$$

where each component of the required vector δ in (12.1.25) is $\delta_k \in (0,1]$ and where, once again, we emphasize that $\kappa_{\nabla,\max}$ is typically unknown to the algorithm. We shall also make the most of the opportunity to approximate $\overline{\phi}_{f,j}^{\delta_k}(x_k)$ rather than compute it exactly, just as we did in the TRq algorithm. In particular, all that we require is that we can find a $d_{k,j} \in \mathcal{B}_{\delta_k}$ for which $\varsigma \overline{\phi}_{f,j}^{\delta_k}(x_k) \leq \overline{\Delta T}_{f,j}(x_k, d_{k,j})$ for some specified $\varsigma \in (0,1]$.

Our aim is thus to ensure that (13.1.12) holds, and we shall achieve this using the trust-region algorithm TRqIDA as follows.

Algorithm 13.1.1: Trust-region algorithm with implicit dynamic accuracy (TRqIDA)

Step 0: Initialization. A criticality order q, a starting point x_0, and an initial trust-region radius Δ_0 are given, as well as desired optimality tolerances $\epsilon \in (0,1)^q$ and an initial set of requested accuracies $\{h_{i,0}\}_{i=1}^q$. The constants ϑ, γ_h, κ_s, κ_h, η_1, η_2, γ_1, γ_2, γ_3, ς, and $\Delta_{\max}$ are also given and satisfy

$$\vartheta \in \left[\min_{j\in[1:q]} \epsilon_j, 1\right], \ \vartheta \leq \Delta_0 \leq \Delta_{\max}, \ 0 < \eta_1 \leq \eta_2 < 1, \ 0 < \gamma_1 < 1 < \gamma_2 < \gamma_3,$$
$$\varsigma \in (0,1], \ \ \gamma_h \in (0,1), \ \ \kappa_s > 0, \ \text{ and } \ 0 < h_{i,0} \leq \kappa_h \ \text{ for } \ i \in [1:q].$$

Compute $f(x_0)$ and set $k = 0$ and $i_h = 0$.

Step 1: Optimality measure computation. Set $\delta_k = \min[\Delta_k, \vartheta]$. For $j = 1, \ldots, q$:

Step 1.1: Compute $\overline{\nabla_x^j f(x_k)}$ to satisfy (13.1.2) and $d_{k,j} \in \mathcal{B}_{\delta_k}$ so that

$$\varsigma \overline{\phi}_{f,j}^{\delta_k}(x_k) \leq \overline{\Delta T}_{f,j}(x_k, d_{k,j}). \tag{13.1.13}$$

Step 1.2: If

$$\overline{\Delta T}_{f,j}(x_k, d_{k,j}) > \tfrac{1}{2}\varsigma \epsilon_j \frac{\delta_k^j}{j!}, \tag{13.1.14}$$

go to Step 2 with the index $j_k = j$ and $d_k = d_{k,j}$.

Step 1.3: Otherwise, consider the next value of j.

If the loop on j finishes, i.e., (13.1.13) holds but (13.1.14) fails for all $j \in [1:q]$, go to Step 5.

Step 2: Step computation. Compute a step $s_k \in \mathcal{B}_{\Delta_k}$ for which

$$\overline{\Delta T}_{f,j_k}(x_k, s_k) \geq \overline{\Delta T}_{f,j_k}(x_k, d_k). \tag{13.1.15}$$

If (13.1.9) holds for $i \in [1:q]$, go to Step 3. Otherwise go to Step 5.

Step 3: Accept the new iterate. Compute $f(x_k + s_k)$ and

$$\rho_k = \frac{f(x_k) - f(x_k + s_k)}{\overline{\Delta T}_{f,j_k}(x_k, s_k)}. \tag{13.1.16}$$

If $\rho_k \geq \eta_1$, then set $x_{k+1} = x_k + s_k$. Otherwise set $x_{k+1} = x_k$.

Step 4: Update the trust-region radius. Set

$$\Delta_{k+1} \in \begin{cases} [\gamma_1 \Delta_k, \gamma_2 \Delta_k] & \text{if } \rho_k < \eta_1, \\ [\gamma_2 \Delta_k, \Delta_k] & \text{if } \rho_k \in [\eta_1, \eta_2), \\ [\Delta_k, \min(\Delta_{\max}, \gamma_3 \Delta_k)] & \text{if } \rho_k \geq \eta_2. \end{cases} \tag{13.1.17}$$

Set $h_{k+1,i} = h_{k,i}$ for $i \in [1:q]$, increment k by one, and go to Step 2 with $d_{k+1} = d_k$ if $x_{k+1} = x_k$ and $\Delta_{k+1} \geq \vartheta$ or to Step 1 otherwise.

Step 5: Improve accuracy. For $i \in [1:q]$, set

$$h_{k+1,i} = \gamma_h h_{k,i}, \tag{13.1.18}$$

increment k and i_h by one, and go to Step 1 with $x_{k+1} = x_k$ and $\Delta_{k+1} = \Delta_k$.

At variance with all algorithms studied so far in this book, the TRqIDA algorithm does not include an explicit termination rule. Providing such a rule that ensures (12.1.25) is impossible in the IDA framework because, as we have seen, verifying approximate optimality with inexact derivatives is challenging, at least without knowing the values $\kappa_{\nabla,i}$ that themselves depend on problem Lipschitz constants. However, as we will see, this does not prevent us from counting the number of evaluations required to guarantee approximate optimality even if we cannot observe this directly. We also note that the choice $\vartheta = 1$ is acceptable since we have required that $\epsilon_j \leq 1$ for all $j \in [1:q]$.

As usual for trust-region algorithms, iteration k is said to be successful when Step 3 is executed, $\rho_k \geq \eta_1$, and $x_{k+1} = x_k + s_k$, and we define sets of *successful*, *unsuccessful*, and *accuracy-improving* iterations, respectively, as

$$\begin{aligned} \mathcal{S} &\stackrel{\text{def}}{=} \{k \in \mathbf{N} \mid \text{Step 5 is not executed and } \rho_k \geq \eta_1\}, \\ \mathcal{U} &\stackrel{\text{def}}{=} \{k \in \mathbf{N} \mid \text{Step 5 is not executed and } \rho_k < \eta_1\}, \quad \text{and} \\ \mathcal{A} &\stackrel{\text{def}}{=} \{k \in \mathbf{N} \mid \text{Step 5 is executed}\}. \end{aligned} \tag{13.1.19}$$

If $\mathcal{T} \stackrel{\text{def}}{=} \mathcal{S} \cup \mathcal{U}$, we also define

$$\begin{aligned} \mathcal{S}_k &\stackrel{\text{def}}{=} \mathcal{S} \cap [0:k], & \mathcal{U}_k &\stackrel{\text{def}}{=} \mathcal{U} \cap [0:k], \\ \mathcal{A}_k &\stackrel{\text{def}}{=} \mathcal{A} \cap [0:k], & \text{and } \mathcal{T}_k &\stackrel{\text{def}}{=} \mathcal{T}_k \cap [0:k], \end{aligned} \tag{13.1.20}$$

the corresponding sets up to iteration k. Notice that $x_{k+1} = x_k + s_k$ for $k \in \mathcal{S}$, while $x_{k+1} = x_k$ for $k \in \mathcal{U} \cup \mathcal{A}$. Moreover,

$$\mathcal{T}_k \subseteq \{i \in [0:k] \mid \text{(13.1.14) (with } i \text{ replacing } k\text{) holds for some } j \in [1:q]\}. \quad (13.1.21)$$

Note also that the objective function is evaluated once per successful or unsuccessful iteration (i.e., once for every $k \in \mathcal{T}$) and that derivatives are estimated once per successful or accuracy-improving iteration (i.e., once for every $k \in \mathcal{S} \cup \mathcal{A}$).

For future reference, we state a property of the TRqIDA algorithm, a direct generalization of Lemma 2.3.1 (p. 36) to the current framework.

Lemma 13.1.1: Successful and unsuccessful iterations (TRqIDA)

Suppose that the TRqIDA algorithm is used and that $\Delta_k \geq \Delta_{\min}$ for some $\Delta_{\min} \in (0, \Delta_0]$. Then

$$|\mathcal{T}_k| \leq |\mathcal{S}_k| \left(1 + \frac{\log \gamma_3}{|\log \gamma_2|}\right) + \frac{1}{|\log \gamma_2|} \left|\log\left(\frac{\Delta_{\min}}{\Delta_0}\right)\right|. \quad (13.1.22)$$

Proof. This is given in Appendix B (p. 484) for the sake of completeness. □

13.1.3.2 Evaluation complexity for the TRqIDA algorithm

We are now ready to analyze the complexity of the TRqIDA algorithm. Our analysis will be conducted under assumptions **f.D0qL**—that the derivatives of degree zero (i.e., the objective function itself) to q are Lipschitz continuous—and **f.Bb**—that the objective function f is bounded below. The first of these is given formally by the following.

f.D0qL

f is q-times continuously differentiable in $\mathbb{R}^n$ and, for all $j \in [0:q]$, the jth derivative of f is Lipschitz continuous with Lipschitz constant $L_{f,j}$; that is, there exist constants $L_{f,j} \geq 1$ such that

$$\|\nabla_x^j f(x) - \nabla_x^j f(y)\| \leq L_{f,j} \|x - y\|$$

for all $x, y \in \mathbb{R}^n$ and all $j \in [0:q]$.

This is similar to **f.D1qL**, except that now we assume additionally that the objective function itself is Lipschitz continuous (i.e., when $j = 0$). We then redefine

$$L_f \stackrel{\text{def}}{=} \max_{j \in [0:q]} L_{f,j} \geq 1. \quad (13.1.23)$$

We also note that the decreasing nature of each requested accuracy $h_{k,i}$ that is guaranteed by the mechanism of Step 5, Theorem A.8.5 (p. 438), and the definition of κ_h imply that

$$\begin{aligned} \|\overline{\nabla_x^i f}(x_k)\| &\leq \|\nabla_x^i f(x_k)\| + \|\overline{\nabla_x^i f}(x_k) - \nabla_x^i f(x_k)\| \leq L_{f,i-1} + \kappa_{\nabla,i} h_{k,i} \\ &\leq L_f + \kappa_{\nabla,\max} \kappa_h \\ &\stackrel{\text{def}}{=} \overline{L}_f \end{aligned} \quad (13.1.24)$$

for all $k \geq 0$ and each $i \in [1:q]$.

We start our analysis with a simple observation, along the lines of Lemma 12.2.1 (p. 294).

Lemma 13.1.2: A useful lower bound for TRqIDA

Consider an iteration $k \in \mathcal{T}$ of the TRqIDA *algorithm, and define*

$$\widehat{\phi}_{f,k} \stackrel{\rm def}{=} \frac{j_k!\,\overline{\Delta T}_{f,j_k}(x_k,d_k)}{\delta_k^{j_k}}, \tag{13.1.25}$$

where j_k is the index j and the direction $d_k = d_{k,j}$ for which (13.1.14) holds in Step 1 of Algorithm 13.1.1. Then

$$\min_{i\in\mathcal{T}_k} \widehat{\phi}_{f,i} \geq \tfrac{1}{2}\varsigma\epsilon_{\min}, \tag{13.1.26}$$

where $\epsilon_{\min} = \min_{j\in[1:q]} \epsilon_j$. Moreover,

$$\overline{\Delta T}_{f,j_k}(x_k,s_k) \geq \widehat{\phi}_{f,k}\frac{\delta_k^{j_k}}{j_k!}. \tag{13.1.27}$$

Proof. This is an obvious variation on Lemma 12.2.1 (p. 294) where k is restricted to the set $\mathcal{T}$ and inexact derivatives are used instead of exact ones. □

We next derive an "inexact" variant of the condition that ensures that an iteration is very successful.

Lemma 13.1.3: Condition for success (TRqIDA)

Suppose that **f.D0qL** *holds, that L_f is given by (13.1.23), and that $\widehat{\phi}_{f,k}$ is defined by (13.1.25). Suppose also that*

$$\Delta_k \leq \min\left\{\vartheta, \frac{(1-\eta_2)}{L_f + 4(q+1)!\,\kappa_s\kappa_{\nabla,\max}}\widehat{\phi}_{f,k}\right\} \tag{13.1.28}$$

at iteration $k \in \mathcal{T}$ of the TRqIDA *algorithm. Then $\rho_k \geq \eta_2$, iteration k is very successful, and $\Delta_{k+1} \geq \Delta_k$.*

Proof. We first note that (13.1.28) trivially implies that

$$\Delta_k \leq \min\left\{\vartheta, \frac{(j_k+1)(1-\eta_2)}{L_f + 2(j_k+1)!\,\kappa_s\kappa_{\nabla,\max}}\widehat{\phi}_{f,k}\right\} \tag{13.1.29}$$

for $k \in \mathcal{T}$ and that $\delta_k = \min[\vartheta, \Delta_k] = \Delta_k$. Moreover, $\delta_k \leq \vartheta \leq 1$. Then we may use (13.1.16), (13.1.7), and the triangle inequality to deduce that

$$\begin{aligned}|\rho_k - 1| &\leq \frac{|f(x_k+s_k) - \overline{T}_{f,j_k}(x_k,s_k)|}{\overline{\Delta T}_{f,j_k}(x_k,s_k)}\\ &\leq \frac{1}{\overline{\Delta T}_{f,j_k}(x_k,s_k)}\Big(|f(x_k+s_k) - T_{f,j_k}(x_k,s_k)|\\ &\qquad + |\overline{T}_{f,j_k}(x_k,s_k) - T_{f,j_k}(x_k,s_k)|\Big).\end{aligned}$$

Invoking (A.19) in Corollary A.8.4 (p. 437), the bounds $\|s_k\| \le \Delta_k = \delta_k$, (13.1.23), (13.1.27), (13.1.11), and (13.1.29), we deduce that

$$\begin{aligned} |\rho_k - 1| &\le \frac{j_k!}{\widehat{\phi}_{f,k}\delta_k^{j_k}} \left(\frac{L_f \delta_k^{j_k+1}}{(j_k+1)!} + 2\kappa_s\kappa_{\nabla,\max}\delta_k^{q+1} \right) \\ &\le \frac{\delta_k}{\widehat{\phi}_{f,k}} \left(\frac{L_f + 2(j_k+1)!\kappa_s\kappa_{\nabla,\max}}{j_k+1} \right) \end{aligned}$$

and thus[86] that $\rho_k \ge \eta_2$. Then iteration k is very successful and (13.1.17) then yields that $\Delta_{k+1} \ge \Delta_k$. □

This allows us to derive lower bounds on the trust-region radius and the model decrease.

Lemma 13.1.4: Lower bound on the trust-region radius (TRqIDA)

Suppose that ***f.D0qL*** *holds. Then the* TRqIDA *algorithm generates*

$$\Delta_k \ge \begin{cases} \min\left\{\gamma_1\vartheta, \kappa_\Delta \min_{i\in\mathcal{T}_k} \widehat{\phi}_{f,i}\right\} & \text{if } \mathcal{T}_k \ne \emptyset, \\ \gamma_1\vartheta & \text{otherwise} \end{cases} \tag{13.1.30}$$

for all $k \ge 0$, where $\widehat{\phi}_{f,i}$ is defined as in (13.1.25) and

$$\begin{aligned} \kappa_\Delta \stackrel{\text{def}}{=} \min&\left[1, \frac{\gamma_1(1-\eta_2)}{L_f + 4(q+1)!\,\kappa_s\kappa_{\nabla,\max}}\right] \\ &\times \min\left[\vartheta, \frac{\Delta_0 \min_{j\in[1:q]} \delta_0^j}{2q(\max_{i\in[1:q]} \|\nabla_x^i f(x_0)\| + \kappa_{\nabla,\max}\kappa_h)}\right]. \end{aligned} \tag{13.1.31}$$

Also, for all $k \ge 0$, we have

$$\delta_k \ge \varsigma\kappa_\delta\epsilon_{\min}, \tag{13.1.32}$$

where

$$\kappa_\delta \stackrel{\text{def}}{=} \tfrac{1}{2}\kappa_\Delta. \tag{13.1.33}$$

Proof. This is a variation on the proof of Lemma 12.2.3 (p. 295), in which inexact derivatives are used and where their upper bound is now given by (13.1.24). The details are given in Appendix B (p. 484). □

We may then deduce a crucial lower bound on the achieved model decrease.

Lemma 13.1.5: Model decrease (TRqIDA)

Suppose that ***f.D0qL*** *holds. Then the* TRqIDA *algorithm ensures that*

$$\overline{\Delta T}_{f,j_k}(x_k, s_k) \ge \frac{(\varsigma\kappa_\delta)^{q+1}}{q!}\,\epsilon_{\min}^{q+1} \tag{13.1.34}$$

[86] We note that the right-hand side of (13.1.28) is slightly larger than needed here. This is done in order to align this result with the corresponding property for the trust-region variant in the forthcoming Explicit Dynamic Accuracy (EDA) framework.

for every iteration $k \in \mathcal{T}$ of algorithm TRqIDA *before reaching approximate optimality, where κ_δ is defined as in (13.1.33). If, in addition, $k \in \mathcal{S}$, then*

$$f(x_k) - f(x_{k+1}) \geq \frac{\eta_1 (\varsigma \kappa_\delta)^{q+1}}{2q!} \, \epsilon_{\min}^{q+1}. \tag{13.1.35}$$

Proof. The proof is very similar to that of Lemma 12.2.4 (p. 297) but is given in Appendix B (p. 486) for completeness.[87] □

The last ingredients before deriving our ultimate result for the TRqIDA algorithm are to make explicit a condition that prevents an infinite loop between Steps 2 and 5 and to ensure that the loop between Steps 1 and 5 leads to approximate optimality.

Lemma 13.1.6: Finite loop in Step 2 of the TRqIDA algorithm

*Suppose that **f.D0qL** holds. Then branching from Step 2 to Step 5 of the* TRqIDA *algorithm cannot occur if*

$$h_{k,\max} \leq \kappa_s \left(\frac{\varsigma^{q+1} \kappa_\delta^q}{4q! \overline{L}_f} \right)^q \epsilon_{\min}^{q(q+1)}. \tag{13.1.36}$$

Proof. Suppose first that $\|s_k\| \leq 1$. From (13.1.14), (13.1.15), and **f.D0qL**, itself implying (13.1.24), and the sum bound (A.36), we see that

$$\tfrac{1}{2} \varsigma \epsilon_{\min} \frac{\delta_k^{j_k}}{j_k!} \leq \overline{\Delta T}_{f,j_k}(x_k, s_k) \leq \sum_{\ell=1}^{j_k} \|\overline{\nabla_x^i f}(x_k)\| \frac{\|s_k\|^\ell}{\ell!} \leq 2\overline{L}_f \|s_k\|$$

if Step 2 is executed. Using (13.1.32), we then deduce that

$$\|s_k\| \geq \frac{\varsigma^{j_k+1} \kappa_\delta^{j_k}}{4 j_k! \overline{L}_f} \epsilon_{\min}^{j_k+1},$$

Then, irrespective of $\|s_k\|$, we have

$$\|s_k\| \geq \min\left[1, \frac{\varsigma^{j_k+1} \kappa_\delta^{j_k}}{4 j_k! \overline{L}_f} \epsilon_{\min}^{j_k+1}\right] = \frac{\varsigma^{j_k+1} \kappa_\delta^{j_k}}{4 j_k! \overline{L}_f} \epsilon_{\min}^{j_k+1} \geq \frac{\varsigma^{q+1} \kappa_\delta^{q}}{4 q! \overline{L}_f} \epsilon_{\min}^{q+1}.$$

This bound implies that for all $i \in [1 : q]$,

$$\|s_k\|^{q-i+1} \geq \left(\frac{\varsigma^{q+1} \kappa_\delta^q}{4q! \overline{L}_f} \epsilon_{\min}^{q+1} \right)^{q-i+1} \geq \left(\frac{\varsigma^{q+1} \kappa_\delta^q}{4q! \overline{L}_f} \epsilon_{\min}^{q+1} \right)^{q} \geq \frac{h_{k,\max}}{\kappa_s} \geq \frac{h_{k,i}}{\kappa_s},$$

where in the penultimate inequality we used (13.1.36) and in the last one we used the definition of $h_{k,\max}$. Thus (13.1.9) holds and the branching to Step 5 does not occur at the end of Step 2. □

If the approximate derivatives are sufficiently accurate (specifically, the accuracies $h_{k,i}$ are suitably small), then the potentially infinite loop between Steps 1 and 5 leads to increasingly closer proximity to a qth-order local minimizer.

[87] The right-hand side of (13.1.35) has a factor of $\frac{1}{2}$ that is not strictly needed here but will be useful later when considering the EDA variant of TRq.

Lemma 13.1.7: The loop in Step 1 of the TRqIDA algorithm

Suppose that **f.D0qL** *holds. If*

$$h_{k,\max} \le \frac{\varsigma^{q+1}\kappa_\delta^q}{4q!\kappa_{\nabla,\max}}\,\epsilon_{\min}^{q+1}, \tag{13.1.37}$$

then branching from Step 1 to Step 5 of the TRqIDA *algorithm achieves the desired approximate optimality condition,* $\phi_{f,j}^{\delta_k}(x_k) \le \epsilon_j \frac{\delta_k^j}{j!}$ *for all* $j \in [1:q]$.

Proof. When branching to Step 5 occurs at the end of Step 1, (13.1.13) and the converse of (13.1.14) together imply that $\overline{\phi}_{f,j}^{\delta_k}(x_k) \le \frac{1}{2}\epsilon_j\delta_k^j/j!$ for all $j \in [1:q]$. From (13.1.6), (13.1.37), and (13.1.32), we then deduce that

$$\phi_{f,j}^{\delta_k}(x_k) \le \tfrac{1}{2}\epsilon_j\frac{\delta_k^j}{j!} + \frac{(\varsigma\kappa_\delta)^q}{2q!}\,\epsilon_{\min}^{q+1} \le \tfrac{1}{2}\epsilon_j\frac{\delta_k^j}{j!} + \tfrac{1}{2}\epsilon_{\min}\frac{\delta_k^j}{j!} \le \epsilon_j\frac{\delta_k^j}{j!}$$

as required, where to obtain the second inequality we also used[88] $\varsigma \le 1$. □

Finally, we arrive at our complexity bounds for the TRqIDA algorithm.

Theorem 13.1.8: qth-order complexity for the TRqIDA algorithm

Suppose that **f.D0qL** *and* **f.Bb** *hold. Then there exist positive constants* κ_{TRqIDA}^A, κ_{TRqIDA}^B, κ_{TRqIDA}^C, κ_{TRqIDA}^E, *and* κ_{TRqIDA}^F *such that, for any* $\epsilon \in (0,1]^q$, *the* TRqIDA *algorithm (p. 329) requires at most*

$$\begin{aligned}\#_{\text{TrqIDA}}^F &\stackrel{\text{def}}{=} \kappa_{\text{TRqIDA}}^A \frac{f(x_0)-f_{\text{low}}}{\min\limits_{j\in[1:q]}\epsilon_j^{q+1}} + \kappa_{\text{TRqIDA}}^B\left|\log\left(\min_{j\in[1:q]}\epsilon_j\right)\right| + \kappa_{\text{TRqIDA}}^C \\ &= \mathcal{O}\left(\max_{j\in[1:q]}\epsilon_j^{-(q+1)}\right)\end{aligned} \tag{13.1.38}$$

evaluations of f *and at most*

$$\begin{aligned}\#_{\text{TRqIDA}}^D &\stackrel{\text{def}}{=} \kappa_{\text{TRqIDA}}^A \frac{f(x_0)-f_{\text{low}}}{\min\limits_{j\in[1:q]}\epsilon_j^{q+1}} + \kappa_{\text{TRqIDA}}^E\left|\log\left(\min_{j\in[1:q]}\epsilon_j\right)\right| + \kappa_{\text{TRqIDA}}^F \\ &= \mathcal{O}\left(\max_{j\in[1:q]}\epsilon_j^{-(q+1)}\right)\end{aligned} \tag{13.1.39}$$

evaluations of the inexact derivatives $\overline{\nabla_x^i f}$ *of order* i *from one to* q *to produce an iterate* x_ϵ *and an optimality radius* $\delta_\epsilon \in (0,1]$ *such that* $\phi_{f,q}^{\delta_\epsilon}(x_\epsilon) \le \epsilon_j\delta_\epsilon^j/j!$ *for all* $j \in [1:q]$.

[88] We have introduced an additional multiplicative factor ς on the right-hand side of (13.1.37) that is not necessary here but which will be needed when we come to the explicit dynamic accuracy case of TRq in Section 13.3.

Proof. Let k be the index of an arbitrary iteration before approximate optimality is achieved. Let $i \in \mathcal{S}$ with $i \le k$. Then (13.1.35) holds at iteration i. Using **f.Bb**, $f(x_i) = f(x_{i+1})$ on iterations that are not successful, and (13.1.35), we deduce that

$$f(x_0) - f_{\rm low} \ge f(x_0) - f(x_{k+1}) = \sum_{i \in \mathcal{S}_k} [f(x_i) - f(x_{i+1})] \ge |\mathcal{S}_k| \left[\kappa^{\rm S}_{\rm TRqIDA}\right]^{-1} \epsilon_{\min}^{q+1},$$

where

$$\kappa^{\rm S}_{\rm TRqIDA} = \frac{2q!}{\eta_1 (\varsigma \kappa_\delta)^{q+1}} \tag{13.1.40}$$

and κ_δ is given by (13.1.33), and thus that the total number of *successful iterations* before achieving approximate optimality is given by

$$|\mathcal{S}_k| \le \kappa^{\rm S}_{\rm TRqIDA} \frac{f(x_0) - f_{\rm low}}{\epsilon_{\min}^{q+1}}.$$

Using the lower bound (13.1.32) on δ_k, we deduce that

$$\Delta_k \ge \delta_k = \min(\vartheta, \Delta_k) \ge \varsigma \kappa_\delta \epsilon_{\min}.$$

We may now invoke Lemma 13.1.1 to deduce that the total number of successful and unsuccessful iterations required, and hence the total number of function evaluations, is bounded by

$$|\mathcal{S}_k| \left(1 + \frac{\log \gamma_3}{|\log \gamma_2|}\right) + \frac{1}{|\log \gamma_2|} \left(|\log(\epsilon_{\min})| + \left| \log\left(\frac{\varsigma \kappa_\delta}{\Delta_0}\right)\right| \right) + 1, \tag{13.1.41}$$

which yields (13.1.38) with the coefficients (13.1.40),

$$\kappa^{\rm A}_{\rm TRqIDA} \stackrel{\rm def}{=} \kappa^{\rm S}_{\rm TRqIDA} \left(1 + \frac{\log \gamma_3}{|\log \gamma_2|}\right), \quad \kappa^{\rm B}_{\rm TRqIDA} \stackrel{\rm def}{=} \frac{1}{|\log \gamma_2|}, \tag{13.1.42}$$

and

$$\kappa^{\rm C}_{\rm TRqIDA} \stackrel{\rm def}{=} \frac{1}{|\log \gamma_2|} \left| \log\left(\frac{\varsigma \kappa_\delta}{\Delta_0}\right)\right| + 1. \tag{13.1.43}$$

In order to derive an upper bound on the number of evaluations of the inexact derivatives, we now need to bound the number of accuracy-improving iterations. The first case where such an iteration may occur is when branching to Step 5 happens at the end of Step 1, and then Lemma 13.1.7 applies. Combining (13.1.37) with the bound (13.1.36) that occurs if branching to Step 5 happens in Step 2, we see that an accuracy-improving iteration before approximate optimality is achieved is impossible if

$$h_{k,\max} \le \min\left[\frac{\varsigma^{q+1}\kappa_\delta^q}{4q!\kappa_{\nabla,\max}}, \kappa_s \left(\frac{\varsigma^{q+1}\kappa_\delta^q}{4q!\overline{L}_f}\right)^q\right] \epsilon_{\min}^{q(q+1)}.$$

But the mechanism of Step 5 ensures that the requested accuracies $\{h_{k,i}\}_{i=1}^q$ decrease linearly by successive multiplications by the factor γ_h, starting from the initial values $\{h_{0,i}\}_{i=1}^q$, each of which is in turn bounded by κ_h. Thus the number of such reductions, which is the same as the number of accuracy-improving iterations, $|\mathcal{A}_k|$, that can occur before the algorithm achieves approximate optimality, is bounded above by the root $i_{h,\epsilon}$ of the equation

$$\gamma_h^{i_{h,\epsilon}} \kappa_h = \kappa_{\rm acc} \epsilon_{\min}^{q(q+1)}, \quad \text{where } \kappa_{\rm acc} \stackrel{\rm def}{=} \min\left[\frac{\varsigma^{q+1}\kappa_\delta^q}{4q!\kappa_{\nabla,\max}}, \kappa_s \left(\frac{\varsigma^{q+1}\kappa_\delta^q}{4q!\overline{L}_f}\right)^q\right],$$

which is given by

$$\begin{aligned} i_{h,\epsilon} &= \frac{1}{\log(\gamma_h)}\left\{q(q+1)\log(\epsilon_{\min}) + \log\left(\frac{\kappa_{\rm acc}}{\kappa_h}\right)\right\} \\ &\leq \frac{1}{|\log(\gamma_h)|}\left\{q(q+1)|\log(\epsilon_{\min})| + \left|\log\left(\frac{\kappa_{\rm acc}}{\kappa_h}\right)\right|\right\}. \end{aligned}$$

Since the inexact derivatives of the objective function are only evaluated at each successful or accuracy-improving iteration, we deduce that the number of these evaluations before achieving approximate optimality is bounded above by the maximum number of successful iterations plus $i_{h,\epsilon}+1$. This yields (13.1.39) with

$$\kappa^{\rm E}_{\sf TRqIDA} = \frac{q(q+1)}{|\log(\gamma_h)|} \quad \text{and} \quad \kappa^{\rm F}_{\sf TRqIDA} = \frac{1}{|\log(\gamma_h)|}\left|\log\left(\frac{\kappa_{\rm acc}}{\kappa_h}\right)\right|. \tag{13.1.44}$$

□

Since, as we have seen in Theorems 12.1.3 and 12.1.6, there is a strong connection between standard first- and second-order criticality measures and $\phi^{\delta}_{f,q}$ for $q=1$ and 2, we anticipate that variants of TRqIDA that use traditional first- and second-order criticality measures can be devised, with similar outcomes and complexity bounds.

13.1.4 The ARqpIDA algorithm and its complexity

13.1.4.1 The regularized model and its associated optimality measures

Having considered a trust-region method that uses implicit dynamic accuracy to control inexact derivative values, we now turn to its regularization-based counterpart. Once again, this is designed to find a qth-order (ϵ,δ)-approximate minimizer and is strongly inspired by the ARqp algorithm from Section 12.2.2. The hope is that the evaluation-complexity advantages we saw in Section 12.2.2.3 when comparing ARqp with its trust-region sibling TRq carry over to the IDA framework here.

As the reader will surely anticipate by now, such an algorithm uses a regularized, *inexact* pth-degree Taylor-series model

$$\overline{m}_k(s) \stackrel{\rm def}{=} \overline{T}_{f,p}(x_k,s) + \frac{\sigma_k}{(p+1)!}\|s\|^{p+1} \tag{13.1.45}$$

at iteration k, where $\overline{T}_{f,p}(x,s)$ is as described in (13.1.1). Thus this model uses $p>0$ inexact derivatives, each of which is required to satisfy bounds of the form (13.1.2), and where the index j in Section 13.1.2 now ranges from one to p. Our adaptive-regularization algorithm imposes a slightly stronger condition on the requested accuracy $h_{k,i}$ in (13.1.2) than the one used by the TRqIDA algorithm. In particular, we now require that

$$h_{k,i} \leq \kappa_s \|s_k\|^{p-i+1} \quad \text{for all } i \in [1:p], \tag{13.1.46}$$

where κ_s is, as before, a positive constant, before we contemplate taking the step s_k. Consequently, the bound (13.1.11) becomes

$$|T_{f,j}(x_k,s_k) - \overline{T}_{f,j}(x_k,s_k)| \leq 2\kappa_s\kappa_{\nabla,\max}\|s_k\|^{p+1}, \tag{13.1.47}$$

where now $\kappa_{\nabla,\max} \stackrel{\rm def}{=} \max_{j\in[1:p]}\kappa_{\nabla,j}$.

As ever, the model (13.1.45) is approximately minimized, which is to say that the model decrease

$$\Delta\overline{m}_k(s) \stackrel{\text{def}}{=} \overline{m}_k(0) - \overline{m}_k(s)$$

is approximately maximized, and the resulting trial point is accepted or rejected depending on whether or not it produces a significant decrease in the exactly computed objective function. While such a mechanism gives some predictable advantages in terms of derived complexity bounds, it unfortunately also makes the algorithm and its analysis somewhat more involved. Indeed, as well as needing to use separate order-dependent optimality radii $\delta_k \in (0,1]^q$, the fact that "model minimization" involves the optimality measure applied to the model also means that we need to ensure that inaccuracies in the derivatives do not alter this optimality measure sufficiently to create difficulties in the resulting analysis.

The definition of the optimality measure for the model involves its Taylor expansion

$$T_{\overline{m}_k,j}(s_k,d) = \overline{m}_k(s_k) + \sum_{\ell=1}^{j} \frac{1}{\ell!} \sum_{i=\ell}^{p} \frac{1}{(i-\ell)!} \overline{\nabla_x^i f(x_k)}[s_k]^{i-\ell}[d]^\ell,$$

so that

$$|\Delta T_{\overline{m}_k,j}(s_k,d) - \Delta T_{m_k,j}(s_k,d)| \le \sum_{\ell=1}^{j} \frac{\delta^\ell}{\ell!} \sum_{i=\ell}^{p} \frac{1}{(i-\ell)!} \kappa_{\nabla,i} h_{k,i} \|s_k\|^{i-\ell} \tag{13.1.48}$$

when $\|d\| \le \delta$, where $\Delta T_{\overline{m}_k,j}(s_k,d) \stackrel{\text{def}}{=} T_{\overline{m}_k,j}(s_k,0) - T_{\overline{m}_k,j}(s_k,d)$. As a consequence,

$$\begin{aligned}
\phi^\delta_{m_k,j}(s_k) &\le \max_{\|d\|\le\delta} \left(\Delta T_{\overline{m}_k,j}(s_k,d) + |\Delta T_{\overline{m}_k,j}(s_k,d) - \Delta T_{m_k,j}(s_k,d)|\right)\\
&\le \phi^\delta_{\overline{m}_k,j}(s_k) + \sum_{\ell=1}^{j} \frac{\delta^\ell}{\ell!} \sum_{i=\ell}^{p} \frac{1}{(i-\ell)!} \kappa_{\nabla,i} h_{k,i} \|s_k\|^{i-\ell}\\
&\le \phi^\delta_{\overline{m}_k,j}(s_k) + \kappa_{\nabla,\max} \sum_{\ell=1}^{j} \frac{\delta^\ell}{\ell!} \left(\sum_{i=\ell}^{p} \frac{1}{(i-\ell)!} \kappa_s \|s_k\|^{p-\ell+1}\right)\\
&\le \phi^\delta_{\overline{m}_k,j}(s_k) + 3\kappa_{\nabla,\max}\kappa_s \sum_{\ell=1}^{j} \frac{\delta^\ell}{\ell!} \|s_k\|^{p-\ell+1}
\end{aligned} \tag{13.1.49}$$

using (12.1.10) (p. 278), the triangle inequality, (13.1.48), (13.1.46), and the sum bound (A.36). Moreover, if $\|s_k\| \le 1$ and $h_{k,\max} = \max_{i\in[1:p]} h_{k,i}$, we find that

$$\begin{aligned}
\phi^\delta_{m_k,j}(s_k) &\le \max_{\|d\|\le\delta} \left(\Delta T_{\overline{m}_k,j}(s_k,d) + |\Delta T_{\overline{m}_k,j}(s_k,d) - \Delta T_{m_k,j}(s_k,d)|\right)\\
&\le \phi^\delta_{\overline{m}_k,j}(s_k) + \kappa_{\nabla,\max} h_{k,\max} \sum_{\ell=1}^{j} \frac{\delta^\ell}{\ell!} \sum_{i=\ell}^{p} \frac{1}{(i-\ell)!}\\
&\le \phi^\delta_{\overline{m}_k,j}(s_k) + 6\kappa_{\nabla,\max} h_{k,\max}
\end{aligned} \tag{13.1.50}$$

using the triangle inequality, (13.1.48) once again, and the requirement $\delta \in (0,1]$. The bounds (13.1.49) and (13.1.50) are crucial for the analysis of the adaptive-regularization algorithm ARqpIDA, on the facing page.

Algorithm 13.1.2: Adaptive-regularization algorithm with implicit dynamic accuracy (ARqpIDA)

Step 0: Initialization. A criticality order $q \le p$, an initial point $x_0 \in \mathbb{R}^n$, and an initial regularization parameter $\sigma_0 > 0$ are given, as well as accuracy levels $\epsilon \in (0,1)^q$ and an initial set of requested accuracies $\{h_{0,i}\}_{i=1}^p$. The constants κ_h, κ_s, δ_0, θ, η_1, η_2, γ_1, γ_2, γ_3, γ_h, and $\sigma_{\min}$ are also given and satisfy $\kappa_h \ge 0$, $\kappa_s > 0$, $\gamma_h \in (0,1)$, $\theta \in (0,1)$, $\delta_0 \in (0,1]^q$, $\sigma_{\min} \in (0, \sigma_0]$,

$$\begin{aligned} 0 < \eta_1 \le \eta_2 < 1, \quad 0 < \gamma_1 < 1 < \gamma_2 < \gamma_3, \\ h_{0,i} \le \kappa_h \quad (i \in [1:p]), \text{ and } \epsilon_j \le \delta_{0,j} \quad (j \in [1:q]). \end{aligned} \tag{13.1.51}$$

Compute $f(x_0)$, and set $k = 0$ and $i_h = 0$.

Step 1: Compute the optimality measures. Set $\delta_k^{(0)} = \delta_k$ and recompute (if necessary) $\{\overline{\nabla_x^i f(x_k)}\}_{i=1}^p$ to satisfy (13.1.2).
For $j = 1, \dots, q$:

Step 1.1: Compute $d_{k,j} \in \mathcal{B}_{\delta_{k,j}}$ such that

$$\varsigma \overline{\phi}_{f,j}^{\delta_{k,j}}(x_k) \le \overline{\Delta T}_{f,j}(x_k, d_{k,j}). \tag{13.1.52}$$

Step 1.2: If

$$\overline{\Delta T}_{f,j}(x_k, d_{k,j}) \le \tfrac{1}{2} \varsigma \epsilon_j \frac{\delta_{k,j}^j}{j!}, \tag{13.1.53}$$

consider the next value of j.

Step 1.3: Otherwise, if

$$\Delta \overline{m}_k(d_{k,j}) \ge \tfrac{1}{4} \varsigma \epsilon_j \frac{\delta_{k,j}^j}{j!}, \tag{13.1.54}$$

go to Step 2 with the index $j_k = j$, and $\delta_k^{(1)} = \delta_k$ and $d_k = d_{k,j}$.

Step 1.4: Otherwise set $\delta_{k,j} = \frac{1}{2}\delta_{k,j}$ and return to Step 1.1.

If the loop on j finishes, i.e., (13.1.53) holds for all $j \in [1:q]$, go to Step 5.

Step 2: Step calculation. Compute a step s_k and optimality radii $\delta_{s_k} \in (0,1]^q$ by approximately minimizing the model $\overline{m}_k(s)$ in (13.1.45) in the sense that

$$\Delta \overline{m}_k(s_k) \ge \Delta \overline{m}_k(d_k) \tag{13.1.55}$$

and either

$$\|s_k\| \ge 1 \text{ with } \delta_{s_k} = \delta_k^{(1)} \tag{13.1.56}$$

or

$$\varsigma \phi_{\overline{m}_k,\ell}^{\delta_{s_k,\ell}}(s_k) \le \Delta T_{\overline{m}_k,\ell}(s_k, d_{s_k,\ell}^m) \le \varsigma \theta \epsilon_\ell \frac{\delta_{s_k,\ell}^\ell}{\ell!} \text{ for } \ell \in [1:q] \tag{13.1.57}$$

for some radii $\delta_{s_k} \in (0,1]^q$ and some displacements $d_{s_k,\ell}^m \in \mathcal{B}_{\delta_{s_k,\ell}}$. If $\|s_k\| < 1$ and (13.1.46) fails for some $i \in [1:q]$, go to Step 5.

Step 3: Acceptance of the trial point. Compute $f(x_k + s_k)$ and set

$$\rho_k = \frac{f(x_k) - f(x_k + s_k)}{\overline{\Delta T}_{f,p}(x_k, s_k)}. \tag{13.1.58}$$

If $\rho_k \geq \eta_1$, then define $x_{k+1} = x_k + s_k$ and $\delta_{k+1} = \delta_{s_k}$ if $\|s_k\| < 1$ or $\delta_{k+1} = \delta_k^{(1)}$ otherwise. If $\rho_k < \eta_1$, define $x_{k+1} = x_k$ and $\delta_{k+1} = \delta_k^{(1)}$.

Step 4: Regularization parameter update. Set

$$\sigma_{k+1} \in \begin{cases} [\max(\sigma_{\min}, \gamma_1 \sigma_k), \sigma_k] & \text{if } \rho_k \geq \eta_2, \\ [\sigma_k, \gamma_2 \sigma_k] & \text{if } \rho_k \in [\eta_1, \eta_2), \\ [\gamma_2 \sigma_k, \gamma_3 \sigma_k] & \text{if } \rho_k < \eta_1. \end{cases} \tag{13.1.59}$$

Increment k by one and go to Step 1.

Step 5: Improve accuracy. For $i \in [1 : p]$, set

$$h_{k+1,i} = \gamma_h h_{k,i},$$

increment k and i_h by one, and go to Step 1 with $x_{k+1} = x_k$, $\delta_{k+1} = \delta_k^{(0)}$, and $\sigma_{k+1} = \sigma_k$.

It would be trivial to modify the algorithm to permit exactly computed derivatives of certain degrees, but we have again avoided the temptation for the sake of simplicity. As for the TRqIDA, there is no termination rule, because specifying a rule guaranteeing exact approximate optimality in this inexact setting is impossible unless stronger assumptions are made. We note that Step 1 specifies the smallest index $j \in [1 : q]$ for which the jth-order approximate criticality condition (13.1.53) fails. We also recall the notation from (12.2.32) in which $\delta_{k,j}$ and $\delta_{s_k,j}$ (in Step 2) are the jth components of the vectors of radii δ_k and δ_{s_k}, respectively. It is helpful for what follows to distinguish the radii

$$\delta_k^{(0)} = \delta_k \text{ at the start of Step 1,} \tag{13.1.60}$$

and inherited from iteration $k-1$, from

$$\delta_k^{(1)} = \delta_k \text{ at the end of Step 1,} \tag{13.1.61}$$

after possible reductions within that step. Clearly, componentwise $\delta_k^{(1)} \leq \delta_k^{(0)}$.

Step 2 starts exactly as for the step computation in the ARqp algorithm, except that now (13.1.54) guarantees that a strict reduction of the model is possible. Just as for the TRqIDA algorithm, failure to satisfy the requirement (13.1.46) is countered by branching to Step 5 to improve the accuracy. Once again, approximate global optimization is allowed to estimate the criticality measures $\overline{\phi}_{f,j}^{\delta_{k,j}}(x_k)$ and $\phi_{\overline{m}_k,\ell}^{\delta_{s_k,\ell}}(s_k)$ in Steps 1.1 and 2. Step 3 is, by now, quite familiar—we may apply Algorithm 12.2.3 (p. 303) to $\overline{m}_k(s)$ to find a step s_k with associated optimality radii in order to satisfy (13.1.55)–(13.1.57). Step 4 is the standard regularization parameter update, and Step 5 is very similar to that in the TRqIDA algorithm.

We define $\mathcal{S}, \mathcal{U}, \mathcal{A}, \mathcal{T}, \mathcal{S}_k, \mathcal{U}_k, \mathcal{A}_k$, and $\mathcal{T}_k$ as in (13.1.19) and (13.1.20). Just as for the TRqIDA algorithm, the objective function is evaluated once per successful or unsuccessful iteration (i.e., once for every $k \in \mathcal{T}$), and derivatives are evaluated once per successful or accuracy-improving iteration (i.e., once for every $k \in \mathcal{S} \cup \mathcal{A}$). The "regularization" variant of Lemma 13.1.1 has the following form.

Lemma 13.1.9: Successful and unsuccessful iterations (AR*qp*IDA)

Suppose that the AR*qp*IDA *algorithm is used and that* $\sigma_k \leq \sigma_{\max}$ *for some* $\sigma_{\max} > 0$. *Then*

$$|\mathcal{T}_k| \leq |\mathcal{S}_k| \left(1 + \frac{|\log \gamma_1|}{\log \gamma_2}\right) + \frac{1}{\log \gamma_2} \log\left(\frac{\sigma_{\max}}{\sigma_0}\right). \tag{13.1.62}$$

Proof. The proof is a direct generalization of that of Lemma 2.4.1 (p. 44) restricted to $\mathcal{T}$. It is given in Appendix B (p. 486) for completeness. □

As for the TR*q*IDA algorithm, our analysis will be carried out under the assumption that the function value and all derivatives are Lipschitz continuous.

f.D0pL

f is p-times continuously differentiable in $\mathbb{R}^n$, and, for all $j \in [0 : p]$, the jth derivative of f is Lipschitz continuous with Lipschitz constant $L_{f,j}$; that is, there exist constants $L_{f,j} \geq 1$ such that

$$\|\nabla_x^j f(x) - \nabla_x^j f(y)\| \leq L_{f,j} \|x - y\|$$

for all $x, y \in \mathbb{R}^n$ and all $j \in [0 : p]$.

We shall require the obvious analogue

$$L_f \stackrel{\text{def}}{=} \max_{j \in [0:p]} L_{f,j} \geq 1$$

of (13.1.23). This, Theorem A.8.5 (p. 438), and the fact that the accuracy tolerances $\{h_{k,i}\}_{i=1}^p$ only decrease in the course of the AR*qp*IDA algorithm imply that

$$\|\overline{\nabla_x^i f(x_k)}\| \leq L_f + \kappa_{\nabla,\max} \kappa_h \stackrel{\text{def}}{=} \overline{L}_f \tag{13.1.63}$$

for each $k \geq 0$ and each $i \in [1 : p]$.

13.1.4.1.1 The outcomes of Step 1: We start by considering the result of performing Step 1 and show that the loop-reducing $\delta_{k,j}$ generated by the possible return to Step 1.1 from Step 1.4 is finite. This is done in two stages: we start by expressing a general property of the value of the model compared to the truncated Taylor series, which we will subsequently apply to obtain the desired conclusion.

Lemma 13.1.10: Relationship between model and Taylor decrements

Suppose that **f.D0pL** *holds, that* $\overline{m}_k(s)$ *is the inexact model (13.1.45) corresponding to some approximate derivative values* $\{\overline{\nabla_x^\ell f(x_k)}\}_{\ell=1}^p$, *and that* $\overline{L}_f$ *is given by (13.1.63). Given* $\alpha > 0$ *and*

$$\delta \in \left(0, \min\left(1, \frac{\alpha}{4\max[\overline{L}_f, \sigma_k]}\right)\right], \tag{13.1.64}$$

suppose that there is a displacement $d \in \mathcal{B}_\delta$ *for which*

$$\overline{\Delta T}_{f,j}(x_k, d) \geq \alpha \frac{\delta^j}{j!}$$

for some $j \in [1:q]$. *Then*

$$\Delta \overline{m}_k(d) \geq \tfrac{1}{2}\alpha \frac{\delta^j}{j!}. \tag{13.1.65}$$

Proof. The proof is essentially that of Lemma 12.2.8 (p. 305), but built on the sequence of inequalities

$$\begin{aligned}
\Delta \overline{m}_k(d_k) &= \overline{\Delta T}_{f,j}(x_k, d) + \sum_{\ell=j+1}^{p} \frac{1}{\ell!}\overline{\nabla_x^\ell f}(x_k)[d]^\ell + \frac{\sigma_k}{(p+1)!}\|d\|^{p+1} \\
&\geq \overline{\Delta T}_{f,j}(x_k, d) - \sum_{\ell=j+1}^{p} \frac{\delta^\ell}{\ell!}\|\overline{\nabla_x^\ell f}(x_k)\| - \frac{\sigma_k}{(p+1)!}\delta^{p+1} \\
&\geq \alpha \frac{\delta^j}{j!} - \max[\overline{L}_f, \sigma_k] \sum_{\ell=j+1}^{p+1} \frac{\delta^\ell}{\ell!} \\
&\geq \alpha \frac{\delta^j}{j!} - 2\max[\overline{L}_f, \sigma_k] \frac{\delta^{j+1}}{j!} \\
&= \frac{\delta^j}{j!}\left(\alpha - 2\max[\overline{L}_f, \sigma_k]\delta\right)
\end{aligned}$$

that arise from the triangle inequality and because of the assumptions made. The required bound (13.1.65) then follows because of (13.1.64). □

We may then summarize the outcome of Step 1 as follows.

Lemma 13.1.11: Outcome of Step 1 of the ARqpIDA algorithm

Suppose that **f.D0pL** *holds and that* $\overline{L}_f$ *is given by (13.1.63). Then the loop on* $\delta_{k,j}$ *in Step 1 is finite (i.e., Step 1.4 cannot be performed* ad infinitum*), and*

$$\delta_{k,j}^{(0)} \geq \delta_{k,j} \geq \delta_{k,j}^{(1)} \geq \min\left[\frac{\varsigma\epsilon_j}{16\max[\overline{L}_f, \sigma_k]}, \delta_{k,j}^{(0)}\right]. \tag{13.1.66}$$

Proof. Firstly, observe that the mechanism of Step 1 ensures that the first inequality in (13.1.66) is satisfied. Moreover, if the loop on a particular $j \in [1:q]$ does not finish and branching to Step 5 does not happen, it must be that (13.1.53) is violated, i.e.,

$$\overline{\phi}_{f,j}^{\delta_{k,j}}(x_k) = \overline{\Delta T}_{f,j}(x_k, d_k) > \tfrac{1}{2}\varsigma\epsilon_j \frac{\delta_{k,j}^j}{j!}, \tag{13.1.67}$$

and the test (13.1.54) is reached. Suppose now that

$$\delta_{k,j} \leq \frac{\varsigma\epsilon_j}{8\max[\overline{L}_f, \sigma_k]}. \tag{13.1.68}$$

In that case, it follows from Lemma 13.1.10 with $\alpha = \frac{1}{2}\varsigma\epsilon_j$ that (13.1.54) holds, and thus Step 1.3 will branch to Step 2. Thus the number of times that (13.1.54) is violated—and hence $\delta_{k,j}$ is halved—is finite since (13.1.54) must hold as soon as (13.1.67) holds. At this stage, if one or more halvings happen, the resulting value of $\delta_{k,j}$ cannot be smaller than half of the right-hand side of (13.1.68), and so (13.1.66) also holds. If no halving of $\delta_{k,j}$ occurs, $\delta_{k,j}^{(1)} = \delta_{k,j}^{(0)}$ and (13.1.66) obviously holds. □

Another potential outcome of Step 1 is that it branches to Step 5 when (13.1.53) holds for all $j \in [1:q]$. We show in Lemma 13.1.15 below that this possibly infinite loop between Steps 1 and 5 leads us ever closer to a qth-order local minimizer.

13.1.4.1.2 The outcomes of Step 2: We next investigate Step 2 in more detail. Specifically, we shall apply Algorithm 12.2.3 (p. 303) to the inexact model $\overline{m}_k(s)$ defined in (13.1.45). We recall that branching to Step 5 does not occur when $k \in \mathcal{T}$.

Lemma 13.1.12: Outcomes of Step 2 of the ARqpIDA algorithm

Suppose that **f.D0pL** *holds and that $\overline{L}_f$ is given by (13.1.63). Then Step 2 of the* ARqpIDA *algorithm, using Algorithm 12.2.3 (p. 303), is well defined and produces a pair (s_k, δ_{s_k}) such that $s_k \neq 0$ and either*

(i) $\|s_k\| \geq 1$ *and* $\delta_{s_k} = \delta_k$, *or*

(ii) *the second inequality in (13.1.57) holds for some δ_{s_k} such that*

$$\delta_{s_k,j} = 1 \quad \text{for } j \in [1:\min(2,q)] \tag{13.1.69}$$

and

$$\delta_{s_k,j} \geq \frac{\varsigma\theta}{24\max[\overline{L}_f, \sigma_k]}\,\epsilon_j \quad \text{for} \quad j \in [3:q]. \tag{13.1.70}$$

Moreover,

$$\delta_{k,j}^{(0)} \geq \delta_{k,j} \geq \delta_{k,j}^{(1)} \geq \kappa_\delta^{(1)}(\sigma_{k,\max})\epsilon_j \tag{13.1.71}$$

for all $k \geq 0$ and $j \in [1:q]$, where $\sigma_{k,\max} = \max_{i\in[0:k]} \sigma_i$ and

$$\kappa_\delta^{(1)}(\sigma) \stackrel{\text{def}}{=} \frac{\varsigma\theta}{24\max[\overline{L}_f, \sigma]}. \tag{13.1.72}$$

Finally, $k \in \mathcal{T}$ whenever

$$h_{k,\max} \leq \kappa_s \left(\frac{\varsigma\left(\kappa_\delta^{(1)}(\sigma_{k,\max})\right)^q}{12q!\,\max[\overline{L}_f, \sigma_{k,\max}]} \right)^p \epsilon_{\min}^{p(q+1)}. \tag{13.1.73}$$

Proof. We first verify that Step 2 is well defined. The problem is almost identical to that of proving that a step can be found in the ARqp algorithm (p. 300), the only difference being that the model now uses approximate derivatives of f at x_k and thus that **f.D1pL** can no longer be invoked directly to ensure that the derivatives of the model of degree two to p are bounded. Fortunately, (13.1.63) provides such a bound. As a consequence, a straightforward variant of Lemma 12.2.9 (p. 306) still holds with L_f replaced by $\overline{L}_f$. In particular, this lemma ensures that, when $\|s_k\| < 1$, $\delta_{s_k} = 1$

for $j \leq \min[q, 2]$ and

$$\delta_{s_k,j} \geq \frac{\varsigma\theta}{8(3\overline{L}_f + \sigma_k)}\,\epsilon_j \geq \frac{\varsigma\theta}{24\max[\overline{L}_f, \sigma_k]}\,\epsilon_j$$

and (13.1.70) holds for $j \in [3:q]$. Thus Step 2 is well defined, and either outcome (i) or outcome (ii) occurs. In addition,

$$\Delta\overline{m}_k(s_k) \geq \Delta\overline{m}_k(d_k) \geq \frac{\varsigma\epsilon_{j_k}}{4}\,\frac{\left(\delta^{(1)}_{k,j_k}\right)^{j_k}}{j_k!} > 0 \tag{13.1.74}$$

from (13.1.55) and (13.1.54), and thus $s_k \neq 0$.

We now consider how the vectors of radii δ_k evolve. Flow from Step 1 in the algorithm is either to Step 4 (via Steps 2 and 3) or Step 5 (either directly or via Step 2). The radii at the start of Step 1 following an iteration via Step 5 are unchanged, i.e., $\delta^{(0)}_k = \delta^{(0)}_{k-1}$ for all $k-1 \in \mathcal{A}$.

Suppose that iterations $0, \ldots, k-1$ have all passed to Step 5 from Step 1 and that iteration k is the first that passes to Step 2. Then $\delta^{(0)}_{k,j} = \delta^{(0)}_{0;j} \geq \epsilon_j$ because of (13.1.51), and thus (13.1.66) gives

$$\delta^{(1)}_{k,j} \geq \min\left[\frac{\varsigma\epsilon_j}{16\max[\overline{L}_f, \sigma_k]}, \epsilon_j\right] = \frac{\varsigma\epsilon_j}{16\max[\overline{L}_f, \sigma_k]} \geq \frac{\varsigma\theta\epsilon_j}{24\max[\overline{L}_f, \sigma_k]}$$

because $\theta \leq 1$, giving us (13.1.71). This pattern then repeats. Suppose that iteration k_1 passed from Step 1 to Step 2 and that (13.1.71) holds with $k = k_1$. As a result, regardless of whether control then passes to Step 3 (and hence Step 4) or to Step 5, the algorithm ensures that

$$\delta^{(0)}_{k_1+1,j} = \delta_{k_1+1,j} \geq \min\left[\frac{\varsigma\epsilon_j}{16\max[\overline{L}_f, \sigma_{k_1}]}, \frac{\varsigma\theta\epsilon_j}{24\max[\overline{L}_f, \sigma_{k_1}]}, 1\right] \geq \frac{\varsigma\theta\epsilon_j}{24\max[\overline{L}_f, \sigma_{k_1,\max}]}. \tag{13.1.75}$$

Now suppose that iterations k_1+1 until $k-1$ pass from Step 1 to Step 5, but iteration k passes to Step 3. Then

$$\delta^{(0)}_k = \delta^{(0)}_{k_1+1} \geq \frac{\varsigma\theta\epsilon_j}{24\max[\overline{L}_f, \sigma_{k_1,\max}]} \geq \frac{\varsigma\theta\epsilon_j}{24\max[\overline{L}_f, \sigma_{k,\max}]}$$

because of (13.1.75), and thus (13.1.71) follows from (13.1.66).

Finally, when $\|s_k\| < 1$, we have that

$$\begin{aligned}
\frac{\varsigma\epsilon_{\min}}{4}\,\frac{\left(\delta^{(1)}_{k,j_k}\right)^{j_k}}{j_k!} &\leq \overline{\Delta T}_{f,p}(x_k, s_k) + \frac{\sigma_k}{(p+1)!}\|s_k\|^{p+1} \\
&\leq \sum_{\ell=1}^{p} \|\overline{\nabla^i_x f}(x_k)\|\frac{\|s_k\|^\ell}{\ell!} + \frac{\sigma_k}{(p+1)!}\|s_k\|^{p+1} \\
&\leq (2\overline{L}_f + \sigma_k)\|s_k\| \\
&\leq 3\max[\overline{L}_f, \sigma_k]\|s_k\|
\end{aligned}$$

from (13.1.74), (13.1.63), and the sum bound (A.34). But (13.1.71) gives

$$\left(\delta^{(1)}_{k,j_k}\right)^{j_k} \geq \left(\delta^{(1)}_{k,j_k}\right)^{q} \geq \left(\kappa^{(1)}_\delta(\sigma_k)\epsilon_{\min}\right)^{q}$$

because we have passed to Step 2, and hence

$$\|s_k\| \geq \frac{\varsigma \left(\kappa_\delta^{(1)}(\sigma_k)\right)^q \epsilon_{\min}^{q+1}}{12q! \max[\overline{L}_f, \sigma_k]} \geq \frac{\varsigma \left(\kappa_\delta^{(1)}(\sigma_{k,\max})\right)^q \epsilon_{\min}^{q+1}}{12q! \max[\overline{L}_f, \sigma_{k,\max}]}.$$

Thus (13.1.73) ensures (13.1.46), which prevents branching from Step 2 to Step 5. □

13.1.4.2 Evaluation complexity of the ARqpIDA algorithm

We are now poised to consider the evaluation-complexity analysis of the ARqpIDA algorithm. We start our analysis by stating a suitable variant of the standard bound on model decrease, which now involves the inexact Taylor series $\overline{T}_{f,p}(x_k, s_k)$ of degree p.

Lemma 13.1.13: Model decrease for the ARqpIDA algorithm

The mechanism of the ARqpIDA algorithm guarantees that

$$\overline{\Delta T}_{f,p}(x_k, s_k) \geq \frac{\sigma_k}{(p+1)!}\|s_k\|^{p+1} \tag{13.1.76}$$

for all $k \in \mathcal{T}$ and that (13.1.58) is well defined.

Proof. Since $k \in \mathcal{T}$, both (13.1.54) and (13.1.55) must hold at iteration k for some $j \in [1:q]$. We then have that

$$0 < \frac{\varsigma\epsilon_j}{4} \cdot \frac{\delta_{k,j}^j}{j!} \leq \Delta\overline{m}_k(d_{k,j}) \leq \Delta\overline{m}_k(s_k) = \overline{\Delta T}_{f,p}(x_k, s_k) - \frac{\sigma_k}{(p+1)!}\|s_k\|^{p+1},$$

using (13.1.74), which ensures both that $s_k \neq 0$ and that (13.1.76) holds. □

We next show that the regularization parameter σ_k has to remain bounded, even when the derivatives of f are computed inexactly.

Lemma 13.1.14: Condition for success (ARqpIDA)

Suppose that **f.D0pL** *holds. Then*

$$\sigma_k \leq \sigma_{\max} \stackrel{\text{def}}{=} \max\left[\sigma_0, \gamma_3 \frac{2L_{f,p} + 4(p+1)!\,\kappa_s\kappa_{\nabla,\max}}{1-\eta_2}\right] \tag{13.1.77}$$

for all k.

Proof. Suppose that $k \in \mathcal{T}$ and that[89]

$$\sigma_k \geq \frac{2L_{f,p} + 4(p+1)!\,\kappa_s\kappa_{\nabla,\max}}{1-\eta_2} \geq \frac{L_{f,p} + 2(p+1)!\,\kappa_s\kappa_{\nabla,\max}}{1-\eta_2}.$$

[89] Note that we are allowing a currently unnecessary factor of 2 in (13.1.77) to address the forthcoming EDA framework in Section 13.3.

Hence

$$\begin{aligned}
|\rho_k - 1| &\le \frac{|f(x_{k+1}) - \overline{T}_{f,p}(x_k, s_k)|}{\overline{\Delta T}_{f,p}(x_k, s_k)} \\
&\le \frac{1}{\overline{\Delta T}_{f,p}(x_k, s_k)} \left[|f(x_{k+1}) - T_{f,p}(x_k, s_k)| + |T_{f,p}(x_k, s_k) - \overline{T}_{f,p}(x_k s_k)| \right] \\
&\le \frac{(p+1)!}{\sigma_k \|s_k\|^{p+1}} \left[\frac{L_{f,p}}{(p+1)!} + 2\kappa_s \kappa_{\nabla,\max} \right] \|s_k\|^{p+1} \\
&\le 1 - \eta_2
\end{aligned}$$

using (13.1.7), the triangle inequality, (13.1.76), (13.1.47), and (A.19) in Corollary A.8.4 (p. 437). Thus iteration k is successful, and $\sigma_{k+1} \le \sigma_k$. The conclusion then follows from the mechanism of Step 4 (for $k \in \mathcal{T}$) and the fact that $\sigma_{k+1} = \sigma_k$ for $k \in \mathcal{A}$. □

If the approximate derivatives are sufficiently accurate (specifically, the accuracies $h_{k,i}$ are suitably small), then the potentially infinite loop between Steps 1 and 5 leads us progressively closer to a qth-order local minimizer.[90]

Lemma 13.1.15: The optimality loop in Step 1 of the ARqpIDA algorithm

Suppose that **f.D0pL** *holds. If*

$$h_{k,\max} \le \varsigma \frac{\left(\kappa_\delta^{(1)}(\sigma_{\max})\right)^q}{4q!\kappa_{\nabla,\max}} \epsilon_{\min}^{q+1}, \tag{13.1.78}$$

then branching from Step 1 to Step 5 of the ARqpIDA *algorithm achieves the desired approximate optimality condition,* $\phi_{f,j}^{\delta_k}(x_k) \le \epsilon_j \frac{\delta_k^j}{j!}$, *for all* $j \in [1:q]$.

Proof. An accuracy-improving iteration may occur at the end of Step 1, where branching to Step 5 happens because $\overline{\phi}_{f,j}^{\delta_{k,j}}(x_k) \le \frac{1}{2}\epsilon_j \delta_{k,j}^j / j!$ for all $j \in [1:q]$, which follows from (13.1.52) and (13.1.53). In this case, if we assume that (13.1.78) holds, and using that $\varsigma \le 1$, we obtain from (13.1.6) that

$$\begin{aligned}
\phi_{f,j}^{\delta_{k,j}}(x_k) &\le \tfrac{1}{2}\epsilon_j \frac{\delta_{k,j}^j}{j!} + \frac{\left(\kappa_\delta^{(1)}(\sigma_{\max})\right)^q}{2q!} \epsilon_{\min}^{q+1} \\
&\le \tfrac{1}{2}\epsilon_j \frac{\delta_{k,j}^j}{j!} + \frac{\left(\kappa_\delta^{(1)}(\sigma_{k,\max})\right)^q}{2q!} \epsilon_{\min}^{q+1} \\
&\le \tfrac{1}{2}\epsilon_j \frac{\delta_{k,j}^j}{j!} + \tfrac{1}{2}\epsilon_{\min} \frac{\delta_{k,j}^j}{j!}
\end{aligned}$$

for all $j \in [1:q]$, where in the second and third inequalities we used (13.1.77) and (13.1.71); this yields the required result. □

[90] The right-hand side of (13.1.78) has an additional ς factor that is not strictly needed here but will be useful in the EDA framework of Section 13.3.

We now follow the tracks of the ARqp algorithm with necessary adaptations for the IDA framework.

Lemma 13.1.16: A useful inequality for the ARqpIDA algorithm

Suppose that **f.D0pL** *holds. Suppose also that iteration* $k \in \mathcal{T}$ *is successful, that approximate optimality is not achieved at iteration* $k+1$ *(i.e., (12.1.25) (p. 285) does not hold), and that* $\|s_k\| < 1$. *Then there exists* $j \in [1:q]$ *such that*

$$(1-\theta)\,\epsilon_j \,\frac{(\delta^{(0)}_{k+1,j})^j}{j!} \le (L_{f,p} + \sigma_{\max} + 3\kappa_{\nabla,\max}\kappa_s) \sum_{\ell=1}^{j} \frac{(\delta^{(0)}_{k+1,j})^\ell}{\ell!} \|s_k\|^{p-\ell+1}, \tag{13.1.79}$$

where $\sigma_{\max}$ *is given by (13.1.77).*

Proof. The proof of this result essentially follows that of Lemma 12.2.11 (p. 308), with minor adaptations. We first note that our assumption that approximate optimality is not achieved at iteration $k+1$ imposes that

$$\phi_{f,j}^{\delta^{(0)}_{k+1,j}}(x_{k+1}) > \epsilon_j \,\frac{\left(\delta^{(0)}_{k+1,j}\right)^j}{j!}$$

for some $j \in [1:q]$. Consider such a j, and let d be the argument of the minimization in the definition of $\phi_{f,j}^{\delta^{(0)}_{k+1,j}}(x_{k+1})$. Then $\|d\| \le \delta_{k+1,j} \le 1$ and we can follow the proof of Lemma 12.2.11 with $\overline{L}_f$ replacing L_f up to the inequality (12.2.56).

The proof is now concluded as in Lemma 12.2.11 (p. 308) by replacing (12.2.55) with the inequalities

$$\begin{aligned}
-\sum_{\ell=1}^{j} \frac{1}{\ell!} \nabla_s^\ell T_{f,p}(x_k,s_k)[d]^\ell - \sum_{\ell=1}^{j} \frac{\sigma_k}{\ell!(p-\ell+1)!} \|s_k\|^{p-\ell+1} (\delta^{(0)}_{k+1,j})^\ell \hspace{-6em}& \\
&\le \phi_{m_k,j}^{\delta_{s_k,j}}(s_k) \\
&\le \phi_{\overline{m}_k,j}^{\delta_{s_k,j}}(s_k) + 3\kappa_{\nabla,\max}\kappa_s \sum_{\ell=1}^{j} \frac{(\delta_{s_k,j})^\ell}{\ell!} \|s_k\|^{p-\ell+1} \\
&\le \theta\epsilon_j \,\frac{\delta^j_{s_k,j}}{j!} + 3\kappa_{\nabla,\max}\kappa_s \sum_{\ell=1}^{j} \frac{(\delta_{s_k,j})^\ell}{\ell!} \|s_k\|^{p-\ell+1} \\
&= \theta\epsilon_j \,\frac{(\delta^{(0)}_{k+1,j})^j}{j!} + 3\kappa_{\nabla,\max}\kappa_s \sum_{\ell=1}^{j} \frac{(\delta^{(0)}_{k+1,j})^\ell}{\ell!} \|s_k\|^{p-\ell+1},
\end{aligned}$$

where the second inequality follows from (13.1.49) and the third from (13.1.57), and the last equality is derived using the fact that $\delta_{s_k,j} = \delta^{(0)}_{k+1,j}$ because iteration k is successful. □

Lemma 13.1.17: Lower bound on the length of the step for ARqpIDA

*Suppose that **f.DOpL** holds, that iteration k is successful, and that approximate optimality is not achieved at iteration $k+1$. Suppose also that the algorithm ensures that (13.1.69) and (13.1.70) hold. Then there exists a $j \in [1:q]$ such that*

$$\|s_k\| \geq \begin{cases} \left(\dfrac{1-\theta}{2j!(L_{f,p}+\sigma_{\max}+3\kappa_{\nabla,\max}\kappa_s)}\right)^{\frac{1}{p-j+1}} \epsilon_j^{\frac{1}{p-j+1}} & \text{if } q \in \{1,2\}, \\ \left(\dfrac{(1-\theta)\kappa_{\delta,\min}^{j-1}}{2j!(L_{f,p}+\sigma_{\max}+3\kappa_{\nabla,\max}\kappa_s)}\right)^{\frac{1}{p}} \epsilon_j^{\frac{j}{p}} & \text{if } q > 2, \end{cases} \tag{13.1.80}$$

where

$$\kappa_{\delta,\min} \stackrel{\text{def}}{=} \frac{\varsigma\theta}{24\max[\overline{L}_f,\sigma_{\max}]} \tag{13.1.81}$$

and $\sigma_{\max}$ is given by (13.1.77).

Proof. The proof is identical to that of Lemma 12.2.12 (p. 309), except that (13.1.70) is used instead of (12.2.51) to conclude the proof. □

As for the ARqp algorithm, we now consolidate the above results by stating lower bounds on the minimal model and function decreases at successful iterations.

Lemma 13.1.18: Model and function decreases (ARqpIDA)

*Suppose that **f.DOpL** holds. Then*

$$\overline{\Delta T}_{f,p}(x_k,s_k) \geq \kappa_{\Delta m} \min_{j\in[1:q]} \epsilon_j^{\pi_j} \tag{13.1.82}$$

for every iteration $k \in \mathcal{T}$ of algorithm ARqpIDA before reaching approximate optimality, where

$$\kappa_{\Delta m} \stackrel{\text{def}}{=} \begin{cases} \dfrac{\sigma_{\min}}{(p+1)!}\left(\dfrac{1-\theta}{2q!(L_{f,p}+\sigma_{\max}+3\kappa_{\nabla,\max}\kappa_s)}\right)^{\frac{p+1}{p-q+1}} & \text{if } q \in \{1,2\}, \\ \dfrac{\sigma_{\min}}{(p+1)!}\left(\dfrac{(1-\theta)\kappa_{\delta,\min}^{q-1}}{2q!(L_{f,p}+\sigma_{\max}+3\kappa_{\nabla,\max}\kappa_s)}\right)^{\frac{q(p+1)}{p}} & \text{if } q > 2, \end{cases}$$

with $\kappa_{\delta,\min}$ being given by (13.1.81) and $\sigma_{\max}$ by (13.1.77), and

$$\pi_j \stackrel{\text{def}}{=} \begin{cases} \dfrac{p+1}{p-j+1} & \text{if } q \in \{1,2\}, \\ \dfrac{j(p+1)}{p} & \text{if } q > 2. \end{cases}$$

If, in addition, $k \in \mathcal{S}$, then

$$f(x_k) - f(x_{k+1}) \geq \tfrac{1}{2}\eta_1\kappa_{\Delta m} \min_{j\in[1:q]} \epsilon_j^{\pi_j}. \tag{13.1.83}$$

Proof. Lemma 13.1.13 (p. 345), (13.1.59), and Lemma 13.1.17 together imply that for every $k \in \mathcal{T}$ there is a $j \in [1:q]$ such that

$$\overline{\Delta T}_{f,p}(x_k, s_k) \geq \begin{cases} \frac{\sigma_{\min}}{(p+1)!}\left(\frac{1-\theta}{2j!(L_{f,p}+\sigma_{\max}+3\kappa_{\nabla,\max}\kappa_s)}\right)^{\frac{p+1}{p-j+1}} \epsilon_j^{\frac{p+1}{p-j+1}} & \text{if } q \in \{1,2\}, \\ \frac{\sigma_{\min}}{(p+1)!}\left(\frac{(1-\theta)\kappa_{\delta,\min}^{q-1}}{2j!(L_{f,p}+\sigma_{\max}+3\kappa_{\nabla,\max}\kappa_s)}\right)^{\frac{j(p+1)}{p}} \epsilon_j^{\frac{j(p+1)}{p}} & \text{if } q > 2, \end{cases}$$

and (13.1.82) follows. If $k \in \mathcal{S}$, then $\rho_k \geq \eta_1$, and (13.1.83) results from (13.1.82) and (13.1.58).[91] □

We are now in a position to formally state a bound on the worst-case evaluation complexity of the ARqpIDA algorithm.

Theorem 13.1.19: qth-order complexity for the ARqpIDA algorithm

Suppose that **f.D0pL** *and* **f.Bb** *hold. Suppose also that the algorithm ensures that (13.1.69) and (13.1.70) hold.*

1. *If $q \in \{1,2\}$, then there exist positive constants $\kappa^S_{\text{ARqpIDA,1}}$, $\kappa^A_{\text{ARqpIDA,1}}$, $\kappa^C_{\text{ARqpIDA}}$, $\kappa^E_{\text{ARqpIDA}}$, and $\kappa^F_{\text{ARqpIDA}}$ such that, for any $\epsilon \in (0,1]^q$, the ARqpIDA algorithm requires at most*

$$\#^F_{\text{ARqpIDA,1}} \stackrel{\text{def}}{=} \kappa^A_{\text{ARqpIDA,1}} \frac{f(x_0)-f_{\text{low}}}{\min_{j\in[1:q]} \epsilon_j^{\frac{p+1}{p-j+1}}} + \kappa^C_{\text{ARqpIDA}} = \mathcal{O}\left(\max_{j\in[1:q]} \epsilon_j^{-\frac{p+1}{p-j+1}}\right) \tag{13.1.84}$$

evaluations of f and at most

$$\#^D_{\text{ARqpIDA,1}} \stackrel{\text{def}}{=} \kappa^S_{\text{ARqpIDA,1}} \frac{f(x_0)-f_{\text{low}}}{\min_{j\in[1:q]} \epsilon_j^{\frac{p+1}{p-j+1}}} + \kappa^E_{\text{ARqpIDA}} \left|\log\left(\min_{j\in[1:q]} \epsilon_j\right)\right| + \kappa^F_{\text{ARqpIDA}} = \mathcal{O}\left(\max_{j\in[1:q]} \epsilon_j^{-\frac{p+1}{p-j+1}}\right) \tag{13.1.85}$$

evaluations of the inexact derivatives $\overline{\nabla^i_x f}$ of order i from one to p to produce an iterate x_ϵ such that $\phi^1_{f,j}(x_\epsilon) \leq \epsilon_j/j!$ for all $j \in [1:q]$.

2. *If $q > 2$, then there exist positive constants $\kappa^S_{\text{ARqpIDA,2}}$, $\kappa^A_{\text{ARqpIDA,2}}$, $\kappa^C_{\text{ARqpIDA}}$, $\kappa^E_{\text{ARqpIDA}}$, and $\kappa^F_{\text{ARqpIDA}}$ such that, for any $\epsilon \in (0,1]^q$, the ARqpIDA algorithm requires at most*

$$\#^F_{\text{ARqIDAp,2}} \stackrel{\text{def}}{=} \kappa^A_{\text{ARqpIDA,2}} \frac{f(x_0)-f_{\text{low}}}{\min_{j\in[1:q]} \epsilon_j^{\frac{j(p+1)}{p}}} + \kappa^C_{\text{ARqpIDA}} = \mathcal{O}\left(\max_{j\in[1:q]} \epsilon_j^{-\frac{j(p+1)}{p}}\right) \tag{13.1.86}$$

[91] Once again, the $\frac{1}{2}$ factor is not needed here—it is added to simplify the analysis in Section 13.3.

evaluations of f and at most

$$\begin{aligned}\#^{D}_{\text{ARqlDAp,2}} &\stackrel{\text{def}}{=} \kappa^{S}_{\text{ARqpDA,2}} \frac{f(x_0) - f_{\text{low}}}{\min_{j\in[1:q]} \epsilon_j^{\frac{j(p+1)}{p}}} + \kappa^{E}_{\text{ARqpIDA}} \left|\log\left(\min_{j\in[1:q]} \epsilon_j\right)\right| + \kappa^{F}_{\text{ARqpIDA}} \\ &= \mathcal{O}\left(\max_{j\in[1:q]} \epsilon_j^{-\frac{j(p+1)}{p}}\right)\end{aligned} \tag{13.1.87}$$

evaluations of the inexact derivatives $\overline{\nabla_x^i f}$ of order i from one to p to produce an iterate x_ϵ such that $\phi_{f,j}^{\delta_{\epsilon,j}}(x_\epsilon) \leq \epsilon_j\, \delta_{\epsilon,j}^j/j!$ for some $\delta_\epsilon \in (0,1]^q$ and all $j \in [1:q]$.

Proof. Let k_ϵ be the index of the first iteration for which (12.1.25) occurs with $x_\epsilon = x_{k_\epsilon}$ and $\delta = \delta_{k_\epsilon}$, and let iteration $k \leq k_\epsilon$ be successful. Using the fact that $\{f(x_k)\}$ decreases monotonically and (13.1.83) in Lemma 13.1.18, we deduce that

$$f(x_0) - f(x_{k+1}) \geq \tfrac{1}{2}\eta_1\kappa_{\Delta m}\,|\mathcal{S}_{k-1}| \min_{j\in[1:q]} \epsilon_j^{\pi_j}$$

for $k < k_\epsilon$. We conclude from **f.Bb** and using $|\mathcal{S}_k| \leq |\mathcal{S}_{k-1}| + 1$ that

$$|\mathcal{S}_k| \leq \left(\tfrac{1}{2}\eta_1\kappa_{\Delta m}\right)^{-1} \frac{f(x_0) - f_{\text{low}}}{\min_{j\in[1:q]} \epsilon_j^{\pi_j}} + 1$$

before approximate optimality is reached. We may then invoke Lemma 13.1.9 to deduce that the number of iterations in $\mathcal{T}_k$ is bounded by

$$|\mathcal{T}_k| \leq |\mathcal{S}_k| \left(1 + \frac{|\log\gamma_1|}{\log\gamma_2}\right) + \frac{1}{\log\gamma_2}\log\left(\frac{\sigma_{\max}}{\sigma_0}\right) + 1.$$

Since each iteration in $\mathcal{T}_k$ involves one evaluation of f, the bounds (13.1.84) and (13.1.86) then follow, with the constants

$$\kappa^{s,1}_{\text{ARqpIDA}} \stackrel{\text{def}}{=} \frac{2(p+1)!}{\eta_1\sigma_{\min}} \left(\frac{2j!(L_{f,p} + \sigma_{\max} + 3\kappa_{\nabla,\max}\kappa_s)}{1-\theta}\right)^{\frac{p+1}{p-j+1}}$$

if $q \in \{1,2\}$ and

$$\kappa^{s,2}_{\text{ARqpIDA}} \stackrel{\text{def}}{=} \frac{2(p+1)!}{\eta_1\sigma_{\min}} \left(\frac{2j!(L_{f,p} + \sigma_{\max} + 3\kappa_{\nabla,\max}\kappa_s)}{(1-\theta)\kappa_{\delta,\min}^{j-1}}\right)^{\frac{p+1}{p-j+1}}$$

if $q > 2$, and so

$$\kappa^{A}_{\text{ARqpIDA,1}} = \kappa^{s,1}_{\text{ARqpIDA}} \left(1 + \frac{|\log\gamma_1|}{\log\gamma_2}\right), \quad \kappa^{A}_{\text{ARqpIDA,2}} = \kappa^{s,2}_{\text{ARqpIDA}} \left(1 + \frac{|\log\gamma_1|}{\log\gamma_2}\right), \tag{13.1.88}$$

and

$$\kappa^{C}_{\text{ARqpIDA}} = \frac{1}{\log\gamma_2}\log\left(\frac{\sigma_{\max}}{\sigma_0}\right) + 1. \tag{13.1.89}$$

As in the proof of Theorem 13.1.8, and in order to derive an upper bound on the number of evaluations of the inexact derivatives, we now have to count the number of additional derivative evaluations caused by the need to approximate them to the desired accuracy, and we may "copy-and-paste" the argument there with simple adaptations.

Combining (13.1.78), (13.1.73), and (13.1.77), we see that no further accuracy-improving iterations are required before approximate optimality is reached if

$$h_{k,\max} \leq \min\left[\varsigma\frac{\left(\kappa_\delta^{(1)}(\sigma_{\max})\right)^q}{4q!\kappa_{\nabla,\max}}, \kappa_s\left(\frac{\varsigma\left(\kappa_\delta^{(1)}(\sigma_{\max})\right)^q}{12q!\max[\overline{L}_f,\sigma_{\max}]}\right)^p\right]\epsilon_{\min}^{p(q+1)} \stackrel{\text{def}}{=} \kappa_{\text{acc}}\epsilon_{\min}^{p(q+1)}.$$

Notice that the mechanism of Step 5 ensures that the $\{h_{k,i}\}_{i=1}^q$ decrease linearly by successive multiplications with the factor γ_h, starting from initial values $\{h_{0,i}\}_{i=1}^q$ that are all bounded by κ_h. Thus the number of such multiplications—which is the same as the number of accuracy-improving iterations $|\mathcal{A}_{k_\epsilon}|$—that can happen before achieving approximate optimality is bounded above by the root $i_{h,\epsilon}$ of the equation

$$\gamma_h^{i_{h,\epsilon}}\kappa_h = \kappa_{\text{acc}}\epsilon_{\min}^{p(q+1)},$$

and this is given by

$$\begin{aligned} i_{h,\epsilon} &= \frac{1}{\log(\gamma_h)}\left\{p(q+1)\log(\epsilon_{\min}) + \log\left(\frac{\kappa_{\text{acc}}}{\kappa_h}\right)\right\} \\ &\leq \frac{1}{|\log(\gamma_h)|}\left\{p(q+1)|\log(\epsilon_{\min})| + \left|\log\left(\frac{\kappa_{\text{acc}}}{\kappa_h}\right)\right|\right\}. \end{aligned}$$

Since an evaluation of the inexact derivatives of the objective function may occur at each successful or accuracy-improving iteration, we deduce that the number of these evaluations before achieving approximate optimality is bounded above by the maximum number of successful iterations plus $i_{h,\epsilon}+1$. This yields the required bounds (13.1.85) and (13.1.87) with

$$\kappa_{\text{ARqpIDA}}^{\text{E}} = \frac{p(q+1)}{|\log(\gamma_h)|} \quad \text{and} \quad \kappa_{\text{ARqpIDA}}^{\text{F}} = \frac{1}{|\log(\gamma_h)|}\left|\log\left(\frac{\kappa_{\text{acc}}}{\kappa_h}\right)\right|. \tag{13.1.90}$$

□

We conclude this section with some general comments regarding the IDA framework that apply to both its trust-region and its adaptive-regularization variants. Firstly, inexact evaluations of the value of the objective function might also be allowed in the IDA methodology, under the assumption that (13.1.2) holds additionally for $i = 0$ and that both (13.1.9) and (13.1.46) are also valid for $i = 0$, meaning that the error in f is of the order of the $(q+1)$st or $(p+1)$st power of the norm of the step; other conditions on the accuracy of the function evaluations are also possible and sometimes beneficial, as illustrated in the next section. However, the significant downside is that the constant $\kappa_{\nabla,0}$ that would be present in (13.1.9) and (13.1.46) with $i = 0$ needs to be known a priori, and so this aspect does not fit with the methodological assumptions of the IDA technique, but it does prove a match for the explicit dynamic accuracy framework.

In order to avoid overloading notation and overcomplicating arguments, we have made a couple of simplifying assumptions in the formal description of the IDA algorithms. The first is that, when accuracy is tightened in Step 5, we have stipulated a uniform improvement for all derivatives of orders one to q. A more refined version of these algorithms is obviously possible in which the need to improve accuracy for each derivative is considered separately, which requires sufficient accuracy on each of the approximate derivatives $\{\overline{\nabla_x^i f(x_k)}\}_{j=1}^q$.

A second simplifying feature of these algorithms relates to the insistence that the accuracies $\{h_{k,i}\}_{i=1}^q$ are initialized in Step 0 once and for all. As a consequence, the accuracy requirements can only become more severe as the iteration proceeds. This might well be viewed as inefficient because the true need for accurate derivatives depends more on their values at an individual iterate rather than the evolving set of iterates. Versions of these algorithms for which the accuracies $\{h_{k,i}\}_{i=1}^q$ are reinitialized at every successful iterate are of course possible, at a moderate increase in the overall complexity bound. Indeed, in such a case, the number of "additional" derivative evaluations (in the proofs of Theorems 13.3.4 and 13.1.19) would no longer need to cover all iterations, but only what happens at a single iterate. Thus the logarithmic term in $\epsilon_{\min}$ is no longer added to the number of successful iterations, but multiplies it, and the worst-case evaluation complexity for the modified algorithms becomes $\mathcal{O}(|\log(\epsilon_{\min})|\epsilon_{\min}^{-(q+1)})$ for TRqIDA, and $\mathcal{O}(\epsilon_{\min}^{-(p+1)/(p-q+1)}|\log(\epsilon_{\min})|)$ if q is one or two or $\mathcal{O}(\epsilon_{\min}^{-q(p+1)/p}|\log(\epsilon_{\min})|)$ if $q \geq 3$ for ARqpIDA.

Notes and references for Section 13.1

The material presented in this section is a significant extension of earlier work [93], in which implementable first-order and derivative-free algorithms are described that use finite differences to build local models. The mechanism for monotonically increasing the desired accuracy of this model is essentially the same as that used in the EDA algorithms that we will encounter in Section 13.3. Similar ingredients (but no complexity analysis) have also been used to handle PDE-constrained optimization problems under uncertainty [259].

13.2 Fast convergence using low-degree oracles

We follow our development and analysis of inexact algorithms within the IDA framework with a crucial observation. This concerns the requirements placed upon evaluation oracles within these algorithms. As we have seen in both the TRqIDA and ARqpIDA algorithms, the worst-case evaluation complexity depends on the order of approximate minimization required and the degree of the inexact Taylor series used. The only certain *oracle* requirement is that exact function values are used (and, as we noted, even this may be relaxed); everything else depends on the inexact-derivative scheme employed. A natural question then is what degree of oracles is actually required to ensure the worst-case bounds we have derived?

We start by recalling that high-degree derivatives can be approximated by low-degree ones and giving two well-known examples of schemes that allow such approximations. The first is to use finite differences, where, for instance, gradients and Hessian values may be approximated componentwise by formulae such as

$$[\overline{\nabla_x^1 f(x)}]_i = \frac{f(x+h_1 e_i) - f(x)}{h} = [\nabla_x^1 f(x)]_i + \mathcal{O}(h)$$

or

$$[\overline{\nabla_x^1 f(x)}]_i = \frac{f(x+he_i) - f(x-he_i)}{2h} = [\nabla_x^1 f(x)]_i + \mathcal{O}\left(h^2\right)$$

and

$$[\overline{\nabla_x^2 f(x)}]_{i,j} = \frac{f(x+he_i+he_j) - f(x+he_i) - f(x+he_j) + f(x)}{h^2} = [\nabla_x^2 f(x)]_{i,j} + \mathcal{O}(h)$$

or

$$\overline{[\nabla_x^2 f(x)]_{i,j}} = \frac{[\nabla_x^1 f(x+he_j)]_i - [\nabla_x^1 f(x)]_i}{h} = [\nabla_x^2 f(x)]_{i,j} + \mathcal{O}(h),$$

where $h \in (0,1]$ is a suitably small stepsize. Such formulae can be extended to cover the componentwise approximation of derivative tensors $\nabla_x^i f(x)$ of degree i, using exact derivatives of degree[92] up to o for any $0 \le o < i$. In particular,[93] if $j_k \in [1:n]$ for $k \in [1:p]$,

$$\begin{aligned} \overline{\nabla_x^i f(x)}[e_{j_1},\ldots,e_{j_i}] &= \sum_{j\in\mathcal{S}_i} c_{i,j} \nabla_x^o f(x+s_{i,j})[e_{i_1},\ldots,e_{j_o}] \\ &= \nabla_x^i f(x)[e_{j_1},\ldots,e_{j_i}] + \mathcal{O}\left(h^{p-i+1}\right), \end{aligned} \tag{13.2.1}$$

where the dimension-dependent $\mathcal{S}_i$ are the indices of a finite set of coefficients $c_{i,j}$ and stencil points $s_{i,j} = \sum_{\ell=1}^n h_{i,j,\ell} e_\ell$ and each $h_{i,j,\ell}$ is a (small) integer multiple of h. Thus the approximation error satisfies a bound exactly of the form (13.1.2) for identical $h_i = h$, $i \in [1:p]$, where the constants $\kappa_{\nabla,i}$ directly involve the (unknown) norm of the ith derivative of f—of course, $L_{f,i-1}$ provides such a bound. We refer the reader to [266, Chap. 1] for a useful discussion of finite-difference techniques and their associated error bounds. We should not underestimate the cost involved, particularly if i is much larger than j. Indeed up to $\mathcal{O}(n^{i-j})$ judicious evaluations of derivatives of degree j may be required to estimate those of order up to i, although the cost may be significantly less if the function has sparse or structured derivatives or if evaluations in parallel are possible.

The second technique is multivariate interpolation, in which a polynomial model of degree p is built as a linear combination of multivariate Lagrange polynomials $\{\ell_i(x)\}_{i=0}^d$ of degree p defined on a set $\mathcal{Y} = \{y_0,\ldots,y_d\}$ of interpolation points[94] by the relations

$$\ell_i(y_j) = \delta_{ij}, \quad i,j \in [0:d], \text{ where } d = \sum_{i=0}^p \binom{n+i-1}{i} \tag{13.2.2}$$

is a sum of binomial coefficients—note that the ith-derivative tensor $\nabla_x^i f(x)$ has $\binom{n+i-1}{i}$ independent entries to reproduce because of symmetry [125, Prop. 3.4], and by convention $\binom{n-1}{0} = 1$. The interpolation model is then given by

$$w(x) = \sum_{j=0}^d f(y_j)\ell_j(x). \tag{13.2.3}$$

It is known [118], [132, Sect. 3.2] that, for $i \in [0:p]$,

$$\|\nabla_x^i w(x) - \nabla_x^i f(x)\| \le \frac{L_{f,p}}{(p+1)!} \sum_{j=0}^d \|y_j - x\|^{p+1} \|\nabla_x^i \ell_j(x)\|. \tag{13.2.4}$$

In order to understand why this bound is of the form (13.1.2), we must step back a little. Firstly, assume without loss of generality that $x = 0$ and consider the set $\hat{\mathcal{Y}} \stackrel{\text{def}}{=}$

[92] Recall that, by convention, we refer to function values as derivatives of degree zero.

[93] We were unable to find precisely this result in the literature, but a proof that generalizes one- and two-dimensional arguments [169, 170], [266, Chap. 1] and uses Taylor's theorem seems plausible, if notationally very complicated.

[94] One can choose a different interpolation set for each i, although this is not commonly done.

$\{\hat{y}_0, \ldots, \hat{y}_d\}$ where $\hat{y}_i = \alpha y_i$ and where we choose α so that the smallest ball in $\mathbb{R}^n$ centered at $x = 0$ and containing all the $\hat{y}_0, \ldots, \hat{y}_d$ is the unit sphere $\mathcal{B}(0,1)$—thus at least one of the $\hat{y}_i$ has norm one. Then let the $\hat{\ell}_j$ be the Lagrange polynomials defined by (13.2.2) using the $\hat{y}_i$. As it turns out, the values of these Lagrange polynomials only depend on the geometry of the $\hat{y}_i$ in $\mathcal{B}(0,1)$ and are invariant under scaling and translation [132, Lems. 3.8 and 3.9]; that is, $\ell_j(v) = \hat{\ell}_j(\alpha v)$ for all v. As a consequence, if $\hat{v} = \alpha v$, then $\nabla_x^i \ell_j(v) = \alpha^i \nabla_x^i \hat{\ell}_j(\hat{v})$ for $i \in [0:p]$. If we now let $\delta_{\mathcal{Y}} = 1/\alpha$ be the radius of the smallest ball centered at x and containing $\{y_0, \ldots, y_d\}$, we then deduce that

$$\|\overline{\nabla_x^i f}(x) - \nabla_x^i f(x)\| \leq \frac{L_{f,p}(d+1)}{(p+1)!} \Lambda_i \delta_{\mathcal{Y}}^{p-i+1}$$

from (13.2.4), where

$$\Lambda_i \stackrel{\text{def}}{=} \max_{j \in [0:d]} \max_{\hat{v} \in \mathcal{B}(0,1)} \|\nabla_x^i \hat{\ell}_j(\hat{v})\|$$

is a constant depending only on the distribution of the scaled interpolation points $\hat{y}_i$ in the unit sphere. Thus, if x is given, one can choose the radius $\delta_{\mathcal{Y}}$, a set of points $\hat{\mathcal{Y}} = \{x + \hat{y}_0, \ldots, x + \hat{y}_d\}$ contained in $x + \mathcal{B}(0,1)$—the latter determining the constant Λ_i via the Lagrange polynomials $\hat{\ell}_j(x)$—and the corresponding interpolation set $\mathcal{Y} = \{x + y_0, \ldots, x + y_d\} = \{x + \delta_{\mathcal{Y}} \hat{y}_0, \ldots, x + \delta_{\mathcal{Y}} \hat{y}_d\}$. Noting that $\ell_j(v) = \hat{\ell}_j(v/\delta_{\mathcal{Y}})$ for all j, we can then build the interpolation model (13.2.3) and define $\overline{\nabla_x^i f}(x) = \nabla_x^i w(x)$. The formula (13.2.4) then yields an error of the form (13.1.2), where now

$$\kappa_{\nabla,i} = \frac{L_{f,p}(d+1)}{(p+1)!} \Lambda_i \quad \text{and} \quad h_i = \delta_{\mathcal{Y}}^{p-i+1}.$$

This approach has been widely used in finite-element methods for partial differential equations [118, 259] and has also been instrumental in the development of early "model-based derivative-free" algorithms for low-order optimization [321, 365, 366], themselves the object of much further research (see [132] for an introduction) and extended to structured large problems [121, 122, 123, 319].

In many algorithms for derivative-free optimization (DFO) that follow the multivariate interpolation approach, points at which the objective function's value is known are used to build interpolation sets $\mathcal{Y}$ from which a (typically linear or quadratic) local model $w(x)$ of the objective function is constructed, and convergence is then ensured by deriving and using conditions under which the model is "fully linear" or "fully quadratic."[95] In particular, if Λ_0 is not too large—one then says that the interpolation set is "Λ_0-poised"—then such models can be constructed and an adequate prediction of the function decrease is guaranteed by (13.2.4). Thus, since the number of such function evaluations required to obtain a pth-degree polynomial model is d, *Theorems 13.1.8 and 13.1.19 provide worst-case evaluation-complexity bounds for model-based trust-region or adaptive-regularization DFO algorithms using models of arbitrary degree.*

This discussion has far-reaching consequences. Our comments on applying TRqIDA- or ARqpIDA-like algorithms to the derivative-free case is just a special instance of a more general framework in which derivatives of a given degree i are approximated, admittedly at some additional cost, by suitable evaluations of those of degrees $0 \leq j < i$. Indeed, an

[95] Note that such model conditions are variants of (13.1.2) with $h_{k,i}$ replaced by some $\Delta_k > 0$ representing either a trust-region or a sampling-type radius [113, 132], which can in turn be viewed as a reformulation of the "conditional-model" conditions [131, (9.1.3) and (9.3.2)]. It is then possible to show that if the DFO method is of trust-region type, the additional control (13.1.9) is no longer necessary.

algorithm of this sort that uses finite differences of function or gradient values to build a local quadratic model and that inherits the worst-case complexity of adaptive cubic regularization has been known for some time [93].

The techniques mentioned above illustrate the use of what one may call a *finite* derivative-approximation scheme. In this, one assumes that derivatives of degree $[0:o]$ may be computed from oracles for some $o \geq 0$. Inexact derivatives of degree $[o+1:q]$ are then estimated (in Step 1 of the appropriate algorithm) by evaluating oracles of degree at most o by a finite process that guarantees accuracy bounds of the form (13.1.2).

The use of such schemes clearly blurs the standard distinction between first-, second-, and higher-order algorithms, and we see that the following conclusion holds.

Theorem 13.2.1: Complexity using low-order derivatives

Suppose that **f.D0pL** *and* **f.Bb** *hold, and that derivative-evaluation oracles are available for all degrees up to o for a given $o \in [0:q]$. Then the* TRqIDA *algorithm, using inexact derivatives $\{\overline{\nabla_x^i f(x)}\}_{i=o+1}^q$ in Step 1 obtained by some finite derivative-approximation scheme, will compute an unconstrained qth-order (ϵ, δ)-approximate minimizer of $f(x)$ in at most*

$$\mathcal{O}\left(\min_{j\in[1:q]} \epsilon_j^{-(q+1)}\right) \tag{13.2.5}$$

calls to oracles of degree up to o. If the ARqpIDA *algorithm with inexact derivatives $\{\overline{\nabla_x^i f(x)}\}_{i=o+1}^p$ obtained as above is used instead, the bound becomes*

$$\mathcal{O}\left(\min_{j\in[1:q]} \epsilon_j^{-\frac{p+1}{p-q+1}}\right) \quad \text{for} \quad q \leq 2 \tag{13.2.6}$$

or

$$\mathcal{O}\left(\min_{j\in[1:q]} \epsilon_j^{-\frac{q(p+1)}{p}}\right) \quad \text{for} \quad q > 2. \tag{13.2.7}$$

Proof. We simply apply Theorem 13.1.8 (for TRqIDA) or 13.1.19 (for ARqpIDA) to the particular implementation of the relevant algorithm using a finite derivative-approximation scheme to compute $\{\overline{\nabla_x^i f(x)}\}_{i=o+1}^p$ in Step 1. The conclusion follows from the fact that the accuracy requested by the algorithm on these higher-order derivatives can be obtained by progressively tightening (in Step 5) the accuracy parameters $h_{k,i}$. The mechanism of the algorithm shows that this can be done in $\mathcal{O}\left(|\log(\min_{j\in[1:q]}\epsilon_j)|\right)$ evaluations of $\{\overline{\nabla_x^i f(x)}\}_{i=0}^o$, a number that remains small compared to (13.2.5)–(13.2.7). □

This very general theorem obviously subsumes our comment about DFO methods and demonstrates the remarkable fact that *"complexitywise fast" convergence can be achieved using derivatives of low degree.* Note, however, that this important conclusion comes with an associated caveat, namely, that the constants hidden within the $\mathcal{O}$ notation may get significantly larger when o departs significantly from q *due to dimension dependencies*[96]—and, furthermore, we have already observed that Lipschitz constants may not be immune to dimension. Observe that we have mentioned finite differences and polynomial interpolation as possible approximation schemes, but these also include finite-element methods [118, 259], and this is obviously not exhaustive.

[96] As illustrated by $|\mathcal{S}_i|$ in (13.2.1) and d in (13.2.2).

Notes and references for Section 13.2

The complexity result of [93] in effect states an early version of Theorem 13.2.1 and, to our knowledge, is the earliest proof that faster than $\mathcal{O}(\epsilon^{-2})$ worst-case complexity can be achieved just using function values (and gradients if available)—both first-order and derivative-free variants are mentioned.

Another very appealing suggestion along similar lines was made later by Carmon et al. [75], using the well-known idea that a fast but potentially unreliable algorithm can be made convergent by inserting steps of a converging slower one. More specifically, if we know a very fast algorithm for a class of problems but are uncertain whether the problem at hand lies in that class, we might nevertheless tentatively apply this fast algorithm so long as we find no evidence to the contrary. Should we be wrong, and we determine that the problem does not belong to the class, an iteration of a slower but safer method should be performed, and the process repeated. In this context, every decrease in the objective function obtained by the fast "tentative" method is exploited, reducing the gap to the solution, thereby also improving the worst-case complexity of the slower "safe" algorithm. Applying this idea is possible if two conditions are fulfilled. Firstly, the transition from the tentative algorithm to the safe one (when the former is judged inappropriate) should not involve an amount of work that would offset the advantage of using the tentative one. Secondly, the decision that the tentative algorithm is inappropriate should be implementable with the information available at decision time. In [75], the authors use the class of strongly convex problems and its associated algorithm of optimal complexity (see Section 1.2.5) as "tentative." They then check the rate of decrease in gradient norms to decide if the problem is effectively strongly convex (hence the name "convex until proved guilty" used in their paper) and show that, if this not the case, a direction of negative curvature can be found by analyzing the history of the fast method (at the potential cost of rerunning it). This analysis then shows that our first condition is satisfied and yields an evaluation-complexity bound of $\mathcal{O}(\epsilon_1^{-7/4}|\log(\epsilon_1)|)$ for finding an ϵ_1-approximate first-order point using only function and gradient values (assuming not only a Lipschitz-continuous gradient but also a Lipschitz-continuous Hessian). Interestingly, this bound can be improved to $\mathcal{O}(\epsilon_1^{-5/3}|\log(\epsilon_1)|)$ if one additionally assumes the existence of a third-order Lipschitz-continuous derivative. However, it seems that the estimation of the rate of decrease expected for a convex problem (the key decision factor) needs the knowledge of the gradient's Lipschitz constant, an assumption that we have deliberately avoided in this book. Whether a method using an algorithmically derived approximate value of this constant can be found to maintain this improved complexity remains an open question.

The general Theorem 13.2.1 covers a very wide range of potential algorithms. This ties in with recent developments in the convex-optimization world in which algorithms with low-degree oracles are able to match worst-case complexity bounds normally associated with higher-degree evaluation [304].

The evaluation complexity of derivative-free methods has been considered by several authors besides [93]. In particular, model-based trust-region DFO methods, which are the class of methods most related to those developed in the section above, were analyzed by Garmanjani, Júdice, and Vicente [174] and Roberts [328]. The bounds obtained there require $\mathcal{O}(n^2\epsilon_1^{-2})$ function evaluations to achieve (standard) first-order criticality and $\mathcal{O}(n^5\epsilon_2^{-3})$ evaluations to ensure second-order criticality. Such orders are no better, and sometimes worse, in terms of appropriate ϵ-dependency than the bounds presented in Theorem 13.2.1. An extensive—but not comprehensive—list of best-known worst-case complexity bounds for various classes of DFO methods (including direct-search and probabilistic approaches) can be found in the survey article by Larson, Menickelly, and Wild [262, Tab. A.1]. For example, one finds that the evaluation bound to achieve first-order criticality using function values alone and a model-based DFO trust-region method is $\mathcal{O}(n^2\epsilon_1^{-2})$. By contrast, the approach in the preceding section guarantees a deterministic $\mathcal{O}(n\epsilon_1^{-3/2})$ bound when second derivatives are sparse and estimated using sparse finite-difference formulae [120, 322], and more generally at most $\mathcal{O}(n^2\epsilon_1^{-3/2})$ evaluations are required using methods of the kind typified by Theorem 13.2.1 [93, Thm. 4.7]. Recently, [202] proposed a second-order cubic-regularization algorithmic variant for which the Hessian is approximated by forward finite differences whose accuracy is required to satisfy a step condition based on the *previous* step s_{k-1} and where nonmonotonic iterates are acceptable; the resulting complexity for achieving standard first- and second-order criticality is $\mathcal{O}(n\max[\epsilon_1^{-3/2}, \epsilon_2^{-3}])$, but note that this method uses exact gradient values. A fully derivative-free model-based cubic-regularization variant with separable regularization terms is proposed in [142], with ensuing complexity of $\mathcal{O}(n^4\epsilon_1^{-3/2})$ for achieving approximate first-order standard criticality.

13.3 Inexact algorithms using explicit dynamic accuracy

In Section 13.1.2, we provided results for optimization in the IDA framework. Although these cover important applications, they are somewhat unsatisfactory, primarily because there is no obvious way to recognize that approximate optimality has been achieved. We now investigate the Explicit Dynamic Accuracy (EDA) framework in which this is possible.

In this new framework, we shall still tolerate inexact derivatives $\{\overline{\nabla_x^i f(x)}\}$ of the objective function satisfying (13.1.2), but we now also permit inexact approximations $\bar{f}(x)$ of $f(x)$ itself[97] that achieve the bounds

$$|\bar{f}(x_k) - f(x_k)| \leq h_{k,0} \tag{13.3.1}$$

and

$$|\bar{f}(x_k + s_k) - f(x_k + s_k)| \leq h_{k,0} \tag{13.3.2}$$

for all x_k and all steps s_k of interest, and for some *requested accuracy* $h_{k,0}$. Indeed, in the EDA methodology, *the accuracy constants $\kappa_{\nabla,i}$ in the error bounds (13.1.2) on the derivatives are assumed to be known.* Thus they can be regarded simply as scalings of the accuracies $h_{k,i}$ and subsumed into $h_{k,i}$,[98] and so without loss of generality, we therefore let

$$\kappa_{\nabla,i} = \kappa_{\nabla,\max} \stackrel{\text{def}}{=} 1 \text{ for } i \in [1:j] \text{ and } j \in \{q,p\}. \tag{13.3.3}$$

Thus error bounds of the form (13.3.1), (13.3.2), and

$$\| \overline{\nabla_x^i f(x_k)} - \nabla_x^i f(x_k)\| \leq h_{k,i} \quad (i \in [1:j]) \tag{13.3.4}$$

are now *enforced* when demanded by the algorithm.

Typical examples of problems falling within the EDA framework are those for which the value and derivative of the objective function are obtained by a controllable iterative procedure. An important case of interest, and of growing relevance in high-performance computing, is the emerging field of "variable-accuracy" or "multiprecision" optimization [16,27,214,269], where the computation of the objective function and its derivative(s) is intentionally truncated to significantly fewer digits than would be required for an accurate calculation. Doing so allows the use of specialized processing units whose chip surface, and hence power consumption, is much less than is needed for standard double-precision arithmetic [171]. Another example is when one considers an objective function of the form

$$f(x) = h(Ax) = h(y),$$

[97]Recall our comment in Subsection 13.1.4.2 (p. 351), which briefly explains why inexact function values are difficult/impossible to include in the IDA framework without knowing the constants involved.

[98]This is also why no (known) constant $\kappa_{\nabla,0}$ was introduced into (13.3.1) and (13.3.2).

yielding

$$\nabla_x^1 f(x) = A^T \nabla_y^1 h(y) \text{ and } \nabla_x^2 f(x) = A^T \nabla_y^2 h(y) A,$$

where A is a matrix so large/dense/complicated that the product Ax is performed inexactly but with controllable accuracy. Such situations occur for instance in data assimilation for weather forecasting [61, 134, 216], when A is the inverse of another complicated matrix.

13.3.1 The TRqEDA algorithm and its complexity

In what follows, we shall consider a trust-region optimization algorithm named TRqEDA (the EDA suffix refers to the explicit dynamic accuracy framework), whose purpose is to find a vector $x = x_\epsilon$ which is a qth-order (ϵ, δ)-approximate minimizer for given $\epsilon \in (0, 1]^q$ and some vector of optimality radii $\delta \in (0, 1]^q$. The TRqEDA algorithm is presented below, and again uses approximate global minimization to estimate the (inexact) optimality measure. The algorithm is a slight modification of the TRqIDA one but crucially presumes that (13.3.1), (13.3.2), and (13.3.4) can be enforced explicitly. This knowledge allows us to certify the progress and approximate optimality of candidate iterates—by contrast, as the reader will recall, the TRqIDA algorithm did not effectively have termination criteria. The conditions on the accuracies $h_{k,i}$, $i \in [1 : q]$, will remain the same as in (13.1.9), but a different relative condition will be imposed on the accuracy $h_{k,0}$ required on the approximate function values.

Throughout this section we maintain all the definitions, constants, and notation in Section 13.1.3 unless otherwise specified or redefined.

Algorithm 13.3.1: Trust-region algorithm with explicit dynamic accuracy (TRqEDA)

Apply the TRqIDA algorithm (p. 329) with the following modifications:

- In Step 0 of the TRqIDA algorithm, add to the list of parameters:

Step 0: Set the parameter $\kappa_s > 0$ such that

$$\kappa_s \le \frac{1}{4} \min[\eta_1, 1 - \eta_2], \tag{13.3.5}$$

and exclude the computation of $f(x_0)$.

- Replace Step 1 of the TRqIDA algorithm as follows:

Step 1: Optimality measure computation. Set $\delta_k = \min[\Delta_k, \vartheta]$.
For $j = 1, \ldots, q$:

Step 1.1: Compute $\overline{\nabla_x^j f}(x_k)$ to satisfy (13.3.4) and $d_{k,j} \in \mathcal{B}_{\delta_k}$ so that

$$\varsigma \overline{\phi}_{f,j}^{\delta_k}(x_k) \le \overline{\Delta T}_{f,j}(x_k, d_{k,j}). \tag{13.3.6}$$

Step 1.2: If

$$\overline{\Delta T}_{f,j}(x_k, d_{k,j}) > \tfrac{1}{2} \varsigma \epsilon_j \frac{\delta_k^j}{j!}, \tag{13.3.7}$$

go to Step 2 with the index $j_k = j$ and $d_k = d_{k,j}$.

Step 1.3: Otherwise, consider the next value of j.

Step 1.4: If the loop on j finishes, i.e., (13.3.6) holds but (13.3.7) fails for all $j \in [1:q]$, then complete the following steps:

Step 1.4.1: If

$$\sum_{i=1}^{q} h_{k,i} \frac{\delta_k^i}{i!} \leq \tfrac{1}{2}\varsigma \epsilon_{\min} \frac{\delta_k^q}{q!}, \tag{13.3.8}$$

where $h_{k,i}$ are the derivative accuracies in (13.3.4) and $\epsilon_{\min} \stackrel{\rm def}{=} \min_{i\in[1:q]} \epsilon_i$, then terminate with the qth-order approximate minimizer $x_\epsilon = x_k$ and associated radius $\delta_\epsilon = \delta_k$.

Step 1.4.2: Otherwise, go to Step 5.

- Replace Step 3 of the TRqIDA algorithm as follows:

Step 3: Accept the new iterate. Set

$$0 < h_{k,0} \leq \kappa_s \overline{\Delta T}_{f,j_k}(x_k, s_k), \tag{13.3.9}$$

and compute $\overline{f}(x_k + s_k)$ to ensure (13.3.2). Also ensure, by either setting $\overline{f}(x_k) = \overline{f}(x_{k-1} + s_{k-1})$ $(k \geq 1)$ or recomputing $\overline{f}(x_k)$ $(k \geq 0)$, that (13.3.1) holds with $h_{k,0}$ in (13.3.9). Then compute

$$\rho_k = \frac{\overline{f}(x_k) - \overline{f}(x_k + s_k)}{\overline{\Delta T}_{f,j_k}(x_k, s_k)}. \tag{13.3.10}$$

If $\rho_k \geq \eta_1$, then set $x_{k+1} = x_k + s_k$. Otherwise set $x_{k+1} = x_k$.

In fact Steps 1.1–1.3 and 1.4.2 are the same as for the TRqIDA algorithm, and only Step 1.4.1 is new. Condition (13.3.8) could be replaced with a stronger requirement on $h_{k,\max}$ similar to the second inequality in (13.1.12), and comparable results would ensue.

While Step 1 is intended to secure approximate optimality, the requirement (13.3.5) and the new Step 3 account for inexact function values. Indeed, Steps 1 and 2 do not depend on the value of $\overline{f}(x_k)$ since they only involve $\overline{\Delta T}_{i,j}(x_k, \cdot)$, which is why we do not need to compute $\overline{f}(x_k)$—or $h_{k,0}$—before Step 3. At the start of Step 3 of iteration $k \geq 0$, the value $h_{k,0}$ can be set without recourse to Step 5. Furthermore, when $k > 1$, a new value of $\overline{f}(x_k)$ has to be computed to ensure (13.3.1) in Step 3 only when $\overline{\Delta T}_{f,j_k}(x_{k-1}, s_{k-1}) > \overline{\Delta T}_{f,j_k}(x_k, s_k)$. If this is the case, an (inexact) function value is computed twice, not once, for that iteration. In addition, we again stress that although (13.3.1), (13.3.2), and (13.3.4) might suggest that we need to know the true f and its derivatives, this is not the case; rather we simply need some mechanism to ensure that x_k and $x_k + s_k$ satisfy the required bounds. Of course, Step 3 simplifies if the function values are exact, since then $\overline{f}(x_k) = f(x_k)$ and $\overline{f}(x_k + s_k) = f(x_k + s_k)$ suffice. In this case, the only difference between TRqEDA and TRqIDA is in Step 1, for which the former can validate termination.

We also note that the condition (13.3.9) on the required accuracy of the function values is different from (13.1.9), which is the corresponding required accuracy on the derivatives. A condition of the form (13.1.9) with $i = 0$ might be used instead, but it yields a more restrictive range for κ_s, which depends on $\epsilon_{\min}$, than does (13.3.5).

Though counting the numbers of iteration and derivative evaluations by means of the sizes of the sets (13.1.19), (13.1.20), and (13.1.21) remains valid for the TRqEDA algorithm, we must remember that the objective function is evaluated at least once—and at most twice—for each successful or unsuccessful iteration (i.e., for every $k \in \mathcal{T}$).

Because the TRqEDA algorithm is a minor modification of its TRqIDA predecessor, the majority of the results obtained for the former continue to hold in the current setting with minor modifications. We only need concern ourselves with those involving the objective-function values and the algorithm termination. In particular, a useful consequence when the conditions (13.1.9) are met is that the error bound (13.1.10) in the Taylor decrements continues to hold (with $\kappa_{\nabla,\max} = 1$) since it is unaffected by the inexact function values, while it follows from (13.1.4), (13.3.1), and (13.3.3) that the Taylor-expansion error now becomes

$$|T_{f,j}(x_k, s_k) - \overline{T}_{f,j}(x_k, s_k)| \leq h_{k,0} + 2\kappa_s \|s_k\|^{q+1}. \tag{13.3.11}$$

The direction of development parallels that in Section 13.1.3.2 for the TRqIDA algorithm; indeed some results are the same. Lemma 13.1.2 may be used without change because (13.1.14) in TRqIDA is (13.3.7) in TRqEDA. However, a new variant of Lemma 13.1.3 is required as follows.

Lemma 13.3.1: Condition for success (TRqEDA)

Suppose that **f.D0qL** *holds, that L_f is given by (13.1.23), and that $\widehat{\phi}_{f,k}$ is defined by (13.1.25). Suppose also that (13.1.28) and (13.3.3) hold at iteration $k \in \mathcal{T}$ of the* TRqEDA *algorithm. Then $\rho_k \geq \eta_2$, iteration k is very successful, and $\Delta_{k+1} \geq \Delta_k$.*

Proof. We first note that (13.1.28) trivially implies that

$$\Delta_k \leq \min\left\{\vartheta, \frac{(j_k+1)(1-\eta_2)}{2L_f + 4(j_k+1)!\,\kappa_s}\,\widehat{\phi}_{f,k}\right\} \tag{13.3.12}$$

for $k \in \mathcal{T}$, and that $\delta_k = \min[\vartheta, \Delta_k] = \Delta_k$. Moreover, $\delta_k \leq \vartheta \leq 1$. Then we may use (13.3.10), the fact that $\overline{\Delta T}_{f,j_k}(x_k, s_k)$ is independent of the value of $\overline{T}(x_k, 0)$, and the triangle inequality to deduce that

$$\begin{aligned}
|\rho_k - 1| &= \frac{|\overline{f}(x_k+s_k) - \overline{T}_{f,j_k}(x_k,s_k)|}{\overline{\Delta T}_{f,j_k}(x_k,s_k)} \\
&\leq \frac{1}{\overline{\Delta T}_{f,j_k}(x_k,s_k)}\Big(|\overline{f}(x_k+s_k) - f(x_k+s_k)| + |f(x_k+s_k) - T_{f,j_k}(x_k,s_k)| \\
&\qquad + |\Delta T_{f,j_k}(x_k,s_k) - \overline{\Delta T}_{f,j_k}(x_k,s_k)| + |f(x_k) - \overline{f}(x_k)|\Big).
\end{aligned}$$

Invoking (A.19) in Corollary A.8.4 (p. 437), the bounds $\|s_k\| \leq \Delta_k = \delta_k$, (13.1.23), (13.1.11), (13.3.11), (13.3.9), (13.3.1), (13.3.2), and (13.3.12), we deduce that

$$\begin{aligned}
|\rho_k - 1| &\leq \frac{1}{\overline{\Delta T}_{f,j_k}(x_k,s_k)}\left(\frac{L_f\delta_k^{j_k+1}}{(j_k+1)!} + 2\kappa_s\delta_k^{q+1} + 2\kappa_s\overline{\Delta T}_{f,j_k}(x_k,s_k)\right) \\
&\leq \frac{\delta_k^{j_k+1}}{\overline{\Delta T}_{f,j_k}(x_k,s_k)}\left(\frac{L_f}{(j_k+1)!} + 2\kappa_s\right) + 2\kappa_s \\
&\leq \frac{\delta_k}{\widehat{\phi}_{f,k}}\left(\frac{L_f}{j_k+1} + 2j_k!\kappa_s\right) + 2\kappa_s \leq 1 - \eta_2,
\end{aligned}$$

and thus that $\rho_k \geq \eta_2$, where in the penultimate inequality we used (13.1.27) and in the last inequality we used the choice $\kappa_s \leq \frac{1}{4}(1-\eta_2)$. Then iteration k is very successful and (13.1.17) then yields that $\Delta_{k+1} \geq \Delta_k$. □

Lemma 13.1.4—with a simplified value of κ_δ resulting from (13.3.3)—follows directly because Lemma 13.1.3 ensures that (13.1.28) holds once again. This leads to a refined version of Lemma 13.1.5, namely, of (13.1.35), which now depends on inexact function values.

Lemma 13.3.2: Model decrease (TRqEDA)

Suppose that **f.D0qL** *holds. Then the* TRqEDA *algorithm ensures that (13.1.34), holds for every iteration* $k \in \mathcal{T}$ *of algorithm* TRqEDA *before termination, where* κ_δ *is defined as in (13.1.33). If, in addition,* $k \in \mathcal{S}$*, then (13.1.35) holds.*

Proof. The proof of (13.1.34) is identical to that for TRqIDA when (13.3.3) holds.

To prove (13.1.35) here, let $k \in \mathcal{S}$. Then (13.3.9), (13.3.1), and (13.3.2) imply

$$\begin{aligned} f(x_k) - f(x_{k+1}) &= f(x_k) - \bar{f}(x_k) + \bar{f}(x_k) - \bar{f}(x_k+s_k) + \bar{f}(x_k+s_k) - f(x_k+s_k) \\ &\geq -2\kappa_s \overline{\Delta T}_{f,j_k}(x_k,s_k) + \bar{f}(x_k) - \bar{f}(x_k+s_k) \\ &\geq [-2\kappa_s + \eta_1]\, \overline{\Delta T}_{f,j_k}(x_k,s_k) \\ &\geq \tfrac{1}{2}\eta_1 \overline{\Delta T}_{f,j_k}(x_k,s_k), \end{aligned}$$

where in the last inequality we used $\kappa_s \leq \frac{1}{4}\eta_1$. Now (13.1.35) follows from (13.1.34). □

Lemma 13.1.6 also holds since it only concerns the unchanged Step 2. But Step 1 now has an extra, fortunately possible, outcome, and we may replace Lemma 13.1.7 by the following.

Lemma 13.3.3: The outcomes of Step 1 of the TRqEDA algorithm

(i) *Suppose that (13.3.8) holds. Then* x_k *is an* (ϵ, δ_k)*-approximate* q*th-order minimizer.*

(ii) *Additionally, suppose that* **f.D0qL** *holds. If (13.1.37) holds with the choice (13.3.3), then the* TRqEDA *algorithm terminates in Step 1.4.1 rather than branching to Step 5 via Step 1.4.2.*

Proof. (i) From (13.3.8), (13.3.3), and the definition of $\epsilon_{\min}$, we obtain that for each $j \in [1:q]$,

$$\sum_{i=1}^{j} h_{k,i}\frac{\delta_k^i}{i!} \leq \sum_{i=1}^{q} h_{k,i}\frac{\delta_k^i}{i!} \leq \tfrac{1}{2}\varsigma\epsilon_{\min}\frac{\delta_k^q}{q!} \leq \tfrac{1}{2}\varsigma\epsilon_j\frac{\delta_k^j}{j!}.$$

Combining this with (13.1.3) when $s = d$, (13.3.6), and the failure of (13.3.7), we deduce that

$$\begin{aligned}
\varsigma\phi_{f,j}^{\delta_k}(x_k) &\leq \varsigma \max_{\|d\|\leq\delta_k} \left(\overline{\Delta T}_{f,j}(x_k,d) + |\overline{\Delta T}_{f,j}(x_k,d) - \Delta T_{f,j}(x_k,d)|\right)\\
&\leq \varsigma\overline{\phi}_{f,j}^{\delta_k}(x_k) + \varsigma\sum_{i=1}^{j} h_{k,i}\frac{\|d_{k,j}\|^i}{i!}\\
&\leq \varsigma\overline{\phi}_{f,j}^{\delta_k}(x_k) + \varsigma\sum_{i=1}^{j} h_{k,i}\frac{\delta_k^i}{i!}\\
&\leq \overline{\Delta T}_{f,j}(x_k,d_{k,j}) + \tfrac{1}{2}\varsigma\epsilon_j\frac{\delta_k^j}{j!}\\
&\leq \varsigma\epsilon_j\frac{\delta_k^j}{j!}
\end{aligned}$$

for all $j \in [1:q]$. It then follows immediately from (12.1.25) that x_k is an (ϵ,δ_k)-approximate qth-order minimizer.

(ii) The definition of $h_{k,\max}$, (13.3.3), and (A.34) provide the first inequality below,

$$\sum_{i=1}^{q} h_{k,i}\frac{\delta^i}{i!} \leq 2h_{k,\max} \leq 2\frac{\varsigma^{q+1}\kappa_\delta^q}{4q!}\epsilon_{\min}^{q+1} \leq \frac{\varsigma^{q+1}\kappa_\delta^q}{2q!}\epsilon_{\min}\frac{\delta_k^q}{(\varsigma\kappa_\delta)^q},$$

while the second inequality follows from (13.1.37), and the third from (13.1.32). Thus the condition (13.3.8) for termination in Step 1.4.1 follows. □

Finally, just as in Section 13.1.3.2, this is then enough to give the desired evaluation-complexity result for the TRqEDA algorithm.

Theorem 13.3.4: qth-order complexity for the TRqEDA algorithm

Suppose that **f.D0qL** *and* **f.Bb** *hold. Then there exist positive constants $\kappa_{\text{TRqEDA}}^{A}$, $\kappa_{\text{TRqEDA}}^{B}$, $\kappa_{\text{TRqEDA}}^{C}$, $\kappa_{\text{TRqEDA}}^{E}$, and $\kappa_{\text{TRqEDA}}^{F}$ such that, for any $\epsilon \in (0,1]^q$, the TRqEDA algorithm (p. 358) requires at most*

$$\begin{aligned}
\#_{\text{TrqEDA}}^{F} &\stackrel{\text{def}}{=} \kappa_{\text{TRqEDA}}^{A}\frac{f(x_0)-f_{\text{low}}}{\min_{j\in[1:q]}\epsilon_j^{q+1}} + \kappa_{\text{TRqEDA}}^{B}\left|\log\left(\min_{j\in[1:q]}\epsilon_j\right)\right| + \kappa_{\text{TRqEDA}}^{C}\\
&= \mathcal{O}\left(\max_{j\in[1:q]}\epsilon_j^{-(q+1)}\right)
\end{aligned}$$

evaluations of $\overline{f}$ and at most

$$\begin{aligned}
\#_{\text{TRqEDA}}^{D} &\stackrel{\text{def}}{=} \kappa_{\text{TRqEDA}}^{A}\frac{f(x_0)-f_{\text{low}}}{\min_{j\in[1:q]}\epsilon_j^{q+1}} + \kappa_{\text{TRqEDA}}^{E}\left|\log\left(\min_{j\in[1:q]}\epsilon_j\right)\right| + \kappa_{\text{TRqEDA}}^{F}\\
&= \mathcal{O}\left(\max_{j\in[1:q]}\epsilon_j^{-(q+1)}\right)
\end{aligned}$$

evaluations of the inexact derivatives $\overline{\nabla_x^i f}$ of order i from one to q to terminate with an iterate x_ϵ and an optimality radius $\delta_\epsilon \in (0,1]$ such that $\phi_{f,q}^{\delta_\epsilon}(x_\epsilon) \leq \epsilon_j\delta_\epsilon^j/j!$ for all $j \in [1:q]$.

Proof. The proof is identical to that of Theorem 13.1.8 (p. 335) with the choice (13.3.3), and the constants remain unchanged, apart from a multiplicative factor of 2 for the total function-evaluation count (13.1.41) (which is the same as the total number of successful and unsuccessful iterations). This doubling is due to the potential additional (re)evaluation of $\overline{f}(x_k)$ in Step 3. In particular, $\kappa^{\text{A}}_{\text{TRqEDA}} \stackrel{\text{def}}{=} 2\kappa^{\text{A}}_{\text{TRqIDA}}$ as defined in (13.1.40) and (13.1.42), $\kappa^{\text{B}}_{\text{TRqEDA}} \stackrel{\text{def}}{=} 2\kappa^{\text{B}}_{\text{TRqIDA}}$ in (13.1.42), $\kappa^{\text{C}}_{\text{TRqEDA}} \stackrel{\text{def}}{=} 2\kappa^{\text{C}}_{\text{TRqIDA}}$ in (13.1.43), and $\kappa^{\text{E}}_{\text{TRqEDA}} \stackrel{\text{def}}{=} \kappa^{\text{E}}_{\text{TRqIDA}}$ and $\kappa^{\text{F}}_{\text{TRqEDA}} \stackrel{\text{def}}{=} \kappa^{\text{F}}_{\text{TRqIDA}}$ are given in (13.1.44). □

Crucially, although there are differences in the constants involved, the ϵ-order in the worst-case bound given in Theorem 13.3.4 for the TRqEDA algorithm is no worse than that in Theorem 12.2.5 (p. 297) for the TRq algorithm. Thus we see once more that using approximate rather than exact derivatives has no significant adverse effect, so long as the approximations are carefully controlled. It also follows from Theorem 12.2.6 (p. 299) that the evaluation bounds of Theorem 13.3.4 are sharp in order.

13.3.2 The ARqpEDA algorithm and its complexity

Confidently armed with a trust-region algorithm that uses EDA to control inexact function and derivative values, we now follow our usual pattern and examine a regularization-method counterpart. The setup required is a reassuringly familiar mixture of ingredients we have seen in the ARqpIDA and TRqEDA algorithms in previous sections. We apply our by now usual approximate global-minimization technique when needed and use the regularized, inexact Taylor-series model defined in (13.1.45), which we approximately minimize. Success or failure is assessed by comparing the reduction in the Taylor series this gives compared to the inexact function value at the resulting trial point. Complications arise since optimality can only be assessed using inexact problem data and because we need the inexact function values and derivatives to maintain appropriate coherence with their unknown true values. As with ARqpIDA, this makes both the algorithm and its analysis more involved.

Without further ado, here is the detailed algorithm, which is, as to be expected, a modification of ARqpIDA. Throughout this section we maintain all the definitions, constants, and notations in Section 13.1.4 unless otherwise specified or redefined. We also assume that (13.3.3) holds.

Algorithm 13.3.2: Adaptive-regularization algorithm with explicit dynamic accuracy (ARqpEDA)

Apply Algorithm 13.1.2 (ARqpIDA (p. 339)) with the following modifications:

- In Step 0 of the ARqpIDA algorithm, add the following to the list of parameters:

Step 0: Choose the parameter $\kappa_s > 0$ such that

$$\kappa_s \le \frac{1}{4}\min[\eta_1, 1-\eta_2], \tag{13.3.13}$$

and exclude the computation of $f(x_0)$.

- Replace Step 1 of the ARqpIDA algorithm as follows:

Step 1: Compute the optimality measures. Set $\delta_k^{(0)} = \delta_k$, and recompute (if necessary) $\{\overline{\nabla_x^i f(x_k)}\}_{i=1}^p$ to satisfy (13.1.2).

For $j = 1, \dots, q$:

Step 1.1: Compute $d_{k,j} \in \mathcal{B}_{\delta_{k,j}}$ such that

$$\varsigma \overline{\phi}_{f,j}^{\delta_{k,j}}(x_k) \le \overline{\Delta T}_{f,j}(x_k, d_{k,j}). \tag{13.3.14}$$

Step 1.2: If

$$\overline{\Delta T}_{f,j}(x_k, d_{k,j}) \le \tfrac{1}{2}\varsigma\epsilon_j \frac{\delta_{k,j}^j}{j!}, \tag{13.3.15}$$

consider the next value of j.

Step 1.3: Otherwise, if

$$\Delta \overline{m}_k(d_{k,j}) \ge \tfrac{1}{4}\varsigma\epsilon_j \frac{\delta_{k,j}^j}{j!}, \tag{13.3.16}$$

go to Step 2 with the index $j_k = j$, and set $\delta_k^{(1)} = \delta_k$ and $d_k = d_{k,j}$.

Step 1.4: Otherwise (i.e., (13.3.16) fails) set $\delta_{k,j} = \frac{1}{2}\delta_{k,j}$ and return to Step 1.1.

Step 1.5: If the loop on j finishes, i.e., (13.3.14) and (13.3.15) hold for all $j \in [1:q]$, then do the following:

Step 1.5.1: If, for each $j \in [1:q]$,

$$\sum_{i=1}^{j} h_{k,i} \frac{\delta_{k,j}^i}{i!} \le \tfrac{1}{2}\varsigma\epsilon_j \frac{\delta_{k,j}^j}{j!}, \tag{13.3.17}$$

terminate with the qth-order approximate minimizer $x_\epsilon = x_k$ and associated radius $\delta_\epsilon = \delta_k$.

Step 1.5.2: Otherwise, go to Step 5.

- Replace Step 3 of the ARqpIDA algorithm as follows:

Step 3: Accept the new iterate. Set

$$0 < h_{k,0} \le \kappa_s \overline{\Delta T}_{f,p}(x_k, s_k), \tag{13.3.18}$$

and compute $\overline{f}(x_k + s_k)$ to ensure (13.3.2). Also ensure, either by setting $\overline{f}(x_k) = \overline{f}(x_{k-1} + s_{k-1})$ $(k \ge 1)$ or recomputing $\overline{f}(x_k)$ $(k \ge 0)$, that (13.3.1) holds with $h_{k,0}$ in (13.3.18). Then compute

$$\rho_k = \frac{\overline{f}(x_k) - \overline{f}(x_k + s_k)}{\overline{\Delta T}_{f,p}(x_k, s_k)}. \tag{13.3.19}$$

If $\rho_k \ge \eta_1$, set $x_{k+1} = x_k + s_k$, and define $\delta_{k+1} = \delta_{s_k}$ if $\|s_k\| < 1$ or $\delta_{k+1} = \delta_k^{(1)}$ otherwise. If $\rho_k < \eta_1$, set $x_{k+1} = x_k$ and $\delta_{k+1} = \delta_k^{(1)}$.

The comments we made concerning the construction of the TRqEDA algorithm apply equally here. In summary, Step 1 is intended to secure approximate optimality by using the requirement (13.3.13) and the new Step 3 to take account of inexact function values.

No exact function and derivative values are required, merely approximations that satisfy (13.3.1), (13.3.2), and (13.3.4). The values of $\overline{f}(x_k)$ and $h_{k,0}$ are not needed until Step 3, and the latter may be computed without visiting Step 5. Step 3 simplifies if the function values are exact, and then the only difference between ARqpEDA and ARqpIDA is that the former can then validate termination in Step 1. Otherwise, a new value of $\overline{f}(x_k)$ needs be computed to ensure (13.3.1) with $h_{k,0}$ in (13.3.18) in Step 3 only when $\overline{\Delta T}_{f,p}(x_{k-1}, s_{k-1}) > \overline{\Delta T}_{f,p}(x_k, s_k)$, and in this case two inexact function values are computed for such iterations. The condition (13.3.18) on the requested function-value accuracy closely resembles the corresponding (13.3.9) for the TRqEDA algorithm, and once again a variant of (13.1.46) with $i = 0$ would be possible.[99] Once again, counting the numbers of iteration and derivative evaluations by means of the sizes of the sets (13.1.19), (13.1.20), and (13.1.21) is still valid for the ARqpEDA algorithm, and we must remember that the objective function is evaluated at least once—and at most twice—for each successful or unsuccessful iteration (i.e., for every $k \in \mathcal{T}$).

If conditions (13.1.46) are met for $i \in [1 : p]$ with the choice (13.3.3), the error bound in the Taylor decrements (13.1.3) for $j = p$ is unaffected by inexact function values and becomes

$$|\overline{\Delta T}_{f,p}(x_k, s_k) - \Delta T_{f,p}(x_k, s_k)| \le 2\kappa_s \|s_k\|^{p+1}.$$

Moreover, (13.1.4) and (13.3.1) imply that the Taylor-expansion error now is

$$|T_{f,p}(x_k, s_k) - \overline{T}_{f,p}(x_k, s_k)| \le h_{k,0} + 2\kappa_s \|s_k\|^{p+1},$$

and furthermore

$$|T_{f,p}(x_k, s_k) - \overline{T}_{f,p}(x_k, s_k)| \le \kappa_s \overline{\Delta T}_{f,p}(x_k, s_k) + 2\kappa_s \|s_k\|^{p+1} \tag{13.3.20}$$

whenever (13.3.18) also holds.

Since the ARqpEDA algorithm is a minor modification of the ARqpIDA one, the majority of the results from Section 13.1.4 continue to hold here precisely as given. Again, we need only concern ourselves with those involving the objective values and the algorithm termination. We proceed by mirroring the results in Section 13.1.4. The outcome of Step 1 in Lemma 13.1.11 remains true since Steps 1.1–1.4 of the ARqpIDA algorithm coincide with the corresponding ones in its counterpart ARqpEDA. Similarly, since Step 2 is the same in the ARqpEDA and ARqpIDA algorithms, Lemma 13.1.12 is satisfied here too.

Considering in detail the results from Section 13.1.4.2, Lemma 13.1.13 is unchanged since the model-minimization conditions in Step 2 of the two algorithms are identical. However, a new version of Lemma 13.1.14 that shows that the regularization parameter stays bounded is required because the (inexact) function values play a role in deciding the success or otherwise of an iteration, and this in turn impacts the change in the value of the regularization parameter.

Lemma 13.3.5: Condition for success (ARqpEDA)

Suppose that **f.D0pL** *holds. Then (13.1.77) with the choice (13.3.3) holds for all k.*

[99] An analysis similar to the one we shall shortly provide indicates that the main conclusions hold, provided that κ_s is smaller than a suitable multiple of $\sigma_{\min}$.

Proof. Suppose that $k \in \mathcal{T}$ and that

$$\sigma_k \geq \frac{2L_{f,p} + 4(p+1)!\,\kappa_s}{1-\eta_2}. \tag{13.3.21}$$

Then we may use (13.3.19), the fact that $\overline{\Delta T}_{f,p}(x_k, s_k)$ is independent of the value of $\overline{T}_{f,p}(x_k, 0)$, and the triangle inequality to deduce that

$$\begin{aligned}|\rho_k - 1| &= \frac{|\bar{f}(x_k+s_k) - \overline{T}_{f,p}(x_k,s_k)|}{\overline{\Delta T}_{f,p}(x_k,s_k)}\\ &\leq \frac{1}{\overline{\Delta T}_{f,p}(x_k,s_k)}\Big(|\bar{f}(x_k+s_k) - f(x_k+s_k)| + |f(x_k+s_k) - T_{f,p}(x_k,s_k)|\\ &\qquad\qquad + |T_{f,p}(x_k,s_k) - \overline{T}_{f,p}(x_k,s_k)|\Big).\end{aligned}$$

Invoking (A.19) in Corollary A.8.4 (p. 437), (13.3.2), (13.3.18), and (13.3.20), we have

$$\begin{aligned}|\rho_k - 1| &\leq \frac{1}{\overline{\Delta T}_{f,p}(x_k,s_k)}\left(\frac{L_{f,p}}{(p+1)!}\|s_k\|^{p+1} + 2\kappa_s\|s_k\|^{p+1} + 2\kappa_s\overline{\Delta T}_{f,p}(x_k,s_k)\right)\\ &\leq 2\kappa_s + \left(2\kappa_s + \frac{L_{f,p}}{(p+1)!}\right)\frac{\|s_k\|^{p+1}}{\overline{\Delta T}_{f,p}(x_k,s_k)}\\ &\leq 2\kappa_s + \left(2\kappa_s + \frac{L_{f,p}}{(p+1)!}\right)\frac{(p+1)!}{\sigma_k}\\ &\leq \tfrac{1}{2}(1-\eta_2) + \tfrac{1}{2}(1-\eta_2) = 1-\eta_2,\end{aligned}$$

where we also used (13.1.76) to derive the third inequality and (13.3.21) and (13.3.13) to derive the fourth one. Thus iteration k is successful, and $\sigma_{k+1} \leq \sigma_k$. The conclusion then follows from the mechanism of Step 4 (for $k \in \mathcal{T}$) and the fact that $\sigma_{k+1} = \sigma_k$ for $k \in \mathcal{A}$. □

The next lemma addresses the optimality check in Step 1 and also shows that the loop between Steps 1.5.2 and Step 5 is finite. While Lemma 13.1.15 shows that the possibly infinite loop between Step 1 and Step 5 in the ARqpIDA algorithm brings us ever closer to optimality, here we show instead the much needed guarantee that the ARqpEDA algorithm terminates finitely, despite using inexact function values and derivatives.

Lemma 13.3.6: The outcome of Step 1.5 of the ARqpEDA algorithm

(i) *Suppose that (13.3.17) holds. Then x_k is an (ϵ, δ_k)-approximate qth-order minimizer.*

(ii) *Additionally, suppose that* ***f.D0pL*** *holds. If (13.1.78) holds with the choice (13.3.3), $\sigma_{\max}$ as defined in (13.1.77), and $\kappa_\delta^{(1)}(\sigma)$ as in (13.1.72), then the ARqpEDA algorithm terminates in Step 1.5.1 rather than branching to Step 5 via Step 1.5.2.*

Proof. (i) Let $j \in [1:q]$. Combining (13.3.17) with (13.1.3) when $s = d$, (13.3.14), and (13.3.15), we deduce that

$$\begin{aligned}
\varsigma\phi_{f,j}^{\delta_{k,j}}(x_k) &\leq \varsigma \max_{\|d\|\leq\delta_{k,j}} \left(\overline{\Delta T}_{f,j}(x_k,d) + |\overline{\Delta T}_{f,j}(x_k,d) - \Delta T_{f,j}(x_k,d)|\right) \\
&\leq \varsigma\overline{\phi}_{f,j}^{\delta_{k,j}}(x_k) + \varsigma\sum_{i=1}^{j} h_{k,i}\frac{\|d_{k,j}\|^i}{i!} \\
&\leq \varsigma\overline{\phi}_{f,j}^{\delta_{k,j}}(x_k) + \varsigma\sum_{i=1}^{j} h_{k,i}\frac{\delta_{k,j}^i}{i!} \\
&\leq \overline{\Delta T}_{f,j}(x_k,d_{k,j}) + \tfrac{1}{2}\varsigma\epsilon_j\frac{\delta_{k,j}^j}{j!} \\
&\leq \varsigma\epsilon_j\frac{\delta_{k,j}^j}{j!}
\end{aligned}$$

for all $j \in [1:q]$. It then follows immediately from (12.1.25) that x_k is an (ϵ, δ_k)-approximate qth-order minimizer.

(ii) Let $j \in [1:q]$. The definition of $h_{k,\max}$ and (A.34) provide the first inequality below,

$$\begin{aligned}
\textstyle\sum_{i=1}^{j} h_{k,i}\frac{\delta_{k,j}^i}{i!} &\leq 2h_{k,\max} \leq 2\frac{\varsigma\left(\kappa_\delta^{(1)}(\sigma_{\max})\right)^q}{4q!}\epsilon_{\min}^{q+1} \\
&\leq \varsigma\frac{\left(\kappa_\delta^{(1)}(\sigma_{\max})\right)^q}{2q!}\epsilon_{\min}^{q+1} \leq \varsigma\frac{\left(\kappa_\delta^{(1)}(\sigma_{\max,k})\epsilon_j\right)^j}{2j!}\epsilon_j \leq \varsigma\frac{\delta_{k,j}^j}{2}\epsilon_j,
\end{aligned}$$

where we also used (13.1.78), (13.1.71), and (13.1.77). Thus the condition (13.3.17) for termination in Step 1.5.1 follows. □

As we are so used to by now, a crucial ingredient for obtaining complexity bounds is the derivation of a lower bound on the length of the step. Since the trial step is calculated in Step 2, which is identical for the ARqpEDA and ARqpIDA algorithms, the results that provide this lower bound, namely, Lemmas 13.1.16 and 13.1.17 for the ARqpIDA algorithm, continue to hold for ARqpEDA, and so we do not repeat them here.

The last issue remaining before the complexity bounds is to evaluate the model and function decreases in ARqpEDA, which need to be recalculated.

Lemma 13.3.7: Model and function decreases (ARqpEDA)

Suppose that **f.D0pL** *holds. Then the model decrease (13.1.82) holds, with the choice (13.3.3), for every iteration $k \in \mathcal{T}$ of the* ARqpEDA *algorithm before termination. If, in addition, $k \in \mathcal{S}$, then (13.1.83) holds as well.*

Proof. The proof of (13.1.82) is identical to that for the ARqpIDA algorithm in Lemma 13.1.18 (p. 348) by making the choice (13.3.3). To show (13.1.83), let $k \in \mathcal{S}$. Then (13.3.18), (13.3.1), and (13.3.2) imply

$$\begin{aligned}
f(x_k) - f(x_{k+1}) &= f(x_k) - \overline{f}(x_k) + \overline{f}(x_k) - \overline{f}(x_k+s_k) + \overline{f}(x_k+s_k) - f(x_k+s_k) \\
&\geq -2\kappa_s\overline{\Delta T}_{f,p}(x_k,s_k) + \overline{f}(x_k) - \overline{f}(x_k+s_k) \\
&\geq [-2\kappa_s + \eta_1]\,\overline{\Delta T}_{f,p}(x_k,s_k) \\
&\geq \tfrac{1}{2}\eta_1\overline{\Delta T}_{f,p}(x_k,s_k),
\end{aligned}$$

where in the last inequality we used (13.3.13) and (13.3.19). Now (13.1.83) follows from the lower bound (13.1.82) on $\overline{\Delta T}_{f,p}(x_k, s_k)$ derived in Lemma 13.1.18 (p. 348). □

We now have all of the ingredients in place to deduce a complexity bound for the ARqEDA algorithm.

Theorem 13.3.8: qth-order complexity for the ARqpEDA algorithm

Suppose that **f.D0pL** *and* **f.Bb** *hold. Suppose also that the algorithm ensures that (13.1.69) and (13.1.70) hold.*

1. *If $q \in \{1,2\}$, then there exist positive constants $\kappa^S_{\rm ARqpEDA,1}$, $\kappa^A_{\rm ARqpEDA,1}$, $\kappa^C_{\rm ARqpEDA}$, $\kappa^E_{\rm ARqpEDA}$, and $\kappa^F_{\rm ARqpEDA}$ such that, for any $\epsilon \in (0,1]^q$, the ARqpEDA algorithm requires at most*

$$\begin{aligned}\#^F_{\rm ARqpEDA,1} &\stackrel{\rm def}{=} \kappa^A_{\rm ARqpEDA,1} \frac{f(x_0) - f_{\rm low}}{\min_{j\in[1:q]} \epsilon_j^{\frac{p+1}{p-j+1}}} + \kappa^C_{\rm ARqpEDA}\\ &= \mathcal{O}\left(\max_{j\in[1:q]} \epsilon_j^{-\frac{p+1}{p-j+1}}\right)\end{aligned}$$

evaluations of $\overline{f}$ and at most

$$\begin{aligned}\#^D_{\rm ARqpEDA,1} &\stackrel{\rm def}{=} \kappa^S_{\rm ARqpEDA,1} \frac{f(x_0) - f_{\rm low}}{\min_{j\in[1:q]} \epsilon_j^{\frac{p+1}{p-j+1}}} + \kappa^E_{\rm ARqpEDA} \left|\log\left(\min_{j\in[1:q]} \epsilon_j\right)\right| + \kappa^F_{\rm ARqpEDA}\\ &= \mathcal{O}\left(\max_{j\in[1:q]} \epsilon_j^{-\frac{p+1}{p-j+1}}\right)\end{aligned}$$

evaluations of the inexact derivatives $\overline{\nabla^i_x f}$ of order i from one to p to terminate with an iterate x_ϵ such that $\phi^1_{f,j}(x_\epsilon) \le \epsilon_j/j!$ for all $j \in [1:q]$.

2. *If $q > 2$, then there exist positive constants $\kappa^S_{\rm ARqpEDA,2}$, $\kappa^A_{\rm ARqpEDA,2}$, $\kappa^C_{\rm ARqpEDA}$, $\kappa^E_{\rm ARqpEDA}$, and $\kappa^F_{\rm ARqpEDA}$ such that, for any $\epsilon \in (0,1]^q$, the ARqpEDA algorithm requires at most*

$$\begin{aligned}\#^F_{\rm ARqpEDA,2} &\stackrel{\rm def}{=} \kappa^A_{\rm ARqpEDA,2} \frac{f(x_0) - f_{\rm low}}{\min_{j\in[1:q]} \epsilon_j^{\frac{j(p+1)}{p}}} + \kappa^C_{\rm ARqpEDA}\\ &= \mathcal{O}\left(\max_{j\in[1:q]} \epsilon_j^{-\frac{j(p+1)}{p}}\right)\end{aligned}$$

evaluations of $\overline{f}$ and at most

$$\begin{aligned}\#^D_{\rm ARpqEDA,2} &\stackrel{\rm def}{=} \kappa^S_{\rm ARqpEDA,2} \frac{f(x_0) - f_{\rm low}}{\min_{j\in[1:q]} \epsilon_j^{\frac{j(p+1)}{p}}} + \kappa^E_{\rm ARqpEDA} \left|\log\left(\min_{j\in[1:q]} \epsilon_j\right)\right| + \kappa^F_{\rm ARqpEDA}\\ &= \mathcal{O}\left(\max_{j\in[1:q]} \epsilon_j^{-\frac{j(p+1)}{p}}\right)\end{aligned}$$

evaluations of the inexact derivatives $\overline{\nabla^i_x f}$ of order i from one to p to terminate with an iterate x_ϵ such that $\phi^{\delta_{\epsilon,j}}_{f,j}(x_\epsilon) \le \epsilon_j\, \delta^j_{\epsilon,j}/j!$ for some $\delta_\epsilon \in (0,1]^q$ and for all $j \in [1:q]$.

Proof. The proof is identical to that of Theorem 13.1.19 (p. 349) with the choice (13.3.3), and the constants remain unchanged apart from a multiplicative factor of two for the total function-evaluation count (13.1.84) ($q = 1, 2$) and (13.1.86) ($q > 2$) (which is the same as the total number of successful and unsuccessful iterations). This doubling is due to the potential additional (re)evaluation of $\overline{f}(x_k)$ in Step 3 of ARqpEDA. In particular, $\kappa^{\rm A}_{\rm ARqpEDA,1} \stackrel{\rm def}{=} 2\kappa^{\rm A}_{\rm ARqpIDA,1}$ and $\kappa^{\rm A}_{\rm ARqpEDA,2} \stackrel{\rm def}{=} 2\kappa^{\rm A}_{\rm ARqpIDA,2}$, both of which are defined in (13.1.88); $\kappa^{\rm C}_{\rm ARqpEDA} \stackrel{\rm def}{=} 2\kappa^{\rm C}_{\rm ARqpIDA}$, as given in (13.1.89); and $\kappa^{\rm E}_{\rm ARqpEDA} \stackrel{\rm def}{=} \kappa^{\rm E}_{\rm ARqpIDA}$ and $\kappa^{\rm F}_{\rm ARqpEDA} \stackrel{\rm def}{=} \kappa^{\rm F}_{\rm ARqpIDA}$, as defined in (13.1.90). □

Since the orders in $\epsilon_{\min}$ are the same as those derived for the ARqp algorithm using exact evaluations and because these were proved to be sharp (see Section 12.2.2.4), the same conclusion obviously holds for the ARqpEDA algorithm using inexact evaluations.

Finally, the better complexity of regularization methods—in comparison with trust region–based ones—strengthens the conclusions of the important Theorem 13.2.1 and the discussions in Section 13.2 on obtaining complexitywise fast algorithms using derivatives of low degree or no derivatives at all.

Notes and references for Section 13.3

In this section, we aimed for our exposition to be as simple as possible, reusing much of the material developed in the IDA section that imposed accuracy conditions on the derivatives of the form (13.1.9) or (13.1.46). As we already remarked earlier, (13.3.1) with (13.3.9) and (13.3.18) use the value of the decrease in the inexact model, $\overline{\Delta T}_{f,p}(x_k, s_k)$, instead of $\kappa_s \|s_k\|^{p+1}$, which would be more coherent with (13.1.46). This latter choice is also possible, although it leads to tighter accuracy restrictions on algorithm parameters. Interestingly, it is also possible to obtain consistency by replacing (13.1.9) and (13.1.46) by conditions based on inexact model decrease. This is the approach taken originally in [28] (where weak approximate minimizers are considered), as well as in [110, 111], albeit at the cost of significantly more complicated algorithms. Whether accuracy conditions on derivatives are better stated in the space of variables (i.e., via (13.1.9) or (13.1.46) and a step norm condition for the accuracy of the objective-function values) or in the space of model values (as in those technical reports, and in (13.3.9) and (13.3.18)) is likely to depend on the scaling of the problem at hand.

First-order trust-region methods with inexact evaluations and explicit dynamic accuracy are described in [79], [131, Sect. 10.6]. The earliest complexity analysis for the first-order case is discussed in [214]. The extension in this section to arbitrary orders in the form of the TRqEDA algorithm is new.

The idea of inexact models has also generated a significant interest in the setting where inexactness is random and cannot be controlled dynamically, for instance, in the case where the objective function is composed of a large number of terms, and functions as well as derivatives are sampled rather than fully computed. This is of special interest in the fast-developing area of machine learning and deep neural networks. Although this is beyond the scope of our book, it is interesting to note that worst-case evaluation bounds in expectation or in probability can still be derived in such a framework. The reader is for instance referred to [28,29,51,58,141,154,276,314] and the references therein for examples.

Finally, the analysis of inexact optimization methods carried out in this chapter can shed some light on the effect of numerical round-off error in the values of the function and its derivatives [30].

Chapter 14

Composite and Constrained Problems

Chapter summary

The ideas in Chapter 12 can be extended in other ways. We discuss three of them here, the first being where the "composite" objective function is the sum of a smooth term and a (possibly) nonsmooth one. We introduce the concept of high-order approximate minimizers for this case and show how evaluation complexity may be affected by the nonsmoothness. We then introduce the class of inexpensive constraints and discuss how remarkably simple it is to generalize the results derived in Chapters 12 and 13 for the unconstrained ARqp *and* TRq *algorithms to this class.*

Turning to expensive constraints, we revisit, and extend, the "penalty-like" approach we developed in Chapter 7 for the problem at hand. We may use a method of our choice to deal with inexpensive side constraints and aim to find an ϵ-approximate qth-order point for $q \leq 3$ when possible, although we have to allow for local infeasibility. The evaluation complexity of the whole process will of course depend on that of the method chosen to solve the inexpensively constrained problem. If this complexity (as given by Chapters 6 and 12) is $\mathcal{O}(\epsilon^{-\pi})$, the resulting method will, under reasonable assumptions, need at most $\mathcal{O}(\epsilon^{1-2\pi})$ evaluations of the objective and expensive constraint functions (and their relevant derivatives) to find an ϵ-approximate qth-order minimizer.

Along the way, we discuss why finding approximate minimizers of order larger than three is a significant challenge for constrained problems.

14.1 Composite optimization problems

14.1.1 The problem and associated optimality conditions

We start by considering "composite" optimization problems. For these, we aim to find (ϵ, δ)-approximate minimizers of arbitrary order for the problem

$$\min_{x\in\mathbb{R}^n} w(x) \stackrel{\rm def}{=} f(x) + h(c(x)), \tag{14.1.1}$$

where c is a smooth function from $\mathbb{R}^n$ into $\mathbb{R}^m$ and h is a potentially nonsmooth function from $\mathbb{R}^m$ into $\mathbb{R}$. To do so, we generalize the ARqp algorithm, and for this we shall need to strengthen our assumptions, particularly so that we can analyze the complexity of finding an (ϵ, δ)-approximate qth-order minimizer. Unsurprisingly, we require that all derivatives involved be Lipschitz continuous. Thus, in addition to **f.D1pL** (p. 299), we require the following.

c.D1pL

Each component of c is p-times continuously differentiable in $\mathbb{R}^n$, and, for $j \in [1:p]$, the jth derivative of each component of c is Lipschitz continuous with Lipschitz constant $L_{c,j} \geq 0$, in that

$$\|\nabla_x^j c_i(x) - \nabla_x^j c_i(y)\| \leq L_{c,j}\|x - y\|$$

for all $x, y \in \mathbb{R}^n$, $j \in [1:p]$, and $i \in [1:m]$.

In view of Theorem A.8.5 (p. 438), this assumption implies, in particular, that

$$\|\nabla_x^j c(x)\| \leq L_{c,j-1} \quad \text{for} \quad j \in [2:p]. \tag{14.1.2}$$

Crucially, we shall not assume that the function h from $\mathbb{R}^m$ to $\mathbb{R}$ is everywhere differentiable but merely that it is Lipschitz continuous.

h.L

h is Lipschitz continuous in $\mathbb{R}^m$ with constant $L_{0,h} \geq 0$, in that

$$|h(x) - h(y)| \leq L_{0,h}\|x - y\|$$

for all $x, y \in \mathbb{R}^m$.

Given **f.D1pL**, **h.L**, and **c.D1pL**, we define

$$L_w \stackrel{\text{def}}{=} L_f + \sqrt{m} L_{h,0} L_c, \quad \text{where} \quad L_f \stackrel{\text{def}}{=} \max_{j\in[1:p]} L_{f,j} \quad \text{and} \quad L_c \stackrel{\text{def}}{=} \max_{j\in[1:p]} L_{c,j}. \tag{14.1.3}$$

We also assume that h is subadditive and zero at zero.

h.S0

h is *subadditive* in $\mathbb{R}$; that is,

$$h(x + y) \leq h(x) + h(y)$$

for all $x, y \in \mathbb{R}$. Moreover, $h(0) = 0$.

Observe that **h.L** and **h.S0** together imply that, for all $x \in \mathbb{R}$,

$$|h(x)| = |h(x) - h(0)| \leq L_{h,0}\|x - 0\| = L_{h,0}\|x\|. \tag{14.1.4}$$

Finally, we assume that problem (14.1.1) is well defined in that its objective function is bounded below.

w.Bb

There is a constant w_{low} such that $w(x) \geq w_{\text{low}}$ for all $x \in \mathbb{R}^n$.

Assumptions **h.L** and **h.S0** allow a fairly general class of composition functions. Examples include the popular $\|\cdot\|_1$-, $\|\cdot\|$-, and $\|\cdot\|_\infty$-norms; concave functions vanishing at zero; and, in the unidimensional case, the ReLu function $\max[0,\cdot]$ and the periodic $|\sin(\cdot)|$. As these examples show, nonconvexity and nondifferentiability are allowed (but not necessary). Trivially, finite sums of functions that satisfy these assumptions also satisfy them. However, just because h is subadditive, it does not follow that h^α is for $\alpha \geq 1$ ($h(c) = c$ is, but $(h(c))^2$ is not), nor that it is concave. If h is odd, i.e., $h(-c) = -h(c)$, equality holds in **h.S0**. Indeed, $h(-x-y) \leq h(-x) + h(-y)$ and thus, since h is odd, $-h(x+y) \leq -h(x) - h(y)$, which, combined with **h.S0**, gives that $h(x+y) = h(x) + h(y)$.

When h is smooth, problem (14.1.1) can be viewed as either composite or noncomposite. Does the composite view present any advantage in this case? The answer is that the assumptions needed on h in the composite case are weaker, in that Lipschitz continuity is only required for h itself, not for its derivatives of orders 1 to p. If any of these derivatives are costly, unbounded, or nonexistent, this can be a significant advantage. However, as we will see shortly (in Theorem 14.1.10), this comes at the price of an inferior worst-case evaluation-complexity bound. For example, the case of linear h is simple to assess, since in that case $h(c)$ amounts to a linear combination of the c_i, and there is obviously no costly or unbounded derivative involved: a noncomposite approach is therefore, from a complexity perspective, preferable to the composite framework we develop in this section. We tacitly assume throughout that $h \neq 0$. We will also see that knowledge of the convexity of h, a very frequent case in applications involving norms, is also an advantage, although it is not mandatory.

For future reference, and abusing notation slightly, we write

$$T_{w,j}(x,s) \stackrel{\text{def}}{=} T_{f,j}(x,s) + h\big(T_{c,j}(x,s)\big), \tag{14.1.5}$$

where $T_{c,j}(x,s)$ is the vector-valued function whose ith component is $T_{c_i,j}(x,s)$ for $i \in [1:m]$ and $j \in [1:p]$; in particular, be aware that $T_{w,j}(x,s)$ is *not* a Taylor expansion. We then define

$$\begin{aligned} \phi^\delta_{w,j}(x) &\stackrel{\text{def}}{=} w(x) - \min_{\|d\|\leq\delta} \left[T_{f,j}(x_k,s) + h(T_{c,j}(x_k,s))\right] \\ &= w(x) - \min_{\|d\|\leq\delta} T_{w,j}(x_k,s) \end{aligned} \tag{14.1.6}$$

for $j \in [1:p]$, by analogy with (12.1.10) (p. 278). This definition allows us to consider (approximate) high-order minimizers of w even when h is nonsmooth, because we have retained h in the optimality measure (14.1.6), rather than resorting to a Taylor expansion.

There is a simple first-order necessary optimality condition for composite problems of the form (14.1.1).

Lemma 14.1.1: Convex-composite first-order necessary optimality

*Suppose that f and c are continuously differentiable and that **h.L** and **h.S0** hold. Suppose in addition that h is convex and that x_* is a global minimizer of w. Then the origin is a global minimizer of $T_{w,1}(x_*,s)$ and $\phi^\delta_{w,1}(x_*) = 0$ for all $\delta > 0$.*

Proof. Suppose that the origin is not a global minimizer of $T_{w,1}(x_*,s)$ but that there exists an $s_1 \neq 0$ with $T_{w,1}(x_*,s_1) < T_{w,1}(x_*,0) = w(x_*)$. By Taylor's theorem, we obtain that, for $\alpha \in [0,1]$,

$$f(x_* + \alpha s_1) = T_{f,1}(x_*,\alpha s_1) + o(\alpha) \tag{14.1.7}$$

and, using **h.S0** and (14.1.4),

$$\begin{aligned} h(c(x_* + \alpha s_1)) &= h\big(T_{c,1}(x_*, \alpha s_1) + o(\alpha\|s_1\|)\big) \\ &\le h\big(T_{c,1}(x_*, \alpha s_1)\big) + h\big(o(\alpha)\|s_1\|\big) \\ &\le h\big(T_{c,1}(x_*, \alpha s_1)\big) + o(\alpha)L_{h,0}\|s_1\| \\ &= h\big(T_{c,1}(x_*, \alpha s_1)\big) + o(\alpha). \end{aligned} \tag{14.1.8}$$

Now note that the convexity of h and the linearity of $T_{f,1}(x_*, s)$ and $T_{c,1}(x_*, s)$ imply that $T_{w,1}(x_*, s)$ is convex and thus that

$$T_{w,1}(x_*, \alpha s_1) - w(x_*) \le \alpha[T_{w,1}(x_*, s_1) - w(x_*)].$$

Hence, using (14.1.7) and (14.1.8), we deduce that

$$\begin{aligned} 0 \le w(x_* + \alpha s_1) - w(x_*) &\le T_{w,1}(x_*, \alpha s_1) - w(x_*) + o(\alpha) \\ &\le \alpha[T_{w,1}(x_*, s_1) - w(x_*)] + o(\alpha), \end{aligned}$$

which is impossible for α sufficiently small since $T_{w,1}(x_*, s_1) - w(x_*) < 0$. As a consequence, the origin must be a global minimizer of the convex $T_{w,1}(x_*, s)$ and therefore $\phi^{\delta}_{w,1}(x_*) = 0$ for all $\delta > 0$. □

Unfortunately, this result does not extend to the case where $q = 2$, as is shown by the following example.

Example 14.1.1 Consider the univariate $w(x) = -\frac{2}{5}x + |x - x^2 + 2x^3|$, where h is the (convex) absolute-value function satisfying **h.L** and **h.S0**. Then $x_* = 0$ is a global minimizer of w (plotted in blue in Figure 14.1.1), and yet

$$\begin{aligned} T_{w,2}(x_*, s) &= T_{f,2}(x_*, s) + |T_{c,2}(x_*, s)| \\ &= 0 - \tfrac{2}{5}s + |0 + (1 - 2x_* + 6x_*^2)s + \tfrac{1}{2}(-2 + 12x_*)s^2| \\ &= -\tfrac{2}{5}s + |s - s^2| \end{aligned}$$

(plotted in red in the figure) admits a global minimum for $s = 1$ whose value $(-\frac{2}{5})$ is smaller than $w(x_*) = 0$. Thus $\phi^{1}_{w,2}(x_*) > 0$ despite x_* being a global minimizer. But it is clear in the figure that $\phi^{\delta}_{w,2}(x_*) = 0$ for sufficiently small δ (smaller than $\frac{1}{2}$, say).

14.1.2 The ARqpC algorithm and its complexity

14.1.2.1 The ARqpC algorithm for composite problems

Returning, once again, to one of the main themes of this book, we now consider an adaptive-regularization algorithm to search for an (ϵ, δ)-approximate qth-order minimizer for problem (14.1.1), that is, a point $x_\epsilon \in \mathbb{R}^n$ such that

$$\phi^{\delta_j}_{w,j}(x_\epsilon) \le \epsilon_j \frac{\delta_j^j}{j!} \quad \text{for} \quad j \in [1:q], \tag{14.1.9}$$

where $\phi^{\delta}_{w,j}(x)$ is defined as in (14.1.6). As for the ARqp algorithm, these inequalities will be checked at every iterate of the algorithm (in order to declare termination) without

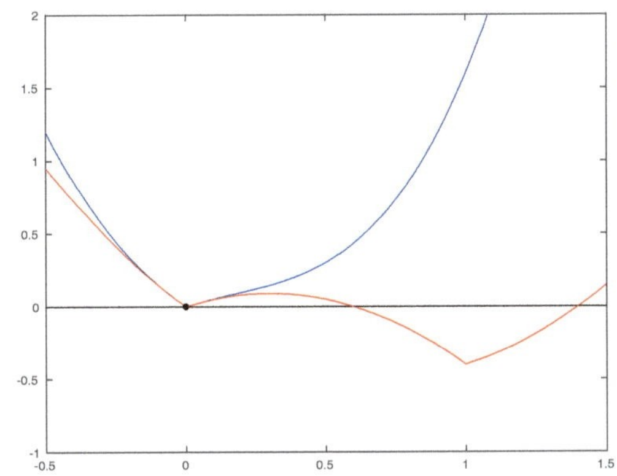

Figure 14.1.1. $w(x)$ (in blue) and $T_{w,2}(x_*, s) = T_{f,2}(x_*, s) + |T_{c,2}(x_*, s)|$ (in red) for Example 14.1.1.

exactly solving the global optimization problem involved in the definition of $\phi_{w,j}^{\delta_{k,j}}(x_k)$ for each $j \in [1:q]$, but instead by checking the approximate variant

$$\varsigma \phi_{w,j}^{\delta_{k,j}}(x_k) \leq \Delta T_{w,j}(x_k, d_{k,j}) \leq \varsigma \epsilon_j \frac{\delta_j^j}{j!}, \tag{14.1.10}$$

involving the decrement

$$\Delta T_{w,j}(x, d) \stackrel{\text{def}}{=} T_{w,j}(x, 0) - T_{w,j}(x, d),$$

for $j \in [1:q]$ and $d_{k,j} \in \mathcal{B}_{\delta_{k,j}}$. At each iteration, our variant of the ARqp algorithm seeks an approximate minimizer of the (possibly nonsmooth) regularized model

$$\begin{aligned} \widehat{m}_k(s) &\stackrel{\text{def}}{=} T_{f,p}(x_k, s) + h\big(T_{c,p}(x_k, s)\big) + \frac{\sigma_k}{(p+1)!}\|s\|^{p+1} \\ &= T_{w,p}(x_k, s) + \frac{\sigma_k}{(p+1)!}\|s\|^{p+1}, \end{aligned} \tag{14.1.11}$$

and, as in Step 2 of the ARqp algorithm (p. 300), this process is allowed to terminate with a step s_k whenever

$$\widehat{m}_k(s_k) \leq \widehat{m}_k(0) = w(x_k) \tag{14.1.12}$$

and either $\|s_k\| \geq 1$ or

$$\varsigma \phi_{\widehat{m}_k,j}^{\delta_{s,j}}(s_k) \leq \Delta T_{\widehat{m}_k,j}(s_k, d_{k,j}^m) \leq \varsigma \theta \epsilon_j \frac{\delta_s^j}{j!} \tag{14.1.13}$$

for some $d_{k,j}^m \in \mathcal{B}_{\delta_{s,j}}$ and each $j \in [1:q]$, where now

$$\begin{aligned} \phi_{\widehat{m}_k,j}^{\delta_{s,j}}(s) = \max_{\|d\| \leq \delta_{s,j}} \Bigg[&-\sum_{\ell=1}^{j} \frac{1}{\ell!} \nabla_s^\ell T_{f,p}(x_k, s)[d]^\ell + h\big(T_{c,p}(x_k, s)\big) \\ &-h\left(\sum_{\ell=0}^{j} \frac{1}{\ell!} \nabla_s^\ell T_{c,p}(x_k, s)[d]^\ell\right) - \frac{\sigma_k}{(p+1)!} \sum_{\ell=1}^{j} \frac{1}{\ell!} \nabla_s^\ell \|s\|^{p+1}[d]^\ell \Bigg] \end{aligned} \tag{14.1.14}$$

and

$$\Delta T_{\widehat{m}_k,j}(s,d) \stackrel{\text{def}}{=} T_{\widehat{m}_k,j}(s,0) - T_{\widehat{m}_k,j}(s,d),$$

exactly as in (12.2.27)–(12.2.29). Observe that $\widehat{m}_k(s)$ is bounded below because (14.1.4) guarantees that the regularization term still dominates for large steps. Obviously, by including h in the definition of the model (14.1.11) we are implicitly assuming that, as is common, the cost of evaluation h is small compared with that of evaluating f or c. It also implies that computing $\Delta T_{w,j}(x_k, d_{k,j})$ and $\Delta T_{\widehat{m}_k,j}(s_k, d^m_{k,j})$ is potentially more complicated than in the noncomposite case—the semidefinite optimization approach we mentioned in Section 12.1.6 for the TRq and ARqp algorithms is no longer directly applicable. The rest of the algorithm is essentially identical to the ARqp algorithm and is stated as follows.

Algorithm 14.1.1: ARqpC algorithm for finding (ϵ,δ)-approximate qth-order minimizers of composite functions

Step 0: Initialization. A criticality order q, a degree $p \geq q$, an initial point x_0, and an initial regularization parameter $\sigma_0 > 0$ are given, as well as accuracy levels $\epsilon \in (0,1)^q$ and optimality radii $\delta_0 \in (0,1]^q$. The constants $\delta_0 \in (0,1]^q$, θ, ς, η_1, η_2, γ_1, γ_2, γ_3, and $\sigma_{\min}$ are also given and satisfy

$$\theta \in (0,1), \quad \varsigma \in (0,1], \quad \sigma_{\min} \in (0,\sigma_0], \quad 0 < \eta_1 \leq \eta_2 < 1, \\ \text{and} \;\; 0 < \gamma_1 < 1 < \gamma_2 < \gamma_3.$$

Compute $w(x_0)$, and set $k = 0$.

Step 1: Test for termination. Evaluate $\{\nabla_x^i f(x_k)\}_{i=1}^q$ and $\{\nabla_x^i c(x_k)\}_{i=1}^q$. If (14.1.10) holds with $\delta = \delta_k$, terminate with the approximate solution $x_\epsilon = x_k$. Otherwise compute $\{\nabla_x^i f(x_k)\}_{i=q+1}^p$ and $\{\nabla_x^i c(x_k)\}_{i=q+1}^p$.

Step 2: Step calculation. Compute an approximate minimizer s_k of model $\widehat{m}_k(s)$ given in (14.1.11) and an optimality radius $\delta_{s_k} \in (0,1]^q$ so that (14.1.12) holds and either $\|s_k\| \geq 1$ or (14.1.13) holds for $j \in [1:q]$. If $\widehat{m}_k(s_k) = \widehat{m}_k(0)$, terminate with the approximate solution $x_\epsilon = x_k$ and $\delta_\epsilon = \delta_{s_k}$.

Step 3: Acceptance of the trial point. Compute $w(x_k + s_k)$ and define

$$\rho_k = \frac{w(x_k) - w(x_k + s_k)}{\Delta T_{w,p}(x_k, s_k)}. \tag{14.1.15}$$

If $\rho_k \geq \eta_1$, then define $x_{k+1} = x_k + s_k$ and $\delta_{k+1} = \delta_{s_k}$. Otherwise define $x_{k+1} = x_k$ and $\delta_{k+1} = \delta_k$.

Step 4: Regularization parameter update. Set

$$\sigma_{k+1} \in \begin{cases} [\max(\sigma_{\min}, \gamma_1\sigma_k), \sigma_k] & \text{if } \rho_k \geq \eta_2, \\ [\sigma_k, \gamma_2\sigma_k] & \text{if } \rho_k \in [\eta_1, \eta_2), \\ [\gamma_2\sigma_k, \gamma_3\sigma_k] & \text{if } \rho_k < \eta_1. \end{cases} \tag{14.1.16}$$

Increment k by one and go to Step 1 if $\rho_k \geq \eta_1$ or to Step 2 otherwise.

As usual, Lemma 2.4.1 (p. 44) is valid for the ARqpC algorithm. However, the existence of a step needs to be reexamined because the new composite model (14.1.11) differs from that in (12.2.23). We noted in Lemma 12.2.9 (p. 306) that, for the ARqp algorithm, one can always choose $\delta_{s_k,j} = 1$ for all k and all j whenever $q = 1$ or $q = 2$. The same conclusion obviously holds here when the composition function $h = 0$, but the situation is more complicated when this is not the case.

Remarkably, Algorithm 12.2.3 (p. 303) remains essentially valid as written, although the meaning of the notation has to be modified, in accordance with our previous discussion, to reflect the composite nature of the problem. The only necessary adaptation is that, because of Example 14.1.1, Step 2.4.3 must now be executed if $j \geq 2$ or h is nonconvex (instead of $j \geq 3$). We repeat the algorithm here for completeness.

Algorithm 14.1.2: A detailed Step 2 for the ARqpC algorithm

Step 2: Step calculation. Suppose that the model $\widehat{m}_k(s)$ from (14.1.11) and the constants $\varsigma \in (0,1]$ and $\theta \in (0,1)$ are given.

Step 2.1: Initialize $\mu \in (0,1]$, and set $\delta_{s_k,j} = 1$ for all $j \in [1:q]$.

Step 2.2: Find s_μ such that

$$\mu(\widehat{m}_k(0) - \widehat{m}_k^*) \leq \Delta\widehat{m}_k(s_\mu),$$

where $\widehat{m}_k^*$ is the (unknown) globally minimal value of $\widehat{m}_k(s)$.

Step 2.3: If $\|s_\mu\| \geq 1$, return the pair (s_k, δ_{s_k}), where $s_k = s_\mu$.

Step 2.4: For $j \in [1:q]$:

Step 2.4.1: Compute $d_{\mu,j}$ such that $\|d_{\mu,j}\| \leq \delta_{s_k,j}$ and

$$\varsigma\phi_{\widehat{m}_k,j}^{\delta_{s_k,j}}(s_\mu) \leq \Delta T_{\widehat{m}_k,j}(s_\mu, d_{\mu,j}).$$

Step 2.4.2: If

$$\Delta T_{\widehat{m}_k,j}(s_\mu, d_{\mu,j}) \leq \varsigma\theta\epsilon_j \frac{\delta_{s_k,j}^j}{j!},$$

consider the next j (if $j < q$) or go to Step 2.5 (if $j = q$).

Step 2.4.3: Otherwise, if either $j \geq 2$ or h is nonconvex and if

$$\widehat{m}_k(s_\mu) - \widehat{m}_k(s_\mu + d_{\mu,j}) < \tfrac{1}{2}\varsigma\theta\epsilon_j \frac{\delta_{s_k,j}^j}{j!},$$

set $\delta_{s_k,j} = \frac{1}{2}\delta_{s_k,j}$ and go back to Step 2.4.1.

Step 2.4.4: Otherwise set $\mu = \frac{1}{2}(1+\mu)$ and go back to Step 2.2.

Step 2.5: Return the pair (s_k, δ_{s_k}), where $s_k = s_\mu$.

We note that the test at the beginning of Step 2.4.3 presumes that we know whether h is convex or not. If this information is unavailable, the algorithm must proceed as if h is nonconvex, but at the price of affecting the final worst-case evaluation-complexity bound (see Theorem 14.1.10 below).

It remains for us to verify that Algorithm 14.1.2 produces the desired step s_k and vector of radius δ_{s_k}. Since the model (14.1.11) is potentially nonsmooth, deriving a generalization of Lemma 12.2.7 for the composite case is impossible, and we are thus led directly to discussing the extension of Lemma 12.2.8 to the composite framework.

Lemma 14.1.2: Relationship between composite model and generalized Taylor decrements

Suppose that **f.D1pL**, **c.D1pL**, **h.L**, *and* **h.S0** *hold, that* $s \in \mathcal{B}_1$, *and that* L_w *is given by (14.1.3). Given* $\alpha > 0$ *and*

$$\delta \in \left(0, \min\left(1, \frac{\alpha}{4(3L_w + \sigma_k)}\right)\right], \tag{14.1.17}$$

suppose that there is a displacement $d \in \mathcal{B}_\delta$ *for which*

$$\Delta T_{\widehat{m}_k,j}(s,d) \geq \alpha \frac{\delta^j}{j!} \tag{14.1.18}$$

for some $j \in [1:q]$. *Then*

$$\widehat{m}_k(s) - \widehat{m}_k(s+d) \geq \tfrac{1}{2}\alpha \frac{\delta^j}{j!}. \tag{14.1.19}$$

Proof. Consider arbitrary $j \in [1:q]$, $s \in \mathcal{B}_1$, and $d \in \mathcal{B}_\delta$. Using the Taylor expansions of f and c (in addition to the regularization term), we have that

$$\begin{aligned}\widehat{m}_k(s) - \widehat{m}_k(s+d) = & -\sum_{\ell=1}^{p} \frac{1}{\ell!} \nabla_s^\ell T_{f,p}(x_k,s)[d]^\ell - h\left(\sum_{\ell=0}^{p} \frac{1}{\ell!} \nabla_s^\ell T_{c,p}(x_k,s)[d]^\ell\right) \\ & + h\big(T_{c,p}(x_k,s)\big) - \frac{\sigma_k}{(p+1)!}\left[\|s+d\|^{p+1} - \|s\|^{p+1}\right].\end{aligned} \tag{14.1.20}$$

Furthermore, it follows from (B.4.3) in Lemma B.4.1 (p. 461) that $\|\nabla_s^{p+1}\|s\|^{p+1}\| = (p+1)!$ and thus from Theorem A.8.5 that $\nabla_s^{p+1}\|s\|^{p+1}$ is Lipschitz continuous with Lipschitz constant $(p+1)!$. We then deduce from (A.19) in Taylor's theorem (Corollary A.8.4 on p. 437) applied to $\|s\|^{p+1}$ that

$$\left| \|s+d\|^{p+1} - \|s\|^{p+1} + \sum_{\ell=1}^{p} \frac{1}{\ell!} \nabla_s^\ell \|s\|^{p+1}[d]^\ell \right| \leq \|d\|^{p+1}$$

and, combining this with (14.1.20), that

$$\begin{aligned}\widehat{m}_k(s) - \widehat{m}_k(s+d) \geq & -\sum_{\ell=1}^{p} \frac{1}{\ell!} \nabla_s^\ell T_{f,p}(x_k,s)[d]^\ell - h\left(\sum_{\ell=0}^{p} \frac{1}{\ell!} \nabla_s^\ell T_{c,p}(x_k,s)[d]^\ell\right) \\ & + h\big(T_{c,p}(x_k,s)\big) - \frac{\sigma_k}{(p+1)!}\left[\sum_{\ell=1}^{p} \frac{1}{\ell!} \nabla_s^\ell \|s\|^{p+1}[d]^\ell + \|d\|^{p+1}\right].\end{aligned} \tag{14.1.21}$$

Now, using the subadditivity of h ensured by **h.S0**, we have that, for all $j \in [1:q]$,

$$h\left(\sum_{\ell=0}^{p} \frac{1}{\ell!} \nabla_s^\ell T_{c,p}(x_k,s)[d]^\ell\right) \leq h\left(\sum_{\ell=0}^{j} \frac{1}{\ell!} \nabla_s^\ell T_{c,p}(x_k,s)[d]^\ell\right) + h\left(\sum_{\ell=j+1}^{p} \frac{1}{\ell!} \nabla_s^\ell T_{c,p}(x_k,s)[d]^\ell\right),$$

and thus

$$\begin{aligned}\widehat{m}_k(s) - \widehat{m}_k(s+d) \geq & -\sum_{\ell=1}^{j} \frac{1}{\ell!} \nabla_s^\ell T_{f,p}(x_k,s)[d]^\ell - \sum_{\ell=j+1}^{p} \frac{1}{\ell!} \nabla_s^\ell T_{f,p}(x_k,s)[d]^\ell \\ & -h\left(\sum_{\ell=0}^{j} \frac{1}{\ell!} \nabla_s^\ell T_{c,p}(x_k,s)[d]^\ell\right) - h\left(\sum_{\ell=j+1}^{p} \frac{1}{\ell!} \nabla_s^\ell T_{c,p}(x_k,s)[d]^\ell\right) \\ & +h(T_{c,p}(x_k,s)) - \frac{\sigma_k}{(p+1)!}\left[\sum_{\ell=1}^{p} \frac{1}{\ell!} \nabla_s^\ell \|s\|^{p+1}[d]^\ell + \|d\|^{p+1}\right].\end{aligned} \tag{14.1.22}$$

But, from the definition (14.1.5),

$$\begin{aligned}\Delta T_{\widehat{m}_k,j}(s,d) = & -\sum_{\ell=1}^{j} \frac{1}{\ell!} \nabla_s^\ell T_{f,p}(x_k,s)[d]^\ell + h(T_{c,p}(x_k,s)) \\ & -h\left(\sum_{\ell=0}^{j} \frac{1}{\ell!} \nabla_s^\ell T_{c,p}(x_k,s)[d]^\ell\right) - \frac{\sigma_k}{(p+1)!}\left[\sum_{\ell=1}^{j} \frac{1}{\ell!} \nabla_s^\ell \|s\|^{p+1}[d]^\ell\right].\end{aligned}$$

Substituting this relation in (14.1.22) then ensures that, for all $s \in \mathcal{B}_1$ and $d \in \mathcal{B}_\delta$,

$$\begin{aligned}\widehat{m}_k(s) - \widehat{m}_k(s+d) \geq & \ \Delta T_{\widehat{m}_k,j}(s,d) \\ & -\sum_{\ell=j+1}^{p} \frac{1}{\ell!} \nabla_s^\ell T_{f,p}(x_k,s)[d]^\ell - h\left(\sum_{\ell=j+1}^{p} \frac{1}{\ell!} \nabla_s^\ell T_{c,p}(x_k,s)[d]^\ell\right) \\ & -\frac{\sigma_k}{(p+1)!}\left[\sum_{\ell=j+1}^{p} \frac{1}{\ell!} \nabla_s^\ell \|s\|^{p+1}[d]^\ell + \|d\|^{p+1}\right],\end{aligned}$$

and, since $d \in \mathcal{B}_\delta$, we deduce that

$$\begin{aligned}\widehat{m}_k(s) - \widehat{m}_k(s+d) \geq & \ \Delta T_{\widehat{m}_k,j}(s,d) \\ & -\sum_{\ell=j+1}^{p} \frac{\delta^\ell}{\ell!} \|\nabla_s^\ell T_{f,p}(x_k,s)\| - L_{h,0}\left(\sum_{\ell=j+1}^{p} \frac{\delta^\ell}{\ell!} \|\nabla_s^\ell T_{c,p}(x_k,s)\|\right) \\ & -\frac{\sigma_k}{(p+1)!}\left[\sum_{\ell=j+1}^{p} \frac{\delta^\ell}{\ell!} \|\nabla_s^\ell \|s\|^{p+1}\| + \|d\|^{p+1}\right]\end{aligned} \tag{14.1.23}$$

using (14.1.4). We also have, by definition and the triangle inequality,

$$\|\nabla_s^\ell T_{f,p}(x_k,s)\| \le \sum_{j=\ell}^{p} \frac{1}{(j-\ell)!}\|\nabla_x^j f(x_k)\|\,\|s\|^{j-\ell} \le 3L_f,$$

$$\|\nabla_s^\ell T_{c,p}(x_k,s)\| \le \sum_{j=\ell}^{p} \frac{1}{(j-\ell)!}\|\nabla_x^j c(x_k)\|\,\|s\|^{j-\ell} \le 3\sqrt{m}L_c,$$

and

$$\|\nabla_s^\ell \|s\|^{p+1}\| \le \frac{(p+1)!}{(p-\ell+1)!}\|s\|^{p-\ell+1} \le (p+1)!$$

for $\ell \in [q+1:p]$, where we used (A.33) in Lemma A.10.1 (p. 442), Lemma B.4.1 (p. 461), and the fact that $s \in \mathcal{B}_1$. Combining these inequalities with (14.1.23) and using (14.1.3) then gives us that

$$\widehat{m}_k(s) - \widehat{m}_k(s+d) \ge \Delta T_{\widehat{m}_k,j}(s,d) - \sum_{\ell=j+1}^{p} \frac{\delta^\ell}{\ell!}[3L_w + \sigma_k] - \sigma_k \frac{\delta^{p+1}}{(p+1)!}.$$

Using the fact that $\delta \le 1$, (A.33), and (14.1.18), we deduce that

$$\widehat{m}_k(s) - \widehat{m}_k(s+d) \ge \alpha\frac{\delta^j}{j!} - 2(3L_w + \sigma_k)\frac{\delta^{j+1}}{j!} = \frac{\delta^j}{j!}(\alpha - 2(3L_w+\sigma_k)\delta),$$

and (14.1.19) then follows from (14.1.17). □

Notice that, although the details of the proof are different and reflect the composite nature of the current problem, the statement of this lemma is formally identical[100] to that of Lemma 12.2.8 (p. 305). It therefore comes as no surprise that Lemma 12.2.9 (p. 306) is easy to generalize to the composite case.

Lemma 14.1.3: Existence of a step (ARqpC)

Suppose that **f.D1pL**, **c.D1pL**, **h.L**, *and* **h.S0** *hold and that L_w is given by (14.1.3). Then a pair (s_k, δ_{s_k}) always exists such that (14.1.12) and either $\|s_k\| > 1$ or (14.1.13) hold. Moreover, it is always possible to ensure that*

$$\delta_{s_k,j} = 1 \quad \textit{when} \quad j = 1 \quad \textit{and } h \textit{ is convex}$$

and, if either $q \ge 2$ or h is nonconvex and $\|s_k\| \le 1$, then

$$\delta_{s_k,j} \ge \kappa_\delta(\sigma_k)\epsilon_j \quad \textit{for} \quad j \in [2:q],$$

where

$$\kappa_\delta(\sigma) \stackrel{\rm def}{=} \frac{\varsigma\theta}{8(3L_w+\sigma)}. \tag{14.1.24}$$

Proof. The proof is a simple variant of that of Lemma 12.2.9 (p. 306) in which the cases where $j \le 2$ and $j > 2$ are replaced by the cases $j = 1$ with h convex and $j > 1$, and where Lemma 14.1.1 is now used instead of Theorem 12.1.4 in the argument leading to (12.2.48). □

[100] Aside from the more detailed assumptions and the adapted Lipschitz constant.

The consequences of terminating in Step 2 of the ARqpC algorithm, just as they were for the ARqp case, also need to be reassessed.

Lemma 14.1.4: Termination in Step 2 of ARqpC

Termination cannot occur in Step 2 of the ARqpC algorithm if $q = 1$ and h is convex. In other cases, if the ARqpC algorithm terminates in Step 2 of iteration k with $x_\epsilon = x_k$, then there exists a $\delta \in (0,1]^q$ such that (14.1.9) holds for $x = x_\epsilon$ and x_ϵ is an (ϵ, δ)-approximate qth-order minimizer.

Proof. See Appendix B (p. 487). As for Lemma 14.1.2, subadditivity is the key to handling the distinction between terms of the Taylor series of degree at most q and those of higher-order Taylor series. □

Thus the ARqpC algorithm is well-defined and, when it terminates, produces an (ϵ, δ)-approximate qth-order minimizer.

14.1.2.2 Evaluation complexity for the ARqpC algorithm

To analyze the evaluation complexity of our algorithm, we first derive the expected decrease in the unregularized model from (14.1.11).

Lemma 14.1.5: Taylor decrease for the ARqpC algorithm

At every iteration k of the ARqpC algorithm, one has that

$$\Delta T_{w,p}(x_k, s_k) = T_{w,p}(x_k, 0) - T_{w,p}(x_k, s_k) \geq \frac{\sigma_k}{(p+1)!}\|s_k\|^{p+1}. \tag{14.1.25}$$

Proof. This is immediate from (14.1.11) and (14.1.5), the fact that $\widehat{m}_k(0) = w(x_k) = T_{w,p}(x_k, 0)$, and (14.1.12). □

Next, we establish the existence of an upper bound on the regularization parameter for the structured composite problem.

Lemma 14.1.6: Upper bound on the regularization parameter for the ARqpC algorithm

Suppose that **f.DpL**, **c.DpL**, *and* **h.L** *hold. Then, for all $k \geq 0$,*

$$\sigma_k \leq \sigma_{\max} \stackrel{\rm def}{=} \max\left[\sigma_0, \frac{\gamma_3 L_{w,p}}{1-\eta_2}\right], \tag{14.1.26}$$

where

$$L_{w,p} \stackrel{\rm def}{=} L_{f,p} + \sqrt{m} L_{h,0} L_{c,p}. \tag{14.1.27}$$

Proof. Just as for the proofs of similar results, the proof of (14.1.26) hinges on the fact that, once the regularization parameter σ_k exceeds the relevant Lipschitz constant ($L_{w,p}$ here), there is no need to increase it any further because the model then overestimates the objective function. Details can be found in Appendix B (p. 488). □

As was the case for the ARqp algorithm, we use this result and Lemma 14.1.3 to presume, without loss of generality, that for each $k \geq 0$,

$$\text{either} \quad \|s_k\| > 1 \quad \text{or} \quad \delta_{s_k,j} \geq \kappa_{\delta,\min}\epsilon_j \tag{14.1.28}$$

for $j \geq 2$, where, from (14.1.24),

$$\kappa_{\delta,\min} \stackrel{\text{def}}{=} \kappa_\delta(\sigma_{\max}) = \frac{\varsigma\theta}{8(3L_w + \sigma_{\max})} \leq \kappa_\delta(\sigma_k), \tag{14.1.29}$$

while $\delta_{s_k,1} = 1$.

Our next step is to derive the appropriate variant of Lemma 12.2.11 (p. 308) for the composite case.

Lemma 14.1.7: A useful inequality (for the ARqpC algorithm)

Suppose that **f.D1pL**, **c.D1pL**, **h.L**, *and* **h.S0** *hold. Suppose also that iteration k is successful and that the ARqpC algorithm does not terminate at iteration $k+1$. Then there exists a $j \in [1:q]$ such that*

$$(1-\theta)\,\epsilon_j\,\frac{\delta_{k+1,j}^j}{j!} \leq (L_{w,p}+\sigma_{\max})\sum_{\ell=1}^{j}\frac{\delta_{k+1,j}^\ell}{\ell!}\|s_k\|^{p-\ell+1} + \frac{\sqrt{m}L_{h,0}L_{c,p}}{(p+1)!}\|s_k\|^{p+1}, \tag{14.1.30}$$

where $\sigma_{\max}$ is given by (14.1.26).

Proof. Notice that (14.1.30) differs from (12.2.52) (p. 308) in that a term in $\|s_k\|^{p+1}$ appears on the right-hand side. This difference is significant (as we will discover) and occurs because the terms in $h(T_{c,p}(x_k,0))$ do not cancel in (14.1.14) when h is nonlinear. The details of the proof, once again strongly hinging on **h.L** and **h.S0**, can be found in Appendix B (p. 489). □

The following lemma then exploits (14.1.30) to derive a lower bound on the step norm.

Lemma 14.1.8: Lower bound on the step for the ARqpC algorithm

Suppose that **f.D1pL**, **c.D1pL**, **h.L**, *and* **h.S0** *hold, that iteration k is successful, and that the ARqpC algorithm does not terminate at iteration $k+1$. In addition, suppose the algorithm ensures for each k that either $\delta_{s_k,1} = 1$, if $q = 1$ and h is convex, or otherwise (14.1.28) holds for all $j \in [1:q]$ (as is possible, according to Lemma 14.1.3). Then there exists a $j \in [1:q]$ such that*

$$\|s_k\| \geq \begin{cases} \left(\dfrac{1-\theta}{3j!(L_{w,p}+\sigma_{\max})}\right)^{\frac{1}{p}} \epsilon_j^{\frac{1}{p}} & \text{if } q = 1 \text{ and } h \text{ is convex,} \\[2ex] \left(\dfrac{(1-\theta)\kappa_{\delta,\min}^j}{3j!(L_{w,p}+\sigma_{\max})}\right)^{\frac{1}{p-j+1}} \epsilon_j^{\frac{j+1}{p+1}} & \text{otherwise,} \end{cases} \tag{14.1.31}$$

where $\kappa_{\delta,\min}$ is defined by (14.1.29), $L_{w,p}$ by (14.1.27), and $\sigma_{\max}$ by (14.1.26).

Proof. The proof essentially parallels that of Lemma 12.2.12 (p. 309) but differs in that the new term in $\|s_k\|^{p+1}$ on the right-hand side of (14.1.30) now dominates the case where $\delta_{k,j} < \|s_k\|$, which happens when $q > 1$ or h is nonconvex. Details are available in Appendix B (p. 490). □

As usual, we now state lower bounds on the minimal model and function decreases at successful iterations.

Lemma 14.1.9: Model and function decreases (ARqpC)

Suppose that **f.D1qL** *holds. Then*

$$\Delta T_{w,p}(x_k, s_k) \geq \kappa_{\Delta m} \min_{j\in[1:q]} \epsilon_j^{\pi_j} \tag{14.1.32}$$

for every iteration of algorithm ARqpC *before termination, where*

$$\kappa_{\Delta m} \stackrel{\text{def}}{=} \begin{cases} \dfrac{\sigma_{\min}}{(p+1)!}\left(\dfrac{1-\theta}{3q!(L_{f,p}+\sigma_{\max})}\right)^{\frac{p+1}{p}} & \text{if } q = 1 \text{ and } h \text{ is convex,} \\ \dfrac{\sigma_{\min}}{(p+1)!}\left(\dfrac{(1-\theta)\kappa_{\delta,\min}^q}{3q!(L_{f,p}+\sigma_{\max})}\right)^{\frac{p+1}{p-q+1}} & \text{otherwise,} \end{cases}$$

with $\kappa_{\delta,\min}$ *given by (14.1.29),* $\sigma_{\max}$ *given by (14.1.26), and*

$$\pi_j \stackrel{\text{def}}{=} \begin{cases} \dfrac{p+1}{p} & \text{if } q = 1 \text{ and } h \text{ is convex,} \\ j+1 & \text{otherwise.} \end{cases}$$

If, in addition, $k \in \mathcal{S}$,

$$f(x_k) - f(x_{k+1}) \geq \eta_1 \kappa_{\Delta m} \min_{j\in[1:q]} \epsilon_j^{\pi_j}. \tag{14.1.33}$$

Proof. Lemma 14.1.5, (14.1.16), and Lemma 14.1.8 (p. 382) together imply that for every $k \in \mathcal{S}$, there exists a $j \in [1:q]$ such that

$$\Delta T_{w,p}(x_k, s_k) \geq \begin{cases} \dfrac{\sigma_{\min}}{(p+1)!}\left(\dfrac{1-\theta}{3j!(L_{f,p}+\sigma_{\max})}\right)^{\frac{p+1}{p}} \epsilon_j^{\frac{p+1}{p}} \\ \qquad\qquad \text{if } q = 1 \text{ and } h \text{ is convex,} \\ \dfrac{\sigma_{\min}}{(p+1)!}\left(\dfrac{(1-\theta)\kappa_{\delta,\min}^j}{3j!(L_{f,p}+\sigma_{\max})}\right)^{\frac{p+1}{p-j+1}} \epsilon_j^{j+1} \\ \qquad\qquad \text{otherwise,} \end{cases}$$

and (14.1.32) follows. If $k \in \mathcal{S}$, then $\rho_k \geq \eta_1$ and (14.1.33) follows from (14.1.32) and (14.1.15). □

We may then derive the expected evaluation-complexity result.

Theorem 14.1.10: qth-order complexity for the ARqpC algorithm

*Suppose that **f.D1pL**, **c.D1pL**, **h.L**, **h.S0**, and **w.Bb** hold. Suppose also that the* ARqpC *(p. 376) algorithm ensures, for each k, that either $\delta_{s_k,1} = 1$ if $q = 1$ and h is convex or otherwise (14.1.28) holds for all $j \in [1:q]$.*

1. *Suppose $q = 1$ and h is convex. Then there exist positive constants $\kappa^{s,1}_{\text{ARqpC}}$, $\kappa^{a,1}_{\text{ARqpC}}$, and $\kappa^{c}_{\text{ARqpC}}$ such that, for any $\epsilon \in (0,1]$, the* ARqpC *algorithm requires at most*

$$\#^F_{\text{ARqpC}} \stackrel{\text{def}}{=} \kappa^{a,1}_{\text{ARqpC}} \frac{w(x_0) - w_{\text{low}}}{\epsilon^{\frac{p+1}{p}}} + \kappa^{c,1}_{\text{ARqpC}} = \mathcal{O}\left(\epsilon^{-\frac{p+1}{p}}\right) \tag{14.1.34}$$

evaluations of f and c and at most

$$\#^D_{\text{ARqpC}} \stackrel{\text{def}}{=} \kappa^{s,1}_{\text{ARqpC}} \frac{w(x_0) - w_{\text{low}}}{\epsilon^{\frac{p+1}{p}}} + 2 = \mathcal{O}\left(\epsilon^{-\frac{p+1}{p}}\right) \tag{14.1.35}$$

evaluations of the derivatives of f and c of orders one to p to produce an iterate x_ϵ such that $\phi^1_{w,j}(x_\epsilon) \le \epsilon$ for all $j \in [1:q]$.

2. *Suppose that h is nonconvex or that $q > 1$. Then there exist positive constants $\kappa^{s,2}_{\text{ARqp}}$, $\kappa^{a,2}_{\text{ARqp}}$, and κ^{c}_{ARqp} such that, for any $\epsilon \in (0,1]^q$, the* ARqpC *algorithm requires at most*

$$\#^F_{\text{ARqpC}} \stackrel{\text{def}}{=} \kappa^{a,2}_{\text{ARqpC}} \frac{w(x_0) - w_{\text{low}}}{\min\limits_{j\in[1:q]} \epsilon_j^{j+1}} + \kappa^{c}_{\text{ARqpC}} = \mathcal{O}\left(\max_{j\in[1:q]} \epsilon_j^{-(j+1)}\right) \tag{14.1.36}$$

evaluations of f and c and at most

$$\#^D_{\text{ARqpC}} \stackrel{\text{def}}{=} \kappa^{s,2}_{\text{ARqpC}} \frac{w(x_0) - w_{\text{low}}}{\min\limits_{j\in[1:q]} \epsilon_j^{j+1}} + 2 = \mathcal{O}\left(\max_{j\in[1:q]} \epsilon_j^{-(j+1)}\right) \tag{14.1.37}$$

evaluations of the derivatives of f and c of orders one to p to produce an iterate x_ϵ such that $\phi^{\delta_\epsilon}_{w,j}(x_\epsilon) \le \epsilon_j \, \delta^j_{\epsilon,j}/j!$ for some $\delta_\epsilon \in (0,1]^q$ and all $j \in [1:q]$.

Proof. Given Lemmas 14.1.5, 14.1.6, and 14.1.8, the familiar argument can be extended without difficulty: the proof is essentially identical to that of Theorem 12.2.14 (p. 311) with f replaced by w, f_{low} replaced by w_{low}, (12.2.57) replaced by (14.1.31), and

$$\pi_j \stackrel{\text{def}}{=} \begin{cases} \dfrac{p+1}{p} & \text{if } q = 1 \text{ and } h \text{ is convex,} \\ j+1 & \text{otherwise.} \end{cases}$$

The constants are then given by

$$\kappa^{s,1}_{\text{ARqpC}} \stackrel{\text{def}}{=} \frac{(p+1)!}{\eta_1 \sigma_{\min}} \left(\frac{3j!(L_{w,p} + \sigma_{\max})}{1-\theta}\right)^{\frac{p+1}{p}} \quad \text{if } q = 1 \text{ and } h \text{ is convex,}$$

$$\kappa^{s,2}_{\text{ARqpC}} \stackrel{\text{def}}{=} \frac{(p+1)!}{\eta_1 \sigma_{\min}} \left(\frac{3j!(L_{w,p} + \sigma_{\max})}{(1-\theta)\kappa^q_{\delta,\min}}\right)^{\frac{p+1}{p-q+1}} \quad \text{otherwise,}$$

where $\kappa_{\delta,\min}$ is given by (14.1.29),

$$\kappa_{\text{ARqpC}}^{a,1} \stackrel{\text{def}}{=} \kappa_{\text{ARqpC}}^{s,1}\left(1+\frac{|\log\gamma_1|}{\log\gamma_2}\right), \qquad \kappa_{\text{ARqpC}}^{a,2} \stackrel{\text{def}}{=} \kappa_{\text{ARqpC}}^{s,2}\left(1+\frac{|\log\gamma_1|}{\log\gamma_2}\right),$$

and

$$\kappa_{\text{ARqpC}}^{c} \stackrel{\text{def}}{=} \frac{1}{\log\gamma_2}\left\{|\log\gamma_1|+\log\left(\frac{\sigma_{\max}}{\sigma_0}\right)\right\}+2,$$

with $\sigma_{\max} = \max\left[\sigma_0, \frac{\gamma_3 L_{w,p}}{1-\eta_2}\right]$. □

The bounds (14.1.36) and (14.1.37) are identical in order to those obtained in Theorem 12.2.5 (p. 297) for the TRq algorithm for the noncomposite case.

As for the noncomposite case, the bounds (14.1.34) and (14.1.35) are sharp.

Corollary 14.1.11: Slow convergence of the ARqpC algorithm for $q = 1$ and h convex

Suppose that $q = 1$ and that h is convex. Then the ARqpC algorithm applied to minimize w may require

$$\epsilon_1^{-\frac{p+1}{p}}$$

iterations and evaluations of f and c and of their derivatives of order one up to p to produce a point x_ϵ such that $\phi_{w,1}^{1}(x_\epsilon) \leq \epsilon_1$.

Proof. It is enough to consider the problem where $w = h(c(x))$ with $h(x) = |x|$ and c is the positive function f constructed in the proof of Theorem 12.2.15. □

Interestingly, it is not known at this stage whether the bounds (14.1.36) and (14.1.37) are sharp. The difficulty of building a suitable example is again caused by the presence of the last term (the multiple of $\|s_k\|^{p+1}$) on the right-hand side of (14.1.30), and it may be that the bounding process leading to this inequality is too loose.

Notes and references for Section 14.1

Composite problems of the form considered in this section occur in a variety of contexts, such as the LASSO[101] method in computational statistics [345], Tikhonov regularization of underdetermined estimation problems [225], compressed sensing [152], artificial intelligence [263], penalty or projection methods for constrained optimization [86], least Euclidean distance and continuous location problems [153], reduced-precision deep learning [361], image processing [24], and others [268]. In many of these applications, the function h is known to be convex, it is cheap to compute,[102] and its Lipschitz constant is known.

Our presentation in this section is extracted from [109] (which considers the more general simply constrained case) and extends earlier ideas presented for adaptive regularization [86, 108, 213]. The introduction of $\phi_{w,j}^{\delta}(x)$ in (14.1.6) as a way to define high-order approximate minimizers for nonsmooth problems is a feature of [109], as are the introduction of **h.S0** (in order to guarantee that the algorithm is well defined for $q < p$) and the general complexity results for the composite case.

Of course, problems of the form (14.1.1) can also be considered as general nonsmooth optimization problems with Lipschitz objective functions, and first-order algorithms for this general

[101] Least absolute shrinkage and selection operator, a regression method that involves penalizing the sum of absolute sizes of the regression variables.

[102] For example, if $h(x)$ is the Euclidean, ℓ_1-, or ℓ_∞-norm.

class of problems, such as proximal-gradient and bundle methods, might also be considered; see [227, 325] for recent references on these topics. Complexity bounds for the former can be found, for example, in [20, Chap. 10], but we are not aware of existing complexity results for the latter methods.

14.2 Problems with simple constraints

In Chapters 6 and 7, we discussed algorithms for finding low-order approximate minimizers of problems subject to the constraint that the vector of variables x must belong to a set $\mathcal{C}$, which we assumed to be convex. In this section, we exploit the similarity[103] between the optimality conditions (6.1.4) and (12.1.25) to extend our analysis to high orders while, at the same time, considering a less-restrictive definition of the feasible set. To set the scene, we now consider applying one of the high-order-capable algorithms of Chapters 12 and 13 to the problem

$$\underset{x \in \mathcal{X}}{\text{minimize}}\ f(x), \tag{14.2.1}$$

which looks suspiciously similar to problem (6.1.1) (p. 177), except that we have swapped the convex feasible set $\mathcal{C}$ for the conceivably more general $\mathcal{X}$. For the problem to make sense, we will assume the following.

X.CNe	The constraint set $\mathcal{X}$ is a closed and nonempty subset of $\mathbb{R}^n$.

As it turns out, minimal changes to our minimization algorithms are needed to handle the more general feasible set $\mathcal{X}$, so long as we use slightly modified variants of the optimality measures $\phi^\delta_{f,j}(x)$ that we defined in (12.1.10). More precisely, adapted variants of the ARqp, ARqpC, ARqpIDA, ARqpEDA, TRq, TRqIDA, and TRqEDA algorithms (which all allow the search for qth-order (ϵ,δ)-approximate minimizers) may be defined and involve only two easily stated modifications.

1. The first necessary modification is to alter the definition of the optimality measure to take the constraints into account. This is achieved by extending the definition (7.1.2) (p. 195) and defining $\phi^\delta_{f,\mathcal{X}(x),j}(x)$, a constrained variant of $\phi^\delta_{f,j}(x)$, as

   $$\phi^\delta_{f,\mathcal{X}(x),j}(x) \stackrel{\text{def}}{=} f(x) - \min_{\substack{\|d\|\leq\delta \\ d\in\mathcal{X}(x)}} T_{f,j}(x,d), \tag{14.2.2}$$

   where

   $$\mathcal{X}(x) \stackrel{\text{def}}{=} \{d \in \mathbb{R}^n \mid x + d \in \mathcal{X}\} \tag{14.2.3}$$

   is the set of moves from x that are feasible for the constraints defined by $\mathcal{X}$. In particular, when $j = 1$ and $\mathcal{X}$ is convex, this coincides with the measure $\omega_{f,\mathcal{C},1}(x)$ employed in (6.1.5) when $\delta = 1$, and $\phi^\delta_{f,\mathcal{X}(x),2}(x)$ will be zero when both $\omega_{f,\mathcal{C},1}(x) = 0$ and $\omega_{f,\mathcal{C},2}(x)$ from (6.1.7) is nonnegative. Similarly, when the optimality measure is applied to the model, it is also redefined to be

   $$\phi^\delta_{m_k,\mathcal{X}(x_k+s),j}(s) \stackrel{\text{def}}{=} f(x) - \min_{\substack{\|d\|\leq\delta \\ d\in\mathcal{X}(x_k+s)}} T_{m_k,j}(s,d). \tag{14.2.4}$$

[103] Not entirely accidental, of course.

2. The second modification occurs in the step computation, where one requires (in addition to the other conditions imposed by the particular algorithm) that
$$x_0 \in \mathcal{X} \quad \text{and} \quad s_k \in \mathcal{X}(x_k) \text{ for all } k, \tag{14.2.5}$$
which ensures that all iterates remain feasible.

Warning lights should start flashing at this point. Clearly, using and requiring (14.2.2)–(14.2.5) for nonconvex $\mathcal{X}$ may not be an issue in theory, but it obviously might make the computation of $\phi^{\delta}_{f,\mathcal{X}(x),j}(x)$, $\phi^{\delta}_{m_k,\mathcal{X}(x_k+s),j}(s)$, and s_k significantly more complicated than if it had been convex. Since we have so far made no assumption to the contrary, $\mathcal{X}$ might not be convex—or even connected—meaning that the measure $\phi^{\delta}_{f,\mathcal{X}(x),j}(x)$ is no longer automatically continuous. While this might be a concern if we were to consider $\phi^{\delta}_{f,\mathcal{X}(x),j}(x)$ as a guaranteed measure of optimality, we have already noticed for $\phi^{\delta}_{f,j}(x)$ in Example 12.2.2 that this is not necessarily our intention. In particular, when δ is too large, the associated optimality displacement might suggest a step that results in an objective-function value smaller than that at x, even when x is a local (but not global) minimizer. Again, this feature should be considered as an advantage if our aim is to minimize the objective function, rather than simply to attain local optimality. Nonetheless, $\phi^{\delta}_{f,\mathcal{X}(x),j}(x)$ retains the crucial property that a very small value of the measure implies a very small local decrease of f, as may be stated in a variant of Theorem 12.1.8 (p. 290) that we now derive. We shall take this opportunity to revise the bounds in Theorem 12.1.8 slightly, a modification that will be useful later.

Theorem 14.2.1: Stability of approximate constrained minimizers

Suppose that f is j-times continuously differentiable and that, for some $\eta \geq 0$,
$$\phi^{\delta_j}_{f,\mathcal{X}(x_\epsilon),j}(x_\epsilon) \leq \epsilon_j \frac{\delta_j^j}{j!} + \eta$$
for some $\eta \geq 0$, $\delta_j \in (0,1]$, and $\epsilon_j \in (0,1]$, where $\phi^{\delta_j}_{f,\mathcal{X}(x),j}(x)$ is defined by (14.2.2) for some closed nonempty feasible set $\mathcal{X}$. Suppose also that $\nabla^j_x f$ is Lipschitz continuous with constant $L_{f,j}$ in an open neighbourhood of a point x_ϵ of radius larger than δ_j. Then, if $L_{f,j} > 0$,
$$f(x_\epsilon + d) \geq f(x_\epsilon) - 2\epsilon_j \frac{\delta^j}{j!} - \eta \quad \text{for all } d \in \mathcal{X}(x_\epsilon) \text{ with } \|d\| \leq \min\left[\delta_j, \frac{(j+1)\,\epsilon_j}{L_{f,j}}\right], \tag{14.2.6}$$
and
$$f(x_\epsilon + d) \geq f(x_\epsilon) - \epsilon_j \frac{\delta_j^j}{j!} - \eta \quad \text{for all } d \in \mathcal{X}(x_\epsilon) \text{ with } \|d\| \leq \delta_j \tag{14.2.7}$$
otherwise.

Proof. The proof is identical to that of Theorem 12.1.8 (p. 290) with $\phi^{\delta}_{f,\mathcal{X}(x),j}(x)$ replacing $\phi^{\delta}_{f,j}(x)$, except that a term $-\eta$ now appears on the left-hand side of (12.1.38) and (12.1.39). Note that (14.2.6) and (14.2.7) describe the behaviour of f in a neighbourhood of x_ϵ, and thus the fact that this neighbourhood is potentially restricted to its feasible part does not affect the outcome. □

As a consequence, algorithms whose aim is to make $\phi^{\delta_j}_{f,\mathcal{X}(x),j}(x)$ small are meaningful for optimization over $\mathcal{X}$. But is it enough for our purposes to simply assume that $\mathcal{X}$ is closed and nonempty? Not really, because we must also make sure that handling the modified subproblems (14.2.2) and (14.2.5) does not affect our oracle-based notion of evaluation complexity. Thus we also need to assume that if $f(x_k)$ and $\{\nabla^\ell_x f(x_k)\}_{\ell=1}^p$ have been evaluated (and counted as oracle calls in our estimate of complexity), no further call to any oracle is needed to solve (14.2.2) or to ensure (14.2.5) when computing the step s_k. This, of course, would not be the case in the context of Chapter 7 because, should we define $\mathcal{X} = \{x \in \mathcal{C} \mid c(x) = 0\}$, this would require evaluation of the oracle $c(x)$. Thus we are led to the following definition.

Key Definition

Simple constraints are constraints such that the associated feasible set is closed and nonempty and the solution of the subproblem defining the optimality measure (14.2.2) and the computation of the step s_k satisfying (14.2.5) do not require any oracle evaluation.

In line with our formal attitude concerning evaluation complexity so far, the computational cost of solving for the optimality measure or the step does not affect the above definition. However, it is abundantly clear that our interest in practice is with "simple" problems for which this cost is small, at least compared to that of evaluating f or its derivatives. Section 14.2.2 will review some motivating examples for which this is the case.

For the sake of conciseness, we do not restate the relevant variants of the ARqp, ARqpC, ARqpIDA, ARqpEDA, TRq, TRqIDA, and TRqEDA algorithms in detail, since they only differ from those of the corresponding unconstrained version by the introduction of (14.2.2)–(14.2.5). We will append the suffix -SC (for "simple constraints") to the names to refer to the modified algorithms, and thus ARqp-SC is the variant of ARqp using $\phi^{\delta}_{f,\mathcal{X}(x),j}(x)$ instead of $\phi^{\delta}_{f,j}(x)$ and guaranteeing (14.2.5).

14.2.1 Evaluation complexity for optimization with simple constraints

Of course, the presence of constraints requires a corresponding adaptation of our stated assumptions: we will only require suitable smoothness and boundedness properties to hold in a open, convex set containing $\mathcal{X}$. The theoretical framework we have relied on thus far has been broad and designed for simplicity of exposition rather than tightness. In the interests of what is to come, at this stage we need to slightly refine our assumptions. Indeed, readers may refer back to the previous chapters and convince themselves that the results we have derived remain true under these new, slightly more technical requirements.

We suppose that we start the algorithm under consideration from a given $x_0 \in \mathcal{X}$ and define the sublevel set

$$\mathcal{L}_f(x_0) \stackrel{\text{def}}{=} \{x \in \mathcal{X} \mid f(x) \leq f(x_0)\}. \tag{14.2.8}$$

We shall then assume[104] the following.

f.D1qLaB.X

f is q-times continuously differentiable in an open, convex set $\mathcal{X}_0$ containing $\mathcal{X}$, and there are constants $L_{f,j} \geq 1$ and $B_{f,j} \geq 1$ such that

$$\|\nabla_x^j f(x) - \nabla_x^j f(y)\| \leq L_{f,j}\|x - y\| \tag{14.2.9}$$

and

$$\|\nabla_x^j f(x)\| \leq B_{f,j} \tag{14.2.10}$$

for any $x \in \mathcal{L}_f(x_0)$, any $y \in \mathcal{X}_0$, and all $j \in [1 : q]$.

That such an assumption suffices rather than a more general one involving Lipschitz continuity and derivative boundedness is perhaps unsurprising. The objective values for successful iterates x_k generated by our algorithms decrease monotonically, and thus each such x_k lies in $\mathcal{L}_f(x_0)$. We only build models at such points, and the boundedness of these follow from (14.2.10). Unsuccessful iterates require that we investigate the function at points outside of $\mathcal{L}_f(x_0)$, but we can relate these to the successful iterates via Taylor's theorem (Corollary A.8.4 (p. 437)) and the *anchored-Lipschitz* condition (14.2.9) anchored at $x = x_k$ (hence the "La" mnemonic in the assumption's name). Our earlier proofs rely on the equivalence of assumed Lipschitz continuity and derivative boundedness throughout $\mathcal{X}_0$, but now the boundedness is explicit in (14.2.10); the theoretical bounds that we have derived earlier involved $L_{f,j}$, but as a careful reader will see, some of these can be replaced by a $B_{f,j}$. Since there is no convexity or connectivity requirement in the definition of simple constraints, the reader may wonder why **f.D1qLaB.X** requires $\mathcal{X}_0$ to be convex, but this is purely so that we may apply Taylor's theorem as needed. We also note in passing that the anchored-Lipschitz requirement (14.2.9) is automatic if the required derivatives of f are Lipschitz over $\mathcal{X}_0$, while (14.2.10) also follows from Theorem A.8.5 if additionally f is Lipschitz over the same set.

By analogy, we will refer to the assumption **f.D1pLaB.X** as being precisely as in **f.D1qLaB.X** but with the index $j \in [1 : p]$ rather than $j \in [1 : q]$. Furthermore, we shall use assumptions **f.D0qLaB.X** and **f.D0pLaB.X**, which are the obvious analogues, when the indices in (14.2.9) run over $j \in [0 : q]$ and $j \in [0 : p]$.

We shall also require that the objective function be bounded below on the feasible region.

f.bB.X

$f(x)$ is bounded below on $\mathcal{X}$; that is, there exists a constant f_{low} such that, for all $x \in \mathcal{X}$,

$$f(x) \geq f_{\text{low}}.$$

Remarkably, all results valid for the ARqp and TRq (general unconstrained case), ARqpC (composite unconstrained case), ARqpIDA, ARqpEDA, TRqIDA, and TRqEDA (unconstrained case with IDA and EDA) algorithms extend in an entirely straightforward manner, with just a single caveat. This is that the choice $\delta_{k,j} = 1$ may no longer be possible when $q = 2$ in Step 2 of ARqp, ARqpIDA, and ARqpEDA, because the optimality conditions of

[104] Strictly, the requirement (14.2.10) is not necessary for $j \in [2 : q]$ since, as the reader may verify, it is implied by (14.2.9) for $j \in [1 : q-1]$ using essentially the same proof as for Theorem A.8.5 (p. 438) but restricting x to $\mathcal{L}_f(x_0)$; we include it simply as a reminder to the reader. Moreover, (14.2.10) may be removed altogether if (14.2.9) is required for $j \in [0 : q]$.

the problem of minimizing the model in $\mathcal{X}$ no longer imply that the approximating quadratic is convex. Hence the argument in the proof of Theorem 12.1.4 fails for $j = 2$ unless $\mathcal{X} = \mathbb{R}^n$, but note that it remains valid for $q = 1$ when the constraints are convex.

We first provide a result that states the objective-function decrease that is guaranteed after successful iterations of the SC algorithms.

Lemma 14.2.2: Function decrease at successful iterations (in the presence of simple constraints)

*Suppose that **f.D1qLaB.X** (for TR algorithms) or **f.D1pLaB.X** (for AR algorithms) holds. Then there exist algorithm-dependent constants κ_{decr} so that*

$$f(x_k) - f(x_{k+1}) \geq \kappa_{\text{decr}} \min_{j\in[1:q]} \epsilon_j^{\pi_j} \tag{14.2.11}$$

for each $k \in \mathcal{S}$, where

$$\pi_j = \begin{cases} \dfrac{p+1}{p-j+1} & \text{for AR}qp\text{-SC} \quad \text{for } q=1,\ \mathcal{X} \text{ convex} \\ & \qquad\qquad\quad\ \text{or } q=2,\ \mathcal{X}=\mathbb{R}^n; \\ \dfrac{j(p+1)}{p} & \text{for AR}qp\text{-SC} \quad \text{in all other cases;} \\ \dfrac{p+1}{p-j+1} & \text{for AR}qp\text{C-SC} \quad \text{for } q=1,\ h, \mathcal{X} \text{ convex;} \\ j+1 & \text{for AR}qp\text{C-SC} \quad \text{in all other cases;} \\ q+1 & \text{for TR}q\text{-SC}. \end{cases} \tag{14.2.12}$$

*Suppose instead that **f.D0qLaB.X** (for TR algorithms) or **f.D0pLaB.X** (for AR algorithms) holds. Then the guaranteed decreases for the TRqIDA-SC and TRqEDA-SC algorithms are identical to those stated for TRq-SC, and the decreases for the ARqpIDA-SC and ARqpEDA-SC algorithms are identical to those stated for ARqp-SC.*

Proof. This is just a compilation of the results of Lemmas 12.2.13 (p. 310) (for ARqp), 13.1.18 (p. 348) (for ARqpIDA), 13.3.7 (p. 367) (for ARqpEDA), 14.1.9 (p. 383) (for ARqpC), 12.2.4 (p. 297) (for TRq), 13.1.5 (p. 333) (for TRqIDA), and 13.3.2 (p. 361) (for TRqEDA), taking into account the comment we just made on the choice of $\delta_{k,j}$ in ARqp-SC, ARqpC-SC, ARqpIDA-SC, and ARqpEDA-SC. The reader is referred to the appropriate lemma for the explicit definition of each κ_{decr} constant. □

Another core ingredient is our capacity to bound the total number of iterations by a multiple of that of successful ones.

Lemma 14.2.3: Count of successful iterations

*Suppose that the TRq-SC algorithm is applied and **f.D1qLaB.X** holds or that the ARqp-SC algorithm is applied and **f.D1pLaB.X** holds. Then there are algorithm-dependent constants κ_{uns} and κ_{unc} for which*

$$k \leq \kappa_{\text{uns}} |\mathcal{S}_k| + \kappa_{\text{unc}}, \tag{14.2.13}$$

*where $\mathcal{S}_k = \mathcal{S} \cap [0:k]$ and $\mathcal{S}$ is the index set of successful iterations. The same is true for the TRqIDA-SC and TRqEDA-SC algorithms under **f.D0qLaB.X** and the ARqpIDA-SC and ARqpEDA-SC algorithms under **f.D0pLaB.X**.*

Proof. The constants $\kappa_{\text{uns}}, \kappa_{\text{unc}} > 0$ depend on the details of the algorithm under consideration. Such an inequality is guaranteed by Lemma 2.3.1 (p. 36) for the exact-evaluation trust-region algorithms and by Lemma 2.4.1 (p. 44) for the regularization ones. The justification is slightly more involved for the algorithms that use inexact evaluations: Lemmas 13.1.1 (p. 331) and 13.1.9 (p. 341) provide an upper bound on $|\mathcal{T}_k|$, the number of unsuccessful and successful iterations up to k, ignoring the accuracy-improving iterations in $\mathcal{A}_k$. However, we argue in the proofs of Theorems 13.1.8 (p. 335) and 13.1.19 (p. 349)—and by analogy those of Theorems 13.3.4 (p. 362) and 13.3.8 (p. 368)—that the number of such iterations is small compared with $|\mathcal{S}_k|$, so that a bound of the form (14.2.13) also holds for these cases. □

We may now state our combined complexity result.

Theorem 14.2.4: Complexity in the presence of simple constraints

Suppose that the feasible set $\mathcal{X}$ corresponds to simple constraints, that **X.CNe**, **f.bB.X**, *and* **f.D0pLaB.X** *(or* **f.D0qLaB.X** *in the trust-region case) hold, and that $\phi^{\delta_j}_{f,\mathcal{X}(x),j}(x)$ is as defined by (14.2.2). Suppose also that, for all iterations k of regularization algorithms, either $\delta_{k,j} = 1$ for $j \in [1:q]$ if $q = 1$ and $\mathcal{X}$ is convex or $q = 2$ and $\mathcal{X} = \mathbb{R}^n$ (as discussed above) or that (12.2.51) holds (as allowed by Lemma 12.2.9). Then the complexity results for the* ARqp, ARqpC, ARqpIDA, ARqpEDA, TRq, TRqIDA, *and* TRqEDA *algorithms extend to their variants for simple constraints, with the aim of finding an $x_\epsilon \in \mathcal{X}$ for which*

$$\phi^{\delta_j}_{f,\mathcal{X}(x_\epsilon),j}(x_\epsilon) \le \epsilon_j \frac{\delta_j^j}{j!} \quad \text{for all} \quad j \in [1:q],$$

as indicated in the following table.

Adapted algorithm	*Applicable result*	*Evaluation-complexity order with simple constraints*	
ARqp-SC	*Thm. 12.2.14 (p. 311)*	$\mathcal{O}\left(\max_{j\in[1:q]} \epsilon_j^{-\frac{p+1}{p-j+1}}\right)$	*($q = 1$; $\mathcal{X}$ convex or $q = 2$; $\mathcal{X} = \mathbb{R}^n$)*
		$\mathcal{O}\left(\max_{j\in[1:q]} \epsilon_j^{-\frac{j(p+1)}{p}}\right)$	*(otherwise)*
ARqpC-SC	*Thm. 14.1.10 (p. 384)*	$\mathcal{O}\left(\max_{j\in[1:q]} \epsilon_j^{-\frac{p+1}{p-j+1}}\right)$	*($q = 1$; $h, \mathcal{X}$ convex)*
		$\mathcal{O}\left(\max_{j\in[1:q]} \epsilon_j^{-(j+1)}\right)$	*(otherwise)*
ARqpIDA-SC	*Thm. 13.1.19 (p. 349)*	$\mathcal{O}\left(\max_{j\in[1:q]} \epsilon_j^{-\frac{p+1}{p-j+1}}\right)$	*($q = 1$; $\mathcal{X}$ convex or $q = 2$; $\mathcal{X} = \mathbb{R}^n$)*
		$\mathcal{O}\left(\max_{j\in[1:q]} \epsilon_j^{-\frac{j(p+1)}{p}}\right)$	*(otherwise)*

(continued)

Adapted algorithm	*Applicable result*	*Evaluation-complexity order with simple constraints*
ARqpEDA-SC	*Thm. 13.3.8 (p. 368)*	$\mathcal{O}\left(\max_{j\in[1:q]} \epsilon_j^{-\frac{p+1}{p-j+1}}\right)$ ($q=1$; $\mathcal{X}$ convex or $q=2$; $\mathcal{X}=\mathbb{R}^n$) $\mathcal{O}\left(\max_{j\in[1:q]} \epsilon_j^{-\frac{j(p+1)}{p}}\right)$ (otherwise)
TRq-SC	*Thm. 12.2.5 (p. 297)*	$\mathcal{O}\left(\max_{j\in[1:q]} \epsilon_j^{-(q+1)}\right)$
TRqIDA-SC	*Thm. 13.1.8 (p. 335)*	$\mathcal{O}\left(\max_{j\in[1:q]} \epsilon_j^{-(q+1)}\right)$
TRqEDA-SC	*Thm. 13.3.4 (p. 362)*	$\mathcal{O}\left(\max_{j\in[1:q]} \epsilon_j^{-(q+1)}\right)$

Moreover, the orders of complexity mentioned in the last column of the table are sharp.

Proof. The validity of the quoted theorems results from the argument above, leading to Lemma 14.2.2. The sharpness results immediately follow from the sharpness of the unconstrained bounds and the fact that unconstrained problems are a special case of simply constrained ones ($\mathcal{X} = \mathbb{R}^n$). □

14.2.2 Examples of simple constraints

We now discuss some interesting examples of simple constraints.

14.2.2.1 Convex constraints.

Clearly, many problems with convex constraints belong to the class of simply constrained problems, such as those involving bound constraints or constraints for which the projection onto the feasible set can be computed purely internally—for example, simple sets such as boxes, cylinders, and spheres [131, Sect. 12.1], the ordered simplex [37, 68, 69, 70, 244], and more general polyhedra [272]. We have already seen this for low-order approximate minimizers in Section 6.1. For the purpose of the discussion which follows, we note that a workhorse for optimization over convex constraints is the "projected-gradient-path" technique in which minimization of (a model of) the objective function is carried along the path $P(x_k - t\nabla_x^1 f(x_k))$, where $t \geq 0$ and P is the projection onto the feasible set. This is particularly useful for trust-region methods (see [131, Chap. 12], for instance), where the solution of the first-order trust-region subproblem can be obtained by following this projected-gradient path as far as possible within the trust region [131, Sect. 12.1.3].

14.2.2.2 Sparsity constraints.

A fascinating class of simply constrained problems are those involving a *sparsity constraint* [21, 152], that is, problems in which the number of nonzeros in the vector representing an approximate minimizer cannot exceed a given threshold. This constraint is often written as

$$\|x\|_0 \leq n_{\rm nz}, \tag{14.2.14}$$

where $\|x\|_0$ is the number of nonzero entries in x and $n_{\rm nz} \leq n$ is the threshold—be aware though that the use of the "norm" symbol here is misleading since $\|\cdot\|_0$ is not a norm; indeed the set (14.2.14) is not convex. We denote the set of indices of the nonzero components of x by

$$\mathcal{A}(x) \stackrel{\rm def}{=} \{i \in [1:n] \mid x_i \neq 0\}, \tag{14.2.15}$$

which is sometimes called the *support* of x; clearly, $\mathcal{A}(x)$ contains at most $n_{\rm nz}$ indices if x is feasible. Checking that the constraint (14.2.14) holds is trivial, and a Euclidean projection onto the feasible set is easily computed using the *hard-thresholding* operator, i.e., selecting the $n_{\rm nz}$ components of x whose absolute value is largest, and setting the remainder to zero. Of course, this projection isn't unique due to potentially equal components in x, but simple rule-breaking ties can be introduced to ensure uniqueness [114].

We contend that these nonconvex sparsity constraints are simple in the sense we have defined above. To justify this, we develop a first-order trust-region framework—we shall refer to this as the TR1-SP algorithm, an instance of the TRq class—for solving the "sparsity-constrained" problem

$$\min_{x\in\mathbb{R}^n,\ \|x\|_0\leq n_{\rm nz}} f(x). \tag{14.2.16}$$

While finding a global solution to (14.2.16) is likely of most interest in practice [114,381], this is clearly beyond our current capabilities in most cases. Thus, as always, we reduce our aim to that of finding a local minimizer. Superficially, one might believe that simply picking any set $\mathcal{A}_0$ of $n_{\rm nz}$ distinct indices in $[1:n]$, assigning variables $x_i = 0$ for all $i \notin \mathcal{A}_0$, and (locally) minimizing $f(x)$ over the remaining variables $x_i \in \mathcal{A}_0$ will be enough. In most cases, none of the $x_i \in \mathcal{A}_0$ so found will be zero, and in this case, the resulting x will indeed be a local minimizer of (14.2.16). However, this offers no guarantee. What if some of the $x_i \in \mathcal{A}_0$ are zero?

Our aim here is to show that methods within the simple-constraints framework that we have just analyzed do not suffer from such a defect and are capable of finding at least a first-order approximate minimizer. Indeed, we argue that if a first-order approximate minimizer is sought, problem (14.2.16) fits the "simply constrained" class that is our focus with the feasibility set

$$\mathcal{X} = \{x \in \mathbb{R}^n \mid \|x\|_0 \leq n_{\rm nz}\}.$$

We remarked at the start of Section 14.2 that such a set needs to play a significant role in two parts of our tentative TR1-SP algorithm, and the first of these is the computation of an optimality measure at $x \in \mathcal{X}$. At slight variance with (14.2.2), we define this to be

$$\phi^1_{f,\mathcal{X}(x),1}(x) \stackrel{\rm def}{=} f(x) - \min_{\substack{\|d\|_\infty\leq\delta\\ d\in\mathcal{X}(x)}} T_{f,1}(x,d), \tag{14.2.17}$$

where we use the ℓ_∞-norm to restrict the range of possible optimality displacements rather than the Euclidean one, and we recollect the definition $\mathcal{X}(x) = \{s \in \mathbb{R}^n \mid \|x+s\|_0 \leq n_{\rm nz}\}$ in (14.2.3)—recall that we commented at the end of Section 6.1.1 on the possibility of using non-Euclidean norms to suit the geometry of the problem at hand, which will be the case here. If we can compute a global solution of the problem involved in (14.2.17), that is, the problem given by

$$\max_{\substack{\|d\|_\infty \le \delta \\ \|x+d\|_0 \le n_{\rm nz}}} \quad \ell(d) \stackrel{\rm def}{=} \langle \nabla_x^1 f(x), d \rangle, \tag{14.2.18}$$

then not only do we obtain the required value of the measure $\phi^\delta_{f,\mathcal{X}(x),j}(x)$ but the resulting minimizing displacement d_k may also be used to define the step—refer to Step 2 of the TRq algorithm (p. 292). Thus we have in place all the pieces necessary to build the TR1-SP algorithm.

We now focus on the crucial linear problem (14.2.18). A first observation is that, because of the nonconvexity of $\mathcal{X}$, the gradient-path technique that helped so much in the convex context can no longer be applied here, and moreover this is not due to the change from the Euclidean norm to the ℓ_∞ one but to the possibly arbitrary value of δ. Indeed, one may also consider the Euclidean variant

$$\max_{\substack{\|d\|_2 \le \delta \\ \|x+d\|_0 \le n_{\rm nz}}} \quad \ell(d) \tag{14.2.19}$$

and examine Figure 14.2.1 on the current page, where we attempt to solve problem (14.2.18) with $n = 2$, $n_{\rm nz} = 1$, $x = (\frac{1}{2}, 0)$, $\nabla_x^1 f(x) = (-\frac{1}{3}, -1)$, and $\delta = \frac{1}{2}$ (left panel), or problem (14.2.19) with the same n, $n_{\rm nz}$, x, and $\nabla_x^1 f(x)$ but $\delta = \frac{65}{100}$ (right panel). In both panels, the steepest-descent direction originating from x is shown in cyan and its (now discontinuous!) projection on $\mathcal{X}$ in blue. The boundary of the ℓ_∞ (left) or Euclidean (right) ball centered at x and of radius δ is shown in dotted red lines, while level lines for the linear objective function are parallel to the black line orthogonal to the steepest-descent direction. In both cases, the solution of the problem is the top extremity of the vertical green line, which does not lie on the projected gradient path, in contrast with the results of [131, Sect. 12.1.3] for convex $\mathcal{X}$.

Thus we are condemned to abandon projected-gradient-path techniques if we wish to solve (14.2.18) for general δ and have to find another approach, which we now outline. It is based on the special structure of the solution of problem (14.2.18)

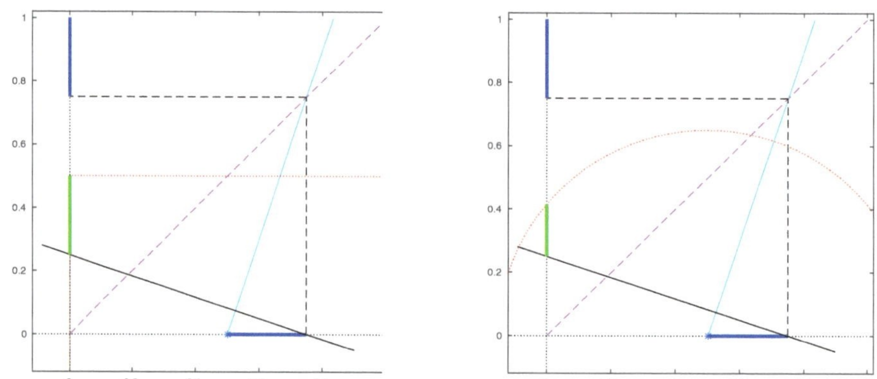

Figure 14.2.1. Examples showing that the solution (top of the vertical green line) of problems (14.2.18) (left) and (14.2.19) (right) may not lie on the projected-gradient path (in blue) for every value of δ ($\frac{1}{2}$ on the left, $\frac{65}{100}$ on the right). The dotted red line is the boundary of $\|d\| \le \delta$.

Lemma 14.2.5: Optimal sparse solution of problem (14.2.18)

Suppose that $x \in \mathcal{X}$. Then the solution to problem (14.2.18) is given by

$$d^* = -\delta \sum_{i\in\mathcal{A}_\delta(x)} \text{sign}([g]_i)e_i - \sum_{i\in\mathcal{A}(x)\setminus\mathcal{A}_\delta(x)} [x]_i e_i - \sum_{i\in\mathcal{A}_\delta^c(x)} \tau_i\Big(\delta\,\text{sign}([g]_i) - [x]_i\Big)e_i, \tag{14.2.20}$$

where $g = \nabla_x^1 f(x)$, $\mathcal{A}(x)$ is defined as in (14.2.15), $\mathcal{A}_\delta(x) \stackrel{\text{def}}{=} \{i \in [1:n] \mid |[x]_i| > \delta\}$, $\mathcal{A}_\delta^c(x) \stackrel{\text{def}}{=} [1:n] \setminus \mathcal{A}_\delta(x)$, and the binary variables τ_i solve the 0-1 knapsack problem[105] given by

$$\max_{\tau\in\{0,1\}^{n-|\mathcal{A}_\delta(x)|}} \sum_{i\in\mathcal{A}_\delta^c(x)} \tau_i\Big(\delta\,|[g]_i| - [x]_i[g]_i\Big) \text{ such that } \sum_{i\in\mathcal{A}_\delta^c(x)} \tau_i \le n_{\text{nz}} - |\mathcal{A}_\delta(x)|. \tag{14.2.21}$$

Proof. Consider first the case where $n_{\text{nz}} = n$, that is, when the sparsity constraint is, in effect, inactive. Because the problem is linear and the step only constrained by the ℓ_∞ ball of radius δ, we have that its optimal value is given by

$$\ell(d^*) = -\delta\sum_{i=1}^n |[g]_i| \text{ at } d^* = -\delta\sum_{i=1}^n \text{sign}([g]_i)e_i;$$

the ith component of the minimizer can be any value in $[-\delta,\delta]$ if $[g]_i = 0$.

On reintroducing the sparsity constraint by letting $n_{\text{nz}} < n$, such a solution may no longer be feasible since, if $\|g\|_0 > n_{\text{nz}}$, the sparsity constraint can be violated at $x + d^*$. Observe that the objective function of problem (14.2.18) is separable in its variables, and we may thus consider solving separately in the two subspaces

$$\mathcal{M}(x) \stackrel{\text{def}}{=} \operatorname*{span}_{i\,\in\,\mathcal{A}_\delta(x)}\{e_i\} \text{ and } \mathcal{N}(x) \stackrel{\text{def}}{=} \mathcal{M}(x)^\perp = \operatorname*{span}_{i\,\in\,\mathcal{A}_\delta^c(x)}\{e_i\}.$$

If $i \in \mathcal{A}_\delta(x)$, zeroing this component would violate the constraint that $|d_i| \le \delta$. As a consequence, the index i must belong to the support $\mathcal{A}(x+d^*)$ and the sparsity constraints are inactive in $\mathcal{M}(x)$. Thus the solution of the problem in this subspace is given by

$$d^*_{\mathcal{M}(x)} \stackrel{\text{def}}{=} -\delta\sum_{i\in\mathcal{A}_\delta(x)} \text{sign}([g]_i)e_i. \tag{14.2.22}$$

To obtain the solution in $\mathcal{N}(x)$, we consider first a step from the projection of x onto $\mathcal{N}(x)$, $[x]_{\mathcal{N}(x)}$, to zero, at which point contributions along any coordinate direction e_i for $i \in \mathcal{A}_\delta^c(x)$ may be considered, subject to the sparsity constraint. However, the origin is not at the center of the ℓ_∞ ball of radius δ in $\mathcal{N}(x)$ centered at $[x]_{\mathcal{N}(x)}$ and thus the step that is made in direction e_i ($i \in \mathcal{N}(x)$) is not $-\delta\,\text{sign}([g]_i)e_i$ but $\delta\,\text{sign}([g]_i) - [x]_i$. Thus the solution in $\mathcal{N}(x)$ can be written as

$$d^*_{\mathcal{N}(x)} = -[x]_{\mathcal{N}(x)} - \sum_{i\in\mathcal{A}_\delta^c(x)} \tau_i\Big(\delta\,\text{sign}([g]_i) - [x]_i\Big)e_i, \tag{14.2.23}$$

[105] This well-studied combinatorial problem takes its name from the dilemma of the hiker preparing a backpack: what items to choose to maximize their usefulness on the trek while still fitting in the bag. We refer the reader to [127, 283, 367] for in-depth coverage of the knapsack problem.

where τ_i is a binary variable which is one if the ith component occurs in $\mathcal{A}(x+d_*)$ and zero otherwise. We then obtain from (14.2.22) and (14.2.23) that the optimal value of the objective function for problem (14.2.18) is

$$\begin{aligned}\ell(d^*) &= -\delta \sum_{i\in\mathcal{A}_\delta(x)} |[g]_i| - \sum_{i\in\mathcal{A}_\delta^c(x)} [x]_i[g]_i - \sum_{i\in\mathcal{A}_\delta^c(x)} \tau_i\Big(\delta\,|[g]_i| - [x]_i[g]_i\Big)\\ &= -\delta \sum_{i\in\mathcal{A}_\delta(x)} |[g]_i| - \sum_{i\in\mathcal{A}(x)\setminus\mathcal{A}_\delta(x)} [x]_i[g]_i - \sum_{i\in\mathcal{A}_\delta^c(x)} \tau_i\Big(\delta\,|[g]_i| - [x]_i[g]_i\Big).\end{aligned}$$

To make $\ell(d^*)$ minimal, it is thus enough to consider the values of the binary variables τ_i which solve the discrete maximization problem (14.2.21), and we obtain (14.2.20) from (14.2.22), (14.2.23), and the fact that $d^* = d^*_{\mathcal{M}(x)}+d^*_{\mathcal{N}(x)}$. It is then easy to verify that $\|d^*\|_\infty \le \delta$ and, since $\|[x]_{\mathcal{M}(x)} + d^*_{\mathcal{M}(x)}\|_0 \le |\mathcal{A}_\delta(x)|$ and $\|[x]_{\mathcal{N}(x)} + d^*_{\mathcal{N}(x)}\|_0 \le \sum_{i\in\mathcal{A}_\delta^c(x)} \tau_i$, that

$$\|x+d^*\|_0 \le |\mathcal{A}_\delta(x)| + \sum_{i\in\mathcal{A}_\delta^c(x)} \tau_i \le |\mathcal{A}_\delta(x)| + n_{\rm nz} - |\mathcal{A}_\delta(x)| = n_{\rm nz},$$

as desired. □

Thus the solution of problem (14.2.18) may be obtained from (14.2.20)–(14.2.21) by computations that remain purely internal to the TR1-SP minimization algorithm, without any call to an external oracle.

Since we can solve problem (14.2.18) without recourse to external oracles, designing the TR1-SP algorithm, a first-order trust-region method for problem (14.2.16), on the basis of the TRq algorithm (p. 292) with $q=1$ is straightforward. It suffices to use the knapsack subproblem solver (14.2.21)–(14.2.20) both to compute (14.2.17) as a replacement for $\phi_{f,1}^{\delta_k}(x_k)$ in Step 1 and to compute the step s_k that solves

$$\min_{s\in\mathbb{R}^n} T_{f,1}(x_k,s) = \langle g_k, s\rangle \;\text{ subject to }\; \|s\|_\infty \le \Delta_k \;\text{ and }\; \|x_k+s\|_0 \le n_{\rm nz},$$

along with the required $\phi_{m_k,1}^{\delta_k}(s_k)$, in Step 2, the rest of the algorithm being unchanged. As a consequence we obtain a practical algorithm for finding a first-order approximate local minimizer of the sparse-constrained optimization problem that requires at most $\mathcal{O}(\epsilon_1^{-2})$ evaluations of f and its derivatives by virtue of Theorem 14.2.4. Note that no claim of global optimality or "guaranteed recovery"[106] is made here.

We also note that the technique we have described easily extends to quadratic models whose Hessian is diagonal, since a simple modification of Lemma 14.2.5 holds for this case. Due to the geometry of the ℓ_∞ trust region, bounds on the variables are also easy to include in this approach.

Obtaining a regularization algorithm AR1-SP by adapting the ARqp algorithm (p. 300) with $q=1$ to handle the sparsity constraint is also possible. As for the trust-region case, the knapsack subproblem solver (14.2.20)–(14.2.21) can be applied to check the termination condition (for $q=p=1$) in Steps 1 and 2. For the solution of the step defining the subproblem

$$\min_{s\in\mathbb{R}^n} T_{f,1}(x_k,s) + \tfrac{1}{2}\sigma_k\|s\|^2 = \langle g_k,s\rangle + \tfrac{1}{2}\sigma_k\|s\|^2 \;\text{ subject to }\; \|x_k+s\|_0 \le n_{\rm nz}$$

[106] By "guaranteed recovery" we mean the ability of the algorithm to recover the global, sparsest solution of (14.2.16) when the objective f in (14.2.16) is a linear least-squares problem, as is common in compressed sensing [152].

in Step 2, we may use the extension of the knapsack-based approach hinted at in the previous paragraph, or even local first-order methods since only local solutions of this subproblem are needed (due to the local approximate termination conditions in the step calculation). The resulting AR1-SP algorithm then finds a first-order approximate minimizer of problem (14.2.16) in $\mathcal{O}(\epsilon_1^{-2})$ evaluations of the objective function and its gradient (see Theorem 14.2.4).

14.2.2.3 Manifold optimization.

Another example of nonconvex simply constrained problems with nonconvex feasible sets can be found within the class of *problems defined on smooth manifolds.* Again, either verification is simple or tools have been developed for it that do not involve oracle calls. Furthermore, under suitable conditions, projections (sometimes called retractions) of tangent vectors on the manifold may also be computed without reevaluating the nonlinear function defining the manifold. Simple examples include minimization on a sphere (as is the case in Rayleigh-quotient minimization for computing eigenvalues) or on the Stiefel manifold of orthogonal matrices. Even when there is an associated computational cost to ensuring feasibility (irrelevant for our current purposes, as we have discussed), the data science and information processing communities have found the "manifold toolbox" computationally very useful.

To be more specific, consider the case of the Stiefel manifold $\mathcal{M}$ of orthonormal matrices, and note that it is known [271, Lem. 1] that this manifold is included in the boundary of the convex set $\mathcal{H} \stackrel{\rm def}{=} \{M \in \mathbb{R}^{n\times n} \mid \|M\|_2 \leq 1\}$.[107] Thus the retraction onto $\mathcal{M}$ of any tangent vector—these do not belong in the interior of $\mathcal{H}$ because $\mathcal{H}$ is convex—is given by the projection of this vector onto the convex $\mathcal{H}$, an operation involving only standard linear algebra and no calls to the external function defining $\mathcal{M}$. So long as we compute the objective function only at feasible points, which is the case in the simply constrained framework, problem (14.2.1) with $\mathcal{X} = \mathcal{M}$ can be viewed as the minimization of the function $f_{\mathcal{M}}$ defined as the restriction of f to $\mathcal{M}$, and whose Riemannian gradient[108] at $x \in \mathcal{M}$ is the projection of $\nabla_x^1 f(x)$ onto the tangent space to $\mathcal{M}$ at x, that is, the tangent space to $\mathcal{H}$ at x. Because no point in this tangent plane belongs to the interior of $\mathcal{H}$, the retraction of points in the tangent space to $\mathcal{M}$ is then given by the projection onto $\mathcal{H}$. Then the computation of $\phi^1_{f_{\mathcal{M}},\mathcal{M}(x),1}(x)$ is equivalent to that of $\phi^1_{f_{\mathcal{M}},\mathcal{H}(x),1}(x)$, and the latter can be obtained using techniques we covered above for convexly constrained problems. The solution of the trust-region subproblem for the specialized TRq algorithm on the Stiefel manifold with $q = 1$ can be computed in the same way, and a complete TR1-SC algorithm is thus well defined. As a consequence, problems for which the manifold $\mathcal{M}$ can be viewed as part of the boundary of a convex set $\mathcal{H}$ in the "ambient" space $\mathbb{R}^n$ and for which the projection/retraction of the tangent space onto $\mathcal{H}$ belongs to $\mathcal{M}$ and does not require any oracle call belong to the class of simply constrained problems, and the evaluation complexity of finding a first-order approximate minimizer of $f_{\mathcal{M}}$ is described by Theorem 14.2.4.

Many other examples of simply constrained problems exist, including feasible sets with potentially complicated geometries—and there is no requirement for the feasible set to be connected—but we leave them to the imagination of the reader. The take-away message of this section, then, is that the class of simply constrained problems encompasses a number of cases of significant practical interest.

[107] Note that there are matrices on the boundary of $\mathcal{H}$ that do not belong to the Stiefel manifold.

[108] Which can also be viewed as the gradient of the Lagrangian of the problem (considered as equality constrained in $\mathbb{R}^n$) for a suitable choice of the Lagrange multiplier.

Notes and references for Section 14.2

Simple constraints were originally introduced under the name "inexpensive constraints" in the context of the analysis of the convergence of regularization methods to weak minimizers for general unconstrained problems [108], possibly involving inexact evaluations [28]. They have also been considered for strong minimizers of unconstrained composite problems in [109]. The extension to exact and inexact trust-region methods is new. We now prefer to call them "simple constraints" in order to avoid the potential confusion between the cost of calling an oracle (which is what evaluation complexity is about) and that of performing internal computations. This terminology is also used elsewhere, for example, for nonfunctional convex constraints for which gradient mappings can be computed [299, Sect. 2.2.4].

The knapsack technique discussed for building the TR1-SP and AR1-SP algorithms is novel, to the best of our knowledge. We also note that the cost of computing (14.2.20) is $\mathcal{O}(n+n|\mathcal{B}(x)|) = \mathcal{O}(nn_{\rm nz})$ floating-point operations and is dominated by that of solving (14.2.21) using dynamic programming. If a greedy approximation d_a guaranteeing that $\ell(d_a) \leq \ell(d_*)/2$ is used,[109] then this cost is reduced to $\mathcal{O}(n\log(n))$, which makes it computationally very attractive. We refer the reader to [127, 283, 367] for more coverage of the knapsack problem and algorithms. We note that (other) integer-programming formulations for sparsity-constrained problems have been proposed, for example, in [36, 165].

A much-discussed extension of the sparsity-constrained framework is provided by the class of "group sparsity-constrained" problems (also called problems with structured sparsity or group LASSO): in this setting, the components of x are divided in $|G|$ groups, with the gth group containing components $\{x_{j,g}\}_{j=1}^{|G_g|}$. The constraint is then reformulated as

$$\sum_{g=1}^{|G|} \|x_g\|_g \leq t, \quad \text{where } \|x_g\|_g^2 = \sum_{j=1}^{|G_g|} x_{j,g}^2.$$

Several variations of the concept are possible, depending, for example, on whether or not groups of entries may overlap. Group sparsity-constrained problems and their complexity have generated a substantial literature in areas such as signal processing (compressed sensing) and neural networks (see, for instance, [22, 23, 181, 223, 344] and the references therein), which are beyond our scope here.

There is a substantial literature on manifold-constrained optimization, often focusing on using the geometry of the manifold itself, rather than considering its embedding in the ambient space, to develop practical algorithms. Because this is beyond our present scope, we refer the reader to the textbooks [1, 59], as well as the related complexity papers for first- and second-order methods [3, 60].

14.3 High-order criticality for nonconvex constraints

In this section, we return to the general constrained problem

$$\underset{x\in\mathcal{X}}{\text{minimize}} \;\; f(x) \;\text{ subject to }\; c(x) = 0, \tag{14.3.1}$$

which involves m explicit constraints $c(x) = 0$, some of which may be nonconvex. At variance with (7.1.1), we no longer presume that x is chosen in a convex set, but we assume instead that the closed, nonempty set $\mathcal{X}$ is defined by inexpensive constraints (in the sense of Section 14.2 above).

[109] This then requires halving ϵ_1 to maintain the same approximate optimality guarantee.

14.3.1 High-order necessary conditions for constrained problems

Our first task is to clarify what we mean by a qth-order critical point of (14.3.1). As we have already observed in Section 12.1 for the unconstrained case, this may be very difficult in general, but, as in Section 12.1.2, we start by considering curve-based necessary optimality conditions. Let x_* be a local minimizer of (14.3.1), and consider paths $x_*(\alpha)$ of the form

$$x_*(\alpha) = x_* + \sum_{i=1}^{j} \alpha^i s_i + o(\alpha^j) \tag{14.3.2}$$

for $j \in [1:q]$ and where $\alpha > 0$. We say that the path $x_*(\alpha)$ is *$\mathcal{X}$-feasible* if $x_*(\alpha) \in \mathcal{X}$ for all α sufficiently small. Using this notion, we then introduce the *jth-order descriptor set of $\mathcal{X}$ at x_** as

$$\mathcal{D}_{\mathcal{X}}^j \stackrel{\text{def}}{=} \left\{(s_1,\ldots,s_j) \in \mathbb{R}^{n\times j} \mid x_*(\alpha) \text{ is } \mathcal{X}\text{-feasible}\right\}.$$

As for the unconstrained case, we now exploit (14.3.2) to derive the desired necessary conditions: the idea is again to state necessary optimality conditions for the unidimensional problem of minimizing $f(x_*(\alpha))$ subject to $c(x_*(\alpha)) = 0$, where $x_*(\alpha)$ is an arbitrary $\mathcal{X}$-feasible path. Substituting (14.3.2) for any $j \in [1:q]$ into the expression $c(x_*(\alpha)) = 0$ and collecting terms of equal degree in α, we find that

$$0 = c(x_*(\alpha)) = \sum_{i=1}^{j} \alpha^i \sum_{k=1}^{i} \frac{1}{k!}\left(\sum_{(\ell_1,\ldots,\ell_k)\in\mathcal{P}(i,k)} \nabla_x^k c(x_*)[s_{\ell_1},\ldots,s_{\ell_k}]\right) + o(\alpha^j)$$

for all $j \in [1:q]$ and for all sufficiently small α, where $\mathcal{P}(i,k)$ is defined by (12.1.5) (p. 275). This requires that the coefficient of each power of α be identically zero, which is to say that

$$\sum_{k=1}^{i} \frac{1}{k!}\left(\sum_{(\ell_1,\ldots,\ell_k)\in\mathcal{P}(i,k)} \nabla_x^k c(x_*)[s_{\ell_1},\ldots,s_{\ell_k}]\right) = 0 \text{ for all } i \in [1:q]. \tag{14.3.3}$$

In order to state the desired necessary optimality conditions in this context, we must ensure that there are feasible paths of the form (14.3.2) satisfying (14.3.3). This is the purpose of the following *constraint qualification.*

X.CqQ If $x_* \in \mathcal{X}$, the set of nonzero matrices $\{(s_1,\ldots,s_j)\} \in \mathcal{D}_{\mathcal{X}}^j$ for which (14.3.3) holds is nonempty for each $j \in [1:q]$.

When $q = 1$ and $\mathcal{X}$ is described by a set of inequalities, **X.CQ** is implied by the LICQ first-order constraint qualification.[110] This can be shown using the implicit-function theorem: assuming that the set $\mathcal{X}$ is defined by a set of inequalities, defining $c_{\mathcal{A}}(x)$ to be the subset of equality and inequality constraints active at x, and given s_1 such that $\nabla_x^1 c_{\mathcal{A}}(x_*)[s_1] = 0$, there exists a feasible path with tangent s_1.

The form of the necessary conditions we are about to state involves $\mathcal{L}(x,y) = f(x) - \langle y, c(x)\rangle$, the Lagrangian function as defined in (1.1.1) on p. 2 (where $y \in \mathbb{R}^m$).

[110] Stating that the normals to active constraints are linearly independent.

Theorem 14.3.1: Curve-based necessary conditions for a constrained minimum

[104, Thm. 2.2]: Suppose that f and each of the $\{c_i\}_{i=1}^m$ are q-times continuously differentiable in an open set containing $\mathcal{X}$ and that x_* is a local minimizer for problem (14.3.1) at which **X.CQ** *holds. Then we have that $c(x_*) = 0$ and, for all $y_* \in \mathbb{R}^m$ and all $j \in [1:q]$,*

$$\sum_{k=1}^{j} \frac{1}{k!} \left(\sum_{(\ell_1,\ldots,\ell_k)\in\mathcal{P}(j,k)} \nabla_x^k \mathcal{L}(x_*, y_*)[s_{\ell_1}, \ldots, s_{\ell_k}] \right) \geq 0 \tag{14.3.4}$$

for all $(s_1, \ldots, s_j) \in \mathcal{D}_{\mathcal{X}}^j$ and such that

$$\sum_{k=1}^{i} \frac{1}{k!} \left(\sum_{(\ell_1,\ldots,\ell_k)\in\mathcal{P}(i,k)} \nabla_x^k \mathcal{L}(x_*, y_*)[s_{\ell_1}, \ldots, s_{\ell_k}] \right) = 0 \quad (i \in [1:j-1]), \tag{14.3.5}$$

and (14.3.3) hold for $j \in [1:q]$.

Recall that in Example 12.1.2 (p. 277) we observed that the necessary optimality condition for unconstrained optimization involves a combination of more than a single derivative tensor with the s_i for orders four and above. If equality constraints are present, this unfortunate possibility occurs for order three (and above).

Indeed, suppose that $\mathcal{X}$ has an interior, suppose that x_* lies there, and consider the necessary conditions (14.3.3)–(14.3.5) for $q = 3$. We observe that (14.3.4) for $j = 1$, the fact that $\mathcal{P}(1,1) = \{1\}$ (see Table 12.1.1 on p. 276), and the fact that the sign of s_1 can be chosen arbitrarily without altering feasibility with respect to $\mathcal{X}$ (since x_* is interior) imply that $\nabla_x^1 \mathcal{L}(x_*, y_*) = 0$. Likewise, for $j = 2$, $\nabla_x^2 \mathcal{L}(x_*, y_*)[s_1]^2 \geq 0$ for all s_1 such that $\nabla_x^1 c(x_*)[s_1] = 0$, which is the familiar result from Theorem 1.1.3. But, for $j = q = 3$, conditions (14.3.4) and (14.3.5) can be written as

$$\nabla_x^2 \mathcal{L}(x_*, y_*)[s_1, s_2] + \tfrac{1}{6} \nabla_x^3 \mathcal{L}(x_*, y_*)[s_1]^3 = 0 \tag{14.3.6}$$

for all $s_1 \in \ker^1[\nabla_x^1 \mathcal{L}(x_*, y_*)] \cap \ker^2[\nabla_x^2 \mathcal{L}(x_*, y_*)]$, and (14.3.3) can be written as

$$\begin{aligned} \nabla_x^1 c(x_*)[s_2] + \tfrac{1}{2} \nabla_x^2 c(x_*)[s_1]^2 = 0 \quad \text{and} \\ \nabla_x^1 c(x_*)[s_3] + \nabla_x^2 c(x_*)[s_1, s_2] + \tfrac{1}{6} \nabla_x^3 c(x_*)[s_1]^3 = 0, \end{aligned} \tag{14.3.7}$$

where as before we recall more generally that $\ker^p[T]$ is the p-kernel of the multilinear p-form $T[v]^p$ defined by (A.8). These conditions do not require that the second term on the left-hand side of (14.3.6) vanish but instead that it balance the first term. This differs from what happens in the unconstrained case, since second-order necessary conditions then guarantee that $\nabla_x^2 \mathcal{L}(x_*, y_*) = \nabla_x^2 f(x_*)$ is positive semidefinite on $\mathbb{R}^n$ and therefore admits a square root. As a consequence, in the unconstrained case,

$$\nabla_x^2 \mathcal{L}(x_*, y_*)[s_1, s_2] = [\nabla_x^2 \mathcal{L}(x_*, y_*)^{\frac{1}{2}} s_2]^T [\nabla_x^2 \mathcal{L}(x_*, y_*)^{\frac{1}{2}} s_1] = 0$$

since s_1 must belong to $\ker^2[\nabla_x^2 \mathcal{L}(x_*, y_*)]$. However, such an argument fails when equality constraints are present because $\nabla_x^2 \mathcal{L}(x_*, y_*)$ is only positive semidefinite on a strict subspace of $\mathbb{R}^n$ and the square root may fail to exist. We illustrate this as follows.

Example 14.3.1 Consider the equality-constrained problem

$$\min_{x\in\mathbb{R}^3} x_1 + x_2^2 + x_2^3 - x_3 \quad \text{subject to} \quad c(x) = \begin{pmatrix} -x_1 - x_2^2 + x_1x_2 + x_3 \\ x_1 + x_2^2 + x_1x_2 + x_3 \end{pmatrix} = 0,$$

for which the origin is a high-order saddle point.
Comparing the constraints' expression with (14.3.2) for $q = 3$, we see that

$$s_1 = e_2, \quad s_2 = -e_1, \quad \text{and} \quad s_3 = e_3$$

define a feasible path for all α since then

$$x_*(\alpha) = \begin{pmatrix} -\alpha^2 \\ \alpha \\ \alpha^3 \end{pmatrix} \quad \text{and} \quad c(x_*(\alpha)) = \begin{pmatrix} \alpha^2 - \alpha^2 - \alpha^3 + \alpha^3 \\ -\alpha^2 + \alpha^2 - \alpha^3 + \alpha^3 \end{pmatrix} = 0.$$

Now,

$$\nabla_x^1 f(x) = \begin{pmatrix} 1 \\ 2x_2 + 3x_2^2 \\ -1 \end{pmatrix}, \quad \nabla_x^2 f(x) = \begin{pmatrix} 0 & 0 & 0 \\ 0 & 2+6x_2 & 0 \\ 0 & 0 & 0 \end{pmatrix},$$

and $[\nabla_x^3 f(x)]_{2,2,2} = 6$. In addition,

$$\nabla_x^1 c(x) = \begin{pmatrix} -1 + x_2 & -2x_2 + x_1 & 1 \\ 1 + x_2 & 2x_2 + x_1 & 1 \end{pmatrix}, \quad \nabla_x^2 c_1(x) = \begin{pmatrix} 0 & 1 & 0 \\ 1 & -2 & 0 \\ 0 & 0 & 0 \end{pmatrix},$$

and $\nabla_x^3 c_1(x) = 0$, while

$$\nabla_x^2 c_2(x) = \begin{pmatrix} 0 & 1 & 0 \\ 1 & 2 & 0 \\ 0 & 0 & 0 \end{pmatrix} \quad \text{and} \quad \nabla_x^3 c_2(x) = 0.$$

Moreover,

$$\begin{aligned} &\nabla_x^1\, c(0)[s_2] + \tfrac{1}{2}\nabla_x^2 c(0)[s_1]^2 \\ &= -\begin{pmatrix} -1 & 0 & 1 \\ 1 & 0 & 1 \end{pmatrix} e_1 + \tfrac{1}{2}\left[e_2^T \begin{pmatrix} 0 & 1 & 0 \\ 1 & -2 & 0 \\ 0 & 0 & 0 \end{pmatrix} e_2 \right] e_1 \\ &\quad + \tfrac{1}{2}\left[e_2^T \begin{pmatrix} 0 & 1 & 0 \\ 1 & 2 & 0 \\ 0 & 0 & 0 \end{pmatrix} e_2 \right] e_2 \\ &= 0 \end{aligned}$$

and

$$\begin{aligned} &\nabla_x^1\, c(0)[s_3] + \nabla_x^2 c(0)[s_1, s_2] + \tfrac{1}{6}\nabla_x^3 c(0)[s_1]^3 \\ &= \begin{pmatrix} -1 & 0 & 1 \\ 1 & 0 & 1 \end{pmatrix} e_3 - \left[e_2^T \begin{pmatrix} 0 & 1 & 0 \\ 1 & -2 & 0 \\ 0 & 0 & 0 \end{pmatrix} e_1 \right] e_1 \\ &\quad - \left[e_2^T \begin{pmatrix} 0 & 1 & 0 \\ 1 & 2 & 0 \\ 0 & 0 & 0 \end{pmatrix} e_1 \right] e_2 - \tfrac{1}{6} 0^T [e_1]^3 \\ &= 0. \end{aligned}$$

Thus (14.3.7) holds. From the values of $\nabla_x^1 f(0)$ and $\nabla_x^1 c(0)$, we have that $\nabla_x^1 \mathcal{L}(0, y_0) = 0$ when $y_0 = (1,0)^T$. Moreover, we have that

$$\ker^1[\nabla_x^1 c(0)] = \ker^1\left[\begin{pmatrix} -1 & 0 & 1 \\ 1 & 0 & 1 \end{pmatrix}\right] = \operatorname{span}\{e_2\}, \quad \nabla_x^2 \mathcal{L}(0, y_0) = \begin{pmatrix} 0 & 1 & 0 \\ 1 & 0 & 0 \\ 0 & 0 & 0 \end{pmatrix},$$

and the only nonzero component of $\nabla_x^3 \mathcal{L}(0, y_0)$ is its $(2,2,2)$ element, which is 6. Thus (14.3.5) also holds for $i = 2$. In addition, it is easy to check that the third-order necessary condition (14.3.6) holds with

$$\nabla_x^2 \mathcal{L}(0, y_0)[s_1, s_2] = -1 \quad \text{and} \quad \nabla_x^3 \mathcal{L}(0, y_0)[s_1]^3 = 6.$$

This shows that the term involving $\nabla_x^3 \mathcal{L}(0, y_0)[s_1]^3$ is not the only one occurring in the third-order necessary condition for this example, the second derivative term $\nabla^2 \mathcal{L}(0, y_0)[s_1, s_2]$ being equally important, as we said.

Thus, the third-order necessary condition must consider both terms in (14.3.6) and cannot simply rely only on the third derivative of the Lagrangian along a well-chosen direction or subspace. In general, the qth-order necessary conditions in (14.3.4) involve a mix of the qth-derivative tensor of the Lagrangian and other lower-order derivative terms applied to vectors s_i for $i > 1$, where the s_i depend on the geometry of the set of feasible arcs. At this stage, for lack of a suitable formal understanding of this geometry, unsurprisingly conditions (14.3.3)–(14.3.5) remain very difficult to interpret or check.

Because of this difficulty, we turn instead to consider a neighbourhood-based technique inspired by the material from Section 12.1.3. We shall use the $\phi_{\cdot,j}^\delta$ optimality measure defined in (14.2.2) but now applied to the Lagrangian $\mathcal{L}(x, y)$ considered as a function of its first argument. More specifically, we consider

$$T_{\mathcal{L},j}(x, y, s) \stackrel{\text{def}}{=} T_{f,j}(x, s) - \langle y, T_{c,j}(x, s) \rangle,$$

the Taylor expansion of the Lagrangian function $\mathcal{L}(x, y)$ as a function of x, and the quantity

$$\phi_{\mathcal{L},\mathcal{V}(x),j}^\delta(x, y) = \mathcal{L}(x, y) - \min_{\substack{\|s\| \le \delta \\ s \in \mathcal{V}(x)}} T_{\mathcal{L},j}(x, y, s) = \max_{\substack{\|s\| \le \delta \\ s \in \mathcal{V}(x)}} \Delta T_{\mathcal{L},j}(x, y, s),$$

where, as usual, $\delta \in (0, 1]$, $\Delta T_{\mathcal{L},j}(x, y, s)$ is now the Taylor decrement—recall the terminology from Section 13.1.2—of the Lagrangian with respect to its first argument, and $\mathcal{V}(x)$ is a suitable subset of

$$\mathcal{X}(x) \stackrel{\text{def}}{=} \{s \in \mathbb{R}^n \mid x + s \in \mathcal{X}\}.$$

Thus $\phi_{\mathcal{L},\mathcal{V}(x),j}^\delta(x, y)$ measures how much decrease in the jth-degree Taylor-series approximation of the Lagrangian can be achieved, for fixed y, in the intersection of a subset of feasible perturbations $\mathcal{V}(x)$ around x with a neighbourhood of x of diameter δ. We shall be concerned with the possibilities $\mathcal{V}(x) = \mathcal{X}(x)$,

$$\mathcal{V}(x) = \mathcal{X}(x) \cap \ker^1[\nabla_x^1 c(x)] \quad \text{and} \quad \mathcal{V}(x) = \mathcal{X}(x) \cap \ker^1[\nabla_x^1 c(x)] \cap \ker^1[\nabla_x^1 f(x)].$$

Notice that we recover the definition (14.2.2) when $\mathcal{V}(x) = \mathcal{X}(x)$, while the other two choices ensure additionally that c does not change, up to first order, in a feasible neighbourhood of x.

We start by providing a generalization of Theorem 7.1.1 (p. 194), showing that the role of the Lagrangian function $\mathcal{L}(x, y)$ extends beyond first-order optimality.

Theorem 14.3.2: Local sufficiency of the Lagrangian for approximate minimizers

Suppose that the set $\mathcal{X} \subseteq \mathbb{R}^n$ is closed and nonempty, that $x_\epsilon \in \mathcal{X}$, and that f and c are j-times continuously differentiable and have Lipschitz-continuous jth derivatives ($j \in \{1,2,3\}$) with constants $L_{f,j}$ and $L_{c,j}$, respectively, in a neighbourhood of x_ϵ of diameter $\delta_j > 0$. Suppose also that $\mathcal{V}(x_\epsilon) \subseteq \mathcal{X}(x_\epsilon)$ is given and that

$$\|c(x_\epsilon)\| \leq \epsilon_0. \tag{14.3.8}$$

Finally, suppose that there exists a vector of Lagrange multipliers $y_\epsilon \in \mathbb{R}^m$ and an $\eta \geq 0$ such that

$$\phi^{\delta_j}_{\mathcal{L},\mathcal{V}(x_\epsilon),j}(x_\epsilon, y_\epsilon) \leq \epsilon_j \|(1, y_\epsilon^T)\| \frac{\delta_j^j}{j!} + \eta, \quad j \in \{1,2,3\}. \tag{14.3.9}$$

Then

$$f(x_\epsilon + d) \geq f(x_\epsilon) - 2\epsilon_0 \|y_\epsilon\| - 2\epsilon_j \|(1, y_\epsilon^T)\| \frac{\delta_j^j}{j!} - \eta, \quad j \in \{1,2,3\}, \tag{14.3.10}$$

for all $d \in \mathcal{V}(x_\epsilon)$ such that $\|c(x_\epsilon + d)\| \leq \epsilon_0$ and[111]

$$\|d\| \leq \min\left[\delta_j, \frac{(j+1)\,\epsilon_j}{\sqrt{2}\max[L_{f,j}, L_{c,j}]}\right]. \tag{14.3.11}$$

In addition, if either $c(x_\epsilon + s) = 0$ and $\langle y_\epsilon, c(x_\epsilon)\rangle \leq 0$ or $c(x_\epsilon + s) = c(x_\epsilon)$, then the bound (14.3.10) simplifies to

$$f(x_\epsilon + d) \geq f(x_\epsilon) - 2\epsilon_j \|(1, y_\epsilon^T)\| \frac{\delta_j^j}{j!} - \eta, \quad j \in \{1,2,3\}. \tag{14.3.12}$$

Proof. Let $j \in \{1,2,3\}$. Using (14.3.9) and applying Theorem 14.2.1 to the function $\mathcal{L}(\cdot, y_\epsilon)$ (with order-j Lipschitz constant $L_{f,j} + \|y_\epsilon\| L_{c,j}$), we deduce that

$$\mathcal{L}(x_\epsilon + d, y_\epsilon) \geq \mathcal{L}(x_\epsilon, y_\epsilon) - 2\epsilon_j \|(1, y_\epsilon^T)\| \frac{\delta_j^j}{j!} - \eta \tag{14.3.13}$$

for all $d \in \mathcal{V}(x_\epsilon)$ such that

$$\|d\| \leq \min\left[\delta_j, \frac{(j+1)\,\epsilon_j \|(1, y_\epsilon^T)\|}{L_{f,j} + \|y_\epsilon\| L_{c,j}}\right]. \tag{14.3.14}$$

[111] If $L_{f,j} = L_{c,j} = 0$, the second term in the minimum disappears.

Since $\sqrt{2(1+\alpha^2)} \geq 1+\alpha$ for any $\alpha > 0$, we see that

$$\|(1, y_\epsilon^T)\| \geq \frac{1+\|y_\epsilon\|}{\sqrt{2}},$$

and thus if d satisfies (14.3.11), (14.3.14) holds. Thus we deduce from (14.3.13) that

$$f(x_\epsilon + d) - \langle y_\epsilon, c(x_\epsilon + d)\rangle \geq f(x_\epsilon) - \langle y_\epsilon, c(x_\epsilon)\rangle - 2\epsilon_j \|(1, y_\epsilon^T)\| \frac{\delta_j^j}{j!} - \eta \quad (14.3.15)$$

for all d satisfying (14.3.11). It then follows immediately from the Cauchy–Schwarz inequality that

$$f(x_\epsilon + d) \geq f(x_\epsilon) - \|y_\epsilon\| \, \|c(x_\epsilon) - c(x_\epsilon + d)\| - 2\epsilon_j \|(1, y_\epsilon^T)\| \frac{\delta_j^j}{j!} - \eta.$$

Since $\|c(x_\epsilon + d)\| \leq \epsilon_0$, the required bound (14.3.10) then follows from the inequality $\|c(x_\epsilon) - c(x_\epsilon + d)\| \leq \|c(x_\epsilon)\| + \|c(x_\epsilon + d)\| \leq 2\epsilon_0$. If, in addition, $c(x_\epsilon + d) = c(x_\epsilon)$, we obtain (14.3.12). If instead $c(x_\epsilon + s) = 0$ and $\langle y_\epsilon, c(x_\epsilon)\rangle \leq 0$, (14.3.12) follows from (14.3.15). □

As was the case for Theorem 7.1.1 (p. 194) and assuming η is suitably small, this theorem provides conditions—namely, (14.3.9)—that guarantee that the objective function cannot decrease (much) in a (nearly) feasible neighbourhood of x_ϵ. That it can decrease at all when feasibility remains within the requested accuracy ϵ_0 obviously results from the fact that (14.3.9) is designed to allow approximate local minimizers, not just exact ones.[112]

In fact, the conditions (14.3.9) for $j = 1$ and 2 restate small variations on familiar themes. Suppose that $\mathcal{X} = \mathbb{R}^n$. Notice that the definition of $\phi_{\mathcal{L},\mathcal{X},j}^{\delta_j}(x, y)$ in (14.2.2) and (14.3.9) for $j = 1$ then implies that

$$\|\nabla_x^1 \mathcal{L}(x_\epsilon, y_\epsilon)\| \leq \epsilon_1 \|(1, y_\epsilon^T)\|$$

if $\eta = 0$ for all $\delta_1 > 0$, which is a perturbation of the usual approximate Lagrangian-based first-order condition (1.1.13) (p. 5) whenever $\|c(x_\epsilon)\| \leq \epsilon_0$. If (14.3.9) holds for $\mathcal{V}(x_\epsilon) = \mathcal{X}(x_\epsilon) \cap \ker^1[\nabla_x^1 c(x_\epsilon)]$ and $\nabla_x^1 \mathcal{L}(x_\epsilon, y_\epsilon) = 0$, then

$$\lambda_{\min}[\nabla_x^2 \mathcal{L}^{\mathcal{V}}(x_\epsilon, y_\epsilon)] \geq -\epsilon_2 \|(1, y_\epsilon^T)\|,$$

where $\nabla_x^2 \mathcal{L}^{\mathcal{V}}(x_\epsilon, y_\epsilon)$ is the restriction of $\nabla_x^2 \mathcal{L}(x_\epsilon, y_\epsilon)$ to $\ker^1[\nabla_x^1 c(x_\epsilon)]$. In other words, the Hessian of the Lagrangian is (nearly) positive semidefinite on the space tangent to the explicit constraints (see (1.1.15) on p. 5). In Section 14.3.2.3, we will return to consider what happens if the gradient of the Lagrangian is small, but not zero.

In conclusion, Theorem 14.3.2 is important because it justifies considering algorithms whose purpose is to compute a pair (x_ϵ, y_ϵ) for which (14.3.8) and (14.3.9) hold.

14.3.2 The TP*qp*-GC algorithm and its complexity

We have already addressed the complexity of finding first-order critical points of (14.3.1) in Chapter 7 (p. 193) using the two-phase algorithm TP*p*-GC. Here, we consider a similar

[112] Obviously, Theorem 7.1.1 remains valid. To recover it from the above result requires that we replace $\phi_{\mathcal{L},\mathcal{V}(x),j}^{\delta_1}$ by $\phi_{\mathcal{L},\mathcal{V}(x),\infty}^{\delta_1} \stackrel{\text{def}}{=} \mathcal{L}(x,y) - \min_{\substack{\|s\| \leq \delta \\ s \in \mathcal{V}(x)}} \mathcal{L}(x+s, y)$.

type of framework, but for finding first-, second-, and third-order[113] approximate minimizers in the sense that they satisfy conditions (14.3.8) and (14.3.9). We will not fully specify which algorithm to use when solving the subproblems but will assume that their worst-case complexity is known so as to derive an evaluation-complexity bound for the complete algorithm. In keeping with our previous conventions for naming algorithms, the resulting two-phase algorithm will be called the TPqp-GC algorithm.

14.3.2.1 Solving nonlinear least-squares subproblems

The two-phase algorithm we are going to consider relies on the solution of nonlinear least-squares subproblems of the form

$$\underset{x\in\mathcal{X}}{\text{minimize}} \quad \Phi_2(x) = \tfrac{1}{2}\|r(x)\|^2 \tag{14.3.16}$$

(as in (5.1.2) on p. 120), where $r(x)$ is a smooth function from $\mathbb{R}^n$ to $\mathbb{R}^m$. We actually need this as a surrogate for the equivalent least-Euclidean-norm problem

$$\underset{x\in\mathcal{X}}{\text{minimize}} \quad \Phi_1(x) = \|r(x)\| \tag{14.3.17}$$

since, as we shall see, it is (14.3.16) that will enable us to tie optimality conditions of (14.3.17) to that of the general constrained problem (14.3.1).

Following ideas in Section 5.1.1, we define an ϵ-approximate qth-order critical point for (14.3.17) as a point x at which

$$\|r(x_\epsilon)\| \le \epsilon_0 \quad \text{or} \quad \phi^{\delta_j}_{\Phi_2,\mathcal{X}(x_\epsilon),j}(x_\epsilon) \le \epsilon_j \|r(x_\epsilon)\| \, \frac{\delta_j^j}{j!} \quad \text{for } j\in[1:q] \tag{14.3.18}$$

and some $\delta \in (0,1]^q$, where $\phi^{\delta_j}_{\Phi_2,\mathcal{X}(x),j}(x)$ is defined by (14.2.2). However, at variance with the methods we encountered in Sections 5.1 and 6.3, we shall not use special-purpose least-Euclidean-norm methods here but rather rely on generic methods for simply constrained problems, such as those we considered in Section 14.2, applied in the least-squares (14.3.16) case.

We consider a nonspecific inner-iteration algorithm, INNER$(f,\mathcal{X},q,x^{\rm s},\epsilon)$, whose aim is to find a point $x_\epsilon \in \mathcal{X}$ and associated radii $\delta \in (0,1]^q$ for which

$$\phi^{\delta_j}_{f,\mathcal{X}(x_\epsilon),j}(x_\epsilon) \le \epsilon_j \, \frac{\delta_j^j}{j!} \quad \text{for all} \quad j\in[1:q] \tag{14.3.19}$$

starting from $x = x^{\rm s} \in \mathcal{X}$. We have already described many candidates, and Theorem 14.2.4 indicates the worst-case evaluation bounds for such methods under appropriate variations of assumption **f.D1qLaB.X**. Generically, in view of this (and by studying the many previous results that culminate in this theorem), we shall assume that INNER$(f,\mathcal{X},q,x^{\rm s},\epsilon)$ requires at most[114]

$$\#^{\rm F}_{\sf INNER} = \kappa^{\rm F}_{\sf INNER}\left(f(x^{\rm s}) - f_{\rm low}\right) \max_{j\in[1:q]} \epsilon_j^{-\pi_j} \tag{14.3.20}$$

[113] The limitation to third order arises because of the complexity of the TPp-GC algorithm (indeed we will keep q general in its analysis) but because the results of running this algorithm do not seem to relate to optimality conditions for higher order (see Lemma 14.3.5 below).

[114] Strictly, we have ignored less-important $\mathcal{O}(\log \epsilon_j)$ terms that are present in some of our previous bounds.

(exact or inexact, as appropriate) evaluations of $f(x)$ and at most

$$\#^{\rm D}_{\sf INNER} = \kappa^{\rm D}_{\sf INNER}\left(f(x^{\rm S}) - f_{\rm low}\right) \max_{j\in[1:q]} \epsilon_j^{-\pi_j} \tag{14.3.21}$$

(exact or inexact) derivative evaluations to find a point $x_\epsilon \in \mathcal{X}$ that satisfies (14.3.19). Here $\kappa^{\rm F}_{\sf INNER}$ and $\kappa^{\rm D}_{\sf INNER}$ depend on problem characteristics—primarily Lipschitz constants—and algorithmic parameters. In addition, $f_{\rm low}$ is an assumed lower bound on f for $x \in \mathcal{X}$, and in the least-squares case of interest here, we may reasonably simply choose $f_{\rm low} = 0$.

Our intention is to apply the INNER algorithm to $\Phi_2(x)$ to approximately solve (14.3.16). If we do so, it follows from (14.3.19) that we will find

$$\phi^{\delta_j}_{\Phi_2,\mathcal{X}(x_\epsilon),j}(x_\epsilon) \le \epsilon_j \frac{\delta_j^j}{j!} \quad \text{for all} \quad j \in [1:q]$$

rather than the often more exacting (14.3.18). We could instead apply INNER($\Phi_2, \mathcal{X}, q, x^{\rm S}, \epsilon_0\epsilon$) using tighter tolerances $\epsilon_0\epsilon_j$ for all $j \in [1:q]$, and it is easy to see that this guarantees (14.3.18). However, the complexity bounds (14.3.20) and (14.3.21) then become $\mathcal{O}(\epsilon_0^{-\pi_j}\epsilon_{\rm min}^{-\pi_j})$, which is considerably worse than $\mathcal{O}(\epsilon_{\rm min}^{-\pi_j})$. Fortunately, it is possible to use the INNER algorithm to ensure a better rate. We do so in the following LENq-SC algorithm.

Algorithm 14.3.1: LENq-SC: An algorithm for least-Euclidean-norm minimization subject to simple constraints

An initial point $x_0 \in \mathcal{X}$, a criticality order q, primal and dual optimality tolerances $\epsilon_j \in (0,1)$ for $j \in [0:q]$, and a constant[115] $\beta \in (0, \min(1, 1/\|r(x_0)\|))$ are given.

If $\|r(x_0)\| \le \epsilon_0$, stop with $x_\epsilon = x_0$.

Iterate, for $k = 1, 2, \dots$:

Compute x_k and $\delta \in (0,1]^q$ using INNER($\Phi_2, \mathcal{X}, q, x_{k-1}, \beta\|r(x_{k-1})\|\epsilon$) so that

$$\phi^{\delta_j}_{\Phi_2,\mathcal{X}(x_k),j}(x_k) \le \beta\|r(x_{k-1})\|\epsilon_j \frac{\delta_j^j}{j!} \quad \text{for all} \quad j \in [1:q]. \tag{14.3.22}$$

If

$$\|r(x_k)\| \le \epsilon_0 \quad \text{or} \quad \phi^{\delta_j}_{\Phi_2,\mathcal{X}(x_k),j}(x_k) \le \epsilon_j\|r(x_k)\| \frac{\delta_j^j}{j!} \text{ for } j \in [1:q], \tag{14.3.23}$$

stop with $x_\epsilon = x_k$.

Although most INNER algorithms do not, as a matter of course, monitor objective-function values, it is trivial to modify them to do so. In particular, a check to see whether $\|r(x_k)\| \le \epsilon_0$ required in (14.3.23) need not wait until other derivative-based tests have completed.

It is very easy to show that the algorithm must converge to a suitable point.

[115] The requirement $\beta \le 1/\|r(x_0)\|$ is made to ensure that any stopping tolerances passed to the INNER algorithm lie in $(0,1]$.

Theorem 14.3.3: Convergence of the LENq-SC algorithm

The LENq-SC algorithm terminates at a point satisfying (14.3.18) after at most $\#^{I}_{\text{LENq-SC}} \stackrel{\text{def}}{=} \max(0, \log(\epsilon_0/\|r(x_0)\|)/\log\beta)$ *iterations.*

Proof. The algorithm terminates immediately if $\|r(x_0)\| \le \epsilon_0$, so suppose otherwise that $\|r(x_0)\| > \epsilon_0$, that the algorithm has reached iteration $k > 0$, and that (14.3.23) is not true. Then

$$\|r(x_k)\| > \epsilon_0 \tag{14.3.24}$$

and

$$\phi^{\delta_j}_{\Phi_2,\mathcal{X}(x_k),j}(x_k) > \epsilon_j \|r(x_k)\| \frac{\delta_j^j}{j!} \quad \text{for some } j \in [1:q]. \tag{14.3.25}$$

Then, combining (14.3.22) and (14.3.25), we have that

$$\epsilon_j \|r(x_k)\| \frac{\delta_j^j}{j!} < \phi^{\delta_j}_{\Phi_2,\mathcal{X}(x_k),j}(x_k) \le \beta\|r(x_{k-1})\|\epsilon_j \frac{\delta_j^j}{j!}$$

for this j, and thus

$$\|r(x_k)\| < \beta\|r(x_{k-1})\| < \beta^k\|r(x_0)\|,$$

where the second inequality occurs by applying the first one recursively for iterations 1 to $k-1$. But this contradicts (14.3.24) once

$$\beta^k < \frac{\epsilon_0}{\|r(x_0)\|} < 1,$$

that is, if $k > \log(\epsilon_0/\|r(x_0)\|)/\log\beta$ iterations are required. □

It is also possible to compute the worst-case evaluation complexity of LENq-SC. Since the algorithm depends upon applying the INNER algorithm to $\Phi_2(x)$, if INNER is of trust-region type, we shall rely on the following property.

Φ_2.D1qLaB.X

Φ_2 is q-times continuously differentiable in an open, convex set $\mathcal{X}_0$ containing $\mathcal{X}$, and there are constants $L_{\Phi_2,j} \ge 1$ and $B_{\Phi_2,j} \ge 1$ such that

$$\|\nabla_x^j \Phi_2(x) - \nabla_x^j \Phi_2(y)\| \le L_{\Phi_2,j}\|x - y\| \tag{14.3.26}$$

and

$$\|\nabla_x^j \Phi_2(x)\| \le B_{\Phi_2,j}$$

for any $x \in \mathcal{L}_{\Phi_2}(x_0)$, any $y \in \mathcal{X}_0$, and all $j \in [1:q]$.

Here, just as in (14.2.8), $\mathcal{L}_{\phi_2}(x_0) \stackrel{\text{def}}{=} \{x \in \mathcal{X} \mid \Phi_2(x) \le \Phi_2(x_0)\}$ and x_0 is the starting point for the LENq-SC algorithm.

By analogy, if INNER is of regularization type, we shall need the stronger Φ_2**.D1pLaB.X**, which is simply Φ_2**.D1qLaB.X** when $q = p$.

Theorem 14.3.4: Evaluation complexity of the LENq-SC algorithm

Suppose that Φ_2***.D1qLaB.X*** *(for trust-region* INNER *algorithms) or* Φ_2***.D1pLaB.X*** *(for regularization* INNER *algorithms) holds, and that each call to the* INNER *algorithm satisfies the evaluation bounds (14.3.20) and (14.3.21). Then the* LENq-SC *algorithm requires at most*[116]

$$\#^{F}_{\text{LENq-SC}} = \tfrac{1}{2}\kappa^{F}_{\text{INNER}}\|r(x_0)\| \max_{j\in[1:q]} \frac{\beta^{1-2\pi_j}}{\beta^{1-\pi_j}-1}\epsilon_0^{1-\pi_j}\epsilon_j^{-\pi_j} \tag{14.3.27}$$

evaluations of $\Phi_2(x)$ *and at most*

$$\#^{D}_{\text{LENq-SC}} = \tfrac{1}{2}\kappa^{D}_{\text{INNER}}\|r(x_0)\| \max_{j\in[1:q]} \frac{\beta^{1-2\pi_j}}{\beta^{1-\pi_j}-1}\epsilon_0^{1-\pi_j}\epsilon_j^{-\pi_j} \tag{14.3.28}$$

evaluations of its derivatives to compute an iterate x_k *satisfying (14.3.23).*

Proof. Suppose that the LENq-SC algorithm terminates at iteration $k_{\max} \leq \#^{\text{I}}_{\text{LENq-SC}}$. Then just as in the previous proof, it must be that

$$\|r(x_{k_{\max}-1})\| > \epsilon_0$$

and, for all $0 \leq k < k_{\max}$,

$$\|r(x_{k_{\max}-1})\| < \beta\|r(x_{k_{\max}-2})\| < \beta^{k_{\max}-k}\|r(x_{k-1})\|,$$

in which case

$$\|r(x_{k-1})\| > \beta^{k-k_{\max}}\epsilon_0. \tag{14.3.29}$$

Applying assumption (14.3.20) with $f(x) = \frac{1}{2}\|r(x)\|^2$ and $f_{\text{low}} = 0$, and invoking the bound (14.3.29), the fact that $\pi_j > 1$, and the decreasing nature of $\{\|r(x_k)\|\}$, iteration k of the LENq-SC algorithm requires at most

$$\begin{aligned}
&\tfrac{1}{2}\kappa^{\text{F}}_{\text{INNER}}\ \|r(x_{k-1})\|^2 \max_{j\in[1:q]}\left[\beta\|r(x_{k-1})\|\epsilon_j\right]^{-\pi_j}\\
&\quad= \tfrac{1}{2}\kappa^{\text{F}}_{\text{INNER}}\|r(x_{k-1})\| \max_{j\in[1:q]}\left[\beta^{-\pi_j}\|r(x_{k-1})\|^{1-\pi_j}\epsilon_j^{-\pi_j}\right]\\
&\quad\leq \tfrac{1}{2}\kappa^{\text{F}}_{\text{INNER}}\|r(x_0)\| \max_{j\in[1:q]}\left[\beta^{-\pi_j}(\beta^{k-k_{\max}}\epsilon_0)^{1-\pi_j}\epsilon_j^{-\pi_j}\right]\\
&\quad= \tfrac{1}{2}\kappa^{\text{F}}_{\text{INNER}}\|r(x_0)\| \max_{j\in[1:q]}\left[\beta^{-\pi_j}(\beta^{1-\pi_j})^{k-k_{\max}}\epsilon_0^{1-\pi_j}\epsilon_j^{-\pi_j}\right]
\end{aligned}$$

evaluations of Φ_2. Thus the total number of evaluations is the sum of these terms from $k = 1$ to $k_{\max}$ and, using the well-known formula for the sum of a geometric

[116] The exponent $1-\pi_j$ on ϵ_0 that results from the trivial identity $2-\pi_j = 1+(1-\pi_j)$ could be replaced by the smaller $1-\delta-\pi_j$ for some $\delta \in [1,\pi_j]$, but at the expense of larger constants.

series,

$$
\begin{aligned}
\sum_{k=1}^{k_{\max}} \tfrac{1}{2}\kappa_{\text{INNER}}^{\text{F}} \,\|r(x_{k-1})\|^2 &\max_{j\in[1:q]} \left[\beta\|r(x_{k-1})\|\epsilon_j\right]^{-\pi_j} \\
&\leq \tfrac{1}{2}\kappa_{\text{INNER}}^{\text{F}}\|r(x_0)\| \max_{j\in[1:q]} \left[\beta^{-\pi_j}\epsilon_0^{1-\pi_j}\epsilon_j^{-\pi_j}\sum_{k=1}^{k_{\max}}(\beta^{1-\pi_j})^{k-k_{\max}}\right] \\
&= \tfrac{1}{2}\kappa_{\text{INNER}}^{\text{F}}\|r(x_0)\| \max_{j\in[1:q]} \left[\beta^{-\pi_j}\epsilon_0^{1-\pi_j}\epsilon_j^{-\pi_j}\beta^{1-\pi_j}\frac{1-(\beta^{1-\pi_j})^{-k_{\max}}}{\beta^{1-\pi_j}-1}\right] \\
&\leq \tfrac{1}{2}\kappa_{\text{INNER}}^{\text{F}}\|r(x_0)\| \max_{j\in[1:q]} \left[\beta^{-\pi_j}\epsilon_0^{1-\pi_j}\epsilon_j^{-\pi_j}\frac{\beta^{1-\pi_j}}{\beta^{1-\pi_j}-1}\right],
\end{aligned}
$$

which is equivalent to (14.3.27). Identical reasoning using (14.3.21) instead of (14.3.20) leads to (14.3.28). □

It is worth remarking that the bounds obtained are similar, but not identical, to those derived for the specialized methods considered in Sections 5.1 and 6.3. In particular, comparing the result above with the evaluation-complexity bounds in Theorems 5.1.7 and 6.3.1, we observe a slight additional dependence on ϵ_0 in the new result, but of course more generality when $q > 3$.

14.3.2.2 The TP*qp*-GC algorithm

Just as in Section 7.2.1, we shall apply a two-phase algorithm—in this case the TP*qp*-GC algorithm—to solve problem (14.3.1). Our method relies heavily on the INNER and LEN*q*-SC algorithms that we encountered in the previous section.[117] The first phase aims to find an approximate feasible point, if at all possible, by solving the feasibility subproblem

$$\min_{x\in\mathcal{X}} \nu(x) \stackrel{\text{def}}{=} \tfrac{1}{2}\|c(x)\|^2,$$

which is trivially of the form (14.3.16), using the LEN*q*-SC algorithm. If one is found, the INNER algorithm is then applied repeatedly in a second phase to approximately solve a series of improvement problems of the form (14.3.16),

$$\min_{x\in\mathcal{X}} \mu(x,t_k) \stackrel{\text{def}}{=} \tfrac{1}{2}\|r(x,t_k)\|^2, \text{ where } r(x,t) \stackrel{\text{def}}{=} \begin{pmatrix} c(x) \\ f(x)-t \end{pmatrix}, \tag{14.3.30}$$

for some monotonically decreasing sequence of *targets* t_k (for $k \geq 1$). The resulting algorithm, which the reader will recognize as a straightforward extension of the TP*p*-GC algorithm (p. 198), is described on the following page. Observe that the recomputation of $\phi_{\mu,\mathcal{X},j}^{\delta_{k,j}}(x_{k+1},t_{k+1})$ ($j \in [1:q]$) required to check (14.3.32) in Step 1 of Phase 2 does not require the reevaluation of $f(x_{k+1})$, $c(x_{k+1})$, or any of their derivatives because these will have already been found in the course of the LEN*q*-SC and INNER algorithms.

[117] Or any other method with similar guarantees.

Algorithm 14.3.2: TPqp-GC: A two-phase algorithm for constrained optimization

Initialization: An initial point $x_0 \in \mathcal{X}$, a criticality order $q \in \{1, 2, 3\}$ (for both the feasibility phase and the optimization phase), and a constant $\theta \in (0, 1)$ are given. The primal and dual tolerances $\epsilon_j \in (0, 1]$ for $j \in [0 : q]$ are also given.

Phase 1: Feasibility phase: Starting from x_0, apply the LENq-SC algorithm with some $\beta \in (0, \min(1, 1/\|c(x_0)\|))$ to approximately minimize $\nu(x) = \frac{1}{2}\|c(x)\|^2$ subject to $x \in \mathcal{X}$ to find a point $x_1 \in \mathcal{X}$ with associated optimality radii $\delta_1 \in (0, 1]^q$ for which

$$\|c(x_1)\| < \theta\epsilon_0 \quad \text{or} \quad \phi^{\delta_{1,j}}_{\nu,\mathcal{X}(x_1),j}(x_1) \le \epsilon_j \|c(x_1)\| \frac{\delta^j_{1,j}}{j!} \quad (j \in [1 : q]). \tag{14.3.31}$$

If $\|c(x_1)\| > \theta\epsilon_0$, terminate with $x_\epsilon = x_1$.

Phase 2: Optimality phase:

Step 0: Initialization. Set $t_1 = f(x_1) - \sqrt{\epsilon_0^2 - \|c(x_1)\|^2}$ and $k = 1$.

Step 1: Inner iteration to compute an improved objective value. Compute $x_{k+1} \in \mathcal{X}$ and optimality radii $\delta_{k+1} \in (0, 1]^q$ to satisfy

$$\|r(x_{k+1}, t_k)\| < \theta\epsilon_0 \quad \text{or} \quad f(x_{k+1}) < t_k$$
$$\text{or} \quad \phi^{\delta_{k+1,j}}_{\mu,\mathcal{X}(x_{k+1}),j}(x_{k+1}, t_k) \le \epsilon_j \|r(x_{k+1}, t_k)\| \frac{\delta^j_{k+1,j}}{j!} \quad (j \in [1 : q]) \tag{14.3.32}$$

using INNER($\mu(x, t_k), \mathcal{X}, q, x_k, \beta\|r(x_k, t_k)\|\epsilon$) for some $\beta \in (0, \theta]$.

Step 2: Decrease the target and test for termination.

(a) If $\|r(x_{k+1}, t_k)\| < \theta\epsilon_0$, set

$$t_{k+1} = f(x_{k+1}) - \sqrt{\epsilon_0^2 - \|c(x_{k+1})\|^2} \tag{14.3.33}$$

and terminate with $(x_\epsilon, t_\epsilon) = (x_{k+1}, t_{k+1})$ if

$$\phi^{\delta_{k+1,j}}_{\mu,\mathcal{X}(x_{k+1}),j}(x_{k+1}, t_{k+1}) \le \epsilon_j \|r(x_{k+1}, t_{k+1})\| \frac{\delta^j_{k+1,j}}{j!} \quad \text{for } j \in [1 : q]. \tag{14.3.34}$$

(b) Otherwise, if $\|r(x_{k+1}, t_k)\| \ge \theta\epsilon_0$ and $f(x_{k+1}) < t_k$, set

$$t_{k+1} = 2f(x_{k+1}) - t_k, \tag{14.3.35}$$

and terminate with $(x_\epsilon, t_\epsilon) = (x_{k+1}, t_{k+1})$ if (14.3.34) holds.

(c) Otherwise, if $\|r(x_{k+1}, t_k)\| \ge \theta\epsilon_0$ and $f(x_{k+1}) \ge t_k$, terminate with $(x_\epsilon, t_\epsilon) = (x_{k+1}, t_k)$.

Step 3: Move to the next iteration. Set $k = k + 1$, and go to Step 1.

The similarities with the TPp-GC algorithm are obvious: the differences occur uniquely in the choice of the LENq-SC algorithm for reducing infeasibilities in Phase 1 and the INNER algorithm with appropriate parameters for solving the subproblems in Phase 2, in the modified conditions (14.3.31), (14.3.32), and (14.3.34), now involving $\phi^{\delta_{k+1,j}}_{\mu,\mathcal{X}(x_k),j}$ (and its associated optimality radius $\delta_{k+1,j}$), and in particular applying for optimality orders j from one to three rather than simply one. Thus, reassuringly, we remain in familiar territory and will be able to follow the tracks of Section 7.2.2, reusing the results derived there when possible. In particular, it is easy to verify that *the conclusions of Lemmas 7.2.3 (p. 202) and 7.2.4 (p. 203) remain valid for the* TPqp-GC *algorithm*, where we now partition the indices of iterates into

$$\mathcal{K}_a \stackrel{\rm def}{=} \{k \geq 0 \mid \|r(x_{k+1}, t_k)\| < \theta\epsilon_0 \text{ and (14.3.33) is applied}\} \tag{14.3.36}$$

and

$$\mathcal{K}_b \stackrel{\rm def}{=} \{k \geq 0 \mid \|r(x_{k+1}, t_k)\| \geq \theta\epsilon_0 \text{ and (14.3.35) is applied}\}.$$

Note that the requirement that $x_0 \in \mathcal{X}$ is reasonable for simple constraints, since the definition of such constraints implies that x_0 can be derived cheaply from any starting guess x_{-1}. We need to check that applying the INNER algorithm in Step 1 of Phase 2 does indeed reveal an $x_{k+1} \in \mathcal{X}$ satisfying (14.3.32). By construction, the algorithm will terminate with an x_{k+1} and radii $\delta_{k+1} \in (0,1]^q$ for which

$$\phi^{\delta_{k+1,j}}_{\mu,\mathcal{X}(x_{k+1}),j}(x_{k+1}, t_k) \leq \epsilon_j \beta \|r(x_k, t_k)\| \frac{\delta^j_{k+1,j}}{j!} \quad (j \in [1:q]).$$

If $\beta\|r(x_k, t_k)\| \leq \|r(x_{k+1}, t_k)\|$, then trivially the third possibility in (14.3.32) occurs. Otherwise,

$$\|r(x_{k+1}, t_k)\| < \beta\|r(x_k, t_k)\| \leq \theta\epsilon_0,$$

since $\beta \leq \theta$ and $\|r(x_k, t_k)\| \leq \epsilon_0$ because of (7.2.27) and (7.2.28) in Lemma 7.2.3 (p. 202), and thus the first possibility in (14.3.32) occurs. Our interest in the TPqp-GC algorithm, just as with its predecessor TPp-GC, is essentially theoretical, since it is unlikely that it will perform as well in practice as, for example, the augmented-Lagrangian, interior-point, or SQP methods.

As for the TPp-GC algorithm, we start by showing that it terminates at a point which is an ϵ-approximate infeasible local minimizer of $\|c(x)\|$ or at an approximate local constrained minimizer in the sense of (14.3.8) and (14.3.9). Understandably, this requires that we make our assumptions specific to problem (14.3.1).

f.D1pLB.X	The objective function $f(x)$ is p-times continuously differentiable in an open, convex set $\mathcal{X}_0$ containing $\mathcal{X}$, and its derivative tensors $\nabla^j_x f(x)$, $j \in [1:p]$, are Lipschitz continuous and bounded in $\mathcal{X}_0$.

c.D1pLB.X	The constraint functions $c_i(x)$, $i \in [1:m]$, are p-times continuously differentiable in an open, convex set $\mathcal{X}_0$ containing $\mathcal{X}$, and their derivative tensors $\nabla^j_x c_i(x)$, $j \in [1:p]$, are Lipschitz continuous and bounded in $\mathcal{X}_0$.

We shall also require that f satisfy **f.B.Fb** (p. 198), where $\mathcal{F}_\beta$ is now suitably redefined to be

$$\mathcal{F}_\beta \stackrel{\text{def}}{=} \{x \in \mathcal{X} \mid \|c(x)\| \leq \beta\}.$$

As was the case in in Section 14.2.1, note that **c.D1pLB.X** and **f.D1pLB.X** can be replaced by **c.D1qLB.X** and **f.D1qLB.Fb** if the INNER algorithm employed by the LENq-SC algorithm is of trust-region, rather than regularization, type.

For future reference, the first four derivatives of $\mu(x,t)$ as a function of x are

$$\nabla_x^1 \mu(x,t) = \sum_{i=1}^m c_i(x)\nabla_x^1 c_i(x) + (f(x)-t)\nabla_x^1 f(x), \tag{14.3.37}$$

$$\begin{aligned}\nabla_x^2 \mu(x,t) = \sum_{i=1}^m \Big[\nabla_x^1 c_i(x) \otimes \nabla_x^1 c_i(x) + c_i(x)\nabla_x^2 c_i(x)\Big] + \nabla_x^1 f(x) \otimes \nabla_x^1 f(x)\\ + (f(x)-t)\nabla_x^2 f(x),\end{aligned} \tag{14.3.38}$$

$$\begin{aligned}\nabla_x^3 \mu(x,t) = \sum_{i=1}^m \Big[3\,\nabla_x^2 c_i(x) \otimes \nabla_x^1 c_i(x) + c_i(x)\nabla_x^3 c_i(x)\Big]\\ + 3\,\nabla_x^2 f(x) \otimes \nabla_x^1 f(x) + (f(x)-t)\nabla_x^3 f(x), \quad \text{and}\end{aligned} \tag{14.3.39}$$

$$\begin{aligned}\nabla_x^4 \mu(x,t) = \sum_{i=1}^m \Big[4\,\nabla_x^3 c_i(x) \otimes \nabla_x^1 c_i(x) + 3\,\nabla_x^2 c_i(x) \otimes \nabla_x^2 c_i(x) + c_i(x)\nabla_x^4 c_i(x)\Big]\\ + 4\,\nabla_x^3 f(x) \otimes \nabla_x^1 f(x) + 3\nabla_x^2 f(x) \otimes \nabla_x^2 f(x) + (f(x)-t)\nabla_x^4 f(x),\end{aligned} \tag{14.3.40}$$

where $\otimes$ denotes the tensor product.

We may now derive a version of Lemma 7.2.1 (p. 199) that considers orders one to three.

Lemma 14.3.5: Consequences of the termination of the TPqp-GC algorithm

Suppose that ***X.CNe***, ***c.D1pLB.X***, ***f.D1pLB.X***, *and* ***f.B.Fb*** *hold. Suppose also that the* TPqp-GC *algorithm terminates at iteration* k_ϵ. *Then one of the two following situations will occur.*

1. Infeasible approximate local minimizer of the constraint violation:

$$\|c(x_\epsilon)\| > \theta\epsilon_0 \quad \text{and} \quad \phi_{\nu,\mathcal{X}(x_\epsilon),j}^{\delta_{k_\epsilon,j}}(x_\epsilon) \leq \epsilon_j \|c(x_\epsilon)\| \, \frac{\delta_{k_\epsilon,j}^j}{j!}. \tag{14.3.41}$$

2. Approximate problem solution:

$$\|c(x_\epsilon)\| \leq \epsilon_0 \quad \text{and} \quad \phi_{\mathcal{L},\mathcal{X}(x_\epsilon),1}^{\delta_{k_\epsilon,1}}(x_\epsilon, y_\epsilon) \leq \epsilon_1 \|(1, y_\epsilon^T)\| \, \delta_{k_\epsilon,1} \tag{14.3.42}$$

for some $y_\epsilon \in \mathbb{R}^m$. *Moreover, if* $q > 1$, *then*

$$\phi_{\mathcal{L},\mathcal{T}(x_\epsilon),2}^{\delta_{k_\epsilon,2}}(x_\epsilon, y_\epsilon) \leq \tfrac{1}{2}\Big[\epsilon_2 \|(1, y_\epsilon^T)\| \, \delta_{k_\epsilon,2}^2 + \phi_{\mathcal{L},\mathcal{X}(x_\epsilon),1}^{\delta_{k_\epsilon,2}}(x_\epsilon, y_\epsilon)\|\nabla_x^1 f(x_\epsilon)\| \, \delta_{k_\epsilon,2}\Big] \tag{14.3.43}$$

and

$$\phi_{\mathcal{L},\mathcal{M}(x_\epsilon),2}^{\delta_{k_\epsilon,2}}(x_\epsilon, y_\epsilon) \leq \tfrac{1}{2}\epsilon_2 \|(1, y_\epsilon^T)\| \, \delta_{k_\epsilon,2}^2, \tag{14.3.44}$$

where

$$\mathcal{T}(x) \stackrel{\rm def}{=} \mathcal{X}(x) \cap \ker^1[\nabla_x^1 c(x)] \quad \text{and} \quad \mathcal{M}(x) \stackrel{\rm def}{=} \mathcal{T}(x) \cap \ker^1[\nabla_x^1 f(x)]. \tag{14.3.45}$$

In addition, if $q = 3$,

$$\phi_{\mathcal{L},\mathcal{T}(x_\epsilon),3}^{\delta_{k_\epsilon,3}}(x_\epsilon, y_\epsilon) \leq \tfrac{1}{6}\Big[\epsilon_3\|(1, y_\epsilon^T)\|\,\delta_{k_\epsilon,3}^3 \\ +3\phi_{\mathcal{L},\mathcal{X}(x_\epsilon),1}^{\delta_{k_\epsilon,3}}(x_\epsilon, y_\epsilon)\left(\|\nabla_x^1 f(x_\epsilon)\|\,\delta_{k_\epsilon,3} + \|\nabla_x^2 f(x_\epsilon)\|\,\delta_{k_\epsilon,3}^2\right)\Big] \tag{14.3.46}$$

and

$$\phi_{\mathcal{L},\mathcal{M}(x_\epsilon),3}^{\delta_{k_\epsilon,3}}(x_\epsilon, y_\epsilon) \leq \tfrac{1}{6}\epsilon_3\|(1, y_\epsilon^T)\|\,\delta_{k_\epsilon,3}^3. \tag{14.3.47}$$

Proof. If $\|c(x_\epsilon)\| > \theta\epsilon_0$ at the end of Phase 1, we obtain (14.3.41) directly from the requirement (14.3.31). Assume therefore that termination occurs in Phase 2.

To start, we follow the proof of Lemma 7.2.1 (p. 199) closely, making adaptations when necessary. We first show that, on termination,

$$\|r(x_\epsilon, t_\epsilon)\| \geq \theta\epsilon_0, \quad f(x_\epsilon) \geq t_\epsilon, \tag{14.3.48}$$

and

$$\phi_{\mu,\mathcal{X}(x_\epsilon),j}^{\delta_{k_\epsilon,j}}(x_\epsilon, t_\epsilon) \leq \epsilon_j\|r(x_\epsilon, t_\epsilon)\|\,\frac{\delta_{k_\epsilon,j}^j}{j!} \quad (j \in [1:q]). \tag{14.3.49}$$

To prove these inequalities, we consider the three possible cases for which termination occurs in Phase 2. The first is in Step 2(a), where $(x_\epsilon, t_\epsilon) = (x_{k+1}, t_{k+1})$ and (14.3.33) holds and thus $\|r(x_{k+1}, t_{k+1})\| = \epsilon_0 > \theta\epsilon_0$. Since $\|r(x_{k+1}, t_k)\| < \theta\epsilon_0$, we have that

$$\|c(x_{k+1})\|^2 \leq \|r(x_{k+1}, t_k)\|^2 \leq \theta^2\epsilon_0^2 < \epsilon_0^2,$$

and therefore $f(x_{k+1}) > t_{k+1}$ from (14.3.33), which gives (14.3.48). Since termination occurs, it must be because (14.3.34) holds, and thus so does (14.3.49). The second terminal case is in Step 2(b), where once again $(x_\epsilon, t_\epsilon) = (x_{k+1}, t_{k+1})$, and (14.3.35) gives that

$$t_{k+1} - f(x_{k+1}) = f(x_{k+1}) - t_k < 0, \tag{14.3.50}$$

and therefore

$$\|r(x_{k+1}, t_{k+1})\| = \|r(x_{k+1}, t_k)\|. \tag{14.3.51}$$

But if Step 2(b) is reached and we take (14.3.34), (14.3.50), and (14.3.51) into account, we deduce that $\|r(x_{k+1}, t_{k+1})\| \geq \theta\epsilon_0$, which again implies that (14.3.48) and (14.3.49) hold. The last case for which termination can occur is Step 2(c), when $\|r(x_{k+1}, t_k)\| \geq \theta\epsilon_0$ and $f(x_{k+1}) \geq t_k$. But then the third inequality in (14.3.32) must hold, again implying (14.3.49), this time at $(x_\epsilon, t_\epsilon) = (x_{k+1}, t_k)$.

It remains to show that (14.3.48) and (14.3.49) imply the desired conclusions. In particular, we have that

$$\epsilon_0 \geq \|r(x_\epsilon, t_\epsilon)\| \geq \theta\epsilon_0, \tag{14.3.52}$$

where the first inequality results from the fact that $\|r(x_k, t_k)\|$ decreases during Steps 1 and 2(b) and is reset to ϵ_0 in Step 2(a). If $f(x_\epsilon) = t_\epsilon$, then

$$\phi^{\delta_{k_\epsilon,j}}_{\nu,\mathcal{X}(x_\epsilon),j}(x_\epsilon) = \phi^{\delta_{k_\epsilon,j}}_{\mu,\mathcal{X}(x_\epsilon),j}(x_\epsilon, t_\epsilon) \leq \epsilon_j \|c(x_\epsilon)\| \frac{\delta^j_{k_\epsilon,j}}{j!} \quad (j \in [1:q])$$

from the definition of $r(x,t)$ in (14.3.30) and (14.3.49), and (14.3.41) follows since (14.3.52) ensures that $\|c(x_\epsilon)\| = \|r(x_\epsilon, t_\epsilon)\| \geq \theta\epsilon_0$.

So suppose instead that $f(x_\epsilon) \neq t_\epsilon$, and observe that $f(x_\epsilon) < t_\epsilon$ is excluded by (14.3.48), and, as a consequence, $f(x_\epsilon) > t_\epsilon$. To establish the desired conclusions when $j = 1$, define

$$y_\epsilon \stackrel{\text{def}}{=} -\frac{c(x_\epsilon)}{f(x_\epsilon) - t_\epsilon}. \tag{14.3.53}$$

Since

$$\nabla^1_x \mu(x_\epsilon, t_\epsilon) = (f(x_\epsilon) - t_\epsilon)\nabla^1_x f(x_\epsilon) + [\nabla^1_x c(x_\epsilon)]^T c(x_\epsilon),$$

it follows that

$$\nabla^1_x \mathcal{L}(x_\epsilon, y_\epsilon) = \nabla^1_x f(x_\epsilon) - [\nabla^1_x c(x_\epsilon)]^T y_\epsilon = \frac{\nabla^1_x \mu(x_\epsilon, t_\epsilon)}{f(x_\epsilon) - t_\epsilon}. \tag{14.3.54}$$

Using the inequality $f(x_\epsilon) > t_\epsilon$, (14.3.37), the termination condition (14.3.49) for $j = 1$, and the definition (14.3.53), we deduce that

$$\phi^{\delta_{k_\epsilon,1}}_{\mathcal{L},\mathcal{X}(x_\epsilon),1}(x_\epsilon, y_\epsilon) = \frac{\phi^{\delta_{k_\epsilon,1}}_{\mu,\mathcal{X}(x_\epsilon),1}(x_\epsilon, t_\epsilon)}{f(x_\epsilon) - t_\epsilon} \leq \epsilon_1 \frac{\|r(x_\epsilon, t_\epsilon)\|}{f(x_\epsilon) - t_\epsilon} \delta_{k_\epsilon,1} = \epsilon_1 \|(y_\epsilon, 1)\| \, \delta_{k_\epsilon,1}.$$

Thus (14.3.42) holds because $\|c(x_\epsilon)\| \leq \|r(x_\epsilon, t_\epsilon)\| \leq \epsilon_0$, which follows from (14.3.52). When $j = 2$, (14.3.38), (14.3.53), and the definition of $\mathcal{T}(x)$ in (14.3.45) give that

$$\nabla^2_x \mathcal{L}(x_\epsilon, y_\epsilon)[d]^2 = \frac{1}{f(x_\epsilon) - t_\epsilon} \left[\nabla^2_x \mu(x_\epsilon, t_\epsilon)[d]^2 - \left(\nabla^1_x f(x_\epsilon)[d] \right)^2 \right] \quad \text{for all } d \in \mathcal{T}(x_\epsilon). \tag{14.3.55}$$

We also know from (14.3.49) for $j = 2$ that

$$\nabla^1_x \mu(x_\epsilon, t_\epsilon)[d] + \tfrac{1}{2}\nabla^2_x \mu(x_\epsilon, t_\epsilon)[d]^2 \geq -\tfrac{1}{2}\epsilon_2 \|r(x_\epsilon, t_\epsilon)\| \, \delta^2_{k_\epsilon,2}$$

for all $d \in \mathcal{T}(x_\epsilon)$. Dividing this inequality by $f(x_\epsilon) - t_\epsilon$ and using (14.3.53)–(14.3.55) then gives that

$$\nabla^1_x \mathcal{L}(x_\epsilon, y_\epsilon)[d] + \tfrac{1}{2}\nabla^2_x \mathcal{L}(x_\epsilon, y_\epsilon)[d]^2 \geq -\tfrac{1}{2}\epsilon_2 \|(1, y_\epsilon^T)\| \, \delta^2_{k_\epsilon,2} - \tfrac{1}{2}|\nabla^1_x f(x_\epsilon)[d]| \frac{|\nabla^1_x f(x_\epsilon)[d]|}{f(x_\epsilon) - t_\epsilon} \tag{14.3.56}$$

for all $d \in \mathcal{T}(x_\epsilon)$, Moreover, (14.3.54) and the definitions of the Lagrangian and $\phi^{\delta_{k_\epsilon,1}}_{\mathcal{L},\mathcal{X}(x_\epsilon),2}(x_\epsilon, y_\epsilon)$ imply that

$$\frac{|\nabla^1_x f(x_\epsilon)[d]|}{f(x_\epsilon) - t_\epsilon} = |\nabla^1_x \mathcal{L}(x_\epsilon, y_\epsilon)[d]| \leq \phi^{\delta}_{\mathcal{L},\mathcal{X}(x_\epsilon),1}(x_\epsilon, y_\epsilon) \tag{14.3.57}$$

for all $d \in \mathcal{T}(x_\epsilon)$ when $\|d\| \leq \delta$. Substituting this bound into (14.3.56) for $\|d\| \leq \delta_{k_\epsilon,2}$ and using the Cauchy–Schwarz inequality gives (14.3.43). If d is further restricted to be in $\mathcal{M}(x_\epsilon) \subseteq \mathcal{T}(x_\epsilon)$, then the last term on the right-hand side of (14.3.56) vanishes, yielding (14.3.44).

The reasoning is similar when $j = 3$. It follows from (14.3.39) that

$$\nabla_x^3 \mathcal{L}(x_\epsilon, y_\epsilon)[d]^3 = \frac{1}{f(x_\epsilon) - t_\epsilon} \left[\nabla_x^3 \mu(x_\epsilon, t_\epsilon)[d]^3 - 3(\nabla_x^2 f(x_\epsilon)[d]^2)(\nabla_x^1 f(x_\epsilon)[d]) \right] \tag{14.3.58}$$

when $d \in \mathcal{T}(x_\epsilon)$. Furthermore, (14.3.49) for $j = 3$ gives

$$\nabla_x^1 \mu(x_\epsilon, t_\epsilon)[d] + \tfrac{1}{2} \nabla_x^2 \mu(x_\epsilon, t_\epsilon)[d]^2 + \tfrac{1}{6} \nabla_x^3 \mu(x_\epsilon, t_\epsilon)[d]^3 \geq -\tfrac{1}{6} \epsilon_3 \|r(x_\epsilon, t_\epsilon)\| \, \delta_{k_\epsilon,3}^3$$

for all $d \in \mathcal{T}(x_\epsilon)$. Dividing this inequality by $f(x_\epsilon) - t_\epsilon$ and using (14.3.53)–(14.3.55) and (14.3.58), we find that

$$\begin{aligned} &\nabla_x^1 \mathcal{L}(x_\epsilon, y_\epsilon)[d] + \tfrac{1}{2} \nabla_x^2 \mathcal{L}(x_\epsilon, y_\epsilon)[d]^2 + \tfrac{1}{6} \nabla_x^3 \mathcal{L}(x_\epsilon, y_\epsilon)[d]^3 \\ &\geq -\tfrac{1}{6} \epsilon_3 \|(1, y_\epsilon^T)\| \, \delta_{k_\epsilon,3}^3 - \tfrac{1}{2} \left(|\nabla_x^1 f(x_\epsilon)[d]| + |\nabla_x^2 f(x_\epsilon)[d]^2| \right) \frac{|\nabla_x^1 f(x_\epsilon)[d]|}{f(x_\epsilon) - t_\epsilon}. \end{aligned} \tag{14.3.59}$$

Substituting the bound (14.3.57) for $\|d\| \leq \delta_{k_\epsilon,3}$ into (14.3.59) and using the Cauchy–Schwarz inequality gives (14.3.46). If d is further restricted to be in $\mathcal{M}(x_\epsilon) \subseteq \mathcal{T}(x_\epsilon)$, then the last term on the right-hand side of (14.3.59) vanishes, yielding (14.3.47). □

Observe that if $\delta_{k_\epsilon,2} \leq \delta_{k_\epsilon,1}$, then clearly

$$\phi_{\mathcal{L},\mathcal{X}(x_\epsilon),1}^{\delta_{k_\epsilon,2}}(x_\epsilon, y_\epsilon) \leq \phi_{\mathcal{L},\mathcal{X}(x_\epsilon),1}^{\delta_{k_\epsilon,1}}(x_\epsilon, y_\epsilon) \leq \epsilon_1 \|(1, y_\epsilon^T)\| \, \delta_{k_\epsilon,1}$$

using (14.3.42), and substituting this into (14.3.43) reveals that

$$\phi_{\mathcal{L},\mathcal{T}(x_\epsilon),2}^{\delta_{k_\epsilon,2}}(x_\epsilon) \leq \tfrac{1}{2} \|(1, y_\epsilon^T)\| \left[\epsilon_2 \delta_{k_\epsilon,2}^2 + \epsilon_1 \|\nabla_x^1 f(x_\epsilon)\| \, \delta_{k_\epsilon,1} \delta_{k_\epsilon,2} \right].$$

Similarly, if $\delta_{k_\epsilon,3} \leq \delta_{k_\epsilon,1}$, then (14.3.46) becomes

$$\phi_{\mathcal{L},\mathcal{T}(x_\epsilon),3}^{\delta_{k_\epsilon,3}}(x_\epsilon) \leq \tfrac{1}{6} \|(1, y_\epsilon^T)\| \left[\epsilon_3 \delta_{k_\epsilon,3}^3 + 3 \epsilon_1 \delta_{k_\epsilon,1} \left(\|\nabla_x^1 f(x_\epsilon)\| \, \delta_{k_\epsilon,3} + \|\nabla_x^2 f(x_\epsilon)\| \, \delta_{k_\epsilon,3}^2 \right) \right].$$

It is also important to note that each of the bounds (14.3.41)–(14.3.47) is of the form (14.3.9) (when one chooses η appropriately) and thus that Theorem 14.3.2 may be used to derive specific lower bounds on the local decrease in the objective function for each of the cases under consideration.

Observe also that the second part of (14.3.48) and (14.3.53) imply that $\langle y_\epsilon, c(x_\epsilon) \rangle = -\|c(x_\epsilon)\|^2 / (f(x_\epsilon) - t_\epsilon) \leq 0$, a property we have seen in Theorem 14.3.2 that can be used to derive the improved bound (14.3.12) on the objective function decrease in a feasible neighbourhood of x_ϵ.

14.3.2.3 Evaluation complexity of the TPqp-GC algorithm

Before starting the analysis proper, we must first verify that the smoothness assumptions that we have made on c and f translate appropriately to the vital auxiliary functions ν and μ that the TPqp-GC algorithm relies upon.

Lemma 14.3.6: Smoothness of ν and μ

(i) *Suppose that* **c.D1pLB.X** *holds. Given* $x_0 \in \mathcal{X}$, *let* $\mathcal{L}_\nu(x_0) \stackrel{\text{def}}{=} \{x \in \mathcal{X} \mid \|c(x)\| \leq \|c(x_0)\|\}$. *Then* $\nu(x)$ *is p-times continuously differentiable in an open, convex set* $\mathcal{X}_0$ *containing* $\mathcal{X}$, *and there are constants* $L^A_{\nu,j} \geq 1$, $L^B_{\nu,j} \geq 1$, $B^A_{\nu,j} \geq 1$, *and* $B^B_{\nu,j} \geq 1$ *such that*

$$\|\nabla^j_x \nu(x) - \nabla^j_x \nu(y)\| \leq \left(L^A_{\nu,j} + L^B_{\nu,j}\|c(x_0)\|\right)\|x - y\| \tag{14.3.60}$$

and

$$\|\nabla^j_x \nu(x)\| \leq B^A_{\nu,j} + B^B_{\nu,j}\|c(x_0)\| \tag{14.3.61}$$

for all $x \in \mathcal{L}_\nu(x_0)$, *all* $y \in \mathcal{X}_0$, *and every* $j \in [1:p]$.

(ii) *Suppose that* **c.D1pLB.X** *and* **f.D1pLB.X** *hold. Let* $\mathcal{L}_\mu \stackrel{\text{def}}{=} \{x \in \mathcal{X} \mid \|r(x,t)\| \leq 1|\}$. *Then* $\mu(x,t)$ *is p-times continuously differentiable in an open, convex set* $\mathcal{X}_0$ *containing* $\mathcal{X}$, *and there are constants* $L_{\mu,j} \geq 1$ *and* $B_{\mu,j} \geq 1$, *independent of t, such that*

$$\|\nabla^j_x \mu(x,t) - \nabla^j_x \mu(y,t)\| \leq L_{\mu,j}\|x - y\| \tag{14.3.62}$$

and

$$\|\nabla^j_x \nu(x)\| \leq B_{\mu,j} \tag{14.3.63}$$

for all $x \in \mathcal{L}_\mu$, *all* $y \in \mathcal{X}_0$, *and every* $j \in [1:p]$.

Proof. The technical proof is given in Appendix B (p. 491). □

The importance of this result is that by assuming **c.D1pLB.X** and **f.D1pLB.X**, we ensure that **f.D1qLaB.X** holds for both the Phase 1 objective function, $\nu(x)$, and the Phase 2 one, $\mu(x,t)$. Consequently, we may apply Lemmas 14.2.2 and 14.2.3 and Theorem 14.2.4 as required throughout the course of the analysis of the TPqp-GC algorithm.

Because the constants κ^μ_{decr} and κ^μ_{uns} in (14.2.11) and (14.2.13) for the specific INNER algorithm used to minimize $\mu(x,t)$ depend directly on the Lipschitz constants of the derivatives of μ with respect to x, the independence of these with respect to t ensures that κ^μ_{decr} and κ^μ_{uns} are also independent of t. By contrast, the Lipschitz constants and derivative bounds for ν illuminated by (14.3.62) and (14.3.63) depend linearly on the value of the norm of constraints at the initial point, and this will permeate into $\kappa^{\text{F}}_{\text{INNER}}$ and $\kappa^{\text{D}}_{\text{INNER}}$ in (14.3.20) and (14.3.21) for ν.

We start our analysis by providing an upper bound on the evaluation complexity of the initial feasibility phase.

Lemma 14.3.7: Evaluation complexity of Phase 1 of the TPqp-GC algorithm

Suppose that **X.CNe** *and* **c.D1pLB.X** *hold. Then there are constants* $\kappa^{\text{F},\nu}_{\text{INNER}}$ *and* $\kappa^{\text{D},\nu}_{\text{INNER}}$ *depending on* $\|c(x_0)\|$ *such that Phase 1 of the* TPqp-GC *algorithm terminates at a point* x_1 *for which either*

$$\|c(x_1)\| \leq \theta\epsilon_0 \quad \text{or} \quad \phi^{\delta_{k_1,j}(x_1)}_{\nu,\mathcal{X}(x_1),j} \leq \epsilon_j \|c(x_1)\| \frac{\delta^j_{k,j}}{j!} \tag{14.3.64}$$

in at most

$$\begin{aligned}\#^{F}_{\text{INNER},1} &\stackrel{\text{def}}{=} \tfrac{1}{2}\kappa^{F,\nu}_{\text{INNER}}(\|c(x_0)\|)\|c(x_0)\| \max_{j\in[1:q]} \frac{\beta^{1-2\pi_j}}{\beta^{1-\pi_j}-1}\theta^{1-\pi_j} \max_{j\in[1:q]} \epsilon_0^{1-\pi_j}\epsilon_j^{-\pi_j}\\ &= \mathcal{O}\left(\max_{j\in[1:q]} \epsilon_0^{1-\pi_j}\epsilon_j^{-\pi_j}\right)\end{aligned} \tag{14.3.65}$$

evaluations of c and at most

$$\begin{aligned}\#^{D}_{\text{INNER},1} &\stackrel{\text{def}}{=} \tfrac{1}{2}\kappa^{D,\nu}_{\text{INNER}}(\|c(x_0)\|)\|c(x_0)\| \max_{j\in[1:q]} \frac{\beta^{1-2\pi_j}}{\beta^{1-\pi_j}-1}\theta^{1-\pi_j} \max_{j\in[1:q]} \epsilon_0^{1-\pi_j}\epsilon_j^{-\pi_j}\\ &= \mathcal{O}\left(\max_{j\in[1:q]} \epsilon_0^{1-\pi_j}\epsilon_j^{-\pi_j}\right)\end{aligned} \tag{14.3.66}$$

evaluations of its derivatives.

Proof. We apply Theorem 14.3.4 to $\nu(x)$ and deduce that the LENq-SC algorithm requires at most

$$\tfrac{1}{2}\kappa^{\text{F},\nu}_{\text{INNER}}(\|c(x_0)\|)\|c(x_0)\| \max_{j\in[1:q]} \frac{\beta^{1-2\pi_j}}{\beta^{1-\pi_j}-1}\theta^{1-\pi_j}\epsilon_0^{1-\pi_j}\epsilon_j^{-\pi_j} \tag{14.3.67}$$

evaluations of $\Phi_2(x)$ and at most

$$\tfrac{1}{2}\kappa^{\text{D},\nu}_{\text{INNER}}(\|c(x_0)\|)\|c(x_0)\| \max_{j\in[1:q]} \frac{\beta^{1-2\pi_j}}{\beta^{1-\pi_j}-1}\theta^{1-\pi_j}\epsilon_0^{1-\pi_j}\epsilon_j^{-\pi_j} \tag{14.3.68}$$

evaluations of its derivatives to compute an iterate x_1 satisfying (14.3.64). Here $\kappa^{\text{F},\nu}_{\text{INNER}}(\|c(x_0)\|)$ and $\kappa^{\text{D},\nu}_{\text{INNER}}(\|c(x_0)\|)$ reflect the dependence of the relevant constants in (14.3.27) and (14.3.28) for $\nu(x)$, implied by Lemma 14.3.6(i), on the relevant Lipschitz constants of $c(x)$ but more particularly on the value of $\|c(x_0)\|$. The bounds (14.3.65) and (14.3.66) then follow directly from (14.3.67) and (14.3.68). □

Having analyzed the complexity of Phase 1 and verified that the problem assumptions are suited to applying the INNER algorithm to Step 1 of Phase 2, we now turn our attention to the sequence of iterations in Step 2 of Phase 2. Our development is actually very close to that we employed in Section 7.2.2.2 for the TPp-GC algorithm, but, since that was a couple hundred pages ago, we do not hesitate to repeat the essence here.

In particular, for any iteration k, we define $\ell(k)$ so that $k+\ell(k)$ is the first iteration after k whose index is in the set $\mathcal{K}_a$, defined in (14.3.36). If $\ell(k)>1$, iterations $[k+1 : k+\ell(k)-1]$ all belong to $\mathcal{K}_b$. Following (7.2.35) (p. 204) and (7.2.36), we now define

$$\mathcal{I}_k \stackrel{\text{def}}{=} \{(j,i) \mid i\in[0:e_j), j\in[k+1:k+\ell(k)-1]\},$$

the set of all inner iterations (during Step 1) necessary to find $x_{k+\ell(k)}$ from x_k. We also define

$$\mathcal{I}_k^{\text{S}} \stackrel{\text{def}}{=} \{(j,i) \mid i=0 \text{ or iteration } i \text{ is successful}\}.$$

These definitions imply that the set of all inner iterations before termination is given by $\cup_{k\in\mathcal{K}_a}\mathcal{I}_k$, and thus that the TP$qp$-GC algorithm needs at most

$$\left|\bigcup_{k\in\mathcal{K}_a}\mathcal{I}_k\right| + 1 \le \max_{k\in\mathcal{K}_a}|\mathcal{I}_k| \frac{f_{\text{up}}-f_{\text{low}}-1}{1-\theta}\epsilon_0^{-1}+1 \tag{14.3.69}$$

evaluations of f and c in Phase 2, where we added one to take the final evaluation into account and where the inequality results from applying Lemma 7.2.4 (p. 203).[118] Similarly, the number of successful iterations before termination is $\cup_{k\in\mathcal{K}_a}\mathcal{I}_k^{\rm S}$, and thus the algorithm needs at most

$$\Big|\bigcup_{k\in\mathcal{K}_a}\mathcal{I}_k^{\rm S}\Big| + 1 \leq \max_{k\in\mathcal{K}_a}|\mathcal{I}_k^{\rm S}|\,\frac{f_{\rm up}-f_{\rm low}-1}{1-\theta}\,\epsilon_0^{-1}+1 \tag{14.3.70}$$

evaluations of the derivatives of f and c in Phase 2.

The next step in our analysis is thus to obtain upper bounds on $|\mathcal{I}_k|$ and $|\mathcal{I}_k^{\rm S}|$ by deriving evaluation-complexity bounds for the INNER algorithm during Step 1 of Phase 2.

Lemma 14.3.8: Bounds on the size of $\mathcal{I}_k$ and $\mathcal{I}_k^{\rm S}$

Suppose that ***X.CNe***, ***c.D1pLB.X***, ***f.D1pLB.X***, *and* ***f.B.Fb*** *hold. Then there are constants $\kappa^{S}_{\rm INNER,2}$, $\kappa^{A}_{\rm INNER,2}$, and $\kappa^{C}_{\rm INNER,2}$ such that, for each $k\in\mathcal{K}_a$ before termination,*

$$|\mathcal{I}_k^{\rm S}| \leq \kappa^{S}_{\rm INNER,2}(1-\theta)\max_{j\in[1:q]}\epsilon_0^{2-\pi_j}\epsilon_j^{-\pi_j} \tag{14.3.71}$$

and

$$|\mathcal{I}_k| \leq \kappa^{A}_{\rm INNER,2}(1-\theta)\max_{j\in[1:q]}\epsilon_0^{2-\pi_j}\epsilon_j^{-\pi_j} + \kappa^{C}_{\rm INNER,2}. \tag{14.3.72}$$

Proof. The proof entirely parallels that of Lemma 7.2.6 (p. 204), where one now uses (14.3.20) and (14.3.21) for $\Phi_2=\mu$, giving the constants

$$\kappa^{\rm S}_{\rm INNER,2} \stackrel{\rm def}{=} [\kappa^{\mu}_{\rm decr}]^{-1}(\beta\theta)^{-\pi_{\max}},\ \kappa^{\rm A}_{\rm INNER,2} \stackrel{\rm def}{=} \kappa^{\mu}_{\rm uns}\kappa^{\rm S}_{\rm INNER,2},\ \text{ and }\ \kappa^{\rm C}_{\rm INNER,2} \stackrel{\rm def}{=} \kappa^{\mu}_{\rm unc},$$

with $\pi_{\max}=\max_{j\in[1:q]}\pi_j$. The proof is given in Appendix B (p. 494) for completeness. □

We finally state our evaluation-complexity result for the TPqp-GC algorithm.

Theorem 14.3.9: Complexity of finding qth-order approximate minimizers by the TPqp-GC algorithm

Suppose that the INNER *algorithm is of regularization type and* ***X.CNe***, ***c.D1pLB.X***, ***f.D1pLB.X***, *and* ***f.B.Fb*** *hold, or that the* INNER *algorithm is of trust-region type and* ***X.CNe***, ***c.D1qLB.X***, ***f.D1qLB.Fb***, *and* ***f.B.Fb*** *hold. Then there exist constants $\kappa^{S}_{\rm INNER,2}$, $\kappa^{A}_{\rm INNER,2}$, and $\kappa^{C}_{\rm INNER,2}$ such that, for any $\epsilon_0\in(0,1]$ and $\epsilon\in(0,1]^q$, the* TPqp-GC *algorithm requires at most*

$$\#^{F}_{\rm TPqpGC@TP\mathit{qp}GC} \stackrel{\rm def}{=} \#^{F}_{\rm INNER,1}+\#^{F}_{\rm INNER,2} = \mathcal{O}\left(\max\left[\epsilon_0^{-1},\ \max_{j\in[1:q]}\epsilon_0^{1-\pi_j}\epsilon_j^{-\pi_j}\right]\right)$$

evaluations of f and c and at most

$$\#^{D}_{\rm TPqpGC@TP\mathit{qp}GC} \stackrel{\rm def}{=} \#^{D}_{\rm INNER,1}+\#^{D}_{\rm INNER,2} = \mathcal{O}\left(\max_{j\in[1:q]}\epsilon_0^{1-\pi_j}\epsilon_j^{-\pi_j}\right)$$

evaluations of their derivatives, where the $\{\pi_j\}_{j=1}^q$ are defined in (14.2.12), $\#^{F}_{\rm INNER,1}$

[118] We noted above that this lemma remains valid for the TPqp-GC algorithm.

and $\#^D_{\text{INNER},1}$ *are given by (14.3.65) and (14.3.66), and*

$$\#^F_{\text{INNER},2} \stackrel{\text{def}}{=} (f_{\text{up}} - f_{\text{low}} - 1)\left(\kappa^A_{\text{INNER},2} \max_{j\in[1:q]} \epsilon_0^{1-\pi_j}\epsilon_j^{-\pi_j} + \kappa^C_{\text{INNER},2}\epsilon_0^{-1}\right) + 1$$

and

$$\#^D_{\text{INNER},2} \stackrel{\text{def}}{=} \kappa^S_{\text{INNER},2}(f_{\text{up}} - f_{\text{low}} - 1) \max_{j\in[1:q]} \epsilon_0^{1-\pi_j}\epsilon_j^{-\pi_j} + 1,$$

to compute a point x_ϵ *such that either (14.3.41) holds, or (14.3.42) holds for some* y_ϵ *together with (14.3.43) and (14.3.44) if* $q > 1$, *and (14.3.46) and (14.3.47) if* $q = 3$.

Proof. If the TPqp-GC algorithm terminates in Phase 1, then the bounds $\#^F_{\text{INNER},1}$ and $\#^D_{\text{INNER},1}$ immediately result from Lemmas 14.3.5 and 14.3.7. Otherwise, the algorithm terminates in Phase 2. We then deduce the upper bound $\#^D_{\text{INNER},2}$ on its number of successful inner iterations (and hence derivative evaluations) in this phase from (14.3.69), (14.3.71), and Lemma 14.3.5. Similarly, the bound $\#^F_{\text{INNER},2}$ results from (14.3.70), (14.3.72), and Lemma 14.3.5. Summing up the numbers of evaluations for both phases gives the requested bounds. □

The attentive reader will have noticed that we haven't mentioned any specific result for q exceeding three. This is not because the TPqp-GC algorithm will not terminate in this case, but rather because it is far from obvious how to extend the conclusions of Lemma 14.3.5 beyond the third order. Indeed, the proof of this lemma is strongly dependent on the observation that, in the subspace $\mathcal{M}(x)$,

$$\frac{1}{f(x_\epsilon) - t_\epsilon}\nabla_x^j \mu(x_\epsilon, t_\epsilon) = \nabla_x^j \mathcal{L}(x_\epsilon, y_\epsilon) \quad (j = 1, 2, 3)$$

(see (14.3.54), (14.3.55), and (14.3.58)), creating a strong link between the minimization of μ and the optimality conditions involving the Lagrangian. For orders four and beyond, the link between $\nabla_x^j \mu(x_\epsilon, t_\epsilon)$ and $\nabla_x^j \mathcal{L}(x_\epsilon, y_\epsilon)$ when j grows is at best tenuous. This is apparent when considering the expressions (14.3.37)–(14.3.40). Indeed, the terms

$$\begin{aligned}\frac{3}{f(x_\epsilon) - t_\epsilon}&\left[\sum_{i=1}^m (\nabla_x^2 c_i(x) \otimes \nabla_x^2 c_i(x))[d]^4 + (\nabla_x^2 f(x) \otimes \nabla_x^2 f(x))[d]^4\right] \\ &= \frac{3}{f(x_\epsilon) - t_\epsilon}\left[\sum_{i=1}^m \left(\nabla_x^2 c_i(x)[d]^2\right)^2 + \left(\nabla_x^2 f(x)[d]^2\right)^2\right]\end{aligned} \tag{14.3.73}$$

in (14.3.40) would vanish in general only if $d \in \ker^2[\nabla_x^2 f(x)] \cap \ker^2[\nabla_x^2 c(x)]$. Although this is formally reminiscent of the definition of $\mathcal{M}(x)$ in (14.3.45), this crucial inclusion now no longer follows from lower-order conditions—remember Example 12.1.2 (p. 277).

This has wider implications. Indeed the first of the problematic terms in (14.3.73) occurs not only in the function $\mu(x, t)$ used above, but also when employing quadratic, ℓ_1- and ℓ_∞-penalty, or augmented-Lagrangian approaches to tackle problem (14.3.1), or even sequential quadratic programming methods that use a merit function that depends on such penalty terms. The same difficulty may also occur if more general penalizations of the type $p(\nu(x))$, for some increasing smooth function p from $\mathbb{R}^+$ to $\mathbb{R}^+$, are employed.

Indeed, consider the derivatives of $p(\nu(x))$. One finds that

$$\begin{aligned}\nabla_x^4 p(\nu(x)) =& \, p_\nu^{(4)}(\nu(x))[\nabla_x^1\nu(x)]^{4\otimes} + 6p_\nu^{(3)}(\nu(x))\nabla_x^2\nu(x) \otimes [\nabla_x^1\nu(x)]^{2\otimes} \\ &+ 4p_\nu^{(2)}(\nu(x))\nabla_x^3\nu(x) \otimes \nabla_x^1\nu(x) + 3p_\nu^{(2)}(\nu(x))[\nabla_x^2\nu(x)]^{2\otimes} \\ &+ p_\nu^{(1)}(\nu(x))\nabla_x^4\nu(x),\end{aligned}$$

whose last term, together with

$$\nabla_x^4\nu(x) = \sum_{i=1}^m \left[4\,\nabla_x^3 c_i(x) \otimes \nabla_x^1 c_i(x) + 3\,[\nabla_x^2 c_i(x)]^{2\otimes} + c_i(x)\nabla_x^4 c_i(x)\right],$$

indicates that the troublesome terms involving $[\nabla_x^2 c_i(x)]^{2\otimes}$ will not vanish unless $p_\nu^{(1)}(\nu(x))$ also vanishes with $\nu(x)$.

To summarize, *none of the standard linear or quadratic penalization approaches can be expected to produce critical points of orders four or more.* Thus, innovative approaches will be required if one is interested in a (theoretical) procedure for finding approximate minimizers of (14.3.1) of higher order.

Notes and references for Section 14.3

The material in this section is adapted from [105] to parallel Section 7.2.2 as much as possible.

Part III

Conclusions and Perspectives

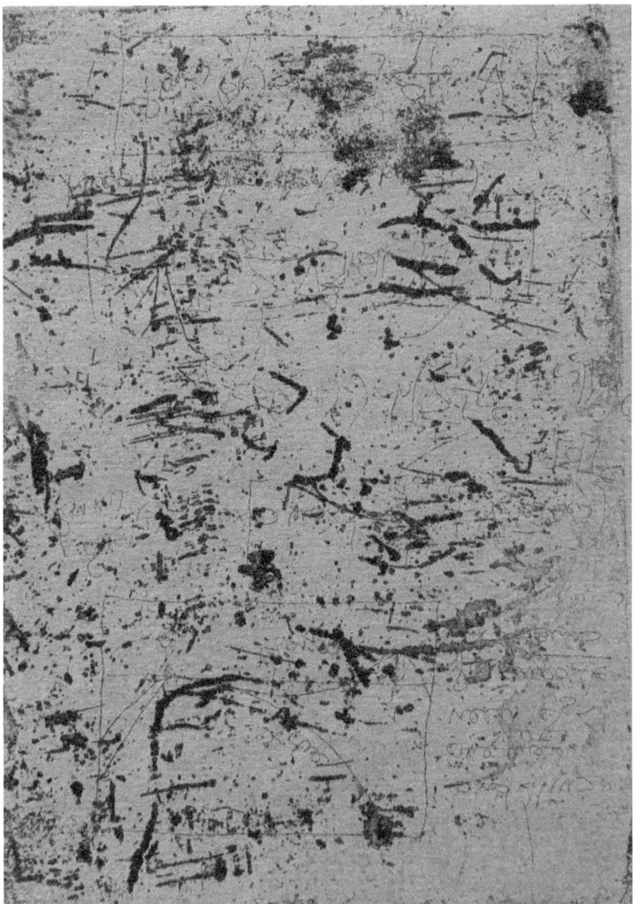

And so we say la revedere, farewell, au revoir, ...

Previous page: From the combook 4273, by Philippe Toint, 2020

What we hope we have achieved

Although we have reached the end of our current trip through the many themes of evaluation-based complexity, the journey is far from over. After a decade or two of emphasis on the convex case, the focus in optimization is now shifting firmly back to nonconvexity, driven by applications. Thus, while we have tried to be as topical as possible, it seems inevitable that there will soon be new results that will complement and even surpass what we already know. Nonetheless we hope that what we have presented here will provide a solid foundation for what is still to come.

The reader will already be aware of the three strands of what we have provided. The first covers by now standard aspects of evaluation complexity using tools that are familiar to those who have studied conventional topics such as global and local convergence of nonconvex-optimization algorithms. We have attempted to provide evaluation-complexity bounds—worst-case, sharp, and sometimes even optimal—for many of the methods that practitioners use today. There have been some surprises along the way. Our second strand concerns the unsung details that convert these theoretical algorithms into practical ones—unashamedly, we believe that optimization is not for spectators; indeed all of the technicalities that we have described have been converted into software. Our final contribution is more revolutionary, but with good reason. As we show, traditional measures of optimality run out of steam once higher-order derivatives are available. Thus we are forced to design optimality measures that can cope. This is not without a downside, since computing these measures may often be beyond our current capabilities. However, within the rules of evaluation complexity—that evaluating values and derivatives is the overbearing cost—these new measures, and the resulting algorithms that aim to enforce them, provide a significant improvement over traditional approaches. This part of our book then looks to the future.

We are not foolish enough to believe that our job is done. Our intention is to keep this book "alive," to fix the inevitable typos and more significant errors, and to grow with the field as it evolves. Version "1.0" is available from

`https://bookstore.siam.org/mo30/bonus`,

and updates will be incorporated as and when they arise.

What we haven't covered

What, the reader might ask, can we possibly not have covered in a book of this length? In truth, a great deal, and we don't simply mean details.

To start, we have focused primarily on deterministic problems for which both the objective function and the constraints are at least continuous and may be evaluated exactly. As such, we have avoided the vast area of *stochastic optimization*, with both its long-standing and recent developments, where the evaluations of the problem functions are stochastic or corrupted by (stochastic) noise even when the underlying "true" function may be smooth. This is a vibrant area of research [29, 51, 141, 154, 257, 276, 314, 325, 370, 371, 373], most especially in machine learning [58, 140, 248, 342], but one that we have purposely avoided because even many of the ground rules—and the computational feasibility of some of the proposed algorithms with the best complexity guarantees—are still being debated.

A related topic concerns the use of randomness within algorithms for deterministic problems. This has long been a theme of global optimization, and more recently has been used very effectively for convex optimization. It may manifest itself in at least two ways.

For very large problems, particularly those for which the objective function comprises a huge sums of subfunctions, e.g.,

$$f(x) = \sum_{i=1}^{m} f_i(x),$$

where each $f_i(x)$ is inexpensive and m is currently in the tens of millions, local sampling, in which f is replaced by

$$f^{\mathrm{S}}(x) = \sum_{i \in \mathcal{S}} f_i(x)$$

for a succession of random subsets $\mathcal{S}$ of $[1:m]$ [58], has been a major theme; readers are almost certainly aware of the celebrated stochastic gradient-descent algorithm [327], in which one applies gradient descent to the subsampled function. Algorithms based on such ideas are usually assessed by their ability to converge in expectation or probability. Within the framework of our book, for example, one might be interested in the question of how many evaluations are required to reduce the norm of $\nabla_x^1 f(x)$ below ϵ with a certain probability, and certainly there have been recent analyses on this theme.

Randomness may also be present when solving subproblems. In particular, we have already mentioned that only probabilistic guarantees are possible [72, 334] when minimizing the trust-region and regularization subproblems by iterative methods in exceptional ("hard") cases. And of course the approximate-minimization criteria that we have championed in Part II involve global optimization, and as we have just mentioned randomization plays an important role for this.

Another important issue in practice is the effects of finite-precision computation, which can often be viewed as deterministic noise. In almost all that we have said—apart from function and derivative evaluations in Chapter 13—our analysis presumes that our algorithms compute exactly. But as numerical analysts, we know full well that this is a convenient fiction. Our algorithms are predicated on many decisions in which we compare numbers; for example, we decide when an iteration is successful or not on this basis. We demonstrate examples of slow convergence, but even the tiniest rounding may well perturb the ensuing future iterates so that the bad behaviour vanishes, or equally might be worse. Our optimistic analysis of the cost of iterative methods for solving subproblems is frequently undeserved, since the computed iterates soon diverge from their exact counterparts; even the solution by "direct" methods may be affected by (preventable) poor numerical methods or (inevitable) bad conditioning. We optimistically and unreasonably expect that we can compute a residual to be smaller than some growing power of a small ϵ. Indeed, the reader might wonder what, if anything, we have said is true or relevant! Our response is simple: what we have described is both an aspiration and a hint of what might (and often does) happen. We can imagine forthcoming computers with larger wordlengths for which more accurate computation, particularly for key components, is possible. We can anticipate algorithms that are aware of current deficiencies and that by cunning aim to avoid them. But most important of all, we can imagine future algorithms and analyses that take it as read that computation is inexact (but nonetheless obeys reasonable rules). This is a core preoccupation in numerical analysis for many simpler tasks, such as solving linear equations or approximating low-dimensional functions, and there is no reason not to expect the same for our problem. And we should of course mention that there is an obvious relationship between computer rounding and randomization, so one can predict a profitable future interaction between the two preoccupations.

We have also (purposely) only touched very briefly on the dimension dependency of algorithms. This is a core component in many disciplines and, as mentioned in our introduction, defines to an extent other aspects of complexity (and sometimes paints a very pessimistic picture, indeed!). At the outset, evaluation complexity tried to avoid the issue by its fundamental premise that function evaluation dominates all, but of course dimension comes into the definition of functions, and potentially even more so into that of their derivatives (notably if sparsity or structure is ignored). More recently there has been a trend to bring some relevant aspects of dimension back into the analysis but at the same time ignoring others (such as how dimension may impinge on the vital Lipschitz constants).[119] This is clearly an issue that will become more relevant in the future, especially given the desire to solve huge nonconvex problems from machine learning and related areas.

Other topics that we have not covered are (non-model-based) methods for derivative-free optimization, infinite-dimensional (functional) optimization,[120] and global optimization alongside its twin, (mixed) integer programming. Each of these could easily fill its own book, and all we can say to our colleagues is that we very much look forward to reading them all.

Open questions and perspectives

We hope that we have been quite open that the worst-case complexity bounds that we have exhibited for nonconvex optimization bear no resemblance to what is observed in practice, especially if second-degree or higher models are employed. Indeed, had this not been the case, it is unlikely that people would have any faith in using current optimization methods at all. This then leads us to why this should be so. Are problems that one encounters in practice somehow special? For almost all of the algorithms we have considered, we have provided examples that match their theoretical worst case (in order). Thus one is tempted to believe that examples of this sort are rare.

Such questions are not restricted to optimization. When solving linear systems of equations, the most common method used, at least for small instances, is Gaussian elimination. But this is known to be unstable, or at least there are examples for which huge growth can occur. However, this seems never to happen in practice. Although this hasn't been fully explained, there are convincing arguments that bad behaviour is extremely rare if the matrices involved are suitably random [352]. Likewise, in linear optimization, the simplex method for linear programming has an exponential worst-case computational-complexity bound with common pivot rules [256], but in practice it behaves perfectly reasonably (often as if the bound were a linear function of its input). This may be explained to a certain extent, using so-called *average-case complexity* analyses, by examining problems whose data is from common random distributions [55], or small random perturbations of general input [341]. Thus it is natural to ask whether one might say something similar for the problem class that has been our focus. The main obstacle is, of course, what one might mean by a "random function." Unlike for the linear-system and optimization cases, generic optimization functions are not simply expressed in terms of matrices and vectors on which we might impose sensible probability distributions. Thus, at the moment, average-case evaluation complexity for generic nonconvex optimization seems a distant dream.

Of course much more may be said if we restrict the class of problems we consider. We have already seen this for least-squares and partially separable problems, indeed more

[119] The notes at the end of Section 4.2 (on page 117) detail some of these aspects and apparent contradictions.

[120] Except for first-order points; see the last paragraph in the notes and references for Section 4.1.2 on page 110 and [208].

generically if we distinguish unconstrained and simply constrained problems from generally constrained ones. It is also being observed in practice in areas like machine learning where hidden (low-rank) structure prevents overparameterization from being the obstacle it might seem. Furthermore, in applications such as phase retrieval and compressed sensing in signal processing, and others, it has been observed computationally—and sometimes also shown theoretically—that the nonconvex landscape has innocuous saddle points and a unique local minimizer—which is thus also global—that has the desired low-rank structure. This may then be found by means of (local) second-order optimization methods such as those described in this book. These are early days, but we strongly believe that the major future advances will come through discovery and exploitation of generic structure.

Another topic that we have touched on, but is far from fully explored, is that of phases of progress for algorithms. In the early days of optimization, research focused on global issues like convergence and local concerns such as asymptotic rates without really connecting the two. Evaluation complexity, at least of the relatively crude kind that has been our main focus in this book, addresses this to some extent, but anyone who has ever monitored a good optimization algorithm in practice will have seen phases where rapid progress is made and others where semistagnation occurs. We expect brisk performance on functions that are locally strongly convex or have significant negative curvature once we enter such neighbourhoods, but what if we are in a neighbourhood for which the objective hovers on the convex-nonconvex border? The best we currently have is the regional-complexity analysis of Curtis and Robinson [135] that ties progress to local (regional) function behaviour. This seems to us a good future paradigm, but are there others?

Finally, of the problem classes covered here, by far the least developed evaluation-complexity theory is that for problems with explicit nonlinear constraints. To a certain extent, by repute, this is true of the algorithms for this class themselves, since in practice they are not seen to be as reliable as those for simpler problems. In particular, failures do occur when convenient assumptions are violated. In part this stems from the need to balance objective reduction with constraint infeasibility, an issue that does not arise for feasible-point methods. Many penalty-based methods do this balancing explicitly, which is not without difficulties since the weight placed on the constraints may never (globally) overcome the downward pull of the objective—just imagine using the quadratic or augmented-Lagrangian function to try to minimize the concave objective $-x^4$ subject to the linear constraint $x = 1$. This is less of an issue for methods that require feasible points throughout, but it does make an a priori assumption that there are feasible points, or, more importantly, that one can be found (which is itself a global-optimization problem).

Thus, although we have exhibited a current champion, it uses a very cautious approach in which first feasibility is sought, and thereafter objective-function reduction is encouraged without losing feasibility. Thus failure of feasibility is assessed even before the more challenging optimization aspects occur. This is not how the most useful (interior-point or SQP) methods work in practice. Indeed it is usual for feasibility to occur only in the limit, while the ability to take "shortcuts" across the infeasible domain is exploited to the full. As we have described in Chapters 7 and 14, some progress has been made in analyzing more traditional methods, and one can hope—even predict—that this will remain an active area for the foreseeable future.

Appendices

This is where we put the material that we hope you already know, and where we put the necessary but often less-exciting technical details.

Previous page: From the combook 4284, by Philippe Toint, 2020

Appendix A

Background and General Technical Results

We have delayed important background material to this appendix, so as not to disturb the flow of the main elements of evaluation complexity as we encountered them.

A.1 Indices, subsets, and the Kronecker delta

If $\mathcal{I}$ is a set of indices, we will write the vector made up from these as $[v]_{\mathcal{I}}$ or, more conveniently, $v_{\mathcal{I}}$. If $\mathcal{J}$ is another such set, $[A]_{\mathcal{I},\mathcal{J}}$ or, more conveniently, $A_{\mathcal{I},\mathcal{J}}$, is the subset of A with row indices in $\mathcal{I}$ and column indices in $\mathcal{J}$.

The Kronecker delta $\delta_{ij} = 1$ when the indices i and j are equal and is 0 otherwise. The unit vectors $e_i \in \mathbb{R}^n$ are such that $[e_i]_j = \delta_{ij}$ for $i, j = 1, \ldots, n$, while the identity matrix $I_n \in \mathbb{R}^{n \times n}$, often abbreviated to I when the dimension is clear from the context, has entries $[I_n]_{i,j} = \delta_{ij}$.

A.2 Inner products and norms

A basic requirement in all that we have studied is that we can measure the size of things, both absolutely and relative to each other. To do so, we suppose that our quantities of interest lie in a real vector space $\mathcal{V}$, and we have a function $\|\cdot\| : \mathcal{V} \rightarrow \mathbb{R}$ for which $\|x\| \geq 0$ (with equality if and only if x is the zero in $\mathcal{V}$), the (triangle) inequality $\|x + y\| \leq \|x\| + \|y\|$ holds, and $\|\alpha X\| = |\alpha|\|x\|$ for all x and y in $\mathcal{V}$ and real scalars α. Any such function is known as a *norm* and clearly satisfies reasonable notions of size. While the particular choice of norm is made for convenience, a highly relevant result is that if $\mathcal{V}$ is finite-dimensional, any pair of norms $\|\cdot\|$ and $\|\cdot\|_N$ are equivalent in the sense that $\kappa_{\rm l}\|x\|_N \leq \|x\| \leq \kappa_{\rm u}\|x\|_N$ for all x and some constants $\kappa_{\rm l}$ and $\kappa_{\rm u}$. Thus, in finite dimensions, it really doesn't matter (from a theoretical perspective) which norm we choose, provided we are not overly concerned with the constant factors $\kappa_{\rm l}$ and $\kappa_{\rm u}$.

In our vector space $\mathcal{V}$, any positive-definite symmetric bilinear form $\langle \cdot, \cdot \rangle$ is said to be an *inner product*. Given a norm $\|\cdot\|$, we may define its *dual norm* via

$$\|x\|_{\rm D} = \sup_{\|y\| \leq 1} |\langle y, x \rangle| \equiv \sup_{y \neq 0} \frac{|\langle y, x \rangle|}{\|y\|},$$

from which we easily deduce the valuable *Hölder inequality*

$$-\|x\|\|y\|_{\rm D} \leq \langle x, y\rangle \leq \|x\|\|y\|_{\rm D}.$$

We may always define a norm directly from an inner product via $\|x\| = \sqrt{\langle x, x\rangle}$.

The most popular norms in $\mathbb{R}^n$ are the ℓ_2-norm, $\|x\|_2 = \sqrt{\sum_{i=1}^n x_i^2}$, the ℓ_1-norm, $\|x\|_1 = \sum_{i=1}^n |x_i|$, and the ℓ_∞-norm, $\|x\|_\infty = \max_{i=1,\ldots,n} |x_i|$, while the most-used inner product is the Euclidean one, $\langle x, y\rangle = \sum_{i=1}^n x_i y_i$; the relationships

$$\|x\|_\infty \leq \|x\|_2 \leq \|x\|_1 \leq \sqrt{n}\|x\|_2 \leq n\|x\|_\infty$$

are often useful. The ℓ_2-norm is self-dual in the Euclidean inner product; the ℓ_1- and ℓ_∞-norms are dual to each other; and the Hölder inequality leads directly to the famous *Cauchy–Schwarz inequality*,

$$-\|x\|_2\|y\|_2 \leq \langle x, y\rangle \leq \|x\|_2\|y\|_2.$$

A.3 Convergence of sequences

A major concern in more traditional numerical analysis is the *rate of convergence* of a sequence to a limit. For simplicity, let $\{a_k\}$, $k \geq 0$, be a sequence of strictly positive real numbers whose limit is zero, and suppose that

$$\lim_{k\to\infty} \frac{a_{k+1}}{a_k} = \rho.$$

Then we say that $\{a_k\}$ converges *Q-superlinearly*[121] if $\rho = 0$, *Q-linearly* if $\rho \in (0, 1)$, and *Q-sublinearly* if $\rho = 1$. We will usually omit the prefix "*Q*-." Linear and superlinear convergence is often distinguished by its *order*. In particular, if

$$\lim_{k\to\infty} \frac{a_{k+1}}{a_k^q} \leq o$$

for some order $q \geq 1$ and $o > 0$, the convergence is superlinear if and only if $q > 1$, and particularly it is *Q-quadratic* if $q = 2$.

Notice that there is no presumption here that $\{a_k\}$ is monotonic; indeed, for many of the quantities measured in this book (such as norms of gradients for unconstrained optimization, or violations of constraints), this is often not the case. In some circumstances, it is more important to assess *global* or *local* rates of convergence. For instance, $\{a_k\}$ is *globally linearly convergent* if $a_{k+1} \leq \rho a_k$ for some $\rho < 1$ and all k and *locally quadratic* if $a_{k+1} \leq \rho a_k^2$ for all k sufficiently large that $\rho a_k < 1$.

A.4 Infinite-dimensional spaces

We only mention infinite-dimensional (function) spaces in passing in this book, but for completeness we briefly recall the main ingredients.

[121] The "*Q*" here stands for quotient. There are other measures of convergence, with "R" or root convergence, in which $a_k \leq b_k$ where $\{b_k\}$ converges, being the most common.

A *vector space* over a field $\mathcal{F}$ (for our needs, invariably $\mathcal{F} = \mathbb{R}$) is a finite- or infinite-dimensional set $\mathcal{V}$ for which the quantities $v_1 + v_2$ and αv_1 lie in $\mathcal{V}$ for any elements $v_1, v_2 \in \mathcal{V}$ and $\alpha \in \mathcal{F}$. A *normed* vector space is a vector space that is equipped with a norm $\|\cdot\|$ (see Section A.2) on its elements, and the norm induces a *metric* $d(v_1, v_2) = \|v_1 - v_2\|$ for each pair of elements $v_1, v_2 \in \mathcal{V}$, which makes $\mathcal{V}$ a *metric space*.

A sequence of vectors $\{v_i\}$, $i \geq 0$, in a metric space $\mathcal{V}$ is a *Cauchy sequence* if for any $\epsilon > 0$, there is an index k for which $d(v_i, v_j) < \epsilon$ for all indices $i, j > k$. A metric space $\mathcal{V}$ is *complete* if every Cauchy sequence in $\mathcal{V}$ has a limit in $\mathcal{V}$. A *Banach* space is a complete, normed vector space. Commonly occurring Banach spaces include the infinite-dimensional L_p-spaces of (Lebesgue-μ-measurable) functions over a domain $\mathcal{D}$ for which

$$\|v\|_p = \left(\int_{\mathcal{D}} |v|^p d\mu\right)^{1/p} < \infty.$$

An *inner-product* space is a vector space $\mathcal{V}$ over the field $\mathcal{F}$ that is equipped with an inner product $\langle\cdot,\cdot\rangle \in \mathcal{F}$ (see Section A.2) on its elements. A *Hilbert* space is an inner-product space $\mathcal{V}$ over a field $\mathcal{F}$ that is also a complete metric space. Obviously, Hilbert spaces are Banach spaces using the norm $\|v\| = \sqrt{\langle v, v\rangle}$. Perhaps the best-known infinite-dimensional Hilbert space is the Banach space L_2, for which the appropriate inner product is

$$\langle v_1, v_2\rangle = \int_{\mathcal{D}} v_1 v_2 d\mu.$$

The development of the rich theory of Banach, Sobolev,[122] and Hilbert spaces is one of the triumphs of early 20th century mathematics. There are many books devoted to such topics, some of which have a particular optimization focus [17, 56, 278].

A.5 Matrices, eigenvalues, and eigenvectors

We encounter several special matrices in this book. We have already seen the identity matrix I, which is a very special diagonal matrix. In general a matrix D is *diagonal* if all of its off-diagonal entries $d_{i,j} = 0$ for all $i \neq j$, and we write $D = \text{diag}(d_{i,i})$. The matrix L is *lower triangular* if $l_{i,j} = 0$ for all $i < j$, and it is *unit* lower triangular if additionally all $l_{i,i} = 1$. Similarly, U is *(unit) upper triangular* if U^T is (unit) lower triangular. Finally, a *banded* matrix B, with *semibandwidth* m, has $b_{i,j} = 0$ for all $|i - j| > m$; in particular T is *tridiagonal* if its semibandwidth is one.

The majority of the matrices A that arise in this book are real and *symmetric*; that is, $a_{i,j} = a_{j,i}$ for each i, j. If $H \in \mathbb{R}^{n\times n}$ is such a matrix, it has a set of n real *eigenvalues* λ_i and mutually orthogonal nonzero *eigenvectors* $v_i \in \mathbb{R}^n$ for which $Hv_i = \lambda_i v_i$, and we arrange these so that

$$\lambda_1 \leq \lambda_i \leq \lambda_{i+1} \leq \lambda_n \tag{A.1}$$

for $i = 2, \ldots, n-2$. We refer to (λ_i, v_i) as an eigenpair of H. For convenience we denote the leftmost and rightmost eigenvalues by $\lambda_{\min}[H] \stackrel{\text{def}}{=} \lambda_1$ and $\lambda_{\max}[H] \stackrel{\text{def}}{=} \lambda_n$. If $\lambda_{\min}[H] \geq 0$, we say that H is *positive semidefinite*, and if $\lambda_{\min}[H] > 0$, it is *positive definite*. When $\lambda_{\min}[H] < 0$ and $\lambda_{\max}[H] > 0$, the matrix is *indefinite*, while if $\lambda_i = 0$ for any i, it is *singular*. When H is positive definite, we frequently refer to the *spectral*

[122] Sobolev spaces are particular Banach spaces that arise from combinations of L_p-norms of functions and their derivatives and that naturally occur in optimal control, solutions of partial differential equations, and related areas.

condition number as the ratio $\kappa(H) \stackrel{\text{def}}{=} \lambda_{\max}[H]/\lambda_{\min}[H] \geq 1$. The *spectral decomposition* $H = V\Lambda V^T$ with $V^TV = VV^T = I$ is a compact representation of the above; here $V \in \mathbb{R}^{n\times n}$ has columns v_i, $\Lambda \stackrel{\text{def}}{=} \text{diag}(\lambda_i) \in \mathbb{R}^{n\times n}$, and $\text{diag}(d_i) \in \mathbb{R}^{n\times n}$ is diagonal with entries d_i, for $i = 1,\ldots,n$. We also refer to the *generalized inverse* $H^+ \stackrel{\text{def}}{=} V\Lambda^+V^T$ of H, where the diagonal matrix Λ^+ has entries

$$[\Lambda^+]_{i,i} \stackrel{\text{def}}{=} \begin{cases} 1/\lambda_i & \text{if } \lambda_i \neq 0, \text{ and} \\ 0 & \text{if } \lambda_i = 0; \end{cases}$$

notice that $H^+ = H^{-1}$ whenever H is nonsingular.

More generally, if M is symmetric and additionally positive definite, there is a set of *generalized eigenpairs* (λ_i, v_i) with real λ_i for which $Hv_i = \lambda_i M v_i$ and the nonzero v_i are M-orthogonal, i.e., for the Euclidean inner product, $\langle v_i, Mv_j\rangle = \delta_{ij}$. Once again we assume that the generalized eigenvalues are ordered so that (A.1) holds and accordingly define $\lambda_{\min}[H,M] \stackrel{\text{def}}{=} \lambda_1$ and $\lambda_{\max}[H,M] \stackrel{\text{def}}{=} \lambda_n$. The above may be written more compactly as $HV = MV\Lambda$ with $V^TMV = I$. We refer to the form $H - \lambda M$ as a *matrix pencil* and denote this by (H,M).

For H and M as above, we make considerable use of the *generalized Rayleigh quotient*

$$\rho_{H,M}(x) \stackrel{\text{def}}{=} \frac{\langle x, Hx\rangle}{\langle x, Mx\rangle}, \tag{A.2}$$

again using the Euclidean inner product. This is particularly important since, for any real x, we have the *Rayleigh-quotient inequality*, namely, that

$$\lambda_{\min}[H,M] \leq \rho_{H,M}(x) \leq \lambda_{\max}[H,M]. \tag{A.3}$$

For the special case when $M = I$, we simply refer to $\rho_H(x) \stackrel{\text{def}}{=} \rho_{H,I}(x)$ as the *Rayleigh quotient*.

If $A \in \mathbb{R}^{m\times n}$ is unsymmetric—perhaps not even square—it is often singular values rather than eigenvalues that are relevant. The set of $\min(m,n)$ *singular values* $\sigma_i \geq 0$, and (mutually orthogonal) right- and left-*singular vectors* $v_i \in \mathbb{R}^n$ and $u_i \in \mathbb{R}^m$, for which $Av_i = \sigma u_i$ and $A^Tu_i = \sigma v_i$, are collectively known as the *singular-triples* of A. As before, it is convenient to presume that the singular values are ordered so that $\sigma_{\min}(A) \stackrel{\text{def}}{=} \sigma_1 \leq \sigma_i \leq \sigma_{i+1} \leq \sigma_{\max}(A) \stackrel{\text{def}}{=} \sigma_{\min(m,n)}$. The singular values of a symmetric matrix are the absolute values of its eigenvalues, while both left- and right-singular vectors are its eigenvectors. The *rank* of A is the number of nonzero singular values, and A is full rank if its rank is $\min(m,n)$. Once again there is a more compact representation, the *singular-value decomposition* (SVD), $A = U\Sigma V^T$, where $U \in \mathbb{R}^{m\times\min(m,n)}$ and $V \in \mathbb{R}^{n\times\min(m,n)}$ have columns u_i and v_i, respectively, the diagonal $\Sigma \in \mathbb{R}^{\min(m,n)\times\min(m,n)}$ has entries σ_i, and $U^TU = V^TV = I$.

It is straightforward to define *(subordinate) matrix norms*

$$\|A\| = \sup_{\|x\|\leq 1} \|Ax\|$$

from their vector equivalents.[123] The three most frequently used subordinate matrix norms are the matrix ℓ_1-, ℓ_2-, and ℓ_∞-*norms*, which have the simplified forms

$$\|A\|_1 = \max_{1\leq i\leq n} \|Ae_i\|_1, \quad \|A\|_2 = \sigma_{\max}[A], \quad \text{and} \quad \|A\|_\infty = \max_{1\leq i\leq m} \|A^Te_i\|_1,$$

[123] Indeed we might use different norms for $\|x\|$ and $\|Ax\|$.

where $A \in \mathbb{R}^{m\times n}$. Thus the matrix ℓ_1- and ℓ_∞-norms are merely the maximum ℓ_1-norms of the columns and rows of A, while the matrix ℓ_2-norm is its largest singular value. Once again, the simple relationship

$$\|A\|_\infty \leq \sqrt{n}\|A\|_2 \leq n\|A\|_1 \leq n\sqrt{m}\|A\|_2 \leq nm\|A\|_\infty$$

exists between these norms. A fourth matrix norm, the *Frobenius norm* $\|A\|_{\text{F}}$, for which

$$\|A\|_{\text{F}} = \sqrt{\sum_{i=1}^{m}\sum_{j=1}^{n} a_{i,j}^2},$$

is also useful, and for this we have the connection

$$\|A\|_2 \leq \|A\|_{\text{F}} \leq \sqrt{n}\|A\|_2$$

between the Frobenius and matrix ℓ_2-norms. If H is symmetric, it is easy to see that its Rayleigh quotient satisfies the inequality

$$-\|H\|_2 \leq \rho_H(x) \leq \|H\|_2. \tag{A.4}$$

An unrelated but equally useful result is that if $H_1, H_2 \in \mathbb{R}^{n\times n}$ are symmetric,

$$|\lambda_{\min}[H_1] - \lambda_{\min}[H_2]| \leq \|H_1 - H_2\|_2 \tag{A.5}$$

[185, Cor. 8.1.6].

A.6 Tensors

A tensor T of order p may be viewed as a multidimensional array $[T]_{i_1,i_2,\ldots,i_p}$, where each $i_j \in [1 : n_j]$ for $j = 1,\ldots,p$. It is cubical if $n_j = n$ for all $j \in [1 : p]$ and symmetric if it is invariant under all permutations of its indices [125, 258]. Of particular significance for our purpose, continuous derivative tensors of multivariate functions are both cubical and symmetric.

For a cubical, symmetric tensor T of order p, the multilinear form

$$T[v_1,\ldots,v_p] \stackrel{\text{def}}{=} \sum_{i_1=1}^{n}\cdots\sum_{i_p=1}^{n}[T]_{i_1,i_2,\ldots,i_p}[v_1]_{i_1}\cdots[v_p]_{i_p}$$

is the result of applying T to the vectors $v_1,\ldots,v_p$, the special case

$$T[v]^p \stackrel{\text{def}}{=} \sum_{i_1=1}^{n}\cdots\sum_{i_p=1}^{n}[T]_{i_1,i_2,\ldots,i_p}[v]_{i_1}\cdots[v]_{i_p} \tag{A.6}$$

is the result of applying T to p copies of the vector v, and

$$\|T\|_{[p]} \stackrel{\text{def}}{=} \max_{\|v\|=1}|T[v]^p| = \max_{\|v_1\|=\cdots=\|v_p\|=1}|T[v_1,\ldots,v_p]| \tag{A.7}$$

is the associated induced norm for such tensors; for vectors g and symmetric matrices H, (A.6) gives the familiar

$$g[v]^1 = \langle g, v\rangle \text{ and } H[v]^2 = \langle v, Hv\rangle,$$

while the second equality in (A.7) results from [379, Thm. 2.1]. Observe that the definition (A.7) implies that

$$\begin{aligned}
\|T[v_1,\ldots,v_j]\|_{[p-j]} &= \max_{\|w_1\|=\cdots=\|w_{p-j}\|=1} T[v_1,\ldots,v_j][w_1,\ldots,w_{p-j}] \\
&= \left(\max_{\|w_1\|=\cdots=\|w_{p-j}\|=1} T\left[\frac{v_1}{\|v_1\|},\ldots,\frac{v_j}{\|v_j\|},w_1,\ldots,w_{p-j}\right]\right)\left(\prod_{i=1}^{j}\|v_i\|\right) \\
&\leq \left(\max_{\|w_1\|=\cdots=\|w_p\|=1} T[w_1,\ldots,w_p]\right)\left(\prod_{i=1}^{j}\|v_i\|\right) \\
&= \|T\|_{[p]}\cdot\prod_{i=1}^{j}\|v_i\|,
\end{aligned}$$

a simple generalization of the standard Cauchy–Schwarz inequality for order-1 tensors (vectors) and of $\|Mv\| \leq \|M\|\,\|v\|$, which is valid for induced norms of matrices (order-2 tensors). Note also that perturbation theory (see [238, Thm. 7]) implies that $\|T\|_{[p]}$ is continuous as a function of T. For symmetric tensors, we often omit the subscript $_{[p]}$ to avoid notational clutter.

If T_1 and T_2 are tensors, $T_1 \otimes T_2$ is their tensor product (i.e., the tensor formed by multiplying each element of the order-p_1 tensor T_1 by the entire order-p_2 tensor T_2 and assembling the result as a $p_1 \times p_2$ tensor—this is a generalization of the matrix Kronecker product) and $T_1^{k\otimes}$ is the product of T_1 k times with itself. If u and v are vectors, the outer product $uv^T \equiv u \otimes v$.

If T is a cubical, symmetric tensor of order p, we define the p-kernel of the multilinear p-form $T[v]^p$ as

$$\ker^p[T] \stackrel{\text{def}}{=} \{v \in \mathbb{R}^n \mid T[v]^p = 0\} \tag{A.8}$$

[63, 64]. Note that, in general, $\ker^p[T]$ is a union of cones. Interestingly, p-kernels are also subspaces when $p = 1$, but this is not true for general p-kernels, since both $(0,1)^T$ and $(1,0)^T$ belong to the 2-kernel of the cubical, symmetric 2-form x_1x_2 on $\mathbb{R}^2$, but not their sum.

A.7 Images and kernels of functions

We denote the image of c over a domain $\mathcal{D}$ as

$$\mathrm{Im}(c,\mathcal{D}) \stackrel{\text{def}}{=} \{y : y = c(x) \text{ for some } x \in \mathcal{D}\},$$

and the kernel of c over the same set is

$$\mathrm{Ker}(c,\mathcal{D}) \stackrel{\text{def}}{=} \{x \in \mathcal{D} : c(x) = 0\}.$$

A.8 Taylor's theorem

The most useful results in real analysis as far as we are concerned in this book are those collectively known as Taylor's theorem. The simplest occurs for differentiable univariate functions and may be proved using the fundamental theorem of calculus and integration by parts.

Theorem A.8.1: Taylor's theorem for a univariate function

Suppose that $\psi : \mathbb{R} \rightarrow \mathbb{R}$ is a k-times continuously differentiable function on the interval $[t, t+h]$. Let

$$\tau_k(t,h) = \psi(t) + \sum_{i=1}^{k} \frac{1}{i!}\psi^{(i)}(t)h^i$$

be the kth-degree Taylor-series approximation of $\psi(t+h)$ about t expressed in terms of the value ψ and its ith derivatives $\psi^{(i)}$, $i = 1, \ldots, k$. Then

$$\psi(t+h) - \tau_k(t,h) = \frac{1}{(k-1)!}\int_0^h (h-\xi)^{k-1}[\psi^{(k)}(t+\xi) - \psi^{(k)}(t)]\,d\xi. \tag{A.9}$$

Proof. It is well known (e.g., [18, Thm. 8.25]) that, under the stated assumptions,

$$\psi(t+h) - \tau_{k-1}(t,h) = \frac{1}{(k-1)!}\int_0^h (h-\xi)^{k-1}\psi^{(k)}(t+\xi)\,d\xi.$$

Subtracting $\psi^{(k)}(t)h^k/k!$ from both sides of this identity and using the fact that $h^k = k\int_0^h (h-\xi)^{k-1}\,d\xi$ then yields (A.9). □

A useful variant of this result is the following Taylor mean-value theorem, which is often known as the Lagrange form.

Theorem A.8.2: Univariate Taylor's theorem (Lagrange form)

[18, Thm. 8.23]: Suppose that $\psi : \mathbb{R} \rightarrow \mathbb{R}$ is a $(k+1)$-times continuously differentiable function on the interval $[t, t+h]$. Then there exists a ξ between t and $t+h$ such that

$$\psi(t+h) - \tau_k(t,h) = \psi^{(k+1)}(\xi)\frac{h^{k+1}}{(k+1)!}.$$

In the multivariate case, let $T_{f,p}(x,s)$ be the standard pth-degree Taylor-series expansion of f about x computed for the increment s, that is,

$$T_{f,p}(x,s) \stackrel{\text{def}}{=} f(x) + \sum_{\ell=1}^{p} \frac{1}{\ell!}\nabla_x^\ell f(x)[s]^\ell. \tag{A.10}$$

Simple differentiation reveals that

$$\nabla_s^j T_{f,p}(x,s) = \sum_{\ell=j}^{p} \frac{1}{(\ell-j)!}\nabla_x^\ell f(x)[s]^{\ell-j}. \tag{A.11}$$

It is well known that if f is p-times differentiable at x, then

$$|f(x+s) - T_{f,p}(x,s)| = o(\|s\|^p). \tag{A.12}$$

The following result provides crucial, more precise, approximation bounds. For an integer k and a real $\beta \in [0,1]$, we define $(k+\beta)! \stackrel{\text{def}}{=} \prod_{\ell=1}^{k}(\ell+\beta)$ (this coincides with the standard factorial if $\beta = 1$) and $\beta! = 1$. As is usual, we also define $0! = 1$.

Theorem A.8.3: Taylor's theorem for a multivariate Hölder function

Let $\mathcal{C}$ be a convex subset of $\mathbb{R}^n$, let $\mathcal{D}$ be any open set containing $\mathcal{C}$, and suppose that $f : \mathcal{D} \to \mathbb{R}$ is a p-times continuously differentiable function for which $\nabla_x^p f(x)$ is Hölder continuous on $\mathcal{C}$ with Hölder constant $H_{f,p}$ and exponent $\beta_p \in [0,1]$, i.e.,

$$\|\nabla_x^p f(x) - \nabla_x^p f(y)\|_{[p]} \le H_{f,p}\|x-y\|^{\beta_p} \tag{A.13}$$

for all $x, y \in \mathcal{C}$, where the tensor norm $\|\cdot\|_{[p]}$ is defined in (A.7). Given $x, x+s \in \mathcal{C}$, let $T_{f,p}(x,s)$ be the pth-degree Taylor approximation of $f(x+s)$ about x given by (A.10). Then

$$|f(x+s) - T_{f,p}(x,s)| \le \frac{H_{f,p}}{(p+\beta_p)!}\|s\|^{p+\beta_p} \quad \text{and} \tag{A.14}$$

$$\|\nabla_x^j f(x+s) - \nabla_s^j T_{f,p}(x,s)\|_{[j]} \le \frac{H_{f,p}}{(p-j+\beta_p)!}\|s\|^{p-j+\beta_p} \quad (j \in [1:p]). \tag{A.15}$$

Proof. We first establish the identity

$$I_{k-1,\beta_p} \stackrel{\text{def}}{=} \int_0^1 \xi^{\beta_p}(1-\xi)^{k-1}\,d\xi = \frac{(k-1)!}{(k+\beta_p)!}. \tag{A.16}$$

For $k = 1$, the result follows from simple integration. To see (A.16) for $k > 1$, integrating by parts, we have that

$$I_{k-1,\beta_p} = \left[\frac{\xi^{1+\beta_p}}{1+\beta_p}(1-\xi)^{k-1}\right]_0^1 + \frac{(k-1)}{1+\beta_p}\int_0^1 \xi^{1+\beta_p}(1-\xi)^{k-2}\,d\xi = \frac{(k-1)}{1+\beta_p}I_{k-2,1+\beta_p}$$

and thus, recursively, that

$$I_{k-1,\beta_p} = \frac{(k-1)!}{(k-1+\beta_p)!}I_{0,k-1+\beta_p} = \frac{(k-1)!}{(k-1+\beta_p)!}\int_0^1 \xi^{k-1+\beta_p}\,d\xi = \frac{(k-1)!}{(k+\beta_p)!}.$$

Then, for given $x, x+s \in \mathcal{C}$ and $x+ts \in \mathcal{C}$ for all $t \in [0,1]$, and picking $\psi(t) = f(x+ts)$, $t = 0$, $h = 1$, and $k = p$ in Theorem A.8.1, the identity (A.9) and the relationships $\psi^{(p)}(t) = \nabla_x^p f(x+ts)[s]^p$ and $\tau_p(1) = T_{f,p}(x,s)$ give that

$$f(x+s) - T_{f,p}(x,s) = \frac{1}{(p-1)!}\int_0^1 (1-\xi)^{p-1}\left(\nabla_x^p f(x+\xi s) - \nabla_x^p f(x)\right)[s]^p\,d\xi,$$

and thus from the definition of the tensor norm (A.7), the Lipschitz bound (A.13), and the identity (A.16) when $k = p$ that

$$\begin{aligned}
&|f(x+s) - T_{f,p}(x,s)| \\
&\quad\le \frac{1}{(p-1)!}\int_0^1 (1-\xi)^{p-1}\left|\left(\nabla_x^p f(x+\xi s) - \nabla_x^p f(x)\right)\left[\frac{s}{\|s\|}\right]^p\right| \|s\|^p\,d\xi \\
&\quad\le \frac{1}{(p-1)!}\int_0^1 (1-\xi)^{p-1}\max_{\|v\|=1}\left|\left(\nabla_x^p f(x+\xi s) - \nabla_x^p f(x)\right)[v]^p\right| \|s\|^p\,d\xi \\
&\quad= \frac{1}{(p-1)!}\int_0^1 (1-\xi)^{p-1}\|\nabla_x^p f(x+\xi s) - \nabla_x^p f(x)\|_{[p]}d\xi \cdot \|s\|^p \\
&\quad\le \frac{1}{(p-1)!}\int_0^1 \xi^{\beta}(1-\xi)^{p-1}\,d\xi \cdot H_{f,p}\|s\|^{p+\beta_p} = \frac{H_{f,p}}{(p+\beta_p)!}\|s\|^{p+\beta_p}
\end{aligned}$$

for all $x, x+s \in \mathcal{C}$, which is the required (A.14).

Likewise, for arbitrary unit vectors $v_1, \ldots, v_j$, choosing $\psi(t) = \nabla_x^j f(x+ts)[v_1, \ldots, v_j]$ and $k = p - j$, it follows from (A.9) and the relationships

$$\psi^{(p-j)}(t) = \nabla_x^p f(x + ts)[v_1, \ldots, v_j][s]^{p-j}$$

and $\tau_{p-j}(1) = \nabla_s^j T_{f,p}(x, s)$ that

$$\begin{aligned}(\nabla_x^j f(x+s) - \nabla_s^j T_{f,p}(x,s))[v_1, \ldots, v_j] \\ = \frac{1}{(p-j-1)!} \int_0^1 (1-\xi)^{p-j-1} \left(\nabla_x^p f(x+\xi s) - \nabla_x^p f(x)\right)[v_1, \ldots, v_j][s]^{p-j}\, d\xi.\end{aligned} \tag{A.17}$$

Then, picking $v_1, \ldots, v_j$ to maximize the absolute value of the left-hand side of (A.17) and using the tensor norm (A.7), the Hölder bound (A.13), and the identity (A.16) when $k = p - j$, we find that

$$\begin{aligned}&(p-j-1)!\|\nabla_x^j f(x+s) - \nabla_s^j T_{f,p}(x,s)\| \\ &\quad \leq \int_0^1 (1-\xi)^{p-j-1} \left| (\nabla_x^p f(x+\xi s) - \nabla_x^p f(x))[v_1, \ldots, v_j] \left[\frac{s}{\|s\|}\right]^{p-j} \right| \|s\|^{p-j}\, d\xi \\ &\quad \leq \int_0^1 (1-\xi)^{p-j-1} \max_{\|v_1\|=\cdots=\|v_p\|=1} \left| (\nabla_x^p f(x+\xi s) - \nabla_x^p f(x))[v_1, \ldots, v_p] \right| \|s\|^{p-j}\, d\xi \\ &\quad = \int_0^1 (1-\xi)^{p-j-1} \|\nabla_x^p f(x+\xi s) - \nabla_x^p f(x)\|_{[p]}\, d\xi \cdot \|s\|^{p-j} \\ &\quad \leq \int_0^1 \xi^{\beta_p}(1-\xi)^{p-j-1}\, d\xi \cdot H_{f,p}\|s\|^{p-j+\beta_p} = \frac{(p-j+1)!}{(p-j+\beta_p)!} H_{f,p}\|s\|^{p-j+\beta_p}\end{aligned}$$

for all $x, x+s \in \mathcal{C}$, which gives (A.15). □

Frequently in the early part of our book, we apply this and subsequent results to the special case $\mathcal{C} = \mathcal{D} = \mathbb{R}^n$.

A very useful corollary for the Lipschitz case follows immediately.

Corollary A.8.4: Taylor's theorem for a multivariate Lipschitz function

Let $\mathcal{C}$ be a convex subset of $\mathbb{R}^n$, let $\mathcal{D}$ be any open set containing $\mathcal{C}$, and suppose that $f : \mathcal{D} \to \mathbb{R}$ is a p-times continuously differentiable function for which $\nabla_x^p f(x)$ is Lipschitz continuous with Lipschitz constant $L_{f,p}$ on $\mathcal{C}$, i.e.,

$$\|\nabla_x^p f(x) - \nabla_x^p f(y)\|_{[p]} \leq L_{f,p}\|x - y\| \tag{A.18}$$

for all $x, y \in \mathcal{C}$, where the tensor norm $\|\cdot\|_{[p]}$ is defined in (A.7). Given $x, x+s \in \mathcal{C}$, let $T_{f,p}(x, s)$ be the pth-degree Taylor approximation of $f(x + s)$ about x given by (A.10). Then

$$|f(x+s) - T_{f,p}(x,s)| \leq \frac{L_{f,p}}{(p+1)!}\|s\|^{p+1} \quad \text{and} \tag{A.19}$$

$$\|\nabla_x^j f(x+s) - \nabla_s^j T_{f,p}(x,s)\|_{[j]} \leq \frac{L_{f,p}}{(p-j+1)!}\|s\|^{p-j+1} \quad (j \in [1:p]). \tag{A.20}$$

Proof. Apply Theorem A.8.3 with $\beta_p = 1$ and $H_{f,p} = L_{f,p}$. □

Lipschitz continuity and boundedness of derivatives are intimately tied together.

Theorem A.8.5: Lipschitz if and only if bounded derivatives

Let $\mathcal{C}$ be an open, convex subset of $\mathbb{R}^n$, and suppose that $f : \mathcal{C} \to \mathbb{R}$ is a $p+1$-times continuously differentiable function. Then $\nabla_x^p f(x)$ is Lipschitz continuous with Lipschitz constant $L_{f,p}$ throughout, i.e., (A.18) holds, if and only if

$$\|\nabla_x^{p+1} f(x)\|_{[p+1]} \leq L_{f,p}. \tag{A.21}$$

Proof. Let $x \in \mathcal{C}$, suppose that s and v are arbitrary Euclidean unit vectors, and consider the function $\psi(\tau) = \nabla_x^p f(x + \tau s)[v]^p$ for any $\tau \in \mathbb{R}$ for which $x + \tau s \in \mathcal{C}$. Then the fundamental theorem of calculus gives that

$$\begin{aligned} |(\nabla_x^p f(x+\alpha s) - \nabla_x^p f(x))\,[v]^p| &= |\psi(\alpha) - \psi(0)| = \left| \int_0^\alpha \psi'(\tau) d\tau \right| \\ &= \left| \int_0^\alpha \nabla_x^{p+1} f(x+\tau s)[s][v]^p d\tau \right|. \end{aligned} \tag{A.22}$$

More specifically, suppose that $y \neq x \in \mathcal{C}$. In this case $y = x + \alpha s$ for some unit vector s with $\alpha = \|y - x\|$. It then follows from (A.7) that

$$\begin{aligned} \|\nabla_x^p f(y) - \nabla_x^p f(x)\|_{[p]} &= \|\nabla_x^p f(x+\alpha s) - \nabla_x^p f(x)\|_{[p]} \\ &= |(\nabla_x^p f(x+\alpha s) - \nabla_x^p f(x))\,[v]^p| \end{aligned}$$

for another unit vector v, and thus from (A.22) that

$$\|\nabla_x^p f(y) - \nabla_x^p f(x)\|_{[p]} = \left| \int_0^\alpha \nabla_x^{p+1} f(x+\tau s)[s][v]^p d\tau \right|.$$

If (A.21) holds,

$$\left| \int_0^\alpha \nabla_x^{p+1} f(x+\tau s)[s][v]^p d\tau \right| \leq \left\| \int_0^\alpha \nabla_x^{p+1} f(x+\tau s) d\tau \right\|_{[p+1]} \leq \alpha L_{f,p} = L_{f,p}\|y - x\|$$

from (A.7) since $\|s\| = \|v\| = 1$, which gives (A.18). Conversely, if (A.18) holds,

$$|(\nabla_x^p f(x+\alpha s) - \nabla_x^p f(x))\,[s]^p| \leq \|\nabla_x^p f(x+\alpha s) - \nabla_x^p f(x)\|_{[p]} \leq L_{f,p}\alpha\|s\| \leq \alpha L_{f,p}$$

for arbitrary s and all α sufficiently small that $x + \alpha s \in \mathcal{C}$, and thus (A.22) with $v = s$ gives

$$\left| \int_0^\alpha \nabla_x^{p+1} f(x+\tau s)[s]^{p+1} d\tau \right| \leq \alpha L_{f,p}.$$

Dividing by α and taking the limit as α tends to zero, we then have

$$\left| \nabla_x^{p+1} f(x)[s]^{p+1} \right| \leq L_{f,p}.$$

Since this is true for all unit vectors s, (A.7) gives (A.21). $\square$

A.9 Polynomial Hermite interpolation

A vital tool in our analysis is polynomial interpolation. This concerns the interpolation of a given set of data by a polynomial of lowest possible degree that passes through the points of the data set. Our data points will be values of a given univariate function evaluated on a discrete set, and more particularly in our case we also wish the derivatives of the chosen polynomial to match those of the function at the evaluation points. This is known as *Hermite* interpolation.

We present some useful properties of Hermite interpolation. Let us assume that we wish to construct a univariate Hermite interpolant π of degree $2p+1$ of the form

$$\pi(\tau) = \sum_{i=0}^{2p+1} c_i \tau^i \tag{A.23}$$

on the interval $[0, s]$ satisfying the $2(p+1)$ conditions

$$\pi^{(i)}(0) = f_0^{(i)} \quad \text{and} \quad \pi^{(i)}(s) = f_1^{(i)} \quad \text{for } i \in [0:p], \tag{A.24}$$

where $f_0^{(i)}$ and $f_1^{(i)}$ are given. The values of the coefficients $c_0, \ldots, c_p$ may then be obtained by

$$c_i = \frac{f_0^{(i)}}{i!} \quad \text{for } i \in [0:p],$$

while the remainder satisfy the linear system

$$\begin{pmatrix} a_{0,0}s^{p+1} & a_{0,1}s^{p+2} & \cdots & a_{0,p-1}s^{2p} & a_{0,p}s^{2p+1} \\ a_{1,0}s^{p} & a_{1,1}s^{p+1} & \cdots & a_{1,p-1}s^{2p-1} & a_{1,p}s^{2p} \\ \vdots & \vdots & \ddots & \vdots & \vdots \\ a_{p,0}s & a_{p,1}s^{2} & \cdots & a_{p,p-1}s^{p} & a_{p,p}s^{p+1} \end{pmatrix} \begin{pmatrix} c_{p+1} \\ c_{p+2} \\ \vdots \\ c_{2p+1} \end{pmatrix} = \begin{pmatrix} f_1^{(0)} - T_{f,p}^{(0)}(0,s) \\ f_1^{(1)} - T_{f,p}^{(1)}(0,s) \\ \vdots \\ f_1^{(p)} - T_{f,p}^{(p)}(0,s)) \end{pmatrix}, \tag{A.25}$$

where

$$T_{f,p}(0,s) = \sum_{i=0}^{p} \frac{f_0^{(i)}}{i!} s^i \quad \text{and} \quad a_{i,j} = \frac{(p+j+1)!}{(p+j+1-i)!} \quad (i,j \in [0:p]).$$

Observe that (A.25) can be rewritten as

$$\begin{pmatrix} s^p & & & \\ & s^{p-1} & & \\ & & \ddots & \\ & & & 1 \end{pmatrix} A_p \begin{pmatrix} s & & & \\ & s^2 & & \\ & & \ddots & \\ & & & s^{p+1} \end{pmatrix} \begin{pmatrix} c_{p+1} \\ c_{p+2} \\ \vdots \\ c_{2p+1} \end{pmatrix} = \begin{pmatrix} f_1^{(0)} - T_{f,p}^{(0)}(0,s) \\ f_1^{(1)} - T_{f,p}^{(1)}(0,s) \\ \vdots \\ f_1^{(p)} - T_{f,p}^{(p)}(0,s)) \end{pmatrix},$$

where A_p is the matrix whose (i,j)th entry is $a_{i,j}$, which only depends on p. It is shown [102, Appendix] that A_p is nonsingular. Therefore

$$\begin{pmatrix} c_{p+1}\, s \\ c_{p+2}\, s^2 \\ \vdots \\ c_{2p+1}\, s^{p+1} \end{pmatrix} = A_p^{-1} \begin{pmatrix} \frac{1}{s^p}[f_1^{(0)} - T_{f,p}^{(0)}(0,s)] \\ \frac{1}{s^{p-1}}[f_1^{(1)} - T_{f,p}^{(1)}(0,s)] \\ \vdots \\ f_1^{(p)} - T_{f,p}^{(p)}(0,s) \end{pmatrix}. \tag{A.26}$$

We therefore deduce, using the triangle inequality, that, for any $\tau \in [0,s]$ and $j \in [0 : p+1]$,

$$\begin{aligned} |\pi^{(j)}(\tau)| &= \left| \sum_{i=j}^{p} \frac{i!}{(i-j)!} c_i\, \tau^{i-j} + \tau^{p-j} \sum_{i=0}^{p} \frac{(p+1+i)!}{(p+1+i-j)!} c_{p+1+i}\, \tau^{i+1} \right| \\ &\leq \sum_{i=j}^{p} \frac{|f_0^{(i)}|}{(i-j)!} s^{i-j} + s^{p+1-j} \sum_{i=0}^{p} \frac{(p+1+i)!}{(p+1+i-j)!} |c_{p+1+i}\, s^{i+1}| s^{-1} \\ &\leq (p+1)\Big[\max_{i\in[j:p]} |f_0^{(i)}| s^{i-j} \\ &\qquad + (2p+1)! s^{p+1-j} \max_{i\in[0:p]} |c_{p+1+i}\, s^{i+1}| s^{-1} \Big] \\ &\leq (p+1)\left[\max_{i\in[j:p]} |f_0^{(i)}| s^{i-j} \right. \\ &\qquad \left. + (2p+1)! s^{p+1-j} \|A_p^{-1}\|_\infty \max_{j\in[0:p]} \left| \frac{f_1^{(j)} - T_{f,p}^{(j)}(0,s)}{s^{p-j+1}} \right| \right]. \end{aligned} \tag{A.27}$$

In particular,

$$|\pi^{(p+1)}(\tau)| \leq (p+1)(2p+1)! \|A_p^{-1}\|_\infty \max_{j\in[0:p]} \left| \frac{f_1^{(j)} - T_{f,p}^{(j)}(0,s)}{s^{p-j+1}} \right|,$$

and the mean-value theorem then implies that

$$\begin{aligned} \frac{|\pi^{(p)}(\tau_1) - \pi^{(p)}(\tau_0)|}{|\tau_1 - \tau_0|} &= |\pi^{(p+1)}(\xi)| \\ &\leq \max_{\tau\in[0,s]} |\pi^{(p+1)}(\tau)| \\ &\leq \frac{(p+1)(2p+1)!}{p!} \|A_p^{-1}\|_\infty \max_{j\in[0:p]} \left| \frac{f_1^{(j)} - T_{f,p}^{(j)}(0,s)}{s^{p-j+1}} \right| \end{aligned} \tag{A.28}$$

for any $0 \leq \tau_0 \leq \tau_1 \leq s$ and some $\xi \in [\tau_0, \tau_1] \subseteq [0,s]$. For $j \in [0 : p]$, (A.27) gives that

$$\begin{aligned} &|\pi^{(j)}(\tau)| \\ &\leq (p+1)\max[1, s^{p+1}] \left[\max_{i\in[j:p]} |f_0^{(i)}| + (2p+1)! \|A_p^{-1}\|_\infty \max_{j\in[0:p]} \left| \frac{f_1^{(j)} - T_{f,p}^{(j)}(0,s)}{s^{p-j+1}} \right| \right]. \end{aligned} \tag{A.29}$$

This line of development leads to the following result.

Theorem A.9.1: General Hermite interpolation on an interval

Suppose that $\{f_\ell^{(j)}\}$ *are given for* $\ell \in \{0,1\}$ *and* $j \in [0:p]$*. Suppose also that there is a constant* $\kappa_f \geq 0$ *for which*

$$|f_1^{(j)} - T_{f,p}^{(j)}(0,s)| \leq \kappa_f\, s^{p-j+1} \tag{A.30}$$

for all $j \in [0:p]$*. Then the Hermite interpolation polynomial* $\pi(\tau)$ *on* $[0,s]$ *given by (A.23) and satisfying (A.24) admits a Lipschitz-continuous pth derivative on* $[0,s]$*, with Lipschitz constant depending only on* p *and* κ_f*. If, in addition,* κ_f *is large enough to ensure that*

$$|f_0^{(j)}| \leq \kappa_f \quad \text{for} \quad j \in [0:p] \quad \text{and} \quad s \leq \kappa_f, \tag{A.31}$$

then $|\pi^{(j)}(\tau)|$ *is bounded for all* $\tau \in [0,s]$ *and all* $j \in [0:p]$ *by a constant depending only on* p *and* κ_f*.*

Proof. The first result follows directly from Theorem A.8.5 and (A.28), (A.30), (A.29), and (A.31). To derive the final conclusion, we note that, from (A.28),

$$\begin{aligned}\frac{|\pi^{(p)}(\tau_1) - \pi^{(p)}(\tau_0)|}{|\tau_1 - \tau_0|} &\leq (p+1)(2p+1)!\|A_p^{-1}\|_\infty \max_{j\in[0:p]} \left|\frac{f_1^{(j)} - T_{f,p}^{(j)}(0,s)}{s^{p-j+1}}\right| \\ &\leq (p+1)(2p+1)!\|A_p^{-1}\|_\infty \kappa_f,\end{aligned}$$

and, from (A.29),

$$\begin{aligned}|\pi^{(j)}(\tau)| &\leq (p+1)\max[1, s^{p+1}] \left[\max_{i\in[j:p]} |f_0^{(i)}| \right. \\ &\qquad \left. + (2p+1)!\|A_p^{-1}\|_\infty \max_{j\in[0:p]} \left|\frac{f_1^{(j)} - T_{f,p}^{(j)}(0,s)}{s^{p-j+1}}\right| \right] \\ &\leq (p+1)\max[1, \kappa_f^{p+1}] \left[\kappa_f + (2p+1)!\|A_p^{-1}\|_\infty \kappa_f\right].\end{aligned}$$

□

Observe that (A.30) is identical to (A.20) when $n = 1$. This means that the conditions of Theorem A.9.1 automatically hold if the interpolation data $\{f_i^{(j)}\}$ is itself extracted from a function having a Lipschitz-continuous pth derivative.

Applying the above results to several interpolation intervals then yields the existence of a smooth Hermite interpolant on the real line.

Theorem A.9.2: General Hermite interpolation

Suppose that, for some integer $k_e > 0$ *and* $p > 0$*, the data* $\{f_k^{(j)}\}$ *and* $\{x_k\}$ *are given for* $k \in [0:k_e]$ *and* $j \in [0:p]$*. Suppose also that* $s_k = x_{k+1} - x_k > 0$ *for* $k \in [0:k_e]$ *and that*

$$|f_{k+1}^{(j)} - T_{f,p}^{(j)}(x_k, s_k)| \leq \kappa_f\, s_k^{p-j+1} \quad \text{for} \quad j \in [0:p], \tag{A.32}$$

some constant $\kappa_f \geq 0$, and all $k \in [0 : k_e - 1]$, where $T_{k,p}(x_k, s) = \sum_{i=0}^{p} f_k^{(i)} s^i/i!$. Then there exists a p-times continuously differentiable function f from $\mathbb{R}$ to $\mathbb{R}$ with Lipschitz-continuous pth derivative and Lipschitz constant depending only on p and κ_f such that

$$f^{(j)}(x_k) = f_k^{(j)} \quad \text{for} \quad j \in [0 : p].$$

If, in addition, κ_f is large enough to ensure that, for $k \in [0 : k_e - 1]$,

$$|f_k^{(j)}| \leq \kappa_f \quad \text{for} \quad j \in [0 : p] \quad \text{and} \quad s_k \leq \kappa_f,$$

then $|f^{(j)}(x)|$ is bounded by a constant depending only on p and κ_f for all $x \in \mathbb{R}^n$ and all $j \in [0 : p]$.

Proof. We first use Theorem A.9.1 to define a Hermite interpolant $\pi_k(s)$ of the form (A.23) on each interval $[x_k, x_{k+1}] = [x_k, x_k + s_k]$ ($k \in [0 : k_e]$) using $f_0^{(j)} = f_k^{(j)}$ and $f_1^{(j)} = f_{k+1}^{(j)}$ for $j \in [0 : p]$, and then we set

$$f(x_k + s) = \pi_k(s)$$

for any $s \in [0, s_k]$. We may then smoothly prolongate f for $x \in \mathbb{R}$ by defining two additional interpolation intervals $[x_{-1}, x_0] = [-s_{-1}, 0]$ and $[x_{k_e}, x_{k_e} + s_{k_e}]$ with end conditions

$$f_{-1} = f_0^{(0)}, \quad f_{k_e+1} = f_{k_e}^{(0)}, \quad \text{and} \quad f_{-1}^{(j)} = f_{k_e+1}^{(j)} = 0 \quad \text{for} \quad j \in [1 : p],$$

where s_{-1} and s_{k_e} are chosen sufficiently large to ensure that (A.32) also holds on intervals $[x_{-1}, x_0]$ and $[x_{k_e}, x_{k_e+1}]$. We next set

$$f(x) = \begin{cases} f_0^{(0)} & \text{for} \quad x \leq x_{-1}, \\ \pi_k(x - x_k) & \text{for} \quad x \in [x_k, x_{k+1}] \text{ and } k \in [-1 : k_e], \\ f_{k_e}^{(0)} & \text{for} \quad x \geq x_{k_e} + s_{k_e}. \end{cases}$$

□

Note that the conclusions of this theorem imply the existence of a bounded interpolant with Lipschitz-continuous derivatives of degree 1 to p, therefore satisfying both **f.D1pL** and **f.Bb**.

A.10 Simple series bounds

Finally, since we often rely on bounds on finite series, we collect them here for reference.

Lemma A.10.1: Sum bounds

Let $\delta \in [0, 1]$. Then, for any $j \geq 0$ and any $p \geq j$,

$$\sum_{\ell=j}^{p} \frac{\delta^{\ell-j}}{(\ell - j)!} \leq 1 + 2\delta \quad \text{and} \quad \sum_{\ell=j+1}^{p+1} \frac{\delta^\ell}{\ell!} \leq \frac{2\delta^{j+1}}{j!}. \tag{A.33}$$

Proof. The first bound follows immediately from the relationship

$$\sum_{\ell=j}^{p} \frac{\delta^{\ell-j}}{(\ell-j)!} = \sum_{i=0}^{p-j} \frac{\delta^i}{i!} \le \sum_{i=0}^{\infty} \frac{\delta^i}{i!} = \exp(\delta)$$

since $\exp(\delta) \le 1 + 2\delta$ when $\delta \in [0,1]$. To establish the second bound, we have the identity

$$\sum_{\ell=j+1}^{p+1} \frac{\delta^\ell}{\ell!} = \frac{\delta^j}{j!} \sum_{\ell=j+1}^{p+1} \frac{\delta^{\ell-j}}{(\ell-j)!} \cdot \frac{(\ell-j)!j!}{\ell!}.$$

But, since $(\ell-j)!j! \le \ell!$,

$$\sum_{\ell=j+1}^{p+1} \frac{\delta^\ell}{\ell!} \le \frac{\delta^j}{j!} \sum_{\ell=j+1}^{\infty} \frac{\delta^{\ell-j}}{(\ell-j)!} = \frac{\delta^j}{j!}(\exp(\delta) - 1) \le \frac{\delta^j}{j!}(2\delta).$$

□

The lemma then leads directly to the useful bounds

$$\sum_{\ell=1}^{j} \frac{\delta^\ell}{\ell!} \le 2\delta \le 2 < 3 \tag{A.34}$$

for all $\delta \in [0,1]$;

$$\sum_{\ell=1}^{j} \frac{\delta^\ell}{\ell!} \le 2\delta^j \tag{A.35}$$

for all $\delta \ge 1$; and

$$\sum_{\ell=1}^{j} \frac{1}{\ell!} \le 2, \quad \sum_{\ell=j}^{p} \frac{1}{(\ell-j)!} < 3, \quad \text{and} \quad \sum_{\ell=j+1}^{p+1} \frac{1}{\ell!} \le \frac{2}{j!} \tag{A.36}$$

when $\delta = 1$.

Appendix B

Technical Results from Individual Chapters

Many of the more technical results that underpin the analysis we have conducted are collected in this appendix. We also provide detailed proofs of other theorems and the like that we felt were best left to an appendix so as not to disrupt the flow of the substantial text.

Appendix to Chapter 2

Proofs of results from Chapter 2

Proof of Theorem 2.2.5 (p. 32)

The idea behind our worst-case example for the SDEX algorithm (p. 31) is to define $f(x, y)$, the objective function in the example, as the sum of $f_1(x)$ and $f_2(x, y)$. Our development closely follows [92]. Because the proof is long and technical, we divide it into a sequence of steps embodied as six lemmas. The first of these specializes Theorem A.9.1 (p. 441).

Lemma B.2.1: Hermite interpolation on an interval with $p = 2$

Suppose that the real values f_0, g_0, H_0, f_1, g_1, and H_1 are known. Then there exists a fifth-order polynomial $p(t) = c_0 + c_1 t + c2t^2 + c_3 t^3 + c_4 t^4 + c_5 t^5$ for $t \in [0, 1]$, such that

$$\begin{array}{llll} p(0) = f_0, & p^{(1)}(0) = g_0, & \text{and} & p^{(2)}(0) = H_0, \ \text{and} \\ p(1) = f_1, & p^{(1)}(1) = g_1, & \text{and} & p^{(2)}(1) = H_1. \end{array}$$

The coefficients of this polynomial are given by

$$\begin{array}{c} c_0 = f_0, \quad c_1 = g_0, \quad c_2 = \frac{1}{2}H_0, \quad c_3 = 10r_0 - 4r_1 + \frac{1}{2}r_2, \\ c_4 = -15r_0 + 7r_1 - r_2, \quad \text{and} \quad c_5 = 6r_0 - 3r_1 + \frac{1}{2}r_2, \end{array} \tag{B.2.1}$$

where

$$r_0 = f_1 - f_0 - g_0 - \tfrac{1}{2}H_0, \quad r_1 = g_1 - g_0 - H_0, \quad \text{and} \quad r_2 = H_1 - H_0. \tag{B.2.2}$$

Moreover, if there is a nonnegative constant $\kappa_f \geq 1$ such that

$$|r_i| \leq \kappa_f \quad (i \in [0:2]), \tag{B.2.3}$$

then there exists a constant $\kappa_h > 0$ dependent only on κ_f such that $|p^{(i)}(t)| \leq \kappa_h$ for $i \in [0:2]$ and all $t \in [0,1]$.

Proof. The existence of the polynomial and (B.2.1)–(B.2.2) result directly from (A.26) in the arguments that lead to Theorem A.9.1 (p. 441). The rest of the statement follows from the last conclusion of the same theorem with $p = 2$. □

The next statement looks very much like our desired conclusion, except that we will still need to verify that the objective function satisfies **f.Bb** (Lemma B.2.4) and **f.gL** (Lemma B.2.5) and that the method it considers is indeed identical to the SDEX algorithm (Lemma B.2.6).

Lemma B.2.2: Construction of the counterexample

For any given $\epsilon \in (0,1]$, there exists a function $f(x,y)$ from $\mathbb{R}^2$ to $\mathbb{R}$ and a starting point $x_0 \in \mathbb{R}^2$ such that the steepest-descent algorithm with prescribed stepsize applied to $f(x,y)$ starting from x_0 needs exactly

$$k_\epsilon = \epsilon^{-2}$$

iterations to produce an iterate x_k such that $\|\nabla_x^1 f(x_k)\| \leq \epsilon$.

Proof. As in Theorem 2.2.3 (p. 27) with $t_{\text{ini}} = 1$, the construction considers iterations up to k_ϵ and, given

$$\omega_k \stackrel{\text{def}}{=} \epsilon \frac{k_\epsilon - k}{k_\epsilon} \in [0, \epsilon], \tag{B.2.4}$$

defines a first univariate function $f_1(x)$ by piecewise Hermite-polynomial interpolation between the sequence of iterates

$$x_0 = 0, \quad x_{k+1} = x_k + (\epsilon + \omega_k) \; (k \geq 0) \tag{B.2.5}$$

and the values

$$\begin{array}{c} f_1(x_0) = 4, f_1(x_{k+1}) = f_1(x_k) - (\epsilon + \omega_k)^2, \\ f_1^{(1)}(x_k) = -(\epsilon + \omega_k), \quad \text{and} \quad f_1^{(2)}(x_k) = 0. \end{array} \tag{B.2.6}$$

From Lemma B.2.1, it is easy to verify that

$$f_1(x) = f_1(x_k) - (\epsilon + \omega_k)^2 t + (\epsilon + \omega_k)(\omega_k - \omega_{k+1})\big[-4t^3 + 7t^4 - 3t^5\big],$$

$$f_1^{(1)}(x) = -(\epsilon + \omega_k) + (\omega_k - \omega_{k+1})\big[-12t^2 + 28t^3 - 15t^4\big], \tag{B.2.7}$$

and

$$f_1^{(2)}(x) = \frac{\omega_k - \omega_{k+1}}{\epsilon + \omega_k}\big[-24t + 84t^2 - 60t^3\big] \tag{B.2.8}$$

for $x \in [x_k, x_{k+1}]$ and $t = (x - x_k)/(\epsilon + \omega_k)$. Moreover, the second part of the same lemma ensures that $f_1(x)$ is uniformly bounded below. Note that

$$-12t^2 + 28t^3 - 15t^4 = -t^2[12 - 28t + 15t^2] \leq 1$$

for $t \in [0, 1]$, and thus, using (B.2.7),

$$f_1^{(1)}(x) \leq -(\epsilon + \omega_{k+1}) < 0 \quad \text{for all} \quad x \in [x_k, x_{k+1}].$$

In addition, taking into account that

$$0 < \frac{\omega_k - \omega_{k+1}}{\epsilon + \omega_k} < 1 \tag{B.2.9}$$

for $k \geq 0$ and that $t \in [0, 1]$ if $x \in [x_k, x_{k+1}]$, a straightforward majorization in (B.2.8) means that

$$|f_1^{(2)}(x)| < 168 \tag{B.2.10}$$

for $x \in [x_k, x_{k+1}]$, which in turn implies that $f_1^{(2)}(x)$ is uniformly bounded for all $x \geq 0$. The behaviour of $f_1(x)$ and of its first and second derivatives is pictured in Figure B.2.1.

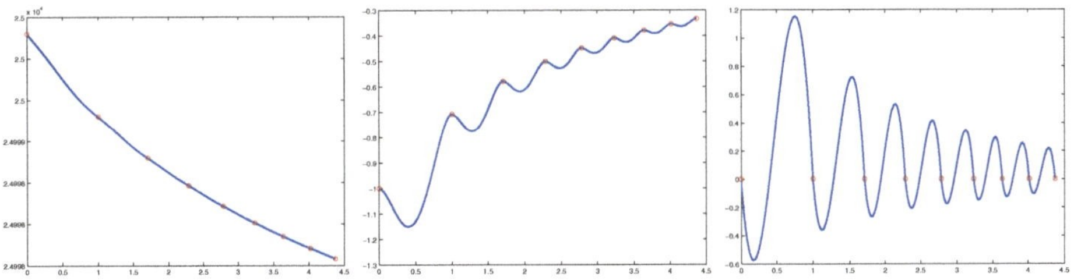

Figure B.2.1. The function $f_1(x)$ and its first two derivatives (from left to right) on the first eight intervals.

We now turn to the specification of the function $f_2(x, y)$, whose role is to limit the iterates in the y-direction to a progressively narrower "corridor," thereby forcing the iteration path to oscillate between its lower and upper limits. We have already prescribed that the x-components of the successive iterates are given by (B.2.5), and $f_2(x, y)$ will be constructed in such a way that its gradient at the iterates is equal in norm to that of $f_1(x)$ but alternating in sign, thus generating the necessary orthogonality conditions and the oscillating iteration path.

More specifically, and in accordance with (B.2.5), let

$$\begin{pmatrix} x_0 \\ y_0 \end{pmatrix} = \begin{pmatrix} 0 \\ 0 \end{pmatrix}, \quad \begin{pmatrix} x_{k+1} \\ y_{k+1} \end{pmatrix} = \begin{pmatrix} x_k \\ y_k \end{pmatrix} + \begin{pmatrix} \epsilon + \omega_k \\ (-1)^k(\epsilon + \omega_k) \end{pmatrix}$$

for all $k \geq 0$, which defines the zig-zagging piecewise-linear iteration path $y(x)$. We impose lower and upper boundaries of the corridor containing the iterates. This is achieved by defining the lower boundary $y_{\text{low}}(x)$ as a twice continuously differentiable curve that interpolates the y-coordinates of the iterates of index $2k$ ($k \geq 0$) and is constant on the intervals $[x_{2k-1}, x_{2k}]$, yielding

$$y_{\text{low}}(x_{2k-1}) = y_{\text{low}}(x_{2k}) = y_{2k}.$$

Polynomial Hermite interpolation (Lemma B.2.1) is then used to twice continuously connect the constant parts. The upper boundary $y_{\text{up}}(x)$ is defined in the same way to interpolate the y-coordinates of the iterates of index $2k+1$ ($k \geq 0$), being constant on $[x_{2k}, x_{2k+1}]$, yielding

$$y_{\text{up}}(x_{2k}) = y_{\text{up}}(x_{2k+1}) = y_{2k+1}.$$

We shall also use the average

$$y_{\text{mid}}(x) \stackrel{\text{def}}{=} \tfrac{1}{2}(y_{\text{low}}(x) + y_{\text{up}})$$

of the two boundaries. We illustrate the path and its corridor in Figure B.2.2.

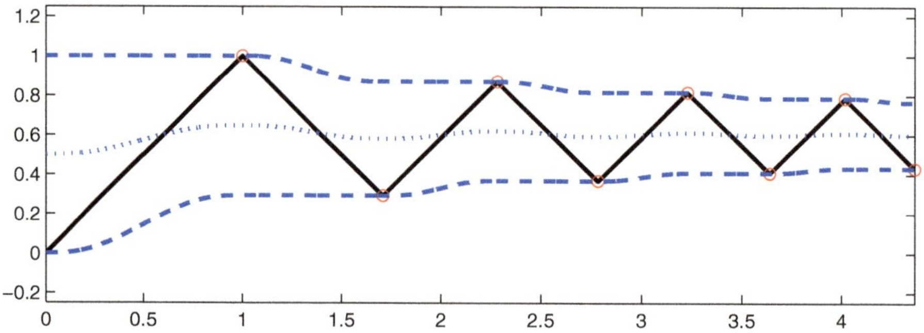

Figure B.2.2. The iteration path $y(x)$ (plain), the upper and lower corridor boundaries $y_{\text{low}}(x)$ and $y_{\text{up}}(x)$ (dashed), and $y_{\text{mid}}(x)$ (dotted) for $k = 1, \ldots, 8$ and $\eta = 10^{-5}$.

If we define

$$\delta(x) \stackrel{\text{def}}{=} y_{\text{up}}(x) - y_{\text{low}}(x) \tag{B.2.11}$$

(the corridor width at x), we note that, by construction,

$$\delta(x_k) = \epsilon + \omega_k \quad \text{for all} \quad k \geq 0. \tag{B.2.12}$$

Moreover, since the interpolation conditions defining $y_{\rm up}(x)$ are given (for $x \in [x_{2k-1}, x_{2k}]$, say) by

$$y_{\rm up}(x_{2k-1}) = \sum_{i=0}^{2k-2}(-1)^i(\epsilon+\omega_i) \quad \text{and} \quad y_{\rm up}(x_{2k}) = y_{\rm up}(x_{2k+1}) = \sum_{i=0}^{2k}(-1)^i(\epsilon+\omega_i),$$
$$y_{\rm up}^{(1)}(x_{2k-1}) = y_{\rm up}^{(1)}(x_{2k}) = 0 \quad \text{and} \quad y_{\rm up}^{(2)}(x_{2k-1}) = y_{\rm up}^{(2)}(x_{2k}) = 0,$$

closer inspection of the interpolating polynomial (see (B.2.1)–(B.2.2)) reveals that, for $x \in [x_{2k-1}, x_{2k}]$,

$$y_{\rm up}(x) = y_{\rm up}(x_{2k-1}) - (\omega_{2k-1} - \omega_{2k})\big[10t^3 - 15t^4 + 6t^5\big], \tag{B.2.13}$$

where $t = (x - x_{2k-1})/(\epsilon + \omega_{2k-1})$. Symmetrically, we have that, for $x \in [x_{2k}, x_{2k+1}]$,

$$y_{\rm low}(x) = y_{\rm low}(x_{2k}) + (\omega_{2k} - \omega_{2k+1})\big[10t^3 - 15t^4 + 6t^5\big], \tag{B.2.14}$$

where $t = (x - x_{2k})/(\epsilon + \omega_{2k})$. We thus obtain from (B.2.13) and (B.2.14), using (B.2.12) and defining $t = (x - x_k)/\sigma_k$, that

$$\delta(x) = \epsilon + \omega_k - (\omega_k - \omega_{k+1})\big[10t^3 - 15t^4 + 6t^5\big] \tag{B.2.15}$$

and

$$y_{\rm mid}(x) = y_{\rm mid}(x_k) + \tfrac{1}{2}(-1)^k(\omega_k - \omega_{k+1})\big[10t^3 - 15t^4 + 6t^5\big]$$

for all $x \in [x_k, x_{k+1}]$. These last two relations yield that

$$\delta^{(1)}(x) = 2(-1)^{k+1}y_{\rm mid}^{(1)}(x) = -30\,\frac{\omega_k - \omega_{k+1}}{\epsilon + \omega_k}\big[t^2 - 2t^3 + t^4\big] \le 0, \tag{B.2.16}$$

and also that

$$\delta^{(2)}(x) = 2(-1)^{k+1}y_{\rm mid}^{(2)}(x) = -60\,\frac{\omega_k - \omega_{k+1}}{(\epsilon + \omega_k)^2}\big[t - 3t^2 + 2t^3\big]. \tag{B.2.17}$$

The inequality in (B.2.16) results from the decreasing nature of ω_k and the fact that $1 - 2t + t^2 = (1-t)^2 \ge 0$ for $t \in [0, 1]$. It immediately implies, with (B.2.12) and (B.2.16), that $\delta(x)$ is nonincreasing and that

$$2 \ge \epsilon + \omega_k = \delta(x_k) \ge \delta(x) \ge \delta(x_{k+1}) = \epsilon + \omega_{k+1} \tag{B.2.18}$$

for all $x \in [x_k, x_{k+1}]$.

The next step is to define, for each x, $f_2(x, y)$ as a twice continuously differentiable function of y whose value is small between $y_{\rm low}(x)$ and $y_{\rm up}(x)$ and that first increases before leveling off when the distance of y to the corridor increases, thereby keeping the iterates within the corridor. The specific $f_2(x, y)$ is given by

$$f_2(x,y) = \begin{cases} 8\,\delta(x)^2 & \text{if } y \le y_{\rm low}(x) - 1, \\ f_2^{\rm L}(x,y) & \text{if } y \in [y_{\rm low}(x) - 1, y_{\rm low}(x)], \\ (y - y_{\rm mid}(x))^2 & \text{if } y \in [y_{\rm low}(x), y_{\rm up}(x)], \\ f_2^{\rm U}(x,y) & \text{if } y \in [y_{\rm up}(x), y_{\rm up}(x) + 1], \\ 8\,\delta(x)^2 & \text{if } y \ge y_{\rm up}(x) + 1, \end{cases} \tag{B.2.19}$$

where Hermite interpolation is once more used to continuously connect the endpoints of the first and third and the third and fifth intervals. Thus, in the second interval, $f_2^{\mathrm{L}}(x, y)$ is defined as a fifth-order polynomial in y translated to $[0, 1]$, with boundary conditions on this interval given by

$$\begin{aligned}
&p(0) = 8\,\delta(x)^2, \quad p^{(1)}(0) = 0, \quad p^{(2)}(0) = 0, \\
&p(1) = (y_{\text{low}}(x) - y_{\text{mid}}(x))^2, \\
&p^{(1)}(1) = 2(y_{\text{low}}(x) - y_{\text{mid}}(x)) = -\delta(x), \text{ and } p^{(2)}(1) = 2.
\end{aligned} \tag{B.2.20}$$

The interpolation conditions for $f_2^{\mathrm{U}}(x, y)$ on the fourth interval are defined symmetrically. Figure B.2.3 shows the shape of $f_2(x, y)$ for fixed x.

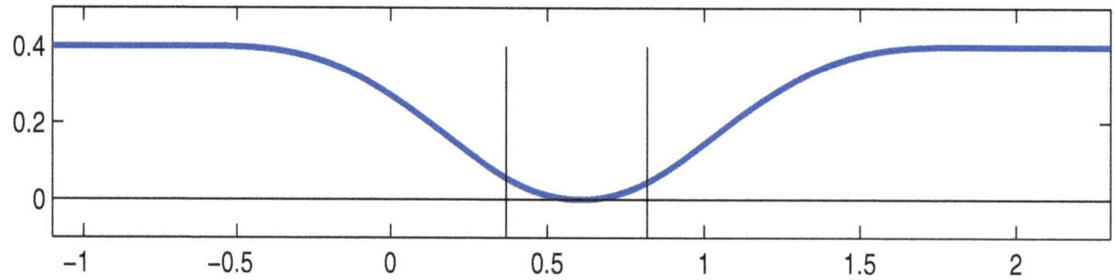

Figure B.2.3. The shape of $f_2(x, y)$ for $x = x_2$ and $\eta = 10^{-5}$, with the vertical lines indicating the values of $y_{\text{low}}(x_2)$ and $y_{\text{up}}(x_2)$.

Note that $f_2(x, y)$ is symmetric in y with respect to $y_{\text{mid}}(x)$ by construction. Note also that, using (B.2.11), the definition of $y_{\text{mid}}(x)$, and (B.2.12),

$$\begin{aligned}
\frac{\partial f_2}{\partial y}(x_{2k}, y_{2k}) &= \frac{\partial f_2}{\partial y}(x_{2k}, y_{\text{low}}(x_{2k})) \\
&= 2(y_{\text{low}}(x_{2k}) - y_{\text{mid}}(x_{2k})) \\
&= -\delta(x_{2k}) \\
&= -(\epsilon + \omega_{2k}),
\end{aligned} \tag{B.2.21}$$

and, similarly,

$$\begin{aligned}
\frac{\partial f_2}{\partial y}(x_{2k+1}, y_{2k+1}) &= \frac{\partial f_2}{\partial y}(x_{2k+1}, y_{\text{up}}(x_{2k+1})) \\
&= 2(y_{\text{up}}(x_{2k+1}) - y_{\text{mid}}(x_{2k+1})) \\
&= \delta(x_{2k+1}) \\
&= \epsilon + \omega_{2k+1}.
\end{aligned} \tag{B.2.22}$$

Note also that, because of (B.2.19) and (B.2.16) taken at $x = x_k$ (i.e., $t = 0$),

$$\frac{\partial f_2}{\partial x}(x_k, y_k) = -2(y_k - y_{\text{mid}}(x_k))y_{\text{mid}}^{(1)}(x_k) = 0. \tag{B.2.23}$$

Finally, we define the objective function of our minimization problem by

$$f_{\text{SDEX}}(x, y) \stackrel{\text{def}}{=} \begin{cases} \dfrac{1}{\sqrt{2}}[f_1(x) + f_2(x, y)] & \text{for } x \geq 0, y \in \mathbb{R}, \\ \dfrac{1}{\sqrt{2}}[f_1(0) + x f_1^{(1)}(0) + f_2(0, y)] & \text{for } x < 0, y \in \mathbb{R}. \end{cases} \tag{B.2.24}$$

We thus obtain, using (B.2.6), (B.2.21)–(B.2.22), and (B.2.23), that

$$\nabla^1 f_{\text{SDEX}}(x_k, y_k) = -\frac{\epsilon + \omega_k}{\sqrt{2}} \begin{pmatrix} 1 \\ (-1)^k \end{pmatrix} \tag{B.2.25}$$

and therefore that

$$\|\nabla^1 f_{\text{SDEX}}(x_k, y_k)\| = \epsilon + \omega_k.$$

Because of the definition of ω_k in (B.2.4), this implies that the steepest-descent algorithm applied to f_{SDEX}, with its stepsize prescribed as above, requires at least ϵ^{-2} iterations to produce an iterate x_k for which $\|\nabla_x^1 f(x_k)\| \leq \epsilon$ for any $\epsilon \in (0,1)$. □

Notice that (B.2.25) implies that $\langle \nabla^1 f_{\text{SDEX}}(x_k), \nabla^1 f_{\text{SDEX}}(x_{k+1})\rangle = 0$ and hence that x_{k+1} is a first-order minimizer of f_{SDEX} along the steepest-descent direction. This will allow us to conclude our main proof provided that we can show that $f_{\text{SDEX}}(x,y)$ both is bounded below and has a globally Lipschitz continuous gradient and that x_{k+1} is the first minimizer along $-\nabla^1 f(x_k, y_k)$. With this in mind, we start by considering the size of the derivatives of f_2.

Lemma B.2.3: Boundedness of the derivatives of f_2

The values of

$$f_2(x,y),\ \frac{\partial f_2}{\partial x}(x,y),\ \frac{\partial f_2}{\partial y}(x,y),\ \frac{\partial^2 f_2}{\partial x^2}(x,y),\ \frac{\partial^2 f_2}{\partial y^2}(x,y),\ \text{and}\ \frac{\partial^2 f_2}{\partial x \partial y}(x,y)$$

are uniformly bounded (in absolute value) for all $x \geq 0$ and all $y \in [y_{\text{low}}(x) - 1, y_{\text{low}}(x)] \cup [y_{\text{up}}(x), y_{\text{up}}(x)+1]$.

Proof. Consider $y \in [y_{\text{low}}(x) - 1, y_{\text{low}}(x)]$ and $x \geq 0$; identical reasoning will hold by symmetry for $y \in [y_{\text{up}}(x), y_{\text{up}}(x)+1]$. Combining (B.2.9) with (B.2.16) and (B.2.17), respectively, we have that

$$\max\left[|\delta^{(1)}(x)|, |\delta^{(2)}(x)|, |y_{\text{mid}}^{(1)}(x)|, |y_{\text{mid}}^{(2)}(x)|\right] \leq 360 \stackrel{\text{def}}{=} \kappa_{\text{dy}} \tag{B.2.26}$$

for all $x \in [x_k, x_{k+1}]$, since then $t \in [0,1]$. Since $f(x,y)$ is a polynomial in y on an interval of length one, we may invoke Lemma B.2.1 to deduce that its values, along with its first and second derivatives with respect to y, are uniformly bounded (in absolute value) provided that its coefficients are also uniformly bounded. This will be the case (see (B.2.3) and (B.2.20)) if the quantities

$$|8\,\delta(x)^2 - (y_{\text{up}}(x) - y_{\text{mid}}(x))^2 - 0 - \tfrac{1}{2}0| \quad \text{and} \quad |\delta(x) - 0 - 0| \tag{B.2.27}$$

are themselves uniformly bounded (the third, $i = 2$, component on the right-hand side of (B.2.3) being identically equal to 2). But this is the case for the first term in (B.2.27) since, using (B.2.18),

$$|8\,\delta(x)^2 - (y_{\text{up}}(x) - y_{\text{mid}}(x))^2| = |8\,\delta(x)^2 - \tfrac{1}{4}\delta(x)^2| < 8\delta(x)^2 \leq 32,$$

and it is also the case for the second because of (B.2.18).

Turning to the derivatives of $f_2(x, y)$ with respect to x, since the function is defined on the interval of concern as a polynomial in y shifted to $[0, 1]$, the dependence on x is entirely captured by the coefficients $c_0, \ldots, c_5$ of this polynomial, and the coefficients themselves depend on the boundary conditions

$$c_0 = 8\,\delta(x)^2, \quad c_1 = 0, \quad \text{and} \quad c_2 = 0$$

and on (B.2.1). The boundedness of the first and second derivatives of $c_0, \ldots, c_5$ (as functions of x) is then implied by (B.2.26) and the boundedness of the two terms in (B.2.27), which we already verified.

Finally, the second derivative of $f_2(x, y)$ with respect to x and y for y in the relevant interval is also a polynomial on a (shifted) interval of length one, obtained by differentiating the coefficients $c_1, \ldots, c_5$ with respect to x in the polynomial corresponding to the derivative of $f_2(x, y)$ with respect to y. But because we just verified that the first derivatives of $c_0, \ldots, c_5$ with respect to x are themselves uniformly bounded in x, this must also be the case for the cross-derivatives of $f_2(x, y)$. □

Lemma B.2.4: Lower bound of f_{SDEX}

*The function $f_{\text{SDEX}}(x, y)$ is uniformly bounded below on $\mathbb{R}^2$, and therefore **f.Bb** holds.*

Proof. This result follows from (B.2.24), the observation that $f_1(x)$ is uniformly bounded below, and Lemma B.2.3. □

To demonstrate that the gradient of $f_{\text{SDEX}}(x, y)$ has a uniform Lipschitz constant requires a more lengthy calculation and is the objective of the next lemma.

Lemma B.2.5: Lipschitz continuity of $\nabla^1 f_{\text{SDEX}}$

*The gradient of function $f_{\text{SDEX}}(x, y)$ is uniformly Lipschitz continuous on $\mathbb{R}^2$, and thus **f.gL** holds.*

Proof. Assume that x is given, and consider the intervals for y given by (B.2.19).

Consider first the case where $y \in [y_{\text{low}}(x), y_{\text{up}}(x)]$. In this interval, we find that

$$\frac{\partial^2 f_2}{\partial y^2}(x, y) = 2 \tag{B.2.28}$$

for $x \in [x_k, x_{k+1}]$. Moreover,

$$\frac{\partial f_2}{\partial x}(x, y) = -2(y - y_{\text{mid}}(x)) y_{\text{mid}}^{(1)}(x) \tag{B.2.29}$$

and thus

$$\frac{\partial^2 f_2}{\partial x^2}(x, y) = 2 y_{\text{mid}}^{(1)}(x)^2 - 2(y - y_{\text{mid}}(x)) y_{\text{mid}}^{(2)}(x).$$

Taking absolute values and noting that, because of the definition of $y_{\rm mid}(x)$ and (B.2.26),

$$|y - y_{\rm mid}(x)| \leq \tfrac{1}{2}\delta(x) \leq \tfrac{1}{2}(\epsilon + \omega_k) \tag{B.2.30}$$

for $x \in [x_k, x_{k+1}]$ and $y \in [y_{\rm low}(x), y_{\rm sup}(x)]$, we deduce that

$$\left|\frac{\partial^2 f_2}{\partial x^2}(x,y)\right| \leq 2\kappa_{\rm dy}^2 + (\epsilon + \omega_k)\kappa_{\rm dy} \leq 2(\kappa_{\rm dy}^2 + \kappa_{\rm dy}) \tag{B.2.31}$$

for x and y in these intervals, where we also used (B.2.26). Finally, for x and y in the same intervals, we have that

$$\left|\frac{\partial^2 f_2}{\partial x\,\partial y}(x,y)\right| = \left|-2y_{\rm mid}^{(1)}(x)\right| \leq 2\kappa_{\rm dy},$$

where we used (B.2.29) and (B.2.26). Considering this last relation together with (B.2.28) and (B.2.31), we thus conclude that the second derivatives of $f_2(x,y)$ are uniformly bounded for all $x \geq 0$ and all $y \in [y_{\rm low}(x), y_{\rm sup}(x)]$.

Next, the case where $y \in [y_{\rm low}(x) - 1, y_{\rm low}(x)] \cup [y_{\rm up}(x), y_{\rm up}(x) + 1]$ is covered by Lemma B.2.3.

To conclude our analysis, we are thus left with checking the boundedness of the second derivatives of $f_2(x,y)$ for $y \geq y_{\rm up} + 1$ and $y \leq y_{\rm low} - 1$. In these intervals, $f_2(x,y) = 8\,\delta(x)^2$, whose second derivatives are bounded because of (B.2.26). We may therefore conclude that the second derivatives of $f_2(x,y)$ are bounded for all $x \geq 0$ and all y.

We may now combine this conclusion with (B.2.10) and (B.2.24) to deduce that $f(x,y)$ has uniformly bounded second derivatives for all $(x,y) \in \mathbb{R}^2$, and the desired Lipschitz continuity of its gradient then follows. □

We complete the construction of our example by verifying that the sequence of iterates may indeed be obtained from the steepest-descent method with exact linesearches, that is, by the SDEX algorithm.

Lemma B.2.6: Linesearches are exact

The iterate (x_{k+1}, y_{k+1}) is the first minimizer along the steepest-descent direction from (x_k, y_k).

Proof. The theorem statement is equivalent to verifying that the slope

$$\begin{aligned}\sigma(x) &= \left\langle \nabla f(x, y(x)), \frac{1}{\sqrt{2}}\begin{pmatrix} 1 \\ (-1)^k \end{pmatrix}\right\rangle \\ &= \tfrac{1}{2}\left[f_1^{(1)}(x) + \frac{\partial f_2}{\partial x}(x,y) + (-1)^k \frac{\partial f_2}{\partial y}(x,y)\right]\end{aligned}$$

of $f(x,y)$ on $[x_k, x_{k+1}]$; that is,

$$\sigma(x) = \tfrac{1}{2} f_1^{(1)}(x) + [y(x) - y_{\rm mid}(x)]\left[(-1)^k - y_{\rm mid}^{(1)}(x)\right] \tag{B.2.32}$$

is always nonpositive and is zero only at the iterates (which are the corners of the trajectory). To prove this property, we first observe that, because of (B.2.16),

$$(-1)^k - y_{\text{mid}}^{(1)}(x) = (-1)^k \left[1 - |y_{\text{mid}}^{(1)}(x)|\right]. \tag{B.2.33}$$

Recalling (B.2.16) and using the fact that $\max_{t\in[0,1]} t^2(1-t)^2 = \frac{1}{16}$, we obtain that, for $x \in [x_{2k-1}, x_{2k}]$,

$$|y_{\text{mid}}^{(1)}(x)| < 15 \max_{t\in[0,1]} t^2(1-t)^2 < \frac{15}{16}. \tag{B.2.34}$$

We deduce similarly that (B.2.34) also holds for $x \in [x_{2k}, x_{2k+1}]$. As a consequence, we obtain that $|y_{\text{mid}}^{(1)}(x)| < 1$ for all $x \geq 0$, and therefore, using (B.2.33),

$$|(-1)^k - y_{\text{mid}}^{(1)}(x)| \leq 1$$

for all $x \in [x_k, x_{k+1}]$, where the inequality is strict except at x_k and x_{k+1} since $y_{\text{mid}}^{(1)}(x_k) = y_{\text{mid}}^{(1)}(x_{k+1}) = 0$. Hence we deduce that

$$\left|[y(x) - y_{\text{mid}}(x)]\,[(-1)^k - y_{\text{mid}}^{(1)}(x)]\right| \leq \tfrac{1}{2}\delta(x) \tag{B.2.35}$$

for all $x \in [x_k, x_{k+1}]$, using (B.2.30). Moreover, since, at the leftmost boundary of $[x_{2k-1}, x_{2k}]$,

$$y(x_{2k-1}) - y_{\text{mid}}(x_{2k-1}) = \tfrac{1}{2}\delta(x_{2k-1}) = \tfrac{1}{2}(\epsilon + \omega_{2k-1})$$

and, at the leftmost boundary of $[x_{2k}, x_{2k+1}]$,

$$y(x_{2k}) - y_{\text{mid}}(x_{2k}) = -\tfrac{1}{2}\delta(x_{2k}) = -\tfrac{1}{2}(\epsilon + \omega_{2k})$$

(where we used (B.2.18)), we deduce from (B.2.33) that the inequality in (B.2.35) can only hold as an equality at x_{k+1}.

Our penultimate step to is note that (B.2.7) and (B.2.15) together give that, for $x \in [x_k, x_{k+1}]$ and $t = (x - x_k)/\sigma_k$,

$$f_1^{(1)}(x) + \delta(x) = (\omega_k - \omega_{k+1})\left[-12\,t^2 + 18\,t^3 - 6\,t^5\right] \leq 0, \tag{B.2.36}$$

where, again, the inequality is strict in the interior of the interval (see Figure B.2.4).

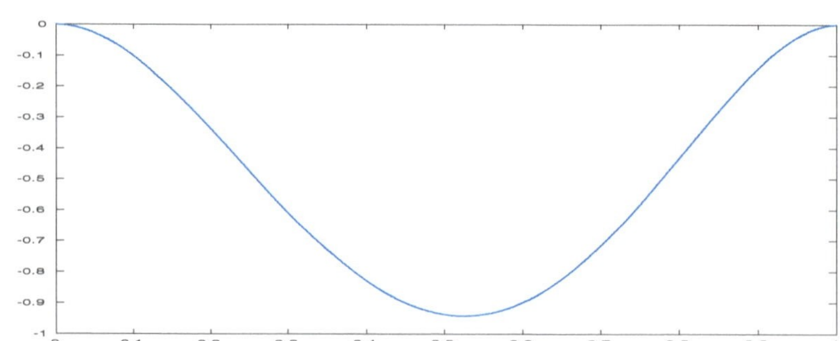

Figure B.2.4. The polynomial $-12\,t^2 + 18\,t^3 - 6\,t^5$ on $[0, 1]$.

Finally, combining (B.2.32), (B.2.35), and (B.2.36), we obtain that, for all $k \geq 0$,

$$\sigma(x) < 0 \text{ for } x \in [x_k, x_{k+1}),$$

and (x_{k+1}, y_{k+1}) is indeed the first local minimizer of $f(x,y)$ along the steepest-descent direction at iterate (x_k, y_k), as is clearly demonstrated in Figure B.2.5, where the values of the objective function along the path of iterates and along the prolongation of each step are shown in blue and red, respectively.

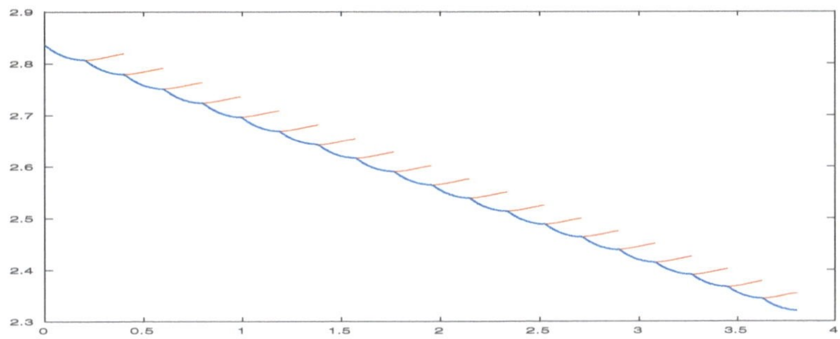

Figure B.2.5. The value of the objective function along the path of iterates (in blue) and the value of the objective function which would be obtained if each step were prolongated (in red).

□

□

Appendix to Chapter 3

We required the following, entirely technical result in the proof of Lemma 3.4.3 (p. 78).

Lemma B.3.1: An algebraic property used in Lemma 3.4.3

[335, Lem. 17, App. A] For $a, b > 0$ and $c \geq 0$, one has that

$$-b + \sqrt{b^2 + ac} \geq \left(-b + \sqrt{b^2 + a}\right) \min[1, c]. \tag{B.3.1}$$

Proof. Consider first the case where $c \geq 1$. Then

$$-b + \sqrt{b^2 + ac} \geq -b + \sqrt{b^2 + a} \geq \left(-b + \sqrt{b^2 + a}\right) \min[1, c],$$

which is the desired result. Suppose now that $c \leq 1$. This implies that

$$\left(a + 2b^2 - 2b\sqrt{b^2 + a}\right) c = \left(-b + \sqrt{b^2 + a}\right)^2 c \geq \left(-b + \sqrt{b^2 + a}\right)^2 c^2,$$

and hence, adding b^2 on the left and right, that

$$b^2 + ac \geq \left(-b + \sqrt{b^2 + a}\right)^2 c^2 + 2b\left(-b + \sqrt{b^2 + a}\right) c + b^2.$$

Now taking the square root on both sides, we obtain that

$$\sqrt{b^2 + ac} \geq \left(-b + \sqrt{b^2 + a}\right) c + b,$$

which yields (B.3.1). □

Proofs of results from Section 3.4.2

Proof of Theorem 3.4.11 (p. 87)

Using our customary telescoping-sum argument, based here on (3.4.35), the fact that $\rho_k \geq \eta$ if $k \in \mathcal{S}$, and **f.Bb**, we have that, for any k for which the iRN algorithm has not terminated before or at iteration $k+1$,

$$f(x_0) - f_{\rm low} \geq f(x_0) - f(x_{k+1}) \geq \eta \sum_{j \in \mathcal{S}_k} \|s_j\|^3, \tag{B.3.2}$$

where $\mathcal{S}_k$ denotes the number of successful iterations up to iteration k. Since the algorithm has not terminated, we must have that $\|\nabla_x^1 f(x_{k+1})\| > \epsilon_1$. Using (B.3.2) and Lemma 3.4.9 for all $j \in \mathcal{S}_{k-1}$, we obtain

$$f(x_0) - f_{\rm low} \geq \eta \sum_{j \in \mathcal{S}_{k-1}} \|s_j\|^3 \geq \frac{2\sqrt{2}\eta\epsilon_1^{3/2}}{(L_{f,2} + 4\sigma_{\max}^{\rm U} + 2\kappa_3)^{3/2}} |\mathcal{S}_{k-1}|.$$

Therefore, since $|\mathcal{S}_k| \leq |\mathcal{S}_{k-1}| + 1$,

$$|\mathcal{S}_k| \leq \frac{(L_{f,2} + 4\sigma_{\max}^{\rm U} + 2\kappa_3)^{3/2}}{2\sqrt{2}\eta} \frac{f(x_0) - f_{\rm low}}{\epsilon_1^{3/2}} + 1,$$

which provides an upper bound on the number of successful iterations and, since the gradient and Hessian are evaluated once for each such iteration, yields the bound (3.4.44) with

$$\kappa_{\rm iRN}^{\rm S} \stackrel{\rm def}{=} \frac{(L_{f,2} + 4\sigma_{\max}^{\rm U} + 2\kappa_3)^{3/2}}{2\sqrt{2}\eta},$$

where we added a 1 to account for the final iteration.
We may now apply Lemma 3.4.10 to deduce an upper bound on the total number of iterations and evaluations of f by multiplying the bound (3.4.44) by the bound (3.4.42) on the number of unsuccessful iterations that can occur between every consecutive pair of successful ones. This leads to the bound (3.4.43), where

$$\kappa_{\rm iRN}^{\rm A} \stackrel{\rm def}{=} \kappa_{\rm iRN}^{\rm S} \left(\frac{1}{\log \gamma_1} \log\left(\frac{\sigma_{\max}^{\rm U}}{\sigma_{\min}}\right) + 1\right)$$

and

$$\kappa_{\rm iRN}^{\rm B} \stackrel{\rm def}{=} 2 \left(\frac{1}{\log \gamma_1} \log\left(\frac{\sigma_{\max}^{\rm U}}{\sigma_{\min}}\right) + 1\right).$$

□

Proof of Theorem 3.4.12 (p. 87)

Let $\mathcal{S}_\epsilon \subseteq \mathcal{S}$ be the set of successful iterations for which either $\|\nabla_x^1 f(x_{j+1})\| > \epsilon_1$ or $\lambda_{\min}[Q_j^T \nabla_x^2 f(x_j) Q_j] < -\epsilon_2$ for $j \in \mathcal{S}_\epsilon$. Then Lemma 3.4.9 and (3.4.35) imply that for each $j \in \mathcal{S}_\epsilon$, we have the function decrease

$$\begin{aligned} f(x_j) - f(x_{j+1}) &\geq \eta \min \left[\frac{2\sqrt{2}}{(L_{f,2} + 4\sigma_{\max}^{\text{U}} + 2\kappa_3)^{3/2}} \epsilon_1^{3/2}, \frac{1}{(\kappa_4 + \sigma_{\max}^{\text{U}})^3} \epsilon_2^3 \right] \\ &\geq \kappa_6 \min \left[\epsilon_1^{3/2}, \epsilon_2^3 \right], \end{aligned}$$

where

$$\kappa_6 \stackrel{\text{def}}{=} \eta \min \left[\frac{2\sqrt{2}}{(L_{f,2} + 4\sigma_{\max}^{\text{U}} + 2\kappa_3)^{3/2}}, \frac{1}{(\kappa_4 + \sigma_{\max}^{\text{U}})^3} \right]. \tag{B.3.3}$$

A standard telescoping-sum argument involving **f.Bb** then leads to

$$f(x_0) - f_{\text{low}} \geq \sum_{j \in \mathcal{S}_\epsilon} \kappa_6 \min \left[\epsilon_1^{3/2}, \epsilon_2^3 \right], \tag{B.3.4}$$

where κ_6 is defined as in (B.3.3). From (B.3.4), we deduce that the number of iterations in $\mathcal{S}_\epsilon$ is finite and

$$|\mathcal{S}_\epsilon| \leq \frac{f(x_0) - f_{\text{low}}}{\kappa_6 \min \left[\epsilon_1^{3/2}, \epsilon_2^3 \right]}.$$

Let k_1 and $k_2 > k_1$ be the first two elements of $\mathcal{S} \setminus \mathcal{S}_\epsilon$. Then, by definition of $\mathcal{S}_\epsilon$, $\|\nabla_x^1 f(x_k))\| \leq \epsilon_1$ for all $k \geq k_1$ and $\lambda_{\min}[Q_k^T \nabla_x^2 f(x_k) Q_k] \geq -\epsilon_2$ for all $k \geq k_2$. Therefore (3.4.34) holds for $k \geq k_2$, and thus at most $|\mathcal{S}_\epsilon| + 2$ successful iterations can occur before termination. This gives the bound (3.4.46), with

$$\kappa_{\text{iRN}}^{\text{T}} \stackrel{\text{def}}{=} \frac{1}{\eta} \max \left[\frac{(L_{f,2} + 4\sigma_{\max}^{\text{U}} + 2\kappa_3)^{3/2}}{2\sqrt{2}}, (\kappa_4 + \sigma_{\max}^{\text{U}})^3 \right],$$

where $\sigma_{\max}^{\text{U}}$, κ_3, and κ_4 are defined as in (3.4.38), (3.4.30), and (3.4.32), respectively. Just as in the proof of Theorem 3.4.11, we use Lemma 3.4.10 to deduce an upper bound on the total number of iterations and evaluations of f by multiplying the bound (3.4.46) by the bound (3.4.42) on the number of unsuccessful iterations that can occur between every two successful ones. This provides the bound (3.4.45), where

$$\kappa_{\text{iRN}}^{\text{C}} \stackrel{\text{def}}{=} \kappa_{\text{iRN}}^{\text{T}} \left(\frac{1}{\log \gamma_1} \log \left(\frac{\sigma_{\max}^{\text{U}}}{\sigma_{\min}} \right) + 1 \right)$$

and

$$\kappa_{\text{iRN}}^{\text{D}} \stackrel{\text{def}}{=} 2 \left(\frac{1}{\log \gamma_1} \log \left(\frac{\sigma_{\max}^{\text{U}}}{\sigma_{\min}} \right) + 1 \right).$$

□

Proofs of results from Section 3.5

Proof of Lemma 3.5.1 (p. 91)

Clearly, (3.5.4) holds when $\lambda_{\min}(\nabla_x^2 f(x_k))+\lambda_k=0$. When $\lambda_{\min}(\nabla_x^2 f(x_k))+\lambda_k>0$ and hence $\nabla_x^2 f(x_k)+\lambda_k I \succ 0$, (3.5.2) implies that

$$\|s_k\| \le \frac{\|\nabla_x^1 f(x_k)\|+\|r_k\|}{\lambda_{\min}(\nabla_x^2 f(x_k))+\lambda_k} \le \frac{(1+\kappa_{rg})\|\nabla_x^1 f(x_k)\|}{\lambda_{\min}(\nabla_x^2 f(x_k))+\lambda_k}.$$

This and (3.5.6) give the inequality

$$\psi(\lambda_k)\le 0 \quad \text{with} \quad \psi(\lambda) \stackrel{\text{def}}{=} \lambda\left[\lambda+\lambda_{\min}(\nabla_x^2 f(x_k))\right]-\overline{\kappa}_\lambda(1+\kappa_{rg})\|\nabla_x^1 f(x_k)\|. \tag{B.3.5}$$

Now note that $\psi(0)=\psi(-\lambda_{\min}(\nabla_x^2 f(x_k)))=-\overline{\kappa}_\lambda(1+\kappa_{rg})\|\nabla_x^1 f(x_k)\|$ and thus

$$\psi(\lambda_{1,k})<0 \quad \text{with} \quad \lambda_{1,k}=\max\{0,-\lambda_{\min}(\nabla_x^2 f(x_k))\}. \tag{B.3.6}$$

Moreover, the form of $\psi(\lambda)$ implies that $\psi(\lambda)$ is strictly increasing for $\lambda\ge\lambda_{1,k}$. Now define

$$\lambda_{2,k} \stackrel{\text{def}}{=} -\lambda_{\min}(\nabla_x^2 f(x_k))+2\max\left\{|\lambda_{\min}(\nabla_x^2 f(x_k))|,|\psi(0)|^{\frac{1}{2}}\right\}, \tag{B.3.7}$$

and note that $\lambda_{2,k}>\lambda_{1,k}$. Suppose first that $\lambda_{\min}(\nabla_x^2 f(x_k))<0$ and $|\lambda_{\min}(\nabla_x^2 f(x_k))|\ge|\psi(0)|^{\frac{1}{2}}$. Then one verifies that $\lambda_{2,k}=3|\lambda_{\min}(H_k)|$ and

$$\begin{aligned}\psi(\lambda_{2,k}) &= (3|\lambda_{\min}(\nabla_x^2 f(x_k))|)^2-(3|\lambda_{\min}(\nabla_x^2 f(x_k))|)|\lambda_{\min}(\nabla_x^2 f(x_k))|-|\psi(0)|\\ &= 2\cdot 3|\lambda_{\min}(\nabla_x^2 f(x_k))|^2-\overline{\kappa}_\lambda(1+\kappa_{rg})\|\nabla_x^1 f(x_k)\|>0.\end{aligned}$$

Suppose that $\lambda_{\min}(\nabla_x^2 f(x_k))\ge 0$ and $|\lambda_{\min}(\nabla_x^2 f(x_k))|\ge|\psi(0)|^{\frac{1}{2}}$. Then $\lambda_{2,k}=\lambda_{\min}(\nabla_x^2 f(x_k))$ and

$$\psi(\lambda_{2,k})=(\lambda_{\min}(\nabla_x^2 f(x_k)))^2+\lambda_{\min}(\nabla_x^2 f(x_k))|\lambda_{\min}(\nabla_x^2 f(x_k))|-|\psi(0)|>0.$$

Thus we deduce that $\psi(\lambda_{2,k})>0$ whenever $|\lambda_{\min}(\nabla_x^2 f(x_k))|\ge|\psi(0)|^{\frac{1}{2}}$. Moreover, the same inequality obviously holds if $|\lambda_{\min}(H_k)|<|\psi(0)|^{\frac{1}{2}}$ because $\psi(\lambda)$ is increasing with λ. As a consequence, $\psi(\lambda_{2,k})>0$ in all cases. We now combine this inequality, (B.3.6), and the monotonicity of $\psi(\lambda)$ for $\lambda\ge\lambda_{1,k}$ to obtain that either $\lambda_k\le\lambda_{1,k}<\lambda_{2,k}$ or $\lambda_k\in[\lambda_{1,k},\lambda_{2,k})$ as a result of (B.3.5). Thus $\lambda_k\le\lambda_{2,k}$, which implies (3.5.4) because of (B.3.7) and $\overline{\kappa}_\lambda>1$. □

Proof of Lemma 3.5.2 (p. 91)

A by now familiar argument gives

$$\|\nabla_x^1 f(x_{k+1})\| \le \tfrac{1}{2}L_{f,2}\|s_k\|^2+\|\nabla_x^1 f(x_k)+\nabla_x^2 f(x_k)s_k\|,$$

which, together with (3.5.2), further becomes

$$\begin{aligned}\|\nabla_x^1 f(x_{k+1})\| &\le \tfrac{1}{2}L_{f,2}\|s_k\|^2 + \lambda_k\|s_k\| + \|r_k\| \\ &\le (\tfrac{1}{2}L_{f,2} + \kappa_{rs})\|s_k\|^2 + \lambda_k\,\|s_k\|.\end{aligned}$$

Using (3.5.6), we deduce that

$$\|\nabla_x^1 f(x_{k+1})\| \le \left[\tfrac{1}{2}L_{f,2} + \kappa_{rs} + \overline{\kappa}_\lambda\right]\|s_k\|^2,$$

which is (3.5.7) with $\overline{\kappa}_s \stackrel{\rm def}{=} \left[\tfrac{1}{2}L_{f,2} + \kappa_{rs} + \overline{\kappa}_\lambda\right]^{-1}$. □

Proof of Lemma 3.5.3 (p. 92)

Let

$$0 \le \alpha_k \stackrel{\rm def}{=} \frac{\|s_k\|}{\sqrt{\|\nabla_x^1 f(x_{k+1})\|}}, \qquad k \ge 0. \tag{B.3.8}$$

From (3.5.9) and the definition of α_k in (B.3.8), we have that, for $k \ge 0$,

$$\begin{aligned}(1-\kappa_{rg})\frac{\|s_k\|}{\alpha_k} &\le \sqrt{\kappa_c}(1-\kappa_{rg})\|\nabla_x^1 f(x_k)\| \le \sqrt{\kappa_c}\|-\nabla_x^1 f(x_k) + r_k\| \\ &= \sqrt{\kappa_c}\|(H_k+\lambda_k I)s_k\| \qquad \le \sqrt{\kappa_c}\|H_k + \lambda_k I\|\cdot\|s_k\|,\end{aligned}$$

where we used (3.5.2) to obtain both the second inequality and the first equality. It follows that

$$\|H_k + \lambda_k I\| \ge \frac{(1-\kappa_{rg})}{\alpha_k\sqrt{\kappa_c}}, \qquad k \ge 0. \tag{B.3.9}$$

From (3.5.8) and **f.gL**, we deduce that $\{H_k + M_k\}$ is uniformly bounded above for all k, namely,

$$\|H_k + \lambda_k I\| \le L_{f,1} + \overline{\kappa}_\lambda, \qquad k \ge 0. \tag{B.3.10}$$

Now (B.3.9) and (B.3.10) give that

$$\alpha_k \ge \frac{1-\kappa_{rg}}{(L_{f,1}+\overline{\kappa}_\lambda)\sqrt{\kappa_c}} \stackrel{\rm def}{=} \overline{\kappa}_s > 0 \;\text{ for all } k \ge 0,$$

and so it follows from (B.3.8) that (3.5.7) holds. □

Proof of Theorem 3.5.4 (p. 94)

Let p_k be the polynomial

$$p_k(s) = c_{0,k} + c_{1,k}s + c_{2,k}s^2 + c_{3,k}s^3 + c_{4,k}s^4 + c_{5,k}s^5,$$

where $s = x - x_k \in [0, s_k]$, with coefficients defined by the interpolation conditions (3.5.21). The existence of the polynomial and the expressions of its coefficients follow from (A.26) (with $p = 2$) in the argument leading to Theorem A.9.1 (p. 441);

we obtain

$$c_{0,k} = f_k, \quad c_{1,k} = g_k, \quad c_{2,k} = \tfrac{1}{2} H_k;$$

$$c_{3,k} = \frac{10}{s_k^3} r_{0,k} - \frac{4}{s_k^2} r_{1,k} + \frac{1}{2 s_k} r_{2,k};$$

$$c_{4,k} = -\frac{15}{s_k^4} r_{0,k} + \frac{7}{s_k^3} r_{1,k} - \frac{1}{s_k^2} r_{2,k};$$

$$c_{5,k} = \frac{6}{s_k^5} r_{0,k} - \frac{3}{s_k^4} r_{1,k} + \frac{1}{2 s_k^3} r_{2,k},$$

where

$$r_{0,k} \stackrel{\rm def}{=} f_{k+1} - f_k - g_k s_k - \tfrac{1}{2} H_k s_k^2, \quad r_{1,k} \stackrel{\rm def}{=} g_{k+1} - g_k - H_k s_k, \quad r_{2,k} \stackrel{\rm def}{=} H_{k+1} - H_k. \tag{B.3.11}$$

Next we show that $f_\epsilon^{\mathcal{M}}$ is bounded in absolute value independent of ϵ and twice continuously differentiable with Lipschitz-continuous gradient and Hessian on the interval

$$[0, x_{k_{\epsilon,\alpha}}] = \bigcup_{k \in \mathcal{K}} [x_k, x_k + s_k].$$

Recall first that (3.5.20) provides that $f_\epsilon^{\mathcal{M}}$ is twice continuously differentiable by construction on $[0, x_{k_{\epsilon,\alpha}}]$. It thus remains to investigate the Lipschitz continuity of the gradient and Hessian, as well as whether $|f_\epsilon^{\mathcal{M}}(x)|$ is bounded on this interval. For this it is sufficient to show that (A.30) holds with $p = 2$, so that we can apply Theorem A.9.1 and then Theorem A.9.2. From (B.3.11), we have

$$\left| \frac{r_{0,k}}{s_k^3} \right| = f_k \frac{|1 - \theta_k|^2}{\theta_k^3} \le \frac{1}{\theta_k^3} \le \kappa_\lambda^3 (1 - \kappa_r)^{-3},$$

$$\left| \frac{r_{1,k}}{s_k^2} \right| = 8 \frac{f_k^2}{\theta_k^2} |f_k - f_{k+1} - \theta_k f_k| \le 8 \kappa_\lambda^2 (1 - \kappa_r)^{-2} (2 + \kappa_r), \quad \text{and}$$

$$\left| \frac{r_{2,k}}{s_k} \right| = \frac{8 f_k^3}{\theta_k} \le 8 \kappa_\lambda (1 - \kappa_r),$$

where we also used (3.5.19) and (3.5.13). The desired conclusions follow from Theorems A.9.1 and A.9.2. □

The constructed function $f_\epsilon^{\mathcal{M}}(x)$, $x \ge 0$, admits an isolated finite global minimizer. Indeed, since $g_{k_\epsilon} < 0$, there must be a value lower than $f(x_{k_\epsilon})$ in the subsequent interval of interpolation, and thus the global minimizer must lie in one of the constructed subintervals in $(-1, x_{k_\epsilon} + 1)$; since $f_\epsilon^{\mathcal{M}}(x)$ is quintic (and not constant) in each of these, the global minimizer must therefore be isolated.

Appendix to Chapter 4

The following technical lemma is extracted from [108, Lem. 2.4] and gives the form and the norm of the jth derivative of the regularization term used in the ARp algorithm. Because it is crucial for the main results of Chapters 4, 5, and 12, we reprove it here. For an integer k and a real $\beta \in (0, 1]$, we define $(k + \beta)! \stackrel{\rm def}{=} \prod_{\ell=1}^k (\beta + \ell)$ (this coincides with the standard factorial if $\beta = 1$) and $\beta! = 1$.

Lemma B.4.1: Derivatives of the regularization term

Let s be a vector of $\mathbb{R}^n$ and $\beta \in (0,1]$. Then

$$\| \nabla_s^j (\|s\|^{p+\beta}) \| = \frac{(p+\beta)!}{(p-j+\beta)!} \|s\|^{p-j+\beta} \quad \text{for } j \in [0:p]; \tag{B.4.1}$$

$$\nabla_s^j (\|s\|^{p+\beta})[s]^j = \frac{(p+\beta)!}{(p-j+\beta)!} \|s\|^{p+\beta} \quad \text{for } j \in [0:p]; \tag{B.4.2}$$

and, whenever $s \neq 0$,

$$\| \nabla_s^{p+1} (\|s\|^{p+\beta}) \| = \beta\,(p+\beta)!\, \|s\|^{\beta-1}. \tag{B.4.3}$$

Moreover, $\nabla_s^j (\|s\|^{p+\beta})[d]^j$ is a convex function of d for any d orthogonal to s. It is also convex for any multiple of s whenever j is even.

Proof. We first observe that $\nabla_s^j (\|s\|^{p+\beta})$ is a jth-order tensor whose norm is defined by

$$\|\nabla_s^j (\|s\|^{p+\beta})\| \stackrel{\text{def}}{=} \max_{\|v\|=1} |\nabla_s^j (\|s\|^{p+\beta})[v]^j| \tag{B.4.4}$$

(see (A.7)). Moreover, using the relationships

$$\nabla_s (\|s\|^\tau) = \tau \|s\|^{\tau-2} s \quad (\tau \in \mathbb{R}) \quad \text{and} \quad \nabla_s (s^{j\otimes}) = j\, s^{(j-1)\otimes} \otimes I \quad (j \in \mathbb{N}), \tag{B.4.5}$$

defining

$$\nu_0 \stackrel{\text{def}}{=} 1 \text{ and } \nu_i \stackrel{\text{def}}{=} \prod_{\ell=1}^{i} (p+2-2\ell),$$

and proceeding by induction, we obtain that, for some $\mu_{j,i} \geq 0$ with $\mu_{1,1} = 1$,

$$\begin{aligned}
&\nabla_s \left[\nabla_s^{j-1} (\|s\|^{p+\beta})\right] \\
&= \nabla_s \left[\sum_{i=2}^{j} \mu_{j-1,i-1} \nu_{i-1} \|s\|^{p+\beta-2(i-1)} \, s^{(2(i-1)-(j-1))\otimes} \otimes I^{((j-1)-(i-1))\otimes} \right] \\
&= \sum_{i=2}^{j} \mu_{j-1,i-1} \nu_{i-1} \Big[(p+\beta-2(i-1)) \|s\|^{p+\beta-2(i-1)-2} \, s^{(2(i-1)-(j-1)+1)\otimes} \otimes I^{(j-i)\otimes} \\
&\qquad + ((2(i-1)-(j-1)) \|s\|^{p+\beta-2(i-1)} \, s^{(2(i-1)-(j-1)-1)\otimes} \otimes I^{((j-1)-(i-1)+1)\otimes} \Big] \\
&= \sum_{i=2}^{j} \mu_{j-1,i-1} \nu_{i-1} \Big[(p+\beta+2-2i) \|s\|^{p+\beta-2i} \, s^{(2i-j)\otimes} \otimes I^{(j-i)\otimes} \\
&\qquad + (2(i-1)-j+1) \|s\|^{p+\beta-2(i-1)} \, s^{(2(i-1)-j)\otimes} \otimes I^{(j-(i-1))\otimes} \Big]
\end{aligned}$$

$$
\begin{aligned}
&= \sum_{i=2}^{j} \mu_{j-1,i-1}\nu_{i-1}(p+\beta+2-2i)\|s\|^{p+\beta-2i}\, s^{(2i-j)\otimes} \otimes I^{(j-i)\otimes} \\
&\quad + \sum_{i=1}^{j-1}(2i-j+1)\mu_{j-1,i}\nu_i\|s\|^{p+\beta-2i}\, s^{(2i-j)\otimes} \otimes I^{(j-i)\otimes} \\
&= \sum_{i=1}^{j} \big((p+\beta+2-2i)\mu_{j-1,i-1}\nu_{i-1} + (2i-j+1)\mu_{j-1,i}\nu_i\big) \\
&\quad \cdot \|s\|^{p+\beta-2i}\, s^{(2i-j)\otimes} \otimes I^{(j-i)\otimes},
\end{aligned}
$$

where the last equation uses the convention that $\mu_{j,0} = 0$ for all j. Thus we may write

$$
\nabla_s^j\big(\|s\|^{p+\beta}\big) = \nabla_s\left[\nabla_s^{j-1}\big(\|s\|^{p+\beta}\big)\right] = \sum_{i=1}^{j} \mu_{j,i}\nu_i\, \|s\|^{p+\beta-2i}\, s^{(2i-j)\otimes} \otimes I^{(j-i)\otimes}, \tag{B.4.6}
$$

with

$$
\begin{aligned}
\mu_{j,i}\nu_i &= (p+\beta+2-2i)\mu_{j-1,i-1}\nu_{i-1} + (2i-j+1)\mu_{j-1,i}\nu_i \\
&= \big[\mu_{j-1,i-1} + (2i-j+1)\mu_{j-1,i}\big]\nu_i,
\end{aligned} \tag{B.4.7}
$$

where we used the identity

$$
\nu_i = (p+\beta+2-2i)\nu_{i-1} \quad \text{for} \quad i = 1,\dots,j \tag{B.4.8}
$$

to deduce the second equality. Now (B.4.6) gives that

$$
\nabla_s^j\big(\|s\|^{p+\beta}\big)[v]^j = \sum_{i=1}^{j} \mu_{j,i}\nu_i\|s\|^{p+\beta-j} \left\langle \frac{s}{\|s\|}, v\right\rangle^{2i-j} \langle v, v\rangle^{j-i}. \tag{B.4.9}
$$

If j is even, then all terms involving v occur with an even power, and thus $\nabla_s^j(\|s\|^{p+\beta})[v]^j$ is a convex function of v. Moreover, $\nabla_s^j(\|s\|^{p+\beta})[v]^j = 0$ for any v orthogonal to s for odd j (in which case no term with $2i = j$ occurs). Thus the last conclusion of the lemma's statement holds. It is also easy to see that the maximum in (B.4.4) is achieved for $v = s/\|s\|$, so that

$$
\|\,\nabla_s^j\big(\|s\|^{p+\beta}\big)\,\| = \left(\sum_{i=1}^{j} \mu_{j,i}\nu_i\right)\|s\|^{p+\beta-j} = \pi_j\|s\|^{p+\beta-j}, \tag{B.4.10}
$$

with

$$
\pi_j \stackrel{\text{def}}{=} \sum_{i=1}^{j} \mu_{j,i}\,\nu_i. \tag{B.4.11}
$$

Furthermore, it follows from (B.4.9) when $v = s$ that

$$
\nabla_s^j\big(\|s\|^{p+\beta}\big)[s]^j = \sum_{i=1}^{j} \mu_{j,i}\nu_i\|s\|^{p+\beta} = \pi_j\|s\|^{p+\beta}. \tag{B.4.12}
$$

Successively using the definition (B.4.11), (B.4.7), (B.4.8) (twice), the identity

$\mu_{j-1,j} = 0$, and (B.4.11) again, we then deduce that

$$\begin{aligned}
\pi_j &= \sum_{i=1}^{j} \mu_{j-1,i-1}\nu_i + \sum_{i=1}^{j}[(2i-j+1)\mu_{j-1,i}]\nu_i \\
&= \sum_{i=1}^{j-1} \mu_{j-1,i}\nu_{i+1} + \sum_{i=1}^{j}[(2i-j+1)\mu_{j-1,i}]\nu_i \\
&= \sum_{i=1}^{j-1} \mu_{j-1,i}\left[\nu_{i+1} + (2i-j+1)\nu_i\right] \qquad \text{(B.4.13)} \\
&= \sum_{i=1}^{j-1} \mu_{j-1,i}\left[(p+\beta+2-2(i+1))\nu_i + (2i-j+1)\nu_i\right] \\
&= (p+\beta+1-j)\sum_{i=1}^{j-1} \mu_{j-1,i}\,\nu_i \\
&= (p+\beta+1-j)\pi_{j-1}.
\end{aligned}$$

Since $\pi_1 = p+\beta$ by taking norms of both sides of the first part of (B.4.5), we obtain that $\pi_j = (p+\beta)!/(p-j+\beta)!$, which, combined with (B.4.10) and (B.4.11), gives (B.4.1), and with (B.4.12) we find (B.4.2). We obtain (B.4.3) from (B.4.10) and (B.4.11), the observation that $\pi_p = (p+\beta)!$, and (B.4.13) for $j = p+1$. □

Appendix to Chapter 5

We need to ensure that certain residual exponents are positive to justify our least-Euclidean-norm complexity bounds.

Lemma B.5.1: Positive exponents

Let $\iota_1(p,\ell,i)$, $\iota_2(p,\ell,i)$, and $\iota_1(p,\ell,i)$ be the exponents given in (5.1.21), (5.1.23), and (5.1.25). Then $\iota_1(p,p,i) > 0$ when p is positive and $i > \log_2 p$, $\iota_2(p,p/2,i) > 0$ when p is a positive multiple of two and $i > \log_2(p-1)$, and $\iota_3(p,p/3,i) > 0$ when p is a positive multiple of three and $i > \log_2(p-2)$.

Proof. We have that $\iota_1(p,p,i) > 0$ if and only if $p(2^i - 1) > (p-1)2^i$ or, rearranging, $p < 2^i$, which implies the first result. Similarly, $\iota_1(p,p/2,i) > 0$ if and only if $(p-1)(2^i-1) > 2^i(p-2)$ or equivalently $p-1 < 2^i$, which gives the second result. Finally, $\iota_1(p,p/3,i) > 0$ if and only if $(p-2)(2^i-1) > 2^i(p-3)$, i.e., $p-2 < 2^i$, as desired. □

Proofs of results from Chapter 5

Proof of Lemma 5.1.3 (p. 125)

The relationship $a \geq b \geq 0$ and the identity

$$a^\ell - b^\ell = \left(\sum_{j=0}^{\ell-1} a^j b^{\ell-j-1}\right)(a-b) \leq \ell a^{\ell-1}(a-b)$$

lead directly to the inequality

$$a - b \geq \frac{a^\ell - b^\ell}{\ell a^{\ell-1}}.$$

Furthermore, a simple induction using the identity $A^2 - B^2 = (A-B)(A+B)$ with $A = a^{1/2^j} > B = b^{1/2^j}$ for increasing $j \leq i$ gives that

$$a^{1/2^i} - b^{1/2^i} \geq \frac{a-b}{2^i a^{\frac{2^i-1}{2^i}}}.$$

Combining the two inequalities gives the required result. □

Proof of Lemma 5.1.5 (p. 127)

If $p = 1$, $p = 1 \cdot 2^0$, so assume inductively that the result is true for all $p \in [1, \ldots, k-1]$. Then if k is odd, then $k = k \cdot 2^0$. By contrast, if k is even, then $k = 2r$ for some $1 \leq r < k$ for which $r = o \cdot 2^j$ for some odd o and $j \in \mathbb{Z}_+$. Hence $k = 2r = o \cdot 2^{j+1}$. Thus the identity holds for $p = k$ and hence for all p. Uniqueness follows from the uniqueness of the prime factorization of p. □

Proof of Theorem 5.3.3 (p. 143)

We first note that the superlinear convergence of the gaps $\{\Gamma_k\}_{k\in\mathcal{S}}$ to zero directly follows from item (i) of Lemma 5.3.1 (p. 141).
It is convenient to rewrite (5.3.11) in terms of a scaled gap ω_k for which

$$\omega_k - \omega_{k+1} \geq \omega_{k+1}^\tau \tag{B.5.1}$$

when $k \in \mathcal{S}$, where $\tau = \nu/\mu$,

$$\omega_k \stackrel{\text{def}}{=} \kappa_\Gamma^{\frac{1}{\alpha}} \Gamma_k \;\text{ with }\; \kappa_\Gamma \stackrel{\text{def}}{=} \frac{\kappa_\nu}{\kappa_\psi^\tau}, \tag{B.5.2}$$

and

$$\alpha = \tau - 1 = \frac{\nu - \mu}{\mu}. \tag{B.5.3}$$

Note that $\{\omega_k\}_{k\in\mathcal{S}}$ is positive and monotonically decreasing because $\{\Gamma_k\}_{k\in\mathcal{S}}$ is, and observe that $\tau < 1$ and $\alpha < 0$ when $\nu < \mu$.
Consider first the case $\omega_0 \geq 1$, and observe that (B.5.1) may be expressed as

$$\frac{\omega_k}{\omega_{k+1}} \geq 1 + \omega_{k+1}^{\tau-1} = 1 + \omega_{k+1}^\alpha$$

because of (B.5.3), and the decreasing nature of the sequence $\{\omega_k\}_{k\in\mathcal{S}}$ gives that

$$\omega_{k+1} \leq \frac{\omega_k}{1 + \omega_{k+1}^\alpha} \leq \frac{\omega_k}{1 + \omega_0^\alpha}$$

for $k \in \mathcal{S}$. Thus it takes at most

$$\frac{\log(\omega_0)}{\log\left(1+\omega_0^\alpha\right)}+1 \tag{B.5.4}$$

successful iterations to find an iteration $k_* \in \mathcal{S}$ for which the scaled gap has a value ω_{k_*+1} smaller than $\omega_* \stackrel{\text{def}}{=} \left(1+\omega_0^\alpha\right)^{-1} < 1$. It then remains to reduce the scaled gap from ω_* to ϵ. The inequality (B.5.1) clearly implies that

$$\omega_k \geq \omega_{k+1}^\tau$$

for $k \in \mathcal{S}$ with $k > k_*$, so that

$$\omega_{k+1} \leq \omega_k^{\frac{1}{\tau}}.$$

As a consequence

$$\omega_{k_j+1} \leq \omega_*^{\left(\frac{1}{\tau}\right)^j},$$

where iteration $k_j \in \mathcal{S}$ is the jth successful iteration after iteration k_*, and additionally we can be sure that

$$\omega_{k_j+1} \leq \epsilon \tag{B.5.5}$$

as soon as

$$\omega_*^{\left(\frac{1}{\tau}\right)^j} \leq \epsilon,$$

that is,

$$\left(\frac{1}{\tau}\right)^j |\log(\omega_*)| \geq |\log(\epsilon)|$$

or, equivalently,

$$j|\log(\tau)| + \log(|\log(\omega_*)|) \geq \log(|\log(\epsilon)|).$$

Thus to guarantee (B.5.5) requires at most

$$\frac{\log(|\log(\epsilon)|) - \log(|\log(\omega_*)|)}{|\log(\tau)|} \tag{B.5.6}$$

successful iterations on top of those that might be required to bring the gap from ω_0 to ω_*. Hence, summing the numbers of successful iterations given by (B.5.4) and (B.5.6), the algorithm requires at most

$$\frac{\log(|\log(\epsilon)|) - \log(|\log(\omega_*)|)}{|\log(\tau)|} + \frac{\log(\omega_0)}{\log\left(1+\omega_0^\alpha\right)} + 1 \tag{B.5.7}$$

successful iterations to compute an iterate x_k for which $\omega_k \leq \epsilon$.
To simplify matters, observe that $|\log(\epsilon)| \geq |\log(\omega_*)|$ since $\epsilon \leq \omega_* < 1$, and thus

$$\begin{aligned}\log(|\log(\epsilon)|) - \log(|\log(\omega_*)|) &= \Big|\log(|\log(\epsilon)|) - \log(|\log(\omega_*)|)\Big| \\ &\leq \Big|\log(|\log(\epsilon)|)\Big| + \Big|\log(|\log(\omega_*)|)\Big|,\end{aligned}$$

so that the bound (B.5.7) is no worse than

$$\frac{|\log(|\log(\epsilon)|) + |\log(|\log(\omega_*)|)}{|\log(\tau)|} + \frac{\log(\omega_0)}{\log\left(1+\omega_0^\alpha\right)} + 1. \tag{B.5.8}$$

If $\omega_0 < 1$, then the bound (B.5.4) is irrelevant, but we need to consider reducing the gap directly from ω_0 to ϵ, which takes at most

$$\frac{\log(|\log(\epsilon)|) - \log(|\log(\omega_0)|)}{|\log(\tau)|} \le \frac{|\log(|\log(\epsilon)|)| + |\log(|\log(\omega_0)|)|}{|\log(\tau)|} \tag{B.5.9}$$

successful iterations. Combining (B.5.8) and (B.5.9) gives that at most

$$\frac{|\log(|\log(\epsilon)|)| + \max\left[|\log(|\log(\omega_*)|)|, |\log(|\log(\omega_0)|)|\right]}{|\log(\tau)|} + \frac{\max[0, \log(\omega_0)]}{\log\left(1+\omega_0^\alpha\right)} + 1$$

successful iterations are needed, and substituting

$$\kappa_\Gamma^{\frac{1}{\alpha}} = \left(\frac{\kappa_\nu}{\kappa_\psi^{\nu/\mu}}\right)^{\frac{\mu}{\nu-\mu}} = \left(\frac{\kappa_\nu^\mu}{\kappa_\psi^\nu}\right)^{\frac{1}{\nu-\mu}}, \quad \omega_0 = \kappa_\Gamma^{\frac{1}{\alpha}}\Gamma_0 = \left(\frac{\kappa_\nu^\mu}{\kappa_\psi^\nu}\right)^{\frac{1}{\nu-\mu}}\Gamma_0, \tag{B.5.10}$$

and $\omega_* = (1+\omega_0^\alpha)^{-1}$ gives that

$$\log(\omega_0) = \frac{1}{\nu-\mu}\log\left(\frac{\kappa_\nu^\mu}{\kappa_\psi^\nu}\right) + \log\left(\Gamma_0\right) \quad \text{and} \quad |\log(\omega_*)| = \log\left(1 + \frac{\kappa_\nu}{\kappa_\psi^{\nu/\mu}}\Gamma_0^{\frac{\nu-\mu}{\mu}}\right).$$

As a consequence, we obtain (5.3.17) with

$$\kappa_{\text{AR,gap,1}}^{\text{E}} \stackrel{\text{def}}{=} \frac{1}{\log(\mu) - \log(\nu)}$$

and

$$\kappa_{\text{AR,gap,1}}^{\text{F}} \stackrel{\text{def}}{=} \frac{\max\left[\left|\log\left(\log\left(1+\frac{\kappa_\nu}{\kappa_\psi^{\nu/\mu}}\Gamma_0^{\frac{\nu-\mu}{\mu}}\right)\right)\right|, \left|\log\left(\left|\frac{1}{\nu-\mu}\log\left(\frac{\kappa_\nu^\mu}{\kappa_\psi^\nu}\right) + \log\left(\Gamma_0\right)\right|\right)\right|\right]}{\log(\mu) - \log(\nu)}$$

$$+ \frac{\max\left[0, \frac{1}{\nu-\mu}\log\left(\frac{\kappa_\nu^\mu}{\kappa_\psi^\nu}\right) + \log\left(\Gamma_0\right)\right]}{\log\left(1+\frac{\kappa_\nu}{\kappa_\psi^{\nu/\mu}}\Gamma_0^\alpha\right)} + 1.$$

The bound (5.3.16) follows from (5.3.9) and (5.3.17), with the appropriate constants

$$\kappa_{\text{AR,gap,1}}^{\text{B}} \stackrel{\text{def}}{=} \kappa_{\text{R}}^{\mathcal{S}}\kappa_{\text{AR,gap,1}}^{\text{E}} \quad \text{and} \quad \kappa_{\text{AR,gap,1}}^{\text{C}} \stackrel{\text{def}}{=} \kappa_{\text{R}}^{\mathcal{S}}\kappa_{\text{AR,gap,1}}^{\text{F}} + \kappa_{\text{A}}^{\mathcal{S}}.$$

Turning now to the analysis of the convergence of the sequence $\{\psi(x_k)\}_{k\in\mathcal{S}}$, we observe also that (5.3.4) and (5.3.5) imply that

$$\Gamma_k \ge f(x_k) - f(x_{k+1}) \ge \kappa_\nu[\psi(x_{k+1})]^\nu$$

and hence that

$$\psi(x_k) \le \epsilon \text{ is satisfied when } \Gamma_{k-1} \le \kappa_\nu\epsilon^\nu. \tag{B.5.11}$$

Thus (5.3.19) and (5.3.18) immediately follow from (5.3.17) and (5.3.16), respectively, with

$$\kappa^{\rm E}_{\sf AR,dom,1} \stackrel{\rm def}{=} \kappa^{\rm E}_{\sf AR,gap,1}, \quad \kappa^{\rm F}_{\sf AR,dom,1} \stackrel{\rm def}{=} \kappa^{\rm F}_{\sf AR,gap,1} + 1,$$

and

$$\kappa^{\rm B}_{\sf AR,dom,1} \stackrel{\rm def}{=} \kappa^{\mathcal{S}}_{\rm R}\kappa^{\rm E}_{\sf AR,dom,1} \text{ and } \kappa^{\rm C}_{\sf AR,dom,1} \stackrel{\rm def}{=} \kappa^{\mathcal{S}}_{\rm R}\kappa^{\rm F}_{\sf AR,dom,1} + \kappa^{\mathcal{S}}_{\rm A}.$$

□

Proof of Theorem 5.3.4 (p. 144)

The linear convergence of the sequence $\{\Gamma_k\}_{k\in\mathcal{S}}$ with factor $1+\kappa_\nu/\kappa_\psi$ follows directly from item (ii) of Lemma 5.3.1 (p. 141). Hence, the algorithm needs at most

$$\frac{\log(\Gamma_0)-\log(\epsilon)}{\log\left(1+\frac{\kappa_\nu}{\kappa_\psi}\right)} = \frac{|\log(\epsilon)|+\log(\Gamma_0)}{\log\left(1+\frac{\kappa_\nu}{\kappa_\psi}\right)}$$

successful iterations to bring the gap below ϵ, giving (5.3.21) with

$$\kappa^{\rm E}_{\sf AR,gap,2} \stackrel{\rm def}{=} \frac{1}{\log\left(1+\frac{\kappa_\nu}{\kappa_\psi}\right)} \text{ and } \kappa^{\rm F}_{\sf AR,gap,2} \stackrel{\rm def}{=} \frac{\log(\Gamma_0)}{\log\left(1+\frac{\kappa_\nu}{\kappa_\psi}\right)} + 1.$$

We then obtain (5.3.20) from (5.3.9) and (5.3.21), where the constants are now

$$\kappa^{\rm B}_{\sf AR,gap,2} \stackrel{\rm def}{=} \kappa^{\mathcal{S}}_{\rm R}\kappa^{\rm E}_{\sf AR,gap,2} \text{ and } \kappa^{\rm C}_{\sf AR,gap,2} \stackrel{\rm def}{=} \kappa^{\mathcal{S}}_{\rm R}\kappa^{\rm F}_{\sf AR,gap,2} + \kappa^{\mathcal{S}}_{\rm A}.$$

Observe also that condition (B.5.11) remains valid when $\nu=\mu$, and hence we immediately see that the algorithm needs at most

$$\kappa^{\rm B}_{\sf AR,gap,2}|\log(\kappa_\nu\epsilon^\nu)| + \kappa^{\rm C}_{\sf AR,gap,2} + 1 = \kappa^{\rm B}_{\sf AR,gap,2}|\log(\kappa_\nu)+\nu\log(\epsilon)| + \kappa^{\rm C}_{\sf AR,gap,2} + 1$$

successful iterations to bring the optimality measure below ϵ, giving (5.3.23) with

$$\kappa^{\rm E}_{\sf AR,dom,2} \stackrel{\rm def}{=} \nu\kappa^{\rm E}_{\sf AR,gap,2} \text{ and } \kappa^{\rm F}_{\sf AR,dom,2} \stackrel{\rm def}{=} \kappa^{\rm E}_{\sf AR,gap,2}|\log(\kappa_\nu)| + \kappa^{\rm F}_{\sf AR,gap,2} + 1.$$

The bounds (5.3.9) and (5.3.23) then give (5.3.22) with

$$\kappa^{\rm B}_{\sf AR,dom,2} \stackrel{\rm def}{=} \kappa^{\mathcal{S}}_{\rm R}\kappa^{\rm E}_{\sf AR,dom,2} \text{ and } \kappa^{\rm C}_{\sf AR,dom,2} \stackrel{\rm def}{=} \kappa^{\mathcal{S}}_{\rm R}\kappa^{\rm F}_{\sf AR,dom,2} + \kappa^{\mathcal{S}}_{\rm A}.$$

□

Proof of Theorem 5.3.5 (p. 144)

Before considering the main result, we first prove the following two useful inequalities:

$$(1-t)^\alpha \leq 1-\alpha t \text{ and } 1-t^\alpha \geq \alpha(1-t). \tag{B.5.12}$$

The first inequality is equivalent to requiring that

$$(1-t) \leq (1-\alpha t)^{\frac{1}{\alpha}}$$

for all $t \in [0,1]$, which is in turn equivalent to requiring that

$$h(t) \stackrel{\text{def}}{=} (1-\alpha t)^{\frac{1}{\alpha}} + t - 1 \geq 0$$

for all $t \in [0,1]$. Now $h(0) = 0$ and

$$h^{(1)}(x) = \frac{1}{\alpha}(1-\alpha t)^{\frac{1}{\alpha}-1}(-\alpha) + 1 = 1 - (1-\alpha t)^{\frac{1}{\alpha}-1} \geq 0$$

for all $t \in [0,1]$ since $(1-\alpha t)^{\frac{1}{\alpha}-1} \in [0,1]$ for all such t. Hence $h(t) \geq 0$ for all $t \in [0,1]$, which proves the first inequality in (B.5.12). The second inequality then follows from the first by replacing t by $1-t$.

We now turn to the proof of Theorem 5.3.5 proper. As in the proof of Theorem 5.3.3 (p. 143) (given on page 464), we consider the scaled inequality (B.5.1) with the definition (B.5.2) and $\tau = \nu/\mu > 1$ and $\alpha > 0$ given by (B.5.3). Now consider bounding

$$\frac{1}{\omega_{k+1}^{\alpha}} - \frac{1}{\omega_k^{\alpha}} = \frac{\omega_k^{\alpha} - \omega_{k+1}^{\alpha}}{\omega_k^{\alpha}\omega_{k+1}^{\alpha}} = \frac{1}{\omega_{k+1}^{\alpha}}\left[1 - \left(\frac{\omega_{k+1}}{\omega_k}\right)^{\alpha}\right].$$

Using the second inequality in (B.5.12), we deduce that

$$\frac{1}{\omega_{k+1}^{\alpha}} - \frac{1}{\omega_k^{\alpha}} \geq \frac{\alpha}{\omega_{k+1}^{\alpha}}\left(1 - \frac{\omega_{k+1}}{\omega_k}\right) = \frac{\alpha}{\omega_{k+1}^{\alpha}\omega_k}(\omega_k - \omega_{k+1}).$$

Taking (B.5.1) into account then gives that

$$\frac{1}{\omega_{k+1}^{\alpha}} - \frac{1}{\omega_k^{\alpha}} \geq \frac{\alpha}{\omega_{k+1}^{\alpha}\omega_k}\,\omega_{k+1}^{\tau} = \alpha\frac{\omega_{k+1}^{\tau-\alpha}}{\omega_k} = \alpha\frac{\omega_{k+1}}{\omega_k}$$

since $\tau - \alpha = 1$ from (B.5.3). We now distinguish two cases. Suppose first that

$$\omega_{k+1} \geq \beta\omega_k, \tag{B.5.13}$$

where β is an arbitrary constant in $(0,1)$, independent of k. Then we obtain that

$$\frac{1}{\omega_{k+1}^{\alpha}} - \frac{1}{\omega_k^{\alpha}} \geq \alpha\beta.$$

If (B.5.13) fails, then we have from the decreasing nature of $\{\omega_k\}$ that

$$\frac{1}{\omega_{k+1}^{\alpha}} - \frac{1}{\omega_k^{\alpha}} \geq \frac{1}{\omega_k^{\alpha}}\left(\frac{1}{\beta^{\alpha}} - 1\right) \geq \frac{1}{\omega_0^{\alpha}}\left(\frac{1}{\beta^{\alpha}} - 1\right).$$

Combining these two cases yields that

$$\frac{1}{\omega_{k+1}^{\alpha}} - \frac{1}{\omega_k^{\alpha}} \geq \min\left[\frac{1}{\omega_0^{\alpha}}\left(\frac{1}{\beta^{\alpha}} - 1\right), \alpha\beta\right] \stackrel{\text{def}}{=} \kappa_{\text{dom}} > 0.$$

If as usual we let $\mathcal{S}_k$ be the index set of the successful iterations up to iteration k, we deduce, by summing over $i \in \mathcal{S}_k$, that

$$\frac{1}{\omega_{k+1}^{\alpha}} \geq \frac{1}{\omega_0^{\alpha}} + \kappa_{\text{dom}}|\mathcal{S}_k|,$$

which is to say that

$$\omega_{k+1} \le \left(\frac{1}{\omega_0^{-\alpha} + \kappa_{\rm dom}|\mathcal{S}_k|} \right)^{\frac{1}{\alpha}}. \tag{B.5.14}$$

Now let k be the largest index in $\mathcal{S}$ for which $\Gamma_{k+1} \ge \epsilon$, that is, $\omega_{k+1} \ge \kappa_\Gamma^{\frac{1}{\alpha}} \epsilon$. Using (B.5.14), this is the case if

$$\kappa_\Gamma \epsilon^\alpha = \left(\kappa_\Gamma^{\frac{1}{\alpha}} \epsilon \right)^\alpha \le \frac{1}{\omega_0^{-\alpha} + \kappa_{\rm dom}|\mathcal{S}_k|},$$

that is, if

$$\omega_0^{-\alpha} + \kappa_{\rm dom}|\mathcal{S}_k| \le \kappa_\Gamma^{-1} \epsilon^{-\alpha},$$

i.e.,

$$|\mathcal{S}_k| \le \frac{1}{\kappa_{\rm dom}} \left[\kappa_\Gamma^{-1} \epsilon^{-\alpha} - \omega_0^{-\alpha} \right] = \frac{1}{\kappa_{\rm dom}\kappa_\Gamma} \left[\epsilon^{-\frac{\nu-\mu}{\mu}} - \Gamma_0^{-\alpha} \right].$$

The constants here are given by (B.5.10) and

$$\begin{aligned} \kappa_{\rm dom}\kappa_\Gamma &= \min \left[\frac{1}{\Gamma_0^\alpha} (\beta^{-\alpha} - 1), \alpha\beta\kappa_\Gamma \right] \\ &= \min \left[\left(\frac{\kappa_\psi^{\nu/\mu}}{\kappa_\nu} \right) \left(\beta^{-\frac{\nu-\mu}{\mu}} - 1 \right), \left(\frac{\nu-\mu}{\mu} \right) \beta \left(\frac{\kappa_\nu}{\kappa_\psi^{\nu/\mu}} \right) \right]. \end{aligned} \tag{B.5.15}$$

Choosing $\beta = \frac{1}{2}$, this implies that (5.3.25) holds with

$$\kappa_{\textsf{AR,gap,3}}^{\rm E} = \frac{1}{\min \left[\left(\frac{\kappa_\psi^{\nu/\mu}}{\kappa_\nu} \right) \left(2^{\frac{\nu-\mu}{\mu}} - 1 \right), \left(\frac{\nu-\mu}{2\mu} \right) \left(\frac{\kappa_\nu}{\kappa_\psi^{\nu/\mu}} \right) \right]}$$

and

$$\kappa_{\textsf{AR,gap,3}}^{\rm F} = -\kappa_{\textsf{AR,gap,3}}^{\rm E} \Gamma_0^{-\frac{\nu-\mu}{\mu}}.$$

The bound (5.3.24) then follows from (5.3.9) and (5.3.25) with

$$\kappa_{\textsf{AR,gap,3}}^{\rm B} \stackrel{\rm def}{=} \kappa_{\rm R}^{\mathcal{S}} \kappa_{\textsf{AR,gap,3}}^{\rm E} \quad \text{and} \quad \kappa_{\textsf{AR,gap,3}}^{\rm C} \stackrel{\rm def}{=} \kappa_{\rm R}^{\mathcal{S}} \kappa_{\textsf{AR,gap,3}}^{\rm F} + \kappa_{\rm A}^{\mathcal{S}}.$$

Finally, consider the convergence of the optimality measures $\{\psi(x_k)\}_{k\in\mathcal{S}}$ to zero, and note that (B.5.14) clearly implies that

$$\kappa_\Gamma^{1/\alpha} \Gamma_{k+1} = \omega_{k+1} \le \left(\frac{1}{\kappa_{\rm dom}|\mathcal{S}_k|} \right)^{\frac{1}{\alpha}}. \tag{B.5.16}$$

Because we know ω_k converges to zero, there is an integer N that is the total number of successful iterations for which $\psi(x_{k+1}) \ge \epsilon$. Suppose first that N is even, and let k_1 be the index of the last iteration in $\mathcal{S}_N$ such that $\mathcal{S}_{N,1}$, the set of successful iterations up to k_1, is at most half the size of $\mathcal{S}_N$. Also let $\mathcal{S}_{N,2} = \mathcal{S}_N \setminus \mathcal{S}_{N,1}$. Thus $\mathcal{S}_N = \mathcal{S}_{N,1} \cup \mathcal{S}_{N,2}$ and $|\mathcal{S}_{N,1}| = |\mathcal{S}_{N,2}|$. As a consequence,

$$|\mathcal{S}_N| = 2|\mathcal{S}_{N,1}|. \tag{B.5.17}$$

From (B.5.16) and (5.3.5), we have that

$$\kappa_\Gamma^{-\frac{1}{\alpha}}\left(\frac{1}{\kappa_{\rm dom}|\mathcal{S}_{N,1}|}\right)^{\frac{1}{\alpha}} \geq \Gamma_{N+1}+\sum_{i=k+1}^{N}[f(x_{i+1})-f(x_i)] \geq \sum_{i=k+1,i\in\mathcal{S}}^{N}\kappa_\nu\epsilon^\nu = \kappa_\nu\epsilon^\nu|\mathcal{S}_{N,1}|,$$

that is,

$$\frac{\kappa_\Gamma^{-\frac{1}{\alpha}}\epsilon^{-\nu}}{\kappa_{\rm dom}^{\frac{1}{\alpha}}\kappa_\nu} \geq |\mathcal{S}_{N,1}|^{1+\frac{1}{\alpha}} = |\mathcal{S}_{N,1}|^{\frac{\nu}{\nu-\mu}},$$

which yields

$$|\mathcal{S}_{N,1}| \leq \left(\frac{\kappa_\Gamma^{-\frac{1}{\alpha}}}{\kappa_{\rm dom}^{\frac{1}{\alpha}}\kappa_\nu}\right)^{\frac{\nu-\mu}{\nu}}\epsilon^{-(\nu-\mu)} = (\kappa_{\rm dom}\kappa_\Gamma)^{-\mu/\nu}\kappa_\nu^{-\frac{\nu-\mu}{\nu}}\epsilon^{-(\nu-\mu)}.$$

Combining this bound with (B.5.17) and (B.5.15) and choosing $\beta = \frac{1}{2}$ then implies (5.3.27) with

$$\kappa^{\rm E}_{\text{AR,dom,3}} \stackrel{\rm def}{=} \frac{\kappa_\nu^{-\frac{\nu-\mu}{\nu}}}{\min\left[\left(\frac{\kappa_\psi^{\nu/\mu}}{\kappa_\nu}\right)\left(\beta^{-\frac{\nu-\mu}{\mu}}-1\right),\left(\frac{\nu-\mu}{\mu}\right)\beta\left(\frac{\kappa_\nu}{\kappa_\psi^{\nu/\mu}}\right)\right]^{\mu/\nu}}$$

and $\kappa^{\rm F}_{\text{AR,dom,3}} = 1$. The case where $|\mathcal{S}_N|$ is odd is taken care of by adding one to the bound obtained for the even case, setting $\kappa^{\rm F}_{\text{AR,dom,3}} = 2$. Finally, (5.3.26) may once again be deduced from (5.3.9), yielding

$$\kappa^{\rm B}_{\text{AR,dom,3}} \stackrel{\rm def}{=} \kappa^{\mathcal{S}}_{\rm R}\kappa^{\rm B}_{\text{AR,dom,2}} \text{ and } \kappa^{\rm C}_{\text{AR,dom,3}} \stackrel{\rm def}{=} \kappa^{\mathcal{S}}_{\rm R}\kappa^{\rm F}_{\text{AR,dom,3}} + \kappa^{\mathcal{S}}_{\rm A}.$$

□

Proof of Theorem 5.3.7 (p. 148)

We showed in Lemma 5.3.6 that $f(x_{k_{\mathcal{S},1}+1}) = f_*$ if $\Gamma_0 = \Gamma_\square$. Otherwise, we have that $\Gamma_0 > \Gamma_\square$ and (5.3.30) holds. Defining

$$\omega_k \stackrel{\rm def}{=} \kappa_\Gamma^{\frac{1}{\alpha}}\Gamma_k \geq 0 \text{ with } \kappa_\Gamma \stackrel{\rm def}{=} \frac{\kappa_\nu}{\kappa_\psi^\tau} \text{ and } \alpha \stackrel{\rm def}{=} \tau - 1 \equiv \frac{\nu-\mu}{\mu} < 0,$$

we obtain from this inequality and the monotonically decreasing nature of the sequence $\{\omega_k\}_{k\in\mathcal{S}}$ that, for $k \in \mathcal{S}$,

$$\omega_{k+1} \leq \omega_k(1-\omega_k^\alpha) \leq \omega_k\left(1-\frac{1}{\omega_0^{|\alpha|}}\right) \leq \omega_0\left(1-\frac{1}{\omega_0^{|\alpha|}}\right)^{|\mathcal{S}_k|}, \tag{B.5.18}$$

where $|\mathcal{S}_k|$ is the number of successful iterations up to iteration k. Thus ω_k decreases at least[124] linearly with factor $(1-1/\omega^{|\alpha|})$ from $\omega_0 > 1$. Moreover, if we let $k_\square$ be the last index in $\mathcal{S}$ such that $\omega_k \geq 1$, we have that $\omega_{k_\square+j} = 0$ for all

[124]We have just seen that the decrease is in fact superlinear.

$j \geq 1$. Because of (B.5.18), we also have that

$$|\mathcal{S}_{k_\square}| \leq |\mathcal{S}_{k_*}|,$$

where k_* is the smallest index such that

$$\omega_0 \left(1 - \frac{1}{\omega_0^{|\alpha|}}\right)^{|\mathcal{S}_{k_*}|} \leq 1,$$

which imposes that

$$|\mathcal{S}_{k_\square}| \leq |\mathcal{S}_{k_*}| \leq \frac{\log(\omega_0)}{|\log(1 - \omega_0^{\alpha})|} + 1.$$

Thus we obtain from the above argument and the definition of ω_k that at most

$$\kappa_{\text{TR},1}^{\text{F}} \stackrel{\text{def}}{=} \frac{\log\left(\kappa_\Gamma^{\frac{\mu}{\nu-\mu}} \Gamma_0\right)}{\left|\log\left(1 - \left(\kappa_\Gamma^{\frac{\mu}{\nu-\mu}} \Gamma_0\right)^{\frac{\nu-\mu}{\mu}}\right)\right|} + 1$$

successful iterations are required to make the gap identically zero, yielding (5.3.36). Using (5.3.9) then gives (5.3.35) with

$$\kappa_{\text{TR},1}^{\text{C}} \stackrel{\text{def}}{=} \kappa_{\text{R}}^{\mathcal{S}} \kappa_{\text{TR},1}^{\text{F}} + \kappa_{\text{A}}^{\mathcal{S}}.$$

Finally, (5.3.37) and (5.3.38) immediately follow from the observation that, because of (5.3.28), $\Gamma_k = 0$ implies that $\psi(x_k) = 0$. □

Proof of Theorem 5.3.8 (p. 149)

We saw in Lemma 5.3.6 (p. 146) that, given that $\mu = \nu$, $\kappa_\nu \leq \kappa_\psi$, and that $f(x_1) = f_*$ if $\kappa_\nu = \kappa_\psi$. Assume therefore that $\kappa_\nu < \kappa_\psi$. The linear convergence of the gaps $\{\Gamma_k\}_{k \in \mathcal{S}}$ to zero with factor $1 - \kappa_\nu / \kappa_\psi$ directly results from Lemma 5.3.6, item (i), and the bound (5.3.40) then follows immediately with

$$\kappa_{\text{TR,gap},2}^{\text{E}} \stackrel{\text{def}}{=} \frac{1}{\left|\log\left(1 - \frac{\kappa_\nu}{\kappa_\psi}\right)\right|} \quad \text{and} \quad \kappa_{\text{TR,gap},2}^{\text{F}} \stackrel{\text{def}}{=} \frac{\log(\Gamma_0)}{\left|\log\left(1 - \frac{\kappa_\nu}{\kappa_\psi}\right)\right|}.$$

We then obtain (5.3.39) from (5.3.9) and (5.3.40), where the constants are now

$$\kappa_{\text{TR,gap},2}^{\text{B}} \stackrel{\text{def}}{=} \kappa_{\text{R}}^{\mathcal{S}} \kappa_{\text{TR,gap},2}^{\text{E}} \quad \text{and} \quad \kappa_{\text{TR,gap},2}^{\text{C}} \stackrel{\text{def}}{=} \kappa_{\text{R}}^{\mathcal{S}} \kappa_{\text{TR,gap},2}^{\text{F}} + \kappa_{\text{A}}^{\mathcal{S}}.$$

The rest of the proof follows exactly that of Theorem 5.3.4 (p. 144) (given on page 467) with the modified convergence factor $1 - \frac{\kappa_\nu}{\kappa_\psi}$ instead of $\left(1 + \frac{\kappa_\nu}{\kappa_\psi}\right)^{-1}$, yielding (5.3.42) and (5.3.41) with

$$\begin{aligned}
\kappa_{\text{TR,dom},2}^{\text{E}} &\stackrel{\text{def}}{=} \nu \kappa_{\text{TR,gap},2}^{\text{E}}, \quad \kappa_{\text{TR,dom},2}^{\text{F}} \stackrel{\text{def}}{=} \kappa_{\text{TR,gap},2}^{\text{E}} |\log(\kappa_\nu)| + \kappa_{\text{TR,gap},2}^{\text{F}} + 1, \\
\kappa_{\text{TR,dom},2}^{\text{B}} &\stackrel{\text{def}}{=} \kappa_{\text{R}}^{\mathcal{S}} \kappa_{\text{TR,dom},2}^{\text{E}}, \quad \text{and} \quad \kappa_{\text{TR,dom},2}^{\text{C}} \stackrel{\text{def}}{=} \kappa_{\text{R}}^{\mathcal{S}} \kappa_{\text{TR,dom},2}^{\text{F}} + \kappa_{\text{A}}^{\mathcal{S}},
\end{aligned}$$

with the latter pair once again using (5.3.9). □

Proof of Theorem 5.3.9 (p. 149)

The proof of (5.3.43) essentially follows that of (5.3.25) in Theorem 5.3.5, whose proof is given on page 467. We again define

$$\omega_k \stackrel{\rm def}{=} \kappa_\Gamma^{\frac{1}{\alpha}} \Gamma_k \geq 0, \quad \text{with} \quad \kappa_\Gamma \stackrel{\rm def}{=} \frac{\kappa_\nu}{\kappa_\psi^\tau} \quad \text{and} \quad \alpha \stackrel{\rm def}{=} \tau - 1 \stackrel{\rm def}{=} \frac{\nu - \mu}{\mu},$$

and obtain from (5.3.30) that

$$\omega_k - \omega_{k+1} \geq \omega_k^{\frac{\nu}{\mu}},$$

in turn leading to

$$\frac{1}{\omega_{k+1}^\alpha} - \frac{1}{\omega_k^\alpha} \geq \frac{\alpha}{\omega_{k+1}^\alpha \omega_k} \omega_k^\tau = \alpha \frac{\omega_k^{\tau-1}}{\omega_{k+1}^\alpha} = \alpha \left(\frac{\omega_k}{\omega_{k+1}} \right)^\alpha > \alpha,$$

where we have used the fact that the sequence $\{\omega_k\}_{k \in \mathcal{S}}$ is monotonically decreasing to obtain the last inequality. Summing over the index set $\mathcal{S}_k$ of the successful iterations up to iteration k, we then deduce that

$$\frac{1}{\omega_{k+1}^\alpha} \geq \frac{1}{\omega_0^\alpha} + \alpha |\mathcal{S}_k|$$

and therefore that

$$\omega_{k+1} \leq \left(\frac{1}{\omega_0^{-\alpha} + \alpha |\mathcal{S}_k|} \right)^{\frac{1}{\alpha}}.$$

This corresponds to inequality (B.5.14) in the proof of Theorem 5.3.5, with $\kappa_{\rm dom} = \alpha$. Modulo that substitution, the rest of the proof is identical to that of Theorem 5.3.5 (page 467), giving the constants

$$\begin{aligned}
\kappa^{\rm E}_{\sf TR,gap,3} &\stackrel{\rm def}{=} \frac{\mu \kappa_\psi^{\mu/\nu}}{(\nu - \mu)\kappa_\nu}, & \kappa^{\rm F}_{\sf TR,gap,3} &\stackrel{\rm def}{=} -\kappa^{\rm B}_{\sf TR,gap,3} \Gamma_0^{-\frac{\nu-\mu}{\mu}}, \\
\kappa^{\rm B}_{\sf TR,gap,3} &\stackrel{\rm def}{=} \kappa_{\rm R}^{\mathcal{S}} \kappa^{\rm E}_{\sf TR,gap,3}, & \kappa^{\rm C}_{\sf TR,gap,3} &\stackrel{\rm def}{=} \kappa_{\rm R}^{\mathcal{S}} \kappa^{\rm F}_{\sf TR,gap,3} + \kappa_{\rm A}^{\mathcal{S}}, \\
\kappa^{\rm E}_{\sf TR,dom,3} &\stackrel{\rm def}{=} \frac{1}{\kappa_\nu} \left(\frac{\nu - \mu}{\mu} \right)^{\mu/\nu} \left(\frac{\kappa_\psi^\nu}{\kappa_\nu^\mu} \right)^{\frac{1}{\nu-\mu}}, & \kappa^{\rm F}_{\sf TR,dom,3} &\stackrel{\rm def}{=} 2, \\
\kappa^{\rm B}_{\sf TR,dom,3} &\stackrel{\rm def}{=} \kappa_{\rm R}^{\mathcal{S}} \kappa^{\rm E}_{\sf TR,dom,3}, \quad \text{and} & \kappa^{\rm C}_{\sf TR,dom,3} &\stackrel{\rm def}{=} \kappa_{\rm R}^{\mathcal{S}} \kappa^{\rm F}_{\sf TR,dom,3} + \kappa_{\rm A}^{\mathcal{S}}.
\end{aligned}$$

□

Appendix to Chapter 7

Yet another, entirely technical, result is needed in the proof of Lemma 7.2.3 (p. 202).

Lemma B.7.1: A global optimization problem used in Lemma 7.2.3

The global minimizer of the two-variable optimization problem

$$\min_{(f,c) \in \mathbb{R}^2} \psi(f,c) \stackrel{\rm def}{=} -f + \sqrt{\epsilon^2 - c^2} \;\; \textit{subject to} \;\; f^2 + c^2 \leq \omega^2, \tag{B.7.1}$$

where $0 < \omega < \epsilon$, *is attained at* $(f_*, c_*) = (\omega, 0)$ *and gives* $\psi(f_*, c_*) = -\omega + \epsilon$.

Proof. We have that $\epsilon^2 - c^2 > 0$ for any feasible (f, c) since $\omega < \epsilon$, and trivially

$$\nabla\psi(f, c) = \left(-1, -\frac{c}{\sqrt{\epsilon^2 - c^2}}\right) \neq 0 \text{ for all } (f, c).$$

Thus the solution of (B.7.1) is attained on the boundary of the feasible region, and hence $f_*^2 + c_*^2 = \omega^2$. Also, (f_*, c_*) satisfies the KKT conditions for (B.7.1), namely,

$$-1 + 2\lambda_* f_* = 0, \quad -\frac{c_*}{\sqrt{\epsilon^2 - c_*^2}} + 2\lambda_* c_* = 0, \quad \lambda_* \geq 0, \quad \text{and} \quad f_*^2 + c_*^2 = \omega^2,$$

where λ_* is the Lagrange multiplier at the minimizer (f_*, c_*). The first condition above implies that $f_* \neq 0$ and $\lambda_* \neq 0$. Thus $\lambda_* = 1/(2f_*)$, and substituting this into the second KKT condition gives

$$c_* \left(\frac{1}{f_*} - \frac{1}{\sqrt{\epsilon^2 - c_*^2}}\right) = 0.$$

But $f_* \neq \sqrt{\epsilon^2 - c_*^2}$ since $f_*^2 + c_*^2 = \omega^2$ and $\omega < \epsilon$. Thus $c_* = 0$ and $f_* = \pm\omega$. Since we are minimizing, the smallest value of $\psi(f, c_*)$ is at $f_* = \omega$. □

Appendix to Chapter 8

The following result is stated, in part, in other sources (e.g., [231, Thm. 4.8]). For completeness, here we state and prove the version we require.

Lemma B.8.1: Convergence rates of one-dimensional Newton-like iteration

[82, Lem. A.1]: Suppose that $\theta : \mathcal{I} \to \mathbb{R}$ is convex (resp., concave), strictly decreasing (resp., strictly increasing), and continuously differentiable on some interval $\mathcal{I} = [\lambda_{\min}, \lambda_{max}] \subseteq \mathbb{R}$, and suppose further that there is a $\lambda_* \in \mathcal{I}$ for which $\theta(\lambda_*) = 0$.
(i) Now suppose that $\theta(\lambda_0) > 0$ for some given $\lambda_0 \in \mathcal{I}$. Then the Newton iterates $\{\lambda_j\}$, where

$$\lambda_{j+1} = \lambda_j - \frac{\theta(\lambda_j)}{\theta'(\lambda_j)} \tag{B.8.1}$$

for $j \geq 0$, all lie in $[\lambda_0, \lambda_*]$ and increase monotonically to λ_*. The convergence is globally Q-linear with factor at least

$$\gamma^{\text{N}} \stackrel{\text{def}}{=} 1 - \frac{\theta'(\lambda_*)}{\theta'(\lambda_0)} < 1$$

and is ultimately Q-superlinear (Q-quadratic if additionally θ' is Lipschitz continuous around λ_*).

(ii) *Suppose that $\theta(\lambda_0)$ and $\theta(\lambda_1) > 0$ for some given $\lambda_0 < \lambda_1 \in \mathcal{I}$. Then the secant iterates $\{\lambda_j\}$, where*

$$\lambda_{j+1} = \lambda_j - \frac{(\lambda_j - \lambda_{j-1})\theta(\lambda_j)}{\theta(\lambda_j) - \theta(\lambda_{j-1})} \tag{B.8.2}$$

for $j \geq 1$, all lie in $[\lambda_0, \lambda_*]$ and increase monotonically to λ_*. The convergence is globally Q-linear with factor at least γ^{N} and is ultimately Q-superlinear.

Proof. We consider the convex case; the concave case then follows directly by considering $-\theta$. The assumptions are such that $\lambda \in \mathcal{I} < \lambda_*$ if and only if $\theta(\lambda) > 0$.
(i) By induction, suppose that $\theta(\lambda_j) > 0$. Since by assumption $\theta'(\lambda_j) < 0$, (B.8.1) shows that $\lambda_{j+1} > \lambda_j$. Additionally, the convexity of θ and (B.8.1) imply that

$$\theta(\lambda_{j+1}) \geq \theta(\lambda_j) + \theta'(\lambda_j)(\lambda_{j+1} - \lambda_j) = 0,$$

and thus $\theta(\lambda_j + 1) > 0$. Convexity also implies that

$$\theta'(\lambda_*)(\lambda_j - \lambda_*) = \theta(\lambda_*) + \theta'(\lambda_*)(\lambda_j - \lambda_*) \geq \theta(\lambda_j), \tag{B.8.3}$$

in which case

$$\lambda_* - \lambda_{j+1} = \lambda_* - \lambda_j + \frac{\theta(\lambda_j)}{\theta'(\lambda_j)} \leq (\lambda_* - \lambda_j)\left(1 - \frac{\theta'(\lambda_*)}{\theta'(\lambda_j)}\right) \leq \gamma^{\mathrm{N}}(\lambda_* - \lambda_j), \tag{B.8.4}$$

which establishes both that $\{\lambda_j\}$ converges to λ_* and that the convergence is at least linear. Ultimate superlinear convergence follows from (B.8.4) since $\theta'(\lambda_j) \to \theta'(\lambda_*)$, while quadratic convergence for Lipschitz-continuous θ' follows since $\theta'(\lambda_*) < 0$ (see Ortega and Rheinboldt [311, Thm. 10.2.2]).
(ii) By induction, suppose that $\lambda_{j-1} < \lambda_j$ and $\theta(\lambda_j) > 0$ (in which case $\theta(\lambda_{j-1}) > \theta(\lambda_j)$). Then it follows directly from (B.8.2) that $\lambda_{j+1} > \lambda_j$. Thus the convexity of θ and (B.8.2) imply that

$$\theta(\lambda_{j+1}) \geq \theta(\lambda_j) + \frac{\lambda_{j+1} - \lambda_j}{\lambda_{j-1} - \lambda_j}(\theta(\lambda_{j-1}) - \theta(\lambda_j)) = 0.$$

Furthermore, the mean-value theorem implies that $\theta(\lambda_j) - \theta(\lambda_{j-1}) = \theta'(\xi_j)(\lambda_j - \lambda_{j-1})$ for some $\xi_j \in (\lambda_{j-1}, \lambda_j)$, and thus from (B.8.2),

$$\lambda_{j+1} = \lambda_j - \frac{\theta(\lambda_j)}{\theta'(\xi_j)}. \tag{B.8.5}$$

Thus, using (B.8.3) and (B.8.5),

$$\lambda_* - \lambda_{j+1} = \lambda_* - \lambda_j + \frac{\theta(\lambda_j)}{\theta'(\xi_j)} \leq (\lambda_* - \lambda_j)\left(1 - \frac{\theta'(\lambda_*)}{\theta'(\xi_j)}\right) \leq \gamma^{\mathrm{N}}(\lambda_* - \lambda_j), \tag{B.8.6}$$

once again establishing both that $\{\lambda_j\}$ converges to λ_* and that the convergence is at least linear. Ultimate superlinear convergence follows from (B.8.6) since $\theta'(\xi_j) \to \theta'(\lambda_*)$; a more precise estimate of the Q-rate may be established if θ' is Lipschitz continuous [311, Thm. 11.2.8]. □

The next, technical, result relates to the size of the step that results from applying Newton's method to the transformed secular equation (8.2.16) for the trust-region subproblem.

Lemma B.8.2: Monotonicity of transformed secular Newton step

Let $\pi(\lambda)$ be the secular function (8.2.4); suppose that $\lambda_C > \lambda_S$, where λ_* is the root of $\pi(\lambda) = \Delta^2$, and let $\Psi_\beta(\lambda) = [\pi(\lambda)]^\beta - \Delta^{2\beta}$. Then the Newton step

$$\delta(\beta) = \left(-\frac{\pi(\lambda_C)}{\pi^{(1)}(\lambda_C)}\right)\frac{1-\mu^\beta}{\beta}, \text{ where } \mu = \frac{\Delta^2}{\pi(\lambda_C)} \equiv \left(\frac{\Delta}{\|s(\lambda_C)\|_M}\right)^2, \tag{B.8.7}$$

for $\Psi_\beta(\lambda)$ from λ_C decreases monotonically as a function of β.

Proof. The Newton step is

$$\delta(\beta) \stackrel{\text{def}}{=} -\Psi_\beta(\lambda_C)/\Psi_\beta^{(1)}(\lambda_C),$$

where

$$\Psi_\beta^{(1)}(\lambda_C) = \beta\pi(\lambda_C)^{\beta-1}\pi^{(1)}(\lambda_C),$$

which leads directly to (B.8.7). But

$$\xi(\beta) \stackrel{\text{def}}{=} \frac{1-\mu^\beta}{\beta}$$

decreases monotonically on $\mathbb{R}$, since

$$\xi'(\beta) = \frac{\exp(\beta\ln\mu)}{\beta^2}[1-\beta\ln\mu - \exp(-\beta\ln\mu)] \leq 0,$$

which follows because $1-t \leq \exp(-t)$ for all t. □

Proofs of results from Chapter 8

Proof of Lemma 8.2.1 (p. 220)

The continuous function $\Psi(\lambda) \stackrel{\text{def}}{=} \pi(\lambda) - \tau(\lambda)$ decreases monotonically throughout (λ_S, ∞) and is strictly positive at λ_S but strictly negative as $\lambda \longrightarrow \infty$. Thus it has a unique root in (λ_S, ∞) by the intermediate-value theorem. □

Proof of Theorem 8.2.2 (p. 221)

It follows trivially from (8.2.7) that the jth derivative, $\pi^{(j)}(\lambda)$, of $\pi(\lambda)$ is

$$\pi^{(j)}(\lambda) = (-1)^j(j+1)!\sum_{i=1}^n \frac{\gamma_i^2}{(\lambda+\lambda_i)^{j+2}}. \tag{B.8.8}$$

Thus Taylor's theorem A.8.2 (p. 435) gives that the error

$$\pi(\lambda_C + \delta) - t_k(\delta) = \frac{1}{(k+1)!}\pi^{(k+1)}(\lambda_C + \xi)\delta^{k+1} \tag{B.8.9}$$

for some ξ strictly between 0 and δ so long as $\delta > \lambda_s - \lambda_C$. But since (B.8.8) shows that even derivatives of $\pi(\lambda)$ are nonnegative and odd derivatives nonpositive for all $\lambda_C > \lambda_S$, (B.8.9) gives (8.2.13) when $\delta > 0$ and

$$\pi(\lambda_C + \delta) \geq t_k(\delta) \quad \text{for all } k$$

when $\lambda_s - \lambda_C < \delta < 0$. This and the relationship

$$t_{k+1}(\delta) - t_k(\delta) = (k+2)(-\delta)^{k+1}\sum_{i=1}^{n}\frac{\gamma_i^2}{(\lambda_C + \lambda_i)^{k+3}},$$

which follows from (B.8.8) and (8.2.12), give (8.2.14) for negative δ.
When $g \neq 0$, at least one of the γ_i in (B.8.8) is nonzero, which is sufficient to ensure that the inequalities that result from (B.8.8) in the above arguments are strict. □

Proof of Lemma 8.2.3 (p. 223)

Differentiating (8.2.7) gives

$$\pi^{(1)}(\lambda) = -2\sum_{i=1}^{n}\frac{\gamma_i^2}{(\lambda+\lambda_i)^3} \quad \text{and} \quad \pi^{(2)}(\lambda) = 6\sum_{i=1}^{n}\frac{\gamma_i^2}{(\lambda+\lambda_i)^4}. \tag{B.8.10}$$

Hence, if $\lambda \in (\lambda_S, \infty)$, (8.2.18) implies that $\pi^{(1)}(\lambda;\beta)$ has the opposite sign to β, since then $\pi(\lambda) > 0$ and $\pi^{(1)}(\lambda) < 0$. Moreover, direct substitution from (B.8.10) and the Cauchy–Schwarz inequality gives

$$\begin{aligned}
&(\beta-1)\,[\pi^{(1)}(\lambda)]^2 + \pi(\lambda)\pi^{(2)}(\lambda) \\
&\quad = 4(\beta-1)\left(\sum_{i=1}^{n}\frac{\gamma_i^2}{(\lambda+\lambda_i)^3}\right)^2 + 6\left(\sum_{i=1}^{n}\frac{\gamma_i^2}{(\lambda+\lambda_i)^4}\right)\left(\sum_{i=1}^{n}\frac{\gamma_i^2}{(\lambda+\lambda_i)^2}\right) \\
&\quad \geq 4(\beta+\tfrac{1}{2})\left(\sum_{i=1}^{n}\frac{\gamma_i^2}{(\lambda+\lambda_i)^3}\right)^2.
\end{aligned}$$

Thus if $\beta > 0$, (8.2.19) implies that $\pi^{(2)}(\lambda;\beta) > 0$, while if $\beta \in [-\frac{1}{2}, 0]$, $\pi^{(2)}(\lambda;\beta) \leq 0$, as required. □

Proof of Theorem 8.2.5 (p. 224)

These results follow directly from the relationship $\pi(\lambda_C + \delta;\beta) = t_k(\delta;\beta) + \pi^{(k+1)}(\lambda_C + \xi_k;\beta)\delta^{k+1}/(k+1)!$ for some ξ_k between 0 and δ (Taylor's theorem), $t_{k+1}(\delta;\beta) = t_k(\delta;\beta) + \pi^{(k+1)}(\lambda_C;\beta)\delta^{k+1}/(k+1)!$ (Taylor series), and Lemma 8.2.3. □

Proof of Theorem 8.2.6 (p. 226)

It follows immediately by differentiating (8.2.2) that $(H + \lambda M)s^{(1)}(\lambda) = -Ms(\lambda)$ and then by induction and continued differentiation that (8.2.29) holds. Now suppose that

$$\pi^{(2k)} = \alpha_k \left\langle s^{(k)}(\lambda), Ms^{(k)}(\lambda) \right\rangle ;$$

this is true for $k = 0$ by definition. Differentiating gives (8.2.27), and a second differentiation reveals

$$\pi^{(2k+2)}(\lambda) = 2\alpha_k \left[\left\langle s^{(k+1)}(\lambda), Ms^{(k+1)}(\lambda) \right\rangle + \left\langle s^{(k)}(\lambda), Ms^{(k+2)}(\lambda) \right\rangle \right]. \quad \text{(B.8.11)}$$

But it follows from (8.2.29) that

$$\begin{aligned}(k+1) \left\langle s^{(k)}(\lambda), Ms^{(k+2)}(\lambda) \right\rangle &= - \left\langle s^{(k+1)}(\lambda), (H + \lambda M)s^{(k+2)}(\lambda) \right\rangle \\ &= (k+2) \left\langle s^{(k+1)}(\lambda), Ms^{(k+1)}(\lambda) \right\rangle ,\end{aligned}$$

and hence (B.8.11) gives

$$\pi^{(2k+2)}(\lambda) = 2\alpha_k \left(1 + \frac{k+2}{k+1}\right) \left\langle s^{(k+1)}(\lambda), Ms^{(k+1)}(\lambda) \right\rangle ,$$

which is (8.2.28) and (8.2.30). □

Appendix to Chapter 9

Proofs of results from Chapter 9

Proof of Corollary 9.1.1 (p. 236)

Since $H + \lambda M = LDL^T$, the definitions of $y^{(k+1)}(\lambda)$, $z^{(k+1)}(\lambda)$, and $s^{(k+1)}(\lambda)$ correspond to solving (8.2.29) by parts. The alternative expression for $\pi^{(2k+1)}(\lambda)$ follows from the identity

$$\begin{aligned}(k+1) \left\langle s^{(k)}(\lambda), Ms^{(k+1)}(\lambda) \right\rangle &= - \left\langle L^T s^{(k+1)}(\lambda), DL^T s^{(k+1)}(\lambda) \right\rangle \\ &= - \left\langle y^{(k+1)}(\lambda), z^{(k+1)}(\lambda) \right\rangle .\end{aligned}$$

The remainder of the result follows immediately from Theorem 8.2.6. □

Proof of Theorem 9.3.2 (p. 243)

Consider $\lambda_C \in \mathcal{F}$ and $s(\lambda)$ from (8.2.2). Then it follows directly from (8.2.2) that

$$\langle g, s(\lambda_C) \rangle = -\langle s(\lambda_C), (H + \lambda_C M)s(\lambda_C) \rangle ,$$

and hence from (8.1.4) we have that

$$\begin{aligned} m^{\mathrm{R}}(s(\lambda_{\mathrm{C}}),\sigma,r) &= \langle g, s(\lambda_{\mathrm{C}})\rangle + \tfrac{1}{2}\langle s(\lambda_{\mathrm{C}}), Hs(\lambda_{\mathrm{C}})\rangle + \tfrac{1}{r}\sigma\|s(\lambda_{\mathrm{C}})\|_M^r \\ &= -\langle s(\lambda_{\mathrm{C}}), (H+\lambda_{\mathrm{C}}M)s(\lambda_{\mathrm{C}})\rangle + \tfrac{1}{2}\langle s(\lambda_{\mathrm{C}}), Hs(\lambda_{\mathrm{C}})\rangle + \tfrac{1}{r}\sigma\|s(\lambda_{\mathrm{C}})\|_M^r \\ &= -\tfrac{1}{2}\langle s(\lambda_{\mathrm{C}}), (H+\lambda_{\mathrm{C}}M)s(\lambda_{\mathrm{C}})\rangle - \tfrac{1}{2}\lambda_{\mathrm{C}}\langle s(\lambda_{\mathrm{C}}), Ms(\lambda_{\mathrm{C}})\rangle + \tfrac{1}{r}\sigma\|s(\lambda)\|_M^r \\ &= -\tfrac{1}{2}\langle s(\lambda_{\mathrm{C}}), (H+\lambda_{\mathrm{C}}M)s(\lambda_{\mathrm{C}})\rangle + \|s(\lambda_{\mathrm{C}})\|_M^2\left(\tfrac{1}{r}\sigma\|s(\lambda_{\mathrm{C}})\|_M^{r-2} - \tfrac{1}{2}\lambda_{\mathrm{C}}\right). \end{aligned}$$

Thus (9.3.14a) is satisfied with $s = s(\lambda_{\mathrm{C}})$ whenever (9.3.15a) holds. Similarly, (8.2.2) gives

$$\begin{aligned} \nabla_s^1 m^{\mathrm{R}}(s(\lambda_{\mathrm{C}}),\sigma,r) &= g + Hs(\lambda_{\mathrm{C}}) + \sigma\|s(\lambda_{\mathrm{C}})\|_M^{r-2}s(\lambda_{\mathrm{C}}) \\ &= g + (H+\lambda_{\mathrm{C}}M)s(\lambda_{\mathrm{C}}) + (\sigma\|s(\lambda_{\mathrm{C}})\|_M^{r-2} - \lambda_{\mathrm{C}})Ms(\lambda_{\mathrm{C}}) \\ &= \left(\sigma\|s(\lambda_{\mathrm{C}})\|_M^{r-2} - \lambda_{\mathrm{C}}\right)Ms(\lambda_{\mathrm{C}}). \end{aligned}$$

Hence

$$\|\nabla_s^1 m^{\mathrm{R}}(s(\lambda_{\mathrm{C}}),\sigma,r)\|_{M^{-1}} = |\sigma\|s(\lambda_{\mathrm{C}})\|_M^{r-2} - \lambda_{\mathrm{C}}|\|s(\lambda_{\mathrm{C}})\|_M,$$

and (9.3.14b) holds at $s = s(\lambda_{\mathrm{C}})$ whenever

$$\left|\sigma\|s(\lambda_{\mathrm{C}})\|_M^{r-2} - \lambda_{\mathrm{C}}\right| \leq \tfrac{1}{2}\theta_1\|s(\lambda_{\mathrm{C}})\|_M,$$

which is equivalent to (9.3.15b). Finally, (8.2.2) gives

$$\begin{aligned} \nabla_s^2 m^{\mathrm{R}}(s(\lambda_{\mathrm{C}}),\sigma,r) &= H + \sigma\|s(\lambda_{\mathrm{C}})\|_M^{r-2}M \\ &\quad + (r-2)\sigma\|s(\lambda_{\mathrm{C}})\|_M^{r-2}\left(\frac{Ms(\lambda_{\mathrm{C}})}{\|s(\lambda_{\mathrm{C}})\|_M}\right)\left(\frac{Ms(\lambda_{\mathrm{C}})}{\|s(\lambda_{\mathrm{C}})\|_M}\right)^T \\ &= H + \lambda_{\mathrm{C}}M + \left(\sigma\|s(\lambda_{\mathrm{C}})\|_M^{r-2} - \lambda_{\mathrm{C}}\right)M \\ &\quad + (r-2)\sigma\|s(\lambda_{\mathrm{C}})\|_M^{r-2}\left(\frac{Ms(\lambda_{\mathrm{C}})}{\|s(\lambda_{\mathrm{C}})\|_M}\right)\left(\frac{Ms(\lambda_{\mathrm{C}})}{\|s(\lambda_{\mathrm{C}})\|_M}\right)^T. \end{aligned}$$

Since $H + \lambda M$ is positive semidefinite for all $\lambda_{\mathrm{C}} \in \mathcal{F}$ and because the trailing rank-one term involving $(Ms(\lambda_{\mathrm{C}}))(Ms(\lambda_{\mathrm{C}}))^T$ shifts eigenvalues to the right, (9.3.14c) holds at $s = s(\lambda_{\mathrm{C}})$ whenever

$$\sigma\|s(\lambda_{\mathrm{C}})\|_M^{r-2} - \lambda_{\mathrm{C}} \geq -\theta_2\|s(\lambda_{\mathrm{C}})\|_M$$

or, equivalently, (9.3.15c) holds. □

Appendix to Chapter 10

We give a useful bound on the entries of the inverse of a positive-definite tridiagonal matrix.

Lemma B.10.1: Inverse of a positive-definite tridiagonal matrix

Let $T \in \mathbb{R}^{m \times m}$ be a symmetric, positive-definite tridiagonal matrix. Then

$$\left|(T^{-1})_{i,j}\right| \leq \frac{2}{\lambda_{\min}(T)} \left(\frac{\sqrt{\kappa(T)}-1}{\sqrt{\kappa(T)}+1}\right)^{|i-j|}$$

for all $i, j = 1, \ldots, m$.

Proof. This follows from a more general result by Demko, Moss, and Smith [145, Thm. 2.4], namely, that if $G \in \mathbb{R}^{m \times m}$ is a symmetric, positive-definite matrix with bandwidth[125] $\beta > 0$, then

$$\left|(G^{-1})_{i,j}\right| \leq \frac{1}{\lambda_{\min}(G)} \max\left(1, \frac{\left(\sqrt{\kappa(G)}+1\right)^2}{2\kappa(G)}\right) \left(\frac{\sqrt{\kappa(G)}-1}{\sqrt{\kappa(G)}+1}\right)^{\frac{|i-j|}{\beta}}$$

for all $i, j = 1, \ldots, m$. The result then follows immediately for the tridiagonal case since then $\beta = 1$ and because $(\sqrt{\kappa(G)}+1)^2 \leq 4\kappa(G)$ since $\kappa(G) \geq 1$. □

Proofs of results from Chapter 10

Proof of Theorem 10.2.2 (p. 255)

[198, Cor. 3.7 when $M = I$]: Since $s_k = Q_k \bar{s}_k$ and $g = My_1 = \gamma_1 M q_1$, we may combine (10.1.4), (10.2.1), and (10.2.2) to deduce that

$$\begin{aligned} \|(H + \lambda_k M)s_k + g\|_{M^{-1}} &= \gamma_{k+1}|\langle e_k, \bar{s}_k\rangle| \\ &= \gamma_{k+1}\|g\|_{M^{-1}}|\langle e_k, (T_k + \lambda_k I_k)^{-1} e_1\rangle| \end{aligned} \tag{B.10.1}$$

[131, Thm. 7.5.10]. It follows immediately from (8.2.34) and (B.10.1) that

$$\|\nabla_s^1 m^{\mathrm{R}}(s_k, \sigma, r)\|_{M^{-1}} = \gamma_{k+1}\|g\|_{M^{-1}}|\langle e_k, (T_k + \lambda_k I_k)^{-1} e_1\rangle|. \tag{B.10.2}$$

Now noting that $\langle e_k, (T_k + \lambda_k I_k)^{-1} e_1\rangle$ is the $(1, k)$th entry of the inverse of $T_k + \lambda_k I_k$, we may invoke Lemma B.10.1 to deduce that

$$|\langle e_k, (T_k + \lambda_k I_k)^{-1} e_1\rangle| \leq \frac{2}{\lambda_{\min}(T_k + \lambda_k I_k)} \left(\frac{\sqrt{\kappa(T_k + \lambda_k I_k)}-1}{\sqrt{\kappa(T_k + \lambda_k I_k)}+1}\right)^{k-1}.$$

Furthermore, it follows from (10.1.4), (10.2.2), the Rayleigh-quotient inequality,

[125] The *bandwidth* of a symmetric matrix is the number of nonzero lower subdiagonals.

and (10.2.3) that

$$\begin{aligned}\|g\|_{M^{-1}} &= \gamma_1 = \|\gamma_1 e_1\| = \|(T_k + \lambda_k I_k)\bar{s}_k\| \leq \|T_k + \lambda_k I_k\| \|\bar{s}_k\| \\ &= \lambda_{\max}[T_k + \lambda_k I_k] \left(\frac{\lambda_k}{\sigma}\right)^{\frac{1}{r-2}}.\end{aligned} \tag{B.10.3}$$

Finally, since M is symmetric and positive definite, it has a symmetric, positive-definite square root $M^{1/2}$ for which $M = M^{T/2}M^{1/2}$. Thus, since

$$T_{k+1} = Q_{k+1}^T H Q_{k+1} = Q_{k+1}^T M^{T/2}(M^{-T/2}HM^{-1/2})M^{1/2}Q_{k+1}$$

from (10.1.3) and γ_{k+1} is the $(k+1,k)$th entry of T_{k+1}, we have

$$\begin{aligned}\gamma_{k+1} &= \langle e_{k+1}, T_{k+1}e_k\rangle \leq \|T_{k+1}\| \leq \|M^{-T/2}HM^{-1/2}\| \|M^{1/2}Q_{k+1}\|^2 \\ &= \|M^{-1}H\|\end{aligned} \tag{B.10.4}$$

from the Cauchy–Schwarz inequality, the M-orthogonality of the columns of Q_{k+1}, and the similarity of $M^{-1}H$ and $M^{-T/2}HM^{-1/2}$. The desired bound (10.2.9) then follows immediately by combining (B.10.2)–(B.10.4). □

Proof of Corollary 10.2.3 (p. 256)

It follows immediately from (10.1.4) and (10.2.2) that

$$\|g\|_{M^{-1}} = \|(T_k + \lambda_k I_k)\bar{s}_k\| \leq \|T_k + \lambda_k I_k\| \|\bar{s}_k\| = (\theta_{k,k} + \lambda_k)\|s_k\|_M.$$

Moreover, the well-known interlacing bound $\theta_{k,k} \leq \lambda_{\max}[H, M]$ [185, Thm. 8.1.7] together with the inequality $\lambda_k \leq \lambda_*$ from Lemma 10.2.1 implies that $\|g\|_{M^{-1}} \leq (\lambda_{\max}[H, M] + \lambda_*)\|s_k\|_M$ and hence that

$$\|s_k\|_M \geq \frac{\|g\|_{M^{-1}}}{\lambda_{\max}[H, M] + \lambda_*}. \tag{B.10.5}$$

Thus if

$$2\|M^{-1}H\|\kappa_* \left(\frac{\lambda_*}{\sigma}\right)^{\frac{1}{r-2}} \left(\frac{\sqrt{\kappa_*}-1}{\sqrt{\kappa_*}+1}\right)^{k-1} \leq \tfrac{1}{2}\theta_1 \left(\frac{\|g\|_{M^{-1}}}{\lambda_{\max}[H, M] + \lambda_*}\right)^2,$$

it immediately follows from (10.2.9), (B.10.5), and the observation that $(\sqrt{\kappa_*} - 1)/(\sqrt{\kappa_*} + 1)$ is monotonically increasing for $\kappa \geq 1$ that the required stopping condition (9.3.14b) holds. This is true when

$$k = \left\lceil \frac{\log\left(\frac{4\|M^{-1}H\|\kappa_*}{\theta_1}\left(\frac{\lambda_*}{\sigma}\right)^{\frac{1}{r-2}}\left(\frac{\lambda_{\max}[H,M]+\lambda_*}{\|g\|_{M^{-1}}}\right)^2\right)}{\log\left(\frac{\sqrt{\kappa_*}+1}{\sqrt{\kappa_*}-1}\right)} \right\rceil + 1,$$

but we should also note that $\nabla_s^1 m^{\mathrm{R}}(s_k, \sigma, r) = 0$, and hence that (9.3.14b) is automatically satisfied, if $k = \text{grade}(M^{-1}H, M^{-1}g)$. □

Proof of Theorem 10.3.2 (p. 257)

[198, Cor. 3.5, when $M = I$]: The proof mirrors that of Theorem 10.2.2 except that now

$$\|g\|_{M^{-1}} \leq \|T_k + \lambda_k I_k\| \|\bar{s}_k\| \leq \|T_k + \lambda_k I_k\| \Delta$$

holds instead of (B.10.3) because of the trust-region constraint. □

Proof of Corollary 10.3.3 (p. 257)

[198, Cor. 3.6, when $M = I$]: The result follows immediately as a result of

$$2\|M^{-1}H\|\kappa_* \Delta \left(\frac{\sqrt{\kappa_*} - 1}{\sqrt{\kappa_*} + 1} \right)^{k-1} \leq \epsilon$$

from Theorem 10.3.2, the monotonic nature of $(\sqrt{\kappa_*} - 1)/(\sqrt{\kappa_*} + 1)$, and the required stopping test (10.3.2). □

Appendix to Chapter 12

The following result justifies the use of the third-order criticality measure $\phi^{\delta}_{f,3}(x)$ in Section 12.1.3.

Lemma B.12.1: $\phi^{\delta}_{f,3}(x) = 0$ implies $\phi^{\delta}_{f,2}(x) = 0$

Given any vector g, symmetric matrix H, and cubically symmetric tensor T, let $\psi_2(s) = \langle g, s \rangle + \frac{1}{2}\langle s, Hs \rangle$ and $\psi_3(s) = \psi_2(s) + \frac{1}{6}T[s]^3$. Suppose that $\min_{s:\|s\| \leq \Delta} \psi_3(s) = 0$. Then $\min_{s:\|s\| \leq \Delta} \psi_2(s) = 0$.

Proof. By assumption we have that

$$\psi_3(s) \geq \min_{s:\|s\| \leq \Delta} \psi_3(s) = 0 \tag{B.12.1}$$

for any s for which $\|s\| \leq \Delta$. Clearly $\min_{s:\|s\| \leq \Delta} \psi_2(s) \leq 0$ since $\psi_2(0) = 0$. Suppose therefore that $s_\Delta = \arg\min_{s:\|s\| \leq \Delta} \psi_2(s)$ and that

$$\psi_2(s_\Delta) < 0. \tag{B.12.2}$$

It then follows from (B.12.1) with $s = s_\Delta$ and (B.12.2) that

$$\begin{aligned} 0 \leq \psi_3(s_\Delta) &= \langle g, s_\Delta \rangle + \tfrac{1}{2}\langle s, Hs_\Delta \rangle + \tfrac{1}{6}T[s_\Delta]^3 \\ &= \psi_2(s_\Delta) + \tfrac{1}{6}T[s_\Delta]^3 < \tfrac{1}{6}T[s_\Delta]^3, \end{aligned} \tag{B.12.3}$$

and thus we have that $T[s_\Delta]^3 > 0$. In addition, using (B.12.1) with $s = -s_\Delta$ gives

$$0 \leq \psi_3(-s_\Delta) = -\langle g, s_\Delta \rangle + \tfrac{1}{2}\langle s, Hs_\Delta \rangle - \tfrac{1}{6}T[s_\Delta]^3,$$

and, adding this to (B.12.3), we find that $\langle s, Hs_\Delta\rangle \geq 0$. Now let

$$\alpha_* \stackrel{\text{def}}{=} \min\left(1, \sqrt{\frac{-3\psi_2(s_\Delta)}{T[s_\Delta]^3}}\right).$$

Then using (B.12.1) with $s = \alpha s_\Delta$ for any $\alpha \in (0, \alpha_*]$ gives

$$\begin{aligned}
0 \leq \frac{\psi_3(\alpha s_\Delta)}{\alpha} &= \langle g, s_\Delta\rangle + \tfrac{1}{2}\alpha\langle s, Hs_\Delta\rangle + \tfrac{1}{6}\alpha^2 T[s_\Delta]^3 \\
&\leq \langle g, s_\Delta\rangle + \tfrac{1}{2}\langle s, Hs_\Delta\rangle + \tfrac{1}{6}\alpha^2 T[s_\Delta]^3 \\
&= \psi_2(s_\Delta) + \tfrac{1}{6}\alpha^2 T[s_\Delta]^3 \\
&\leq \tfrac{1}{2}\psi_2(s_\Delta) \\
&< 0,
\end{aligned}$$

which is a contradiction. Thus (B.12.2) cannot happen. □

Proofs of results from Chapter 12

Proof of Theorem 12.1.7 (p. 287)

Suppose that (12.1.31) does not hold; that is, for some d_{j-1} with $\|d_{j-1}\| \leq \delta_j$,

$$\epsilon_{j-1}\frac{\delta_j^{j-1}}{(j-1)!} < \phi_{f,j-1}^{\delta_j}(x) = f(x) - T_{f,j-1}(x, d_{j-1}) = T_{f,j-1}(x,0) - T_{f,j-1}(x, d_{j-1}).$$

Then, by the definition (12.1.10),

$$\begin{aligned}
\phi_{f,j}^{\delta_j}(x) &\geq T_{f,j}(x,0) - T_{f,j}(x, d_{j-1}) \\
&\geq T_{f,j-1}(x,0) - T_{f,j-1}(x, d_{j-1}) - \frac{\delta_j^j\|\nabla_x^j f(x)\|}{j!} \\
&= \phi_{f,j-1}^{\delta_j}(x) - \frac{\delta_j^j\|\nabla_x^j f(x)\|}{j!} \\
&> \epsilon_{j-1}\frac{\delta_j^{j-1}}{(j-1)!} - \frac{\delta_j^j\|\nabla_x^j f(x)\|}{j!} \\
&\geq \epsilon_j\frac{\delta_j^{j-1}}{(j-1)!} - \frac{\delta_j^j\|\nabla_x^j f(x)\|}{j!}.
\end{aligned}$$

Suppose now that δ_j is small enough to ensure that $\delta_j \leq \min[1, j\epsilon_j/2\|\nabla_x^j f(x)\|]$. Then, using the facts that $\delta_j \leq 1$ and $j \geq 2$,

$$\phi_{f,j}^{\delta_j}(x) > \tfrac{1}{2}\epsilon_j\frac{\delta_j^{j-1}}{(j-1)!} \geq \tfrac{1}{2}j\epsilon_j\frac{\delta_j^j}{j!} \geq \epsilon_j\frac{\delta_j^j}{j!},$$

contradicting (12.1.30), which is impossible. Hence (12.1.31) holds. □

Proof of Theorem 12.2.6 (p. 299)

We achieve the theorem's conclusion by setting $\epsilon_j = j!\,\epsilon$ for $\epsilon \in (0,1]$, $\theta = 1$, and $j \in [1:q]$ and building a sufficiently smooth function $f_{\rm STRq}$ from $\mathbb{R}$ to $\mathbb{R}$ such that, when applied to minimize this function, the TRq algorithm (starting from the origin) needs exactly $\epsilon^{-(q+1)}$ evaluations of f, $\nabla_x^1 f, \ldots, \nabla_x^q f$ to produce an iterate such that (12.1.25) holds. Let us define

$$k_\epsilon = \lceil \epsilon^{-(q+1)} \rceil \tag{B.12.4}$$

and the sequences $\{f_k\}_{k=0}^{k_\epsilon}$ and $\{D_{j,k}\}_{k=0}^{k_\epsilon}$ by

$$f_0 = 1, \quad f_{k+1} \stackrel{\rm def}{=} f_k - \tfrac{1}{2}\epsilon^{q+1}\left(2 - \frac{k}{k_\epsilon}\right), \tag{B.12.5}$$

$$D_{j,k} = 0 \;\text{ for }\; j \in [1:q-1], \quad \text{and} \quad D_{q,k} = -\epsilon\left(2 - \frac{k}{k_\epsilon}\right) q!. \tag{B.12.6}$$

As we have done before, these are intended to specify the objective function and derivative values at successive iterates x_k generated by the TRq algorithm. We also choose $\Delta_k = \epsilon = \min[1, \Delta_k] = \delta_k$ for all $k \in [1:k_\epsilon]$. Immediately observe that, for each $k \in [0:k_\epsilon - 1]$, (B.12.5) and (B.12.6) imply that $s_k = \Delta_k$ and thus that this algorithm will terminate exactly at iteration k_ϵ since $\phi_{f,j}^{\delta_k}(x_k) = 0$ for all $j \in [1:q-1]$ and all $k \in [1:k_\epsilon]$, while

$$\phi_{f,q}^{\delta_k}(x_k) = \frac{|D_{q,k}|}{q!}\delta_k^q > \epsilon\delta_k^q \quad \text{for} \quad k \in [0:k_\epsilon - 1]$$

and $\phi_{f,q}^{\delta_{k_\epsilon}}(x_{k_\epsilon}) = \epsilon\delta_{k_\epsilon}^q$. Also note that $\{f_k\}_{k=0}^{k_\epsilon}$ is monotonically decreasing and that

$$f_{k_\epsilon} = 1 - \sum_{k=0}^{k_\epsilon - 1} \tfrac{1}{2}\epsilon^{q+1}\left(2 - \frac{k}{k_\epsilon}\right) = 1 - \tfrac{1}{2}\epsilon^{q+1}\left(2k_\epsilon - \frac{k_\epsilon - 1}{2}\right) > 1 - \epsilon^{q+1}k_\epsilon \geq 0,$$

where we used the fact that (B.12.4) implies that $\epsilon^{q+1}k_\epsilon \leq 1$ to deduce the last inequality. Hence

$$f_k \in [0,1] \quad \text{for} \quad k \in [0:k_\epsilon].$$

Suppose now that $\delta_0 = \epsilon$. Then $s_1 = \epsilon$ and

$$T_{f,q}(x_0, 0) - T_{f,q}(x_0, s_0) = -\frac{1}{q!}D_{q,0}s_0^q = \epsilon^{q+1} = f_0 - f_1.$$

As a consequence, $\rho_0 = 1$, iteration 0 is successful, $x_1 = \epsilon$, and (3.2.12) allows the choice $\Delta_1 = \Delta_0 = \epsilon = \delta_0$. It is easy to verify that the same situation is reproduced at every iteration and therefore that the sequences $\{f_k\}_{k=0}^{k_\epsilon}$ and $\{\{D_{j,k}\}_{j=1}^q\}_{k=0}^{k_\epsilon}$ defined by (B.12.5)–(B.12.6) and $\{x_k\}_{k=0}^{k_\epsilon}$, $\{s_k\}_{k=0}^{k_\epsilon}$, $\{\Delta_k\}_{k=0}^{k_\epsilon}$, and $\{\delta_k\}_{k=0}^{k_\epsilon}$ defined by

$$x_k = (k-1)\epsilon \quad \text{and} \quad s_k = \Delta_k = \epsilon = \delta_k \quad \text{for} \quad k \in [0:k_\epsilon] \tag{B.12.7}$$

can be interpreted as generated by the TRq algorithm, provided that the values

(B.12.5)–(B.12.6) can be interpolated by a sufficiently smooth function. Observing now that

$$f_{k+1} - T_{f,q}(x_k, s_k) = 0 \tag{B.12.8}$$

and that, for $j \in [1:q]$,

$$|D_{j,k+1} - \nabla_s^j T_{f,j}(x_k, s_k)| = \left|0 - \frac{1}{j!} D_{j,k} s_k^{q-j}\right| \le \epsilon^{q-j+1}\left(2 - \frac{k}{k_\epsilon}\right) \le 2 s_k^{q-j+1}, \tag{B.12.9}$$

the smoothness is guaranteed by Theorem A.9.2 (p. 441) with $\kappa_f = 2$ (see the bounds (B.12.8)–(B.12.9) above, (B.12.5), and (B.12.7)). We therefore deduce that TRq applied on the resulting interpolated function f_{STRq} will stop exactly at iteration k_ϵ. Moreover, this function has bounded range and q first derivatives, as well as Lipschitz-continuous qth derivative. Assumptions **f.Bb** and **f.D1qL** therefore hold, and we may deduce the desired conclusion. □

Appendix to Chapter 13

Proofs of results from Chapter 13

Proof of Lemma 13.1.1 (p. 331)

First note that the trust-region radius is unmodified at accuracy-improving iterations ($k \in \mathcal{A}$). Now (13.1.17) and our assumption imply that

$$\Delta_{i+1} \le \gamma_3 \Delta_i, \quad i \in \mathcal{S}_k, \quad \text{and} \quad \Delta_{i+1} \le \gamma_2 \Delta_i, \quad i \in \mathcal{U}_k.$$

Using our assumption, we thus deduce inductively that

$$\Delta_{\min} \le \Delta_k \le \Delta_0 \gamma_3^{|\mathcal{S}_k|} \gamma_2^{|\mathcal{U}_k|},$$

which gives that

$$\gamma_3^{|\mathcal{S}_k|} \gamma_2^{|\mathcal{U}_k|} \ge \frac{\Delta_{\min}}{\Delta_0},$$

and we obtain inequality (13.1.22) by taking logarithms on both sides and recalling that $\gamma_2 \in (0,1)$ and that, from (13.1.20), $|\mathcal{T}_k| = |\mathcal{S}_k| + |\mathcal{U}_k|$. □

Proof of Lemma 13.1.4 (p. 333)

Let k_0 be the smallest $k \in \mathcal{T}$ (if any), and thus $k \in \mathcal{A}$ and $\mathcal{T}_k = \emptyset$ for all $k \in [0:k_0)$. Consider any iteration $k \in [0:k_0)$. Then the mechanism of the algorithm implies that $\Delta_k = \Delta_0$, and thus (13.1.30) holds for all $k \in [0:k_0)$, since $\Delta_0 \ge \vartheta$ and $\gamma_1 < 1$.
This then proves the lemma if k_0 is infinite, so suppose otherwise and consider iteration k_0. Since all previous iterations passed through Step 5, we have $\Delta_{k_0} = \Delta_0$ and $x_{k_0} = x_0$, and combining this with (13.1.2) gives

$$\|\overline{\nabla_x^i f}(x_{k_0})\| \le \|\nabla_x^i f(x_{k_0})\| + \kappa_{\nabla,i} h_{k,i} \le \|\nabla_x^i f(x_0)\| + \kappa_{\nabla,\max}\kappa_h$$

for all $i \in [1:q]$. Thus, using (13.1.25), (13.1.5), $x_{k_0} = x_0$, and $\delta_{k_0,j} = \delta_{0,j} \le 1$, we have that

$$
\begin{aligned}
\widehat{\phi}_{f,k_0} &\le \max_{j\in[1:q]} \frac{j!\,\overline{\Delta T}_{f,j}(x_{k_0}, d_{k_0,j})}{\delta_{k_0}^j} \\
&\le q \max_{j\in[1:q]} \frac{\max_{i\in[1:j]} \|\overline{\nabla_x^i f}(x_{k_0})\|}{\delta_{k_0}^j} \sum_{i=1}^{j} \frac{\delta_{k_0}^i}{i!} \\
&\le q \,\frac{\max_{i\in[1:q]} \|\overline{\nabla_x^i f}(x_{k_0})\|}{\min_{j\in[1:q]} \delta_{k_0}^j} \left(\exp(\delta_{k_0}) - 1\right) \\
&\le 2q \,\frac{\max_{i\in[1:q]} \|\nabla_x^i f(x_0)\| + \kappa_{\nabla,\max}\kappa_h}{\min_{j\in[1:q]} \delta_0^j}
\end{aligned}
$$

and therefore

$$
\kappa_\Delta \le \min\left[\vartheta, \frac{\Delta_0}{\widehat{\phi}_{f,k_0}}\right] \le \frac{\Delta_0}{\widehat{\phi}_{f,k_0}} = \frac{\Delta_{k_0}}{\widehat{\phi}_{f,k_0}} \tag{B.13.1}
$$

since $\Delta_{k_0} = \Delta_0$. Hence (B.13.1) ensures that (13.1.30) holds for $k = k_0$ since additionally $\Delta_0 \ge \vartheta$ and $\gamma_1 < 1$.
It remains to consider indices $k > k_0$, for which it follows that $\mathcal{T}_k \neq \emptyset$ by definition of k_0. Suppose that a particular $k > k_0$ is the index of the first iteration for which (13.1.30) is violated and thus that

$$
\Delta_k < \kappa_\Delta \min_{i\in\mathcal{T}_k} \widehat{\phi}_{f,i} \;\text{ and }\; \Delta_k < \gamma_1\vartheta. \tag{B.13.2}
$$

Notice that we automatically have that

$$
\min_{i\in\mathcal{T}_k} \widehat{\phi}_{f,i} \le \min_{i\in\mathcal{T}_{k-1}} \widehat{\phi}_{f,i} \le \widehat{\phi}_{f,k-1}. \tag{B.13.3}
$$

If $k-1 \notin \mathcal{T}$, Step 5 of TrqIDA ensures that $\Delta_{k-1} = \Delta_k$, and thus (B.13.3) implies that

$$
\Delta_{k-1} = \Delta_k < \kappa_\Delta \min_{i\in\mathcal{T}_k} \widehat{\phi}_{f,i} \le \kappa_\Delta \min_{i\in\mathcal{T}_{k-1}} \widehat{\phi}_{f,i} \;\text{ and }\; \Delta_{k-1} = \Delta_k < \gamma_1\vartheta
$$

Since this contradicts the assumption that iteration k is the first for which (13.1.30) is violated, it must be that $k-1 \in \mathcal{T}$. Then the updating rule (13.1.17) ensures that

$$
\Delta_{k-1} < \frac{\kappa_\Delta}{\gamma_1} \min_{i\in\mathcal{T}_k} \widehat{\phi}_{f,i} \;\text{ and }\; \Delta_{k-1} \le \vartheta. \tag{B.13.4}
$$

Moreover, since (B.13.3) holds, we deduce that

$$
\frac{\Delta_{k-1}}{\widehat{\phi}_{f,k-1}} \le \frac{\Delta_{k-1}}{\min_{i\in\mathcal{T}_{k-1}} \widehat{\phi}_{f,i}} \le \frac{\Delta_{k-1}}{\min_{i\in\mathcal{T}_k} \widehat{\phi}_{f,i}} < \frac{\kappa_\Delta}{\gamma_1}. \tag{B.13.5}
$$

Lemma 13.1.3 and the second part of (B.13.4) then ensure that $\Delta_{k-1} \le \Delta_k$. Using this bound, the second inequality of (B.13.5), and (B.13.2), we deduce that

$$\frac{\Delta_{k-1}}{\min\limits_{i\in\mathcal{T}_{k-1}} \widehat{\phi}_{f,i}} \leq \frac{\Delta_k}{\min\limits_{i\in\mathcal{T}_k} \widehat{\phi}_{f,i}} < \kappa_\Delta \text{ and } \Delta_{k-1} < \gamma_1\vartheta.$$

As a consequence, (13.1.30) is also violated at iteration $k-1$. But this contradicts the assumption that iteration k is the first such that this inequality fails. This latter assumption is thus impossible, and no such iteration can exist.
We use the above lower bound on Δ_k to derive a lower bound on $\delta_k = \min[\Delta_k, \vartheta]$. If $\mathcal{T}_k = \emptyset$, (13.1.30) implies that $\delta_k \geq \gamma_1\vartheta \geq \gamma_1\epsilon_{\min}$. By contrast, if $\mathcal{T}_k \neq \emptyset$, we combine, respectively, the stipulation $\kappa_\Delta \leq \gamma_1 < 1$, (13.1.30), (13.1.31), (13.1.26), (13.1.33), and the requirement that $\vartheta \geq \epsilon_{\min}$ to deduce that

$$\delta_k \geq \min\left(\gamma_1\vartheta, \kappa_\Delta \min_{i\in\mathcal{T}_k} \widehat{\phi}_{f,i}\right) \geq \varsigma\kappa_\delta\epsilon_{\min}, \tag{B.13.6}$$

which is a lower bound than for the $\mathcal{T}_k = \emptyset$ case. Iterations indexed by $\mathcal{A}_k$ may be ignored since they do not change the radius, and thus (B.13.6) holds for all k. □

Proof of Lemma 13.1.5 (p. 333)

Suppose first that $\Delta_k > \vartheta$ and therefore $\delta_k = \vartheta$. Then (13.1.27), (13.1.26), and the bound $\vartheta \geq \epsilon_{\min}$ give that

$$\overline{\Delta T}_{f,j_k}(x_k, s_k) \geq \frac{\vartheta^{j_k}}{j_k!}\widehat{\phi}_{f,k} \geq \frac{\vartheta^q}{q!}\widehat{\phi}_{f,k} \geq \tfrac{1}{2}\varsigma\frac{\epsilon_{\min}^{q+1}}{q!}$$

since $k \in \mathcal{T}_k$, which yields (13.1.34) since $\varsigma\kappa_\delta < \frac{1}{2}\kappa_\Delta < 1$. By contrast, if $\Delta_k \leq \vartheta$, then $\delta_k = \Delta_k$, and (13.1.30) implies that the first inequality in (B.13.6) holds. Therefore, using (13.1.27) and (13.1.26) once again,

$$\begin{aligned}\overline{\Delta T}_{f,j_k}(x_k, s_k) &\geq \varsigma\frac{1}{j_k!}\widehat{\phi}_{f,k}\min\left\{\gamma_1\vartheta, \kappa_\Delta\min_{i\in\mathcal{T}_k}\widehat{\phi}_{f,i}\right\}^{j_k}\\ &\geq \varsigma\frac{1}{q!}\min\left\{\gamma_1\vartheta, \kappa_\Delta\min_{i\in\mathcal{T}_k}\widehat{\phi}_{f,i}\right\}^{q+1}\\ &\geq \varsigma\frac{1}{q!}\min\left\{\gamma_1\vartheta, \tfrac{1}{2}\varsigma\epsilon_{\min}\kappa_\Delta\right\}^{q+1}\end{aligned} \tag{B.13.7}$$

because $j_k \leq q$, and, since $\kappa_\delta < \gamma_1$, $\vartheta \geq \epsilon_{\min}$ and $\varsigma \leq 1$, (B.13.7) in turn implies (13.1.34). If additionally $k \in \mathcal{S}$, then $\rho_k \geq \eta_1$ and (13.1.35) follows from (13.1.34) and (13.1.16). □

Proof of Lemma 13.1.9 (p. 341)

Observe that $\sigma_{k+1} = \sigma_k$ for $k \in \mathcal{A}$. The regularization parameter update (13.1.59) now gives that, for each k,

$$\gamma_1\sigma_j \leq \max[\gamma_1\sigma_j, \sigma_{\min}] \leq \sigma_{j+1}, \quad j \in \mathcal{S}_k, \quad \text{and} \quad \gamma_2\sigma_j \leq \sigma_{j+1}, \quad j \in \mathcal{U}_k.$$

Thus we deduce inductively that

$$\sigma_0 \gamma_1^{|\mathcal{S}_k|} \gamma_2^{|\mathcal{U}_k|} \le \sigma_k.$$

Therefore, using our assumption that $\sigma_k \le \sigma_{\max}$, we deduce that

$$|\mathcal{S}_k| \log \gamma_1 + |\mathcal{U}_k| \log \gamma_2 \le \log \left(\frac{\sigma_{\max}}{\sigma_0} \right),$$

which then implies that

$$|\mathcal{U}_k| \le -|\mathcal{S}_k| \frac{\log \gamma_1}{\log \gamma_2} + \frac{1}{\log \gamma_2} \log \left(\frac{\sigma_{\max}}{\sigma_0} \right),$$

since $\gamma_2 > 1$. The desired result (13.1.62) then follows from the equality $|\mathcal{T}_k| = |\mathcal{S}_k| + |\mathcal{U}_k|$ and the inequality $\gamma_1 < 1$ given by (13.1.51). □

Appendix to Chapter 14

Proofs of results from Chapter 14

Proof of Lemma 14.1.4 (p. 381)

If the algorithm terminates within Step 2, it must be because every global minimizer s_k^* of $m_k(s)$ is such that $m_k(s_k^*) \ge m_k(0)$. In that case, $s_k^* = 0$ is one such global minimizer. If $q = 1$ and h is convex, Lemma 14.1.1 ensures that termination must have happened in Step 1, and termination in Step 2 is thus impossible. Otherwise, we have that, for any $j \in [1:q]$ and for all d,

$$\begin{aligned}
0 &\le m_k(d) - m_k(0) \\
&= \sum_{\ell=1}^{j} \frac{1}{\ell!} \nabla_x^\ell f(x_k)[d]^\ell + \sum_{\ell=j+1}^{p} \frac{1}{\ell!} \nabla_x^\ell f(x_k)[d]^\ell \\
&\qquad + h\left(c(x_k) + \sum_{\ell=1}^{j} \frac{1}{\ell!} \nabla_x^\ell c(x_k)[d]^\ell + \sum_{\ell=j+1}^{p} \frac{1}{\ell!} \nabla_x^\ell c(x_k)[d]^\ell \right) \\
&\qquad + \frac{\sigma_k}{(p+1)!} \|d\|^{p+1} - h\big(c(x_k)\big) \\
&\le \sum_{\ell=1}^{j} \frac{1}{\ell!} \nabla_x^\ell f(x_k)[d]^\ell + \sum_{\ell=j+1}^{p} \frac{1}{\ell!} \nabla_x^\ell f(x_k)[d]^\ell \\
&\qquad + h\left(\sum_{\ell=1}^{j} \frac{1}{\ell!} \nabla_x^\ell c(x_k)[d]^\ell \right) + h\left(\sum_{\ell=j+1}^{p} \frac{1}{\ell!} \nabla_x^\ell c(x_k)[d]^\ell \right) \\
&\qquad + \frac{\sigma_k}{(p+1)!} \|d\|^{p+1},
\end{aligned}$$

where we used the subadditivity of h (ensured by **h.S0**) to derive the last inequality. Hence

$$-\sum_{\ell=1}^{j}\frac{1}{\ell!}\nabla_x^\ell f(x_k)[d]^\ell - h\left(\sum_{\ell=1}^{j}\frac{1}{\ell!}\nabla_x^\ell c(x_k)[d]^\ell\right)$$
$$\leq \sum_{\ell=j+1}^{p}\frac{1}{\ell!}\nabla_x^\ell f(x_k)[d]^\ell + h\left(\sum_{\ell=j+1}^{p}\frac{1}{\ell!}\nabla_x^\ell c(x_k)[d]^\ell\right) + \frac{\sigma_k}{(p+1)!}\|d\|^{p+1}.$$

Using (14.1.4) (p. 372), we may now choose $\delta \in (0,1]$ small enough to ensure that the absolute value of the right-hand side is at most $\epsilon\|d\| \leq \epsilon\delta_{k,j}^j/j!$ for all d with $\|d\| \leq \delta_{k,j}$, which, in view of (14.1.6), implies (14.1.9). □

Proof of Lemma 14.1.6 (p. 381)

Successively using (14.1.15), Corollary A.8.4 (p. 437) applied to f and c, and (14.1.25), we deduce that

$$\begin{aligned}
|\rho_k - 1| &= \left|\frac{w(x_k) - w(x_k+s_k)}{w(x_k) - T_{f,p}(x_k,s_k) - h(T_{c,p}(x_k,s_k))} - 1\right| \\
&= \left|\frac{w(x_k+s_k) - T_{f,p}(x_k,s_k) - h(T_{c,p}(x_k,s_k))}{w(x_k) - T_{f,p}(x_k,s_k) - h(T_{c,p}(x_k,s_k))}\right| \\
&= \left|\frac{f(x_k+s_k) + h(c(x_k+s_k)) - T_{f,p}(x_k,s_k) - h(T_{c,p}(x_k,s_k))}{w(x_k) - T_{f,p}(x_k,s_k) - h(T_{c,p}(x_k,s_k))}\right| \\
&\leq \left|\frac{\dfrac{L_{f,p}\|s_k\|^{p+1}}{(p+1)!} + L_{h,0}\|c(x_k+s_k) - T_{c,p}(x_k,s_k)\|}{w(x_k) - T_{f,p}(x_k,s_k) - h(T_{c,p}(x_k,s_k))}\right| \\
&\leq \left|\frac{\dfrac{L_{f,p} + \sqrt{m}L_{h,0}L_{c,p}}{(p+1)!}\|s_k\|^{p+1}}{\dfrac{\sigma_k}{(p+1)!}\|s_k\|^{p+1}}\right| \\
&= \left|\frac{L_{f,p} + \sqrt{m}L_{h,0}L_{c,p}}{\sigma_k}\right|
\end{aligned}$$

at iteration k. Thus, if $\sigma_k \geq L_{w,p}$, then iteration k is successful, $x_{k+1} = x_k$, and (14.1.16) implies that $\sigma_{k+1} \leq \sigma_k$. The conclusion then follows from the mechanism of (14.1.16). □

Proof of Lemma 14.1.7 (p. 382)

If the algorithm does not terminate at iteration $k+1$, there must exist a $j \in [1:q]$ such that (14.1.9) fails at order j at iteration $k+1$. Consider such a j, and let d be the argument of the minimization in the definition of $\phi_{w,j}^{\delta_{k+1,j}}(x_{k+1})$, so that $\|d\| \le \delta_{k+1,j} \le 1$. The definition of $\phi_{w,j}^{\delta_{k+1,j}}(x_{k+1})$ in (14.1.6) then gives that

$$
\begin{aligned}
\epsilon_j \frac{\delta_{k+1,j}^j}{j!} &< \phi_{w,j}^{\delta_{k+1,j}}(x_{k+1}) \\
&= -\sum_{\ell=1}^{j} \frac{1}{\ell!} \nabla_x^\ell f(x_{k+1})[d]^\ell + h\big(c(x_{k+1})\big) - h\left(\sum_{\ell=0}^{j} \frac{1}{\ell!} \nabla_x^\ell c(x_{k+1})[d]^\ell\right) \\
&= -\sum_{\ell=1}^{j} \frac{1}{\ell!} \nabla_x^\ell f(x_{k+1})[d]^\ell + \sum_{\ell=1}^{j} \frac{1}{\ell!} \nabla_s^\ell T_{f,p}(x_k, s_k)[d]^\ell \\
&\quad + h\big(c(x_{k+1})\big) - h\big(T_{c,p}(x_k, s_k)\big) \\
&\quad - h\left(\sum_{\ell=0}^{j} \frac{1}{\ell!} \nabla_x^\ell c(x_{k+1})[d]^\ell\right) + h\left(\sum_{\ell=0}^{j} \frac{1}{\ell!} \nabla_s^\ell T_{c,p}(x_k, s_k)[d]^\ell\right) \\
&\quad - \sum_{\ell=1}^{j} \frac{1}{\ell!} \nabla_s^\ell T_{f,p}(x_k, s_k)[d]^\ell + h\big(T_{c,p}(x_k, s_k)\big) \\
&\quad - h\left(\sum_{\ell=0}^{j} \frac{1}{\ell!} \nabla_s^\ell T_{c,p}(x_k, s_k)[d]^\ell\right) - \frac{\sigma_k}{(p+1)!} \sum_{\ell=1}^{j} \frac{1}{\ell!} \nabla_s^\ell \left(\|s_k\|^{p+1}\right) [d]^\ell \\
&\quad + \frac{\sigma_k}{(p+1)!} \sum_{\ell=1}^{j} \frac{1}{\ell!} \nabla_s^\ell \left(\|s_k\|^{p+1}\right) [d]^\ell .
\end{aligned}
\tag{B.14.1}
$$

Now, using Corollary A.8.4 (p. 437),

$$
\begin{aligned}
-\sum_{\ell=1}^{j} \frac{1}{\ell!} \nabla_x^\ell f(x_{k+1})[d]^\ell &+ \sum_{\ell=1}^{j} \frac{1}{\ell!} \nabla_s^\ell T_{f,p}(x_k, s_k)[d]^\ell \\
&\le \sum_{\ell=1}^{j} \frac{\delta_{k+1,j}^\ell}{\ell!} \left\|\nabla_x^\ell f(x_{k+1}) - \nabla_s^\ell T_{f,p}(x_k, s_k)\right\| \\
&\le L_{f,p} \sum_{\ell=1}^{j} \frac{\delta_{k+1,j}^\ell}{\ell!(p-\ell+1)!} \|s_k\|^{p-\ell+1}.
\end{aligned}
\tag{B.14.2}
$$

In the same spirit, also using **h.L** and applying Corollary A.8.4 (p. 437) to c, we obtain that

$$
\begin{aligned}
h\big(c(x_{k+1})\big) - h\big(T_{c,p}(x_k, s_k)\big) &\le L_{h,0} \|c(x_{k+1}) - T_{c,p}(x_k, s_k)\| \\
&\le \frac{\sqrt{m} L_{h,0} L_{c,p}}{(p+1)!} \|s_k\|^{p+1}
\end{aligned}
\tag{B.14.3}
$$

and that

$$\begin{aligned}
-h\left(\sum_{\ell=0}^{j}\frac{1}{\ell!}\nabla_x^\ell c(x_{k+1})[d]^\ell\right) + h\left(\sum_{\ell=0}^{j}\frac{1}{\ell!}\nabla_s^\ell T_{c,p}(x_k,s_k)[d]^\ell\right)& \\
\leq L_{h,0}\left\|\sum_{\ell=0}^{j}\frac{1}{\ell!}\left[\nabla_x^\ell c(x_{k+1}) - \nabla_s^\ell T_{c,p}(x_k,s_k)\right][d]^\ell\right\|& \\
\leq L_{h,0}\sum_{\ell=0}^{j}\frac{\delta_{k+1,j}^\ell}{\ell!}\|\nabla_x^\ell c(x_{k+1}) - \nabla_s^\ell T_{c,p}(x_k,s_k)\|& \\
\leq \sqrt{m}L_{h,0}L_{c,p}\sum_{\ell=0}^{j}\frac{\delta_{k+1,j}^\ell}{\ell!(p-\ell+1)!}\|s_k\|^{p-\ell+1}.&
\end{aligned} \tag{B.14.4}$$

In view of (14.1.11) and (14.1.13),

$$\begin{aligned}
-\sum_{\ell=1}^{j}\frac{1}{\ell!}\nabla_s^\ell T_{f,p}(x_k,s_k)[d]^\ell + h(T_{c,p}(x_k,s_k)) - h\left(\sum_{\ell=0}^{j}\frac{1}{\ell!}\nabla_s^\ell T_{c,p}(x_k,s_k)[d]^\ell\right)& \\
-\frac{\sigma_k}{(p+1)!}\sum_{\ell=1}^{j}\frac{1}{\ell!}\nabla_s^\ell\left(\|s_k\|^{p+1}\right)[d]^\ell& \\
\leq \phi_{m_k,j}^{\delta_{s,j}}(s_k) \leq \theta\epsilon_j\frac{\delta_{s,j}^j}{j!} = \theta\epsilon_j\frac{\delta_{k+1,j}^j}{j!},&
\end{aligned} \tag{B.14.5}$$

where the final equality follows because $\delta_{s,j} = \delta_{k+1,j}$ when iteration k is successful. Moreover, just as in (12.2.56),

$$\frac{\sigma_k}{(p+1)!}\sum_{\ell=1}^{j}\frac{1}{\ell!}\nabla_s^\ell\left(\|s_k\|^{p+1}\right)[d]^\ell \leq \sigma_{\max}\sum_{\ell=1}^{j}\frac{\delta_{k+1,j}^\ell}{\ell!(p-j+1)!}\|s_k\|^{p-j+1} \tag{B.14.6}$$

because of Lemma 14.1.6 (p. 381). Substituting (B.14.2)–(B.14.6) into (B.14.1) and using the inequality $(p-\ell+1)! \geq 1$, we obtain (14.1.30). □

Proof of Lemma 14.1.8 (p. 382)

The proof closely follows the pattern of Lemma 12.2.12 (p. 309). Let

$$\xi \stackrel{\text{def}}{=} \left(\frac{1-\theta}{3j!(L_{w,p}+\sigma_{\max})}\right)^{\frac{1}{p-j+1}} \in (0,1).$$

If $\|s_k\| \geq \xi$, (14.1.31) clearly holds for all $j \in [1:q]$ since $\epsilon \leq 1$ and $\kappa_{\delta,\min} < 1$. Assume therefore that $\|s_k\| < \xi$. Because the algorithm has not terminated, Lemma 14.1.7 ensures that (14.1.30) holds for some $j \in [1:q]$. It is easy to verify that this inequality is equivalent to

$$\alpha\,\epsilon\,\delta_{k+1,j}^j \leq \|s_k\|^{p+1}\chi_j\left(\frac{\delta_{k+1,j}}{\|s_k\|}\right) + \beta\|s_k\|^{p+1}, \tag{B.14.7}$$

where the function

$$\chi_j(t) = \sum_{\ell=1}^{j} \frac{t^\ell}{\ell!}$$

is as defined in (12.2.59) and where we have set

$$\alpha = \frac{1-\theta}{j!(L_{w,p}+\sigma_{\max})} \quad \text{and} \quad \beta = \frac{\sqrt{m}}{(p+1)!}\frac{L_{h,0}L_{c,p}}{(L_{w,p}+\sigma_{\max})} \in [0,1),$$

with the last inclusion resulting because $\sqrt{m}L_{h,0}L_{c,p} \le L_w$ from (14.1.27). In particular, since (A.35) gives that $\chi_j(t) \le 2t^j$ for $t \ge 1$ and since $\beta < 1$, we have that

$$\alpha\,\epsilon \le 2\|s_k\|^{p+1}\left(\frac{1}{\|s_k\|}\right)^j + \left(\frac{\|s_k\|}{\delta_{k+1,j}}\right)^j \|s_k\|^{p-j+1} \le 3\|s_k\|^{p-j+1} \qquad \text{(B.14.8)}$$

when $\|s_k\| \le \delta_{k+1,j}$.
If $q = 1$ and h is convex, then by assumption $j = 1$, $\delta_{k+1,j} = 1$, and $\|s_k\| \le \xi < 1 = \delta_{k+1,j}$. Thus (B.14.8) yields the first case of (14.1.31). Otherwise, $q > 1$ or h is nonconvex. Then our assumptions imply that (14.1.28) holds, and since $\|s_k\| < \xi < 1$, we have that $\delta_{k+1,j} = \delta_{s_k,j} \ge \kappa_{\delta,\min}\epsilon_j$. If $\|s_k\| > \delta_{k+1,j} \ge \kappa_{\delta,\min}\epsilon_j$, then (B.14.7), $\beta < 1$, and $\chi_j(t) < 2$ for $t \in [0,1]$ from (A.34) ensure that

$$\alpha\kappa_{\delta,\min}^j\epsilon_j^{j+1} \le \alpha\,\epsilon\,\delta_{k+1,j}^j \le \|s_k\|^{p+1}\chi_j\left(\frac{\delta_{k+1,j}}{\|s_k\|}\right) + \beta\|s_k\|^{p+1} \le 3\|s_k\|^{p+1},$$

and the second case of (14.1.31) holds since obviously $p+1 \ge p-j+1$. By contrast, if $\|s_k\| \le \delta_{k+1,j}$, we may again deduce from (B.14.8) that

$$\|s_k\| \ge \left(\frac{\alpha}{3}\right)^{1/(p-j+1)} \epsilon_j^{1/(p-j+1)}.$$

But $p-j+1 \ge (p+1)/(j+1)$ for $j \in [0:p]$ since $p-j+1-(p+1)/(j+1) = j(p-j)/(j+1)$, and thus

$$\|s_k\| \ge \left(\frac{\alpha}{3}\right)^{1/(p-j+1)} \epsilon_j^{(j+1)/(p+1)},$$

which once again is the second case of (14.1.31). □

Proof of Lemma 14.3.6 (p. 416)

Denote the Lipschitz constants for $\|\nabla_x^j c_i(x)\|$ and the upper bounds on $\|\nabla_x^j c_i(x)\|$ that are ensured by **c.D1pLB.X** for each $j \in [1:p]$ by $L_{c_i,j}$ and $B_{c_i,j}$, respectively. Since $\nu(x) = \frac{1}{2}\sum_{i=1}^m c_i^2(x)$, ν is clearly as differentiable as c, and it is easy to show by induction that

$$\nabla_x^j \nu(x) = \sum_{i=1}^m \left[\sum_{u,v>0,u+v=j} \alpha_{uv}\nabla_x^u c_i(x) \otimes \nabla_x^v c_i(x) + c_i(x)\nabla_x^j c_i(x)\right], \qquad \text{(B.14.9)}$$

where the α_{uv} are suitable nonnegative finite coefficients. To establish (14.3.60), we examine each of the terms in (B.14.9). The triangle inequality gives that

$$\begin{aligned}&\|\nabla_x^u c_i(x) \otimes \nabla_x^v c_i(x) - \nabla_x^u c_i(y) \otimes \nabla_x^v c_i(y)\| \\ &\quad \leq \|\nabla_x^u c_i(x) - \nabla_x^u c_i(y)\| \|\nabla_x^v c_i(x)\| + \|\nabla_x^u c_i(y)\| \left[\|\nabla_x^v c_i(x) - \nabla_x^v c_i(y)\|\right]\end{aligned}$$

and thus, since the first term on the right-hand side is bounded by $L_{c_i,u}B_{c_i,v}\|x-y\|$ and the second by $B_{c_i,u}L_{c_i,v}\|x-y\|$ because of **c.D1pLB.X**,

$$\begin{aligned}&\|\nabla_x^u c_i(x) \otimes \nabla_x^v c_i(x) - \nabla_x^u c_i(y) \otimes \nabla_x^v c_i(y)\| \\ &\quad \leq (L_{c_i,u}B_{c_i,v} + L_{c_i,v}B_{c_i,u})\|x-y\|. \end{aligned} \tag{B.14.10}$$

Moreover, using the triangle inequality, the mean-value theorem, and **c.D1pLB.X**,

$$\begin{aligned}&\|c_i(x)\nabla_x^j c_i(x) - c_i(y)\nabla_x^j c_i(y)\| \\ &\quad \leq |c_i(x)|(\|\nabla_x^j c_i(x) - \nabla_x^j c_i(y)\|) + |c_i(x) - c_i(y)|\|\nabla_x^j c_i(y)\| \\ &\quad \leq |c_i(x)|(\|\nabla_x^j c_i(x) - \nabla_x^j c_i(y)\|) + \|\nabla_x^1 c_i(z_i)\|\|x-y\|\|\nabla_x^j c_i(x)\| \\ &\quad \leq \Big(|c_i(x)|L_{c_i,j} + B_{c_i,1}B_{c_i,j}\Big)\|x-y\| \\ &\quad \leq \Big(\|c(x)\|L_{c_i,j} + B_{c_i,1}B_{c_i,j}\Big)\|x-y\| \end{aligned} \tag{B.14.11}$$

for some z_i between x and y. But since $x \in \mathcal{L}_\nu(x_0)$, it follows that

$$\|c_i(x)\nabla_x^j c_i(x) - c_i(y)\nabla_x^j c_i(y)\| \leq \Big(B_{c_i,1}B_{c_i,j} + \|c(x_0)\|L_{c_i,j}\Big)\|x-y\|. \tag{B.14.12}$$

Combining (B.14.9), (B.14.10), and (B.14.12), we deduce that (14.3.60) holds with

$$L_{\nu,j}^{\mathrm{A}} \stackrel{\mathrm{def}}{=} \sum_{i=1}^m \left[\sum_{u,v>0,u+v=j} \alpha_{uv}(L_{c_i,u}B_{c_i,v} + L_{c_i,v}B_{c_i,u}) + B_{c_i,1}B_{c_i,j}\right]$$

and

$$L_{\nu,j}^{\mathrm{B}} \stackrel{\mathrm{def}}{=} \sum_{i=1}^m L_{c_i,j}.$$

It also follows immediately from (B.14.9), the triangle inequality, and the bounds on the derivatives that

$$\begin{aligned}\|\nabla_x^j \nu(x)\| &\leq \sum_{i=1}^m \left[\sum_{u,v>0,u+v=j} \alpha_{uv}\|\nabla_x^u c_i(x)\|\|\nabla_x^v c_i(x)\| + |c_i(x)|\|\nabla_x^j c_i(x)\|\right] \\ &\leq \sum_{i=1}^m \left[\sum_{u,v>0,u+v=j} \alpha_{uv}B_{c_i,u}B_{c_i,v} + \|c(x)\|B_{c_i,j}\right]. \end{aligned} \tag{B.14.13}$$

Again, since $x \in \mathcal{L}_\nu(x_0)$, we deduce that

$$\|\nabla_x^j \nu(x)\| \leq \sum_{i=1}^m \left[\sum_{u,v>0,u+v=j} \alpha_{uv}B_{c_i,u}B_{c_i,v} + \|c(x_0)\|B_{c_i,j}\right],$$

which is (14.3.61) with

$$B^{\mathrm{A}}_{\nu,j} \stackrel{\text{def}}{=} \sum_{i=1}^{m} \left[\sum_{u,v>0,u+v=j} \alpha_{uv} B_{c_i,u} B_{c_i,v} \right] \quad \text{and} \quad B^{\mathrm{B}}_{\nu,j} \stackrel{\text{def}}{=} \sum_{i=1}^{m} B_{c_i,j}.$$

In order to establish (14.3.62) and (14.3.63), we note that if $x \in \mathcal{L}_\mu$, then (14.3.30) guarantees that

$$\|c(x)\| \leq 1 \quad \text{and} \quad |f(x) - t| \leq 1. \tag{B.14.14}$$

Moreover, $\nu(x)$ and $\mu(x,t)$ differ by the term $\frac{1}{2}(f(x) - t)^2$, and as before ν is obviously as differentiable as c and f. Just as for (B.14.9), we have that

$$\nabla^j_x[\tfrac{1}{2}(f(x) - t_j)^2] = \sum_{u,v>0,u+v=j} \alpha_{uv} \nabla^u_x f(x) \otimes \nabla^v_x f(x) + (f(x) - t)\nabla^j_x f(x), \tag{B.14.15}$$

and assumption **f.D1pLB.X** gives that

$$\|\nabla^u_x f(x) \otimes \nabla^v_x f(x) - \nabla^u_x f(y) \otimes \nabla^v_x f(y)\| \leq (L_{f,u} B_{f,v} + L_{f,v} B_{f,u})\|x - y\|, \tag{B.14.16}$$

where we have denoted the upper bound on $\|\nabla^j_x f(y)\|$ for each $j \in [1:p]$ that is guaranteed by the assumption by $B_{f,j}$. Moreover, the triangle inequality and the mean-value theorem reveal that

$$\begin{aligned}
&\|(f(x) - t)\nabla^j_x f(x) - (f(y) - t)\nabla^j_x f(y)\| \\
&\quad \leq |f(x) - t|\|\nabla^j_x f(x) - \nabla^j_x f(y)\| + |f(x) - f(y)|\|\nabla^j_x f(y)\| \\
&\quad = |f(x) - t|\|\nabla^j_x f(x) - \nabla^j_x f(y)\| + \|\nabla^1_x f(z)\|\|x - y\|\|\nabla^j_x f(y)\| \\
&\quad \leq \Big(|f(x) - t| L_{f,j} + B_{f,1} B_{f,j}\Big)\|x - y\|
\end{aligned} \tag{B.14.17}$$

for some z between x and y. But then adding (B.14.9) and (B.14.15), invoking the bounds (B.14.10), (B.14.11), (B.14.16), and (B.14.17), and recalling (B.14.14), we recover (14.3.62) with

$$\begin{aligned}
L_{\mu,j} \stackrel{\text{def}}{=} &\sum_{i=1}^{m} \left[\sum_{u,v>0,u+v=j} \alpha_{uv}(L_{c_i,u} B_{c_i,v} + L_{c_i,v} B_{c_i,u}) + B_{c_i,1} B_{c_i,j} + L_{c_i,j} \right] \\
&+ \sum_{u,v>0,u+v=j} \alpha_{uv}(L_{f,u} B_{f,v} + L_{f,v} B_{f,u}) + B_{f,1} B_{f,j} + L_{f,j}.
\end{aligned}$$

In addition, it follows from (B.14.15), the triangle inequality, and the bounds on the derivatives of f that

$$\begin{aligned}
\|\nabla^j_x[\tfrac{1}{2}(f(x) - t_j)^2]\| &\leq \sum_{u,v>0,u+v=j} \alpha_{uv} \|\nabla^u_x f(x)\|\|\nabla^v_x f(x)\| + |f(x) - t|\|\nabla^j_x f(x)\| \\
&\leq \sum_{u,v>0,u+v=j} \alpha_{uv} B_{f,u} B_{f,v} + |f(x) - t| B_{f,j},
\end{aligned} \tag{B.14.18}$$

and, combining this with (B.14.13) when x satisfies (B.14.14), we infer (14.3.63) with

$$B_{\mu,j} \stackrel{\text{def}}{=} \sum_{i=1}^{m} \left[\sum_{u,v>0,u+v=j} \alpha_{uv} B_{c_i,u} B_{c_i,v} + B_{c_i,j} \right] + \sum_{u,v>0,u+v=j} \alpha_{uv} B_{f,u} B_{f,v} + B_{f,j}.$$

□

Proof of Lemma 14.3.8 (p. 418)

Let $k \in \mathcal{K}_a$. Then for each $j \in [k+1 : k+\ell(k)-1]$, the residuals of the inner iterates $\|r(x_{j,i}, t_j)\|$, $i \in [0 : e_j]$, generated by the INNER algorithm in Step 1, decrease monotonically. Moreover, if $j \in [k+1 : k+\ell(k))$, where j lies in $\mathcal{K}_b$, by construction in Step 1 we have $x_{j,e_j} = x_{j+1} = x_{j+1,0}$, and hence, from (7.2.28) in Lemma 7.2.3,

$$\|r(x_{j,e_j}, t_j)\| = \|r(x_{j+1}, t_j)\| = \|r(x_{j+1}, t_{j+1})\| = \|r(x_{j+1,0}, t_{j+1})\|. \quad \text{(B.14.19)}$$

Thus the entire sequence $\{\|r(x_{j,i}, t_j)\|\}_{(j,i)\in\mathcal{I}_k}$ decreases monotonically if $k \in \mathcal{K}_a$. In addition, since $k \in \mathcal{K}_a$, (7.2.27) (again in Lemma 7.2.3) ensures that $\|r(x_{k+1,0}, t_{k+1})\| = \|r(x_{k+1}, t_{k+1})\| = \epsilon_0$, while $\|r(x_{k+\ell(k),e_{k+\ell(k)}}, t_{k+\ell(k)})\| > \theta\epsilon_0$, because otherwise the $(k+\ell(k))$th inner iteration would have stopped at or before its $(e_{k+\ell(k)}-1)$st iteration. Thus each member of the monotonic sequence of residuals lies in the range

$$\theta\epsilon_0 \leq \|r(x_{j,i}, t_j)\| \leq \epsilon_0, \quad \text{(B.14.20)}$$

and the entire sequence of iterates $x_{j,i}$ lies in $\mathcal{F}_\beta$.
Let $\mathcal{I}_k^{\text{S}}$ be the subset of indices of successful iterations in $\mathcal{I}_k$. If $(j,s) \in \mathcal{I}_k^{\text{S}}$, $k \in \mathcal{K}_a$, is the index of a successful inner iteration, then, so long as the third termination test in (14.3.32) has not been satisfied, we may then apply Lemma 14.2.2—with the appropriate $\kappa_{\text{decr}}^{\mu}$ in (14.2.11) that is independent of t_j because of Lemma 14.3.6(ii)—to show that the decrease produced by an iteration of INNER$(\mu(x,t_j), \mathcal{X}, q, x_j, \beta\|r(x_{j,0}, t_j)\|\epsilon)$ satisfies

$$\begin{aligned}
&\|r(x_{j,s}, t_j)\| \left(\|r(x_{j,s}, t_j)\| - \|r(x_{j,s+1}, t_j)\|\right) \\
&\qquad \geq \tfrac{1}{2}\left[\|r(x_{j,s}, t_j)\| + \|r(x_{j,s+1}, t_j)\|\right]\left[\|r(x_{j,s}, t_j)\| - \|r(x_{j,s+1}, t_j)\|\right] \\
&\qquad \geq \tfrac{1}{2}\|r(x_{j,s}, t_j)\|^2 - \tfrac{1}{2}\|r(x_{j,s+1}, t_j)\|^2 \\
&\qquad = \mu(x_{j,s}, t_j) - \mu(x_{j,s+1}, t_j) \\
&\qquad \geq \kappa_{\text{decr}}^{\mu} \min_{l\in[1:q]} [\beta\|r(x_{j,0}, t_j)\|\epsilon_l]^{\pi_l} \\
&\qquad \geq \kappa_{\text{decr}}^{\mu} \beta^{\pi_{\max}} \min_{l\in[1:q]} \|r(x_{j,s+1}, t_j)\|^{\pi_l} \epsilon_l^{\pi_l},
\end{aligned} \quad \text{(B.14.21)}$$

where $\pi_{\max} = \max_{l\in[1:q]} \pi_l$, for $0 \leq s < e_j$; when $s = e_j$, we have instead

$$\|r(x_{j,e_j}, t_j)\| - \|r(x_{j+1,0}, t_{j+1})\| = 0$$

from (B.14.19).

Now consider two possibilities. Firstly, suppose that for some index $(j,s) \in \mathcal{I}_k^{\mathrm{S}}$,

$$\|r(x_{j,s+1},t_j)\| < \theta\|r(x_{j,s},t_j)\|.$$

Then we know from (7.2.27) (p. 202) and the monotonicity of the sequence of residuals that

$$\|r(x_{j,s},t_j)\| \leq \|r(x_{k+1,0},t_{k+1})\| = \epsilon_0,$$

and thus $\|r(x_{j,s+1},t_j)\| < \theta\epsilon_0$. But in this case, the first inner-iteration termination test in (14.3.32) holds, (j,s) is not in $\mathcal{I}_k$, and $j+1 \in \mathcal{K}_a$. Thus, for all $(j,s) \in \mathcal{I}_k^{\mathrm{S}}$, it must be that

$$\|r(x_{j,s+1},t_j)\| \geq \theta\|r(x_{j,s},t_j)\|$$

for all $(j,s) \in \mathcal{I}_k^{\mathrm{S}}$. In this case, (B.14.21) and the lower bound $\theta\epsilon_0$ on the sequence of residuals given by (B.14.20) give that

$$\begin{aligned}
\|r(x_{j,s},t_j)\| - \|r(x_{j,s+1},t_j)\| &\geq \kappa_{\mathrm{decr}}^{\mu}\beta^{\pi_{\max}} \min_{l\in[1:q]} \epsilon_l^{\pi_l}\frac{\|r(x_{j,s+1},t_j)\|^{\pi_l}}{\|r(x_{j,s},t_j)\|} \\
&\geq \kappa_{\mathrm{decr}}^{\mu}\beta^{\pi_{\max}} \min_{l\in[1:q]} \epsilon_l^{\pi_l}\theta^{\pi_l}\|r(x_{j,s},t_j)\|^{\pi_l-1} \\
&\geq \kappa_{\mathrm{decr}}^{\mu}\beta^{\pi_{\max}} \min_{l\in[1:q]} \theta^{\pi_l}\epsilon_0^{\pi_l-1}\epsilon_l^{\pi_l} \\
&\geq \kappa_{\mathrm{decr}}^{\mu}(\beta\theta)^{\pi_{\max}} \min_{l\in[1:q]} \epsilon_0^{\pi_l-1}\epsilon_l^{\pi_l}
\end{aligned}$$

for all $(j,s) \in \mathcal{I}_k^{\mathrm{S}}$. Then, using the standard telescoping-sum argument and the bounds (B.14.20), we have that

$$\begin{aligned}
(1-\theta)\epsilon_0 &\geq \|r(x_{k+1,0},t_{k+1})\| - \|r(x_{j_k,s_k+1},t_{j_k})\| \\
&\geq \sum_{j,s\in\mathcal{I}_k^{\mathrm{S}}} \left(\|r(x_{j,s},t_j)\| - \|r(x_{j,s+1},t_j)\|\right) \\
&\geq \kappa_{\mathrm{decr}}^{\mu}\beta^{\pi_{\max}}\theta^{\pi_{\max}} \min_{l\in[1:q]} \epsilon_0^{\pi_l-1}\epsilon_l^{\pi_l}|\mathcal{I}_k^{\mathrm{S}}|
\end{aligned}$$

and hence that

$$|\mathcal{I}_k^{\mathrm{S}}| \leq \kappa_{\mathsf{INNER},2}^{\mathrm{S}}(1-\theta)\max_{l\in[1:q]} \epsilon_0^{2-\pi_l}\epsilon_l^{-\pi_l}, \quad \text{where } \kappa_{\mathsf{INNER},2}^{\mathrm{S}} \stackrel{\mathrm{def}}{=} \left[\kappa_{\mathrm{decr}}^{\mu}\right]^{-1}(\beta\theta)^{-\pi_{\max}},$$

which is (14.3.71). Invoking (14.2.13) to account for possible unsuccessful inner iterations, we obtain (14.3.72) with $\kappa_{\mathsf{INNER},2}^{\mathrm{A}} \stackrel{\mathrm{def}}{=} \kappa_{\mathrm{uns}}^{\mu}\kappa_{\mathsf{INNER},2}^{\mathrm{S}}$ and $\kappa_{\mathsf{INNER},2}^{\mathrm{C}} \stackrel{\mathrm{def}}{=} \kappa_{\mathrm{unc}}^{\mu}$, where we note that once again $\kappa_{\mathrm{decr}}^{\mu}$ is independent of t_j because of Lemma 14.3.6(ii). $\square$

Bibliography

The following works have been cited in this book. Some are general references to background material, while others are central to the development of the methods we have covered.

[1] P.-A. ABSIL, R. MAHONY, AND R. SEPULCHRE. *Optimization Algorithms on Matrix Manifolds*. Princeton University Press, Princeton, NJ, 2008.

[2] S. ADACHI, S. IWATA, Y. NAKATSUKASA, AND A. TAKEDA. Solving the trust-region subproblem by a generalized eigenvalue problem. *SIAM Journal on Optimization*, 27(1):269–291, 2017. https://doi.org/10.1137/16M1058200.

[3] N. AGARWAL, N. BOUMAL, B. BULLINS, AND C. CARTIS. Adaptive regularization with cubics on manifolds. *Mathematical Programming*, 188:85–134, 2021. https://doi.org/10.1007/s10107-020-01505-1.

[4] M. A. AJIZ AND A. JENNINGS. A robust incomplete Choleski-conjugate gradient algorithm. *International Journal for Numerical Methods in Engineering*, 20(5):949–966, 1984. https://doi.org/10.1002/nme.1620200511.

[5] E. L. ALLGOWER, K. BÖHMER, F. A. POTRA, AND W. C. RHEINBOLDT. A mesh-independence principle for operator equations and their discretizations. *SIAM Journal on Numerical Analysis*, 23(1):160–169, 1986. https://doi.org/10.1137/0723011.

[6] P. AMESTOY, A. BUTTARI, I. S. DUFF, A. GUERMOUCHE, J.-Y. L'EXCELLENT, AND B. UÇAR. MUMPS (MUltifrontal Massively Parallel Solver). In D. Padua, editor, *Encyclopedia of Parallel Computing*, pages 1232–1238. Springer, Heidelberg, Berlin, New York, 2011. https://doi.org/10.1007/978-0-387-09766-4_204.

[7] A. ANANDKUMAR AND R. GE. Efficient approaches for escaping higher order saddle points in non-convex optimization. arXiv e-prints, arXiv:1602.05908, February 2016. https://arxiv.org/abs/1602.05908.

[8] R. ANDREANI, E. G. BIRGIN, J. M. MARTÍNEZ, AND M. L. SCHUVERDT. On augmented Lagrangian methods with general lower-level constraints. *SIAM Journal on Optimization*, 18(4):1286–1309, 2007. https://doi.org/10.1137/060654797.

[9] T. M. APOSTOL. *Mathematical Analysis*. Addison-Wesley, Reading, MA, second edition, 1974.

[10] Y. ARJEVANI, Y. CARMON, J. C. DUCHI, D. J. FOSTER, A. SEKHARI, AND K. SRIDHARAN. Second-order information in non-convex stochastic optimization: Power and limitations. In J. Abernethy and S. Agarwal, editors, *Proceedings of Thirty Third Conference on Learning Theory (COLT)*, volume 125 of *Proceedings of Machine Learning Research*, pages 242–299. PMLR, 9–12 July 2020. http://proceedings.mlr.press/v125/arjevani20a.html.

[11] Y. ARJEVANI, O. SHAMIR, AND R. SHIFF. Oracle complexity of second-order methods for smooth convex optimization. *Mathematical Programming Series A*, 178(1-2):327–360, 2019. https://doi.org/10.1007/s10107-018-1293-1.

[12] L. ARMIJO. Minimization of functions having Lipschitz continuous first partial derivatives. *Pacific Journal of Mathematics*, 16:1–3, 1966. http://projecteuclid.org/euclid.pjm/1102995080.

[13] C. ASHCRAFT, R. G. GRIMES, AND J. G. LEWIS. Accurate symmetric indefinite linear equation solvers. *SIAM Journal on Matrix Analysis and Applications*, 20(2):513–561, 1999. https://doi.org/10.1137/S0895479896296921.

[14] C. Audet and D. Orban. Finding optimal algorithmic parameters using derivative-free optimization. *SIAM Journal on Optimization*, 17(3):642–664, 2006. https://doi.org/10.1137/040620886.

[15] O. Axelsson and L. Yu. Kolotilina. Diagonally compensated reduction and related preconditioning methods. *Numerical Linear Algebra with Applications*, 1(2):155–177, 1994. https://doi.org/10.1002/nla.1680010207.

[16] M. Baboulin, A. Buttari, J. Dongarra, J. Kurzak, J. Langou, P. Luszczek, and S. Tomov. Accelerating scientific computations with mixed precision algorithms. *Computer Physics Communications*, 180(12):2526–2533, 2009. https://doi.org/10.1016/j.cpc.2008.11.005.

[17] V. Barbu and T. Precupanu. *Convexity and Optimization in Banach Spaces*. Springer Monographs in Mathematics. Springer, Dordrecht, Netherlands, fourth edition, 2012.

[18] R. Beals. *Analysis: An Introduction*. Cambridge University Press, Cambridge, UK, 2004.

[19] A. Beck. *Introduction to Nonlinear Optimization*, volume 19 of *MOS-SIAM Series on Optimization*. SIAM, Philadelphia, 2014.

[20] A. Beck. *First-Order Methods in Optimization*, volume 25 of *MOS-SIAM Series on Optimization*. SIAM, Philadelphia, 2017.

[21] A. Beck and Y. Eldar. Sparsity constrained nonlinear optimization: Optimality conditions and algorithms. *SIAM Journal on Optimization*, 23(3):1480–1509, 2013. https://doi.org/10.1137/120869778.

[22] A. Beck and N. Hallak. On the minimization over sparse symmetric sets: Projections, optimality conditions, and algorithms. *Mathematics of Operations Research*, 41(1):196–223, 2016. https://doi.org/10.1287/moor.2015.0722.

[23] A. Beck and N. Hallak. Optimization problems involving group sparsity terms. *Mathematical Programming Series A*, 178(1-2):39–67, 2019. https://doi.org/10.1007/s10107-018-1277-1.

[24] A. Beck and M. Teboulle. A fast iterative shrinkage-thresholding algorithm for linear inverse problems. *SIAM Journal on Imaging Sciences*, 2(1):183–202, 2009. https://doi.org/10.1137/080716542.

[25] A. Beck and Y. Vaisbourd. Globally solving the trust region subproblem using simple first-order methods. *SIAM Journal on Optimization*, 28(3):1951–1967, 2018. https://doi.org/10.1137/16M1150281.

[26] S. Bellavia, C. Cartis, N. I. M. Gould, B. Morini, and Ph. L. Toint. Convergence of a regularized Euclidean residual algorithm for nonlinear least-squares. *SIAM Journal on Numerical Analysis*, 48(1):1–29, 2010. https://doi.org/10.1137/080732432.

[27] S. Bellavia, S. Gratton, and E. Riccietti. A Levenberg-Marquardt method for large nonlinear least-squares problems with dynamic accuracy in functions and gradients. *Numerische Mathematik*, 140(3):791–825, 2018. https://doi.org/10.1007/s00211-018-0977-z.

[28] S. Bellavia, G. Gurioli, B. Morini, and Ph. L. Toint. Adaptive regularization algorithms with inexact evaluations for nonconvex optimization. *SIAM Journal on Optimization*, 29(4):2881–2915, 2019. https://doi.org/10.1137/18M1226282.

[29] S. Bellavia, G. Gurioli, B. Morini, and Ph. L. Toint. Adaptive regularization for nonconvex optimization using inexact function values and randomly perturbed derivatives. *Journal of Complexity*, 68: 101591, 2022. https://arxiv.org/abs/2005.04639.

[30] S. Bellavia, G. Gurioli, B. Morini, and Ph. L. Toint. The impact of noise on evaluation complexity: The deterministic trust-region case. arXiv e-prints, arXiv:2104.02519, April 2021. https://arxiv.org/abs/2104.02519.

[31] A. Ben-Tal and M. Teboulle. Hidden convexity in some nonconvex quadratically constrained quadratic programming. *Mathematical Programming Series A*, 72(1):51–63, 1996. https://doi.org/10.1007/BF02592331.

[32] A. Bensoussan and J. Frehse. *Regularity Results for Nonlinear Elliptic Systems and Applications*, volume 151 of *Applied Mathematical Sciences*. Springer-Verlag, Berlin, 2002.

[33] M. Benzi. Preconditioning techniques for large linear systems: A survey. *Journal of Computational Physics*, 182(2):418–477, 2002. https://doi.org/10.1006/jcph.2002.7176.

[34] E. Bergou, Y. Diouane, and S. Gratton. On the use of the energy norm in trust-region and adaptive cubic regularization subproblems. *Computational Optimization and Applications*, 68(3):533–554, 2017. https://doi.org/10.1007/s10589-017-9929-2.

[35] D. P. BERTSEKAS. On penalty and multiplier methods for constrained minimization. *SIAM Journal on Control and Optimization*, 14(2):216–235, 1976. https://doi.org/10.1137/0314017.

[36] D. BERTSIMAS AND B. VAN PARYS. Sparse high-dimensional regression: Exact scalable algorithms and phase transitions. *Annals of Statistics*, 48(1):300–323, 2020. https://doi.org/10.1214/18-AOS1804.

[37] M. J. BEST AND N. CHAKRAVARTI. Active set algorithms for isotonic regression; A unifying framework. *Mathematical Programming Series A*, 47(3):425–439, 1990. https://doi.org/10.1007/BF01580873.

[38] W. BIAN AND X. CHEN. Worst-case complexity of smoothing quadratic regularization methods for non-Lipschitzian optimization. *SIAM Journal on Optimization*, 23(3):1718–1741, 2013. https://doi.org/10.1137/120864908.

[39] W. BIAN AND X. CHEN. Linearly constrained non-Lipschitz optimization for image restoration. *SIAM Journal on Imaging Sciences*, 8(4):2294–2322, 2015. https://doi.org/10.1137/140985639.

[40] W. BIAN, X. CHEN, AND Y. YE. Complexity analysis of interior point algorithms for non-Lipschitz and nonconvex minimization. *Mathematical Programming Series A*, 149(1-2):301–327, 2015. https://doi.org/10.1007/s10107-014-0753-5.

[41] T. BIANCONCINI, G. LIUZZI, B. MORINI, AND M. SCIANDRONE. On the use of iterative methods in cubic regularization for unconstrained optimization. *Computational Optimization and Applications*, 60(1):35–57, 2015. https://doi.org/10.1007/s10589-014-9672-x.

[42] T. BIANCONCINI AND M. SCIANDRONE. A cubic regularization algorithm for unconstrained optimization using line search and nonmonotone techniques. *Optimization Methods & Software*, 31(5):1008–1035, 2016. https://doi.org/10.1080/10556788.2016.1155213.

[43] E. G. BIRGIN, J. L. GARDENGHI, J. M. MARTÍNEZ, AND S. A. SANTOS. On the use of third-order models with fourth-order regularization for unconstrained optimization. *Optimization Letters*, 14(4):815–838, 2020. https://doi.org/10.1007/s11590-019-01395-z.

[44] E. G. BIRGIN, J. L. GARDENGHI, J. M. MARTÍNEZ, S. A. SANTOS, AND PH. L. TOINT. Evaluation complexity for nonlinear constrained optimization using unscaled KKT conditions and high-order models. *SIAM Journal on Optimization*, 26(2):951–967, 2016. https://doi.org/10.1137/15M1031631.

[45] E. G. BIRGIN, J. L. GARDENGHI, J. M. MARTÍNEZ, S. A. SANTOS, AND PH. L. TOINT. Worst-case evaluation complexity for unconstrained nonlinear optimization using high-order regularized models. *Mathematical Programming Series A*, 163(1-2):359–368, 2017. https://doi.org/10.1007/s10107-016-1065-8.

[46] E. G. BIRGIN AND J. M. MARTÍNEZ. *Practical Augmented Lagrangian Methods for Constrained Optimization*, volume 10 of *Fundamentals of Algorithms*. SIAM, Philadelphia, 2014.

[47] E. G. BIRGIN AND J. M. MARTÍNEZ. The use of quadratic regularization with a cubic descent condition for unconstrained optimization. *SIAM Journal on Optimization*, 27(2):1049–1074, 2017. https://doi.org/10.1137/16M110280X.

[48] E. G. BIRGIN AND J. M. MARTÍNEZ. On regularization and active-set methods with complexity for constrained optimization. *SIAM Journal on Optimization*, 28(2):1367–1395, 2018. https://doi.org/10.1137/17M1127107.

[49] E. G. BIRGIN AND J. M. MARTÍNEZ. Complexity and performance of an augmented Lagrangian algorithm. arXiv e-prints, arXiv:1907.024018, July 2019. https://arxiv.org/abs/1907.02401.

[50] E. G. BIRGIN AND J. M. MARTÍNEZ. A Newton-like method with mixed factorizations and cubic regularization for unconstrained minimization. *Computational Optimization and Applications*, 73(3):707–753, 2019. https://doi.org/10.1007/s10589-019-00089-7.

[51] J. BLANCHET, C. CARTIS, M. MENICKELLY, AND K. SCHEINBERG. Convergence rate analysis of a stochastic trust region method via submartingales. *INFORMS Journal on Optimization*, 1(2):92–119, 2019. https://doi.org/10.1287/ijoo.2019.0016.

[52] P. T. BOGGS AND J. W. TOLLE. Sequential quadratic programming. *Acta Numerica*, 4:1–51, 1995. https://doi.org/10.1017/S0962492900002518.

[53] J. BOLTE, A. DANILIDIS, O. LEY, AND L. MAZET. Characterizations of Łojasiewicz inequalities: Subgradient flows, talweg, convexity. *Transactions of the American Mathematical Society*, 362(6):3319–3363, 2010. https://doi.org/10.1090/S0002-9947-09-05048-X.

[54] J. F. Bonnans, R. Cominetti, and A. Shapiro. Second order optimality conditions based on parabolic second order tangent sets. *SIAM Journal on Optimization*, 9(2):466–492, 1999. `https://doi.org/10.1137/S1052623496306760`.

[55] H.-K. Borgwardt. *The Simplex Method: A Probabilistic Analysis*, volume 1 of *Algorithms and Combinatorics: Study and Research Texts*. Springer-Verlag, Berlin, 1987.

[56] F. Botelho. *Functional Analysis and Applied Optimization in Banach Spaces*. Springer, Cham, 2014.

[57] C. A. Botsaris and D. H. Jacobson. A Newton-type curvilinear search method for optimization. *Journal of Mathematical Analysis and Applications*, 54(1):217–229, 1976. `https://doi.org/10.1016/0022-247X(76)90246-8`.

[58] L. Bottou, F. E. Curtis, and J. Nocedal. Optimization methods for large-scale machine learning. *SIAM Review*, 60(2):223–311, 2018. `https://doi.org/10.1137/16M1080173`.

[59] N. Boumal. *An Introduction to Optimization on Smooth Manifolds*. Available online, November 2020. `http://www.nicolasboumal.net/book`.

[60] N. Boumal, P.-A. Absil, and C. Cartis. Global rates of convergence for nonconvex optimization on manifolds. *IMA Journal of Numerical Analysis*, 39(1):1–33, 2019. `https://doi.org/10.1093/imanum/drx080`. (Erratum in 40(4):2940, 2020. `https://doi.org/10.1093/imanum/drz064`).

[61] F. Bouttier and Ph. Courtier. Data assimilation concepts and methods. Technical report, ECMWF, Reading, UK, 1999. ECMWF Meteorological Training Course Lecture Series. `https://www.ecmwf.int/node/16928`.

[62] S. Boyd and L. Vandenberghe. *Convex Optimization*. Cambridge University Press, Cambridge, UK, 2004.

[63] O. A. Brezhneva and A. A. Tret'yakov. Optimality conditions for degenerate extremum problems with equality constraints. *SIAM Journal on Control and Optimization*, 42(2):729–745, 2003. `https://doi.org/10.1137/S0363012901388488`.

[64] O. A. Brezhneva and A. A. Tret'yakov. The pth-order optimality conditions for inequality constrained optimization problems. *Nonlinear Analysis: Theory, Methods & Applications*, 63(5):e1357–e1366, 2005. `https://doi.org/10.1016/j.na.2005.02.022`.

[65] S. Bubeck, Q. Jiang, Y. T. Lee, Y. Li, and A. Sidford. Near-optimal method for highly smooth convex optimization. In A. Beygelzimer and D. Hsu, editors, *Proceedings of the Thirty-Second Conference on Learning Theory*, volume 99 of *Proceedings of Machine Learning Research*, pages 492–507, Phoenix, AZ, 25–28 June 2019. PMLR. `http://proceedings.mlr.press/v99/bubeck19a.html`.

[66] J. R. Bunch and L. C. Kaufman. Some stable methods for calculating inertia and solving symmetric linear systems. *Mathematics of Computation*, 31(137):163–179, 1977. `https://doi.org/10.2307/2005787`.

[67] J. R. Bunch and B. N. Parlett. Direct methods for solving symmetric indefinite systems of linear equations. *SIAM Journal on Numerical Analysis*, 8:639–655, 1971. `https://doi.org/10.1137/0708060`.

[68] O. Burdakov, A. Grimvall, and O. Sysoev. Data preordering in generalized PAV algorithm for monotonic regression. *Journal of Computational Mathematics*, 24(6):771–790, 2006. `https://www.jstor.org/stable/43693336`.

[69] O. Burdakov and O. Sysoev. A dual active-set algorithm for regularized monotonic regression. *Journal of Optimization Theory and Applications*, 172(3):929–949, 2017. `https://doi.org/10.1007/s10957-017-1060-0`.

[70] O. Burdakov, O. Sysoev, A. Grimvall, and M. Hussian. An $\mathcal{O}(n^2)$ algorithm for isotonic regression. In G. Di Pillo and M. Roma, editors, *Large-Scale Nonlinear Optimization. Nonconvex Optimization and Its Applications*, pages 25–33. Springer, New York, 2006. `https://doi.org/10.1007/0-387-30065-1_3`.

[71] H. Calandra, S. Gratton, E. Riccietti, and X. Vasseur. On high-order multilevel optimization strategies. arXiv e-prints, arXiv:1904.04692, April 2019. `https://arxiv.org/abs/1904.04692`.

[72] Y. Carmon and J. C. Duchi. Gradient descent efficiently finds the cubic-regularized non-convex Newton step. arXiv e-prints, arXiv:1612.00547, December 2016. `https://arxiv.org/abs/1612.00547`.

[73] Y. Carmon and J. C. Duchi. Analysis of Krylov subspace solutions of regularized non-convex quadratic problems. In S. Bengio, H. M. Wallach, H. Larochelle, K. Grauman, N. Cesa-Bianchi, and R. Garnett, editors, *Advances in Neural Information Processing Systems 31*, pages 10705–10715, 2018. https://proceedings.neurips.cc/paper/8269-analysis-of-krylov-subspace-solutions-of-regularized-non-convex-quadratic-problems.pdf.

[74] Y. Carmon and J. C. Duchi. First-order methods for nonconvex quadratic minimization. *SIAM Review*, 62(2):395–436, 2020. https://doi.org/10.1137/20M1321759.

[75] Y. Carmon, J. C. Duchi, O. Hinder, and A. Sidford. "Convex until proven guilty": Dimension-free acceleration of gradient descent on non-convex functions. In *Proceedings of the 34th International Conference on Machine Learning*, volume 70, pages 654–663, 2017. http://proceedings.mlr.press/v70/carmon17a.html.

[76] Y. Carmon, J. C. Duchi, O. Hinder, and A. Sidford. Accelerated methods for nonconvex optimization. *SIAM Journal on Optimization*, 28(2):1751–1772, 2018. https://doi.org/10.1137/17M1114296.

[77] Y. Carmon, J. C. Duchi, O. Hinder, and A. Sidford. Lower bounds for finding stationary points I. *Mathematical Programming Series A*, 184(1-2):71–120, 2020. https://doi.org/10.1007/s10107-019-01406-y.

[78] Y. Carmon, J. C. Duchi, O. Hinder, and A. Sidford. Lower bounds for finding stationary points II: First-order methods. *Mathematical Programming Series A*, 185(1-2):315–355, 2021. https://doi.org/10.1007/s10107-019-01431-x.

[79] R. G. Carter. Numerical experience with a class of algorithms for nonlinear optimization using inexact function and gradient information. *SIAM Journal on Scientific Computing*, 14(2):368–388, 1993. https://doi.org/10.1137/0914023.

[80] C. Cartis, J. M. Fowkes, and N. I. M. Gould. Branching and bounding improvements for global optimization algorithms with Lipschitz continuity properties. *Journal of Global Optimization*, 61(3):429–457, 2015. https://doi.org/10.1007/s10898-014-0199-6.

[81] C. Cartis, N. I. M. Gould, and M. Lange. On monotonic estimates of the norm of the minimizers of regularized quadratic functions in Krylov spaces. *BIT. Numerical Mathematics*, 60(3):583–589, 2020. https://doi.org/10.1007/s10543-019-00791-2.

[82] C. Cartis, N. I. M. Gould, and Ph. L. Toint. Trust-region and other regularisations of linear least-squares problems. *BIT. Numerical Mathematics*, 49(1):21–53, 2009. https://doi.org/10.1007/s10543-008-0206-8.

[83] C. Cartis, N. I. M. Gould, and Ph. L. Toint. On the complexity of steepest descent, Newton's and regularized Newton's methods for nonconvex unconstrained optimization problems. *SIAM Journal on Optimization*, 20(6):2833–2852, 2010. https://doi.org/10.1137/090774100.

[84] C. Cartis, N. I. M. Gould, and Ph. L. Toint. Adaptive cubic regularisation methods for unconstrained optimization. Part I: Motivation, convergence and numerical results. *Mathematical Programming Series A*, 127(2):245–295, 2011. https://doi.org/10.1007/s10107-009-0286-5.

[85] C. Cartis, N. I. M. Gould, and Ph. L. Toint. Adaptive cubic regularisation methods for unconstrained optimization. Part II: Worst-case function- and derivative-evaluation complexity. *Mathematical Programming Series A*, 130(2):295–319, 2011. https://doi.org/10.1007/s10107-009-0337-y.

[86] C. Cartis, N. I. M. Gould, and Ph. L. Toint. On the evaluation complexity of composite function minimization with applications to nonconvex nonlinear programming. *SIAM Journal on Optimization*, 21(4):1721–1739, 2011. https://doi.org/10.1137/11082381X.

[87] C. Cartis, N. I. M. Gould, and Ph. L. Toint. Optimal Newton-type methods for nonconvex optimization. Technical Report RAL-TR-2011-011, STFC-Rutherford Appleton Laboratory, Oxfordshire, UK, 2011. http://purl.org/net/epubs/manifestation/6708.

[88] C. Cartis, N. I. M. Gould, and Ph. L. Toint. An adaptive cubic regularization algorithm for nonconvex optimization with convex constraints and its function-evaluation complexity. *IMA Journal of Numerical Analysis*, 32(4):1662–1695, 2012. https://doi.org/10.1093/imanum/drr035.

[89] C. Cartis, N. I. M. Gould, and Ph. L. Toint. Complexity bounds for second-order optimality in unconstrained optimization. *Journal of Complexity*, 28(1):93–108, 2012. https://doi.org/10.1016/j.jco.2011.06.001.

[90] C. Cartis, N. I. M. Gould, and Ph. L. Toint. Evaluation complexity of adaptive cubic regularization methods for convex unconstrained optimization. *Optimization Methods & Software*, 27(2):197–219, 2012. https://doi.org/10.1080/10556788.2011.602076.

[91] C. Cartis, N. I. M. Gould, and Ph. L. Toint. How much patience do you have? A worst-case perspective on smooth nonconvex optimization. *Optima*, 88:1–10, 2012. http://www.mathopt.org/Optima-Issues/optima88.pdf.

[92] C. Cartis, N. I. M. Gould, and Ph. L. Toint. On the complexity of the steepest-descent with exact linesearches. Technical Report RAL-TR-2012-015, STFC-Rutherford Appleton Laboratory, Oxfordshire, UK, 2012. http://purl.org/net/epubs/manifestation/8250.

[93] C. Cartis, N. I. M. Gould, and Ph. L. Toint. On the oracle complexity of first-order and derivative-free algorithms for smooth nonconvex minimization. *SIAM Journal on Optimization*, 22(1):66–86, 2012. https://doi.org/10.1137/100812276.

[94] C. Cartis, N. I. M. Gould, and Ph. L. Toint. An example of slow convergence for Newton's method on a function with globally Lipschitz continuous Hessian. Technical Report RAL-TR-2013-004, STFC-Rutherford Appleton Laboratory, Oxfordshire, UK, 2013. http://purl.org/net/epubs/manifestation/9328.

[95] C. Cartis, N. I. M. Gould, and Ph. L. Toint. A note about the complexity of minimizing Nesterov's smooth Chebyshev–Rosenbrock function. *Optimization Methods & Software*, 28(3):451–457, 2013. https://doi.org/10.1080/10556788.2012.722632.

[96] C. Cartis, N. I. M. Gould, and Ph. L. Toint. On the evaluation complexity of cubic regularization methods for potentially rank-deficient nonlinear least-squares problems and its relevance to constrained nonlinear optimization. *SIAM Journal on Optimization*, 23(3):1553–1574, 2013. https://doi.org/10.1137/120869687.

[97] C. Cartis, N. I. M. Gould, and Ph. L. Toint. Improved worst-case evaluation complexity for potentially rank-deficient nonlinear least-Euclidean-norm problems using higher-order regularized models. Technical Report RAL-TR-2015-011, STFC-Rutherford Appleton Laboratory, Oxfordshire, UK, 2015. http://purl.org/net/epubs/manifestation/27207475.

[98] C. Cartis, N. I. M. Gould, and Ph. L. Toint. On the evaluation complexity of constrained nonlinear least-squares and general constrained nonlinear optimization using second-order methods. *SIAM Journal on Numerical Analysis*, 53(2):836–851, 2015. https://doi.org/10.1137/130915546.

[99] C. Cartis, N. I. M. Gould, and Ph. L. Toint. Corrigendum: On the complexity of finding first-order critical points in constrained nonlinear optimization. *Mathematical Programming Series A*, 161(1-2):611–626, 2017. https://doi.org/10.1007/s10107-016-1016-4.

[100] C. Cartis, N. I. M. Gould, and Ph. L. Toint. Evaluation complexity bounds for smooth constrained nonlinear optimisation using scaled KKT conditions, high-order models and the criticality measure χ. arXiv e-prints, arXiv:1705.04895, May 2017. https://arxiv.org/abs/1705.04895.

[101] C. Cartis, N. I. M. Gould, and Ph. L. Toint. Worst-case evaluation complexity of regularization methods for smooth unconstrained optimization using Hölder continuous gradients. *Optimization Methods & Software*, 32(6):1273–1298, 2017. https://doi.org/10.1080/10556788.2016.1268136.

[102] C. Cartis, N. I. M. Gould, and Ph. L. Toint. Second-order optimality and beyond: Characterization and evaluation complexity in convexly constrained nonlinear optimization. *Foundations of Computational Mathematics*, 18(5):1073–1107, 2018. https://doi.org/10.1007/s10208-017-9363-y.

[103] C. Cartis, N. I. M. Gould, and Ph. L. Toint. Evaluation complexity bounds for smooth constrained nonlinear optimization using scaled KKT conditions and high-order models. In I. Demetriou and P. Pardalos, editors, *Approximation and Optimization: Algorithms, Complexity and Applications*, pages 5–26. Springer International, 2019.

[104] C. Cartis, N. I. M. Gould, and Ph. L. Toint. Optimality of orders one to three and beyond: Characterization and evaluation complexity in constrained nonconvex optimization. *Journal of Complexity*, 53:68–94, 2019. https://doi.org/10.1016/j.jco.2018.11.001.

[105] C. Cartis, N. I. M. Gould, and Ph. L. Toint. Universal regularization methods: Varying the power, the smoothness and the accuracy. *SIAM Journal on Optimization*, 29(1):595–615, 2019. https://doi.org/10.1137/16M1106316.

[106] C. Cartis, N. I. M. Gould, and Ph. L. Toint. Worst-case evaluation complexity and optimality of second-order methods for nonconvex smooth optimization. In P. N. de Souza, B. Sirakov, and M. Viana, editors, *Proceedings of the International Congress of Mathematicians (ICM 2018)*. World Scientific, Singapore, 2019. https://doi.org/10.1142/9789813272880_0198.

[107] C. Cartis, N. I. M. Gould, and Ph. L. Toint. A concise second-order evaluation complexity for unconstrained nonlinear optimization using high-order regularized models. *Optimization Methods & Software*, 35(2):243–256, 2020. https://doi.org/10.1080/10556788.2019.1678033.

[108] C. Cartis, N. I. M. Gould, and Ph. L. Toint. Sharp worst-case evaluation complexity bounds for arbitrary-order nonconvex optimization with inexpensive constraints. *SIAM Journal on Optimization*, 30(1):513–541, 2020. https://doi.org/10.1137/17M1144854.

[109] C. Cartis, N. I. M. Gould, and Ph. L. Toint. Strong evaluation complexity bounds for arbitrary-order optimization of nonconvex nonsmooth composite functions. arXiv e-prints, arXiv:2001.10802, January 2020. https://arxiv.org/abs/2001.10802. Published as part of: The evaluation complexity of finding high-order minimizers of nonconvex optimization. In D. Belyaev and S. Smirnov, editors, *Proceedings of the International Congress of Mathematicians (ICM 2022)*. EMS Press. https://doi.org/10.4171/ICM2022/95.

[110] C. Cartis, N. I. M. Gould, and Ph. L. Toint. An adaptive regularization algorithm for unconstrained optimization with inexact function and derivatives values. arXiv e-prints, arXiv:2111.14098, November 2021. https://arxiv.org/abs/2111.14098.

[111] C. Cartis, N. I. M. Gould, and Ph. L. Toint. Strong evaluation complexity of an inexact trust-region algorithm for arbitrary-order unconstrained nonconvex optimization. arXiv e-prints, arXiv:2011.00854, March 2021. https://arxiv.org/abs/2011.00854.

[112] C. Cartis, Ph. R. Sampaio, and Ph. L. Toint. Worst-case evaluation complexity of non-monotone gradient-related algorithms for unconstrained optimization. *Optimization*, 64(5):1349–1361, 2015. https://doi.org/10.1080/02331934.2013.869809.

[113] C. Cartis and K. Scheinberg. Global convergence rate analysis of unconstrained optimization methods based on probabilistic models. *Mathematical Programming Series A*, 169(2):337–375, 2018. https://doi.org/10.1007/s10107-017-1137-4.

[114] C. Cartis and A. Thompson. A new and improved quantitative recovery analysis for iterative hard thresholding algorithms in compressed sensing. *IEEE Transactions on Information Theory*, 61(4):2019–2042, 2015. https://doi.org/10.1109/TIT.2015.2399919.

[115] A. Cauchy. Méthode générale pour la résolution des systèmes d'équations simultanées. *Comptes Rendus de l'Académie des Sciences*, 25:536–538, 1847.

[116] X. Chen and Ph. L. Toint. High-order evaluation complexity for convexly-constrained optimization with non-Lipschitzian group sparsity terms. *Mathematical Programming*, 187:47–78, 2021. https://doi.org/10.1007/S10107-020-01470-9.

[117] X. Chen, Ph. L. Toint, and H. Wang. Complexity of partially separable convexly constrained optimization with non-Lipschitzian singularities. *SIAM Journal on Optimization*, 29(1):874–903, 2019. https://doi.org/10.1137/18M1166511.

[118] P. G. Ciarlet and P. A. Raviart. General Lagrange and Hermite interpolation in $\mathbf{R}^n$ with applications to finite element methods. *Archive for Rational Mechanics and Analysis*, 46:177–199, 1972. https://doi.org/10.1007/BF00252458.

[119] A. K. Cline, C. B. Moler, G. W. Stewart, and J. H. Wilkinson. An estimate for the condition number of a matrix. *SIAM Journal on Numerical Analysis*, 16(2):368–375, 1979. https://doi.org/10.1137/0716029.

[120] T. F. Coleman and J. J. Moré. Estimation of sparse Hessian matrices and graph coloring problems. *Mathematical Programming*, 28(3):243–270, 1984. https://doi.org/10.1007/BF02612334.

[121] B. Colson and Ph. L. Toint. Exploiting band structure in unconstrained optimization without derivatives. *Optimization and Engineering*, 2(4):349–412, 2001. https://doi.org/10.1023/A:1016090421852.

[122] B. Colson and Ph. L. Toint. A derivative-free algorithm for sparse unconstrained optimization problems. In A. H. Siddiqi and M. Kočvara, editors, *Trends in Industrial and Applied Mathematics*, volume 72 of *Applied Optimization*, pages 131–147. Kluwer Academic, Dordrecht, Netherlands, 2002. https://doi.org/10.1007/978-1-4613-0263-6_7.

[123] B. Colson and Ph. L. Toint. Optimizing partially separable functions without derivatives. *Optimization Methods & Software*, 20(4-5):493–508, 2005. https://doi.org/10.1080/10556780500140227.

[124] R. Cominetti. Metric regularity, tangent sets, and second-order optimality conditions. *Applied Mathematics and Optimization*, 21(3):265–287, 1990. https://doi.org/10.1007/BF01445166.

[125] P. Comon, G. H. Golub, L.-H. Lim, and B. Mourrain. Symmetric tensors and symmetric tensor rank. *SIAM Journal on Matrix Analysis and Applications*, 30(3):1254–1279, 2008. https://doi.org/10.1137/060661569.

[126] D. Conforti and M. Mancini. A curvilinear search algorithm for unconstrained optimization by automatic differentiation. *Optimization Methods & Software*, 15(3-4):283–297, 2001. https://doi.org/10.1080/10556780108805822.

[127] M. Conforti, G. Cornuejols, and G. Zambelli. *Integer Programming*. Springer International, 2014.

[128] A. R. Conn, N. I. M. Gould, A. Sartenaer, and Ph. L. Toint. Convergence properties of minimization algorithms for convex constraints using a structured trust region. *SIAM Journal on Optimization*, 6(4):1059–1086, 1996. https://doi.org/10.1137/S1052623492236481.

[129] A. R. Conn, N. I. M. Gould, and Ph. L. Toint. A globally convergent augmented Lagrangian algorithm for optimization with general constraints and simple bounds. *SIAM Journal on Numerical Analysis*, 28(2):545–572, 1991. https://doi.org/10.1137/0728030.

[130] A. R. Conn, N. I. M. Gould, and Ph. L. Toint. LANCELOT*: A Fortran Package for Large-Scale Nonlinear Optimization (Release A)*, volume 17 of *Springer Series in Computational Mathematics*. Springer-Verlag, Heidelberg, Berlin, 1992.

[131] A. R. Conn, N. I. M. Gould, and Ph. L. Toint. *Trust-Region Methods*. MPS/SIAM Series on Optimization. SIAM, Philadelphia, 2000.

[132] A. R. Conn, K. Scheinberg, and L. N. Vicente. *Introduction to Derivative-Free Optimization*, volume 8 of *MPS/SIAM Series on Optimization*. SIAM, Philadelphia, PA, 2009.

[133] G. M. Constantine and T. H. Savits. A multivariate Faà di Bruno formula with applications. *Transactions of the American Mathematical Society*, 348(2):503–520, 1996. https://doi.org/10.1090/S0002-9947-96-01501-2.

[134] Ph. Courtier. Dual formulation of four-dimensional variational assimilation. *Quarterly Journal of the Royal Meteorological Society*, 123(544):2449–2461, 1997. https://doi.org/10.1002/qj.49712354414.

[135] F. E. Curtis and D. P. Robinson. Regional complexity analysis of algorithms for nonconvex smooth optimization. *Mathematical Programming*, 187(1-2):579–615, 2021. https://doi.org/10.1007/s10107-020-01492-3.

[136] F. E. Curtis, D. P. Robinson, C. W. Royer, and S. J. Wright. Trust-region Newton-CG with strong second-order complexity guarantees for nonconvex optimization. *SIAM Journal on Optimization*, 31(1):518–544, 2021. https://doi.org/10.1137/19M130563X.

[137] F. E. Curtis, D. P. Robinson, and M. Samadi. A trust region algorithm with a worst-case iteration complexity of $\mathcal{O}(\epsilon^{-3/2})$ for nonconvex optimization. *Mathematical Programming Series A*, 162(1-2):1–32, 2017. https://doi.org/10.1007/s10107-016-1026-2.

[138] F. E. Curtis, D. P. Robinson, and M. Samadi. Complexity analysis of a trust funnel algorithm for equality constrained optimization. *SIAM Journal on Optimization*, 28(2):1533–1563, 2018. https://doi.org/10.1137/16M1108650.

[139] F. E. Curtis, D. P. Robinson, and M. Samadi. An inexact regularized Newton framework with a worst-case iteration complexity of $\mathcal{O}(\epsilon^{-3/2})$ for nonconvex optimization. *IMA Journal of Numerical Analysis*, 39(3):1296–1327, 2019. https://doi.org/10.1093/imanum/dry022.

[140] F. E. Curtis and K. Scheinberg. *Optimization Methods for Supervised Machine Learning: From Linear Models to Deep Learning*, Chapter 5, pages 89–114. TutORials in Operations Research. INFORMS, 2017.

[141] F. E. Curtis and K. Scheinberg. Adaptive stochastic optimization. arXiv e-prints, arXiv:2001.06699, January 2020. https://arxiv.org/abs/2001.06699.

[142] A. L. Custódio, R. Garmanjani, and M. Raydan. Derivative-free separable quadratic modeling and cubic regularization for unconstrained optimization. Technical report, optimization online, CMA, Nova University of Lisbon, Portugal, 2021. http://www.optimization-online.org/DB_HTML/2021/07/8506.html.

[143] E. de Klerk and M. Laurent. Worst-case examples for Lasserre's measure-based hierarchy for polynomial optimization on the hypercube. *Mathematics of Operations Research*, 45(1):86–98, 2019. https://doi.org/10.1287/moor.2018.0983.

[144] E. de Klerk and M. Laurent. Convergence analysis of a Lasserre hierarchy of upper bounds for polynomial minimization on the sphere. *Mathematical Programming*, 2020. https://doi.org/10.1007/s10107-019-01465-1.

[145] S. DEMKO, W. F. MOSS, AND P. W. SMITH. Decay rates for inverses of band matrices. *Mathematics of Computation*, 43(168):491–499, 1984. https://doi.org/10.2307/2008290.

[146] J. W. DEMMEL. *Applied Numerical Linear Algebra*. SIAM, Philadelphia, 1997.

[147] J. E. DENNIS, JR., AND R. B. SCHNABEL. *Numerical Methods for Unconstrained Optimization and Nonlinear Equations*. Prentice Hall Series in Computational Mathematics. Prentice Hall, Englewood Cliffs, NJ, 1983. Reprinted as *Classics in Applied Mathematics 16*, SIAM, Philadelphia, 1996.

[148] P. DEUFLHARD. *Newton Methods for Nonlinear Problems*, volume 35 of *Springer Series in Computational Mathematics*. Springer, Heidelberg, 2011.

[149] P. DEUFLHARD AND F. A. POTRA. Asymptotic mesh independence of Newton–Galerkin methods via a refined Mysovskii theorem. *SIAM Journal on Numerical Analysis*, 29(5):1395–1412, 1992. https://doi.org/10.1137/0729080.

[150] N. DOIKOV AND YU. NESTEROV. Local convergence of tensor methods. *Mathematical Programming, Series A*, 2021. https://doi.org/10.1007/s10107-020-01606-x.

[151] H. S. DOLLAR, N. I. M. GOULD, AND D. P. ROBINSON. On solving trust-region and other regularised subproblems in optimization. Technical Report RAL-TR-2009-003, STFC-Rutherford Appleton Laboratory, Oxfordshire, UK, 2009. http://www.numerical.rl.ac.uk/reports/dgrRAL2009003r.pdf.

[152] D. L. DONOHO. Compressed sensing. *Institute of Electrical and Electronics Engineers. Transactions on Information Theory*, 52(4):1289–1306, 2006. https://doi.org/10.1109/TIT.2006.871582.

[153] Z. DREZNER AND H. W. HAMACHER. *Facility Location: Applications and Theory*. Springer-Verlag, Berlin, 2002.

[154] J. C. DUCHI AND F. RUAN. Stochastic methods for composite and weakly convex optimization problems. *SIAM Journal on Optimization*, 28(4):3229–3259, 2018. https://doi.org/10.1137/17M1135086.

[155] I. S. DUFF AND J. K. REID. The multifrontal solution of indefinite sparse symmetric linear equations. *ACM Transactions on Mathematical Software*, 9(3):302–325, 1983. https://doi.org/10.1145/356044.356047.

[156] I. S. DUFF AND J. K. REID. Exploiting zeros on the diagonal in the direct solution of indefinite sparse symmetric linear systems. *ACM Transactions on Mathematical Software*, 22(2):227–257, 1996. https://doi.org/10.1145/229473.229480.

[157] J. P. DUSSAULT. Simple unified convergence proofs for trust-region and a new ARC variant. Technical report, University of Sherbrooke, Sherbrooke, QC, 2015. https://pdfs.semanticscholar.org/58d2/73d0eb62c6b34fc2e62eca6935adc7d6785b.pdf.

[158] J. P. DUSSAULT. $\mathrm{ARC_q}$: A new adaptive regularization by cubics. *Optimization Methods & Software*, 33(2):322–335, 2018. https://doi.org/10.1080/10556788.2017.1322080.

[159] J. P. DUSSAULT AND D. ORBAN. Scalable adaptive cubic regularization methods. Technical Report G-2015-109, GERAD–HEC Montréal, QC, 2017. https://pdfs.semanticscholar.org/80d1/cafede7f35decf5e9b473885cc47c579ae58.pdf.

[160] ENCYCLOPEDIA OF MATHEMATICS. The Łojasiewicz inequality, 2015. https://encyclopediaofmath.org/wiki/Lojasiewicz_inequality.

[161] J. B. ERWAY AND P. E. GILL. A subspace minimization method for the trust-region step. *SIAM Journal on Optimization*, 20(3):1439–1461, 2009. https://doi.org/10.1137/08072440X.

[162] C. F. FAÁ DI BRUNO. Note sur une nouvelle formule du calcul différentiel. *The Quarterly Journal of Pure and Applied Mathematics*, 1:359–360, 1855.

[163] F. FACCHINEI, V. KUNGURTSEV, L. LAMPARIELLO, AND G. SCUTARI. Ghost penalties in nonconvex constrained optimization: Diminishing stepsizes and iteration complexity. arXiv e-prints, arXiv:1709.03384, September 2017. https://arxiv.org/abs/1709.03384.

[164] R. W. FAREBROTHER. *Fitting Linear Relationships: A History of the Calculus of Observations 1750–1900*. Springer Series in Statistics. Perspectives in Statistics. Springer-Verlag, New York, 1999.

[165] M. FENG, J. E. MITCHELL, J.-S. PANG, X. SHEN, AND A. WÄCHTER. Complementarity formulations of ℓ_0-norm optimization. *Pacific Journal of Optimization*, 14(2):273–305, 2018. http://www.optimization-online.org/DB_HTML/2013/09/4053.html.

[166] R. Fletcher. Factorizing symmetric indefinite matrices. *Linear Algebra and Its Applications*, 14(3):257–272, 1976. https://doi.org/10.1016/0024-3795(76)90071-9.

[167] R. Fletcher. Penalty functions. In A. Bachem, B. Korte, and M. Grötschel, editors, *Mathematical Programming: The State of the Art*, pages 87–114. Springer-Verlag, Berlin, 1983. https://doi.org/10.1007/978-3-642-68874-4_5.

[168] R. Fletcher. *Practical Methods of Optimization*. A Wiley-Interscience Publication. John Wiley & Sons, Chichester, second edition, 1987.

[169] B. Fornberg. Generation of finite difference formulas on arbitrarily spaced grids. *Mathematics of Computation*, 51(184):699–706, 1988. https://doi.org/10.2307/2008770.

[170] B. Fornberg. Classroom note: Calculation of weights in finite difference formulas. *SIAM Review*, 40(3):685–691, 1998. https://doi.org/10.1137/S0036144596322507.

[171] S. Galal and M. Horowitz. Energy-efficient floating-point unit design. *IEEE Transactions on Computers*, 60(7):913–922, 2011. https://doi.org/10.1109/TC.2010.121.

[172] A. Galántai. The theory of Newton's method. *Journal of Computational and Applied Mathematics*, 124(1-2):25–44, 2000. https://doi.org/10.1016/S0377-0427(00)00435-0.

[173] M. R. Garey and D. S. Johnson. *Computers and Intractability. A Guide to the Theory of NP-Completeness*, W. H. Freeman and Co., San Francisco, CA, 1979.

[174] R. Garmanjani, D. Júdice, and L. N. Vicente. Trust-region methods without using derivatives: Worst case complexity and the nonsmooth case. *SIAM Journal on Optimization*, 26(4):1987–2011, 2016. https://doi.org/10.1137/151005683.

[175] Gas Processors and Suppliers Association. *Engineering Data Book*, volume 2. GPSA, Tulsa, OK, 1994.

[176] A. Gasnikov, P. Dvurechensky, E. Gorbunov, E. Vorontsova, D. Selikhanovych, and C. A. Uriber. Optimal tensor methods in smooth convex and uniformly convex optimization. In A. Beygelzimer and D. Hsu, editors, *Proceedings of the Thirty-Second Conference on Learning Theory*, volume 99 of *Proceedings of Machine Learning Research*, pages 1374–1391, Phoenix, AZ, 25–28 June 2019. PMLR. http://proceedings.mlr.press/v99/gasnikov19a.html.

[177] A. Gasnikov, P. Dvurechensky, E. Gorbunov, E. Vorontsova, D. Selikhanovych, C. A. Uriber, B. Jiang, H. Wang, S. Zhang, S. Bubeck, Q. Jiang, Y. T. Lee, Y. Li, and A. Sidford. Near optimal methods for minimizing convex functions with Lipschitz p-th derivatives. In A. Beygelzimer and D. Hsu, editors, *Proceedings of the Thirty-Second Conference on Learning Theory*, volume 99 of *Proceedings of Machine Learning Research*, pages 1392–1393, Phoenix, AZ, 25–28 June 2019. PMLR. http://proceedings.mlr.press/v99/gasnikov19b.html.

[178] D. M. Gay. Computing optimal locally constrained steps. *SIAM Journal on Scientific and Statistical Computing*, 2(2):186–197, 1981. https://doi.org/10.1137/0902016.

[179] M. E. Gertz and P. E. Gill. A primal-dual trust region algorithm for nonlinear optimization. *Mathematical Programming Series B*, 100(1):49–94, 2004. https://doi.org/10.1007/s10107-003-0486-3.

[180] M. Ghasemi, J. B. Lasserre, and M. Marshall. Lower bounds on the global minimum of a polynomial. *Computational Optimization and Applications*, 57(2):387–402, 2014. https://doi.org/10.1007/s10589-013-9596-x.

[181] N. Gillis, R. Ohib, S. Plis, and V. Potluru. Grouped sparse projection. arXiv e-prints, arXiv:1912.03896, December 2019. https://arxiv.org/abs/1912.03896.

[182] D. Goldfarb. The use of negative curvature in minimization algorithms. Technical Report TR80-412, Department of Computer Science, Cornell University, Ithaca, NY, 1980. https://hdl.handle.net/1813/6252.

[183] S. M. Goldfeld, R. E. Quandt, and H. F. Trotter. Maximization by quadratic hill climbing. *Econometrica*, 34:541–551, 1966. https://doi.org/10.2307/1909768.

[184] G. H. Golub and W. Kahan. Calculating the singular values and pseudo-inverse of a matrix. *Journal of the Society for Industrial and Applied Mathematics. Series B. Numerical Analysis*, 2:205–224, 1965. https://doi.org/10.1137/0702016.

[185] G. H. Golub and C. F. Van Loan. *Matrix Computations*. Johns Hopkins Studies in the Mathematical Sciences. Johns Hopkins University Press, Baltimore, MD, fourth edition, 2013.

[186] N. I. M. Gould. On the convergence of a sequential penalty function method for constrained minimization. *SIAM Journal on Numerical Analysis*, 26(1):107–128, 1989. https://doi.org/10.1137/0726007.

[187] N. I. M. GOULD, Y. HU, AND J. A. SCOTT. A numerical evaluation of sparse direct solvers for the solution of large sparse symmetric linear systems of equations. *ACM Transactions on Mathematical Software*, 33(2):Article 10, 32, 2007. https://doi.org/10.1145/1236463.1236465.

[188] N. I. M. GOULD AND S. LEYFFER. An introduction to algorithms for nonlinear optimization. In *Frontiers in Numerical Analysis (Durham, 2002)*, Universitext, pages 109–197. Springer, Berlin, 2003.

[189] N. I. M. GOULD, S. LUCIDI, M. ROMA, AND PH. L. TOINT. Solving the trust-region subproblem using the Lanczos method. *SIAM Journal on Optimization*, 9(2):504–525, 1999. https://doi.org/10.1137/S1052623497322735.

[190] N. I. M. GOULD AND J. NOCEDAL. The modified absolute-value factorization norm for trust-region minimization. In R. De Leone, A. Murli, P. M. Pardalos, and G. Toraldo, editors, *High Performance Algorithms and Software in Nonlinear Optimization, Applied Optimization*, volume 24, pages 225–241. Kluwer Academic, Dordrecht, Netherlands, 1998.

[191] N. I. M. GOULD, D. ORBAN, A. SARTENAER, AND PH. L. TOINT. Sensitivity of trust-region algorithms to their parameters. *4OR. A Quarterly Journal of Operations Research*, 3(3):227–241, 2005. https://doi.org/10.1007/s10288-005-0065-y.

[192] N. I. M. GOULD, D. ORBAN, AND PH. L. TOINT. GALAHAD—A library of thread-safe Fortran 90 packages for large-scale nonlinear optimization. *ACM Transactions on Mathematical Software*, 29(4):353–372, 2003. https://doi.org/10.1145/962437.962438.

[193] N. I. M. GOULD, M. PORCELLI, AND PH. L. TOINT. Updating the regularization parameter in the adaptive cubic regularization algorithm. *Computational Optimization and Applications*, 53(1):1–22, 2012. https://doi.org/10.1007/s10589-011-9446-7.

[194] N. I. M. GOULD, T. REES, AND J. A. SCOTT. Convergence and evaluation-complexity analysis of a regularized tensor-Newton method for solving nonlinear least-squares problems. *Computational Optimization and Applications*, 73(1):1–35, 2019. https://doi.org/10.1007/s10589-019-00064-2.

[195] N. I. M. GOULD, T. REES, AND J. A. SCOTT. Convergence and evaluation-complexity analysis of a regularized tensor-Newton method for solving nonlinear least-squares problems subject to convex constraints. Technical Report RAL-TR-2019-001, STFC-Rutherford Appleton Laboratory, Oxfordshire, UK, 2019. http://purl.org/net/epubs/manifestation/40739646.

[196] N. I. M. GOULD, D. P. ROBINSON, AND H. S. THORNE. On solving trust-region and other regularised subproblems in optimization. *Mathematical Programming Computation*, 2(1):21–57, 2010. https://doi.org/10.1007/s12532-010-0011-7.

[197] N. I. M. GOULD AND J. A. SCOTT. A numerical evaluation of HSL packages for the direct solution of large sparse, symmetric linear systems of equations. *ACM Transactions on Mathematical Software*, 30(3):300–325, 2004. https://doi.org/10.1145/1024074.1024077.

[198] N. I. M. GOULD AND V. SIMONCINI. Error estimates for iterative algorithms for minimizing regularized quadratic subproblems. *Optimization Methods & Software*, 35(2):304–328, 2020. https://doi.org/10.1080/10556788.2019.1670177.

[199] N. I. M. GOULD AND PH. L. TOINT. SQP methods for large-scale nonlinear programming. In M. J. D. Powell and S. Scholtes, editors, *System Modelling and Optimization*, pages 149–178, Springer-Verlag, Heidelberg, Berlin, New York, 2000. https://doi.org/10.1007/978-0-387-35514-6_7.

[200] N. I. M. GOULD AND PH. L. TOINT. FILTRANE, a Fortran 95 filter-trust-region package for solving systems of nonlinear equalities, nonlinear inequalities and nonlinear least-squares problems. *ACM Transactions on Mathematical Software*, 33(1):3–25, 2007. https://doi.org/10.1145/1206040.1206043.

[201] N. I. M. GOULD AND PH. L. TOINT. Nonlinear programming without a penalty function or a filter. *Mathematical Programming Series A*, 122(1):155–196, 2010. https://doi.org/10.1007/s10107-008-0244-7.

[202] G. N. GRAPIGLIA, M. L. N. GONĆALVES, AND G. N. SILVA. A cubic regularization of Newton's method with finite-difference Hessian approximations. Technical report, optimization online, Department of Mathematics, Federal University of Parana, Brazil, 2021. http://www.optimization-online.org/DB_HTML/2021/09/8590.html.

[203] G. N. GRAPIGLIA AND YU. NESTEROV. Regularized Newton methods for minimizing functions with Hölder continuous Hessians. *SIAM Journal on Optimization*, 27(1):478–506, 2017. https://doi.org/10.1137/16M1087801.

[204] G. N. GRAPIGLIA, J. YUAN, AND E. W. SACHS. On the worst-case evaluation complexity of non-monotone line search algorithms. *Computational Optimization and Applications*, 68(3):555–577, 2017. https://doi.org/10.1007/s10589-017-9928-3.

[205] G. N. GRAPIGLIA, J. YUAN, AND Y. YUAN. On the convergence and worst-case complexity of trust-region and regularization methods for unconstrained optimization. *Mathematical Programming Series A*, 152(1-2):491–520, 2015. https://doi.org/10.1007/s10107-014-0794-9.

[206] G. N. GRAPIGLIA, J. YUAN, AND Y. YUAN. Nonlinear stepsize control algorithms: Complexity bounds for first- and second-order optimality. *Journal of Optimization Theory and Applications*, 171(3):980–997, 2016. https://doi.org/10.1007/s10957-016-1007-x.

[207] G. N. GRAPIGLIA AND Y. YUAN. On the complexity of an augmented Lagrangian method for nonconvex optimization. arXiv e-prints, arXiv:1906.05622, June 2019. https://arxiv.org/abs/1906.05622.

[208] S. GRATTON, S. JERAD, AND PH. L. TOINT. Hölder gradient descent and adaptive regularization methods in Banach spaces for first-order points. arXiv e-prints, arXiv:2104.02564, April 2021. https://arxiv.org/abs/2104.02564.

[209] S. GRATTON, A. LAWLESS, AND N. K. NICHOLS. Approximate Gauss–Newton methods for nonlinear least squares problems. *SIAM Journal on Optimization*, 18(1):106–132, 2007. https://doi.org/10.1137/050624935.

[210] S. GRATTON, M. MOUFFE, PH. L. TOINT, AND M. WEBER-MENDONÇA. A recursive l_∞-trust-region method for bound-constrained nonlinear optimization. *IMA Journal of Numerical Analysis*, 28(4):827–861, 2008. https://doi.org/10.1093/imanum/drn034.

[211] S. GRATTON, C. W. ROYER, AND L. N. VICENTE. A decoupled first/second-order steps technique for nonconvex nonlinear unconstrained optimization with improved complexity bounds. *Mathematical Programming Series A*, 179(1-2):195–222, 2020. https://doi.org/10.1007/s10107-018-1328-7.

[212] S. GRATTON, A. SARTENAER, AND PH. L. TOINT. Recursive trust-region methods for multiscale nonlinear optimization. *SIAM Journal on Optimization*, 19(1):414–444, 2008. https://doi.org/10.1137/050623012.

[213] S. GRATTON, E. SIMON, AND PH. L. TOINT. Minimization of nonsmooth nonconvex functions using inexact evaluations and its worst-case complexity. *Mathematical Programming Series A*, 187(1–2):1–24, 2020. https://arxiv.org/abs/1902.10406.

[214] S. GRATTON AND PH. L. TOINT. A note on solving nonlinear optimization problems in variable precision. *Computational Optimization and Applications*, 76(3):917–933, 2020. https://doi.org/10.1007/s10589-020-00190-2.

[215] S. GRATTON AND PH. L. TOINT. Adaptive regularization minimization algorithms with nonsmooth norms. *IMA Journal of Numerical Analysis*, drac005, 2022. https://doi.org/10.1093/imanum/drac005.

[216] S. GRATTON, PH. L. TOINT, AND J. TSHIMANGA. Conjugate gradients versus multigrid solvers for diffusion-based correlation models in data assimilation. *Quarterly Journal of the Royal Meteorological Society*, 139(675):1481–1487, 2013. https://doi.org/10.1002/qj.2050.

[217] J. GREENSTADT. On the relative efficiencies of gradient methods. *Mathematics of Computation*, 21:360–367, 1967. https://doi.org/10.2307/2003238.

[218] A. GRIEWANK. The modification of Newton's method for unconstrained optimization by bounding cubic terms. Technical Report NA/12, Department of Applied Mathematics and Theoretical Physics, University of Cambridge, Cambridge, UK, 1981.

[219] A. GRIEWANK AND PH. L. TOINT. On the unconstrained optimization of partially separable functions. In M. J. D. Powell, editor, *Nonlinear Optimization 1981*, NATO Conference Series II: Systems Science, pages 301–312. Academic Press, London, 1982.

[220] L. GRIPPO, F. LAMPARIELLO, AND S. LUCIDI. A nonmonotone line search technique for Newton's method. *SIAM Journal on Numerical Analysis*, 23(4):707–716, 1986. https://doi.org/10.1137/0723046.

[221] G. HAESER, H. LIU, AND Y. YE. Optimality condition and complexity analysis for linearly-constrained optimization without differentiability on the boundary. *Mathematical Programming Series A*, 178(1-2):263–299, 2019. https://doi.org/10.1007/s10107-018-1290-4.

[222] W. W. HAGER. Condition estimates. *SIAM Journal on Scientific and Statistical Computing*, 5(2):311–316, 1984. https://doi.org/10.1137/0905023.

[223] W. W. HAGER, D. T. PHAN, AND J. ZHU. Projection algorithms for nonconvex minimization with application to sparse principal component analysis. *Journal of Global Optimization*, 65(4):657–676, 2016. https://doi.org/10.1007/s10898-016-0402-z.

[224] H. HANCOCK. *The Theory of Maxima and Minima.* The Athenaeum Press, Ginn & Co, New York, 1917.

[225] P. C. HANSEN. *Rank-Deficient and Discrete Ill-Posed Problems.* SIAM Monographs on Mathematical Modeling and Computation. SIAM, Philadelphia, 1998. Numerical aspects of linear inversion.

[226] P. C. HANSEN, V. PEREYRA, AND G. SCHERER. *Least Squares Data Fitting with Applications.* Johns Hopkins University Press, Baltimore, MD, 2013.

[227] W. HARE, C. SAGASTIZÁBAL, AND M. V. SOLODOV. A proximal bundle method for nonsmooth nonconvex functions with inexact information. *Computational Optimization and Applications*, 63(1):1–28, 2016. https://doi.org/10.1007/s10589-015-9762-4.

[228] M. D. HEBDEN. An algorithm for minimization using exact second derivatives. Technical Report T. P. 515, Theoretical Physics Division, A.E.R.E. Harwell Laboratory, Oxfordshire, UK, 1973. http://www.numerical.rl.ac.uk/reports/AERE_TP515.pdf.

[229] L. HEI. A self-adaptive trust region algorithm. *Journal of Computational Mathematics*, 21(2):229–236, 2003. http://www.jstor.org/stable/43693061.

[230] M. HEINKENSCHLOSS. Mesh independence for nonlinear least squares problems with norm constraints. *SIAM Journal on Optimization*, 3(1):81–117, 1993. https://doi.org/10.1137/0803005.

[231] P. HENRICI. *Elements of Numerical Analysis.* John Wiley & Sons, New York, London, Sydney, 1964.

[232] M. R. HESTENES. Multiplier and gradient methods. *Journal of Optimization Theory and Applications*, 4:303–320, 1969. https://doi.org/10.1007/BF00927673.

[233] N. J. HIGHAM. FORTRAN codes for estimating the one-norm of a real or complex matrix, with applications to condition estimation. *ACM Transactions on Mathematical Software*, 14(4):381–396, 1988. https://doi.org/10.1145/50063.214386.

[234] C. J. HILLAR AND L.-H. LIM. Most tensor problems are NP-hard. *Journal of the ACM*, 60(6):Article 45, 39, 2013. https://doi.org/10.1145/2512329.

[235] O. HINDER. Cutting plane methods can be extended into nonconvex optimization. arXiv e-prints, arXiv:1805.08370, May 2018. https://arxiv.org/abs/1805.08370.

[236] O. HINDER AND Y. YE. Worst-case iteration bounds for log barrier methods for problems with nonconvex constraints. arXiv e-prints, arXiv:1807.00404, July 2018. https://arxiv.org/abs/1807.00404.

[237] N. HO-NGUYEN AND F. KILINÇ-KARZAN. A second-order cone based approach for solving the trust-region subproblem and its variants. *SIAM Journal on Optimization*, 27(3):1485–1512, 2017. https://doi.org/10.1137/16M1065197.

[238] W. W. HOGAN. Point-to-set maps in mathematical programming. *SIAM Review*, 15:591–603, 1973. https://doi.org/10.1137/1015073.

[239] J. D. HOGG, E. OVTCHINNIKOV, AND J. A. SCOTT. A sparse symmetric indefinite direct solver for GPU architectures. *ACM Transactions on Mathematical Software*, 42(1):Article 1, 25, 2016. https://doi.org/10.1145/2756548.

[240] J. D. HOGG AND J. A. SCOTT. HSL_MA97: A bit-compatible multifrontal code for sparse symmetric systems. Technical Report RAL-TR-2011-024, STFC-Rutherford Appleton Laboratory, Oxfordshire, UK, 2011. http://purl.org/net/epubs/work/61445.

[241] M. HONG. Decomposing linearly constrained nonconvex problems by a proximal primal dual approach: Algorithms, convergence, and applications. arXiv e-prints, arXiv:1604.00543, April 2016. https://arxiv.org/abs/1604.00543.

[242] K. HOU AND A. M-C. SO. Hardness and approximation results for L_p-ball constrained homogeneous polynomial optimization problems. *Mathematics of Operations Research*, 39(4):1084–1108, 2014. https://doi.org/10.1287/moor.2014.0644.

[243] S. HSIA, R.-L. SHEU, AND Y. YUAN. Theory and application of p-regularized subproblems for $p > 2$. *Optimization Methods & Software*, 32(5):1059–1077, 2017. https://doi.org/10.1080/10556788.2016.1238917.

[244] Y.-Q. HU, C.-L. HAO, AND Y.-H. DAI. Projected gradient algorithms for optimization over order simplices. *Optimization Methods & Software*, 29(5):1090–1117, 2014. https://doi.org/10.1080/10556788.2014.911872.

[245] H.-N. Huang, S. A. M. Marcantognini, and N. J. Young. Chain rules for higher derivatives. *The Mathematical Intelligencer*, 28(2):61–69, 2006. https://doi.org/10.1007/BF02987158.

[246] J. H. Hubbard and B. B. Hubbard. *Vector Calculus, Linear Algebra, and Differential Forms. A Unified Approach.* Prentice Hall, Upper Saddle River, NJ, 1999.

[247] V. V. Ivanov. Optimal algorithms of minimization of certain classes of functions. *Cybernetics*, 8(4):620–634, 1972. https://doi.org/10.1007/BF01068284.

[248] P. Jain and P. Kar. Non-convex optimization for machine learning. *Foundations and Trends in Machine Learning*, 10(3–4):142–336, 2017. https://doi.org/10.1561/2200000058.

[249] F. Jarre. On Nesterov's smooth Chebyshev–Rosenbrock function. *Optimization Methods & Software*, 28(3):478–484, 2013. https://doi.org/10.1080/10556788.2011.638924.

[250] B. Jiang, H. Wang, and S. Zhang. An optimal high-order tensor method for convex optimization. In A. Beygelzimer and D. Hsu, editors, *Proceedings of the Thirty-Second Conference on Learning Theory*, volume 99 of *Proceedings of Machine Learning Research*, pages 1799–1801, Phoenix, AZ, 25–28 June 2019. PMLR. http://proceedings.mlr.press/v99/jiang19a.html.

[251] A. Kamandi, K. Amini, and M. Ahookhosh. An improved adaptive trust-region algorithm. *Optimization Letters*, 11(3):555–569, 2017. https://doi.org/10.1007/s11590-016-1018-4.

[252] H. Karimi, J. Nutini, and M. Schmidt. Linear convergence of gradient and proximal-gradient methods under the Polyak-Łojasiewicz condition. In *Joint European Conference on Machine Learning and Knowledge Discovery in Databases*, page 795–811. Springer, Berlin, 2016. Expanded version arXiv:1608.04636, August 2016. https://arxiv.org/abs/1608.04636.

[253] C. T. Kelley and E. W. Sachs. Quasi-Newton methods and unconstrained optimal control problems. *SIAM Journal on Control and Optimization*, 25(6):1503–1516, 1987. https://doi.org/10.1137/0325083.

[254] D. S. Kershaw. The incomplete Cholesky–conjugate gradient method for the iterative solution of systems of linear equations. *Journal of Computational Physics*, 26(1):43–65, 1978. https://doi.org/10.1016/0021-9991(78)90098-0.

[255] M. Kimiaei and S. Ghaderi. A new restarting adaptive trust-region method for unconstrained optimization. *Journal of the Operations Research Society of China*, 5(4):487–507, 2017. https://doi.org/10.1007/s40305-016-0149-8.

[256] V. Klee and G. J. Minty. How good is the simplex algorithm? In O. Shisha, editor, *Inequalities III (Proceedings of the Third Symposium on Inequalities*, University of California, Los Angeles, California, 1–9 September, 1969, dedicated to the memory of Theodore S. Motzkin), pages 159–175, 1972.

[257] J. M. Kohler and A. Lucchi. Sub-sampled cubic regularization for non-convex optimization. arXiv e-prints, arXiv:705.05933, May 2017. https://arxiv.org/abs/1705.05933.

[258] T. G. Kolda and B. W. Bader. Tensor decompositions and applications. *SIAM Review*, 51(3):455–500, 2009. https://doi.org/10.1137/07070111X.

[259] D. P. Kouri, M. Heinkenschloss, D. Rizdal, and B. G. van Bloemen-Waanders. Inexact objective function evaluations in a trust-region algorithm for PDE-constrained optimization under uncertainty. *SIAM Journal on Scientific Computing*, 36(6):A3011–A3029, 2014. https://doi.org/10.1137/140955665.

[260] J. Kowalik, M. R. Osborne, and D. M. Ryan. A new method for constrained optimization problems. *Operations Research*, 17:973–983, 1969. https://doi.org/10.1287/opre.17.6.973.

[261] K. Kurdyka. On gradients of functions definable in o-minimal structures. *Université de Grenoble. Annales de l'Institut Fourier*, 48(3):769–783, 1998. https://aif.centre-mersenne.org/item/10.5802/aif.1638.pdf.

[262] J. Larson, M. Menickelly, and S. M. Wild. Derivative-free optimization methods. *Acta Numerica*, 28:287–404, 2019. https://doi.org/10.1017/s0962492919000060.

[263] Y. LeCun, L. Bottou, Y. Bengio, and P. Haffner. Gradient-based learning applied to document recognition. *Proceedings of the IEEE*, 86(11):2278–2324, 1998. https://doi.org/10.1109/5.726791.

[264] F. Lenders, C. Kirches, and A. Potschka. trlib: A vector-free implementation of the GLTR method for iterative solution of the trust region problem. *Optimization Methods & Software*, 33(3):420–449, 2018. https://doi.org/10.1080/10556788.2018.1449842.

[265] K. Levenberg. A method for the solution of certain non-linear problems in least-squares. *Quarterly of Applied Mathematics*, 2:164–168, 1944. https://doi.org/10.1090/qam/10666.

[266] R. J. LeVeque. *Finite Difference Methods for Ordinary and Partial Differential Equations: Steady-State and Time-Dependent Problems*. SIAM, Philadelphia, 2007.

[267] A. Lewis, private communication, Cornell University.

[268] A. S. Lewis and S. J. Wright. A proximal method for composite minimization. *Mathematical Programming Series A*, 158(1-2):501–546, 2016. https://doi.org/10.1007/s10107-015-0943-9.

[269] S. Leyffer, S. M. Wild, M. Fagan, M. Snir, K. Palem, K. Yoshii, and H. Finkel. Doing Moore with less—Leapfrogging Moore's law with inexactness for supercomputing. In *Proceedings of PMES 2018: 3rd International Workshop on Post Moore's Era Supercomputing*, 2016. https://arxiv.org/abs/1610.02606.

[270] G. Li and T. K. Pong. Calculus of the exponent of Kurdyka–Łojasiewicz inequality and its applications to linear convergence of first-order methods. *Foundations of Computational Mathematics*, 18(5):1199–1232, 2018. https://doi.org/10.1007/s10208-017-9366-8.

[271] X. Li, S. Chen, Z. Deng, Q. Qu, Z. Zhu, and A. Man Cho So. Weakly convex optimization over Stiefel manifold using Riemannian subgradient-type methods. arXiv e-prints, arXiv:1911.05047v3, February 2020. https://arxiv.org/abs/1911.05047.

[272] X. D. Li, D. F. Sun, and K. C. Toh. On the efficient computation of a generalized Jacobian of the projector over the Birkhoff polytope. *Mathematical Programming Series A*, 179(1-2):419–446, 2020. https://doi.org/10.1007/s10107-018-1342-9.

[273] Z. Li and Y. Xu. Augmented Lagrangian based first-order methods for convex-constrained programs with weakly-convex objective. arXiv e-prints, arXiv:2003.08880, February 2021. https://arxiv.org/abs/2003.08880.

[274] C.-J. Lin and J. J. Moré. Incomplete Cholesky factorizations with limited memory. *SIAM Journal on Scientific Computing*, 21(1):24–45, 1999. https://doi.org/10.1137/S1064827597327334.

[275] S. Łojasiewicz. Ensembles semi-analytiques. Technical report, Institut des Hautes Etudes Scientifiques, Bures-sur-Yvette, France, 1965. http://perso.univ-rennes1.fr/michel.coste/Lojasiewicz.pdf.

[276] A. Lucchi and J. M. Kohler. A stochastic tensor method for non-convex optimization. arXiv e-prints, arXiv:1911.10367, November 2019. https://arxiv.org/abs/1911.10367.

[277] S. Lucidi, L. Palagi, and M. Roma. On some properties of quadratic programs with a convex quadratic constraint. *SIAM Journal on Optimization*, 8(1):105–122, 1998. https://doi.org/10.1137/S1052623494278049.

[278] D. G. Luenberger. *Optimization by Vector Space Methods*. John Wiley & Sons, New York, 1969.

[279] L. Lukšan, C. Matonoha, and J. Vlček. On Lagrange multipliers of trust-region subproblems. *BIT. Numerical Mathematics*, 48(4):763–768, 2008. https://doi.org/10.1007/s10543-008-0197-5.

[280] O. L. Mangasarian. *Nonlinear Programming*. McGraw-Hill, New York, London, Sydney, 1969.

[281] T. A. Manteuffel. An incomplete factorization technique for positive definite linear systems. *Mathematics of Computation*, 34(150):473–497, 1980. https://doi.org/10.2307/2006097.

[282] D. W. Marquardt. An algorithm for least squares estimation of non-linear parameters. *SIAM Journal*, 11:431–441, 1963. https://doi.org/10.1137/0111030.

[283] S. Martello and P. Toth. *Knapsack Problems: Algorithms and Computer Implementations*. Wiley-Interscience Series in Discrete Mathematics and Optimization. John Wiley & Sons, Chichester, UK, 1990.

[284] B. Martinet. Brève communication. Régularisation d'inéquations variationnelles par approximations successives. *Revue Française d'Informatique et de Recherche Opérationnelle*, 4(Sér. R-3):154–158, 1970. https://www.numdam.org/item/M2AN_1970__4_3_154_0.pdf.

[285] J. M. Martínez. Local minimizers of quadratic functions on Euclidean balls and spheres. *SIAM Journal on Optimization*, 4(1):159–176, 1994. https://doi.org/10.1137/0804009.

[286] J. M. Martínez and M. Raydan. Cubic-regularization counterpart of a variable-norm trust-region method for unconstrained minimization. *Journal of Global Optimization*, 68(2):367–385, 2017. https://doi.org/10.1007/s10898-016-0475-8.

[287] J. A. MEIJERINK AND H. A. VAN DER VORST. An iterative solution method for linear systems of which the coefficient matrix is a symmetric M-matrix. *Mathematics of Computation*, 31(137):148–162, 1977. https://doi.org/10.2307/2005786.

[288] A. MOKHTARI, A. OZDAGLAR, AND A. JADBABAIE. Escaping saddle points in constrained optimization. arXiv e-prints, arXiv:1809.02162, September 2018. https://arxiv.org/abs/1809.02162.

[289] R. D. C. MONTEIRO AND B. F. SVAITER. An accelerated hybrid proximal extragradient method for convex optimization and its implications to second-order methods. *SIAM Journal on Optimization*, 23(2):1092–1125, 2013. https://doi.org/10.1137/110833786.

[290] J. L. MORALES AND J. NOCEDAL. Automatic preconditioning by limited memory quasi-Newton updating. *SIAM Journal on Optimization*, 10(4):1079–1096, 2000. https://doi.org/10.1137/S1052623497327854.

[291] J. J. MORÉ AND D. C. SORENSEN. On the use of directions of negative curvature in a modified Newton method. *Mathematical Programming*, 16(1):1–20, 1979. https://doi.org/10.1007/BF01582091.

[292] J. J. MORÉ AND D. C. SORENSEN. Computing a trust region step. *SIAM Journal on Scientific and Statistical Computing*, 4(3):553–572, 1983. https://doi.org/10.1137/0904038.

[293] D. D. MORRISON. Methods for nonlinear least squares problems and convergence proofs. In J. Lorell and F. Yagi, editors, *Proceedings of the Seminar on Tracking Programs and Orbit Determination*, pages 1–9. Jet Propulsion Laboratory, Pasadena, CA, 1960.

[294] D. D. MORRISON. Optimization by least squares. *SIAM Journal on Numerical Analysis*, 5:83–88, 1968. https://doi.org/10.1137/0705006.

[295] K. G. MURTY AND S. N. KABADI. Some NP-complete problems in quadratic and nonlinear programming. *Mathematical Programming*, 39(2):117–129, 1987. https://doi.org/10.1007/BF02592948.

[296] A. S. NEMIROVSKY AND D. B. YUDIN. *Problem Complexity and Method Efficiency in Optimization*. A Wiley-Interscience Publication. Translated from the Russian and with a preface by E. R. Dawson. Wiley-Interscience Series in Discrete Mathematics. John Wiley & Sons, New York, 1983.

[297] YU. NESTEROV. Squared functional systems and optimization problems. In *High Performance Optimization*, volume 33 of *Applied Optimization*, pages 405–440. Kluwer Academic, Dordrecht, Netherlands, 2000.

[298] YU. NESTEROV. Random walk in a simplex and quadratic optimization over convex polytopes. Technical Report 2003/71, CORE, UCL, Louvain-la-Neuve, Belgium, 2003.

[299] YU. NESTEROV. *Introductory Lectures on Convex Optimization: A Basic Course*, volume 87 of *Applied Optimization*. Kluwer Academic, Boston, MA, 2004.

[300] YU. NESTEROV. Modified Gauss–Newton scheme with worst case guarantees for global performance. *Optimization Methods & Software*, 22(3):469–483, 2007. https://doi.org/10.1080/08927020600643812.

[301] YU. NESTEROV. How to make the gradients small. *Optima*, 88:10–11, 2012. http://www.mathopt.org/Optima-Issues/optima88.pdf.

[302] YU. NESTEROV. Universal gradient methods for convex optimization problems. *Mathematical Programming Series A*, 152(1-2):381–404, 2015. https://doi.org/10.1007/s10107-014-0790-0.

[303] YU. NESTEROV. *Lectures on Convex Optimization*, volume 137 of *Springer Optimization and Its Applications*. Springer, Cham, second edition, 2018.

[304] YU. NESTEROV. Implementable tensor methods in unconstrained convex optimization. *Mathematical Programming Series A*, 186(1-2):157–183, 2021. https://doi.org/10.1007/s10107-019-01449-1.

[305] YU. NESTEROV AND B. T. POLYAK. Cubic regularization of Newton method and its global performance. *Mathematical Programming Series A*, 108(1):177–205, 2006. https://doi.org/10.1007/s10107-006-0706-8.

[306] J. NIE. Sum of squares methods for minimizing polynomial forms over spheres and hypersurfaces. *Frontiers of Mathematics in China*, 7(2):321–346, 2012. https://doi.org/10.1007/s11464-012-0187-4.

[307] J. NOCEDAL AND S. J. WRIGHT. *Numerical Optimization*. Springer Series in Operations Research. Springer-Verlag, New York, 1999. Second edition, 2006.

[308] D. Noll and A. Rondepierre. Convergence of linesearch and trust-region methods using the Kurdyka-Łojasiewiecz inequality. In *Computational and Analytical Mathematics*, volume 50 of *Springer Proceedings in Mathematics and Statistics*, pages 593–611. Springer, New York, 2013.

[309] M. Nouiehed, J. Lee, and M. Razaviyayn. Convergence to second-order stationarity for constrained non-convex optimization. arXiv e-prints, arXiv:1810.02024, October 2018. https://arxiv.org/abs/1810.02024.

[310] M. Nouiehed and M. Razaviyayn. A trust region method for finding second-order stationarity in linearly constrained non-convex optimization. arXiv e-prints, arXiv:1904.06784, April 2019. https://arxiv.org/abs/1904.06784.

[311] J. M. Ortega and W. C. Rheinboldt. *Iterative Solution of Nonlinear Equations in Several Variables.* Academic Press, New York, London, 1970.

[312] C. C. Paige and M. A. Saunders. LSQR: An algorithm for sparse linear equations and sparse least squares. *ACM Transactions on Mathematical Software*, 8(1):43–71, 1982. https://doi.org/10.1145/355984.355989.

[313] C. H. Papadimitriou and K. Steiglitz. *Combinatorial Optimization: Algorithms and Complexity.* Prentice-Hall, Englewood Cliffs, NJ, 1982.

[314] C. Paquette and K. Scheinberg. A stochastic line search method with expected complexity analysis. *SIAM Journal on Optimization*, 30(1):349–376, 2020. https://doi.org/10.1137/18M1216250.

[315] B. N. Parlett. *The Symmetric Eigenvalue Problem.* Prentice-Hall Series in Computational Mathematics, Englewood Cliffs, NJ, 1980. Reprinted as volume 20 of *Classics in Applied Mathematics*, SIAM, Philadelphia, 1998.

[316] G. Peano. *Calcolo differentiale e principii di calcolo integrale.* Fratelli Bocca, Roma, 1884.

[317] B. T. Polyak. Gradient methods for minimization of functionals. *Žurnal Vyčislitel' noĭ Matematiki i Matematičeskoĭ Fiziki*, 3:643–653, 1963.

[318] M. Porcelli and Ph. L. Toint. BFO, a trainable derivative-free brute force optimizer for nonlinear bound-constrained optimization and equilibrium computations with continuous and discrete variables. *ACM Transactions on Mathematical Software*, 44(1): Article 6, 2017. https://doi.org/10.1145/3085592.

[319] M. Porcelli and Ph. L. Toint. Exploiting structure in derivative free optimization. *ACM Transactions on Mathemical Software*, 48(1): Article 6, 2022. https://doi.org/10.1145/3474054.

[320] M. J. D. Powell. A method for nonlinear constraints in minimization problems. In R. Fletcher, editor, *Optimization*, pages 283–298. Academic Press, London, 1969.

[321] M. J. D. Powell. A direct search optimization method that models the objective by quadratic interpolation. Presentation at the 5th Stockholm Optimization Days, Stockholm, 1994.

[322] M. J. D. Powell and Ph. L. Toint. On the estimation of sparse Hessian matrices. *SIAM Journal on Numerical Analysis*, 16(6):1060–1074, 1979. https://doi.org/10.1137/0716078.

[323] C. J. Price and Ph. L. Toint. Exploiting problem structure in pattern-search methods for unconstrained optimization. *Optimization Methods & Software*, 21(3):479–491, 2006. https://doi.org/10.1080/10556780500137116.

[324] L. Qi. Eigenvalues of a real supersymmetric tensor. *Journal of Symbolic Computation*, 40(6):1302–1324, 2005. https://doi.org/10.1016/j.jsc.2005.05.007.

[325] S. Reddi, S. Sra, B. Póczos, and A. Smola. Proximal stochastic methods for nonsmooth nonconvex finite-sum optimization. In D. D. Lee, M. Sugiyama, U. V. Luxburg, I. Guyon, and R. Garnett, editors, volume 29 of *Advances in Neural Information Processing Systems (NIPS 2016)*, pages 1145–1153. Curran Associates, 2019. http://papers.nips.cc/paper/6116-proximal-stochastic-methods-for-nonsmooth-nonconvex-finite-sum-optimization.

[326] C. H. Reinsch. Smoothing by spline functions. II. *Numerische Mathematik*, 16(5):451–454, 1971. https://doi.org/10.1007/BF02169154.

[327] H. Robbins and S. Monro. A stochastic approximation method. *The Annals of Mathematical Statistics*, 22(3):400–407, 1951. https://doi.org/10.1214/aoms/1177729586.

[328] L. Roberts. *Derivative-Free Algorithms for Nonlinear Optimization Problems.* PhD thesis, Mathematical Institute, University of Oxford, Oxford, UK, 2019.

[329] R. T. Rockafellar. Augmented Lagrange multiplier functions and duality in nonconvex programming. *SIAM Journal on Control and Optimization*, 12(2):268–285, 1974. https://doi.org/10.1137/0312021.

[330] R. T. ROCKAFELLAR. Augmented Lagrangians and applications of the proximal point algorithm in convex programming. *Mathematics of Operations Research*, 1(2):97–116, 1976. https://doi.org/10.1287/moor.1.2.97.

[331] M. ROJAS, S. A. SANTOS, AND D. C. SORENSEN. A new matrix-free algorithm for the large-scale trust-region subproblem. *SIAM Journal on Optimization*, 11(3):611–646, 2000/01. https://doi.org/10.1137/S105262349928887X.

[332] M. ROJAS, S. A. SANTOS, AND D. C. SORENSEN. Algorithm 873: LSTRS: MATLAB software for large-scale trust-region subproblems and regularization. *ACM Transactions on Mathematical Software*, 34(2):Article 11, 28, 2008. https://doi.org/10.1145/1326548.1326553.

[333] H. H. ROSENBROCK. An automatic method for finding the greatest or least value of a function. *The Computer Journal*, 3:175–184, 1960/61. https://doi.org/10.1093/comjnl/3.3.175.

[334] C. W. ROYER, M. O'NEILL, AND S. J. WRIGHT. A Newton-CG algorithm with complexity guarantees for smooth unconstrained optimization. *Mathematical Programming Series A*, 180(1-2):451–488, 2020. https://doi.org/10.1007/s10107-019-01362-7.

[335] C. W. ROYER AND S. J. WRIGHT. Complexity analysis of second-order line-search algorithms for smooth nonconvex optimization. *SIAM Journal on Optimization*, 28(2):1448–1477, 2018. https://doi.org/10.1137/17M1134329.

[336] H. E. SALZER. A note on the solution of quartic equations. *Mathematics of Computation*, 14:279–281, 1960. https://doi.org/10.1090/S0025-5718-1960-0117882-6.

[337] A. SARTENAER. Automatic determination of an initial trust region in nonlinear programming. *SIAM Journal on Scientific Computing*, 18(6):1788–1803, 1997. https://doi.org/10.1137/S1064827595286955.

[338] J. A. SCOTT AND M. TŮMA. On positive semidefinite modification schemes for incomplete Cholesky factorization. *SIAM Journal on Scientific Computing*, 36(2):A609–A633, 2014. https://doi.org/10.1137/130917582.

[339] L. SLOT AND M. LAURENT. Improved convergence analysis of Lasserre's measure-based upper bounds for polynomial minimization on compact sets. *Mathematical Programming*, 2020. https://doi.org/10.1007/s10107-020-01468-3.

[340] M. V. SOLODOV. Constraint qualifications. In J. J. Cochran, L. A. Cox, P. Keskinocak, J. P. Kharoufeh, and J. C. Smith, editors, *Wiley Encyclopedia of Operations Research and Management Science*, John Wiley & Sons, Hoboken, NJ, 2011. 12 pages. https://doi.org/10.1002/9780470400531.eorms0978.

[341] D. A. SPIELMAN AND S.-H. TENG. Smoothed analysis of algorithms: Why the simplex algorithm usually takes polynomial time. arXiv e-prints, arXiv:cs/0111050, November 2001. https://arxiv.org/abs/cs/0111050.

[342] S. SRA, S. NOWOZIN, AND S. J. WRIGHT. *Optimization for Machine Learning.* Neural information processing series. MIT Press, Cambridge, MA, 2012.

[343] T. STEIHAUG. The conjugate gradient method and trust regions in large scale optimization. *SIAM Journal on Numerical Analysis*, 20(3):626–637, 1983. https://doi.org/10.1137/0720042.

[344] M. THOM, M. RAPP, AND G. PALM. Efficient dictionary learning with sparseness-enforcing projections. *International Journal of Computer Vision*, 114(2-3):168–194, 2015. https://doi.org/10.1007/s11263-015-0799-8.

[345] R. TIBSHIRANI. Regression shrinkage and selection via the lasso. *Journal of the Royal Statistical Society B*, 58(1):267–288, 1996. https://www.jstor.org/stable/2346178.

[346] M. TISMENETSKY. A new preconditioning technique for solving large sparse linear systems. *Linear Algebra and Its Applications*, 154/156:331–353, 1991. https://doi.org/10.1016/0024-3795(91)90383-8.

[347] PH. L. TOINT. Towards an efficient sparsity exploiting Newton method for minimization. In I. S. Duff, editor, *Sparse Matrices and Their Uses*, pages 57–88. Academic Press, London, New York, 1981.

[348] PH. L. TOINT. Global convergence of a class of trust region methods for nonconvex minimization in Hilbert space. *IMA Journal of Numerical Analysis*, 8(2):231–252, 1988. https://doi.org/10.1093/imanum/8.2.231.

[349] PH. L. TOINT. Nonlinear stepsize control, trust regions and regularizations for unconstrained optimization. *Optimization Methods & Software*, 28(1):82–95, 2013. https://doi.org/10.1080/10556788.2011.610458.

[350] J. F. TRAUB. *Iterative Methods for the Solution of Equations.* Prentice-Hall Series in Automatic Computation. Prentice-Hall, Englewood Cliffs, NJ, 1964.

[351] L. N. TREFETHEN AND D. BAU. *Numerical Linear Algebra.* SIAM, Philadelphia, 1997.

[352] L. N. TREFETHEN AND R. S. SCHREIBER. Average-case stability of Gaussian elimination. *SIAM Journal on Matrix Analysis and Applications*, 11(3):335–360, 1990. https://doi.org/10.1137/0611023.

[353] K. UEDA AND N. YAMASHITA. Convergence properties of the regularized Newton method for the unconstrained nonconvex optimization. *Applied Mathematics and Optimization*, 62(1):27–46, 2010. https://doi.org/10.1007/s00245-009-9094-9.

[354] K. UEDA AND N. YAMASHITA. On a global complexity bound of the Levenberg-Marquardt method. *Journal of Optimization Theory and Applications*, 147(3):443–453, 2010. https://doi.org/10.1007/s10957-010-9731-0.

[355] K. UEDA AND N. YAMASHITA. A regularized Newton method without line search for unconstrained optimization. *Computational Optimization and Applications*, 59(1-2):321–351, 2014. https://doi.org/10.1007/s10589-014-9656-x. (Erratum in 66(1):219–222, 2017, https://doi.org/10.1007/s10589-016-9859-4.)

[356] M. ULBRICH AND S. ULBRICH. Superlinear convergence of affine-scaling interior-point Newton methods for infinite-dimensional problems with pointwise bounds. *SIAM Journal on Scientific Computing*, 38(6):1938–1984, 2000. https://doi.org/10.1137/S0363012997325915.

[357] A. VAN DER SLUIS. Condition numbers and equilibration of matrices. *Numerische Mathematik*, 14:14–23, 1969/1970. https://doi.org/10.1007/BF02165096.

[358] S. A. VAVASIS. *Nonlinear Optimization: Complexity Issues*, volume 8 of *International Series of Monographs on Computer Science.* The Clarendon Press, Oxford University Press, Oxford, UK, 1991.

[359] S. A. VAVASIS. Black-box complexity of local minimization. *SIAM Journal on Optimization*, 3(1):60–80, 1993. https://doi.org/10.1137/0803004.

[360] J. M. B. WALMAG AND E. J. M. DELHEZ. A note on trust-region radius update. *SIAM Journal on Optimization*, 16(2):548–562, 2005. https://doi.org/10.1137/030602563.

[361] N. WANG, J. CHOI, D. BRAND, C.-Y. CHEN, AND K. GOPALAKRISHNAN. Training deep neural networks with 8-bit floating point numbers. In S. Bengio, H. M. Wallach, H. Larochelle, K. Grauman, and N. Cesa-Bianchi, editors, *32nd Conference on Neural Information Processing Systems*, NIPS'18, pages 7686–7695. Curran Associates Inc., 2018. https://papers.nips.cc/paper/7994-training-deep-neural-networks-with-8-bit-floating-point-numbers.

[362] Z. WANG, Y. ZHOU, Y. LIANG, AND G. LAN. A note on inexact gradient and Hessian conditions for cubic regularized Newton's method. *Operations Research Letters*, 47(2):146–149, 2019. https://doi.org/10.1016/j.orl.2019.01.009.

[363] A. J. WATHEN. Preconditioning. *Acta Numerica*, 24:329–376, 2015. https://doi.org/10.1017/S0962492915000021.

[364] M. WEISER, P. DEUFLHARD, AND B. ERDMANN. Affine conjugate adaptive Newton methods for nonlinear elastomechanics. *Optimization Methods & Software*, 22(3):413–431, 2007. https://doi.org/10.1080/10556780600605129.

[365] D. WINFIELD. *Function and Functional Optimization by Interpolation in Data Tables.* PhD thesis, Harvard University, Cambridge, MA, 1969.

[366] D. WINFIELD. Function minimization by interpolation in a data table. *Journal of the Institute of Mathematics and Its Applications*, 12:339–347, 1973. https://doi.org/10.1093/imamat/12.3.339.

[367] L. A. WOLSEY. *Integer Programming.* John Wiley & Sons, Chichester, UK, 1998.

[368] Y. XIE AND S. J. WRIGHT. Complexity of proximal augmented Lagrangian for nonconvex optimization with nonlinear equality constraints. arXiv e-prints, arXiv:1908.00131, July 2019. https://arxiv.org/abs/1908.00131.

[369] C. X. XU AND J. Z. ZHANG. Scaled optimal path trust-region algorithm. *Journal of Optimization Theory and Applications*, 102(1):127–146, 1999. https://doi.org/10.1023/A:1021846613163.

[370] P. XU, F. ROOSTA-KHORASAN, AND M. W. MAHONEY. Second-order optimization for non-convex machine learning: An empirical study. arXiv e-prints, arXiv:1708.07827, August 2017. https://arxiv.org/abs/1708.07827.

[371] P. Xu, F. Roosta-Khorasan, and M. W. Mahoney. Newton-type methods for non-convex optimization under inexact Hessian information. *Mathematical Programming Series A*, 184(1-2):35–70, 2020. https://doi.org/10.1007/s10107-019-01405-z.

[372] T. Yamamoto. Historical developments in convergence analysis for Newton's and Newton-like methods. *Journal of Computational and Applied Mathematics*, 124(1-2):1–23, 2000. https://doi.org/10.1016/S0377-0427(00)00417-9.

[373] Z. Yao, P. Xu, F. Roosta-Khorasan, and M. W. Mahoney. Inexact non-convex Newton-type methods. arXiv e-prints, arXiv:1802.06925, February 2018. https://arxiv.org/abs/1802.06925.

[374] Y. Yuan. Recent advances in trust region algorithms. *Mathematical Programming Series B*, 151(1):249–281, 2015. https://doi.org/10.1007/s10107-015-0893-2.

[375] J. Z. Zhang and C. X. Xu. A class of indefinite dogleg path methods for unconstrained minimization. *SIAM Journal on Optimization*, 9(3):646–667, 1999. https://doi.org/10.1137/S105262349627523X.

[376] L.-H. Zhang and C. Shen. A nested Lanczos method for the trust-region subproblem. *SIAM Journal on Scientific Computing*, 40(4):A2005–A2032, 2018. https://doi.org/10.1137/17M1145914.

[377] L.-H. Zhang, C. Shen, and R.-C. Li. On the generalized Lanczos trust-region method. *SIAM Journal on Optimization*, 27(3):2110–2142, 2017. https://doi.org/10.1137/16M1095056.

[378] L.-H. Zhang, W.-H. Yang, C. Shen, and J. Feng. Error bounds of Lanczos approach for trust-region subproblem. *Frontiers of Mathematics in China*, 13(2):459–481, 2018. https://doi.org/10.1007/s11464-018-0687-y.

[379] X. Zhang, C. Ling, and L. Qi. The best rank-1 approximation of a symmetric tensor and related spherical optimization problems. *SIAM Journal on Matrix Analysis and Applications*, 33(3):806–821, 2012. https://doi.org/10.1137/110835335.

[380] Y. Zhou, Z. Wang, and Y. Liang. Convergence of cubic regularization for nonconvex optimization under KŁ property. In *Advances in Neural Information Processing Systems*, NIPS'18, pages 3764–3773. Curran Associates Inc., 2018. http://papers.neurips.cc/paper/7633-convergence-of-cubic-regularization-for-nonconvex-optimization-under-kl-property.pdf.

[381] Z. Zhu, Q. Li, G. Tang, and M. B. Wakin. Global optimality in low-rank matrix optimization. *IEEE Transactions on Signal Processing*, 66(13):3614–3628, 2018. https://doi.org/10.1109/TSP.2018.2835403.

Subject and Notation Index

Note: A boldface number indicates the page on which the subject is introduced. An underlined number indicates that the subject is a primary topic in the section or subsection starting on the page given.

Author Index

Here is a list of other authors whose work is cited in our book.